Biology and Conservation of the European Sturgeon *Acipenser sturio* L. 1758

Patrick Williot • Eric Rochard •
Nathalie Desse-Berset • Frank Kirschbaum •
Jörn Gessner
Editors

Biology and Conservation of the European Sturgeon *Acipenser sturio* L. 1758

The Reunion of the European and Atlantic Sturgeons

Springer

Editors
Dr. Patrick Williot
Sturgeon Consultant
Rue du Pas de Madame 4
33980 Audenge
France
williot.patrick@neuf.fr

Dr. Nathalie Desse-Berset
Université de Nice-Sophia Antipolis
CNRS-CEPAM-UMR 6130, SJA3
Avenue des Diables Bleus 24
06357 Nice cedex 4
France
berset.desse@hotmail.com
berset@cepam.cnrs.fr

Dr. Jörn Gessner
Leibniz-Institut für
Gewässerökologie und
Binnenfischerei
Müggelseedamm 310
12587 Berlin
Germany
sturgeon@igb-berlin.de

Dr. Eric Rochard
Cemagref
Estuarine Ecosystems and Diadromous
Fish Research Unit
Avenue de Verdun 50
33612 Cestas cedex
France
eric.rochard@cemagref.fr

Prof. Frank Kirschbaum
Humboldt-Universität Berlin
Abt. Biologie und Ökologie der
Fische
Philippstr. 13
10115 Berlin
Haus 16
Germany
frank.kirschbaum@staff.hu-berlin.de

ISBN 978-3-642-20610-8 e-ISBN 978-3-642-20611-5
DOI 10.1007/978-3-642-20611-5
Springer Heidelberg Dordrecht London New York

Library of Congress Control Number: 2011935343

Printed on acid-free paper

Springer is part of Springer Science+Business Media (www.springer.com)

Dedication to Dr. Etienne Magnin

Dr. Étienne Magnin (1922–1990) was one of the few contemporary researchers to take an interest in European sturgeon before it became an endangered species. His doctoral thesis and subsequent work on the biology of the species – and, more specifically, of the population of the Gironde basin – on its taxonomic position relative to Atlantic sturgeon and on its status, which was already considered a cause for concern in the 1950s, laid much of the scientific groundwork for a body of knowledge that others continue to build on and that will, we hope, some day make possible the survival of this magnificent fish.

Paradoxically, Dr. Magnin, though he came late to the field of biology, was a pioneer, and not only with respect to his research in France on European sturgeon. After settling in Canada, where he was first a consultant (1962), then a professor in the Department of Biology at the Université de Montréal (1964–1987), he was also a pioneer in the development of knowledge about the aquatic ecosystems of the St. Lawrence River and the James Bay region during the early stages of construction of the large-scale hydroelectric projects in this vast, little-known northern territory. His areas of interest were many. Both an ichthyologist and a limnologist, Dr. Magnin authored or co-authored, with the graduate students and researchers who worked in his laboratory and that of the Société d'énergie de la Baie James where he was scientific director for more than 7 years, 81 scientific papers: 48 on fish (including 16 on European and North American Acipenseridae), 17 on benthic organisms, six on zooplankton, five on phytoplankton and five on various other subjects (such as amphibians). His book on the freshwater ecology of the James Bay region (1977) still remains a major reference work.

Dr. Magnin was also an outstanding teacher. Those, like us, who had the opportunity of taking his zoology and comparative anatomy courses were able to appreciate his rigorous approach, his clarity of thought, his attention to detail, his respect for students and his sense of humour. He was also a mentor, whose knowledge, commitment, enthusiasm and support enabled him to train three PhD students and 28 master's students in a span of 20 years. Many of these went on to their own research careers in academe and public service, contributing, in turn, to improving our understanding and protection of freshwater ecosystems, preparing a new generation of biologists and developing a profession that plays an increasingly integrated role in the decision-making processes of today's society.

Pierre Dumont
Ministère des Ressources naturelles et de la Faune du Québec

Peter Harper
Université de Montréal

Dedication to Dr. Juraj Holčik

RNDr. Juraj Holčík, CSc. – eminent Slovak zoologist was born on October 18, 1934 in Trnava in Western Slovakia. He completed his B.Sc. (1958) and PhD (1966) studies at Charles University in Prague.

Before being appointed Director of the newly established Institute of Zoology and Ecosozology (in 1995 re-named Institute of Zoology), Slovak Academy of Sciences in 1990, Dr. Holčík was research officer and after 1972 the Senior Research Officer of the Institute of Fishery Research and Hydrobiology in Bratislava. He also worked at the Regional Museum at Trnava and the Slovak National Museum in Bratislava as the Curator of Zoology and Curator of Lower Vertebrates respectively.

The scope of his scientific activity was very diverse, and included taxonomy, zoogeography, ecology, population dynamics, limnology, and ecosozology, as well as production, introduction, acclimatization and conservation of fish gene pool (genofond). However, ichthyology was the main focus of his work. He was an internationally renowned expert on Palearctic Petromyzontidae, Salmonidae, and Cyprinidae (especially the subfamilies Acheilognathinae and Percidae). He described six new fish species, three subspecies and two new hybrid forms of fish fauna in Slovakia, Romania, Mongolia, the Balkans and Iran. Of these, *Gymnocephalus baloni* (Holčík and Hensel 1974) is considered to be one of the last new vertebrate taxa to be described on Slovak territory. He contributed to the development of a methodology for studying ecological fish production in open waters. He contributed significantly to the fact that the Slovak–Hungarian section of the Danube river is one of the best studied sections not only of the Danube but also of large rivers overall. He also contributed to our understanding of the importance of floodplains and knowledge of the causal relationships between the density, ecological production and activity of fish populations on one hand, and main abiotic environmental factors on the other hand. Concerning the population dynamics of fish in man-made water reservoirs, he amended principles of stepwise evolution of ichthyofauna. He was an uncompromising advocate of the conservation of the

Danube inland delta below Bratislava which he supported even as a Member of Parliament of the Czecho–Slovak Federative Republic.

During his career he published, either alone or with co-authors, 170 original articles, over 200 popular science articles, and 25 books or book chapters. Some of them were published in several languages and multiple editions. For example, the book Holčík J, Mihlik J and Malý J: Freshwater Fishes was published in four English editions as well as in seven additional editions in five different languages. Dr. Holčík is coauthor of an ongoing book series entitled The Freshwater Fishes of Europe, for which he created the structure of individual book chapters and was the editor of the first part of the series (Vol. 1/I: Petromyzontes, 1986, Vol. 1/II: General introduction to Fishes, Acipenseriformes, 1989, AULA Verlag Wiesbaden) for which he also authored several chapters. It is noteworthy that in 1998 he published the first and still the only textbook of ichthyology in the former Czechoslovakia (Ichtyológia, Vydavateľstvo Príroda, Bratislava). This significant scientific output is a demonstration of his creativity, extraordinary industriousness, and determination.

In the 1960s J. Holčík led a Czechoslovak limnological expedition to Cuba. In addition, he worked as a visiting professor at the University of Waterloo (Canada, 1986–1987; 1992), and as a technical advisor for FAO in Iran (1989–1991) and Ghana (1995–1996). He also supervised and graduated several promising scientists.

Dr. Holčík's publications and his activity in international ichthyological organizations (which was negatively influenced by political discrimination from the 1950s to the 1980s) had a significant impact on his scientific reputation. He was frequently asked to chair congresses and symposia, became a member of editorial boards of five international scientific journals (*Biologia*, *Folia Zoologica*, *Bios*, *Environmental Biology of Fishes*, and *Voprosy ichtiologii*), and was an active board member in many, primarily ichthyological organizations. For his lifetime achievements he was awarded the World Wide Fund for Nature *Award for Conservation Merit* in 1996 and the G.J. Mendel honour medal of the Czech Academy of Sciences in 1998.

He passed away on May 16, 2010.

Kristina Holčik

Preface

General Introduction, Context, Origin and Setting Up of the Project

The European sturgeon, *Acipenser sturio* L. 1758, used to be one of the most widespread sturgeon species (Magnin 1959). However, like most Eurasian sturgeon species, its status has now become critical (Williot et al. 2002), it has been extirpated from the great majority of its natural biotopes, and survives only in the French Atlantic coast façade, the Garonne basin, with a population on the verge of extinction (Rochard et al. 1990). The species was exploited more or less intensively for decades and even centuries, as it was in most European countries bordering the Mediterranean Sea (Italy, France, and Spain), the Eastern Atlantic Ocean (Spain, France), the North Sea (the Netherlands and Germany), and the Baltic Sea (Classen 1944; Holčik 1989).

The species can therefore be looked upon as an emblematic European species.

Compared with terrestrial animals or plants, fish combine several handicaps: they are not visible, they move, sometimes far away from the administrative limits of human societies. Fisheries are under-managed, as can be seen from changes in fish population status, and fish conservation is almost ignored. The sturgeon is no exception. Moreover, the European sturgeon is further handicapped. It is an anadromous species which colonises alternately a great variety of biotopes in fresh, brackish, and marine waters which do not tick the administrative boxes. Other biological characteristics such as longevity, late puberty, and non-yearly oogenesis make them even more susceptible to exploitation in spite of their apparent robustness (Boreman 1997; Jager et al. 2008).

Much has been done in France in the last few decades in favour of the preservation–conservation–restoration of the species, especially by research bodies. However, despite great efforts and protection, the status of the species has deteriorated.

The conservation programme of the European sturgeon in France was begun in the early 1970s. In 2007 for the first time, a breakthrough was achieved with the first reproduction of cultured brood fish (Williot et al. 2009) which made it possible for the restoration programme to start again. This was the opportunity to draw up an extensive assessment of the long-term programme, to update available knowledge, describe the context, and make known some long-term concerns.

The European sturgeon restoration–conservation programme has been a complicated programme, full of pitfalls. We realised that, among other things, besides its unusual scientific and technical aspects, conservation does not only have scientific relevance. Indeed, social, economic, political, financial, and administrative aspects are an integral part of any preservation–conservation programme.

In the field of science, conservation issues are facing many difficulties. Research into biodiversity is favoured but not protection stricto sensu, and conservation actions are not considered with much enthusiasm in France. Thus, grants from this source are for the most part not accessible for concrete actions. In addition, in the absence of training courses in Conservation Biology, most people working in the field are undereducated, at least when they begin their careers. Thus, it is necessary to take time to explain, to discuss, to exchange and finally to get the consensus of all those involved, even though a specific decision might be not considered as the optimal one from scientific and technical points of view. Consensus is a key issue for going further. Many biological sciences are potentially concerned by fish management and conservation, sometimes at both fundamental and applied levels. Unfortunately, it is often difficult to mobilise specialists. One of the great challenges of conservation biology is that the situation of certain species is so critical that waiting for supposedly needed data would inexorably lead to the extinction of a species. The European sturgeon conservation programme was a good example of the difficulties that have been faced. Due to the high depletion status of the species, the decision was taken by a few people in a research institute to acquire experience and set up methods using another sturgeon species as a biological model, the Siberian sturgeon, *Acipenser baerii* Brandt 1869. Not only did the species play its role of model species almost to perfection, but also the further economic development achieved by farming the species enabled the leading French institute to obtain a new research facility. And this proved to be a key point for building the ex situ European sturgeon brood stock. Another fruitful consequence was that conservation and farming sturgeon activities enabled us to hold ACIPENSER, the first International Symposium on sturgeon (ISS1) in Bordeaux in 1989 (Williot (ed) 1991).

The evaluation of the primary project submitted to Springer was rapid, positive and accompanied by two requests. The first was to include the restoration plan of the Atlantic sturgeon (*Acipenser oxyrinchus*) in the US, and the second to update the data regarding the status of the European sturgeon in the River Rioni in Georgia. No recent paper has been published on the status of the species in Georgia since the synthesis by Ninua (1976).

At the time, it had been known for some years that the Atlantic sturgeon (*A. oxyrinchus*) inhabited the Baltic Sea (Ludwig et al. 2002; Tiedemann et al. 2007). Consequently, restoration actions for the species have started in both Poland and Germany. At almost the same time as our book project got underway, a French archaeozoologist was publishing a scoop showing that the Atlantic sturgeon had inhabited the French Atlantic coasts for a long time (Desse-Berset 2009), from 5000 YBP to 300 YAD, which strongly suggested that the two species were sympatric in the area. We therefore seized the opportunity offered by the Springer incentive, and decided to broaden the Atlantic sturgeon issues in this book by introducing several actions in favour of the species, and looking for somebody able to provide the latest available data from the Georgian population. So this book is timely in taking into account the new situation of sturgeons in Western Europe, especially in France. It is worth noting that it was only recently in the 1960s that the two species were recognised as different (Magnin 1962, 1964; Magnin and Beaulieu 1963). And further genetic investigations have constantly pointed out the great proximity of the two species (e.g., Fontana et al. 2008).

Clearly, such a project would have missed its objectives without the participation of co-editors, including the unusual involvement of an archaeozoologist. This illustrates the need to open up conservation biology to new fields, and the need for cooperation, as mentioned above. All of them have brought very decisive inputs to the book, which would have been much less valuable otherwise. I would like to gratefully acknowledge Eric, Nathalie, Jörn and Frank for their enthusiasm, kindness and efficiency in contributing to our work.

The contents of the book were based on the following central themes: (a) providing a book on conservation in practice, with the European sturgeon as an example, (b) updating knowledge in the field, (c) broadening the range of concerned disciplines, (d) highlighting aspects that are not strictly scientific, (e) assuming new issues from the presence of two sturgeon species, the European and the Atlantic sturgeons, and (f) showing internal coherence, illustrated by the numerous cross-references.

The very great majority of those we contacted agreed to contribute. The contributors are greatly acknowledged, all the more so since many of them provided unpublished inputs, either results or analysis.

The book is divided into six unequal parts. The first part deals with all the available biological data on species population. A very brief introduction provides the reader unfamiliar with sturgeon with some basic characteristics, with a special focus on the European sturgeon. In addition to cytogenetic matters, a large section is devoted to geographical extension, present and past, with different approaches, literature, archaeozoology, and palaeogenetics, to the history of populations and fisheries in diverse countries, especially those for which there was no recent synthesis, and to some specific biological traits. As the Atlantic sturgeon was "invited" to participate in the book, two chapters deal with comparisons of the two species in terms of morphology and osteometry of the bones and morphology for the juveniles.

It is noteworthy that most of the physiological functions of the species were, and still are, to a great extent under-documented, e.g., reproduction, osmoregulation, endocrinology, nutrition, and haematology, with the exception of a preliminary investigation on the hydromineral balance (Magnin 1962), and on reproductive endocrinology (Davail-Cuisset et al. 2008).

The second part is an illustrated interlude devoted to part of the iconography of the European sturgeon, i.e., mainly photos and stamps. The sources are limited to France for the photos.

The third part focuses on the restoration–conservation actions that have been undertaken in France and later in Germany. Apart from the two chapters that draw an historic overview of European sturgeon-related management issues in France and Germany, the other chapters can be arranged under three topics dealing with: population, ex situ efforts, and efforts relying rather on non-scientific actions. The chapters reporting on population studies focus on methods and their applications, e.g., age determination, marking and tagging, monitoring of post-release sturgeon, and an attempt to model the future of these released fish. The ex situ actions cover sex determination and maturation staging, reproduction with two alternatives, wild and farmed brood fish because the management logics are different, larval rearing, brood stock building, its genetic variability, and sperm cryoconservation. The third topic in this part deals with the role of non-governmental stakeholders, awareness campaigns among marine professional fishermen, and a synthesis on French–German cooperation.

The fourth part is focused entirely on Atlantic sturgeon management, illustrated by fisheries in Québec (Ca), conservation in the US, and restoration in the Baltic Sea.

Some perspectives and prospective concerns are described in part five. They deal with the in situ life history, the potential effects of climate change, a population viability analysis, some preliminary investigation into cryobanking of somatic cells, and biological variability.

Finally, part six contains some thoughts on the future of the species.

As a preface to part 1, I am very pleased to dedicate the book to the memory of two renowned scientists because the present European sturgeon combined biology–restoration programme benefited from their previous studies and investment in the species. Dr Etienne Magnin started his career in France by publishing the first documented compendium on the biology of the European sturgeon and then went on to work in Québec, Canada where he also spent time working on sturgeon, especially (but not only) the Atlantic sturgeon. He was responsible for the first documented distinction between the two species, the European and Atlantic sturgeons. E. Magnin unfortunately missed our invitation to participate in ACIPENSER, the ISS1 in 1989. The second scientist is Dr Juraj Holčik, well known for many important works, but in particular for publishing The Freshwater Fishes of Europe, focusing on Petromyzontiformes and Acipenseriformes, with the last edition published in 1989. He personally wrote with colleagues the chapter that dealt with the European sturgeon, and continued to be involved in those issues. Due to health problems, he declined our invitation to contribute to the present book. Many thanks are due to Pierre Dumont and Peter Harper for the lines on E. Magnin. I am very grateful to Kristina Holčik for her kind offer of a text on Juraj's activities.

Audenge, France Patrick Williot

References

Boreman J (1997) Sensitivity of North American sturgeons and paddlefish to fishing mortality. Environ Biol Fish 48:399–405

Classen TEA (1944) Estudio bio-estadistico del esturion o sollo del Guadalquivir. Instituto Espanol de Oceanografia, Ministerio de Marina, Trabajos, N 19, 112 p + XVII planches

Davail-Cuisset B, Lacomme S, Viaene E, Williot P, Lepage M, Gonthier P, Davail S, Rouault T (2008) Hormonal profile in adults of Atlantic European sturgeon, *Acipenser sturio*, adapted to hatchery in France. Cybium 32(2 suppl):169–170

Desse-Berset N (2009) First archaeozoological identification of Atlantic sturgeon (*Acipenser oxyrinchus* Mitchill 1815) in France. CR Palevol 8:717–724

Fontana F, Lanfredi M, Kirschbaum F, Garrido-Ramos MA, Robles F, Forlani A, Congiu L (2008) Comparison of karyotypes of *Acipenser oxyrinchus* and *A. sturio* by chromosome banding and fluorescent in situ hybridization. Genetica 1342:281–286

Holčik J, Kinzelbach R, Sokolov LI, Vassilev V P (1989) *Acipenser sturio* Linnaeus, 1758. In: Holčik J (ed) The freshwater fishes of Europe. Aula Verlag, Wiesbaden, pp 367–394

Jager HI, Rose KA, Vila-Gispert A (2008) Life history correlates and extinction risk of capital-breeding fish. In: Dufour S, Prévost E, Rochard E, Williot P (eds) Fish and diadromy in Europe (ecology, management, conservation). Hydrobiologia 602:15–25

Ludwig A, Debus L, Lieckfeldt D, Wirgin I, Benecke N, Jenneckens I, Williot P, Waldman JR, Pitra C (2002) When the American sea sturgeon swam east. Nature 419:447–448

Magnin E (1959) Répartition actuelle des Acipenséridés. Rev Trav Inst Pêches Marit 23(3): 277–285

Magnin E (1962) Recherches sur la systématique et la biologie des Acipenséridés *Acipenser sturio* L., *Acipenser oxyrinchus* Mitchill et *Acipenser fulvescens* Raf. Ann Station Centr Hydrobiol Appl 9:7–242

Magnin E, Beaulieu G (1963) Etude morphométrique comparée de *l'Acipenser oxyrinchus* Mitchill du Saint Laurent et de l'*Acipenser sturio* Linné de la Gironde. Le Naturaliste Canadien XC (1):5–38

Magnin E (1964) Validité d'une distinction spécifique entre les deux Acipenséridés: *Acipenser sturio* L. d'Europe et *Acipenser oxyrinchus* d'Amérique du Nord. Le Naturaliste Canadien XCI (1):5–20

Ninua NSh (1976) Atlantic sturgeon of the Rioni River. Editions Metsniereba, Tbilissi, pp 122 (in Russian)

Rochard E, Castelnaud G, Lepage M (1990) Sturgeons (Pisces: Acipenseridae); threats and prospects. J Fish Biol 37A:123–132

Tiedemann R, Moll K, Paulus KB, Scheer M, Williot P, Bartel R, Gessner J, Kirschbaum F (2007) Atlantic sturgeons (*Acipenser sturio, Acipenser oxyrinchus*): American females successful in Europe. Naturwissenschaften 94:213–217. doi: 10.1007/s0014-006-0175-1

Williot P (ed) (1991) Acipenser. Actes du premier colloque international sur l'esturgeon. Cemagref, Antony, 518p

Williot P, Arlati G, Chebanov M, Gulyas T, Kasimov R, Kirschbaum F, Patriche N, Pavlovskaya L, Poliakova L, Pourkazemi M, Kim Yu, Zhuang P, Zholdasova IM (2002) Status and management of Eurasian sturgeon: an overview. Int Rev Hydrobiol 87:483–506

Williot P, Rouault T, Pelard M, Mercier D, Jacobs L (2009) Artificial reproduction of captive endangered European Atlantic sturgeon, *Acipenser sturio*. Endang Species Res 6:251–257. doi: 10.3354/esr00174

Acknowledgements

We would like to thank all the colleagues at Cemagref who spent time and effort into this long-term programme, and more specifically Chantal Gardes for her decisive help with documentation and Philippe Camoin for his map drawing. I am also thankful to the EPBx research unit of Cemagref for the logistic help. Thanks are due to Antoine Pasqualini and Jean-Denys Strich (CEPAM-CNRS-Nice University) for their inputs in infography and photography, to Tamas Gulyas for his previous support in sperm cryoconservation and androgenesis investigations.

This project would not exist without the efficient and kind help of Verena Penning, Anette Lindqvist, and Athiappan Kumar from Springer.

Contents

Contributors

M.L. Acolas Cemagref, Estuarine Ecosystems and Diadromous Fish Research Unit, 50 Avenue de Verdun, 33612 Cestas Cedex, France, marie-laure.acolas@cemagref.fr

Gerd-Michael Arndt Fisch und Umwelt Mecklenburg-Vorpommern e.V., Fischerweg 408, 18069 Rostock, Germany

Ryszard Bartel Instytut Rybactwa Śródlądowego, Zakład Ryb Wędrownych, ul. Synów Pułku 37, 80-298 Gdańsk, Poland

M. Béguer Cemagref, Estuarine Ecosystems and Diadromous Fish Research Unit, 50 Avenue de Verdun, 33612 Cestas Cedex, France, melanie.beguer@cemagref.fr

Patrick Berrebi Institut des Sciences de l'Evolution, UMR 5554 CNRS/UM2/IRD, Université Montpellier 2, cc065, Place Bataillon, 34095 Montpellier Cedex 05, France, patrick.berrebi@univ-montp2.fr

Paolo Bronzi World Sturgeon Conservation Society, via Piave, 8, 20854 Vedano al Lambro, Italy, aerespe@mac.com

Laurent Brosse Aqua-Logiq, 527 rue Zac Petite Camargue, 34400 Lunel, France, laurent.brosse@aqua-logiq.fr

Rémi Brun Cemagref, Estuarine Ecosystems and Diadromous Fish Research Unit, 50 Avenue de Verdun, 33612 Cestas Cedex, France

Giuseppe Castaldelli Department of Biology and Evolution, University of Ferrara, via L. Borsari, 46, 44121 Ferrara, Italy

Gérard Castelnaud Cemagref, Estuarine Ecosystems and Diadromous Fish Research Unit, 50 Avenue de Verdun, 33612 Cestas Cedex, France, gerard.castelnaud@cemagref.fr

Stefano Cataudella Department of Biology, University of Tor Vergata, via della Ricerca Scientifica, 00173 Rome, Italy

Olivier Chassaing Paléogénétique et Evolution Moléculaire, Institut de Génomique Fonctionnelle de Lyon, Université de Lyon, Université Lyon 1, CNRS, INRA, Ecole Normale Supérieure de Lyon, 46 allée d'Italie, 69364 Lyon Cedex 07, France; Institut des Sciences de l'Evolution, UMR 5554 CNRS/UM2/IRD, Université Montpellier 2, cc065, Place Bataillon, 34095 Montpellier Cedex 05, France

Patrick Chèvre Cemagref, Estuarine Ecosystems and Diadromous Fish Research Unit, Moulin de la logerie, 33660 Saint-Seurin-sur-l'Isle, France; Cemagref, CREA de Saint-Seurin-sur-l'Isle, Moulin de la logerie, 33660 Saint-Seurin-sur-l'Isle, France, patrick.chevre@cemagref.fr

G. Cvijanović Institute for Multidisciplinary Research, Kneza Višeslava 1, 11000 Belgrade, Serbia, mitrandir@imsi.rs

Alexandra Depince INRA, UR1037 SCRIBE, Cryopreservation and Regeneration, F-35000, Rennes, France

Nathalie Desse-Berset Université de Nice-Sophia Antipolis – CNRS-CEPAM-UMR 6130, SJA3, 24 avenue des Diables Bleus, 06357 NICE Cédex 4, France, berset.desse@hotmail.com; berset@cepam.cnrs.fr

Arkadiusz Duda Department of Ichthyology, Inland Fisheries Institute, Oczapowskiego 10, 10-719 Olsztyn-Kortowo, Poland

Marilyne Duffraisse Paléogénétique et Evolution Moléculaire, Institut de Génomique Fonctionnelle de Lyon, Université de Lyon, Université Lyon 1, CNRS, INRA, Ecole Normale Supérieure de Lyon, 46 allée d'Italie, 69364 Lyon Cedex 07, France

Francesco Fontana Department of Biology and Evolution, University of Ferrara, Via L. Borsari 46, 44100 Ferrara, Italy, francesco.fontana@unife.it

Frank Fredrich Leibniz-Institut für Gewässerökologie und Binnenfischerei, Müggelseedamm 310, 12587 Berlin, Germany

Christine Gazeau Cemagref, Estuarine Ecosystems and Diadromous Fish Research Unit, 33612 Cestas Cedex, France

Jörn Gessner Leibniz-Institute of Freshwater Ecology and Inland Fisheries, Müggelseedamm 310, 12587 Berlin, Germany, sturgeon@igb-berlin.de

Catherine Hänni Paléogénétique et Evolution Moléculaire, Institut de Génomique Fonctionnelle de Lyon, Université de Lyon, Université Lyon 1, CNRS, INRA, Ecole Normale Supérieure de Lyon, 46 allée d'Italie, 69364 Lyon Cedex 07, France

Ákos Horváth Department of Aquaculture, Szent István University, 2100 Gödöllő, Páter K. u. 1., Hungary, Horvath.Akos@mkk.szie.hu

Louis Jacobs Cemagref, Estuarine Ecosystems and Diadromous Fish Research Unit, Moulin de Logerie, 33660 Saint-Seurin-sur-l'Isle, France

J. Knežević-Jarić Ecological Society "Endemit", Oračka 42, 11080 Belgrade, Serbia, jelak@isp.b92.net

Ivan Jarići Institute for Multidisciplinary Research, Kneza Višeslava 1, 11000 Belgrade, Serbia, jaric@imsi.rs

Philippe Jatteau Cemagref, Estuarine Ecosystems and Diadromous Fish Research Unit, 33612 Cestas Cedex, France, philippe.jatteau@cemagref.fr

Andrzej Kapusta Department of Ichthyology, Inland Fisheries Institute, Oczapowskiego 10, 10-719 Olsztyn-Kortowo, Poland

Frank Kirschbaum Faculty of Agriculture and Horticulture, Biology and Ecology of Fishes, Humboldt University, Philippstr. 13, Haus 16, 10115 Berlin, Germany, Frank.kirschbaum@staff.hu-berlin.de; frank.kirschbaum@staff.hu-berlin.de

Ryszard Kolman Department of Ichthyology, Inland Fisheries Institute, Oczapowskiego 10, 10-719 Olsztyn-Kortowo, Poland, kolrys@infish.com.pl

Catherine Labbe INRA, UR1037 SCRIBE, Cryopreservation and Regeneration, F-35000, Rennes, France, Catherine.Labbe@rennes.inra.fr

Patrick Lambert Cemagref, Estuarine Ecosystems and Diadromous Fish Research Unit, 50 Avenue de Verdun, 33612 Cestas Cedex, France

Géraldine Lassalle Cemagref, Estuarine Ecosystems and Diadromous Fish Research Unit, 50 avenue de Verdun, 33612 Cestas Cedex, France; UMR 6250 LIENSs, University of La Rochelle – CNRS, Institut du Littoral et de l'Environnement, 2 rue Olympe de Gouges, 17000 La Rochelle, France, geraldine.lassalle@univ-lr.fr

Pierre-Yves Le Bail INRA, UR1037 SCRIBE, Cryopreservation and Regeneration, F-35000, Rennes, France

Mirjana Lenhardt Institute for Biological Research, Despota Stefana 142, 11000 Belgrade, Serbia

Mario Lepage Cemagref, Estuarine Ecosystems and Diadromous Fish Research Unit, 50, Avenue de Verdun, 33612 Cestas Cedex, France, mario.lepage@cemagref.fr

Aude Lochet Cemagref, Estuarine Ecosystems and Diadromous Fish Research Unit, 33612 Cestas Cedex, France

Arne Ludwig Department of Evolutionary Genetics, Institute for Zoo and Wildlife Research, Alfred-Kowalke-Str. 17, 10315 Berlin, Germany, ludwig@izw-berlin.de

Daniel Mercier Cemagref, Estuarine Ecosystems and Diadromous Fish Research Unit, Moulin de logerie, 33660 Saint-Seurin-sur-l'Isle, France

Nicolas Michelet French National Committee for Marine Fisheries and Sea Farming (CNPMEM), 134 Avenue de Malakoff, 75116 Paris, France, nmichelet@comite-peches.fr

Arturo Morales-Muñiz Laboratorio Arqueozoologia, Universidad Autónoma de Madrid, Darwin, 2, 28049 Madrid, Spain

Didier Moreau WWF-France (consultant), 34 rue des Souchères, 26110 Nyons, France, dmoreau@wwf.fr

Jacek Morzuch Department of Ichthyology, Inland Fisheries Institute, Oczapowskiego 10, 10-719 Olsztyn-Kortowo, Poland

Marcel Pelard Cemagref, Estuarine Ecosystems and Diadromous Fish Research Unit, Moulin de logerie, 33660 Saint-Seurin-sur-l'Isle, France

Gaël Piquès Archéologie des sociétés Méditerranéennes, UMR 5140 CNRS, 390 Avenue de Pérols, 34970 Lattes, France

Stefan Reiser Institute for Hydrobiology and Fisheries Science, University of Hamburg, Olbersweg 24, 22767 Hamburg, Germany

Eric Rochard Cemagref, Estuarine Ecosystems and Diadromous Fish Research Unit, 50 Avenue de Verdun, 33612 Cestas Cedex, France, eric.rochard@cemagref.fr

Charles Roqueplo Cemagref, Estuarine Ecosystems and Diadromous Fish Research Unit, 50 Avenue de Verdun, 33612 Cestas Cedex, France

Eufrasia Roselló-Izquierdo Laboratorio Arqueozoologia, Universidad Autónoma de Madrid, Darwin, 2, 28049 Madrid, Spain

Remiggio Rossi Department of Biology and Evolution, University of Ferrara, via L. Borsari, 46, 44121 Ferrara, Italy

Thierry Rouault Cemagref, Estuarine Ecosystems and Diadromous Fish Research Unit, Moulin de logerie, 33660 Saint-Seurin-sur-l'Isle, France

E. Rouleau Cemagref, Estuarine Ecosystems and Diadromous Fish Research Unit, 50 Avenue de Verdun, 33612 Cestas Cedex, France

Julien Saint-Sevin Cemagref, Estuarine Ecosystems and Diadromous Fish Research Unit, Moulin de Logerie, 33660 Saint-Seurin-sur-l'Isle, France

Anja Schneider Unit of Evolutionary Biology/Systematic Zoology, University of Potsdam, Karl-Liebknecht-Straße 24-25 (Haus 26), 14476 Potsdam, Germany

S. Spratte Landesamt für Landwirtschaft, Umwelt und ländliche Räume des Landes Schleswig-Holstein (LLUR), Abt. 3 Fischerei, Dezernat Binnenfischerei und Aquakultur, Hamburger Chaussee 25, 24220 Flintbek, Germany

Ana Stanković Department of Biology, Institute of Genetics and Biotechnology, University of Warsaw and Institute of Biochemistry and Biophysics, Polish Academy of Sciences, Pawińskiego 5A, 02-106 Warsaw, Poland, anams@ibb.waw.pl

Catherine Taverny Cemagref, Estuarine Ecosystems and Diadromous Fish Research Unit, 50 Avenue de Verdun, 33612 Cestas Cedex, France

Ralph Tiedemann Unit of Evolutionary Biology/Systematic Zoology, Institute of Biochemistry and Biology, University of Potsdam, Karl-Liebknecht-Straße 24-25 (Haus 26), 14476 Potsdam, Germany, tiedeman@rz.uni-postdam.de

Guy Trencia Ministère des Ressources Naturelles et de la Faune, 8400 Sous-le-Vent, Charny, QC, Canada, G6X 3S9

Béla Urbányi Department of Aquaculture, Szent István University, 2100 Gödöllő, Páter K. u. 1., Hungary

Guy Verreault Ministère des Ressources Naturelles et de la Faune, 186 rue Fraser, Rivière-du-Loup, QC, Canada, G5R 1C8, Guy.verreault@mrnf.gouv.qc.ca

Henning von Nordheim Fachgebiet Meeres- und Küstennaturschutz, Bundesamt für Naturschutz, Außenstelle Insel Vilm, 18581 Putbus, Germany

John Waldman Biology Department, Queens College, 65–30 Kissena Boulevard, Flushing, New York, NY 11367, USA, john.waldman@qc.cuny.edu

Patrick Williot Sturgeon Consultant, 4 Rue du pas de madame, 33980 Audenge, France, williot.patrick@neuf.fr

Sven Wuertz Leibniz-Institute of Freshwater Ecology and Inland Fisheries, Müggelseedamm 310, 12587 Berlin, Germany, sveno@igb-berlin.de

Part I
Biology, History, Geographical Distribution and Status, from Past to Present

Chapter 1
Brief Introduction to Sturgeon with a Special Focus on the European Sturgeon, *Acipenser sturio* L. 1758

P. Williot, E. Rochard, N. Desse-Berset, J. Gessner, and F. Kirschbaum

1.1 Introduction

Sturgeon comprise a small family of fish with 4 genera and 27 species. The genus *Acipenser* accounts for 17 species. In general, they are known for their large size, their unusual appearance, and their famous fish product, caviar. They are ancient fish ranging only over the northern hemisphere. All Eurasian sturgeon species are currently under threat, while the situation is slightly better in America. The European sturgeon colonized the continent from the Black Sea to the Baltic, through the Mediterranean, the Western Atlantic Ocean, the English Channel, and the North Sea (Fig. 1.1).

The aim of this very brief overview is to provide the reader who may be unfamiliar with sturgeon with a minimum of information about this fish, briefly introducing its morphology, some relevant biological characteristics, its taxonomy and origin, behaviour, reproductive characteristics, and utilization of sturgeon, with

P. Williot (✉)
Sturgeon Consultant, 4 Rue du Pas de madame, 33980 Audenge, France
e-mail: williot.patrick@neuf.fr

E. Rochard
Cemagref, Estuarine Ecosystems and Diadromous Fish Research Unit, 50 Avenue de Verdun, 33612 Cestas Cedex, France

N. Desse-Berset
Université de Nice-Sophia Antipolis – CNRS-CEPAM-UMR 6130, SJA3, 24 avenue des Diables Bleus, 06357 NICE Cédex 4, France

J. Gessner
Leibniz-Institute of Freshwater Ecology and Inland Fisheries, Müggelseedamm 310, 12587 Berlin, Germany

F. Kirschbaum
Faculty of Agriculture and Horticulture, Biology and Ecology of Fishes, Humboldt University, Philippstr. 13, Haus 16, 10115 Berlin, Germany

P. Williot et al. (eds.), *Biology and Conservation of the European Sturgeon Acipenser sturio L. 1758*, DOI 10.1007/978-3-642-20611-5_1,

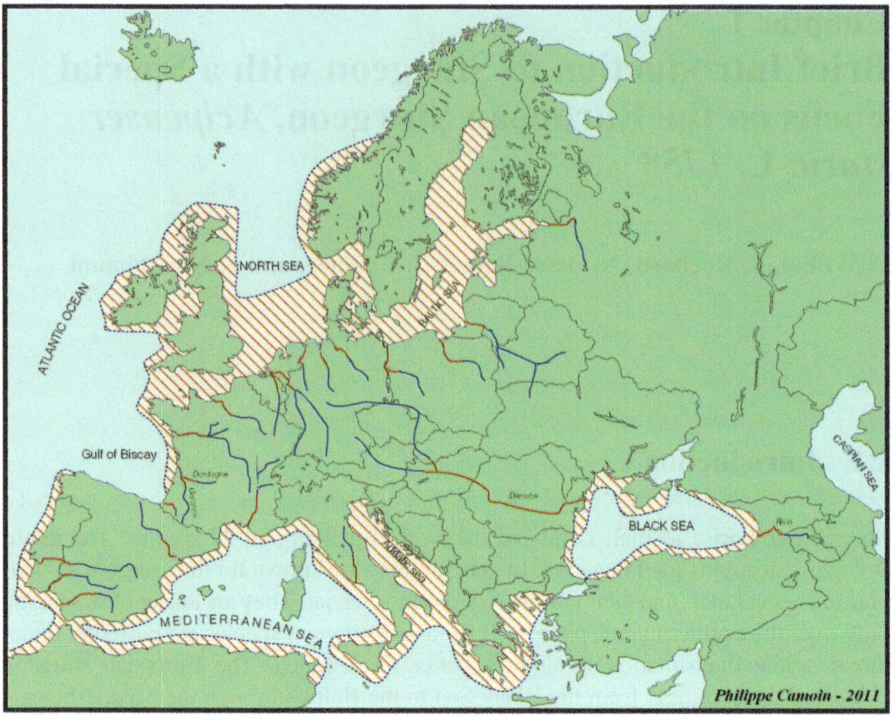

Fig. 1.1 Previous geographical distribution of *Acipenser sturio*

a focus on the European sturgeon. Some characteristics will be covered in more detail and discussed in several chapters in the present volume.

1.2 General Morphological Description

The body is elongated and spindle-shaped. Five longitudinal rows of bony scutes cover the body (one dorsal, two lateral, and two ventral), which constitute the remains of an exoskeleton (Fig. 1.2). In some rare cases, the number of scutes can be a discriminating characteristic between sturgeon species. The rest of the skin is covered by small denticles and platelets. The mouth is positioned ventrally on the underside of the head and is surrounded by fleshy lips. It is protrusible, located behind four barbels which carry mechanoreceptors and taste buds (Fig. 1.2), indicating that most sturgeons are benthic feeders. In two piscivorous species, *Huso huso* (beluga in the Ponto Caspian region) and *Huso dauricus* (kaluga in the River Amur), the mouth is crescent-shaped and can be projected in front of the fish to catch their prey. They have no teeth except at the larval stage. The skull is cartilaginous but covered by bony dermal plates. The snout is elongated, conic, and spatulate, varying

Fig. 1.2 Two spermiating *A. sturio* males. The five rows of scutes are clearly visible in both photographs. The mouth, barbells, snout, and the uro-genital pore between the two pelvic fins are shown on the upper photo. The positions of all fins can be seen in the lower photo, especially the heterocercal caudal fin (Credit Cemagref)

in length in proportion to the length of the head. The caudal fin is heterocercal, with the upper part being elongated. The first ray of the pectoral fins is transformed into a strong spine with a bony structure (Grassé 1958; Sokolov and Berdichevskii 1989).

The size of the adults is species-dependent and ranges widely, with the smallest being the *Pseudoscaphirhynchus* sp. of the Aral Sea system with a maximum length close to 0.5 m, and the largest, *H. huso*, reaching 8.5 m (review in Rochard et al. 1991). The European sturgeon, with a maximum length of 5 m (Laporte 1853), is one of the largest sturgeon species. It is the largest fish species in European fresh waters. The decreasing maximum length observed in the course of the last century, illustrated by the maximum length of ≅255 cm for a female in the very early 1960s (Magnin 1962), is caused by high fishing pressure and hence lower maximum age.

1.3 Taxonomic Position and Origin

The taxonomic position is summarized in Table 1.1, with data from Magnin (1962), Sokolov and Berdichevskii (1989), and Birstein and DeSalle (1998). These last two authors introduced an additional taxon by adding two subfamilies, *Acipenserinae* and *Scaphirynchinae*. The family Acipenseridae is a semi-ossified fish group, and is represented by approximately 25 extant species (Birstein and DeSalle 1998). The status of some species is considered debatable (Birstein and Bemis 1997).

The existence of Acipenseriformes is reported in the lower Jurassic, i.e. 200–175 MYBP, according to Bemis et al. (1997), Peng et al. (2007), and Birstein and DeSalle (1998).

Based on an interpretation of a phylogenetic relationship within the genus *Acipenser*, Bemis et al. (1997) situated the European sturgeon, *A. sturio*, in one clade with the Atlantic sturgeon, *A. oxyrinchus*. In a study on molecular phylogeny of *Acipenserinae*, Birstein and DeSalle (1998), using different nuclear DNA sequences, came to the same conclusion. Using different mtDNA primers, Krieger et al. (2008) clearly identified a common clade for both species. Using synapomorphies, Choudhury and Dick (1998) settled *A. sturio* and *A. oxyrinchus* in the same clade as well. Both species are very similar from the cytogenetic point of view (Fontana et al. 2008). Their external morphology is quite similar too, so this explains why the two species were not distinguished until recently. Indeed, in his review of sturgeon species, Magnin (1959) mentioned the possible presence of *A. sturio* in Northern America. The distinction between the two species was established soon after (Magnin 1962, 1964; Magnin and Beaulieu 1963). It is worth noting that this differentiation is still under debate (Artyukhin and Vecsei 1999). The controversy is not surprising, as the European sturgeon has exhibited a large variability, and there have been suggestions of the former existence of 9–12 possibly reproductively isolated populations with a disrupted gene flow (Holčik 2000).

Peng et al. (2007) suggest that the separation between the two species *A. sturio* and *A. oxyrinchus* occurred at a mean of 57.9 (23.4–112.2) MYBP, while Birstein and DeSalle (1998) determined the origin of the *A. sturio* lineage in the upper cretaceous, i.e. ~95 MYBP.

Table 1.1 Taxonomic position of the European sturgeon, *Acipenser sturio* L, 1758

Taxis	Name	Observations	Geographical area
Super class	Pisces		
Class	Osteichthyes		
Sub-class	Actinopterygii		
Super order	Chondrostei	Semi-ossified fish	
Order	Acipenseriformes	27 species	Worldwide (North)
Family	Acipenseridae	25 species	
Genus	*Acipenser*		Worldwide (North)
Species	*Acipenser sturio* Linnaeus, 1758		Europe

1.4 Some Biological Peculiarities

Most of the biological peculiarities observed in sturgeon are typical for Chondrostei. The skeleton is only partially ossified. The axial skeleton is cartilaginous, and the notochord is persistent. The intestine possesses a spiral valve, which is very easy to see in the larval stage. The swim bladder is connected to the oesophagus (physostoma). The spiracle and the relict gill arch are present, especially in *Acipenserinae*. The anatomy of their reproductive systems is unusual, as both Müllerian and Wolfian ducts are present. Sturgeons are long-lived fish with late puberty (7–22 year-old for *A. sturio*) and non-yearly oogenesis in wild populations.

Despite the fact that Acipenseriformes have many primitive characteristics (see above) which classify them as ancient fish, midway between Chondrichthyes and Teleosts, the group exhibits both secondary-reduced characteristics (reduction in ossification, absence of branchiostegal rays) but also specialized characteristics (snout, ventral mouth, barbels). Therefore, Acipenseriformes should not be considered as a primitive group, but rather as a finished evolutive series (Grassé 1958; Magnin 1962). The current conventional view states that Acipenseriformes evolved from a paleonisciformes ancestor via a paedomorphic reduction of the skeleton and specialization of the feeding system (Bemis et al. 1997).

The sturgeon karyotype presents some peculiarities with a large number of chromosomes ($2n \cong 120$ and $2n \cong 240$, and even $2n \cong 340$ for one species), about half of which are microchromosomes (Fontana 2002). The European sturgeon belongs to the $2n \cong 120$ group (Fontana and Colombo 1974; Tagliavini et al. 1999); however, Tagliavini et al. refute the argument that the species "could be closely related to primitive species within the genus *Acipenser*", at least from the cytogenetic point of view.

1.5 Main Behaviour Patterns

All sturgeons spawn in fresh water. Most are anadromous, i.e. they migrate from freshwater to brackish or salt water during their active growth period, and Bemis and Kynard (1997) placed many species in this group. However, a few of these spend their intense growth period in high salinity water, such as *A. sturio* and its sister species *A. oxyrinchus*, as well as *A. sinensis* and *A. medirostris*, while most anadromous species are confined to either estuaries or low salinity waters, as exampled by the Adriatic sturgeon and the Ponto-Caspian species respectively.

Sturgeons are not known to utilize deep environments while at sea, and they do not in general make extensive offshore migrations (Bemis and Kynard 1997). European sturgeon, and others too, do not utilize waters exceeding 100–200 m in depth (Letaconnoux 1961; Grubisic 1967 in Holčik 1989; Rochard et al. 1997).

As a result of genetic investigations, it is known that homing rates, at least for some species, are very high (>94% per generation) (Waldman et al. 2002; Grunwald et al. 2008; Peterson et al. 2008; Homola et al. 2010). Previous marked–recaptured studies have shown that very few fish from the European French Atlantic population travel very long distances (Castelnaud et al. 1991; Rochard et al. 1997).

Some sturgeons are characterized by their intra-population structure; this concerns anadromous species, especially the Ponto-Caspian species (Kozhin 1964). There are seasonal forms that migrate upstream either in autumn or spring and spawn either the next spring or straight away at migration. Sokolov and Berdichevskii (1989) argue that these characteristics allow the adults and the fingerlings to optimize the use of available spawning grounds and food resources in a river system. There is no support for such an ecological trait in the European sturgeon.

However, the French Atlantic European sturgeon population in the Garonne basin is known to migrate from the adjacent continental shelf of the Gulf of Biscay into the lower part of the Gironde estuary during summer time until autumn during their juvenile phase, i.e. between 3 and 8 years old (Castelnaud et al. 1991).

It is generally observed that at least two sturgeon species are in sympatry, i.e. they cohabit the same river basin. The best examples are the Po River, the Danube River, and the Volga River, where up to 6–7 species have been present simultaneously. Numerous natural hybridizations have been reported (Kozhin 1964; Rochard et al. 1991), without any deleterious effects on each species. Bemis and Kynard (1997) pointed out the peculiarity of the situation, with *A. sturio* being the only sturgeon species in Western Europe, particularly in Western France and the Iberian Peninsula.

At least for France, the situation has potentially changed recently with the discovery of Atlantic sturgeon remains from between 5000 YBP up to 1700 AD on the French Atlantic coast and the Channel (Desse-Berset 2009; Desse-Berset and Williot 2011). This strongly suggests sympatry between the two species in these areas.

1.6 Reproductive Characteristics

Sturgeon are gonochorists. However, there is a disposition to hermaphroditism due to their anatomy. Indeed, functional hermaphroditism has been obtained in farmed sterlet, *Acipenser ruthenus* (Williot et al. 2005). There is no reliable sexual dimorphism. When gametogenesis is complete, the gonads represent, on average, 10–20% of body weight.

Sturgeon are a lithopelagophil species that spawn on rock and gravel bottoms, and their larvae are pelagic (Holčik 1989). The envelope of the fertilized eggs adheres to the gravel or stone (Holčik et al. 1989) allowing the embryogenesis to develop in well-oxygenated conditions. The size of the eggs is species-dependent, ranging from about 2 to 4 mm. European sturgeon eggs currently measure about 2.5 mm. In case of hybridization, the fertility of the progeny depends on the parents having the same ploidy. Puberty is late, and the females do not spawn each year.

1.7 Uses

European sturgeons were exploited for consumables and several other by-products in the countries which had significant sturgeon populations (Poland, Germany, the Netherlands, France, Spain, and Italy). The meat was very much appreciated in countries where there were traditional sturgeon fisheries. It was eaten either fresh, pickled, or smoked. The skin was used as leather. The swim bladder was used to prepare isinglass (fish glue), renowned in cabinet-making as well as being a clarifying agent (Sauvage 1883; Guénaux 1923 among others). Interestingly, these uses are mentioned as early as ancient times (Desse-Berset 1994). Although not the only reason, the lure of caviar was responsible for the dramatic decline in European sturgeon populations. Because of their biological traits in particular, sturgeon species are very sensitive to fishing mortality (Boreman 1997), and this life history, typical of capital-breeding species, meant that sturgeon were at much greater risk of extinction (Jager et al. 2008).

The most expensive product originating from sturgeons is caviar. Due to the collapse of most of the previous sturgeon fisheries, especially in the Ponto-Caspian region (Williot et al. 2002), the great majority of caviar marketed today is from aquaculture.

References

Artyukhin E, Vecsei P (1999) On the status of Atlantic sturgeon: conspecificity of European *Acipenser sturio* and North American *Acipenser oxyrinchus*. J Appl Ichthyol 15:35–37

Bemis WE, Kynard B (1997) Sturgeon rivers: an introduction to Acipenseriform biogeography and life history. Environ Biol Fishes 48:167–183

Bemis WE, Findeis EK, Grande L (1997) An overview of Acipenseriformes. Environ Biol Fishes 48:25–71

Birstein VJ, Bemis WE (1997) How many species are there within the genus *Acipenser*? Environ Biol Fishes 48:157–163

Birstein VJ, DeSalle R (1998) Molecular phylogeny of Acipenserinae. Mol Phylogenet Evol 9:141–155

Boreman J (1997) Sensitivity of North American sturgeons and paddlefish to fishing mortality. Environ Biol Fishes 48:399–405

Castelnaud G, Rochard E, Jatteau P, Lepage M (1991) Données actuelles sur la biologie d'*Acipenser sturio* dans l'estuaire de la Gironde. In: Williot P (ed) Acipenser. Cemagref, Antony, pp 251–275

Choudhury A, Dick TA (1998) The historical biogeography of sturgeons (Osteichtyes: Acipenseridae): a synthesis of phylogenetics, palaeontology and palaeogeography. J Biogeogr 25:623–640

Desse-Berset N (1994) Sturgeon of the Rhône during Protohistory in Arles (6th–2nd century BC). In: Fish exploitation in the past, Proceedings of the 7th meeting of the ICAZ Fish Remains Working Group (Louvain, Sept. 1993). Annales du Musée Royal de l'Afrique Centrale, Tervueren, vol 274, pp 81–90

Desse-Berset N (2009) First archaeozoological identification of Atlantic sturgeon (*Acipenser oxyrinchus* Mitchill 1815) in France. CR Palevol 8:717–724

Desse-Berset N, Williot P (2011) Emerging questions from the discovery of the long term presence of *Acipenser oxyrinchus* in France. J Appl Ichthyol 27:263–268, doi:10.1111/j.1439-0426.2010.01649.x

Fontana F (2002) A cytogenetic approach in the study of taxonomy and evolution in sturgeons. J Appl Ichthyol 18:226–233

Fontana F, Colombo G (1974) The chromosomes of Italian sturgeon. Experientia 30:739–742

Fontana F, Lanfredi M, Kirschbaum F, Garrido-Ramos MA, Robles F, Forlani A, Congiu L (2008) Comparison of karyotypes of *Acipenser oxyrinchus* and *A.sturio* by chromosome banding and fluorescent in situ hybridization. Genetica 1342:281–286

Grassé PP (sous la direction) (1958) Traité de Zoologie. Anatomie, systématique, Biologie. Agnathes et Poissons. Anatomie, Ethologie, systématique, tome XIII, fascicule III. Masson et Cie Editeurs, Paris, pp 1813–2758

Grunwald C, Maceda L, Waldman JR, Stabile J, Wirgin I (2008) Conservation of Atlantic sturgeon *Acipenser oxyrinchus oxyrinchus*: delineation of stock structure and distinct population. Conserv Genet 9:1111–1124

Guénaux P (1923) Les poissons d'eau douce dans leurs rapports avec la pêche et la pisciculture. Librairie Baillière et Fils, Paris, 144 p

Holčik J (2000) Major problems concerning the conservation and recovery of the Atlantic sturgeon *Acipenser sturio* L., 1758. Bol Inst Esp Oceanogr 16(1–4):139–148

Holčik J, Bănărescu P, Evans D (1989) General introduction to fishes. In: Holčik J (ed) The freshwater fishes of Europe. Aula Verlag, Wiesbaden, pp 18–147

Holčik J, Kinzelbach R, Sokolov LI, Vasil'ev VP (1989) Acipenser sturio Linnaeus, 1758. In/The freshwater fishes of Europe. General introduction to fishes Acipenseriformes. Holcik J (ed), AULA-Verlag Wiesbaden pp 367–394

Homola JJ, Scribner KT, Baker EA, Auer NA (2010) Genetic assessment of straying rates of wild and hatchery reared lake sturgeon (*Acipenser fulvescens*) in Lake Superior tributaries. J Great Lakes Res 36:798–802

Jager HI, Rose KA, Vila-Gispert A (2008) Life history correlates and extinction risk of capital-breeding fish. In: Dufour S, Prévost E, Rochard E, Williot P (eds) Fish and diadromy in Europe (ecology, management, conservation). vol 602. Springer, Dordrecht, pp 15–25

Kozhin NI (1964) Sturgeon from USSR and their reproduction. Trudy VNIRO LII:21–58 (in Russian)

Krieger J, Hett AK, Fuerst PA, Artyukhin E, Ludwig A (2008) The molecular phylogeny of the order Acipenseriformes revisited. J Appl Ichthyol 24(suppl 1):36–45

Laporte E (1853) Faune Ichtyologique du département de la Gironde. Actes de la Société Linnéenne de Bordeaux, tome XIX, Deuxième série, tome IX, Bordeaux, chez Th Lafargue, Libraire, 8 rue de Bagne-Cap, pp 157–224

Letaconnoux R (1961) Note sur la fréquence de la distribution des captures d'esturgeons (*Acipenser sturio* L) dans le golfe de Gascogne. Rev Trav Inst Pêches Marit 25:253–261

Magnin E (1959) Répartition actuelle des acipenseridés. Rev Trav Inst Pêches Marit 23(3): 277–285

Magnin E (1962) Recherche sur la systématique et la biologie des Acipenséridés *Acipenser sturio* L., *Acipenser oxyrinchus* Mitchill et *Acipenser fulvescens* Raf. Ann Stat Cent Hydrobiol Appl 9:7–242

Magnin E (1964) Validité d'une distinction spécifique entre les deux acipenséridés: *Acipenser sturio* L. d'Europe et *Acipenser oxyrinchus* d'Amérique du Nord. Le naturaliste Canadien XCI(1):5–20

Magnin E, Beaulieu G (1963) Etude morphométrique comparée de l'*Acipenser oxyrinchus* Mitchill du Saint Laurent et de l'*Acipenser sturio* Linné de la Gironde. Le Naturaliste Canadien XC(1):5–38

Peng Z, Ludwig A, Wang D, Diogo R, Wei Q, He S (2007) Age and biogeography of major clades in sturgeons and paddlefishes. Mol Phylogenet Evol 42:854–862

Peterson DL, Schueller P, De Vries R, Fleming J, Grunwald C, Wirgin I (2008) Annual run size and genetic characteristics of Atlantic sturgeon in the Altamaha River, Georgia. Trans Am Fish Soc 137:393–401

Rochard E, Williot P, Castelnaud G, Lepage M (1991) Eléments de systématique et de biologie des populations sauvages d'esturgeons. In: Williot P (ed) Acipenser. Cemagref, Antony, pp 475–507

Rochard E, Lepage M, Meauze L (1997) Identification et caractérisation de l'aire de répartition marine de l'esturgeon européen *Acipenser sturio* à partir de déclarations de captures. Aquat Living Resour 10:101–109

Sauvage HE (1883) La grande pêche (les poissons). Bibliothèque instructive. Jouvet et Cie, Paris, 315 p

Sokolov LI, Berdichevskii LS (1989) Acipenseridae Bonaparte, 1831. In: Holčik J (ed) The freshwater fishes of Europe. Aula Verlag, Wiesbaden, pp 150–153

Tagliavini J, Williot P, Congiu L, Chicca M, Lanfredi M, Rossi R, Fontana F (1999) Molecular cytogenetic analysis of the karyotype of the European Atlantic sturgeon, *Acipenser sturio*. Heredity 83:520–525

Waldman JR, Grunwald C, Stabile J, Wirgin I (2002) Impacts of life history and biogeography on the genetic stock structure of Atlantic sturgeon *Acipenser oxyrinchus oxyrinchus*, Gulf sturgeon *A. oxyrinchus desotoi*, and shortnose sturgeon *A. brevirostrum*. J Appl Ichthyol 18: 509–518

Williot P, Arlati G, Chebanov M, Gulyas T, Kasimov R, Kirschbaum F, Patriche N, Pavlovskaya L, Poliakova L, Pourkazemi M, Kim Yu, Zhuang P, Zholdasova IM (2002) Status and management of Eurasian sturgeon: an overview. Intern Rev Hydrobiol 87:483–506

Williot P, Brun R, Rouault T, Pelard M, Mercier D, Ludwig A (2005) Artificial spawning in cultured sterlet sturgeon, *Acipenser ruthenus* L., with special emphasis on hermaphrodites. Aquaculture 246:263–273

Chapter 2
Cytogenetics as a Tool for an Exploration of *A. sturio* Status Within Sturgeons

Francesco Fontana

Abstract A karyotype analysis carried out on the European sturgeon, *Acipenser sturio*, revealed a chromosome number of $2n = 122 \pm 3$. A representative karyotype of 122 chromosomes was composed of 66 meta- and submetacentrics, 56 acrocentrics and microchromosomes. The telomeric sequence repeat (TTAGGG)n detected by fluorescent in situ hybridization (FISH) was localized at the telomeres of all chromosomes. The ribosomal DNA (rDNA) genes were detected by FISH with a digoxigenin-labelled probe for 28S rDNA in the telomeric regions of six chromosomes. The 5S rDNA was found in the interstitial region of a small metacentric pair. The close association between the two rDNA families detected by simultaneous two-colour in situ hybridization (sim-FISH) supports the hypothesis that the *A. sturio* karyotype is primitive. The results are discussed in relation to morphophysiological affinities between *A. sturio* and *A. oxyrinchus*.

Glossary

Cytochrome b (cyt b)	A protein component of respiratory chain, involved in electron transport and the production of ATP.
Fluorescent in situ hybridization (FISH)	Cytogenetic technique used to detect and localize the presence of specific DNA sequences on chromosomes. FISH uses fluorescent probes that only bind to the chromosome parts with high sequence similarity.

F. Fontana (✉)
Department of Biology and Evolution, University of Ferrara, Via L. Borsari 46, 44100 Ferrara, Italy
e-mail: francesco.fontana@unife.it

P. Williot et al. (eds.), *Biology and Conservation of the European Sturgeon Acipenser sturio L. 1758*, DOI 10.1007/978-3-642-20611-5_2,

Haplotype	An haplotype is a combination of DNA sequences on one chromosome which tend to be inherited together.
*Hind*III and *Pst*I satellite DNA	Two satellite DNA families (respectively identified by restriction enzymes of *Haemophilus indicus* and *Providencia stuartii*) frequently studied in sturgeons.
Satellite DNA	Short and highly repeated DNA sequences. Most satellite DNA is localized to the telomeric or the centromeric region of the chromosomes. The nucleotide sequence of the repeats is fairly well conserved among species.
sim-FISH	Simultaneous detection of the two gene families by different staining of two different gene probes. The simultaneous visualization of the two colours is obtained with an epifluorescence microscope equipped with special band filters.
Telomeric sequence	A region of repetitive DNA sequences located at the end of a chromosome, which protects the end of the chromosome from being cut at the end of linear DNA replication.
28S, 16S, 5S rDNA	Regions of DNA respectively codifying for 28S, 16S and 5S ribosomal RNAs. They are composed of tandem repeats of sequences.

2.1 Introduction

Within Acipenseridae taxonomy, a highly debated issue is the relationship between the two species *Acipenser sturio* (European sturgeon) and *A. oxyrinchus* (Atlantic sturgeon) (see Magnin 1962). Based on detailed morphological and anatomical studies, Magnin (1962) and Magnin and Beaulieu (1963) stated that the two species, although very close, should be considered as separate ones. Other authors (Artyukhin and Vecsei 1999) concluded on the same bases that the differences were too small to support the status of separate species, and that therefore the two taxa had to be considered as geographically separated populations of the same species.

Recently, several molecular biology data show that *A. sturio* and *A. oxyrinchus* not only are closely related species, but belong to a group significantly separated from the other Acipenseridae, which represent the most ancestral sturgeon group. The phylogenetic data obtained by molecular analyses of partial sequences of three mitochondrial gene regions (respectively the cytochrome b region and fragments of 12S and 16S rDNA from 22 species of Acipenseridae) (Birstein and DeSalle 1998) show that *A. sturio* and *A. oxyrinchus* represent a separate evolutionary lineage within Acipenseridae. These results are supported by analyses of other sequences of mitochondrial genes in a higher number of species (Birstein et al. 2002). Moreover, molecular phylogeny analyses based on cytochrome b sequences support the separation of these two species from the other 13 species belonging to the subfamily Acipenserinae (Ludwig et al. 2001). Also, recent studies on satellite DNA distribution in the same 15 species of Acipenserinae by Southern blot hybridization show that *Pst*I satellite DNA sequences are present in all species analyzed, while *Hind*III satellite DNA sequences are present in all species except *A. sturio* and *A. oxyrinchus* (Robles et al. 2004). A more detailed analysis of *Pst*I satellite DNA in the same 15 species showed a cladistic association of *A. sturio* and *A. oxyrinchus* sequences, and very similar results were also obtained for 5S rDNA (Robles et al. 2005). Apparently, the two species represent a single clade separated from other Acipenserinae species (Robles et al. 2005). A very recent revision of molecular phylogeny of 20 Acipenseriformes species (Krieger et al. 2008), based on combined mitochondrial DNA sequences, shows that *A. sturio* and *A. oxyrinchus* are phylogenetically ancestral species, early branching from the common ancestor of other sturgeons. These data support previous palaeontological findings (Nesov and Kaznyshkin 1983), which consider *A. sturio* and *A. oxyrinchus* as the most primitive species within the genus *Acipenser*.

Within this complex issue, a detailed knowledge of *A. sturio* cytogenetics could be relevant to clarify the evolution and phylogeny of this species in comparison to the other sturgeons.

2.2 Materials and Methods

Tissue fragments of three living individuals of *A. sturio* were sampled in 1998 for cytogenetic analyses from the Gironde estuarine population, and sent to the Cytogenetics Laboratory at the Department of Biology and Evolution, University of Ferrara (Italy) in a refrigerated container. From these fragments, primary cell lines were established from fibroblast fin culture using the technique described by Fontana et al. (1997). The cell lines were maintained in liquid nitrogen at −70°C, to ensure euploidy along time, as verified by regular use in cytogenetic investigations (Lanfredi et al. 2001; Fontana et al. 2003). When required, fibroblasts were recovered from liquid nitrogen, transferred to sterile plastic flasks and allowed to grow to confluence. Cells were then collected, hypotonically treated and fixed for slide preparations of metaphases (Fontana et al. 1997).

Fluorescent in situ hybridization (FISH) was employed on *A. sturio* fibroblasts to localize the telomeric sequence by the deoxynucleotide (TTAGGG)n oligomer, according to standard techniques (Appligene Oncor, Illkirch, France). The two ribosomal DNA probes for FISH were derived from the genomic DNA of *A. naccarii*. The 28S rDNA probe was composed of two different fragments (approximately 400- and 700-bp long) of the coding region, obtained by two sets of previously described primers (Zardoya and Meyer 1996). The 5S probe, derived from genomic DNA of *A. naccarii*, was composed of a 230-bp fragment, cloned in pMOS-blue vector (Amersham Int., USA) and digoxigenin-labelled using the PCR DIG probe Synthesis Kit (Boehringer, Mannheim, Germany). To simultaneously detect the distribution of the two gene families, a simultaneous two-colour FISH (sim-FISH) was performed. The ribosomal sequences 28S probe was labelled with biotin-16-dUTP (Boehringer), the 5S with digoxigenin-11-dUTP (Boehringer), according to the manufacturer's instructions. Slides were examined using a Leitz epifluorescence microscope with triple band filter for simultaneous visualization of the three colours.

The meaning of the most common genetic expressions is given as a glossary at the end of the chapter.

2.3 Results

The mitotic karyotypes of fibroblast cell lines derived from three *A. sturio* individuals of unknown sex were examined. At least 30 metaphase plates were scored for each cell line. Analyses of all data indicate that the mean chromosome number of *A. sturio* is $2n = 122 \pm 3$. Among all available metaphase plates, one was chosen for karyotype reconstruction because it was isolated, round and surrounded by a faint cytoplasmic residue which protected it from loss or gain of chromosomes. This representative plate, with 122 chromosomes, is shown in Fig. 2.1. Because of their gradual size and morphology, the chromosomes are approximately gathered into two groups, arranged in order of decreasing length: the meta- submetacentrics (33 pairs), acrocentrics and microchromosomes (56 in total). The chromosomes belonging to the second group were individually aligned in the karyotype because they can not be paired. The fundamental number (that is the number of chromosome arms) is therefore 188.

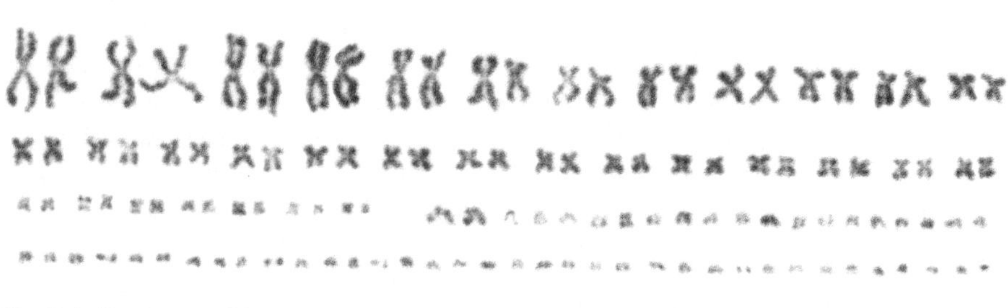

Fig. 2.1 Karyotype of *Acipenser sturio* ($2n = 122$). The meta- and submetacentric chromosomes are aligned in order of decreasing size, followed by acrocentrics and microchromosomes

The telomere signals, detected by FISH as definite spots, are located at both ends of each chromosome arm. No hybridization signal is observed at the interstitial sites (Fig. 2.2a). The *in situ* hybridization with the 28S rDNA probes reveal clear signals of different intensities, interspersed within the telomeric regions of six chromosomes (Fig. 2.2b). After hybridization with the 5S probe, two intense fluorescent signals are observed in the middle region of a pair of small chromosomes (Fig. 2.2c). The results of sim-FISH with biotin-labelled 28S (green hue) and digoxigenin-labelled (red hue) 5S rDNA probes on chromosomes counterstained with DAPI (blue hue) are shown in Fig. 2.2d. The signals of 28S probe, which show different intensities, are localized within the telomeric regions of six chromosomes. There

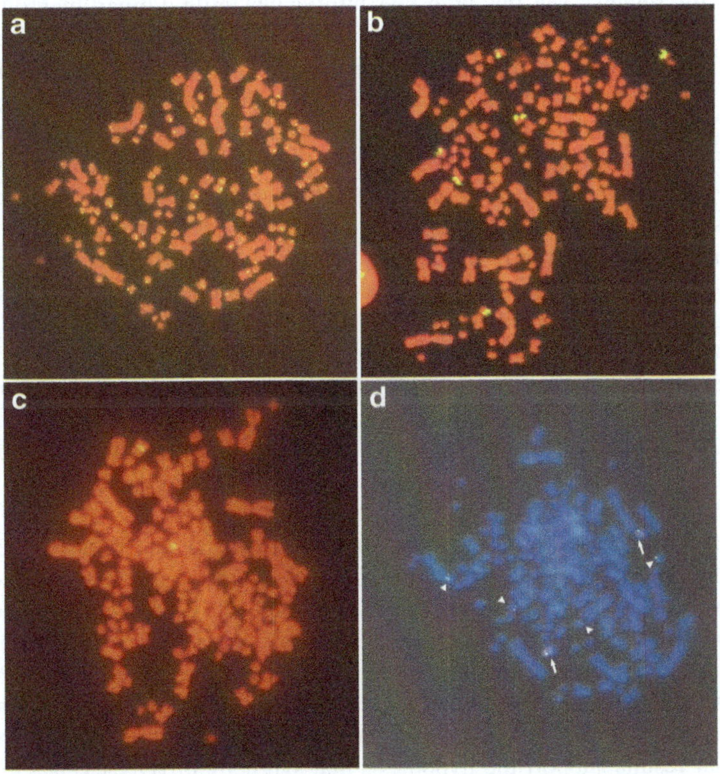

Fig. 2.2 Metaphases of *A. sturio*. **a** FISH with biotin-labelled (TTAGGG)n oligomers, counterstained with propidium iodide. **b** FISH of the 28S rDNA probe detected with digoxigenin and counterstained with propidium iodide. **c** FISH of the 5S rDNA probe detected with digoxigenin and counterstained with propidium iodide. **d** simultaneous two-colour fluorescent in situ hybridization. The micrographs were taken with a triple band filter allowing the simultaneous visualization of the DAPI-stained chromosomes (*blue colour*), the hybridization sites of the 18S–28S (*green*, fluorescein) and the 5S (*red*, Texas Red) rDNA probes. In the loci where the major and minor rDNA signals overlap, the resulting colour is *yellow* (*arrows*). *Arrowheads* show the major rDNA signals

are two 5S-positive regions on one chromosome pair, apparently overlapping two 28S-positive regions.

2.4 Discussion

The karyotype of the Acipenseridae family is characterized by a very high number of chromosomes, about half of which are microchromosomes. The sturgeons may be karyotypically divided into three groups: the first with ~120 chromosomes (from 112 to 146), the second with ~250 chromosomes (from 240 to 270) and the third containing only one species with 372 chromosomes (Fontana et al. 2008a) (for reviews see the Web site http://www.unife.it/dipartimento/biologia-evoluzione/progetti/geneweb). The karyotype of *A. sturio*, the European sturgeon ($2n = 116 \pm 4$), was first described by Fontana and Colombo (1974), and later as ($2n = 121 \pm 3$) by Tagliavini et al. (1999). These data are supported by the present results ($2n = 122 \pm 3$), which confirm that the species belongs to the 120-chromosome group. The telomeric sequence (TTAGGG)n repeats detected by FISH are scattered in the end regions of chromosome arms. The absence of interstitial telomeric signals suggests that no recent chromosome rearrangements have occurred in this species. A similar pattern of telomeric signal distribution was observed in all sturgeon species except *A. gueldenstaedtii*, in which two chromosomes are entirely marked with blocks of repeating telomeric sequences (Fontana et al. 1998). Assuming that the ancestral number of 5S rDNA chromosome pairs is 1, the 120-chromosome group could represent the primitive (plesiomorphic) condition among sturgeons, while the 250-chromosome group and the 372 one (respectively with two and three pairs of 5S rDNA) could represent the derived (or apomorphic) condition. The primitive position of *A. sturio* within the genus *Acipenser* is also supported by the fact that the species belongs to the 120-chromosome group.

The karyotype association (syntheny) between major and minor rDNA genes repeatedly appeared and disappeared during the evolution of eukaryotic genomes. Its widespread presence in lower eukaryotes has been interpreted as an ancestral condition (Drouin and Moniz de Sá 1995; Martins and Galetti 1999), but other authors maintain that in genome evolution the ancestral condition is the separation of the two rDNA gene families on different chromosome pairs (Pendás et al. 1994; Martínez et al. 1996). Given the ancestral position of Acipenseriformes within fishes, our results support the hypothesis that syntheny of rDNA genes represents the ancestral condition within fishes and most probably also in vertebrates.

The main karyotype features, such as chromosome number, distribution of telomeric regions, major (28S) and minor (5S) rDNA, are shared not only by *A. sturio* and *A. oxyrinchus*, but also by other species belonging to the 120-chromosome group, such as *A. ruthenus*, *A. stellatus* and *Huso huso* (Fontana 2002). Thus, the above cytogenetic features are scarcely informative tools to discriminate *A. sturio* and *A. oxyrinchus* from other sturgeon species, but the results obtained by FISH with satellite DNA probes are useful for this purpose. *A. sturio* and *A. oxyrinchus* do not show any signal with *Hind*III satellite DNA probe, contrary to *A. stellatus*, *A. ruthenus* and *Huso huso*, and to 240-chromosome species such as *A. naccarii*, *A. gueldenstaedtii*, *A. baeri* and *A. transmontanus* (Lanfredi et al. 2001; Fontana et al. 2008b). These cytogenetic results are in agreement with previous molecular data obtained by Southern-blot hybridization (Robles et al. 2004). *A. sturio* and *A. oxyrinchus* can also be discriminated from all other sturgeon species by *Pst*I satellite DNA probes: they do not show any hybridization signal, while *A. stellatus* and *A. gueldenstaedtii* show hybridization (Fontana et al. 2008b). The most relevant result is that *A. sturio* and *A. oxyrinchus* react in the same way to both satellite DNA probes. These data support the close association of the two species in a clade that separated early from the other sturgeons (Robles et al. 2004).

Overall, the karyological, cytogenetic and molecular data support the hypothesis that *A. sturio* and *A. oxyrinchus* are very similar and represent a sister clade in comparison to all other sturgeon species. Moreover, the molecular data favour the hypothesis that *A. sturio* and *A. oxyrinchus* are the oldest clade within the genus *Acipenser* (Ludwig et al. 2000). It is reasonable to presume that about 90 M years ago there was an ancestral Atlantic *Acipenser* form (Birstein and Doukakis 2000), which later divided into two forms (presumably *A. sturio* and *A. oxyrinchus*), now surviving on the opposite sides of the North Atlantic: therefore, their interbreeding is presently prevented. Detailed data on geographical distribution of mitochondrial DNA haplotypes show the presence of *A. oxyrinchus* in Northern Europe (Baltic Sea), presumably during the Middle Ages (Ludwig et al. 2002). More recent archaeozoological and paleontological studies witness that *A. oxyrinchus* inhabited the French Atlantic region at the end of the Neolithic Age, about 5,000 years ago (Desse-Berset 2009). With regard to the time elapsed since the separation of the two species, two hypotheses can be advanced. If the separation occurred recently, i.e., during climatic changes in Pliocene and Pleistocene (Choudhury and Dick 1998), then there was not enough time for the two species to significantly differentiate. On the contrary, if the separation occurred a long time ago, their high degree of similarity could be due to the slow genome evolution of Acipenseriformes (Gardiner 1984; Krieger and Fuerst 2002), as also supported by cytogenetic studies on species with the same ploidy degree which show a highly conserved genome (Fontana et al. 2001).

References

Artyukhin E, Vecsei P (1999) On the status of Atlantic sturgeon: conspecificity of European *Acipenser sturio* and the North American *Acipenser oxyrhynchus*. J Appl Ichthyol 15:35–37

Birstein VJ, DeSalle R (1998) Molecular phylogeny of Acipenserinae. Mol Phylogenet Evol 9:141–155

Birstein VJ, Doukakis P (2000) Molecular analysis of *Acipenser sturio*, L., 1758 and *A. oxyrinchus* Mitchill, 1815: a review. Bol Inst Esp Oceanogr 16:61–73

Birstein VJ, Doukakis P, DeSalle R (2002) Molecular phylogeny of Acipenseridae: nonmonophyly of Scaphirhynchidae. Copeia 2:287–301

Choudhury A, Dick TA (1998) The historical biogeography of sturgeons (Osteichthyes: Acipenseridae): a synthesis of phylogenetics, palaeontology and palaeogeography. J Biogeogr 25: 623–640

Desse-Berset N (2009) First archaeozoological identification of Atlantic sturgeon (*Acipenser oxyrinchus* Mitchill 1815) in France. CR Palevol 8:717–724

Drouin G, Moniz de Sá M (1995) The concerted evolution of 5S ribosomal genes linked to the repeat units of other multigene families. Mol Biol Evol 12:481–493

Fontana F (2002) A cytogenetic approach to the study of taxonomy and evolution in sturgeons. J Appl Ichthyol 18:226–233

Fontana F, Colombo G (1974) The chromosomes of Italian sturgeons. Experientia 30:739–742

Fontana F, Rossi R, Lanfredi M, Arlati G, Bronzi P (1997) Cytogenetic characterization of cell lines from three sturgeon species. Caryologia 50:91–95

Fontana F, Lanfredi M, Chicca M, Aiello V, Rossi R (1998) Localization of the repetitive telomeric sequence (TTAGGG)n in four sturgeon species. Chromosome Res 6:303–306

Fontana F, Tagliavini J, Congiu L (2001) Sturgeon genetics and cytogenetics: recent advancements and perspectives. Genetica 111:359–373

Fontana F, Lanfredi M, Congiu L, Leis M, Chicca M, Rossi R (2003) Chromosomal mapping of 18S-28S and 5S rRNA genes by two-colour fluorescent in situ hybridization in six sturgeon species. Genome 46:473–477

Fontana F, Congiu L, Mudrak VA, Quattro JM, Smith TIJ, Ware K, Doroshov SI (2008a) Evidence of hexaploid karyotype in shortnose sturgeon. Genome 51(2):113–119. doi:10.1139/G07-112

Fontana F, Lanfredi M, Kirschbaum F, Garrido-Ramos MA, Robles F, Forlani A, Congiu L (2008b) Comparison of karyotypes of *Acipenser oxyrinchus* and *A. sturio* by chromosome banding and fluorescent in situ hybridisation. Genetica 132:281–286. doi:10.1007/s10709-007-9171-4

Gardiner BG (1984) Sturgeons as living fossils. In: Eldredge N, Stanley SM (eds) Living fossils. Springer, New York, pp 148–152

Krieger J, Fuerst PA (2002) Evidence for a slowed rate of molecular evolution in the Order Acipenseriformes. Mol Biol Evol 19:891–897

Krieger J, Hett AK, Fuerst PA, Artyukhin E, Ludwig A (2008) The molecular phylogeny of the order Acipenseriformes revisited. J Appl Ichthyol 24(suppl 1):36–45

Lanfredi M, Congiu L, Garrido-Ramos MA, De La Herrán R, Leis M, Chicca M, Rossi R, Tagliavini J, Ruiz Rejón C, Ruiz Rejón M, Fontana F (2001) Chromosomal location and evolution of a satellite DNA family in seven sturgeon species. Chromosome Res 9:47–52

Ludwig A, May B, Debus L, Jennekens I (2000) Heteroplasmy in the mtDNA control region of sturgeon (*Acipenser*, *Huso* and *Scaphirhynchus*). Genetics 156:1933–1947

Ludwig A, Belfiore NM, Pitra C, Svirsky V, Jenneckens I (2001) Genome duplication events and functional reduction of ploidy levels in sturgeon (Acipenser, Huso and Scaphirhynchus). Genetics 158:1203–1215

Ludwig A, Debus I, Lieckfeldt D, Wirgin I, Benecke N, Jenneckens I, Williot P, Waldman JR, Pitra C (2002) When the American sea sturgeon swam east. Nature 419:447–448

Magnin E (1962) Recherches sur la systématique et la biologie des Acipenséridés, *Acipenser sturio* L., *Acipenser oxyrhynchus* Mitchill et *Acipenser fulvescens* Raf. Ann Stat Cent Hydrobiol Appl 9:7–242

Magnin E, Beaulieu G (1963) Étude morphométrique comparée de l'*Acipenser oxyrhynchus* Mitchell du Saint-Laurent et l'*Acipenser sturio* Linné de la Gironde. Le Naturaliste Canadien 40:5–38

Martínez JL, Morán P, García-Vázquez E, Pendás AM (1996) Chromosomal localization of the major and 5S rRNA genes in the European eel (*Anguilla anguilla*). Cytogenet Cell Genet 73: 149–152

Martins C, Galetti PM Jr (1999) Chromosomal localization of 5S rDNA genes in *Leporinus* fish (Anostomidae, Characiformes). Chromosome Res 7:363–367

Nesov LA, Kaznyshkin MN (1983) New sturgeons from the Cretaceous and Paleogene of the USSR. In: Menner VV (ed) Contemporary problems of paleoichthyology. Nauka, Moscow, pp 68–76

Pendás AM, Morán P, Freije JP, Garcia-Vazquez E (1994) Chromosomal mapping and nucleotide sequence of two tandem repeats of Atlantic salmon. Cytogenet Cell Genet 67:31–36
Robles F, De La Herrán R, Ludwig A, Ruiz Rejón C, Ruiz Rejón M, Garrido-Ramos MA (2004) Evolution of ancient satellite DNAs in sturgeon genomes. Gene 338:133–142
Robles F, De La Herrán R, Ludwig A, Ruiz Rejón C, Ruiz Rejón M, Garrido-Ramos MA (2005) Genomic organization and evolution of the 5S ribosomal DNA in the ancient fish sturgeon. Genome 48:18–28
Tagliavini J, Williot P, Congiu L, Chicca M, Lanfredi M, Rossi R, Fontana F (1999) Molecular cytogenetic analysis of the karyotype of the European Atlantic sturgeon, Acipenser sturio. Heredity 83:520–525
Zardoya R, Meyer A (1996) Evolutionary relationships of the Coelacanth, Lungfishes, and Tetrapods based on the 28S ribosomal RNA gene. Proc Natl Acad Sci USA 93:5449–5454

Chapter 3
Discrimination of *Acipenser sturio*, *Acipenser oxyrinchus* and *Acipenser naccarii* by Morphology of Bones and Osteometry

Nathalie Desse-Berset

Abstract Studying ancient sturgeon populations according to archaeological remains requires the identification of the species through bone morphology. Osteometry makes it possible to determine the different sizes and age categories of ancient specimens by measuring specific bones which are correlated with the same measurements taken on modern specimens of known total length (TL) and mass. These references have been established for *Acipenser sturio*, *Acipenser oxyrinchus* (both attested in Europe from the Neolithic) and *Acipenser naccarii*, providing a set of morphological and osteometric criteria to discriminate the above-mentioned species.

Through these methods, the size distribution of ancient sturgeons of different periods and locations can be reconstructed.

The frequent presence of large-sized specimens of *A. sturio* and *A. oxyrinchus* on the French Atlantic and North basins strongly suggests their sympatry.

3.1 Introduction

Acipenseridae, primitive fish with partly cartilaginous skeletons, possess a certain number of bony elements (Grassé 1958). Determining a sturgeon in archaeological fauna would seem easy. Its large bony plates, even fragmented, don't go unnoticed and can be rapidly identified even by non-specialists. However, the "sturgeon" determination analysis quickly finds a limit there. If one wishes to go past this stage, one will stumble upon numerous difficulties which have discouraged most of the archaeozoologists. Only a few studies have focused upon sturgeon remains: the European sturgeon *A. sturio* is mentioned in faunal site lists and in local synthesises dealing with fishing or palaeoeconomy (Heinrich 1987; Clavel 2001; Bartosiewicz

N. Desse-Berset (✉)
Université de Nice-Sophia Antipolis – CNRS-CEPAM-UMR 6130, SJA3, 24 avenue des Diables Bleus, 06357 NICE Cédex 4, France
e-mail: berset.desse@hotmail.com; berset@cepam.cnrs.fr

P. Williot et al. (eds.), *Biology and Conservation of the European Sturgeon Acipenser sturio L. 1758*, DOI 10.1007/978-3-642-20611-5_3,

and Bonsall 2008; Makowiecki 2008), but few studies make reference to the morphology or osteometry (we can cite Benecke 1986; Brinkhuizen 1986, 1989, 2006; Desse-Berset 1994, 2009a, b). Some works deal with other species (Bartosiewicz and Takacs 1997) or address specific aspects (Findeis 1997; Hilton and Bemis 1999). Working on animals listed as threatened with extinction and appearing on the red list of endangered species sets the problem of constituting a reference collection of individual skeletons characterizing size, mass, age and origin. These bone references are, however, a necessity for archaeological study, to provide the information about animal remains in excavations. Specific determination is the basis for further knowledge about the history of populations, to allow palaeogeographic synthesises. The only species known to occur in France until recent times was *A. sturio*. The hypothesis of an Adriatic sturgeon presence (*A. naccarii*) along the Mediterranean coasts put forward by Spanish researchers (Garrido-Ramos et al. 1997; De la Herran et al. 2004) has been contested (Elvira et al. 1991; Elvira and Almodovar 2000; Almodovar et al. 2000; Doukakis et al. 2000). A project to reintroduce sturgeon in the Rhône River (south east of France) by the MRM (Migrateurs Rhône–Méditerranée) Association at the beginning of the 2000s made it necessary to verify this question by precisely determining which species lived there in the past (Brosse et al. 2006, 2009).

Recently the study of archaeological sturgeon remains originating from the French Atlantic coast has shown the existence of another species, the Atlantic sturgeon (*A. oxyrinchus*) (Desse-Berset 2009a; Desse-Berset and Williot 2011; see Chap. 7).

The goal of this chapter is to answer the following questions:

1. What are the morphological differences between *A. sturio* and *A. naccarii*, as well as between *A. sturio* and *A. oxyrinchus*?
2. Are there any osteometrical differences?
3. How can we establish relations between some osteometrical criteria and the total length of sturgeons?

All these data provide basic knowledge of ancient fish populations and their history. It is part of the global sturgeon study, and is proving to be all the more precious now that these populations are threatened with extinction. These data are necessary in the developement of reintroduction programmes.

3.2 Material and Methods

3.2.1 What Species in France? Discrimination Between Species

Most of the determination criteria between the sturgeon species, published in various books (Magnin 1962, 1964; Magnin and Beaulieu 1963; Whitehead et al. 1984; Fischer et al. 1987; Lelek 1987; Holcik 1989; CITES Identification Guide

2001) have generally proved to be unusable on archaeozoological material, because most of the discrimination keys are based on whole-fish meristic characteristics and don't apply to archaeological bone remains.

Only exoskeleton surface morphology provides this information about the species, thanks to the external aspect of the bony plates, tubercular for the *A. sturio* and alveolar for the *A. oxyrinchus* (Magnin 1962; see Chap. 7, Fig. 7.1). The biological and biometrical data mentioned in this work (number of scutes, relation TL/age/mass, age/TL for reproduction, etc.) come mainly from Magnin's publications (1962, 1964; Magnin and Beaulieu 1963).

The aim of this topic is not to synthesize all this given data, but to find discrimination criteria between the species on isolated bones by searching for anatomic differences according to archaeo-ichtyological methods (Desse et al. 2002; Desse-Berset 1994).

The anatomic attributions of bones and the reconstitution of identified specimen sizes are added to the data provided by archaeology, such as for instance geographic origin and chronology (dating of these remains is obtained by their stratigraphical environment or by physical–chemical dating methods such as C^{14}), along with the natural or anthropical environmental context.

The large collection of sturgeon bone remains which were discovered on the Arles-Jardin d'Hiver archaeological site (France, sixth to second century BC), and which proves the presence of sturgeon populations in the Rhône River over a period of 5 centuries, was an extraordinary source of information. This collection represented the starting point of the construction of an analysis methodology of this taxis, and has made it possible to answer the question concerning the present species. The constitution of a skeleton reference collection was a necessary stage.

3.2.1.1 Reference Specimens

Every study on bone remains requires modern reference collection and osteometric data. This was non-existent for sturgeon before the year 1990. Specimens preserved in museums of natural history were not usable, because their ossified parts were neither accessible nor measurable. Palaeontological or zoological publications (Goodrich 1930; Gregory 1933; Grassé 1958; CITES Identification Guide 2001) reviewing the complex anatomy of Acipenseriformes, works presenting systematic data for each species (Magnin 1962; Rochard et al. 1991; Williot and Rochard 2007), keys based on meristic criteria for specific determination, as number of dorsal, lateral or ventral scutes (Whitehead et al. 1984; Lelek 1987; Holcik et al. 1989), and morphometric characteristics (percentages of lengths for each body part compared to the total length for example) are very useful with regard to the whole fish, but unfortunately are of limited applicability for fragmented or isolated bones typical for archaeological bone remains.

The reference collection was developed based upon nine specimens from *A. sturio*, either whole fish or their heads (provided by the Cemagref between 1985 and 2009); two *A. naccarii* specimens (given in 2006 by the Giovanini fish farm, Azienda Agricola V.I.P., Italy); one *A. oxyrinchus* skeleton (provided in 2008 by Montreal's university osteotheque); and ten *A. oxyrinchus* specimens whose bones were prepared in 2009 in Canada (provided by the company "Acadian Sturgeon and Caviar Inc.", St John, Nouveau Brunswick, Canada).

3.2.1.2 Anatomy of Sturgeons

Acipenseridae possess a certain number of ossified elements (Fig. 3.1):

- The cranium exoskeleton is made up of ossified plates of dermal origin, with irregular shape and number: the occipital, the rostral bones, the parietal, the post-temporal, the pterotic, and the supratemporal
- The splanchnocranium includes the dentary, the maxillary (or premaxillo-maxillary), the palatopterygoid (or palatoquadrate), the parasphenoid, the subopercular, the hyomandibular, and the branchial archs
- The pectoral girdle include the cleithrum, the supracleithrum, the clavicle, the pectoral spine, and the pectoral fin rays
- The dermal scutes divide up into five rows: a dorsal row, prolonged by the fulcra along the caudal fin, two lateral rows (left and right), and two ventral rows (left and right)

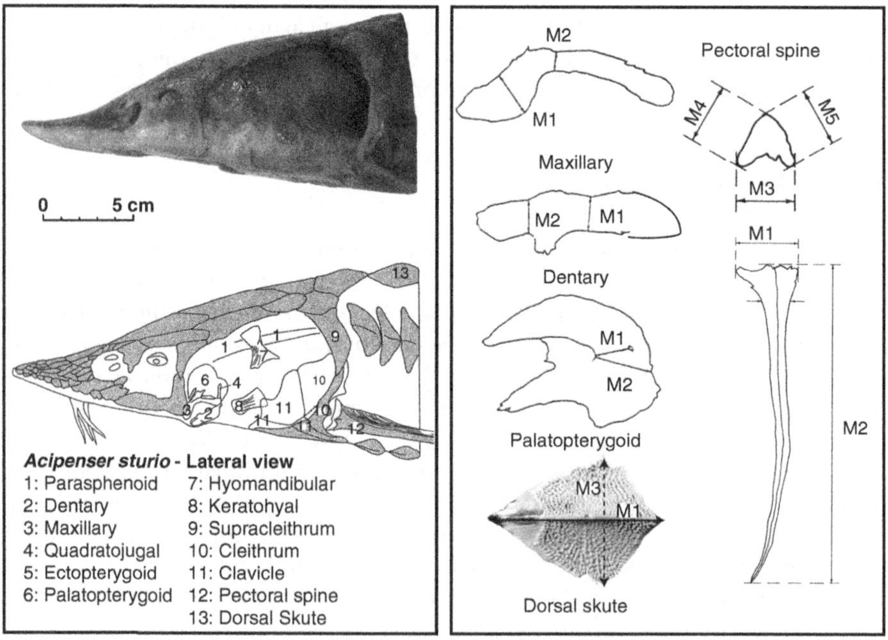

Fig. 3.1 Anatomical position and measurement points selected on various bones of the splanchnocranium (maxillary, dentary, palatopterygoid), of the pectoral spine and of the dorsal scutes (*left side*: after Holmgren and Stensiö 1936; *right side*: after Desse-Berset 1994, Figs. 7–9)

These ossified remains can be found in archaeological excavations, and can be given a specific identification, an estimation of the number of remains and the number of individuals, and a size reconstitution of the fish.

3.2.1.3 Morphological and Osteometrical Study of the Bones

The sturgeon skeleton is complex, and almost no morphological study of the bones as a whole is available and proposes any osteometrical methods for the reconstitution of their sizes (TL) and of their masses. Identification methods based on morphology have thus been defined, along with measuring points chosen according to their conservation expectancy on archaeological bones, often fragmented, according to archaeo-ichtyological methods (Desse-Berset 1994; Desse et al. 2002).

The measuring points are chosen by avoiding the biggest lengths for instance, or the fragile extremities, which are most of the time broken on ancient remains (see Chap. 7, Figs. 7.3–7.6). More compact zones, where the maximum constriction points are favoured, according to higher resistance points or clearer measuring takes (cf. Fig. 3.1). Several measurements have been taken on all the sturgeon scutes in our collection. In the case of dorsal scutes, the relationships M1/M3 (M1: cranio-caudal diameter or length; M3: transverse diameter or width) have been systematically exploited.

The osteometrical analysis of the large Arles Jardin d'Hiver collection (2,500 remains) was done after the morphological study and the bone measurement of modern *A. sturio*, which made it possible to carry out size reconstitutions of sturgeon populations from the Rhône River (Desse-Berset 1994).

Some bones were chosen for this study because of their frequency on the site.

The scutes are the most numerous bony elements which can be found on the site; the number of scutes on a European sturgeon is approximately 105 for each individual, but can vary between 89 and 122 (Magnin 1962). They are always hard to exploit both for the size reconstitution of the fish or for the estimation of the minimum number of individuals:

(i) Their fragmental shape complicates the attribution of their row in cranio-caudal positions, because there are few morphological differences between most scutes of the same row/serie (dorsal, lateral ventral), which go from head to tail. When there is only one scute left, it is often impossible to establish if it is the largest one (which gives us information on the maximum size) or one of the smallest ones of a large individual.

(ii) Another difficulty, when several scute fragments are present, is that it is hard to say whether they belong to the hundreds of scutes of a same individual or whether, on the contrary, the scutes belong to different individuals. Therefore, we will also integrate the archaeological context data (stratigraphy, spatial distribution) in order to evaluate the minimum number of individuals (Desse-Berset 2009b).

Some other bones fit better than scutes for such evaluations. Since the Arles material included anatomic elements permitting the size reconstitution and the number of sturgeons, the splanchnocranium bones (maxillary, dentary, palato-pterygoid) were favoured, both for the counting of the minimum number of individuals and for the total length reconstitutions (TL).

Other bony elements from the pectoral girdle have provided precious data: pectoral spines have made it possible to provide a great deal of information about protohistorical sturgeon populations in the Rhône River (Desse-Berset 1994). To achieve the above-mentioned objectives, the metric data from Arles was merged with the data Magnin had published about sizes correlated with modern sturgeon ages (Magnin 1962, 1964; Magnin and Beaulieu 1963). This data provided the table of the European sturgeon (*A. sturio*) population, still numerous in the Gironde River at that time.

Finally, the morphology of the surface of the exoskeletal dermal plates (cranial dermal plates, dorsal, lateral and ventral scutes), alveolar or tubercular, has proven to be the best discrimination criteria between the species *A. sturio* and *A. oxyrinchus* (Magnin 1962; Desse-Berset 2009a; see Chap. 7, Fig. 7.1).

3.3 Results and Discussion

3.3.1 Discriminating Morphological Characteristics Between A. sturio *and* A. naccarii

The first objective was to search for the morphological differences between the bones of the two species, by eliminating the strictly individual variations. As a matter of fact, sturgeon bones are subject to important variations in number and extension, the welds being frequent between the elements of the cranial exoskeleton (Jarvik 1948; Grassé 1958), and all the authors interested in the morphology of external ossified sturgeon parts have stated that, within the same species, individual variations in number and shape of the ossified parts are very frequent.

The illustrations (Fig. 3.2) show specimens of relatively similar sizes; they have all spent numerous years in captivity in raceways. Their external skeleton may have suffered deterioration; the pectoral fins, for instance, are atrophied and deformed. These bones were not selected for a morphological determination.

The two *A. naccarii* are the same age (more than 30 years old), and have the same sex and size (TL: 1.46 m and 1.48 m). However, they show individual morphological differences (Fig. 3.2). It is a great illustration of the individual variations previously mentioned.

However, some constant morphological differences between the two species can be highlighted. Several discriminating criteria have been defined (Desse-Berset in Brosse et al. 2006).

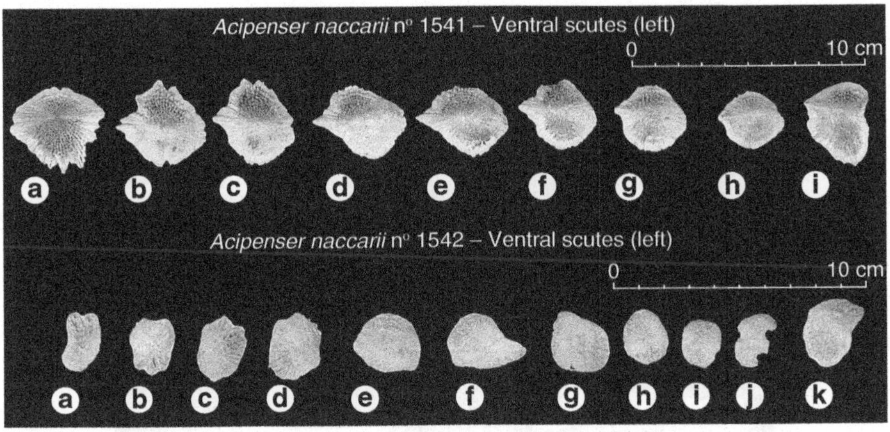

Fig. 3.2 Individual morphological variations on the ventral scutes of two *A. naccarii* (no. 1541: 1.46 m; no. 1542: 1.48 m). These wild-originated specimens have roughly the same size. They have spent several years in captivity in the raceways of the Italian fish farm Azienda Agricola

3.3.1.1 Splanchnocranium Bones (Fig. 3.3)

The splanchnocranium bones seem to be less subject to individual variations than external ossified elements. Morphological differences have been revealed on three splanchnocranium bones: the maxillary, the dentary and the palatopterygoid. They are not always represented in archaeological remains, but are frequent among *A. sturio* remains from Arles.

- Maxillary (or premaxillo-maxillary) (Fig. 3.3a):
  For *A. sturio*, the maxillary has its anterior part clearly individualized and protruding. This part corresponds to 1/3 of the total length of the bone.
  For *A. naccarii*, this part is more antero-posteriorly developed, and the posterior part is proportionally shorter; it corresponds to half of the total length of the bone.
- Dentary (Fig. 3.3b):
  For *A. sturio*, the ascendant process is strongly marked and forms a perpendicular appendix; in internal view, the articular side is slightly inferior to half of the total length of the bone.
  For *A. naccarii*, the ascendant process is almost absent. The articular side is more than half of the total length of the bone (it is closer to 2/3). The general morphology is more elongated.
- Palatopterygoid (or palatoquadrate) (Fig. 3.3c):
  For *A. sturio*, the general shape of the anterior part is generally rounded. In internal view, the insertion pit is strongly marked and hollow.
  For *A. naccarii*, the general shape is elongated; the posterior part of the bone is rectangular. On the internal side, the insertion pit is barely marked.

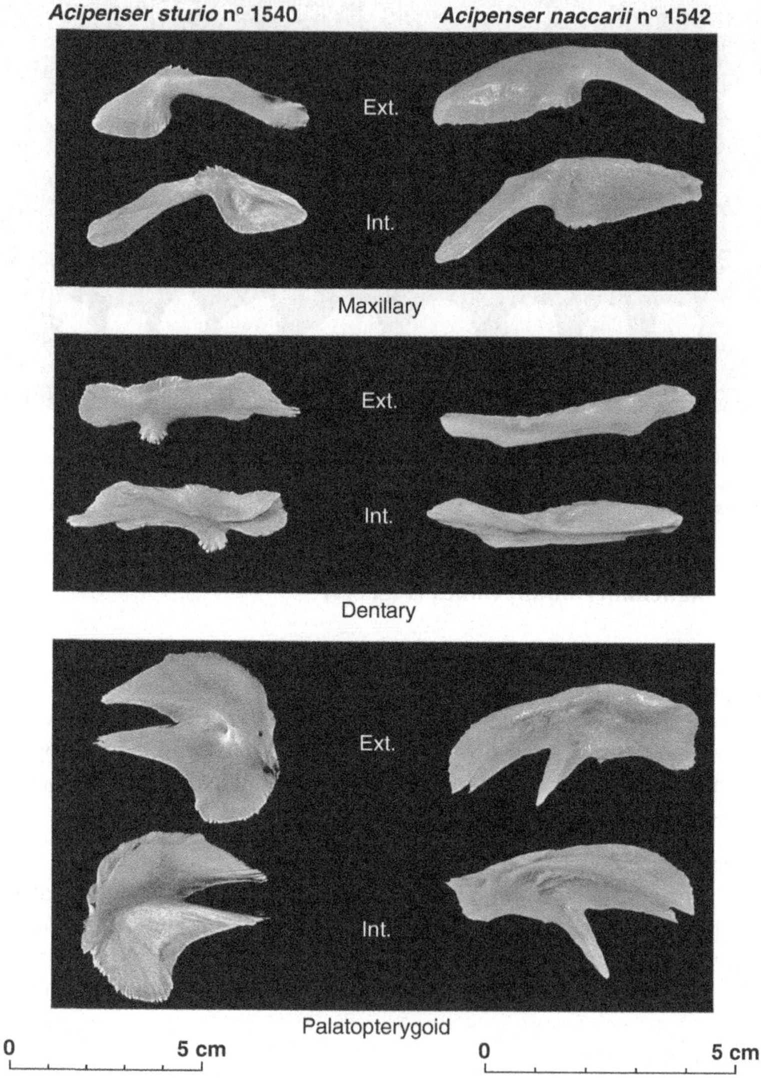

Fig. 3.3 Morphological differences between *A. sturio* and *A. naccarii* on splanchnocranium bones (*external* and *internal view* of the right bones): **a** maxillary; **b** dentary; **c** palatopterygoid

3.3.1.2 Dermal Scutes (Fig. 3.4)

These are ossified elements, the most numerous among all sturgeons; they present frequent individual variations, but also specific differences.

On every *A. naccarii* scute nos. 1541 and 1542 (numbers in our reference collection), one can observe that the thickness of the bones is much inferior to those of *A. sturio*. Their surface is little marked, granuliform, or flat and stellate (detail

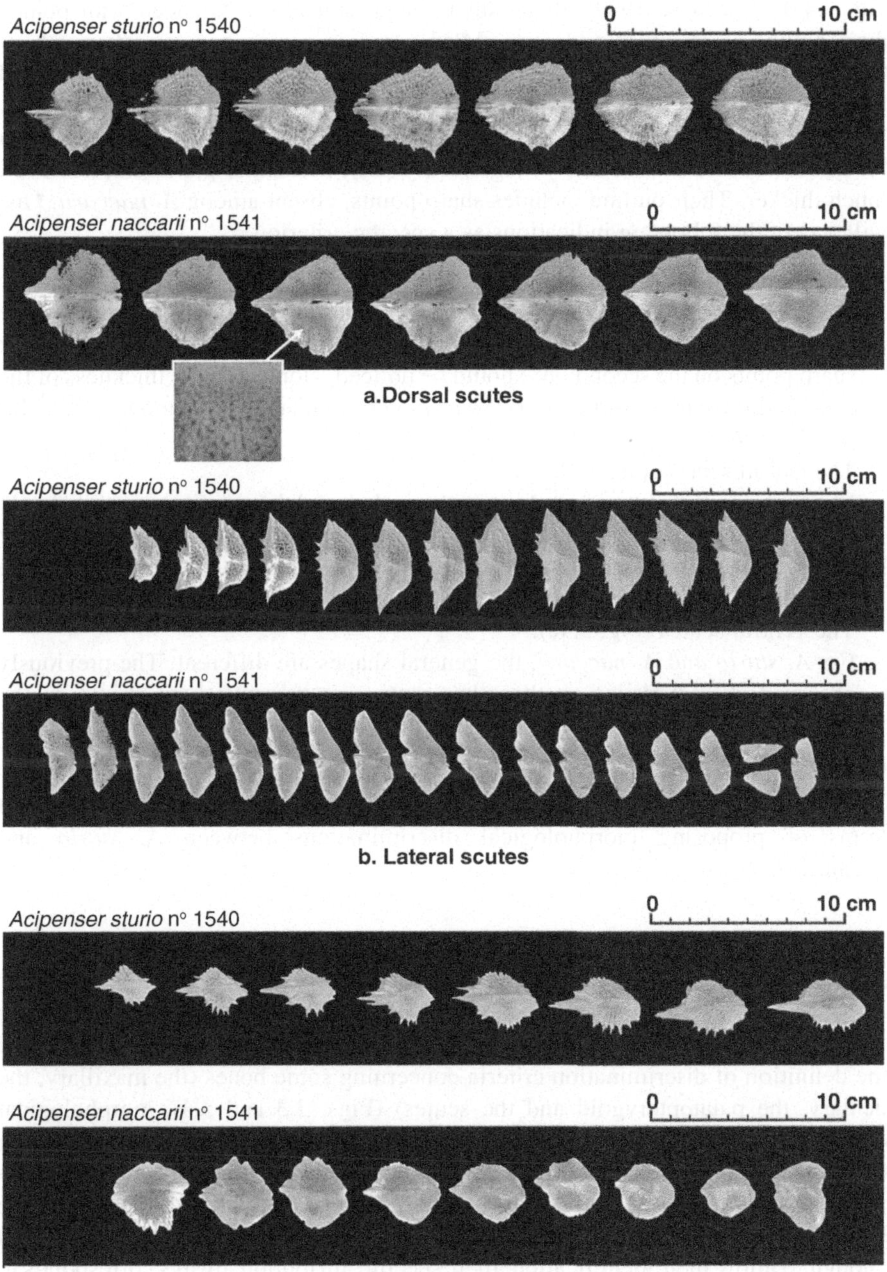

Fig. 3.4 Morphological differences between *A. sturio* and *A. naccarii* on scutes; only a part of the scutes is represented here: **a** dorsal scutes; detail of the surface structure of *A. naccarii* (surface barely marked, granuliform or flat and stellate) (see Chap. 7 and Fig. 7.1 for *A. sturio* and *A. oxyrinchus*); **b** lateral scutes; **c** ventral scutes

Fig. 3.4a) (Tortonese 1989); their hull is blunt, and is not extended with points; their outline is irregular and slim. Maybe one is seeing here an aftermath of their captivity in raceways and their age (about 30 years old) (Sokolov and Berdichevskii 1989).

The *A. sturio* no. 1540 (number in our reference collection), of wild origin, is a bit smaller (TL: 1.27 m), and younger (12 years); its dermal plates are, however, much thicker. Their outline includes sharp points, absent among *A. naccarii*. One will accept the following indications as a specific criterion.

- The dorsal scutes (or dorsomedians) (Fig. 3.4a):
  For *A. sturio* and *A. naccarii*, the general shapes are roughly similar. Only the presence of sharp points on the outline of the first one, and the absence of these sharp points on the second one should be noticed, along with the thickness of the bone and its marked surface for *A. sturio*, very thin and with smooth surface for *A. naccarii*.
- The lateral scutes (Fig. 3.4b):
  For *A. sturio*, the posterior part is more rounded, and the anterior parts fitted with relatively sharp points on the outline, whereas on the *A. naccarii*, the posterior part forms an obtuse angle, and the whole is devoid of points; the bone is also thin and its surface is smooth.
- The ventral scutes (Fig. 3.4c):
  For *A. sturio* and *A. naccarii*, the general shapes are different. The previously observed characteristics on the other scutes are present: sharp points on the outline for *A. sturio*, absent for *A. naccarii*; thick bone, marked relief for *A. sturio*; thin bone, smooth surface and eroded outline for *A. naccarii*.

However, the scutes are less reliable bones than the splanchnocranium bones in terms of proposing morphological discriminations between *A. sturio* and *A. naccarii*.

3.3.1.3 Application to Archaeological Material

After the morphological study comparing reference specimens of both species, and the definition of discrimination criteria concerning some bones (the maxillary, the dentary, the palatopterygoid and the scutes) (Figs. 3.3 and 3.4), morphological differences were proposed for several bones. The re-examination of bones coming from the Arles archaeological collection was carried out using all the remains in which discriminating characteristics had been observed, in order to determine the species present. When the shape of the archaeological bones from Arles was good enough to allow their identification, their specific attribution always corresponds to *A. sturio*.

The analysis of the archaeological material from Arles as a whole was broadened to Montmajour (close to Arles, the Montmajour abbey was located on the site from the fourth until the eighteenth century AD; study still in process). It makes it possible to conclude that *A. sturio* was the only species present in the Rhône valley. The hypothesis of an *A. naccarii* presence was never confirmed in France in archaeological collections.

Other than morphological criteria distinguishing the two species, another element deserves to be considered: the sizes of the sturgeons from Arles (Desse-Berset 1994). They have been reconstituted, and one third of the individuals has a TL much greater than 2 m, which does not correspond to the biometric data documented for *A. naccarii*, whose 2 m TL (for 25 kg) is mentioned in publications as a maximum (Tortonese 1989).

Finally, a palaeogenetic analysis programme carried out on this collection has also confirmed the sole presence of *A. sturio* in the Rhône River (Chassaing 2010; Desse-Berset et al. 2008; Pagès et al. 2009 see Chap. 8).

The absence of *A. naccarii* among identified archaeological remains has justified neither the preparation of other reference specimens of this species, nor the implementation of a set of measurements for an osteometric application of size reconstitution of individuals.

3.3.2 Discriminating Morphological Characteristics Between **A. sturio** *and* **A. oxyrinchus**

Some discriminating morphological criteria applying to the surface of the exoskeleton, already suggested by Tichij (1929, in Debus 1999), have been thoroughly detailed by Magnin (1962). The observations made on modern *A. sturio* and *A. oxyrinchus* skeletons, as well as on various MNHN (Muséum National d'Histoire Naturelle, Paris) specimens, confirm these discriminant characteristics.

The morphological criteria have been recently redefined during the study of a set of Atlantic sturgeons collected and prepared in Canada in 2009. Applied to archaeological material from several sites, they have confirmed their reliability (Desse-Berset 2009a; Desse-Berset and Williot 2011; see Chap. 7).

Except the external aspect of dermal ossified elements, tubercular for *A. sturio* and alveolar for *A. oxyrinchus* (see Chap. 7, Fig. 7.1), other morphological criteria have been researched. We would like to underline the fact that the bone morphologies of the two species are roughly similar, and that the differences are not as evident as those defined between *A. sturio* and *A. naccarii*.

However , some signs can be presented. They concern particularly the bones of the splanchnocranium (Fig. 3.5) and the scutes (Fig. 3.6).

3.3.2.1 Splanchnocranium Bones (Fig. 3.5)

Morphological differences have been revealed on three splanchnocranium bones: the maxillary, the dentary and the palatopterygoid.

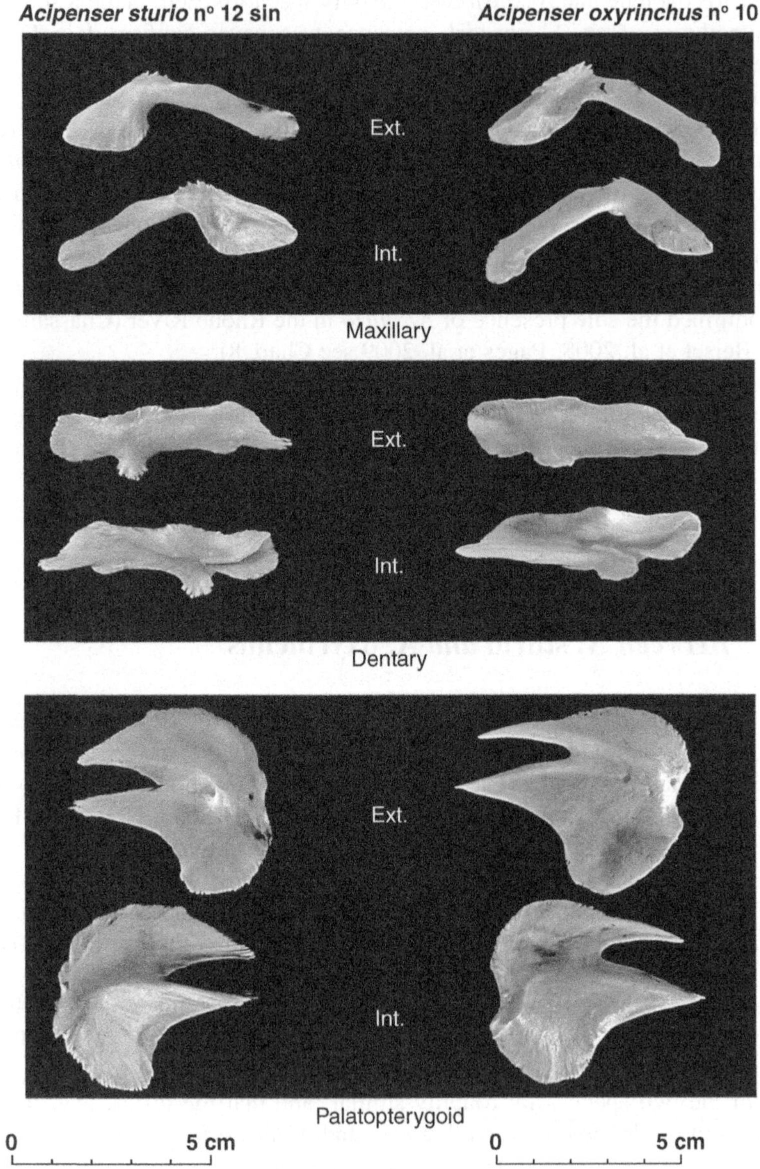

Fig. 3.5 Morphological differences between *A. sturio* and *A. oxyrinchus* on bones of the splanchnocranium (*external* and *internal view*): **a** maxillary; **b**) dentary; **c** palatopterygoid

– Maxillary (or premaxillo-maxillary) (Fig. 3.5a):
  For *A. sturio*, the maxillary has a spade-shaped anterior part markedly individualised and protruding. This part corresponds to 1/3 of the total length of the bone.

For *A. oxyrinchus*, the anterior part is less antero-posteriorly developed, and the posterior part is proportionally longer; it corresponds to more than half of the total length of the bone.

- Dentary (Fig. 3.5b):
  For *A. sturio*, the ascendant process is strongly marked and forms a perpendicular appendix; in internal view; the articular side is slightly inferior to half of the total length of the bone.
  For *A. oxyrinchus*, the ascendant process is little marked. The articular side is slightly superior to half of the total length of the bone. The general morphology is stockier and less sinuous.
- Palatopterygoid (Fig. 3.5c):
  For *A. sturio*, the general shape of the anterior part is most of the time rounded. In internal view, the insertion pit is strongly marked and hollow. The inferior element of both anterior processes is more developed than the superior.
  For *A. oxyrinchus*, the general shape is similar to *A. sturio*; in the anterior part, the inferior element of both anterior processes is narrower than the superior. On the internal side, the insertion pit is less marked.

3.3.2.2 Dorsal Scutes (Fig. 3.6)

For *A. sturio* and *A. oxyrinchus*, the general shapes are quite similar, but for *A. sturio* the dorsal scutes are more numerous (between 13 and 14; Fig 3.8) than for *A. oxyrinchus* (on average 10). This data corresponds to that of Magnin (1962), who reports an average of 13 scutes, and in some extreme cases a number of scutes going from 9 to 16 for *A. sturio*, and an average of ten scutes, and in some extreme cases a number of scutes going from 7 to 13 for *A. oxyrinchus*. In addition, the dorsal scutes for *A. sturio* are smaller compared to the TL of individuals than those for *A. oxyrinchus*.

Thus, *A. oxyrinchus*' dorsal scutes, which are less numerous, are bigger in size compared to the TL of fish; on each studied specimen, the scutes going from the fourth until the seventh row are markedly bigger than the first three and last three rows.

The size reconstitution of the fish from isolated scutes will above all depend on their specific determination. On each scute of the two species, the most obvious specific difference and the least obvious remains the morphology of their surface, with its tubercular patterns for *A. sturio*, and its sharp-sided alveoli for *A. oxyrinchus*. We will accept these indications as a specific criterion.

The dissection of several specimens, and the sampling of all scutes in the cranio-caudal anatomic order, show the frequent morphological variations from an individual to another. More or less elongated, more or less angular or oval dihedron-shaped, the dorsal scutes (which are however more homogenous than lateral or ventral scutes) have a general morphology that varies depending on their row.

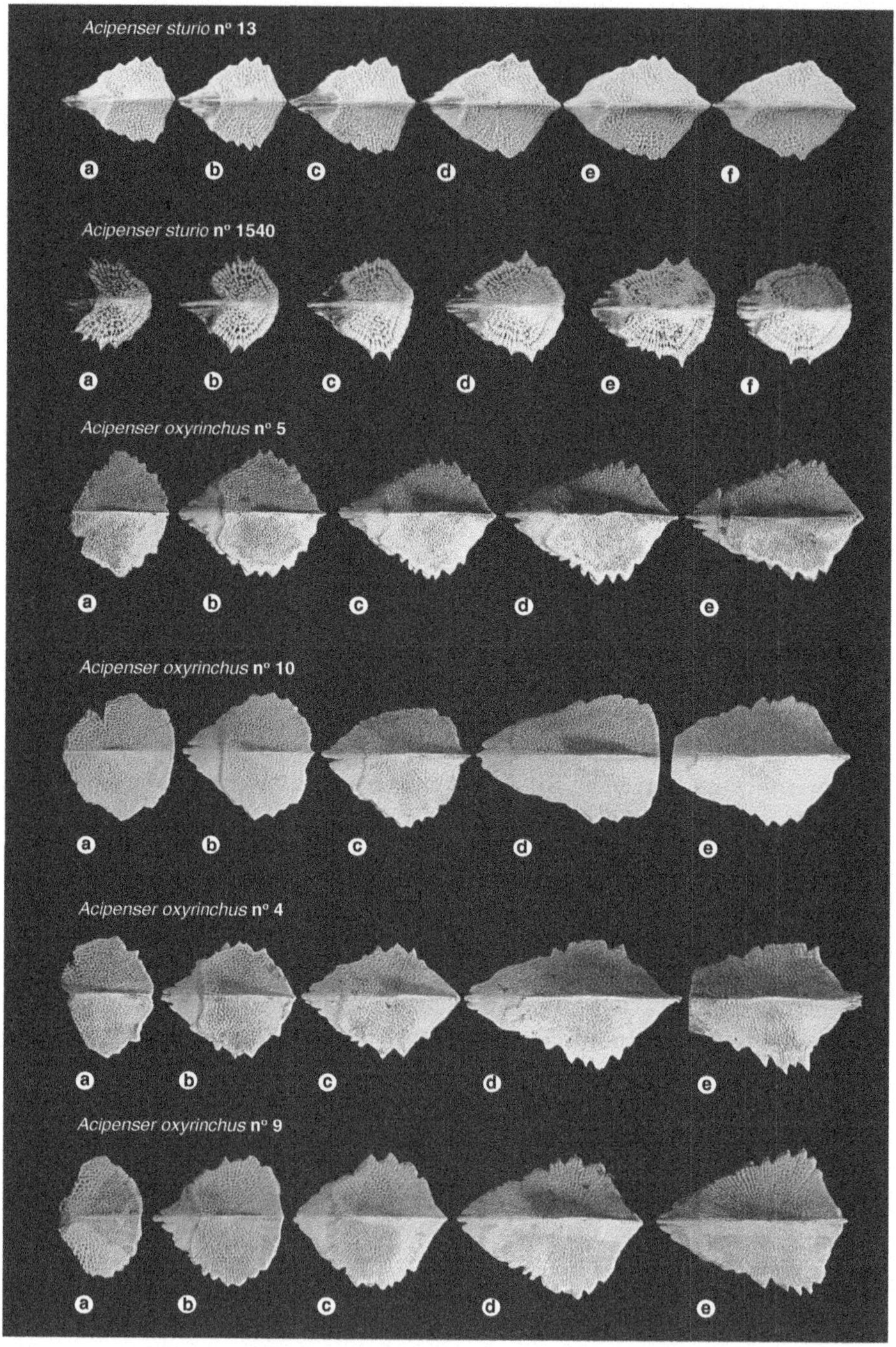

Fig. 3.6 Dorsal scutes, in the cranio-caudal anatomical order: morphological differences between *A. sturio* (first and second line) and *A. oxyrinchus* (third to sixth line). These sets show the individual variations within the same species and their constancy. Some scutes bear the marks of saw cuts because these specimens were collected after the sawing made by the Canadian company "Acadian Sturgeon and Caviar Inc." (third line: i; fourth line: e, h; fifth line: e, h; sixth line: f, i)

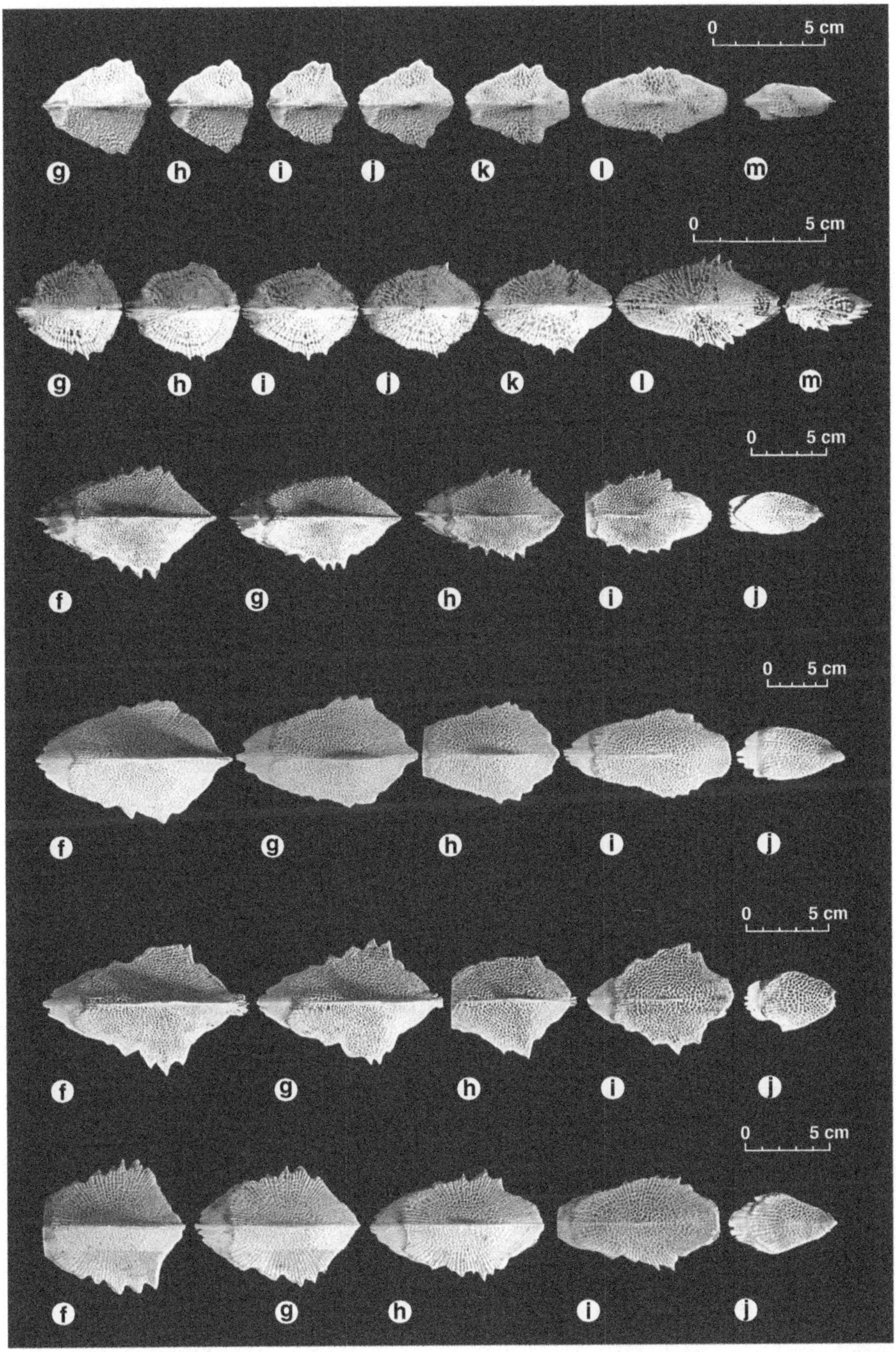

Fig. 3.6 (Continued)

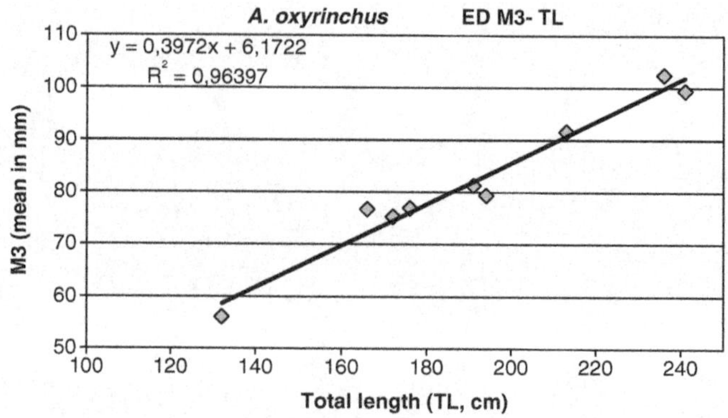

Fig. 3.7 Mean width of dorsal scutes of *A. oxyrinchus* showing an excellent correlation between the relation M3 (width)/TL (Total Length) ($R^2 = 0.964$). The first and the last scutes have not been considered. The size distribution is not significantly different than the normal (Kolmogorov–Smirnov's test, $p = 0.05$)

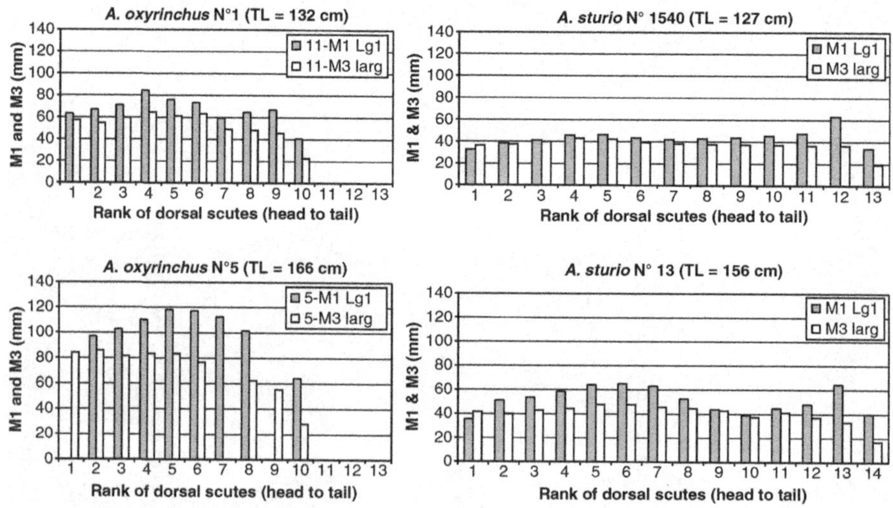

Fig. 3.8 Histograms showing the relations between two average measurements (M1–M3), made on *A. sturio* and *A. oxyrinchus* dorsal scutes in our collection

3.3.3 A. sturio *and* A. oxyrinchus: *Elements of Osteometry*

3.3.3.1 Dorsal Scutes

The dorsal scutes (or dorsomedians) have been favoured, because they fit better than lateral or ventral scutes for size reconstitutions on archaeologic scutes, generally fragmentary. Six measures have been made on each scute: two lengths or antero-posterior diameters (one of the total of the scute = M1, the other without the recovering part = M2), three width or transverse diameters (one of the total = M3, the two others of the left half = M4 and of the right half = M5), and finally the biggest thickness of the scute (= M6). It is thus possible to reconstitute the sizes, even on broken archaeological material.

The measures selected here are M1 and M3 (Fig. 3.1). For the two species, M3 varies little according to the row of the scute and always remains inferior to M1. It is the best measure for TL reconstitutions. The example shown here (Fig. 3.7) proves the excellent correlation ($R^2 = 0.964$) between M3 and TL for *A. oxyrinchus*.

For *A. sturio*, the relationships of the measures M1/M3 on the dorsal scutes show little difference and are relatively regular. The lengths are always very slightly superior to the widths, except for the first scute, whose width is equal or slightly superior to the length. Only the last but one scute systematically distinguishes itself by its length, significantly greater than the others (Fig. 3.8: histograms of two *A. sturio* compared to two *A. oxyrinchus*).

For *A. oxyrinchus*, the dorsal scutes have lengths (M1) always superior to the widths (M3). Their dimension is markedly higher from the fourth until the seventh scute, compared to the three first and to the three last scutes (Fig. 3.9: histograms of ten *A. oxyrinchus*).

Both species *A. sturio* and *A. oxyrinchus* show clear dorsal scute differences between one another. The three or four scutes from the head have for both species the same proportions (equivalent length/width), but the subsequent scute dimensions are very different (Fig. 3.8).

Contrary to what has been written (Vladykov and Greeley 1963, in Ludwig and Gessner 2007), the more or less square or oval shape of the scute depends on its row in the dorsal set, and is not bound to the species (cf. Figs. 3.6, 3.8, and 3.9). Moreover, individual differences are frequent. But the main difference concerns the relationship between the scute measurements (M1: cranio-caudal diameter or length; M3: transverse diameter or width) and the fish TL, which is different between the two species. This observation is an important one for sturgeon size reconstitutions made from isolated scutes. A measurement of identical value made on a bone will noticeably vary, depending on which species it refers to, which is even more explicit as the sizes increase. This highlights the inter-specific differences between the relations of some measurement points of the bones and the fish TL.

The example given by Benecke (1986) and followed by Debus (1999) is a great illustration. From a 121-mm-long lateral scute, originating from the Ralswiek site, Benecke proposed a 5-m-long TL reconstitution. This calculation is based on the largest lateral scute measurement of *A. sturio*. This publication was prior to the knowledge of the presence of a species other than *A. sturio*. And yet according to the surface morphology (visible on Fig. 12 in Debus 1999, as well as in the caption indicating "alveolated structure of scutae dorsale from *Acipenser sturio*"), it is an *A. oxyrinchus* scute. By the way, Debus considers this size reconstitution to be an overestimation: according to his measurements on lateral scutes of "Baltic Sea *Acipenser sturio*", the length calculated from the largest lateral scute was 275 cm

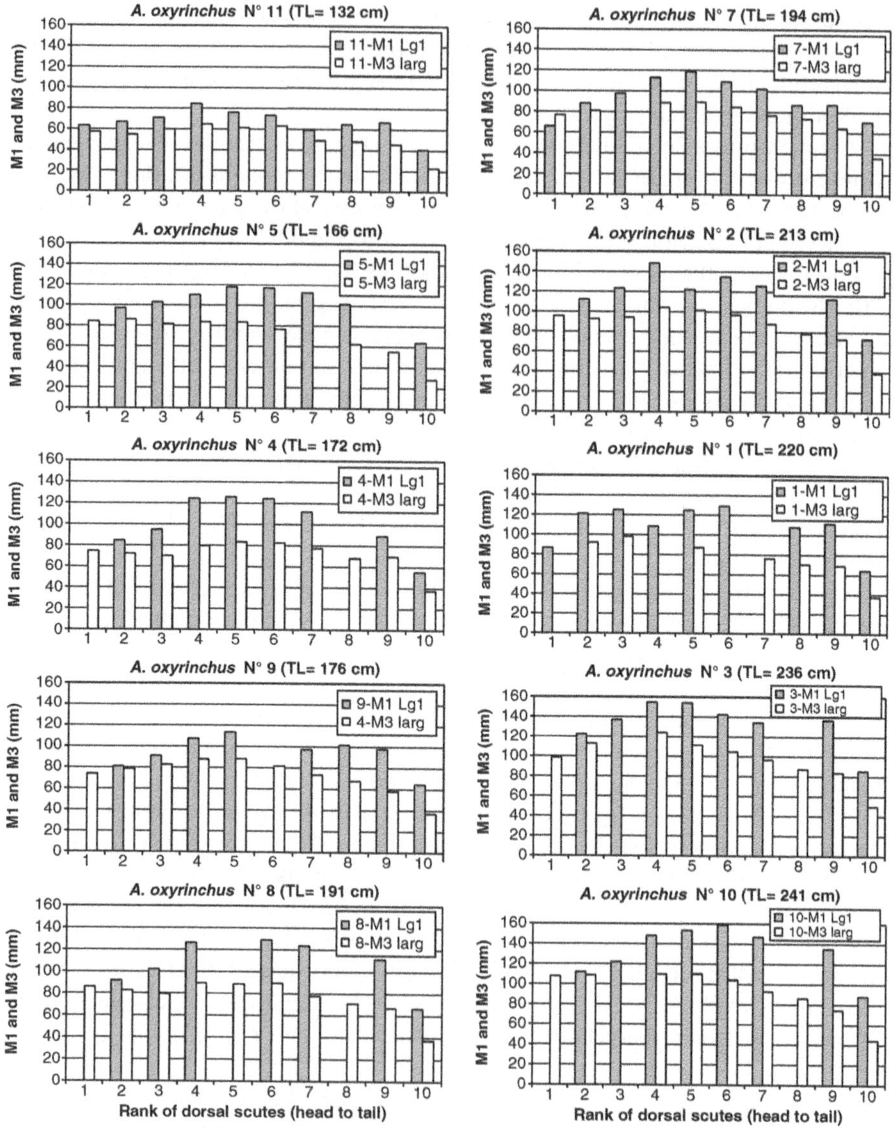

Fig. 3.9 Histograms showing the relations between two average measurements (M1–M3), made on every *A. oxyrinchus* dorsal scute in our collection

(Debus 1999, Fig. 8). "Baltic Sea *Acipenser sturio*" corresponds in reality to *A. oxyrinchus*.

It is therefore necessary to start by determining the species, then the approximate row of the scute in the cranio-caudal order, before calculating the size (TL) according to this data.

3.3.3.2 Splanchnocranium Bones

The splanchnocranium bones, such as the maxillary, the dentary, and the palato-pterygoid, also reveal a good correlation between the measurements made on the bones and the TL.

Thus on the dentary, the measurement M1 is perfectly correlated with the TL for each of the two species: $R^2 = 0.96$ for *A. sturio* and $R^2 = 0.90$ for *A. oxyrinchus* (Fig. 3.10). A statistic comparison under R shows that the slopes of the two regression lines are significantly different.[1]

Thus, a 5-mm-long M1 would correspond to a 1,005 mm LT for *A. sturio*, and a 977 mm LT for *A. oxyrinchus*: the difference would amount to 3%. A 15-mm-long M1 measurement would correspond to a 2,930 mm TL for *A. sturio*, and a 2,217 mm TL for *A. oxyrinchus*: the difference would thus reach 25% (which is important for reconstitution of the sizes of ancient sturgeon).

3.3.3.3 Sizes of Ancient Sturgeon

The main objective of the previous data, after the specific determination, is to recalculate the sizes of ancient sturgeon, originating from precise periods and places. In this book (see Chap. 7), the geographical and chronological distribution of sturgeon remains established in France is presented, along with the distribution of the two species, *A. sturio* and *A. oxyrinchus*, which are determined there.

The size data is essential information, since it makes it possible to give information about the state of a population in a precise location and at a precise time, for a

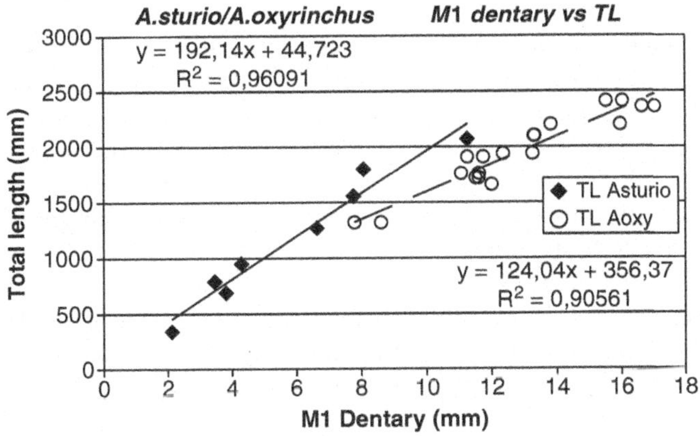

Fig. 3.10 Relationship between TL and M1 dentaries of *A. sturio* and *A. oxyrinchus* in our collection. The *slopes* are significantly different

[1]Corresponds to the slope independence test of a covariance analysis.

group whose special feature is the late age of puberty (at the age of 14 for males and 18 for females before reaching their sexual maturity), and as a result, a length superior to 140 cm for males and 180 cm for females (Magnin 1962).

The size of the biggest individuals during earlier centuries has been passed on to us thanks to textual data, and for recent periods thanks to photographs (see Chap. 19).

The maximum mentioned sizes vary according to naturalists. Sauvage (1883) writes that the common sturgeon can reach a size of 5–6 m long. In the sixteenth century the French naturalist Pierre Belon reports that during King François 1's stay in Montargis, he was shown a 5.40-m-long sturgeon that had been captured in the Loire River. Several authors reported the same information (Larousse 1870; Moreau 1897; Sauvage 1883).

One cannot resist mentioning the empirical size reconstitution method suggested by Jules Verne (1863) in "5 weeks in a Balloon": "After the meeting, the doctor was led to the *Traveller's Club*, in Pall Mall; a magnificent feast was waiting for him there; the size of the served pieces was equivalent to the degree of importance of the personage, and the sturgeon which appeared in this splendid meal measured 3 inches less in length than Samuel Fergusson himself"...). The sturgeon served at this banquet represented a luxury meal offered to Jules Verne's hero, and if the man was 170 cm tall, the fish must have reached 160 cm long, which is very likely to be true (but one must not forget that it is a novel!).

Apart from the reconsideration of some ancient references, one must consider the possible confusion which could have arisen in the past in the determination of the species: in western Europe, each sturgeon until recent times has automatically been considered as *A. sturio*; and yet it is very likely that some *A. oxyrinchus* could have been present (Desse-Berset 2009a, b; Desse-Berset and Williot 2011; see Chap. 7). Some sizes attributed to *A. sturio* could refer to the other species, *A. oxyrinchus*.

The great naturalist G. Rondelet in "L'Histoire entière des Poissons" published in 1553 illustrates his chapter VIII entitled "de l'esturgeon" with a very precise engraving of a specimen on which one can count nine dorsal scutes (Rondelet 1553, re-edition 2002). It would be tempting to interpret this illustration as an *A. oxyrinchus*.[2]

For *A. sturio*, the biggest size ever mentioned is 5.5–6 m (Sauvage 1883; Jenkins 1925; Duncker 1935, in Holcik 1989) and 1,000 kg in weight (Grubisic 1967, in Holcik 1989).

In France, Ernest Laporte saw 5-m-long sturgeons in the Gironde River, and commonly 2-, 3- and even 4-m-long sturgeons (Laporte 1853). Roule (1922) reports

[2]One can also read with interest and attention Rondelet's review on the name attribution of the word *silurus* to the sturgeon in Ausone's poem ("la Moselle"), an attribution which he reconsiders. The word "silurus" has sometimes been thought to be a synonym of sturgeon, but the translation of "Magne Silure", according to Rondelet, refers to *Silurus glanis*, and not to sturgeon (unlike the opinion of his contemporary Paul Jove, 1531), which is also attested by today's specialists. His review is also based on his excellent observation of the customs of these two species.

that some sturgeons reach up to 80–100 kg for lengths going from 2 to 3 m (Roule 1922), some specimens could even reach up to 4–5 m in length and several hundreds of kilos (Roule 1925). For Letaconnoux (1961), the biggest sturgeon unloaded in La Rochelle measured 3.25 m for 215 kg, but the majority of these fish measured less than 2.20 m. The biggest mentioned by Magnin (1962) measured 255 cm long.

For *A. oxyrinchus*, the biggest specimen ever recorded was 4.30 m and 367 kg (Leim and Scott 1966; Scott and Crossman 1973).

3.3.3.4 Archaeological Data

Since textual data is not always reliable, we will rather refer to factual data such as sturgeon bones, which do not lie. The presence of large-sized specimens in the archaeological collections provides proof of the presence of breeders in ancient populations, during defined periods and in defined drainages. The largest sizes will be mentioned, which does not in the end refute the existence of young individuals in these ancient faunas.

According to the methods previously described, sturgeon bones have served for size reconstitutions within the study of archaeological sites from which they originate. Some examples have been chosen to illustrate this point (for more data on species determination, see also Chap. 7).

North Sea, Scheldt basin: the city of Douai (see Chap. 7, Fig. 7.6c, d), located along the Scarpe–Scheldt River (ninth to tenth century AD) delivered, among the 25 differently-sized sturgeon remains, two bones which were part of the pectoral girdle: one clavicle and one very large right cleithrum. These remains may originate from the same individual. It is an *A. oxyrinchus*, whose TL exceeded 2 m.

On the Atlantic coast, several sites prove the presence of large-sized individuals.

The Gallo-Roman site of Le Langon (see Chap. 7, Fig. 7.4c, d) (first century AD) has provided fragments of very large dermal plates (22 remains) (Fig. 7.4c), which refer exclusively to *A. oxyrinchus* (Desse-Berset 2009a). The scutes, often fragmentary, all belong to large-sized individuals. Those which allow a TL estimation show sizes going from 2 m to 2.40 m. They originate from three distinct archaeological places in the settlement, and therefore refer to at least three individuals.

In Nantes (see Chap. 7, Fig. 7.5a, b), along the Loire River, during the Roman period, *A. oxyrinchus* scutes also reached sizes bigger than 2 m (2.10 to 2.20 m).

On the Oléron island, in Ponthezières, the only splanchnocranium bone ever found on the Neolithic site is a dentary one (Desse-Berset 2009b) (Fig. 3.11a). This bone, fully preserved, belongs to a large-sized specimen. Its specific determination is complex because of the erosion of its outline. The dentary of the *A. sturio* and *A. oxyrinchus* species are in fact very similar (Figs. 3.5 and 3.11b, c), and the discrimination criteria between the two species proposed hereafter seem to depend on this large-sized bone: this dentary owns some characteristics from each species. The strongly marked appendix, typical of *A. sturio*, is here eroded, and it is hard to say if it has been damaged by time, or if it was absent from the beginning, which would make the *A. sturio* look like an *A. oxyrinchus*. However, its general

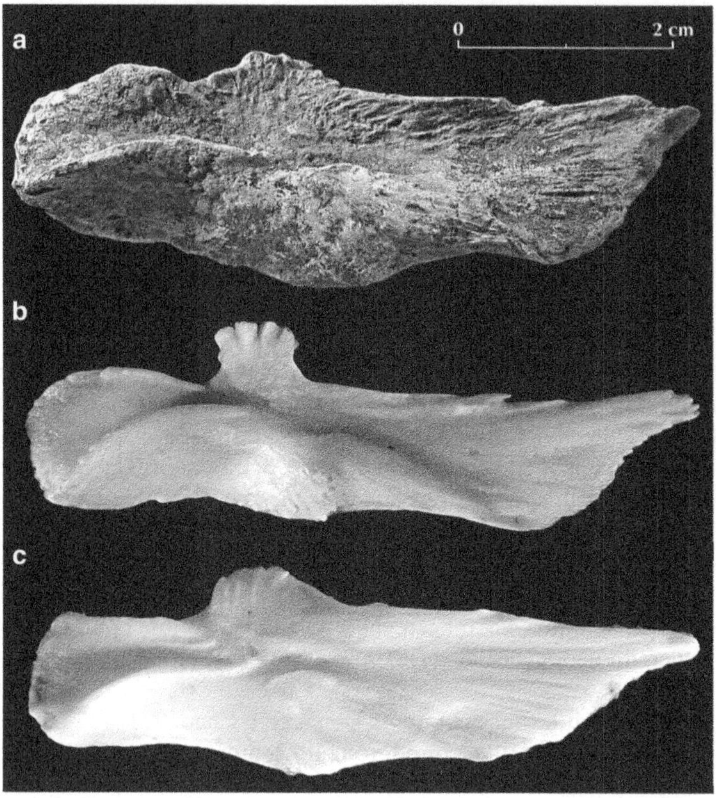

Fig. 3.11 **a** Neolithic dentary from Ponthezières on the Atlantic coast (layer 3, e10 d2): only bone belonging to the internal skeleton found on the site, it is very large and widely exceeds our biggest reference specimen. **b** Modern dentary of *A. sturio* (origin: Gironde Estuary, France) enlarged. **c** Modern dentary of *A. oxyrinchus* (origin: Saint John River, Canada) enlarged

morphology is stocky, and more similar to that of *A. oxyrinchus* (Fig. 3.11c). Maybe this could be a hybrid, which might be clarified by genetics.

Initially, TL was calculated according to measurements made on modern *A. sturio* bones (Desse-Berset 2009b, Figs. 504 and 505). The masses were obtained using the LT/mass relationship, which is a power curve corresponding to: mass $= 7 \times 10^{-07} \times \mathrm{LT}^{3.2848}$ (mass in g and TL in mm). The R^2 obtained here is 0.995. The measured dentary corresponded to a specimen of 3.40 m and a mass of more than 260 kg (~262 kg).

In the meantime, two new large-sized *A. sturio* specimens have been added to our reference collection, thus increasing the data base and slightly modifying the size regression. The estimation of the total length calculated according to the *A. sturio* model gives a 3.06 m TL, which would correspond to a live weight of 202 kg (cf. Fig. 3.10).

If this dentary belonged to an *A. oxyrinchus*, the reconstitution calculation of the TL would have a slightly inferior value, that is to say 230 cm (for a calculated mass of 70 kg).

One can thus notice that the difference between the TL(s) of both species would reach 25%, which is significant.

The other ossified sturgeon remains found on this site are by a majority ossified dermal plates and fragmentary scutes. They almost all belong to large-sized specimens, presenting the alveolar surface morphology typical of *A. oxyrinchus* (Desse-Berset 2009a; see Chap. 7, Fig. 7.4a). Two scutes originating from two different layers (Co 3 and Co 5) correspond to 2.20 m sizes. A subopercular fragmentary is much bigger in size than our largest reference specimen which measures 2.40 m long, and the bone is much thicker.

The only bone originating from a small-sized specimen possesses the typical tubercular morphology of an A. *sturio* (see Chap. 7, Fig. 7.4b).

Mediterranean Façade

Arles (Rhône Valley, France): an ancient population of sturgeons (Fig. 3.12). The numerous remains collected on this site have made it possible to build a size distribution table of this 2,500-year-old population, attested over a period of

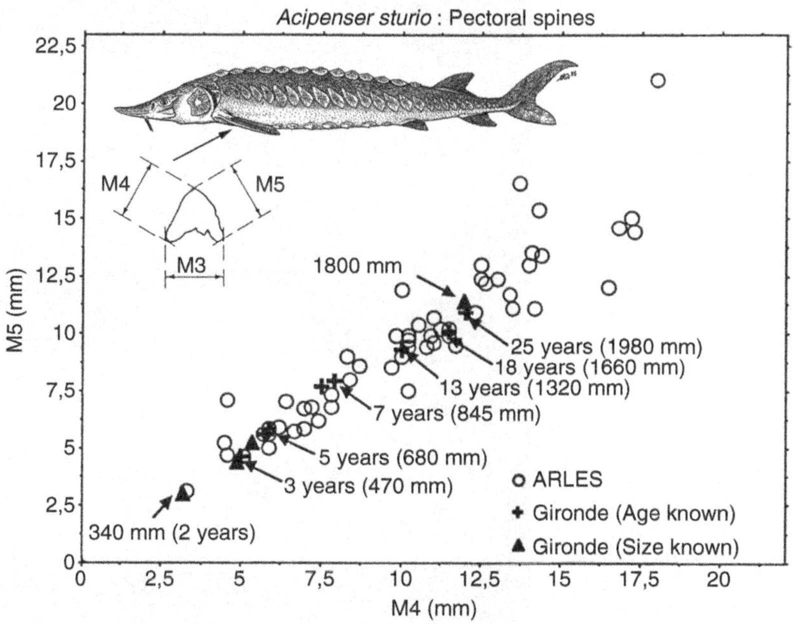

Fig. 3.12 Size distribution of the ancient sturgeon population (*A. sturio*) from Arles based on the following information: two measurements (M4 and M5) of 70 archaeological pectoral spines (*open circles*); the same measurements on some modern specimens from our reference collection of known TL (*dark triangles*) and age (*dark cross*) from the Gironde (from the Cemagref collection) and the age/TL distribution of *A. sturio* from the Gironde, established on ca. 100 sturgeons from Gironde by Magnin (1962) (after Desse-Berset 1994, Fig. 6)

five centuries. The anatomical distribution of all the archaeological remains of this fish is presented as follows: cranium dermal bone (4%); splanchnocranium (17%); pectoral girdle (30%); dermal scutes (49%).

In an earlier publication, histograms illustrated the size distribution of the sturgeons, carried out from sets of measurements made on splanchnocranium bones (maxillary, dentary, palatopterygoid) and also pectoral spines (Desse-Berset 1994, Figs. 7–9).

The pectoral spines from Arles represent 7% of the determined bones. Size estimation of the ancient specimens was obtained using the following information: two measurements taken on the base of each of 70 protohistoric pectoral spines; the same measurements taken from some modern sturgeons of known length and age from the Gironde; and the age/size distribution of *A. sturio* from the Gironde established by Magnin (1962) (Desse-Berset 1994, Fig. 5).

By cross-referencing these different data, they can provide the image of sturgeon populations in the Rhône River, between the sixth and the second century BC. They show that half of the individuals were larger than 145 cm, and one third exceeded a

Fig. 3.13 Fragmentary dentary of an exceptionally large sturgeon (*A. sturio*, Arles JH 058; second half of the fifth century BC), whose TL reach 5.63 m: **a** external side; **b** internal side; one can observe cut marks visible in two places; **c** anatomical positioning of this fragmentary bone on the enlarged photograph of a modern *A. sturio* dentary

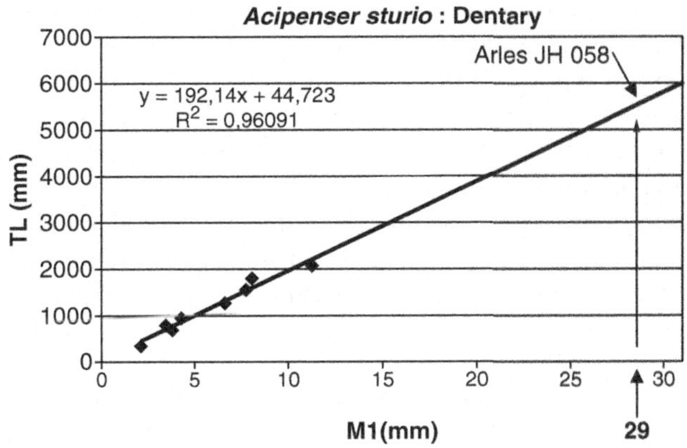

Fig. 3.14 TL reconstitution of the Arles sturgeon (*A. sturio*, Arles JH 058) (cf. Fig. 3.13) according to the M1 (M1 = 29 mm) measurements made on this dentary (TL = 5.63 m for a calculated weight of 1,500 kg)

size of 2 m (Desse-Berset 1994, Fig. 6) (Fig. 3.12). Similar information is obtained from all the bones of the splanchnocranium used for the reconstruction of size.

Some individuals reach exceptional sizes. As was said previously, the site of Arles has only delivered one sturgeon species. Moreover, the archaeozoological discriminations have been confirmed by the palaeogenetic data (Desse-Berset et al. 2008; Pagès et al. 2009). The discrimination between *A. sturio* and *A. oxyrinchus* is no longer a problem here, because only *A. sturio* is present. A very large fragmentary bone has been determined to be a dentary by osteological analysis (Fig. 3.13). Its particularly large measurements made it possible to reconstitute the exceptional size of the fish: its calculated TL measures in fact 5.63 m (and this is not the estimate of a Mediterranean fisherman!) (Fig. 3.14).

Moreover this bone bears marks of deep anthropic cuts, attesting that the animal has been cut up in situ, which is not surprising considering the exceptional weight of the fish: 1,500 kg, according to the formula already used for the dentary from Oléron (LT/mass relationship, which is a power curve corresponding to: mass $= 7 \times 10^{-07} \times \mathrm{LT}^{3.2848}$; the R^2 obtained here is 0.995).

3.4 Conclusion

The results presented here required the dissection of 20 sturgeons from the three species mentioned (*A. sturio*, *A. oxyrinchus* and *A. naccarii*), the preparation of their different bones, the definition of measurement points, and the implementation of thousands of mensurations. The processing of these data provides results that allow researchers who are interested by these species to answer specific discrimination problems and make size reconstitution calculations.

The presented methods have been applied to bony remains of several ancient sturgeon populations of the Mediterranean Sea, of the Atlantic Ocean and of the Channel–North Sea. They inform us that sturgeons of all kinds of sizes, belonging to both the species *A. sturio* and *A. oxyrinchus*, often lived in sympatry, in the drainages of all French rivers for millennia, and that until recent times (with the exception of the Mediterranean catchment, where only *A. sturio* was determined). Their moderate to large size, sometimes very large, attests that they were possible breeders, and that sturgeon populations found suitable environmental and climatic conditions in the past before various consequences of man's action led to their extinction.

Acknowledgements I sincerely thank all those who supplied me with modern sturgeons of *A. sturio*, *A. naccarii* and *A. oxyrinchus*, and particularly the Cemagref, France; the fish farm Giovanini, Azienda Agricola V.I.P., Italy; the Osteotheque of the University of Montreal, Canada; and the company "Acadian Sturgeon and Caviar Inc.", St John, Nouveau Brunswick, Canada.
I also thank JD Strich for the photographs, A Pasqualini for the infography of the illustrations (CEPAM/CNRS-University of Nice–Sophia Antipolis) and JM Paillard ("Starboard Silent Side") for the revision of the English version of this paper.
Finally, I am very grateful to P. Williot for all his help with "my" sturgeons, from Canada to Bordeaux-Audenge.

References

Almodovar A, Machordom A, Suarez J (2000) Preliminary results from characterization of Iberian Peninsula sturgeon based on analysis of the mt-DNA cytochrome b. Bol Inst Esp Oceanogr 16:17–27

Bartosiewicz L, Bonsall C (2008) Complementary taphonomies: Medieval sturgeons from Hungary. In: Béarez P, Grouard S, Clavel B (eds) Archéologie du poisson. Trente ans d'archéo-ichtyologie au CNRS. Hommage aux travaux de Jean Desse et Nathalie Desse-Berset. XVIIIe Rencontres Internationales d'Archéologie et d'Histoire. APDCA, Antibes, pp 35–45

Bartosiewicz L, Takacs I (1997) Osteomorphological studies on the great sturgeon (*Huso huso* Brandt). Archaeofauna 6:9–16

Benecke N (1986) Some remarks on sturgeon fishing in the southern Baltic region in medieval times. In: Brinkhuizen DC, Clason A (eds) Fish and archaeology. Studies in osteometry, taphonomy, seasonality and fishing methods, BAR Intern Series 294. BAR, Oxford, pp 9–17

Brinkhuizen DC (1986) Features observed on the skeletons of some recent European Acipenseridae: their importance for the study of excavated remains of sturgeon. In: Brinkhuizen DC, Clason A (eds) Fish and archaeology. Studies in osteometry, taphonomy, seasonality and fishing methods, BAR Intern Series 294. BAR, Oxford, pp 18–33

Brinkhuizen DC (1989) Ichthio-Archeologisch onderzoek: methoden en toepassing aan de hand van romeins vismaterial uit Velsen (Nederland). Ph. D, Rijksuniversiteit Groningen

Brinkhuizen DC (2006) Fish. In: Kooijmans LL, Jongste PFB (eds) Schipluiden, a Neolithic settlement on the Dutch North Sea Coast c. 3500 cal BC, vol 37/38. Analecta Praehistorica Leidensia, Leiden, pp 449–470

Brosse L, Desse-Berset N, Berrebi P, Lepage M, Menella JY (2006) Travaux préalables à la restauration de l'esturgeon dans le Rhône. Rapport final, Association MRM, Arles

Brosse L, Berrebi P, Desse-Berset N, Lepage M (2009) Sturgeon recovery plan in the Rhône River (France). Preliminary results on species determination and habitat suitability. In: Carmona R, Domezain A, Garcia-Gallego M, Hernando JA, Rodriguez F, Ruiz-Rejon M (eds) Biology, conservation and sustainable development of sturgeons, vol 29, Fish & fisheries series. Springer, New York, pp 403–421, Chapter 25

Chassaing O (2010) Organisation génétique des populations d'esturgeon européen *Acipenser sturio*: passé, présent et futur. PhD dissertation, Université Montpellier 2

Clavel B (2001) L'animal dans l'alimentation médiévale et moderne en France du Nord (XIIe-XVIIe siècles). Revue Archéologique de Picardie N° Spécial 19:204

De La Herran R, Robles F, Lorente JA, Ruiz Rejon C, Garrido-Ramos MA, Ruiz Rejon M (2004) Genetic identification of western Mediterranean sturgeons and its implication for conservation. Conserv Genet 5:545–551

Debus L (1999) Meristic and morphological features of the Baltic sturgeon (*Acipenser sturio* L.). J Appl Ichthyol 15:38–45

Desse J, Desse G, Desse-Berset N (2002) L'Archéoichtyologie. In: Miskovsky JJ (ed) Géologie de la Préhistoire (Réédition). Géopré, Paris, pp 815–822, P 1519

Desse-Berset N (1994) Sturgeons of the Rhône during Protohistory in Arles (6th-2nd century BC). In: Fish exploitation in the past, proceedings of the 7th meeting of the ICAZ Fish Remains Working Group (Louvain, Sept. 1993), vol 274. Annales du Musée royal de l'Afrique centrale, Tervueren, pp 81–90

Desse-Berset N (2009a) First archaeozoological identification of Atlantic sturgeon (*Acipenser oxyrinchus* Mitchill 1815) in France. CR Palevol 8:717–724. doi:10.1016/j.crpv.2009.06.001

Desse-Berset N (2009b) La pêche dans l'économie de subsistance des sites de Ponthezières et de la Perroche (île d'Oléron, Néolithique final). In: Laporte L (ed) Des premiers paysans aux premiers métallurgistes sur la façade atlantique de la France (3500–2000 av JC). Mémoire XXXIII . Association des Publications Chauvinoises, Chauvigny, pp 584–609

Desse-Berset N, Williot P (2011) Emerging questions from the discovery of the long term presence of *A. oxyrinchus* in France. J Appl Ichthyol 27:263–268 Blackwell Verlag, Berlin doi:10.1111/j.1439-0426.2010.01649.x

Desse-Berset N, Pagès M, Brosse L, Tougard C, Chassaing O, Hänni C, Berrebi P (2008) Specific identification of the extinct population of sturgeon from the Rhône River by mtDNA analysis from bone remains (Jardin d'Hiver, Arles, France, 6th to 2nd century BC). In: Béarez P, Grouard S, Clavel B (eds) Archéologie du poisson. Trente ans d'archéo-ichtyologie au CNRS. Hommage aux travaux de Jean Desse et Nathalie Desse-Berset. XVIIIe Rencontres Internationales d'Archéologie et d'Histoire. Éditions APDCA, Antibes, pp 195–200

Doukakis P, Birstein VJ, Desalle R, Ludwig AN, Ludwig A, Machordom A, Almodovar A, Elvira B (2000) Failure to confirm previous identification of two putative museum specimens of the Atlantic sturgeon, *Acipenser sturio*, as the Adriatic sturgeon, *A. naccarii*. Mar Biol 136: 373–377

Duncker G (1935) Die Fische der Nordmark. Lipsius and Tischer, Kiel

Elvira B, Almodovar A (2000) Further observations on the morphological characters of *Acipenser sturio* L., 1758 from the Iberian Peninsula: A comparison with North and Adriatic Sea populations. Bol Inst Esp Oceanogr 16(1–4):89–97

Elvira B, Almodovar A, Lobon Cervia J (1991) Sturgeon (Acipenser sturio L., 1758) in Spain. The population of the River Guadalquivir: a case history and a claim for a restoration programme. In: Williot P (ed) Acipenser. Cemagref, Antony, pp 337–347

Findeis E (1997) Osteology and phylogenetic interrelationships of sturgeons (Acipenseridae). Environ Biol Fishes 48:73–126

Fischer W, Schneider M, Bauchot ML (rédacteurs) (1987) Fiches FAO d'identification des espèces pour les besoins de la pêche (révision 1). Méditerranée et Mer Noire. Zone de pêche 37. Volume II. Vertébrés. Publication préparée par la FAO, résultat d'un accord entre la FAO et la Commission des Communautés Européennes (Projet GCP/INT/422/EEC) financée conjointement par ces deux organisations, vol 2. Rome, FAO, pp 761–1530

Garrido-Ramos MA, Soriguer MC, De La Herran R, Jamilena M, Domezain A, Hernando JA, Ruiz-Rejon M (1997) Morphometric and genetic analysis as proof for the existence of two sturgeon species in the Guadalquivir River. Mar Biol 129:33–39

Goodrich ES (1930) Studies on the structure and development of vertebrates. Macmillan, London

Grassé PP (sous la direction de) (1958) Traité de zoologie. Anatomie, systématique, biologie. Tome III, Agnathes et poissons. 3 fascicules. Masson, Paris, p 2758

Gregory WK (1933) Fish skulls: a study of the evolution of natural mechanisms. Am Philos Soc 23:75–481

Grubisic F (1967) Ribe, rakovi i skoljke Jadrana. Jugoriba, Zagreb

Heinrich D (1987) Untersuchungen an mittelalterlichen Fischresten aus Schleswig. Ausgrabung Schild 1971–1975. Ausgrabungen in Schleswig, Berichte und Studien 6. Karl Wachholtz Verlag, Neumünster

Hilton EJ, Bemis WE (1999) Skeletal variation in shortnose sturgeon (*Acipenser brevirostrum*) from the Connecticut River: implications for comparative osteological studies of fossil and living fishes. In: Arratia G, Schultze HP (eds) Mesozoic Fishes 2 – systematics and fossil record. Verlag Dr. Friedrich Pfeil, München, pp 69–94

Holcik J (1989) The freshwater fishes of Europe. Vol I, Part II: General introduction to fishes, Acipenseriformes. Aula Verlag, Wiesbaden, 471 p

Holcik J, Kinzelbach R, Sokolov LI, Vasil'ev VP (1989) *Acipenser sturio* Linnaeus, 1758. In: Holcik J (ed) The freshwater fishes of Europe. Vol I, Part II: General introduction to fishes, Acipenseriformes. Aula Verlag, Wiesbaden, pp 367–394, 470 p

Holmgren N, Stensiö E (1936) Kranium und Visceralskelett der Akranier, Cyclostomen und Fische. In: Bolk L, Kallius E, Lubosch W (eds) Handbuch der vergleichenden Anatomie der Wirbeltiere, vol IV. Urban & Schwarzenberg, Berlin, pp 233–500

CITES Identification Guide (2001) Sturgeons and paddlefish: guide to the identification of sturgeon and paddlefish species controlled under the Convention on International Trade in Endangered Species of Wild Fauna and Flora

Jarvik E (1948) On the morphology and taxonomy of the Middle Devonian osteolepid fishes of Scotland. Kungliga Svenska Vetenskapsakademiens Handlingar 3(25):1–301

Jenkins JT (1925) The fishes of the British Isles. F Warne, London

Laporte E (1853) Faune Ichtyologique du département de la Gironde. In: Actes de la Société Linnéenne de Bordeaux, tome XIX, Deuxième série, tome IX. Th Lafargue, Bordeaux, pp 157–224

Larousse P (1870) Grand Dictionnaire universel du XIXe siècle, vol 16. Pierre Larousse, Paris

Leim AH, Scott BW (1966) Fishes of the Atlantic coast of Canada. Bull Fish Res Board Can (155):485 DOI, http://dx.doi.org

Lelek A (1987) The freshwater fishes of Europe. Vol 9 Threatened fishes of Europe. Wiesbaden, Aula Verlag (chapitre 3.33 Acipenseridae: 42–57)

Letaconnoux R (1961) Fréquence et distribution des captures d'esturgeons (*Acipenser sturio* LINNE,1758) dans le Golfe de Gascogne. Rev Trav Inst Pêches Marit 25(3):253–261

Ludwig A, Gessner J (2007) What makes the difference? Sea Sturgeon on both sides of the Atlantic Ocean. Am Fish Soc Symp 56:285–300

Magnin E (1962) Recherches sur la systématique et la biologie des Acipenséridés *Acipenser sturio* L, *Acipenser oxyrhynchus* Mitchill, et *Acipenser fulvescens* Raf. Ann Stat Centr Hydrobiol Appl 9:7–242

Magnin E (1964) Validité d'une distinction spécifique entre les deux Acipensérідés: *Acipenser sturio* L. d'Europe et *Acipenser oxyrhynchus* d'Amérique du Nord. Travaux sur les Pêcheries du Québec 1:5–20

Magnin E, Beaulieu G (1963) Etude morphométrique comparée de l'*Acipenser oxyrinchus* Mitchill du Saint Laurent et de l'*Acipenser* sturio Linné de la Gironde. Le Naturaliste Canadien XC(1):5–38

Makowiecki D (2008) Sturgeon fishing on Polish Lowland during Holocene. In: Béarez P, Grouard S, Clavel B (eds) Archéologie du poisson. Trente ans d'archéo-ichtyologie au

CNRS. Hommage aux travaux de Jean Desse et Nathalie Desse-Berset. XVIIIe Rencontres Internationales d'Archéologie et d'Histoire. Éditions APDCA, Antibes, pp 327–339

Moreau E (1897) Bulletin de la Société des Sciences historiques et naturelles de l'Yonne, vol 51. Paris

Pagès M, Desse-Berset N, Tougard C, Brosse L, Hänni C, Berrebi P (2009) Historical presence of the sturgeon *Acipenser sturio* in the Rhône basin determined by the analysis of ancient DNA cytochrome b sequences. Conserv Genet 10:217–224. doi:10.1007/s10592-008-9549-6

Rochard E, Williot P, Castelnaud G, Lepage M (1991) Eléments de systématique et de biologie des populations sauvages d'esturgeons. In: Williot P (ed) Acipenser. Cemagref, Antony, pp 475–507

Rondelet G (2002) L'histoire entière des poissons (traduction française de l'original paru en deux volumes, sous le titre: Libri de piscibus marinis in quibus verae piscium effigies expressae sunt, Lyon, 1554, et Universae aquatilium histotiae pars altera, Lyon, 1555), Lyon, Mace Bonhome, 1558, réimpression par Meunier FJ, avec une préface par Meunier FJ et D'hondt JL. Editions du Comité des Travaux Historiques et Scientifiques, Paris

Roule L (1922) Etude sur l'Esturgeon du Golfe de Gascogne et du Bassin Girondin. Office Scientifique et Technique des Pêches Maritimes, Notes et Mémoires 20, 12 p

Roule L (1925) Les poissons des eaux douces de la France. Presses Universitaires de France, Paris, 228 p

Sauvage HE (1883) La grande pêche. Jouvet, Paris, 315 p

Scott WB and Crossman EJ (1973) Poissons d'eau douce du Canada. Bulletin 184. Ministère de l'Environnement. Service des Pêches et des Sciences de la Mer, Ottawa, p 1026

Sokolov LI, Berdichevskii LS (1989) Acipenseridae Bonaparte, 1831. In: Holčik J (ed) The freshwater fishes of Europe. Aula Verlag, Wiesbaden, pp 150–153

Tortonese E (1989) *Acipenser naccarii* Bonaparte, 1836. In: Holcik J (ed) The freshwater fishes of Europe. General introduction to fishes Acipenseriformes. Aula Verlag, Wiesbaden, pp 285–293, 470 p

Verne J (1863) Cinq semaines en ballon. Hetzel, Paris

Vladykov VD, Greeley JR (1963) Order Acipenseroidei. In: n.I. Memoirs Sears Foundation for Marine Research (ed) Fishes of the Western North Atlantic. Yale University, New Haven, pp 24–60

Whitehead P, Bauchot ML, Bauchot JP, Hureau JC, Nielsen J, Tortonese E (1984) Fishes of the North-eastern Atlantic and the Mediterranean, vol 1. UNESCO, Paris

Williot P, Rochard E (coord) (2007) Biologie, exigences environnementales et éléments d'évaluation a priori du risque relatif à plusieurs espèces d'esturgeons susceptibles d'être élevées en pisciculture en France: phases 1 & 2. Etude Cemagref EPBx pour le Ministère de l'Environnement et du Développement Durable, 231 p

Chapter 4
Morphological Distinction Between Juvenile Stages of the European Sturgeon *Acipenser sturio* and the Atlantic Sturgeon *Acipenser oxyrinchus*

Sven Wuertz, Stefan Reiser, Jörn Gessner, and Frank Kirschbaum

Abstract Despite widespread use of genetic species discrimination, morphological characteristics provide the only means of species identification in the field. In Atlantic and European sturgeon parameters used for species segregation have been established for subadults and adults. So far, with regard to juvenile stages of *A. oxyrinchus* currently used for restocking, diagnostic morphological characteristics allowing the distinction between the two species have not been defined. Here, morphometric and meristic characteristics of early juveniles of *A. oxyrinchus* and *A. sturio* are presented, and their value for the species determination in fish of 4–20 cm is assessed. We outline valid diagnostic characteristics in juveniles (number of dorsal and lateral scutes, surface/type of scutes, presence of fontanelle), giving comments on their usefulness with regard to the diagnostic significance in adults.

S. Wuertz (✉)
Leibniz-Institute of Freshwater Ecology and Inland Fisheries, Müggelseedamm 310, 12587 Berlin, Germany

Gesellschaft für Marine Aquakultur mbH, Hafentörn 3, 25761 Büsum, Germany
e-mail: wuertz@igb-berlin.de

S. Reiser
Institute for Hydrobiology and Fisheries Science, University of Hamburg, Olbersweg 24, 22767 Hamburg, Germany

J. Gessner
Leibniz-Institute of Freshwater Ecology and Inland Fisheries, Müggelseedamm 310, 12587 Berlin, Germany

F. Kirschbaum
Faculty of Agriculture and Horticulture, Biology and Ecology of Fishes, Humboldt University, Philippstr. 13, Haus 16, 10115 Berlin, Germany

P. Williot et al. (eds.), *Biology and Conservation of the European Sturgeon Acipenser sturio L. 1758*, DOI 10.1007/978-3-642-20611-5_4,

4.1 Introduction

The European sturgeon *Acipenser sturio* was once distributed from the Black Sea via the Mediterranean and the Eastern North Atlantic to the North, Baltic and White Sea (Holcik 1989). Today, only a single relict population of *A. sturio* is documented for the Gironde-Garonne-Dordogne basin in France (Rochard et al. 1997; Williot et al. 1997, 2002) and *A. sturio* is one of the most threatened fish species globally. Historically, a Baltic subspecies has been discussed based on distinct morphology (Artyukhin and Vecsei 1999). Analysing 33 subadults and adults (32–262 cm) from museums and comparing the data with previous studies on Atlantic sturgeon *A. oxyrinchus* and *A. sturio*, Debus (1999) reported that Baltic sturgeon shared more morphological characteristics with *A. oxyrinchus* than *A. sturio* from the Gironde (Atlantic) and the Rioni River (Black Sea). Among those features, the number of dorsal and ventral scutes and the scute surface clearly resembled the diagnostic features previously identified by Magnin (1964).

Subsequently, Ludwig et al. (2002) found strong evidence in maternally inherited mtDNA as well as morphologic features, that *A. oxyrinchus* became the dominant sturgeon species in the Baltic Sea, replacing the European sturgeon 800–1,200 years ago, although this view has recently been challenged by Tiedemann et al. (2007) suggesting a sex-biased introgression of *A. oxyrinchus*. Given the fact that the two species are closely related (Artyukhin 2006; Birstein and DeSalle 1998; Ludwig et al. 2001; Robles et al. 2004) and hybridization of sturgeon species is frequently met under sympatry (Holcik 1989; Ludwig et al. 2009; Tranah et al. 2004), this scenario needs to be considered in recent restoration, and genetic analysis has to be supported by sophisticated morphologic characteristics. Furthermore, recent findings have even suggested that *A. oxyrinchus* was once present along the northern Atlantic coast of Europe (Desse-Berset 2009; see Chap. 7).

Increasing interest since the early 1990s culminated in an international sturgeon restoration programme (see Chap. 41) focusing on *A. oxyrinchus* for the tributaries of the Baltic Sea, and *A. sturio* for the North Sea and Atlantic region. The first experimental stocking measures were carried out with *A. oxyrinchus* starting in 2006 (Gessner et al. 2010; see Chap. 43) and with *A. sturio* since 2008 (Chap. 21), involving juveniles between 15–75 cm and 20–35 cm respectively. In France, regular stocking has been taking place in the Garonne and Dordogne Rivers since 2007 (see Chap. 30) using similar sized juveniles (4–40 cm). First recaptures from the Baltic and the North Sea indicate wide dispersal and make it necessary to provide means to distinguish the species to allow correct assessments based on species-specific diagnostic characteristics in juveniles.

In the past, morphometric studies of Magnin (1962, 1964) and Magnin and Beaulieu (1963) focussed on subadults and adults (4–252 cm). The studies revealed considerable variability. Magnin (1962, 1964) reported a larger pectoral fin in *A. sturio* up to 1 m total length (TL), a species-specific characteristic which inversed in bigger sized fish. Debus (1999) rejected the shape of the rostrum suggested by Berg (1948), Mohr (1962) and Sokolov and Berdichevski (1989) as a reliable criterion, since he observed atypic short blunt snouts in large individuals. Furthermore, the number of gill rakers and fulera in adult *A. sturio*, *i* decreases with body length (Ninua 1976). In addition to the mentioned uncertainties, to our knowledge no data on morphological characteristics of earlier juveniles are currently available. Consequently, species distinction in juveniles needs to be addressed.

Here a morphologic study of juveniles is presented, aiming at the distinction between the two species. Furthermore, morphological characteristics for species identification in juveniles are critically reviewed with regard to adults (see also Chaps. 3 and 7). Ultimately, a recommendation on characteristics applicable for species identification in juveniles and adults is provided.

4.2 Material and methods

Sturgeon for this study originated from controlled reproduction. In *A. oxyrinchus* wild caught fish of the St John River (Canada) were used for reproduction in 2006 and 2007 and yolk-sac larvae were transported to Germany for ongrowing. In 2006 three females and five males and, in 2007, four females and six males were used for reproduction. Subsequently, the juveniles were reared for experimental release. *A. sturio* originate from a transfer of 350 fish from the first reproduction of the species from the ex situ stock in St Seurin in 2007 (Williot et al. 2009). Material for the study was obtained by utilizing mortalities during the rearing process (emaciated individuals were not considered).

Upon fixation in 4% phosphate-buffered formalin, 47 juvenile *A. oxyrinchus* (4.75–19.34 cm) and 23 juvenile *A. sturio* (4.16–14.08 cm) were examined (Fig. 4.1). To reduce the impact of fixation, mainly hard structures were considered. Both meristic (number of dorsal, lateral and ventral scutes and gill rakers on the first branchial arch) and metric characteristics [eye diameter, mouth opening, fin length (longest lateral length), horizontal angle of the rostrum, total length TL and standard length SL] were included. As described by Debus (1999), largest and smallest scutes of the ventral, dorsal and lateral line were measured in all specimens. Head length was determined parallel to the axis, from the point of the rostrum to the most posterior margin of the operculum (Holcik 1989) and to the most posterior margin of the branchial cavity (hl according to Magnin 1964). All metric characteristics were determined with a digital calliper gauge (Top Craft, Germany) to the nearest 0.01 mm. The angle of the rostrum did not allow a reproducible determination to the nearest degree and was thus determined to the nearest 2° on a drawing board.

The data on the number of scutes in each lateral line represent the mean ± standard deviation of n replicates. Data were analysed upon test for normal distribution by Shapiro–Wilk test and homogeneity of variances by Levene's test (failed, $p > 0.05$) using non-parametric Mann–Whitney U-Test ($p < 0.001$). To account for the multivariate nature of the morphometric and meristic data and to ultimately determine morphological features for differentiation, linear discriminant analysis (LDA) using 12 morphometric characters (scutes: dorsal, lateral and ventral; head: hl, angle, eye, mouth; fins: dorsalis, pectoralis, analis, ventralis, caudalis)

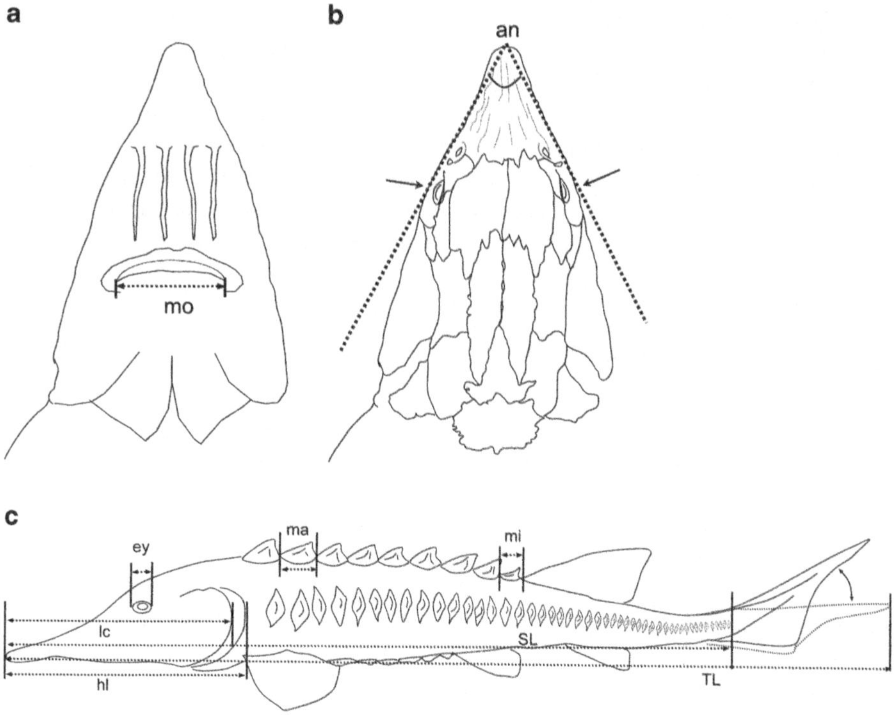

Fig. 4.1 Definition of the morphometric parameters (details see text), ventral (**a**) and dorsal (**b**) view of the head, lateral view (**c**). *mo* mouth opening, *an* horizontal angle of the rostrum, *ey* eye diameter, *TL* total length, *SL* standard length, *ma* largest scutes of the ventral, dorsal and lateral line, *mi* smallest scutes of the ventral, dorsal and lateral line, *hl* head length according to Magnin (1962), *lc* head length to the most posterior margin of the operculum (Holcik 1989)

was carried out. The contribution of each variable to the total sum of Mahalanobis distances was used to identify the most important characteristic in differentiating both species. Morphometric characteristics were divided by TL to correct the data for differences in size and morphometric and meristic data were log- or square root-transformed, respectively, to achieve normality. Prerequisites for LDA were analysed following Zuur et al. (2007). Due to colinearity, lo was excluded from LDA. For the statistical analysis, R 2.11.0 and Brodgar 2.57 (Highland Statistics Ltd) were used.

4.3 Results

SL and TL were highly correlated ($R^2 = 0.99$ *A. oxyrinchus*, $R^2 = 0.97$ *A. sturio*, Fig. 4.2a). Consequently, all figures on allometry (Figs. 4.2 and 4.3) were related to TL here. Regarding the head length, lo and hl revealed isometric scaling when related to TL (Fig. 4.2b, c). When relating *lo* to *hl*, lo in *A. oxyrinchus* seemed to

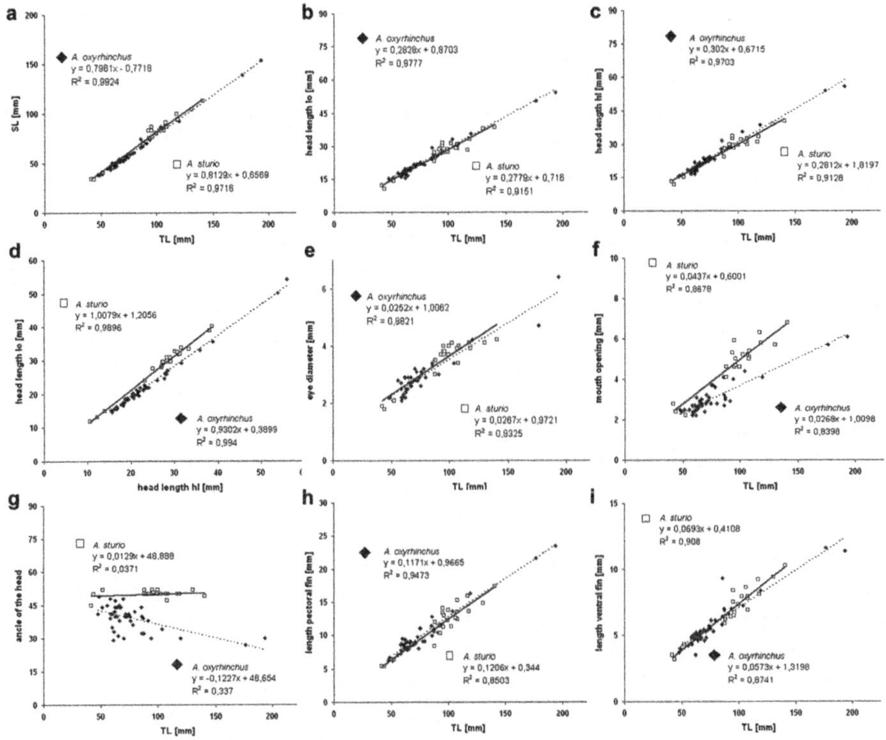

Fig. 4.2 Morphometric characteristics of juvenile *Acipenser oxyrinchus* and *A. sturio*

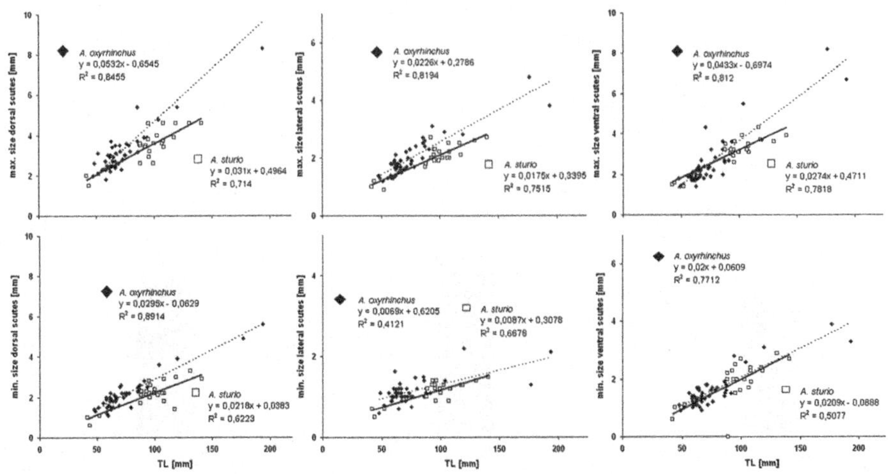

Fig. 4.3 Size of smallest (min size) and biggest scutes (max size) of the dorsal, lateral and ventral line of juvenile *Acipenser oxyrinchus* and *A. sturio*

increase faster than in *A. sturio* (slope 0.93 vs 1.01), suggesting a bigger cleft – not covered by the operculum – in *A. sturio* (Fig. 4.2d).

Eye diameter (Fig. 4.2e), mouth opening (Fig. 4.2f) as well as minimum and maximum size of the scutes (Fig. 4.3) increased isometrically with TL. Minimum and maximum size of the scutes exhibited substantial individual variation, although species-specific pattern was indicated with regard to the biggest scutes but not the smallest (Fig. 4.3). The number of dorsal and lateral scutes was highly species-specific (Fig. 4.4). In *A. oxyrinchus* number of dorsal scutes ranged from 8 to 12 (median 10) and from 18 to 27 in the lateral line (median 23). In *A. sturio* 12–15 scutes were recorded in the lateral line (median 14) and 26–34 in the lateral line (median 30). Numbers of ventral scutes ranging from 7 to 11 (median 8 in *A. oxyrinchus* and 9 in *A. sturio*) were not significantly different in the two species. Linear discriminant analysis identified number of dorsal scutes followed by lateral scutes and mouth opening as the most important characteristic in differentiating the two species (Fig. 4.7) based on the contribution of each variable to the total sum of Mahalanobis distances. These characteristics have not been dropped in the backward selection process using the Mahalanobis distances.

In the juveniles investigated, the surface of the scutes, most prominent in the dorsal ones was smooth in *A. sturio* but spiny in *A. oxyrinchus* (Fig. 4.5). This characteristic was species-specific in all specimens analysed here.

No change was determined in the horizontal angle of the rostrum with increasing TL in *A. sturio* (Fig. 4.2g). In *A. oxyrinchus*, the rostrum appeared slightly more pointed at higher TL. This suggests that angle of the rostrum may be used as a diagnostic characteristic in juveniles at a TL > 10 cm.

The length of the dorsal, pectoral, ventral, anal and caudal fins revealed isometric scaling (as example Fig. 4.2h, i). There was no difference detected between species.

The number of gill rakers was difficult to determine under the stereoscope, varied considerably in successive countings, and consequently represented an ambiguous characteristic in juvenile stages.

In juvenile *A. oxyrinchus*, the parietale and the frontale are clearly separated by a fontanelle (Fig. 4.6). This fontanelle was not found in *A. sturio*.

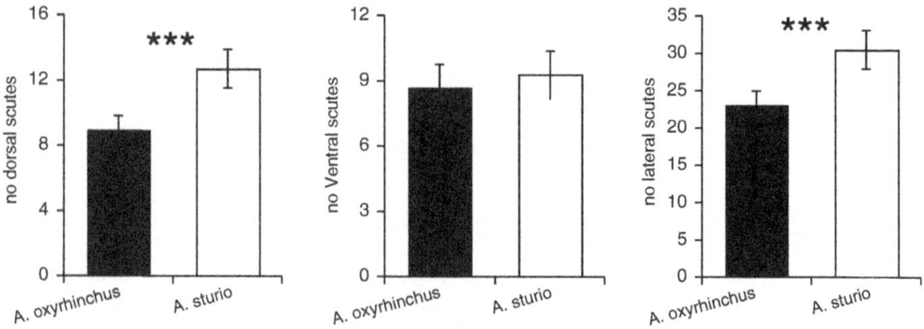

Fig. 4.4 Number of scutes from juvenile *Acipenser oxyrinchus* ($n = 47$) and *A. sturio* ($n = 23$). *** Highly significant ($p < 0.001$, Mann–Whitney rank sum test)

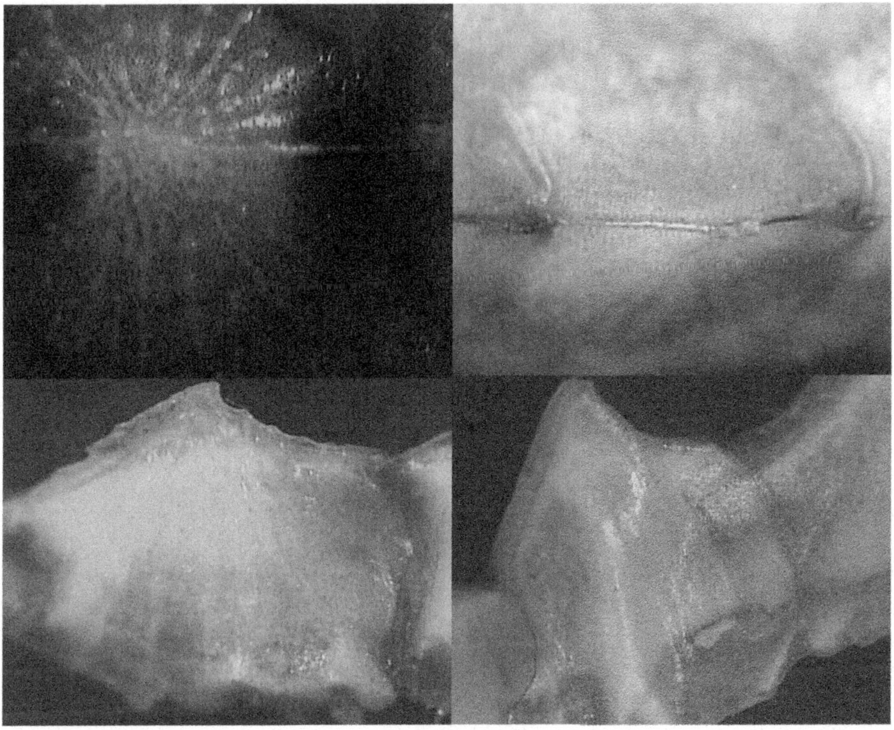

Fig. 4.5 Scutes in lateral (*upper row*) and horizontal view (*lower row*) from juvenile *Acipenser oxyrinchus* (*left*, spiny type) and *A. sturio* (*right*, smoothy type), TL 12.5 and 13.3 cm respectively

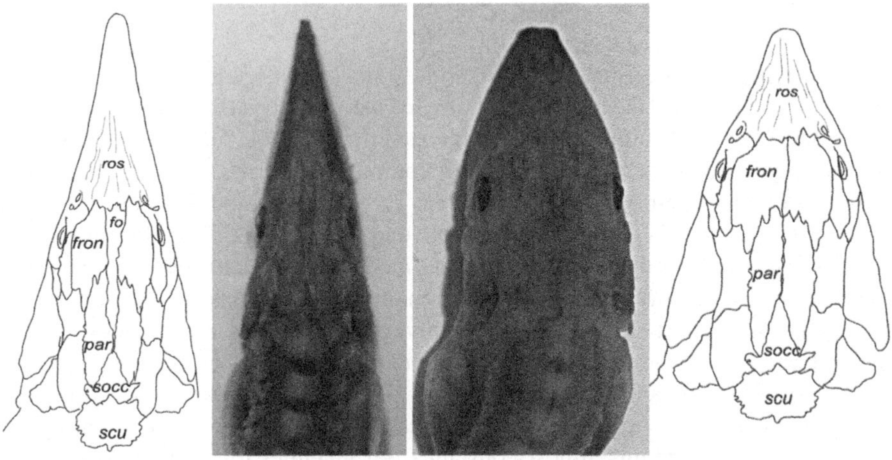

Fig. 4.6 Dermatocranium of the American Atlantic (*left*) and the European sturgeon (*right*). *fo* Fontanelle, *fron* frontale, *par* parietale, *scu* first dorsal scute, *socc* supraoccipitale

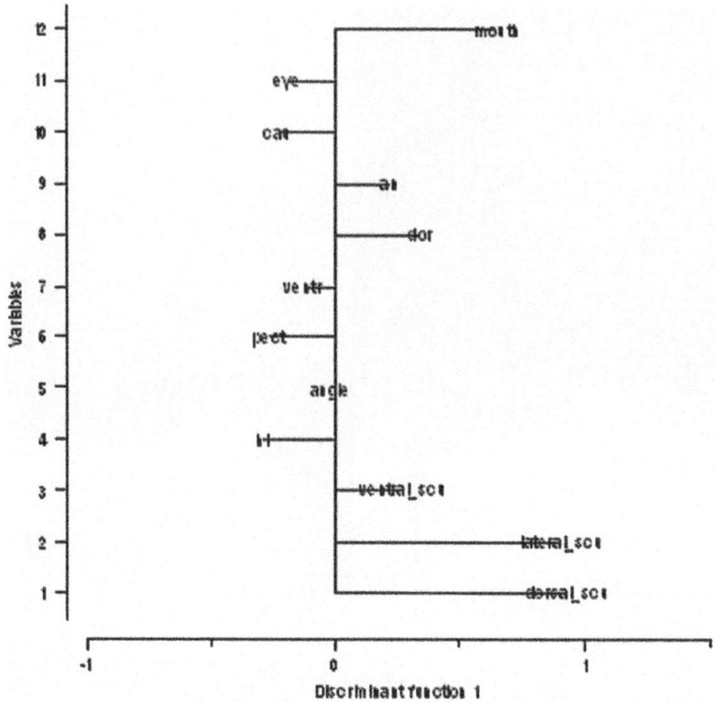

Fig. 4.7 Intraclass correlations between the 12 morphological variables used for linear discriminant analysis (lDA) of *A. sturio* and *A. oxyrinchus* (*mouth* mouth opening, *eye* eye diameter, *cau* caudalis, *an* analis, *dor* dorsalis, *ventr* ventralis, *pect* pectoralis, *hl* head length)

4.4 Discussion

Adult specimen of the European sturgeon *A. sturio* can readily be distinguished from *A. oxyrinchus* by means of morphologic characteristics (Debus 1999; Magnin 1964; Magnin and Beaulieu 1963) as indicated in Table 4.1. The characteristics of juveniles considered here (Table 4.2) are commonly referred to in the identification of adult specimen (Debus 1999; Holcik 1989; Magnin and Beaulieu 1963). In general, morphometric data are less reliable due to a series of difficulties. Changes in the parameters throughout the life cycle, incomparability attributed to differing methods and high individual variation are main reasons (Debus 1999). During the life cycle, some traits change considerably whereas others are diagnostic over a long period. For example, the number of gill rakers or number of fulcra decrease with increasing body length in adults (Debus 1999), and are ambiguous in juveniles as reproducibility of counting revealed mistakes of 20% or more. In contrast, the number of scutes remains almost constant from juveniles to adult spawners. The number of dorsal scutes and number of lateral scutes were the most important diagnostic characteristics differentiating between *A. sturio* and *A. oxyrinchus* in the

Table 4.1 Parameters used for the identification of adults of the Atlantic sturgeon *A. oxyrinchus* and the European sturgeon *A. sturio*. Shaded parameters comprise recommendations based on the current literature

	A. oxyrinchus	*A. sturio*	Diagnostic value
"Pointed long snout" [02]	Ambiguous	Ambiguous	Blunt snout with age [02]
Reticulate arrangement of rombic plates between dorsal and lateral scutes [01–04]	Less developed, rather irregular shape	Pronounced, regular shape	Questioned in [07]
Number of gill rakers 15–31	21.54	20.19	
Shape of gill rakers [09]	Rounded	Pointed	Similar, change during development [01, 02]
Form dorsal scutes [07]	Longer than broad	Square shaped	In *A. o. desotoi* square shaped [07, 08], intraspecific variation?
Scute surface [07]	Alveolar	Tubercular	Absorption dermal ossification in large individuals [07]
Number dorsal/lateral scutes	Diagnostic	Diagnostic	
Neurocranial fontanelle [07]	Present, but closed with age	Absent	[03]

According to [01] Artyukhin (1995), [02] Debus (1999), [03] Magnin and Beaulieu (1963), [04] Mohr (1962), [05] Fitzinger and Heckel (1836), [06] Heckel and Kner (2011), [07] Artyukhin and Vecsei (1999), [08] Vladykov (1955), and [09]

Table 4.2 Evaluation of parameters used for the identification of juvenile stages of the Atlantic sturgeon *A. oxyrinchus* and the European sturgeon *A. sturio*. Shaded parameters comprise recommendations based upon the results of this work (details see text)

	A. oxyrinchus	*A. sturio*	Diagnostic value
"Pointed long snout"	Rather pointed	Rather blunt	
Number of gill rakers	Ambiguous	Ambiguous	Highly variable among individuals low reproducibility
Shape of gill rakers	Rather rounded	Pointed	Similar, change during development
Mouth opening		Wider	Normalisation to TL
Scute surface	Spiny type	Smoothy type	
Number dorsal/lateral scutes	Diagnostic	Diagnostic	
Neurocranial fontanelle	Present	Absent	

linear discriminant analysis accompanying the results of the non-parametric Mann–Whitney rank sum tests. Interestingly, mouth opening was the third important characteristic to differentiate between the two species, whereas the remaining variables were dropped in the full backward selection process using Mahalanobis distances. In the juveniles tested, the number of dorsal scutes was determined as 9.1 ± 0.5 (max 10, $n = 15$) for *A. oxyrinchus* and 12.4 ± 1.2 (min 11, $n = 18$) for *A. sturio*, comparable to earlier reports in adults (Debus 1999; Magnin 1964; Magnin and Beaulieu 1963).

Debus (1999) suggested that body length can be calculated from the scute size in adults, yielding a good approximation. In juveniles, minimum and maximum size of the scutes exhibited substantial individual variation. Still, a species-specific pattern was indicated with regard to the biggest scutes but not the smallest.

Scute surface has been recognised as reliable diagnostic characteristic in adults (Artyukhin and Vecsei 1999; Ludwig et al. 2002; Magnin 1964; Magnin and Beaulieu 1963), described as tubular type in *A. sturio* and alveolar type in *A. oxyrinchus*. In juvenile *A. oxyrinchus* (Fig. 4.5), a prominent spiny surface is observed which develops into the alveolar type at a later stage, presumably due to large scale fusion of protrusions. The scutes of juvenile *A. sturio* are smooth at the comparable size. Still, occasional small spiny-like structures are found in fish exceeding a total length of 14 cm. These are smaller and present only in very low numbers than the prominent structures observed in *A. oxyrinchus*. The fontanelle between the parietale and the frontale was observed only in juvenile *A. oxyrinchus*. Hence the fontanelle is a diagnostic feature in juveniles. In adults, the fontanelle disappears with age (Magnin 1962).

In adults, head length is unreliable since pointed or blunt snout have been described as individual characteristics and head growth decreases with age (Debus 1999). This would require a determination throughout the life cycle and the assessment of the underlying process, its variability and the development of a model to account for the changes over time. In this investigation of juvenile fish, the head in *A. oxyrinchus* juveniles was more pointed (revealed a smaller horizontal angle of rostrum) compared to *A. sturio*. This is congruent to the findings of Magnin and Beaulieu (1963) in subadults (40–252 cm) (Fig. 4.8).

Fig. 4.8 Photo of a juvenile *A. sturio* (*top*) and *A. oxyrinchus* (*bottom*) at approximately 30 cm TL

The sturgeon mouth is unique in lacking a jaw bone and being capable of down and outward projection, which is considered an adaptation to benthic diet and feeding ecology of the species (Vecsei and Peterson 2008). The morphology of the mouth in both species is similar, but the size revealed a larger mouth opening (normalised to TL) in *A. sturio* juveniles.

The gap in the gill opening (hl – lo) was bigger in juvenile *A. sturio* when compared with *A. oxyrinchus*. These two characteristics are hypothesised to be related to a difference in prey items predominantly utilised prey items. *A. oxyrinchus* from the St. Lawrence Estuarine has been reported to feed predominantly on gammarids in juveniles below 20 cm (Guilbardi et al. 2007). In contrast, *A. sturio* mainly feeds on organisms buried in the substrate such as chironomids or polychaetes (Brosse et al. 2000).

So far, species distinction by diagnostic characteristics has only been addressed in adults. The characteristics of juveniles described here allow such an identification using parameters which are easy to access. The most important diagnostic characteristics in juveniles were the number of dorsal and lateral scutes, the presence or absence of the fontanelle and the surface (spiny or smooth) of the scutes.

Acknowledgment This study was supported in part by a grant from the German Ministry of Education and Science (BMBF 0330718) as part of the project "Genetic population structure, breeding plan and artificial reproduction of the European sturgeon". We would like to thank Marcel Jansen and Philipp Lachner for their help in the lab.

References

Artyukhin E (1995) On biogeography and relationships within the genus *Acipenser*. Sturgeon Q 3(2):6–8

Artyukhin E (2006) Morphological phylogeny of the order Acipenseriformes. J Appl Ichthyol 22:66–69

Artyukhin E, Vecsei P (1999) On the status of Atlantic sturgeon: conspecificity of European *Acipenser sturio* and North American *Acipenser oxyrinchus*. J Appl Ichthyol Z Angew Ichthyol 15:35–37C

Berg LS (1948) Acipenseriformes. Ryby presnych vod SSSR i sopredel'nych stran. Izd. AN SSSR, Moskwa Leningrad 1:57–109

Birstein VJ, DeSalle R (1998) Molecular phylogeny of Acipenserinae. Mol Phylogenet Evol 9:141–155

Brosse L, Rochard E, Dumont P, Lepage M (2000) First results on the diet of the European sturgeon, *Acipenser sturio*, in the Gironde estuary. Comparison with the benthic fauna. Cybium 24:49–61

Debus L (1999) Meristic and morphological features of the Baltic sturgeon (*Acipenser sturio* L.). J Appl Ichthyol 15:38–45

Desse-Berset N (2009) First archaeozoological identification of Atlantic sturgeon (*Acipenser oxyrinchus* Mitchill 1815) in France. CR Palevol 8:717–724

Fitzinger LJ, Heckel J (1836) Monographische Darstellung der Gattung *Acipenser*. Annalen des Wiener Museums der Naturgeschichte 1:261–326

Gessner J, Fredrich F, Arndt GM, von Nordheim H (2010) Arterhaltung und Wiedereinbürgerungsversuche für die Atlantischen Störe (*Acipenser sturio* und *A. oxyrinchus*) im Nord- und Ostseeeinzugsgebiet [Conservation and remediation attempts for the Atlantic sturgeons in the North- and Baltic Sea tributaries]. Natur und Landschaft 85(6):514–519

Guilbardi F, Munro J, Dumont P, Hatin D, Fortin R (2007) Feeding ecology of Atlantic sturgeon and Lake sturgeon co-occurring in the St. Lawrence estuarine transition zone. Am Fish Soc Symp 56:85–104

Heckel J, Kner R (2011) Die Süßwasserfische der Oesterreichischen Monarchie. Verlag von Wilhelm Engelmann, Leipzig, 119–123

Holcik J (1989) The freshwater fishes of Europe, vol. 1, part II, general introduction to fishes. AULA-Verlag, Wiesbaden

Ludwig A, Belfiore NM, Pitra C, Svirsky V, Jenneckens I (2001) Genome duplication events and functional reduction of ploidy levels in sturgeon (*Acipenser, Huso* and *Scaphirhynchus*). Genetics 158:1203–1215

Ludwig A, Debus L, Lieckfeldt D, Wirgin I, Benecke N, Jenneckens I, Williot P, Waldman JR, Pitra C (2002) When the American sea sturgeon swam east — a colder Baltic Sea greeted this fish from across the Atlantic Ocean in the Middle Ages. Nature 419:447–448

Ludwig A, Lippold S, Debus L, Reinartz R (2009) First evidence of hybridization between endangered sterlets (*Acipenser ruthenus*) and exotic Siberian sturgeons (*Acipenser baerii*) in the Danube River. Biol Invasions 11:753–760

Magnin E (1962) Recherches sur la systématique et la biologie des Acipenséridés *Acipenser sturio* L., *Acipenser oxyrhynchus* Mitchell et *Acipenser fulvescens* Raf. Ann Stat Cent Hydrobiol Appl 9:7–242

Magnin E (1964) Validité d'une distinction spécifique entre les deux Acipensérides: *Acipenser sturio* L. d'Europe et *Acipenser oxyrinchus* d'Amérique du Nord. Le Naturaliste Canadien 91:5–20

Magnin E, Beaulieu G (1963) Etude morphométrique comparée de l'*Acipenser oxyrhynchus* Mitchell du Saint-Laurent et *Acipenser sturio* Linne de la Gironde. Le Naturaliste Canadien 90:5–38

Mohr E (1962) Ganoiden, Stoere. Handbuch der Binnenfischerei Mitteleuropas 3:235–261

Ninua NS (1976) Atlanticeskij osetr reki rioni [Atlantic sturgeon of the Rioni river]. *Metsniereba Tbilisi*, 122

Robles F, de la Herran R, Ludwig A, Rejon CR, Rejon MR, Garrido-Ramos MA (2004) Evolution of ancient satellite DNAs in sturgeon genomes. Gene 338:133–142

Rochard E, Lepage M, Meauze L (1997) Identification and characterisation of the marine distribution of the European sturgeon *Acipenser sturio*. Aquat Living Resour 10:101–109

Sokolov LI Berdichevski LS (1989) B Systematic part. In: Holcik J (eds) The freshwater fishes of Europe – General introduction to fishes Acipenseriformes. Vol. 1, Part II, pp 201–205

Tiedemann R, Moll K, Paulus KB, Scheer M, Williot P, Bartel R, Gessner J, Kirschbaum F (2007) Atlantic sturgeons (*Acipenser sturio, Acipenser oxyrinchus*): American females successful in Europe. Naturwissenschaften 94:213–217

Tranah G, Campton DE, May B (2004) Genetic evidence for hybridization of pallid and shovelnose sturgeon. J Hered 95:474–480

Vecsei P, Peterson D (2008) Sturgeon ecomorphology. In: Breton GTO, Breamish FWH, Scott RS (eds) Sturgeons and paddlefish of North America, Fish & fisheries series. Kluwer, The Netherlands, pp 101–133

Vladykov VD (1955) A Comparison of the Atlantic Sea sturgeon with a new subspecies from the Gulf of Mexico (Acipenser oxyrhynchus desotoi). Fisheries Research Board of Canada 12:754–761

Williot P, Rochard E, Castelnaud G, Rouault T, Brun R, Lepage M, Elie P (1997) Biological characteristics of European Atlantic sturgeon, *Acipenser sturio*, as the basis for a restoration program in France. Environ Biol Fishes 48:359–372

Williot P, Arlat G, Chebanov M, Gulyas T, Kasimov R, Kirschbaum F, Patriche N, Pavlovskaya LP, Poliakova L, Pourkazemi M, Kim Y, Zhuang P, Zholdasova IM (2002) Conservation and broodstock management. Int Rev Hydrobiol 87:483–506

Williot P, Rouault T, Pelard M, Mercier D, Jacobs L (2009) Artificial reproduction of captive endangered European Atlantic sturgeon, *Acipenser sturio*. Endanger Species Res 6:251–257

Zuur AF, Ieno EN, Smith GM (2007) Analysing Ecological Data. In: Gail M, Krickeberg J, Samet J, Tsiatis A, Wong W (eds) Statistics for Biology and Health 1st Edition, Springer, New York

Chapter 5
Ontogeny of the European Sturgeon, *Acipenser sturio*

Frank Kirschbaum and Patrick Williot

Abstract Early development from late embryo stage [stage 34 (stage numbers correspond to the description of Dettlaff et al. (1993) Sturgeon fishes. Developmental biology and aquaculture. Springer, Berlin)] up to exogenous feeding (stage 44) is described in the European sturgeon, *Acipenser sturio*. The habit is documented in dorsal and lateral views from the larval stage up to adulthood. Special emphasis is given to the morphological development of the dorsal bony scutes and their dorsal spikes. First dorsal scutes occur in 21-day-old larvae, 20 mm long. In juveniles 15 cm long, the dorsal scutes and their dorsal, caudally oriented spikes are very prominent. In older juveniles the spikes regress; in 30 cm long juveniles the spikes have nearly disappeared. The ontogeny of the incomplete cover of the gills by the operculum is shown. The distribution of the ampullary electroreceptors in a 27-day-old larva is described.

5.1 Introduction

Sturgeon as non-teleost chondrostei possess many original features (Marinelli and Strenger 1973; Mills 2003) such as a partly ossified skeleton, thick bony plates covering the skin, ganoid scales on the heterocercal caudal fin, an unrestricted notochord, a spiral intestine, ampullary electroreceptors for the detection of weak electrical fields, and egg development comprising total cleavage and pronounced larval stages (Dettlaff et al. 1993).

F. Kirschbaum (✉)
Faculty of Agriculture and Horticulture, Biology and Ecology of Fishes, Humboldt University, Philippstr. 13, Haus 16, 10115 Berlin, Germany
e-mail: Frank.kirschbaum@staff.hu-berlin.de; frank.kirschbaum@staff.hu-berlin.de

P. Williot
Sturgeon Consultant, 4 Rue du pas de madame, 33980 Audenge, France

P. Williot et al. (eds.), *Biology and Conservation of the European Sturgeon Acipenser sturio L. 1758*, DOI 10.1007/978-3-642-20611-5_5,

The ontogeny of sturgeons has been described in many species (e.g. Kowalewsky et al. 1870; Salensky 1878, 1880, 1881; Sydner 1976; Schmalhausen 1983; Dettlaff et al. 1993; Sydner 1994; Gisbert et al. 1998b; Gisbert 1999; Sydner 2002) because these fish are interesting as basic fishes and also because of their economic value, in particular due to the fact that they deliver the luxury product caviar.

Distinct organ systems have been described in several species: the early ontogeny of the digestive tract (Devitsina and Gadzhieva 1996; Gisbert et al. 1998a, 1999b; Ostos-Garrido et al. 2009), the development of the heart (Icardo et al. 2009) and the brain of *Acipenser naccarii* (Vasquez et al. 2002; Gomez et al. 2009), the ontogeny of the eye and the retina (Loew and Sillman 1993; Chai et al. 2006; Rodriguez and Gisbert 2002), the early ontogeny of *Acipenser naccarii* with special focus on the digestive tract and cutaneous receptors (Boglione et al. 1999, 2006), the development of chemosensory organs (Devitsina and Kazhlayev 1993; Devitsina and Gadzhieva 1996), development of teeth (Iakovleva 1952) and skeleton (Jollie 1980), as well as behavioural ontogenetic aspects (Kasumyan and Kazhlayev 1993; Gisbert and Williot 1997; Gisbert et al. 1999a; Gessner et al. 2009).

The early ontogeny of the European sturgeon, *Acipenser sturio*, was described a long time ago, comprising just a few stages (Ehrenbaum 1894). These data will be shown in this paper together with ontogenetic data which were available due to artificial reproduction in France in 1995, 2007, and 2008, and the subsequent rearing of these fish up to adulthood or the large juvenile stage, respectively (Williot et al. 2000). This has made it possible to describe for the first time the ontogeny of the habit of the European sturgeon, *Acipenser sturio* from embryo stage up to adulthood based on photographic documentation including the description of the development of some specific features such as the dorsal scutes. More detailed description based on fixed material was not possible up to now due to the highly protected status of the species. Nevertheless, we believe that this description of the ontogeny of *A. sturio* is a useful contribution to this monograph because these data have never been published before.

5.2 Material and Methods

The description of the early development *A. sturio* is based on the published data of Ehrenbaum (1894) and the interpretation of these data with regard to staging according to the detailed description of Dettlaff et al. (1993). The free embryo and larval stages described through photographic documentation comprise material from the artificial reproduction performed in France in 2007 (Williot et al. 2009), whereas the larger juveniles represent fish originating from the artificial reproduction obtained in 2008; the large adult fish were born in 1995 (Williot et al. 2000) and reared subsequently (see Chap. 21). The free embryo and larval stages comprise fish which died accidentally and were fixed subsequently in 10% Formalin. The early stages were photographically documented with a digital Leica camera (DFC 420) mounted on a Leica binocular (S6D), whereas the juvenile and adult specimens were documented with a digital Canon camera (EOS 350D). The problem with the documentation of developmental stages of *A. sturio* is based on the protected status of the species; therefore only a few early developmental stages, which died accidentally, were available for the study. The description of the older stages therefore was based on photographic documentation alone.

5.3 Results and Discussion

5.3.1 Early Development [Description After Ehrenbaum (1894) and Dettlaff et al. (1993)]

Egg development takes about 3 days at 20°C. The 2½-day-old embryo, representing stage 34 according to the ontogenetic description of *Acipenser gueldenstaedtii colchicus* (Dettlaff et al. 1993) is clearly visible through the egg shell (Fig. 5.1a). This embryo taken out of the 2.8 mm egg reveals a quite undeveloped stage comprising a large yolk sac and a small embryological fin fold (Fig. 5.1b). At hatching the free embryo (developmental terms after Balon 1975), 9.3 mm long, representing stage 35, shows a small eye, nose openings and the ear structure, about 60 somites, a well-developed embryological fin fold and a large yolk sac (Fig. 5.1c). Interestingly, at this stage, when the gills are not yet covered by an operculum, the free embryo possesses three gill clefts, the anterior one of which is considered the future spiracle. The 6-day-old free embryo, 11 mm long (Fig. 5.1d) possesses the first anlagen of the pectoral and ventral fins, first gill structures and first indication of a dorsal fin development. It is comparable to stage 38 of *Acipenser gueldenstaedtii colchicus*. The 10-day-old free embryo, 13.2 mm long (Fig. 5.1e) has a well-developed eye, four barbels are visible, the two paired fins (pectoral and ventral fins) have developed, in the embryological fin fold dorsal and anal fins start to differentiate and the heterocercal caudal fin emerges; teeth are already present. This free embryo represents stage 44 of *Acipenser gueldenstaedtii colchicus*. Exogenous feeding has not yet started. The 17-day-old larva (Fig. 5.1f), 16.5 mm long, is morphologically much more advanced than the 10-day-old embryo. At 18°C exogenous feeding starts at about day 16. The 17-day-old larva described in Fig. 5.1f, raised at 20°C, has therefore apparently already passed the stage of the beginning of exogenous feeding. However, this larva described by Ehrenbaum (1894) (Fig. 5.1f) could not be fed exogenously (such knowledge was not available at that time) and this larva therefore probably showed some sign of starvation. These early ontogenetic stages of *A. sturio* described here are morphologically similar to the appropriate stages described in *Acipenser gueldenstaedtii colchicus* (Dettlaff et al. 1993). This is interesting as *A. sturio* is considered to be the most primitive of all extant *Acipenser* species (see Chap. 2).

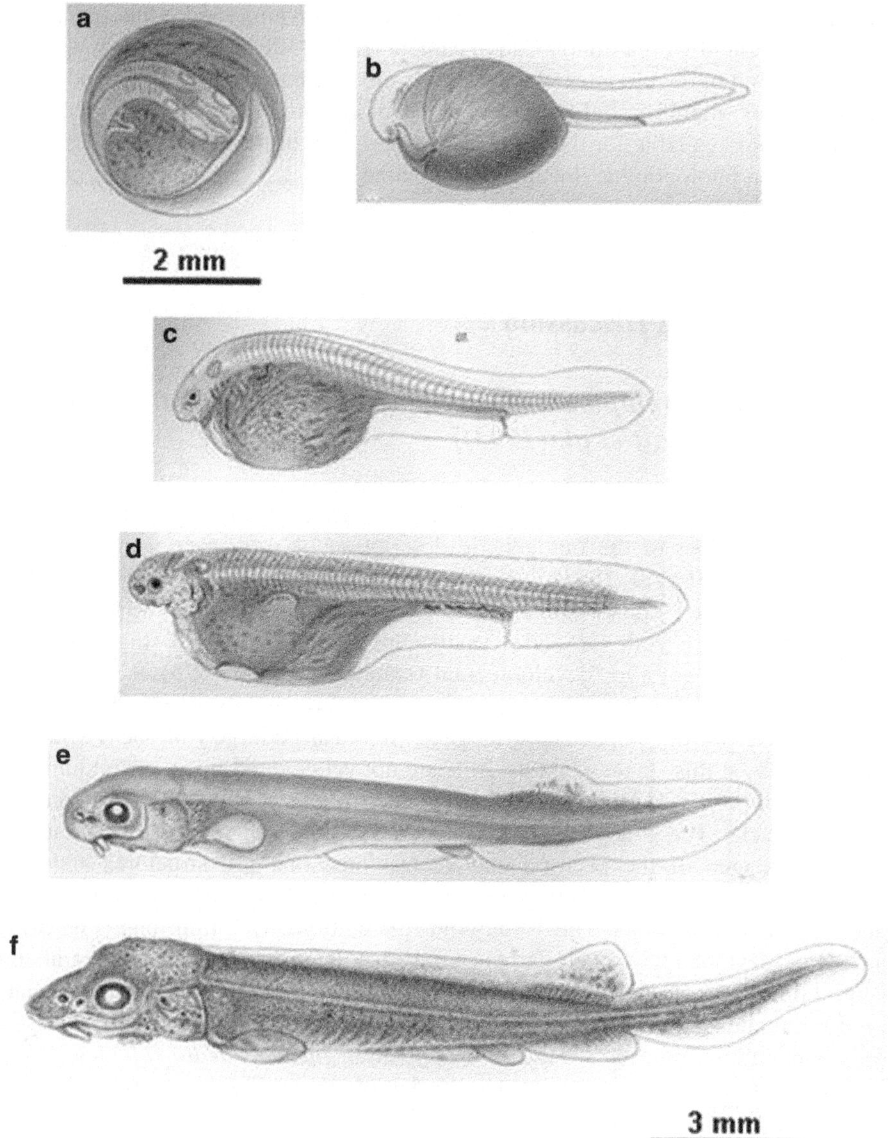

Fig. 5.1 Early ontogeny of *Acipenser sturio*. **a** Embryo, 60 h old, stage 34. **b** Embryo as in **a** taken out of the chorion. **c** Hatched free embryo, 80 h old representing stage 35. **d** Free embryo, 6 days old representing stage 38. **e**) Free embryo, 10 days old, just before exogenous feeding, stage 44. **f** Larva, 17 days old; after transition to exogenous feeding; however, this larva could not be supplied with appropriate food (after Ehrenbaum 1894) and probably has been starving for some time. *Scale bar* (3 mm) applies to **b–f**. Staging according to Dettlaff et al. (1993)

5.3.2 Development of the Habit from Free Embryo Stage Up to Adulthood

5.3.2.1 Dorsal View

The fry raised in 2007 at 18°C started to feed exogenously around day 16; they measured about 15 mm in length. Nineteen-day-old larvae in dorsal view are shown in Fig. 5.2. These larvae possess a broad head, round pectoral fins and a thin body. Head length is less than 1/3 of total length. This corresponds to values measured by Würtz et al. (see Chap. 4), though these authors have started to measure head length from 40-mm-long fish onwards. In 42-day-old larvae (Fig. 5.3) the snout is now slightly pointed, and body width has increased compared to the 19-day-old larva. Bony scutes are now visible (see also Fig. 5.9). Juveniles 107 days old had fact grown to between 6 and 15 cm in length. In a specimen about 7 cm long (Fig. 5.4) the head is more pointed than in the stage before (Fig. 5.3); this feature is even more pronounced in a 13-cm-long specimens (Fig. 5.5). With increasing length, the head becomes (Fig. 5.6) more and more stunted; this is even more pronounced in large adults. This allometric growth in older specimens has been mentioned by Debus

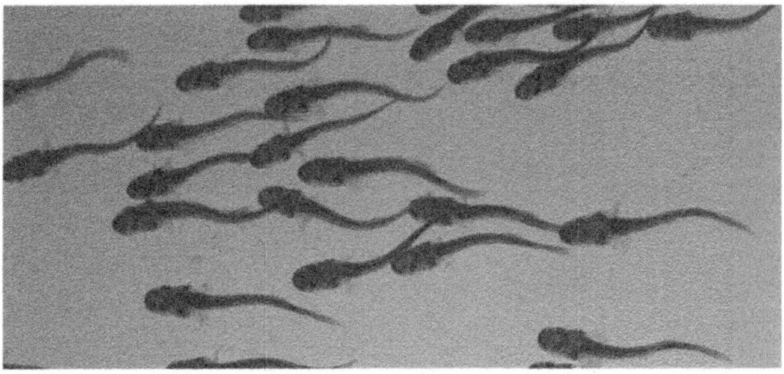

Fig. 5.2 *Acipenser sturio* larvae, 16 days old, 15 mm long, after the transition to exogenous feeding

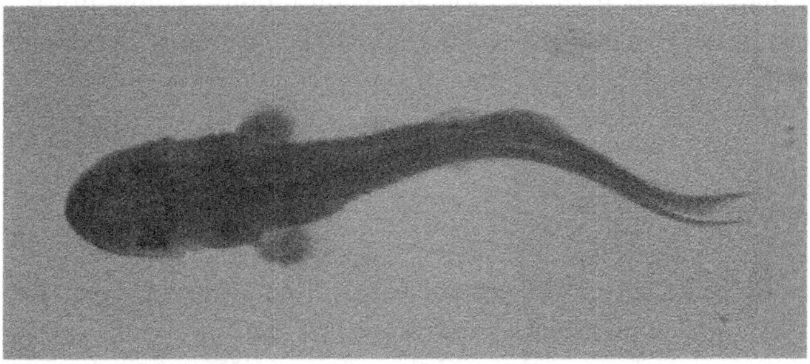

Fig. 5.3 *Acipenser sturio* larva, 42 days old, ca. 35 mm long

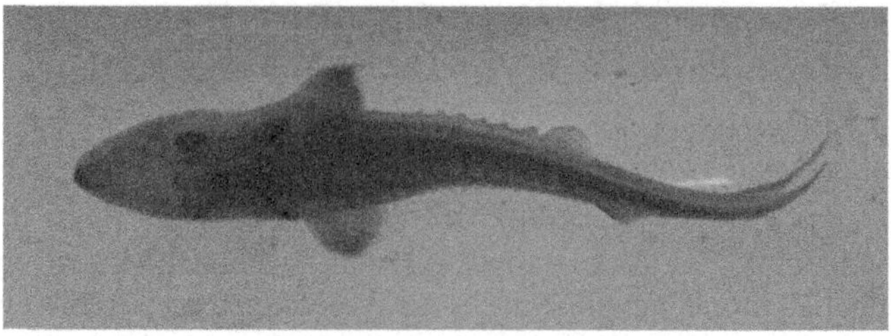

Fig. 5.4 *Acipenser sturio* juvenile 107 days old, about 6 cm long

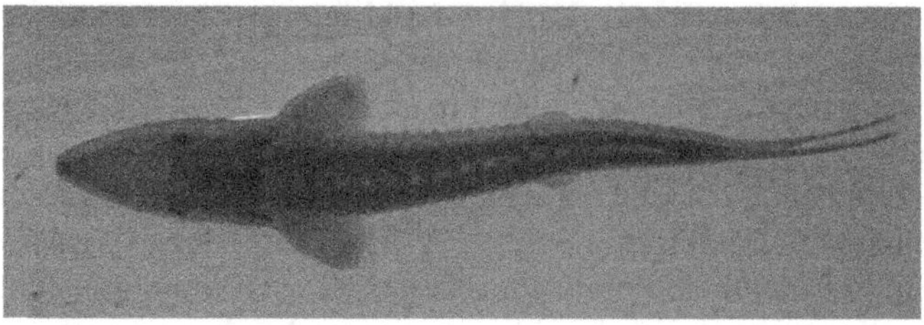

Fig. 5.5 *Acipenser sturio* juvenile 107 days old, about 13 cm long; note the more slender body proportions compared to the smaller juvenile (see Fig. 5.4)

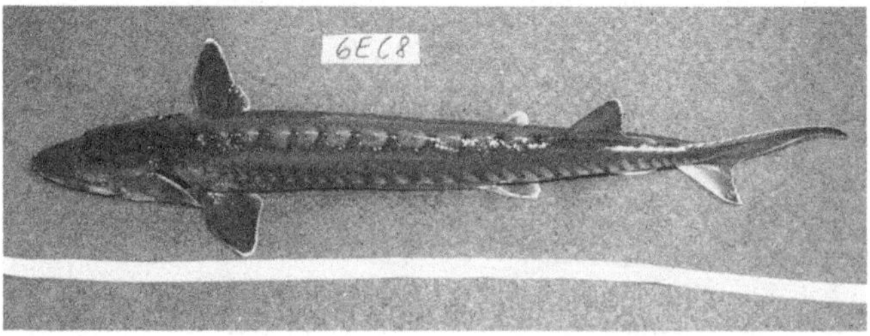

Fig. 5.6 Juvenile *Acipenser sturio*, 50 cm long; the head is less pointed than in younger specimens

(1999), for example. Though head length shows isometric growth related to total length (see Chap. 4), our pictures indicate a continuous change in the shape of the head. This can only be explained on the basis of a continuous change in the shape of the bones of the head during this part of ontogeny.

5.3.2.2 Lateral View

The 15-day-old larva (Fig. 5.7) shows a complete embryological fin fold (EFF), dorsal and anal fins, which start to differentiate, and a caudal fin which already possesses a heterocercal shape. In a 27-day-old larva (Fig. 5.8) dorsal and anal fins have emerged from the EFF and spikes of the dorsal scutes have penetrated the EFF

Fig. 5.7 *Acipenser sturio* larva, 15 days old, 15 mm long, in lateral view

Fig. 5.8 *Acipenser sturio* larva, 27 days old, 27 mm long; spikes of the dorsal scutes have penetrated the embryological fin fold

Fig. 5.9 *Acipenser sturio* larva, 42 days old, ca. 35 mm long; remnants of the embryological fin fold are still present

(for details see Fig. 5.15). During these 13 days of early ontogenetic development the larvae grew 1 mm per day. The 42-day-old larva (Fig. 5.9) is more darkly pigmented than the stage before, dorsal and anal fins are more prominent, and the dorsal scutes are now compact structures. These features described above are more pronounced in the 63-day-old juvenile (Fig. 5.10). We have not been able to determine exactly the transition from larval to juvenile stage. It seems to occur between 40 and 50 mm length. Boglione et al. (1999) state that in *A. naccarii* a 180-day-old fish is still a larva; for example, they still possess teeth. In a 3½-month-old juvenile of *A. sturio* the external structures, in particular the exoskeleton, are further developed (Fig. 5.11). With increasing size (30, 50 cm; Figs. 5.12 and 5.13, respectively) the body gets broader, the fins slightly change their form. The habit of the adult specimen shown in Fig. 5.14 is characterized by fin abnormalities (in particular of the ventral and caudal fins) due to the length of time spent in the restricted space of the rearing tanks.

Fig. 5.10 *Acipenser sturio* juvenile, 63 days old, about 5 cm long

Fig. 5.11 Juvenile of *Acipenser sturio* 3½ months old, 15 cm long

Fig. 5.12 Juvenile of *Acipenser sturio* 1½ years old, 30 cm long

Fig. 5.13 Juvenile of *Acipenser sturio* 1½ years old, 50 cm long

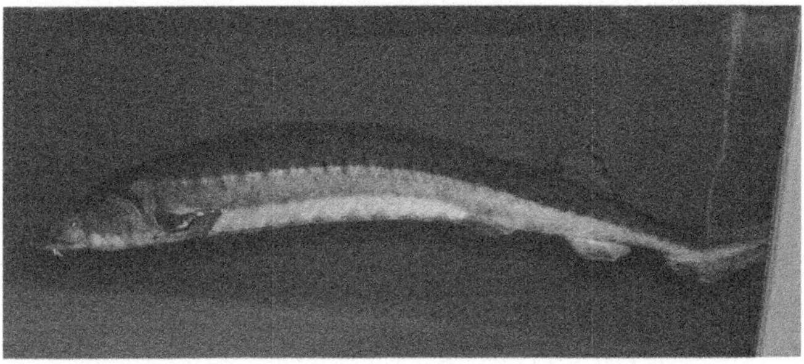

Fig. 5.14 Adult *Acipenser sturio*, 10 years old, 100 cm long

5.3.3 Morphological Development of the Dorsal Scutes

In a larva, 15 days old, the dorsal scutes are not yet visible in the embryological fin fold (Fig. 5.15a). Six days later (larva 20 mm long, 21 days old), small spikes, first indication of the dorsal scutes, have differentiated in the embryological fin fold (Fig. 5.15b). In a 27-day-old larva (26 mm long) the spikes of the dorsal scutes penetrate the embryological fin fold for the first time (Fig. 5.15c). A juvenile, 3½ months old, 15 cm long, possesses dorsal scutes comprising thick bony plates equipped with caudally oriented spikes (Fig. 5.16). Interestingly, there is a anterior–posterior gradient in the development of these spikes. In a juvenile, 1½ years old, 30 cm long, the dorsal scutes represent thick bony plates (see Fig. 5.17); however, the caudally oriented spike have nearly completely disappeared (compare to Fig. 5.16). A 50-cm-long juvenile of the same age possesses dorsal flat scutes without any indication of spikes (Fig. 5.18). An adult *A. sturio*, 10 years old, 100 cm long, has spike-free dorsal scutes, well-isolated from each other (Fig. 5.19). Interestingly, small, caudally oriented small spikes seem to reappear (compare to Fig. 5.17).

Scutes and spikes are very important anti-predator structures. Though we do not know anything about predators of the European sturgeon, we can conclude that the

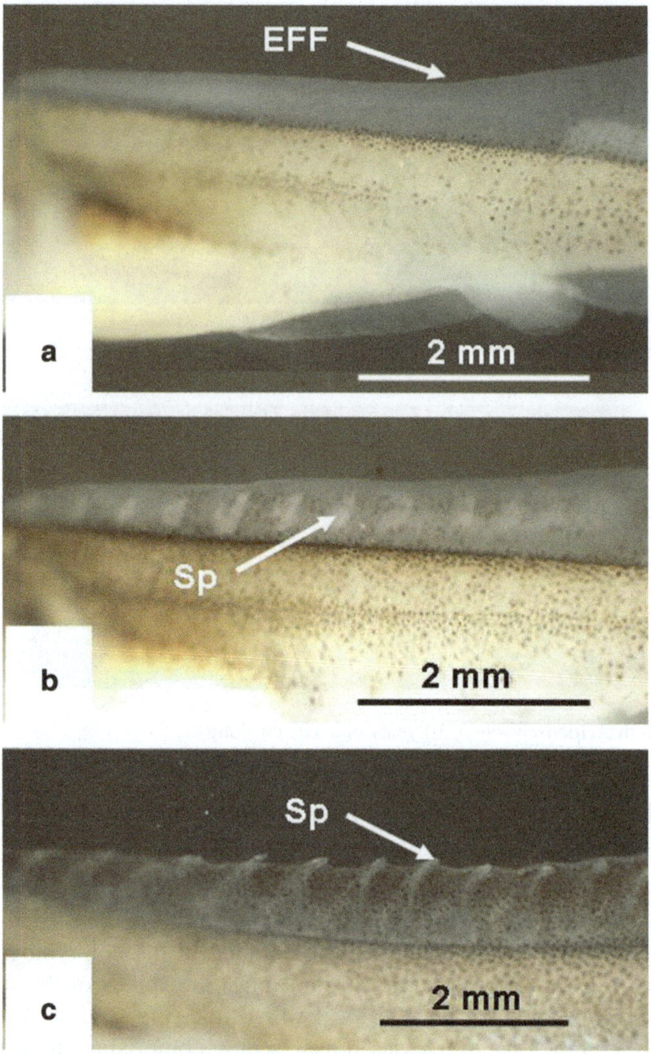

Fig. 5.15 Early development of the dorsal scutes in larvae of *Acipenser sturio*. **a** Larva, 15 days old, 15 mm long; dorsal scutes are not yet visible in the embryological fin fold (*EFF*). **b** Larva, 21 days old, 20 mm long; small spikes (*Sp*), first indication of the dorsal scutes, appear in the EFF. **c** Larva, 27 days old, 26 mm long; the spikes (*Sp*) of the dorsal scutes project from the embryological fin fold for the first time

quick differentiation of the bony scutes (and not only the dorsal ones) comprising the spikes indicates a very effective defence mechanism at this stage of ontogeny. This anti-predator device seems to be less important in older specimens; indeed, few fish species are large enough to be potential predators of large juvenile and

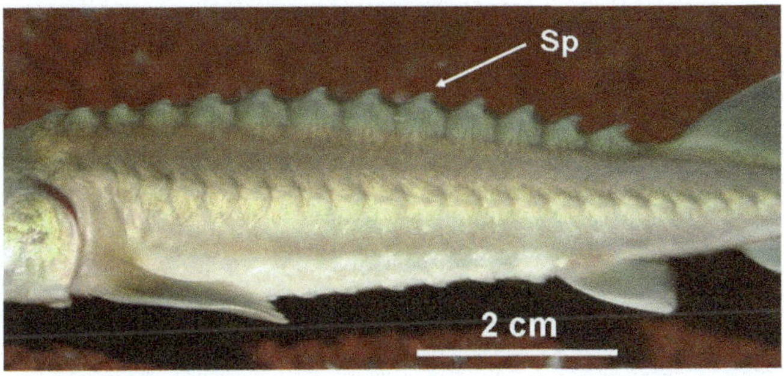

Fig. 5.16 Juvenile of *Acipenser sturio* 3½ months old, 15 cm long; the dorsal scutes now represent thick bony plates with caudally oriented spikes (Sp); there is an anterior–posterior gradient in the development of the spikes

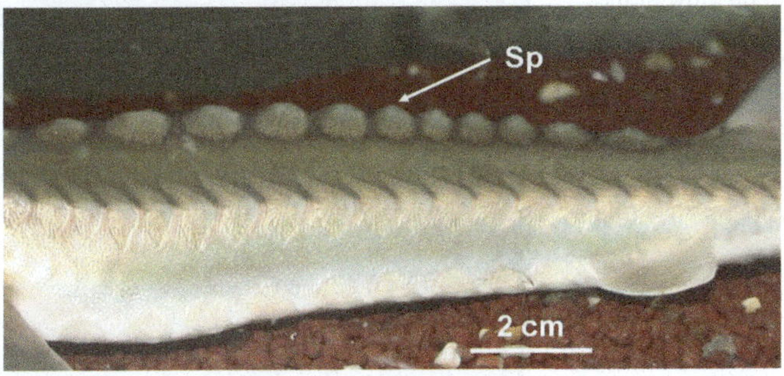

Fig. 5.17 Juvenile of *Acipenser sturio* 1½ years old, 30 cm long; the dorsal scutes now represent thick bony plates (see Fig. 5.15), however, the caudally oriented spikes (Sp) have nearly completely disappeared (compare to Fig. 5.15)

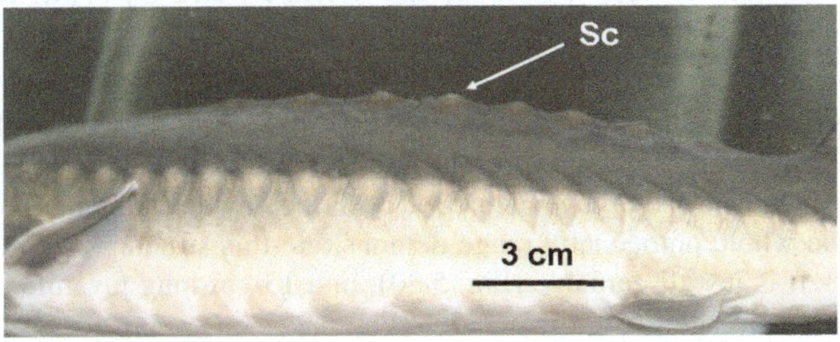

Fig. 5.18 Juvenile of *Acipenser sturio* 1½ years old, 50 cm long. The dorsal scutes (Sc) now are more flat than in the 30 cm long juvenile (compare to Fig. 5.16)

adult sturgeon. The wide space between the individual dorsal bony scutes of the adult specimens indicates an allometric growth comparing scutes and total length (see e.g. Debus 1999, Chap. 3).

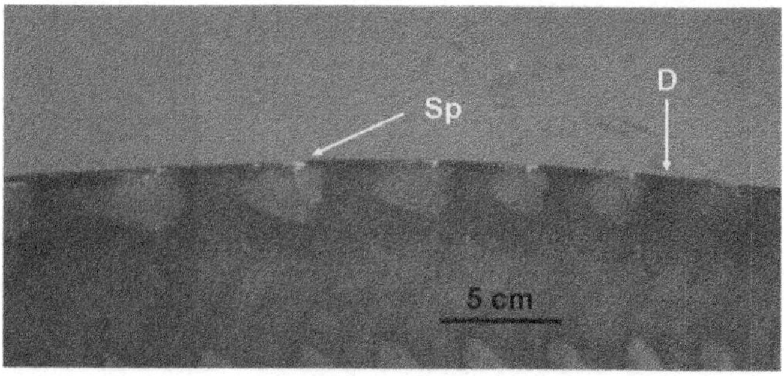

Fig. 5.19 Adult *Acipenser sturio*, 10 years old, 100 cm long. The dorsal scutes are spike-free, well-isolated (D) from each other. Interestingly, small, caudally oriented small spikes (Sp) seem to reappear (compare to Fig. 5.17)

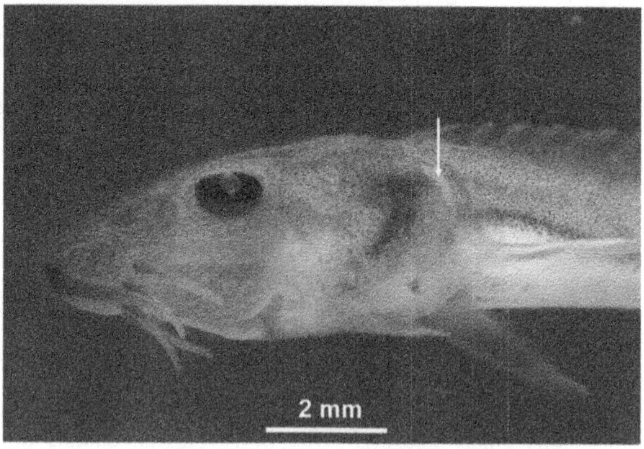

Fig. 5.20 Lateral view of the head of a larva of *Acipenser sturio*, 27 days old, 26 mm long. The operculum nearly completely covers and protects the underlying gills; a small cleft seems to be present (*arrow*)

5.3.4 The Incomplete Cover of the Gills by the Operculum

In adult *A. sturio* as well as in other sturgeon the operculum does not completely cover the gills. This cleft allows the penetration of water towards the gill chamber externally. This cleft is already present in small fish (see Chap. 4) and does not change during the period of juvenile period described by the authors. Three pictures covering the whole ontogenetic range demonstrate this structure. In a 27-day-old larva the cleft is just discernable (Fig. 5.20), in a 15-cm-long juvenile it is visible (Fig. 5.21), and in adult fish it is obvious (Fig. 5.22).

5.3.5 Distribution of the Ampullary Electroreceptors

Ampullary electroreceptors perceive weakly electric fields. They are already present on the ventral part of the lower jaw in 15-day-old larvae. In a 27-day-old larva

Fig. 5.21 Lateral view of the head of a juvenile *Acipenser sturio*, 3½ months old, 15 cm long. The operculum leaves dorsally a space (*arrow*) and does not completely cover the underlying gills; instead water has external access to the gill chamber

Fig. 5.22 Lateral view of the head of an adult *Acipenser sturio*, 10 years old, 136 cm long. Note the space above the operculum (*arrow*) where the gills emerge; this open space allows the penetration of water externally

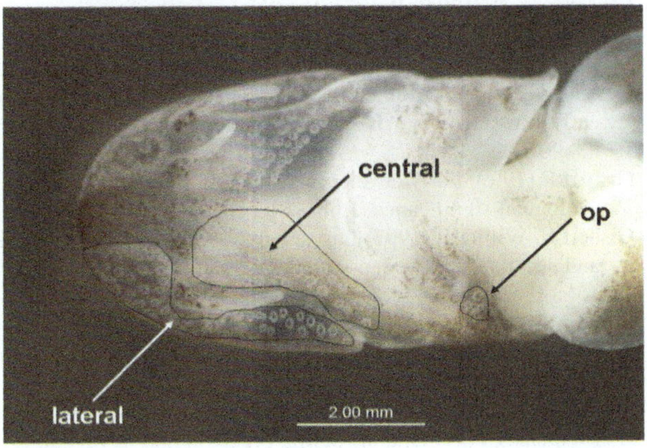

Fig. 5.23 Distribution of the ampullary electroreceptors (*white circles*) in a 27-day-old larva of *Acipenser sturio* in ventral view. In the lateral sensory fields (*lateral*) there are about 70 receptors each, in the two central fields (*central*) about 60 each, and just a few are located on the ventral part of the operculum (op)

(Fig. 5.23) the two lateral sensory fields comprise about 70 receptors each, the central sensory fields about 60 each; in addition, there are some receptors present on the ventral part of the operculum. This arrangement is also seen in larval *A. naccarii* (Boglione et al. 1999). The investigation of the development of ampullary electroreceptors in sturgeon has not attained much interest up to now.

Acknowledgements We thank Yvonne Kirschbaum for technical assistance during the preparation of the manuscript.

References

Balon EK (1975) Terminology of intervals in fish development. J Fish Res Board Can 32: 1663–1670

Boglione C, Bronzi P, Cataldi E, Serra S, Gagliardi F, Cataudella S (1999) Aspect of early development in the Adriatic sturgeon *Acipenser naccarii*. J Appl Ichthyol 15:207–213

Boglione C, Cataldi E, Sighicelli M, Bronzi P, Catadaudella S (2006) A contribution on the trophic ecology of the Adriatic Sturgeon, *Acipenser naccarii*: morphological observations on mouth and head sensorial equipment. J Appl Physiol 22(suppl 1):208–212

Chai Y, Xie C, Wei Q, Chen X, Liu J (2006) The ontogeny of the retina of Chinese sturgeon (*Acipenser sinensis*). J Appl Physiol 22(suppl 1):196–201

Debus L (1999) Meristic and morphological features of the Baltic sturgeon (*Acipenser sturio* L.). J Appl Ichthyol 15:38–45

Dettlaff TA, Ginsburg AS, Schmalhausen OI (1993) Sturgeon fishes. Developmental biology and aquaculture. Springer, Berlin, 300 p

Devitsina GV, Gadzhieva AR (1996) Dynamics of morphological development of gustatory system during the early ontogenesis of two representatives of Acipenserids *Acipenser nudiventri* and *Acipenser persicus*. J Ichthyol 36(8):642–653

Devitsina GV, Kazhlayev AA (1993) Development of chemosensory organs in Siberian sturgeon *Acipenser baeri* and stellate sturgeon, *Acipenser stellatus*. J Ichthyol 33(3):9–19

Ehrenbaum E (1894) Beiträge zur Naturgeschichte einiger Elbfische (*Osmerus eperlanus* L., *Clupea finta* Cuv., *Acerina cernua* L., *Acipenser sturio* L.). Beilage zu den Mitteilungen des Deutschen Seefischereivereins 10:1–49

Gessner J, Würtz S, Kamerichs CM, Kloas W (2009) Substrate related behavioural response in early life stages of American Atlantic sturgeon *A. oxyrinchus*. J Appl Ichthyol 25(suppl 2): 83–90

Gisbert E (1999) Early development and allometric growth patterns in Siberian sturgeon and their ecological significance. J Fish Biol 54:852–862

Gisbert E, Williot P (1997) Larval behaviour and effect of the timing of initial feeding on growth and survival of Siberian sturgeon larvae under small scale hatchery production. Aquaculture 156:63–86

Gisbert E, Rodriguez A, Castelló-Orvay F, Williot P (1998a) A histological study of the development of digestive tract of Siberian sturgeon *Acipenser baerii* during early ontogeny. Aquaculture 167:195–209

Gisbert E, Williot P, Castello-Orvay F (1998b) Morphological development of Siberian sturgeon (*Acipenser baeri* Brandt) during prelarval and larval stages. Riv Ital Acquacolt 33:121–130

Gisbert E, Williot P, Castelló-Orvay F (1999a) Behavioral modifications in the early life stages of Siberian sturgeon *Acipenser baerii*, Brandt. J Appl Ichthyol 15:237–242

Gisbert E, Sarasquette MC, Williot P, Castello-Orvay F (1999b) Histochemistry of the development of the digestive system of Siberian sturgeon during early ontogeny. J Fish Biol 55: 596–616

Gomez A, Durán E, Ocaña FM, Jiménez-Moya F, Broglio C, Domezain A, Salas C, Rodríguez F (2009) Observations on the brain development of the sturgeon *Acipenser naccarii*. In: Carmona R, Domezain A, Garcia Gallego M, Hernando-Casal J, Rodriguez F, Ruiz Rejon M (eds) Biology, conservation and sustainable development of sturgeons, Fish & fisheries series. Springer, New York, pp 155–174

Iakovleva I (1952) Development of teeth in sturgeons with regard to larval stages. Doklady AN SSSR 94(4):775–778

Icardo JM, Guerero A, Durán AC, Colvee E, Domezain A, Sams-Coma V (2009) The developmental anatomy of the heart of the sturgeon *Acipenser naccarii*. In: Carmona R, Domezain A, Garcia Gallego M, Hernando-Casal J, Rodriguez F, Ruiz Rejon M (eds) Biology, conservation and sustainable development of sturgeons, Fish & fisheries series. Springer, New York, pp 137–153

Jollie M (1980) Development of head and pectoral girdle skeleton and scales in *Acipenser*. Copeia 2:226–249

Kasumyan AO, Kazhlayev A (1993) Formation of searching behavioural reaction and olfactory sensitivity to food chemical signals during ontogeny of sturgeons (Acipenseridae). J Ichthyol 33:51–65

Kowalewsky A et al (1870) Die Entwicklungsgeschichte der Störe. Vorläufige Mitteilung. Bull Acad Sci St Petersburg 14:317–325

Loew E, Sillman AJ (1993) Age-related changes in the visual pigments of the white sturgeon *Acipenser transmontanus*. Can J Zool 71:1552–1557

Marinelli W, Strenger A (1973) Vergleichende Anatomie und Morphologie der Wirbeltiere. 4. Lieferung. Franz Deuticke, Wien

Mills KE (2003) Acipenseriformes. In: Gale Group (ed) Grzimek's animal life encyclopedia. Gale Group, Florence, pp 213–220

Ostos-Garrido MV, Llorente JI, Camacho S, García-Gallego M, Sanz A, Domezain A, Carmona R (2009) Histological, histochemical and ultrastructural changes in the digestive tarct of sturgeon *Acipenser naccarii* during early ontogeny. In: Carmona R, Domezain A, Garcia Gallego M, Hernando-Casal J, Rodriguez F, Ruiz Rejon M (eds) Biology, conservation and sustainable development of sturgeons, Fish & fisheries series. Springer, New York, pp 121–136

Rodriguez A, Gisbert E (2002) Eye development and the role of vision during Siberian sturgeon early ontogeny. J Appl Ichthyol 18(4–6):281–285

Salensky VV (1878) Developmental history of the sterlet (*Acipenser ruthenus*) 1. Embryogenesis. Tr Ova Estestvoispytatelei pri Kazanskom Univ 7(3):1–226

Salensky VV (1880) Developmental history of the sterlet (*Acipenser ruthenus*). Postembryonic development of organs. Estestvoispytatelei pri Kazanskom Univ 10(pt 2):227–545

Salensky VV (1881) Recherches sur le développement du sterlet *Acipenser ruthenus*. Arch Biol 2:233–241

Schmalhausen OI (1983) Duration and patterns of development of the giant and stellate sturgeon prelarval development. In: Dettlaff TA, Vassetzky SG (eds) Animal species for developmental studies, vol 2 Vertebrates. Consultants Bureau, New York, pp 67–88

Sydner DE (1976) Terminologies for intervals of larval fish development. In: Boreman J (ed) Great Lakes fish egg and larvae identification. US Fish and Wildlife Service Biological Service Program FWS-OBS (86/23), pp 41–58

Sydner DE (1994) Morphological development and identification of pallid, shovelnose, and hybrid sturgeon larvae. Final report of Colorado State University Larval Fish Lab to US Army Corps of Engineers CEMRO-PD-M, Omaha, NE (paper and fiche copies available via Fish and Wildlife Reference Service, Bethesda, MD, request MIN#809440098)

Sydner DE (2002) Pallid and shovelnose sturgeon larvae — morphological description and identification. J Appl Ichthyol 18(4–6):240–265

Vasquez M, Rodriguez F, Domezain A, Salas C (2002) Development of the brain of the sturgeon *Acipenser naccari*. J Appl Ichthyol 18(4–6):275–279

Williot P, Brun R, Pelard M, Mercier D (2000) Unusual induced maturation and spawning in an incidentally caught pair of adults of the critically endangered European sturgeon *Acipenser sturio* L. J Appl Ichthyol 16(6):279–281

Williot P, Rouault T, Pelard M, Mercier D, Jacobs L (2009) Artificial reproduction of captive endangered European Atlantic sturgeon, *Acipenser sturio*. Endanger Species Res 6:251–257. doi:10.3354/esr00174

Chapter 6
An Overview on Geographical Distribution from Past Descriptions

Géraldine Lassalle, M. Béguer, and E. Rochard

Abstract In the mid-nineteenth century, the European sturgeon (*Acipenser sturio* L, 1758) was proved to have been present in a significant proportion of the Western Palearctic region, i.e., Europe and North Africa. Of the 196 basin units investigated, 24 were selected for spawning by sexually mature individuals, and 44 were frequented by occasional vagrants. Its spatial distribution did not follow the classic "core–periphery" model noted for most European diadromous fish. Spawning basins were generally separated one from another by hundreds or even thousands of kilometers of coast. In these gaps, sturgeon were totally absent from inland waters, or occasionally migrants entered rivers, sometimes traveling far above the tidal limit, for exploration and feeding activities. Because of diverse anthropogenic regional pressures, the sturgeon's range decreased drastically, and population disappearances were reported earlier in the twentieth century in the northern part of its range.

6.1 Introduction

The presence of the European sturgeon (*Acipenser sturio* L, 1758) and its use by human civilizations are demonstrated as far back as Antiquity and the Middle Ages by various source materials: (1) literary references [one of the most famous being Rondelet (1558)], (2) pictorial representations such as manuscript illustrations,

G. Lassalle (✉)
Cemagref, Estuarine Ecosystems and Diadromous Fish Research Unit, 50 avenue de Verdun, 33612 Cestas Cedex, France

UMR 6250 LIENSs, University of La Rochelle – CNRS, Institut du Littoral et de l'Environnement, 2 rue Olympe de Gouges, 17000 La Rochelle, France
e-mail: geraldine.lassalle@univ-lr.fr

M. Béguer • E. Rochard
Cemagref, Estuarine Ecosystems and Diadromous Fish Research Unit, 50 avenue de Verdun, 33612 Cestas Cedex, France

P. Williot et al. (eds.), *Biology and Conservation of the European Sturgeon Acipenser sturio L. 1758*, DOI 10.1007/978-3-642-20611-5_6,

mosaics, sculptures and frescoes, and (3) fish bones and other archaeo-zoological remains, on which both morphological and genetic analysis are performed [Arles area, Rhone basin, Desse-Berset (1994)] (see Chaps. 3, 7–9). This intriguing species was among the most valuable fish in Early Modern Europe, with a great commercial importance (Kinzelbach 1997). Its flesh and eggs, or roe (caviar), were sold for food, and the swim bladder was used in isinglass, a gelatine (see Chaps. 1, 13, 14, 16, 19). Therefore, its distribution has long been studied by biologists, and there has been even more motivation to study this research topic, due to the sharp and generalized decline of the species observed as early as the mid-nineteenth century. It is generally admitted that the European sturgeon was a fairly common species, being found throughout the European continent (Magnin 1959, 1962). However, previous studies did not make the distinction between basins inhabited by occasional vagrants, feeding, or reproducing individuals. This information would be particularly relevant to the restoration of this species, which faces a high risk of becoming extinct in the near future. Furthermore, since the last distribution map was published by Kinzelbach (1997), several studies have clarified the presence and the taxonomic status of sturgeons in certain parts of their range [e.g., Baltic Sea, Ludwig et al. (2002) and Iberian Peninsula, Ludwig et al. (2009)] (see Chaps. 7, 8, 9). Consequently, in this work, precise and updated distributional data were gathered from a large spectrum of literature sources and presented in tables and maps to detail the species distribution and its evolution from 1750 until the present time. This time frame was chosen because in ancient periods, few quantitative data and even fewer time series are available. Added to this, the absence of any systematic surveys of sturgeon occurrences and the taxonomic uncertainties mean that data are not fully reliable, and language particularities make the documents time-consuming to analyze.

6.2 Material and Methods

The study area was defined as covering the entire zone where the European sturgeon was known or reported to occur, according to former authors [e.g., Magnin (1959), Kinzelbach (1997)], and encompassing an entire biogeographical area. Thus, it comprises the Western Palearctic region, including Europe, North Africa, and the Middle East. Based on a list provided by the European Environment Agency (http://www.eea.europa.eu), a total of 196 basins were selected as follows: (1) all basins with a drainage area exceeding 50,000 km^2 were retained, (2) medium-sized basins were chosen, avoiding any obvious geographical bias, and omitting basins with temporary water flow.

Next, the distribution of the European sturgeon was established in terms of presence/absence over three periods: 1750–1850, 1851–1950 and 1951–present time. No additional subdivisions were retained to further detail the evolution of the species distribution, because of the limited amount of data available. References published during each period, recent synthesis papers, international databases and expert consultations were used. Species presence in a system was categorized into three functional groups: (1) spawning basins, where the species reproduced, (2) transitory basins, where the species occurred infrequently and in very low numbers, (3) unexplored basins, where the species was never recorded. For the first group, the upstream migration limit, i.e., how far the species penetrates into the main river and its principal tributaries, was also investigated at that time, as well as the date on which the last individual(s) was captured.

6.3 Results and Discussion

At the turn of the eighteenth century, the European sturgeon was found to have an almost pan-European distribution (Fig. 6.1a). The species was totally absent from the enclosed Caspian environment and from the Eastern Mediterranean region. At the southern and northern edges of the distribution, isolated catches were reported along the Maghreb littoral (Heldt 1934; Furnestin et al. 1958), and single individuals were fished in the waters surrounding Iceland (Saemundsson 1949) and along the Norwegian coast (Collett 1875) as far as the Kola Peninsula (Lagunov and Konstantinov 1954). Compared to diadromous species belonging to other families such as Petromyzontidae, Clupeidae or Salmonidae, the absence of sturgeon species from a given basin can be ascertained with a higher level of confidence, and thus considered as "true absence" from the biogeographers point of view, i.e., because the environment is totally unsuitable (Lütolf et al. 2006). Indeed, it appeared that in regions where the species was not familiar, single-specimen catches were considered as extraordinary events, with pictures being taken and reports published in newspapers (see Chap. 19). This was most probably due to some impressive morphological features of the sturgeon family, i.e., a prehistoric-looking fish characterized by armor-like bony scales and by a remarkably large size, and also the unexpectedly placid nature of these animals and their anadromous behavior, quite unusual for such a large fish.

Forty-four basins were visited by single and immature specimens (Table 6.1). There were two possible circumstances: catches corresponded either (1) to vagrants episodically entering basins for exploration, or (2) to individuals frequenting inland waters once a year for feeding, sometimes far above the tidal limit. The latter case is illustrated by the British catches occurring regularly throughout England, Wales, Scotland and Ireland (Yarrell 1836; Day 1880–1884). The question of whether or not sturgeon bred in British waters has been asked for decades. It appeared that neither specimens smaller than 50 cm nor mature individuals with fully developed gonads were captured in the region (see http://www.glaucus.org.uk/ for complementary discussion). The transitory basins were numerous and spread over the whole study area. This large perimeter, where straying individuals from spawning populations were encountered, demonstrated considerable swimming capabilities as well as a potential developed exploration scheme.

The 196 basins retained for this study encompassed all the basins where the species historically reproduced (Table 6.2). Most of the spawning basins

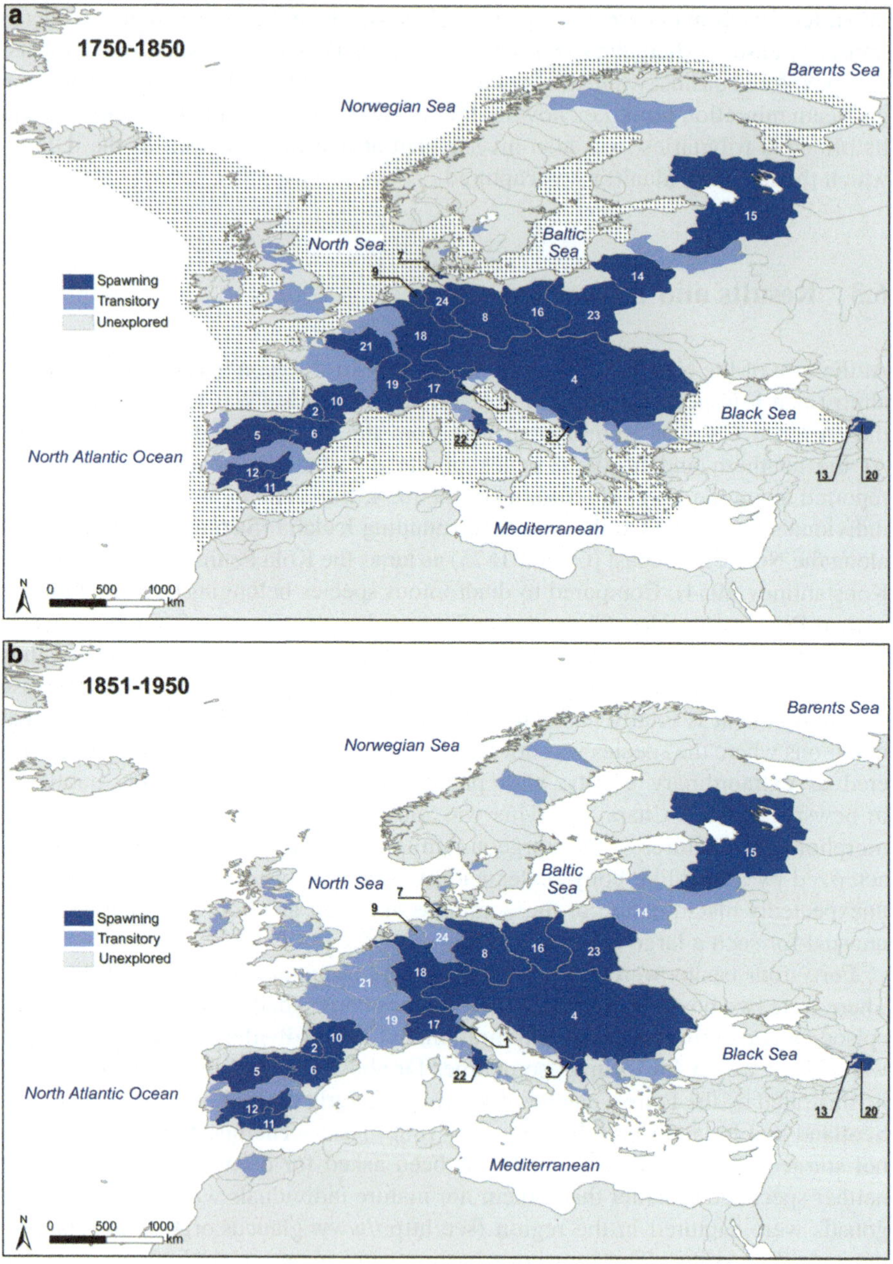

Fig. 6.1 (continued)

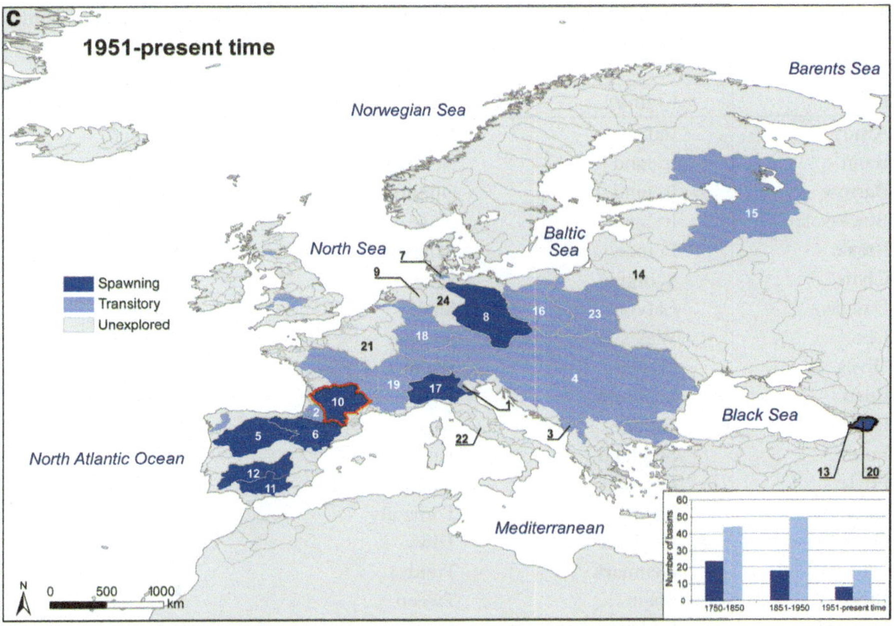

Fig. 6.1 Continental distribution of the European sturgeon for the period 1750–1850 (**a**), 1851–1950 (**b**) and 1951–present time (**c**). The distinction between the three functional groups of basins is made. Numbers refer to supplementary information on spawning basins given in Table 6.2 (name, upstream limit of migration and last catches in the twentieth century). In (**a**), the *dotted area* approximately delineates its marine distribution. In (**c**), the basins where a remaining population occurs (Gironde–Garonne–Dordogne system; 10) or is suspected to occur (Rioni basin; 20) are marked with a *bold red line* and a *bold dark line* respectively. The Nile basin (Egypt) and the Caspian Sea are truncated to offer better visualization of smaller features on the map. In both environments, the European sturgeon was absent

corresponded to major hydrographical units, located in populated areas and submitted to increasing direct human interventions over time (Fig. 6.1a). Consequently, sturgeon populations started to decrease early in the nineteenth century because of overfishing, damming, pollution, poaching, water pumping, and dredging. (Rochard et al. 1990; Williot et al. 2002) (see Chaps. 20 and 34). Twenty-four basins were visited by spawning shoals in 1850; this number decreased to 18 in 1950, and today only one functional population remains in the Gironde–Garonne–Dordogne system in France (Fig. 6.1a–c, Table 6.2, see Chap. 18). Precocious disappearances affected most exclusively those basins located north of the Loire River. Even chronology and causes of sturgeon extinction are unique from one basin to another, intensive coastal fishing of juveniles and overfishing of adults appear to be the major agents of sturgeon decline in the northern territories (Debus 1995). In contrast, the loss of longitudinal connectivity because of the construction of hydraulic structures is pointed out as the recurrent factor in the southern part

Table 6.1 Transitory basins for the period 1750–1850. Basins are given in alphabetical order. Only the country at the outlet is referred to. Sources of the data should be requested from the authors

Basin	Country	Basin (continued)	Country (continued)
Arno	Italy	Minho	Portugal
Bann	Ireland	Neretva	Croatia
Barrow	Ireland	Ombrone	Italy
Blackwater	Ireland	Piave	Italy
Bresle	France	Scheldt	Netherlands
Clyde	Scotland	Severn	Wales
Daugava	Latvia	Slaney	Ireland
Dee	Wales	Somme	France
Dives	France	Struma	Greece
Eden	Scotland	Tagliamento	Italy
Evros	Greece	Tagus	Portugal
Forth	Scotland	Tay	Scotland
Foyle	Ireland	Thames	England
Göta älv	Sweden	Torne älv	Sweden
Great Ouse	England	Towy	Wales
Gudenå	Denmark	Trent	England
Jucar	Spain	Tweed	Scotland
Kemijoki	Finland	Tyne	England
Lima	Portugal	Vjosë	Albania
Loire	France	Vire	France
Mersey	England	Volturno	Italy
Meuse	Netherlands	Wye	Wales

of its range (Nicola et al. 1996). Indeed, during their anadromous migration sturgeons reached upper sections of formerly unregulated rivers (Table 6.2).

When the spatial arrangement of transitory basins with spawning basins was further analyzed, it appeared that the historic sturgeon distribution did not follow the classic "core–periphery" model (Hengeveld and Haeck 1982; Brown 1984). On the contrary, other emblematic European diadromous species (lamprey, shad, salmon, trout, etc.) did have a former distribution that matched the major criteria of this pattern: (1) abundances are highest in the central parts of the species ranges, decreasing towards the range margins, and (2) each species occurs in many sample sites in the center of its range, but becomes much more patchily distributed towards the periphery. Added to this, each species inhabits a geographic range corresponding to a small fraction of the Western Palearctic region. In comparison, spawning basins of the European sturgeon were generally separated one from another by hundreds to thousands of kilometers of coast. In these intervals, sturgeon was totally absent from the inland waters, or sporadically entered numerous basins for feeding or exploration. Consequently, for this species, the ratio between spawning and transitory basins was less than one. Furthermore, a significant proportion of the Western Palearctic region was colonized by this sturgeon. This interspecific

Table 6.2 Spawning basins for the period 1750–1850

Basin (country)	Upstream limit	Last catch
1. Adige (Italy)	Zevio	≥1970s
2. Adour (France)	Peyrehorade	1960s
3. Buna (Montenegro/Albania)	Fierza (Drin River)	1995–1998
4. Danube (Romania/Ukraine)	Delta[a]	Doubtful in 1991
5. Douro (Portugal)	Barca d'Alva[a]	1984
6. Ebro (Spain)	Tudela	1970
7. Eider (Germany)	Rendsburg	1969
8. Elbe (Germany)	Melnik and Litomerice	Late 1980s[b]
9. Ems (Germany/Netherlands)	Rheine	1929
10. Gironde–Garonne–Dordogne (France)	Toulouse and Domme	Still present; probably functional
11. Guadalquivir (Spain)	Cordoba	1992
12. Guadiana (Spain/Portugal)	Merida	Early 1980s
13. Inguri (Georgia)	Dzhavari	Undefined status[c]
14. Neman (Russia/Lithuania)	Druskenik	1939
15. Neva (Russia)	Lake Ladoga	1984
16. Oder (Poland/Germany)	Bohumin	~1950
17. Po (Italy)	Turin	1994
18. Rhine (Netherlands)	Rheinfelden	1942
19. Rhône (France)	Lyon	1974
20. Rioni (Georgia)	Above Kutaisi	Still present?[d]
21. Seine (France)	Bray	1856
22. Tiber (Italy)	Todi	≥1920
23. Vistula (Poland)	Krakow	1965
24. Weser (Germany)	Above Hann	1938

Upstream limit of migration in the main watercourse and last individuals caught are provided. Basins are given in alphabetical order. Only the country at the outlet is referred to. According to Panos Stavros Economidis (personal communication), reproduction of the species in the Evros basin, flowing through Greece, Bulgaria and Turkey, is highly hypothetical. Neva basin includes the Neva River, the Ladoga and Onega lakes and their major tributaries. Buna basin includes the Buna River, Lake Shkoder, the Drin River and Lake Ohrid. Numbers correspond to the basin code used in Fig. 6.1. Sources of the data are presented in supplementary file 1 of Lassalle et al. (2010) for basins where the species historically reproduced.

[a]Upstream limit of migration for the period 1851–1950 is given.

[b]It is not known whether these catches included fish from the Gironde–Garonne–Dordogne population.

[c]No available information since 1991 because of the Georgian–Abkhazian conflict. Zurab Zarkua mentioned that the fish were not observed after the construction of the dam in the 1970s.

[d]In this river, last reproduction occurred in the early 1990s, but one male was captured in 1999, most probably a hybrid *A. sturio* × *A. colchicus*. However, according to Zurab Zarkua, last catch was a 2-meter-long fish in 1991, and last date of reproduction is doubtful. For complementary details, see Chap. 17.

comparison was made possible thanks to the distributional data stored in the Euro-Diad database version 3.2 (Cemagref, Cestas, France, http://www.cemagref.fr). This irregular pattern among diadromous species appears to be partly related to an abrupt change in a single environmental variable, usually either an abiotic factor or the population density of an intensively interacting species (competitor, predator or prey).

Various recent studies have showed the overlapping distributions of two sturgeon sister-species in Europe, the Atlantic sturgeon (*Acipenser oxyrinchus* Mitchill, 1815) and the European sturgeon. According to genetic, morphological and archaeozoological evidence, both species had lived sympatrically in several northern European locations (i.e., Baltic Sea vicinities), where the former is highly suspected of having replaced the latter between 800 and 1,200 years ago (Ludwig et al. 2002), the Bay of l'Aiguillon, the Gironde estuary, and the island of Oléron in France as far back as 5,000 years ago (Desse-Berset 2009) (see Chap. 7). Further information is still needed to disentangle the complex relationships between the two species, particularly elements concerning their potential hybridization and interspecific interactions (e.g., competitive exclusion), as this has already been studied in the Baltic surrounds by Tiedemann et al. (2007). Consequently, according to the current state of our knowledge, all mentions of sturgeon in these regions were considered as European sturgeon. Furthermore, this ambiguity about species identification might be useful in explaining the unusual distributional pattern found for the European sturgeon compared with other diadromous species.

6.4 Conclusion

After the recognition of full species status for both the western and eastern Atlantic sturgeons (Magnin 1964), recent findings concerning their intersected history in Europe demonstrated how far biogeography is a complex science in constant evolution. Indeed, the Atlantic sturgeon is now included in the recently published handbook of European freshwater fish (Kottelat and Freyhof 2007) whereas Baltic sea populations were not considered as belonging to a distinct species in a paper published 10 years earlier (Kottelat 1997). In this era of biological crisis [e.g., scientist warns of sixth mass extinction of wildlife (Thomas et al. 2004)], long-term and updated distributional data could be crucial for restoring freshwater systems. It would help stakeholders and policy-makers in the complicated task of sustaining or reintroducing the "right" species in the "right" places [see Chaps. 20 (Gironde–Garonne–Dordogne system), 34 (Elbe basin), 42 (Oder basin) and 44 (Vistula basin)] (Williot et al. 2009).

Acknowledgments We thank national expert consultants who provided helpful comments on the past and present status of the European sturgeon in their country: Murat Bilecenoglu, Adnan Menderes University, Department of Biology, Turkey; Panos Stavros Economidis, Aristotle University, Department of Zoology, Greece; Tamas Gulyas, Independent Sturgeon Specialist, Hungary; Bella Japoshvili, Institute of Zoology, Laboratory of Hydrobiology and Ichthyology, Georgia; Emmanuil Koutrakis, Fisheries Research Institute, National Agricultural Research Foundation, Greece; Nirgiza Ninua, Georgian National Museum, Collection Zoology, Georgia. This study was carried out with financial support from the French national agency for water and aquatic environments (ONEMA) and the European Environment Agency (EEA).

References

Brown JH (1984) On the relationship between abundance and distribution of species. Am Nat 124: 255–279

Collett R (1875) Norges fiske: med bemaerkninger om deres udbredelse. Brøgger AW, Christiania

Day F (1880–1884) The fishes of Great Britain and Ireland. Williams & Norgate, London

Debus L (1995) Historic and recent distribution of *Acipenser sturio* in the North Sea and Baltic Sea. In: Gershanivich AD, Smith TIJ (eds) Proceedings of the international symposium on sturgeons. VNIRO, Moscow, pp 189–203

Desse-Berset N (1994) Sturgeons of the Rhône during Protohistory in Arles (6th-2nd century BC). In: Van Neer W (ed) Fish exploitation in the past – Proceedings of the 7th meeting of the ICAZ – Fish Remains Working Group, Ann Musee Roy Afr Centr Ser Sci Zool, pp 81–90

Desse-Berset N (2009) First archaeozoological identification of Atlantic sturgeon (*Acipenser oxyrinchus* Mitchill 1815) in France. CR Palevol 8:717–724

Furnestin J, Dardignac C, Maurin C, Vincent A, Coupé R, Boutière H (1958) Données nouvelles sur les poissons du Maroc atlantique. Rev Trav Inst Peches Marit 22:378–493

Heldt MH (1934) Note sur la capture de l'esturgeon (*Acipenser* L.) dans les mers tunisiennes. Comm Int Expl Sci Mer Médit, Rapp et P.V. 8:81–84

Hengeveld R, Haeck J (1982) The distribution of abundance I. Measurements. J Biogeogr 9: 303–316

Kinzelbach R (1997) The sturgeon (*Acipenser sturio* L. 1758) in Europe. Z Oekol Natschutz 6: 129–135

Kottelat M (1997) European freshwater fishes: an heuristic checklist of the freshwater fishes of Europe (exclusive of former USSR), with an introduction for non-systematists and comments on nomenclature and conservation. Biologia (Bratisl) 52:1–271

Kottelat M, Freyhof J (2007) Handbook of European freshwater fishes. Kottelat/Freyhof, Cornol/Berlin

Lagunov II, Konstantinov KG (1954) Baltiyskiy osetr v Belom more (Baltic sturgeon in White Sea). Priroda (Mosc) 3:113–114

Lassalle G, Crouzet P, Gessner J, Rochard E (2010) Global warming impacts and conservation responses for the critically endangered European Atlantic sturgeon. Biol Conserv 143: 2441–2452

Ludwig A, Debus L, Lieckfeldt D, Wirgin I, Benecke N, Jenneckens I, Williot P, Waldman JR, Pitra C (2002) When the American sea sturgeon swam east. Nature 419:447–448

Ludwig A, Arndt U, Debus L, Rosello E, Morales A (2009) Ancient mitochondrial DNA analyses of Iberian sturgeons. J Appl Ichthyol 25:5–9

Lütolf M, Kienast F, Guisan A (2006) The ghost of past species occurrence: improving species distribution models for presence-only data. J Appl Ecol 43:802–815

Magnin E (1959) Répartition actuelle des acipenséridés. Rev Trav Inst Peches Marit 23:277–285

Magnin E (1962) Recherches sur la systématique et la biologie des acipenséridés. Ann Sta Cent Hydrobiol Appl 9:7–242

Magnin E (1964) Validité d'une distinction spécifique entre les deux acipenséridés *Acipenser sturio* L. d'Europe et *Acipenser oxyrhynchus* d'Amérique du Nord. Trav Pech Que 91:5–20

Nicola GG, Elvira B, Almodovar A (1996) Dams and fish passage facilities in the large rivers of Spain: effects on migratory species. Arch Hydrobiol 113:375–379

Rochard E, Castelnaud G, Lepage M (1990) Sturgeons (Pisces: Acipenseridae); threats and prospects. J Fish Biol 37:123–132

Rondelet G (1558) La première partie de l'histoire entière des poissons. Macé Bonhome, Lyon

Saemundsson B (1949) Marine pisces. In: Fridriksson A et al (eds) The zoology of Iceland. Ejnar Munksgaard, Copenhagen, p 149

Thomas JA, Telfer MG, Roy DB, Preston CD, Greenwood JJD, Asher J, Fox R, Clarke RT, Lawton JH (2004) Comparative losses of British butterflies, birds and plants and the global extinction crisis. Science 303:1879–1881

Tiedemann R, Moll K, Paulus K, Scheer M, Williot P, Bartel R, Gessner J, Kirschbaum F (2007) Atlantic sturgeons (*Acipenser sturio, Acipenser oxyrinchus*): American females successful in Europe. Naturwissenschaften 94:213–217

Williot P, Arlat G, Chebanov M, Gulyas T, Kasimov R, Kirschbaum F, Patriche N, Pavlovskaya LP, Poliakova L, Pourkazemi M, Kim Y, Zhuang P, Zholdasova IM (2002) Conservation and broodstock management. Int Rev Hydrobiol 87:483–506

Williot P, Rochard E, Kirschbaum F (2009) Acceptability and prerequisites for the successful introduction of sturgeon species. In: Carmona R, Domezain A, Garcia Gallego M, Hernando Casal J, Rodriguez F, Ruiz Rejon M (eds) Biology, conservation and sustainable development of sturgeons. Springer, pp 369–384

Yarrell W (1836) A history of British fishes. John Van Voorst, London

Chapter 7
Ancient Sturgeon Populations in France Through Archaeozoological Remains, from Prehistoric Time Until the Eighteenth Century

Nathalie Desse-Berset

Abstract The recent discovery in France of *A. oxyrinchus* in archaeological sites along the Atlantic coast has led to the creation of an inventory of bony sturgeon remains all over France, and the review of specific determinations according to morphological criteria. Whilst *A. sturio* remains to this day the only identified species on the Mediterranean side, rivers of the Atlantic coast and the north of France have sheltered by a majority another species, *A. oxyrinchus*, from the Neolithic and at least until the seventeenth century.

This latter species, identified in the Final Neolithic (5,000 years ago) represents the first colonizing by sturgeons of French waters at the end of the last glaciation. More than 30 sites have revealed sturgeon remains, some of which are very high upstream. Both species (*A. sturio* and *A. oxyrinchus*) were present on the Atlantic front and the Channel–North Sea. Today, new questions are raised and discussed.

7.1 Introduction

The existence of sturgeons since the Jurassic era has been proved in the form of fossils at an approximate date of 200 MYBP (Bemis and Kynard 1997; Bemis et al. 1997; Birstein and DeSalle 1998; Choudhury and Dick 1998).

The hypothesis of their colonization of European waters at the end of the last glaciation (about 12,000 years ago) is recurring but does not, however, rely on concrete fossil remains. Where were the sturgeons during the thousand years that separate them from their paleontological origins? When did they colonize West European shores? And which species are we talking about?

N. Desse-Berset
Université de Nice-Sophia Antipolis – CNRS-CEPAM-UMR 6130, SJA3, 24 avenue des Diables Bleus, 06357 NICE Cédex 4, France
e-mail: berset.desse@hotmail.com; berset@cepam.cnrs.fr

P. Williot et al. (eds.), *Biology and Conservation of the European Sturgeon Acipenser sturio L. 1758*, DOI 10.1007/978-3-642-20611-5_7,

The discovery in France of a second sturgeon species, *A. oxyrinchus*, in several sites and at several time periods (Desse-Berset 2009a; Desse-Berset and Williot 2011), has encouraged the author to review all the sturgeon determinations systematically attributed to *A. sturio*, and thus to broaden the inventory.

In this book dedicated to the biology and the conservation of sturgeon, it was important to gather the traces of their past existence through archaeozoological data, and to locate them as precisely as possible in time and space. Thanks to archaeological evidence and faunal analyses, new factual data has allowed us to progress in the knowledge of sturgeon populations during the past thousands of years, and will perhaps allow us to redraw a part of their history.

Lying between Northern and Southern Europe, France is surrounded by various seas and possesses different fronts on the Mediterranean, Atlantic and Channel. All the rivers that flow into these seas were until the beginning of the twentieth century swum by sturgeon (Magnin 1959, 1962).

Sturgeon bone remains dug up in anthropic contexts, from prehistory until the modern era, have made it possible to attest to the presence, number, size of these sturgeon and to determine their species. Archaeological excavation can thus give evidence of their presence – or of their absence – through time and space at precise time periods and places.

The limits are, however, those of human intervention; without human captures attested by bone remains, the existence of sturgeon in French rivers and seas would have been impossible to prove, because no trace is left.

7.2 Material and Methods

7.2.1 Origin of Data

The information presented here comes from an inventory, still in progress, of bony sturgeon remains stemming from archaeological sites mentioned until today. Developing a data set is not an easy thing to do, due to the large number of rescue excavations during which faunas are unfortunately not always studied and published.

As with any archaeological synthesis, this current knowledge report can at any time be completed with new finds and discoveries. The information collected to date concerning these taxon remains is herein gathered (Table 7.1). It is based on: (1) the analyses of the author in numerous sites, published or not (Arles, Montmajour, Ponthezières, La Perroche, Saint Germain d'Esteuil, Le Langon, Paris Grand Louvre, Bennecourt, Douai), and (2) publications and unpublished data communicated directly or via the MNHN inventory (Callou 2009) (Fig. 7.1). In some cases, this data is still incomplete, because we have not yet been able to examine all the bones from some sites. The specific determinations which have always been attributed to *A. sturio* have been reviewed according to our latest work (Desse-Berset 2009a, b). The places where sturgeon remains were discovered are indicated.

Table 7.1 Distribution of archaeological sturgeon remains in France depending on location, chronology and species (revision of specific determinations previously identified as *A. sturio*)

	Geographical area or river drainage	N°	Sites	Chronology	Number of Remains (NR) (in brackets: Bony plates and scutes)[1] [2]	Archaeozoologist (or responsible)	Revision of species (Desse Berset)		
							A.oxyrinchus morphology	*A. sturio morphology*	*A.* sp.
MEDITERRANEAN SEA	Rhône R.	1	Arles JH	6th-2nd cent. BC	2500	NDesse-Berset1994, 2009a	0	2500	
		2	Montmajour	4th-18th cent AD	34	NDesse-Berset	0	34	
	Lez R.	3	Lattes	5th BC-1st cent AD	29	GPiquès-MSternberg	0	29	
	Coast	4	Le Cailar (gard)	5th centBC	1	GPiquès	0	1	
	Coast	6	Aigues Mortes (Psalmodi)	Middle Ages	3	ALocker (pers comm.)	0	3	
	Rhône-Ouvèze R.	5	Vaison-la-Romaine	6th-12th cent AD	41	MSternberg	0	41	
	Rhône R.	7	Lyon	3rd cent AD	2	GPiquès et al 2008	0	2	
ATLANTIC OCEAN	Oléron Island	8	Ponthezières	4th Mill BC	36 (35)	NDesse-Berset 1995, 2009a, b	33	3	
		9	La Perroche	4th Mill BC	2(2)	NDesse-Berset 1995, 2009a, b	2 [3]	0	
	Gironde Estuary	10	St-Germain d'Esteuil	3rd.BC-2nd cent. AD	65(37)	NDesse-Berset 2009b	7	30	
		11	Barzan	1st cent.BC-2nd cent. AD	28 (23)	BEphrem	0	23	5
	Charente R.	12	Andone	10th cent AD	12(12)	LBourgeois-IRodet 2009	-	6	6
	Sèvres Niortaise R.	13	Niort (Donjon)	7th-10th cent AD	11	CVallet T.Cornec	0	11	
	Vendée R.	14	Le Langon	1st cent AD	22(22)	NDesse-Berset 2009a	22	0	
	Loire R.	15	Rezé (2)	1st cent.AD (20-30/40AD)	1(1)	ABorvon	1	0	
		16	Nantes(Poissonnerie)	2nd cent AD	3	MHSantrot Musée Nantes	3	0	
		17	Montsoreau [4]	11th cent. AD	13(9)	ABorvon	9	0	
		18	Angers [5]	10th cent. AD	1(1)	JHYvinec	1	0	
		19	Angers [6] (château)	4th-13th cent AD	7	Inv.MNHN Callou	-	-	
		20	Limoges (musée Evêché)	9th-14th cent. AD	49	CVallet CPironet	0	15	34
	Clain-Vienne-Loire R.	21	Poitiers (les Hospitalières)	5th cent AD	3	CVallet 2007	2	1	
		21b	Poitiers (les Hospitalières)	5th-11th AD	13	CVallet 2007	-	-	13
		22	Poitiers-Vienne	middle12th cent AD	9	ACreuzieux 2008	0	7	2
CHANNEL & NORTH SEA	Seine R.	23	Bennecourt	2nd cent. BC	2(2)	NDesse-Berset 1999	2	0	
		24	Paris Gd Louvre	16th-17th cent. AD	18(18)	JDesse/NDesse-Berset 1992	6	7	5
		25	Vincennes [7]	14th -15th cent. AD	1(1)	BClavel 2001	1	0	
		26	Arnières-sur-Iton	1rst cent AD (Gallo-roman)	1(1)	CBémilli 2008	1	0	
	Aisne R. Oise-Seine	27	Laon	5th-10th cent AD	2?	Callou 2009-BClavel. 2001-2002	-	-	
	Somme R.	28	Boves (château)	11th-13th cent. AD	8	BClavel 2001	8	0	
		29	Amiens (ZAC cathédrale)	?	?	BClavel	-	-	
	Oise R•	30	Compiègne (les Hallettes)	11th cent. AD	1	JHYvinec 1997	0	1	
	Scarpe/Scheldt R.	31	Douai	9th-10th cent. AD	25(24)	NDesse-Berset, BClavel	18	0	6
	Rhin (Bas-Rhin)	32	Andlau (Abbaye)	10th-11th AD	1	O.Putelat	1	0	

Other sources of information have been sought, such as iconographical data (for instance paintings or cave carvings, as well as animal representations in mobiliary art) and then for more recent periods, written sources.

As compared to previous studies in the field where the remains of three (Desse-Berset 2009a) and 12 sites (Desse-Berset and Williot 2011) had been investigated, the present study has been extended to more than 30 archaeological sites.

7.2.2 The Archaeozoological Methods

Bony sturgeon remains come mainly from the exoskeleton, and consist of scutes (dorsal, lateral, ventral) and of dermal bony plates from the cranium. Each individual owns about a hundred, plus those present on the two pectoral spines, which are used to estimate the age of the specimen (see Chap. 23). The internal skeleton is essentially cartilaginous. Only a few elements of the splanchnocranium are ossified, such as the dentary, the maxillary and the palatopterygoid (see Chap. 3). Unfortunately, they are seldom identified by archaeologists. The bone remains revealed during excavation are dated by archaeologists according to their stratigraphical origin, and then reach the archaeozoologist, who determines:

(i) The species, thanks to anatomical attribution and thanks to morphological studies of bones, in particular the aspect of their surface, based on comparison with specimens of modern reference (Magnin 1962, 1964; Desse-Berset 2009a)
(ii) The size reconstruction of individuals thanks to osteometry (Desse and Desse-Berset 1998, 2002)
(iii) The quantification of remains (number of remains (NR), whole or parts of bones; minimal number of individuals: MNI) (Desse et al. 2002)

In the case of the sturgeon, when we started studying specimens from Arles around 1985, these sturgeon-related methods did not exist, and collecting an endangered animal is difficult. Naturalized specimens conserved in museums cannot be used because their bony parts are not accessible or measurable. Palaeontological or zoological publications (Goodrich 1930; Gregory 1933; Grassé 1958) provide syntheses about the complex anatomy of Acipenseriformes; the works of naturalists present for each species systematic data, specific determination criteria, that is to say meristic characteristics (number of dorsal, lateral and ventral scutes for instance) (Magnin 1964; Lelek 1987; Holcik 1989; Whitehead et al. 1984; Rochard et al. 1991) and morphometric characteristics (percentage of length of each body part compared to the rest of the body, for instance) (Magnin 1962, 1964; Magnin and Beaulieu 1963; Debus 1999; Holcik 1989; Ludwig and Gessner 2007), which can be very useful when studying an entire fish, but are unfortunately not very helpful when studying isolated bone fragments generally found in archaeological remains.

7.2.3 Reference Collections

It was essential to constitute a collection of three species: *Acipenser sturio*, *Acipenser oxyrinchus* and *Acipenser naccarii*. Including this latter species originating from the Adriatic Sea was supposed to answer the hypothesis that it existed in the Rhône basin (south of France, Mediterranean coast), following Spanish work reporting its presence in the Iberian peninsula (Garrido-Ramos et al. 1997) or not (Doukakis et al. 2000).

The *A. sturio* specimens have graciously been supplied by the Cemagref since 1985, the *A. naccarii* specimens by the Italian fish farming "Azienda Agricola V.I.P" and the *A. oxyrinchus* specimens by the Ostéothèque of Montreal University and the Canadian company "Acadian Sturgeon and Caviar Inc.". These modern specimens have permitted us to study the morphological and osteometrical of archaeological remains, which most of the time consist of isolated, broken or used bones.

Species discrimination between *A. sturio* and *A. oxyrinchus* can be made by using morphological criteria of external bony plates (cranial and scutes) and splanchnocranium bones (maxillaries, dentaries or palatopterygoids) (see Chap. 3) While those of *A. sturio* show round tubercles, deep alveoli separated by sharp septa are specific to *A. oxyrinchus* according to Magnin's description (1962) (Fig. 7.1). Some of the remains were either too small or too deteriorated to distinguish the species, so some remains are undetermined. Following the pioneering work of

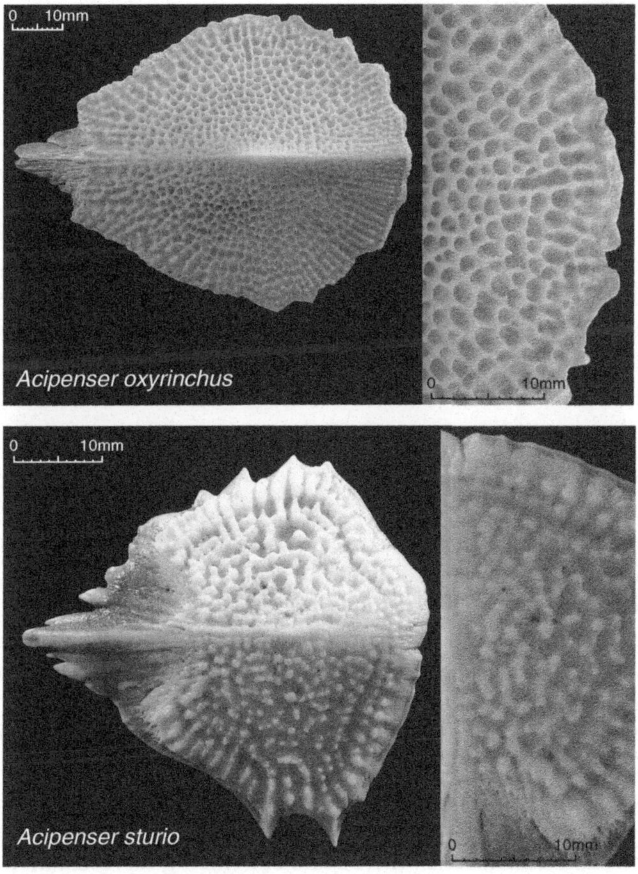

Fig. 7.1 Morphological differences observed on modern dorsal scutes of *Acipenser oxyrinchus* from Canada (**a**) and *Acipenser sturio* from the Gironde estuary (**b**) (reference collection, laboratory of Archaeozoology, Valbonne)

Magnin (1962), the morphological specificity of dermal plates has become a widely spread method used to distinguish between the two species *A. oxyrinchus* and *A. sturio*. Recent examples have been given by Debus (1999) and Artyukhin and Vecsei (1999) to discriminate between the two species in Northern Europe. Even more recently, this method was used to support genetic investigations (Ludwig et al. 2002, 2008; Ludwig and Gessner 2007; see Chap. 9).

Eventually, the connection with natural sturgeon population characteristics has been possible thanks to the works of Magnin (1959, 1962, 1964). Without them, the knowledge about several sturgeon populations would have been forever lost.

When it was possible, the *z*-test was used to compare the relative number of remains coming from both species as well as the number of sites.

7.3 Results and Discussion

7.3.1 Ancientness of Sturgeons in France

We can assume that if the sturgeon was present through all the French rivers, men would have fished it and perhaps pictured it. One of France's most favourable zones with regard to this information research is the Gironde–Garonne–Dordogne basin on the Atlantic coast. This is the last European region to shelter sturgeons: it probably offers prosperous environmental conditions for the fish. This region has also revealed hundreds of prehistoric sites rich in fauna (especially fish) and rock and mobiliary art in the lower basin of the Dordogne.

7.3.1.1 Prehistoric Iconography

Fish representations in prehistoric art are not very numerous (about 20 in parietal art and approximately 250 in all mobiliary art, carved or sculpted in bones or rocks (Cleyet-Merle 1987). Some realistic representations have made it possible to identify salmonids or pike, but most of them are simplified and do not permit the determination of a precise species. The possibility of a sturgeon representation has been suggested by some authors, but most of these authors doubt this assumption (Cleyet-Merle 1987, 1990) (for instance, a carving on deer antlers coming from Gourdan, Haute Garonne marked "Sturgeon?": Museum of National Antiquities, Saint Germain en Laye, Collection no. 47354).

A recent article dedicated to figurative sturgeon representation in prehistoric art has led to the same conclusions: there is no convincing representation designating sturgeon (Citerne 2004) while pike and salmon have been represented in realistic ways. Our own examination of all these elements has disapproved the hypothesis of sturgeon representation in prehistoric iconography.

7.3.1.2 Bone Remains

The bone remains represent the unique source of information with regard to the former presence of sturgeons. In fact, if sturgeons had been captured by men during Prehistory, their bones would have been found: large and resistant dermal plates but also various elements of the internal skeleton would have been conserved and easily identified by specialists who have studied the fauna of these numerous sites.

And yet, for ancient periods (Palaeolithic), no sturgeon remains have been found, while fishing activities have been proved (Cleyet-Merle 1990; Desse-Berset 1994b).

In the following period, which came after the last glaciation (about 20000 BP), the waters were quickly colonized by salmonids that can often be found in archaeological sites from the Upper Palaeolithic but also by shads, eels and pike. Their relatively fragile bones are well-represented in ancient settlements, and identified by archaeo-ichtyologists (Desse and Desse 1976; Le Gall 1984; Cravinho and Desse-Berset 2005). Much more resistant sturgeon remains would not have gone unnoticed. These elements represent significant and negative evidence leading to the suggestion that sturgeons did not live along the French coasts or in French rivers before the Neolithic era. Global warming from the eighth millennium BP might have favoured the colonization of the French coasts by these migratory fish.

And yet in France, the oldest evidence of sturgeon presence only appears in the Final Neolithic, 5,000 years ago, on the Oléron Island. On the Ponthezières and La Perroche sites the author has identified the most ancient factual proofs (Desse-Berset 1995, 2009a, b)

Despite numerous excavated settlements which have provided abundant faunas, no other French coast site (neither on the Atlantic side nor on the Mediterranean side) has confirmed sturgeon presence up to now, for these ancient periods.

One needs to look farther south on the Iberian coast to find sturgeon remains during former periods. Although rare, they have been identified in Portugal in shell middens of the Mesolithic (Lentacker 1994; NR = 4) and in the east of Spain on the settlement of Cueva de Nerja (next to Malaga) on the Mediterranean coast, where sturgeon remains have been found among coastal ichtyofaunas from Upper Palaeolithic levels (Boessneck and von den Driesch 1980; Morales et al. 1994; Ludwig et al. 2009; see Chap. 9).

During protohistory, in the centuries BC, sturgeons appeared progressively in French archaeological fauna tables (Tables 7.1 and 7.2).

Later, through the centuries, these sturgeons appeared during the Middle Ages and during modern periods, but most of the time in a very limited number.

7.3.2 Where? When? How Many Remains?

Sites which have revealed sturgeons are presented in Table 7.1 and in Fig. 7.2.

Table 7.2 Distribution of archaeological sturgeon remains in France depending on location and chronology of species *A. oxyrinchus* (**O**, *black square*), *A. sturio* (**S**, *gray square*); both species (**OS**); *A.* sp. (*dark gray square*)

	Site	3MI	-6	-5	-4	-3	-2	-1	+1	2	3	4	5	6	7	8	9	10	11	12	13	14	15	16	17	18	TOTAL
	1		S	S	S	S	S																				2500
	2											S	S	S				S	S	S						S	34
	3			S	S	S		S	S																		29
M	4			S																							1
	5													S				S	S	S							41
	6													S													3
	7										S																2
	8	OS																									36
	9	O																									2
	10					OS	OS	OS	OS	OS																	65
	11							S	S	S																	28
	12																	S									12
	13														S	S	S	S									7
	14							O																			14
A	15								O																		1
	16									O																	3
	17																		O								13
	18																	O									1
	19																										7
	20																S	S	S	S	S	S					49
	21												OS														16
	22																			S							9
	23						O																				2
	24																							OS	OS		18
	25																					O	O				1
	26								O																		1
N	27																										2?
	28																		O	O	O						8
	29																										?
	30																		S								1
	31																O	O									22
	32																	O	O								1

Mediterranean front (**M**), Atlantic front (**A**), and Channel and North Sea front (**N**). 3 Mi = 3rd Millennium BC; −6 to 18 = 6th cent. BC to 18th centuries AD

7.3.2.1 The Mediterranean Sea

It has been proved that the sturgeon thrived in this area during the following periods: from the sixth century BC until the second century BC, from the fourth century AD until the sixth century AD, from the tenth century AD until the twelfth century AD, and during the eighteenth century AD depending on the location.

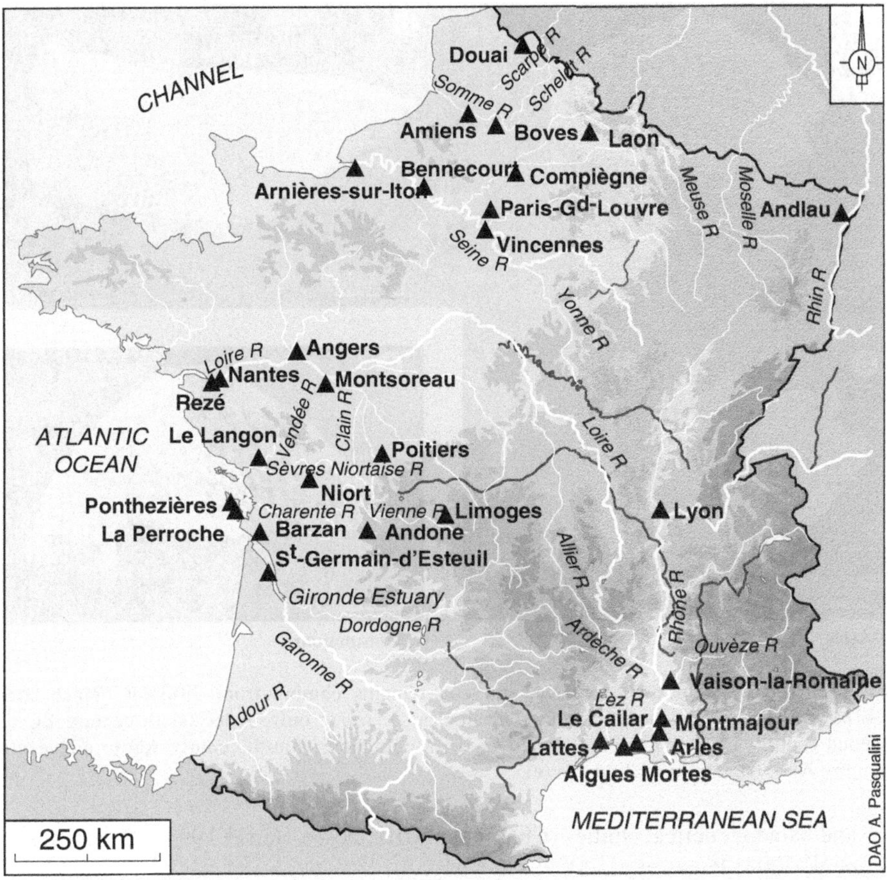

Fig. 7.2 Location of archaeological sites from which sturgeon remains are determined

Rhône River

The Rhône area has always been a privileged region, in which sturgeons reproduced even very recently, because the latest captures date back to the 1970s (Tabardel 1994; Brosse et al. 2005; see Chap. 19). "*Although this magnificent fish can be captured in most French rivers, it is never as common and as big as in the Rhône, where it reigns alone. In other fresh waters, the sturgeon has to fight for royalty with the salmon.*"(De La Blanchère 1868–1869: 396).

– Several sites have revealed sturgeon remains, especially Arles, where the "Jardin d'Hiver" excavation provided a large number of sturgeon remains (NR = 2,500 bones) in areas occupied by the Greeks between the sixth and the second century BC (Fig. 7.3a to d). Thus, this site has become an exceptional source for the study of the species over five centuries. This collection has encouraged archaeozoological

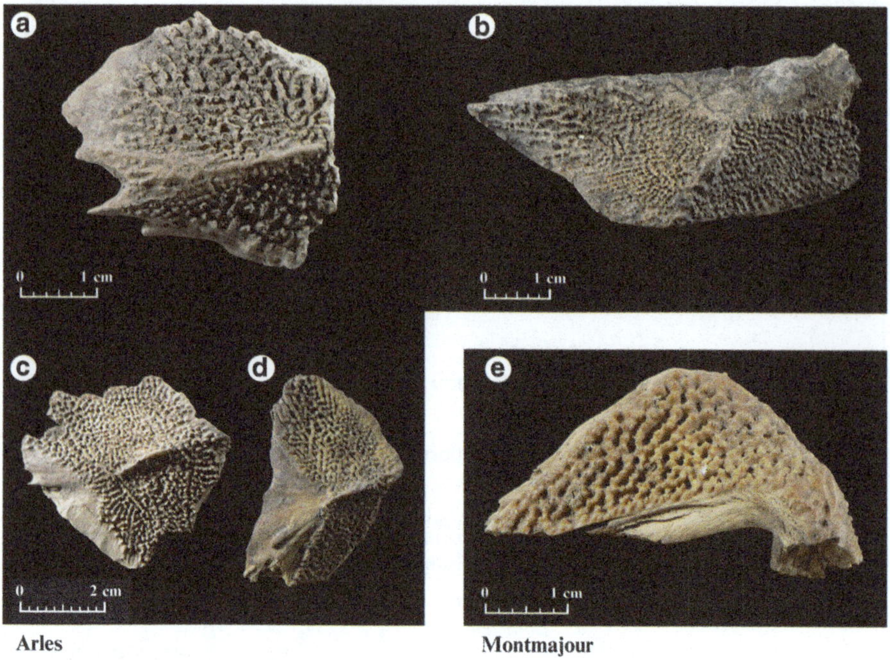

Fig. 7.3 Examples of archaeological sturgeon remains coming from different French sites (*A. oxyrinchus* and *A. sturio*): Mediterranean Sea, *A. sturio* only. Arles (sixth century BC to second century BC): skutes of *A. sturio* (**a–d**); Montmajour (fourth century AD to eighteenth century AD): Cleithrum of *A. sturio* (**e**)

and palaeogenetical studies (Chassaing 2010; Desse-Berset 1994a; Desse-Berset et al. 2008; Pagès et al. 2009; see Chap. 8).

- Other Mediterranean sites have for the same periods revealed a few remains, such as for instance Lattes, a lagunary site next to the little river Lèz (Sternberg 1995; Piquès 2010: fifth century BC until first century AD, NR = 29) and Le Cailar in the Gard (fifth century BC; NR = 1) (Piquès personal communication).
- For more recent periods, the sturgeon was present in several levels of Montmajour Abbey, located on a Rhône islet right next to Arles, between the fourth and the eighteenth century (during the fourth century, the tenth and twelfth centuries and the eighteenth century, NR = 33, Desse-Berset in progress) (Fig. 7.3e). According to some written sources, the monks used to eat sturgeon throughout the centuries, but its capture was regulated. Texts dating from the twelfth century provide information about the right which monks had to be given the first sturgeons captured each year. Thus the first 'eggs sturgeon' (female) and the first 'milk sturgeon' (male) were given to the monastery (Stouff 1984).
- The species is reported in various other sites such as, for instance, Vaison La Romaine along the Ouvèze River, a tributary of the Rhône (Sternberg 1996;

late antiquity — sixth century AD: NR = 18; Middle Ages — tenth to twelfth century: NR = 23) and Aigues-Mortes, where three scute fragments were dug up in the Benedictine abbey of Psalmodi (A. Locker personal communication).
- We may also mention the discovery in Lyon, in the Rhône valley (Saint-Georges Park, third century AD) of a few bones belonging to two pectoral spine fragments which gave birth to a palaeogenetical analysis (Piquès et al. 2008).

7.3.2.2 The Atlantic Ocean

It has been proved that the sturgeon thrived in this area during the following periods: end of the fourth millennium BC until the middle of the third millennium BC, from the third century BC until the second century AD, and from the fourth century AD until the fourteenth century AD depending on the location.

On the Atlantic coast, sturgeons were present in the main basins until the beginning of the twentieth century, and have been confirmed in numerous archaeological sites in the Gironde–Garonne–Dordogne basin, the Loire basin, the Charente basin and the "Sèvres Niortaise" basin (see Chap. 19).

Gironde–Garonne–Dordogne Basin

- On the Oléron island, the Neolithic sites of Ponthezières and La Perroche (Director L. Laporte) have provided the most ancient sturgeon remains (Desse-Berset 1995, 2009a, b): NR = 36 and NR = 2 respectively) (Fig. 7.4a and b). These specimens can be considered as Garonne basin populations (best hypothesis) or of the Charente and "Sèvres Niortaise", although these two rivers are not known for having hosted sturgeon populations.
- In Saint Germain d'Esteuil in the Gironde estuary, the site of Brion (Director P. Garmy) has produced the discovery of numerous remains dated from the third century BC until the second century AD, according to the excavations. They mainly consist of bony plates and pectoral spine elements (NR = 65) (Desse-Berset 2009b; Desse-Berset in progress) (Fig. 7.4e and f).
- Another Gallo-Roman site, Barzan, is likely to have revealed sturgeon remains (NR = 28) (Ephrem, Personal communication).

The small number of sites which have revealed sturgeon remains in this basin during these historical periods can be explained by the lack of faunal analyses.

The future will certainly modify this perspective.

Sèvres Niortaise River

- Niort's Donjon. (NR = 11) Merovingian–Carolingian period (seventh to tenth century AD (C. Vallet personal communication)

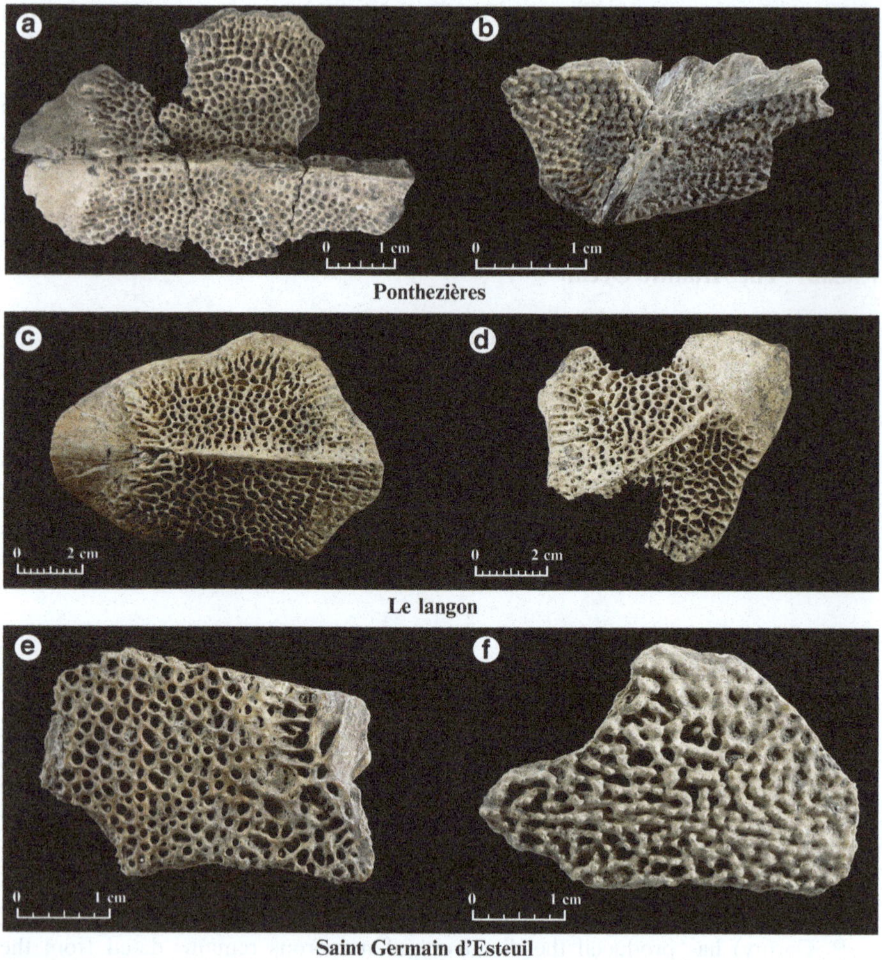

Fig. 7.4 Examples of archaeological sturgeon remains coming from different French sites (*A. oxyrinchus* and *A. sturio*): Atlantic Ocean. Ponthezières (end of fourth millennium BC): dorsal skute of *A. oxyrinchus* (**a**); lateral skute of *A. sturio* (**b**); Le Langon (first century AD): dorsal skutes of *A. oxyrinchus* (**c** and **d**); St Germain d'Esteuil third century BC to second century AD): Skute fragment of *A. oxyrinchus* (**e**); bony plate fragment of *A. sturio* (**f**)

- Le Langon (first century AD) is located in the Vendée estuary tributary of the "Sèvres Niortaise". This Gallo-Roman site is located next to the "Marais Poitevin", which has dried up today. It was an important Gallo-Roman town on the course of a Roman road and a harbour on the Vendée River estuary. It has provided fragments of very large dermal plates (NR = 22) (Desse-Berset 2009a) (Fig. 7.4c and d)
- The Andone Castrum (near the Charente River) in Poitou Charente is a fortified formation dating back to the Gallo-Roman era which was still occupied during the High Middle Ages. It has revealed fragments of sturgeon plates (NR = 15) in levels of the tenth century AD (Bourgeois 2009)

Loire River

Several sites of the Loire basin have revealed sturgeon remains dating from the Roman period to the Middle Age. Remains have been dug up in sites next to the mouth of the Loire River, far upstream, next to tributaries of the Loire River such as, for instance, in Poitiers (Clain River) or Limoges (Vienne River).

- Rezé located on the south bank of the Loire, North West of Nantes, has provided large remains (NR = 1) dating from the Roman period (20–30/40 AD) (A. Borvon personal communication)
- Nantes: bony plates are conserved at the Dobrée Museum in Nantes. Three large scutes found among three tons of smelt bones dating from the Roman period and coming from the Roman Poissonnerie neighbourhood excavation in Nantes are kept at the museum (unpublished excavation) (M. H. Santrot personal communication, curator at the Dobrée Museum Nantes. Field Director Nicolas Rouzeau, 1985) (Fig. 7.5a and b)
- The Montsoreau Castle located on the Loire River (150 km upstream from Nantes) has provided sturgeon fragments (NR = 13) dating from the eleventh century, among them 12 bony plates (more or less complete) and a pectoral spine (Borvon personal communication; Borvon 2006) (Fig. 7.5c)
- Angers (Maine et Loire) (NR = 1) (period V: tenth century AD) (Yvinec personal communication and photo): a large plate (Fig. 7.5d)
- The Castle of Angers (NR 7) (from the fourth until the thirteenth century) (MNHN inventory, Yvinec).
- Poitiers ("Les Hospitalières") (the Clain, a tributary of the Vienne/Loire Rivers (fifth century AD to eleventh century AD; NR = 13) (Fig. 7.5e, f, g) (C. Vallet personal communication)
- Poitiers (NR = 9, from which seven were photographed in Creuzieux 2007) (middle of the twelfth century AD)
- Limoges (the Vienne River is a Loire tributary), (ninth to fourteenth century AD) (NR = 49) (Vallet 2007 personal communication) (Fig. 7.5h to k)

7.3.2.3 The Channel and the North Sea

It has been proved that sturgeon thrived in this area during the following periods: second century BC, first century AD, and from the fifth century AD until seventeenth century AD depending on the location.

Seine Basin

Sturgeon remains were present in the ten sites quoted below, of which Bennecourt (Fig. 7.6a), Arnières-sur-Iton (Fig. 7.6b), Douai (Fig. 7.6c and d), Boves (Fig. 7.6e, f, g), Compiègne (Fig. 7.6h), and Paris Grand Louvre (Fig. 7.6i, j, k).

Fig. 7.5 Examples of archaeological sturgeon remains coming from different French sites (*A. oxyrinchus* and *A. sturio*): Loire basin. Nantes (second century AD): dorsal skutes of *A. oxyrinchus* (crédit photo: copyright Hervé Neveu-Dérotrie, musée Dobrée, Nantes. Dépôt de l'État et Ville de Nantes 2006, no. 2881: dépôt de l'Etat (SRA) au musée Dobrée) (**a** and **b**); Montsoreau Castel (eleventh century AD): lateral skute of *A. oxyrinchus* (**c**); Angers (tenth century AD): lateral skute of *A. oxyrinchus* (**d**); Poitiers (fifth century AD: lateral skute of *A. sturio* (**e**); lateral skutes of *A. oxyrinchus* (**f** and **g**); Limoges (ninth to fourteenth century AD): scutes of *A. sturio* (**h–k**); subopercular of *A. sturio* (**l**)

Seine River

- Bennecourt (Desse-Berset 1999) is a Celto-Roman sanctuary of the second century BC (NR = 2) (Fig. 7.6a).
- Arnières-sur-Iton, Normandy, near the Iton River (Bémilli 2008, Fig. 50, fosse ST 170) first century AD (NR = 1) (Fig. 7.6b).

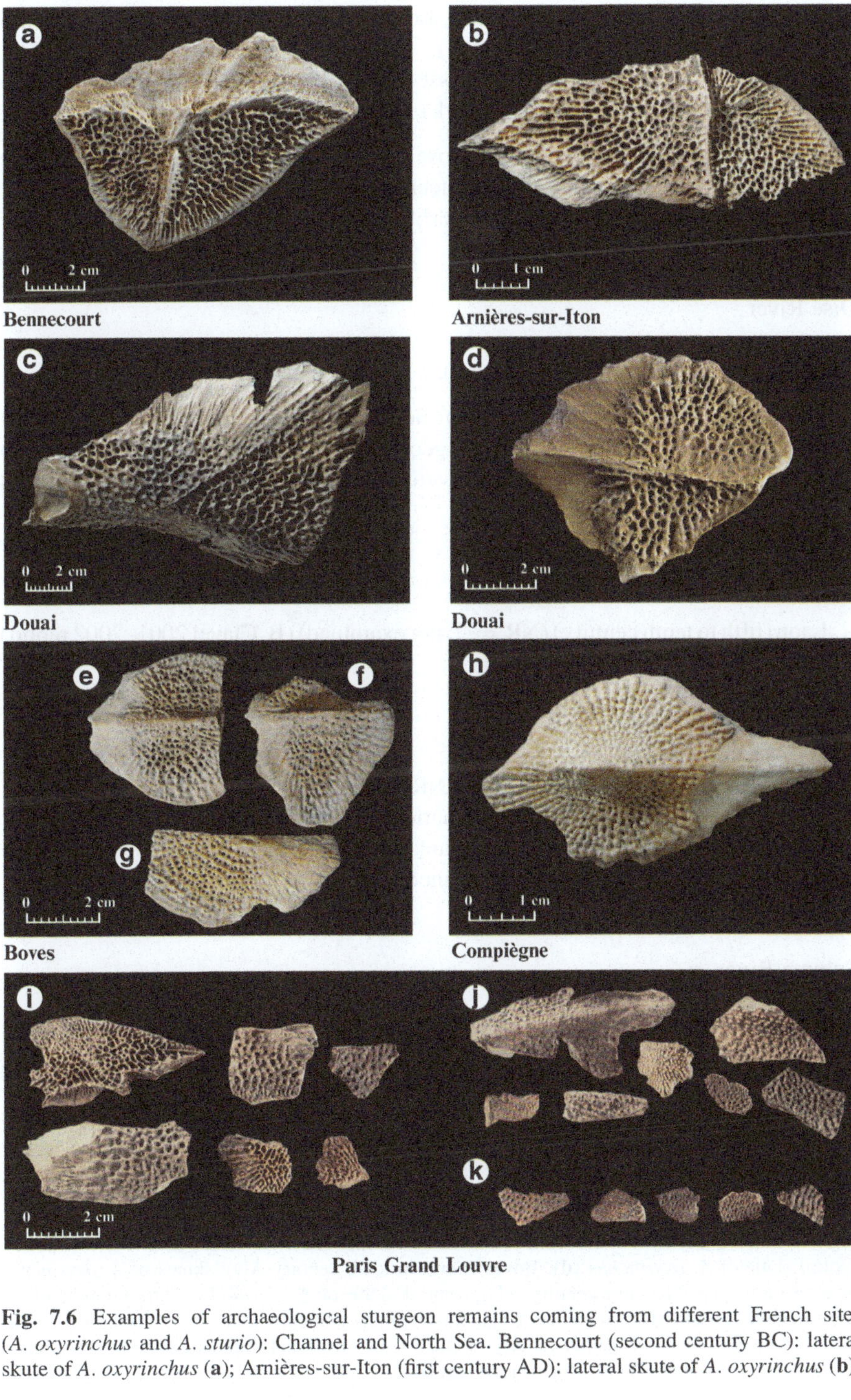

Fig. 7.6 Examples of archaeological sturgeon remains coming from different French sites (*A. oxyrinchus* and *A. sturio*): Channel and North Sea. Bennecourt (second century BC): lateral skute of *A. oxyrinchus* (**a**); Arnières-sur-Iton (first century AD): lateral skute of *A. oxyrinchus* (**b**);

- Vincennes (the Castle of Vincennes, Low Middle Ages, fourteenth to fifteenth century AD, (Clavel 2001) (NR = 1).
- Paris Grand Louvre (Desse and Desse-Berset 1992), sixteenth to seventeenth century AD (NR = 18) (Fig. 7.6i, j, k).

Of the 11,000 fish remains collected close to the Paris Grand Louvre royal palace between the fourteenth and the eighteenth centuries, only 18 sturgeon bone fragments were determined, which underlines the scarcity of the species.

Oise River

(Tributary of the Oise and Aisne rivers)

- Compiègne, Les Hallettes (Picardie). Seigneurial site (tenth to twelfth century AD). The report mentions some sturgeon remains, of which only one has been observed by the author (NR = 1) (eleventh century AD) (Fig. 7.6h) (Yvinec 1997)

Aisne River (Oise/Seine Tributary)

- Laon (fifth to tenth century) (NR = 2; not examined) (B. Clavel 2001–2002 report)

Scarpe–Scheldt River

- Douai (ninth to tenth century AD) (NR = 25) found at the "Puits du Donjon" (unpublished, Desse-Berset determination) (Fig. 7.6c and d).
- Hamage: Benedictine abbey on the banks of the Scarpe River founded during the seventh century (NR = 2; not examined) Period II.

Somme River

- Boves (Boves Castle, junkyard, Picardie, eleventh to thirteenth century, Clavel 2001) (NR = 9). Some large remains were discovered in a medieval garbage dump (Fig. 7.6e, f, g).
- Amiens (ZAC Amiens cathedral, B. Clavel personal communication, not examined).

Fig. 7.6 (continued) Douai (eleventh to thirteenth century AD): clavicular of *A. oxyrinchus* (**c**); ventral skute of *A. oxyrinchus* (**d**); Boves (ninth to tenth century AD): skutes of *A. oxyrinchus* (**e–g**); Compiègne (eleventh century AD): ventral skute of *A. sturio* (**h**); Paris Grand Louvre (seventeenth century AD): only small fragments of skutes and bony plates have been found, all in the same level of the seventeenth century. Some are too small and fragmentary to be determined precisely. *A. oxyrinchus* (**i**); *A. sturio* (**j**); *A.* sp. (**k**)

Lower Rhine River

- Andlau abbey (O. Putelat, second half of the tenth century to first half of the eleventh century, NR = 1)

7.3.3 Sturgeons? Yes, But What Species?

7.3.3.1 Identified Species (Tables 7.1 and 7.2 and Fig. 7.7)

Until recently, the only species considered as indigenous had been the *A. sturio*, and all the sturgeon remains determined in archaeological sites were always attributed to the *A. sturio*. In the Mediterranean, the presence of *A. naccarii* was sought in the collection of 2,500 bony sturgeon remains dug up on the archaeological settlement of Arles (Jardin d'Hiver) located on the Rhône River and occupied between the sixth and the second centuries BC. A first publication of this collection mentioned only the *A. sturio* (Desse-Berset 1994a). A re-examination of the entire material made it possible to declare the specific determination of *A. sturio* based on to a morphological study of current *A. naccarii* specimens compared to modern *A. sturio* specimens (those classified in our skeleton database) (Desse-Berset in Brosse et al. 2005, 2009; See Chap. 3). These results were confirmed by palaeogenetic analyses (Chassaing 2010; Desse-Berset et al. 2008; Pagès et al. 2009, see Chap. 8) (Fig. 7.7).

Yet the identification of another species *A. oxyrinchus* in several sites, thanks to undeniable morphological characteristics, has recently been proved (Desse-Berset 2009a, b; Desse-Berset and Williot 2011).

7.3.4 Spatio-Temporal Distribution of Both Species A. oxyrinchus *and* A. sturio *in France (Table 7.2)*

After the review of the specific determination, the data has been classified in three coastal fronts: the Mediterranean, the Atlantic and the Channel–North Sea.

The inventory of sturgeon remains determined in the French archaeological sites has made it possible to present the results over more than 30 sites. Although the re-examination of all bony remains has not yet been possible, we can already observe some significant characteristics that fully modify our sturgeon knowledge in France. Firstly, the most ancient determined sturgeon vestiges are in majority *A. oxyrinchus*.

On the Atlantic coast in Ponthezières, most of the bony plates refer to the *A. oxyrinchus*, and remains presenting the morphology of *A. sturio* are rare (three

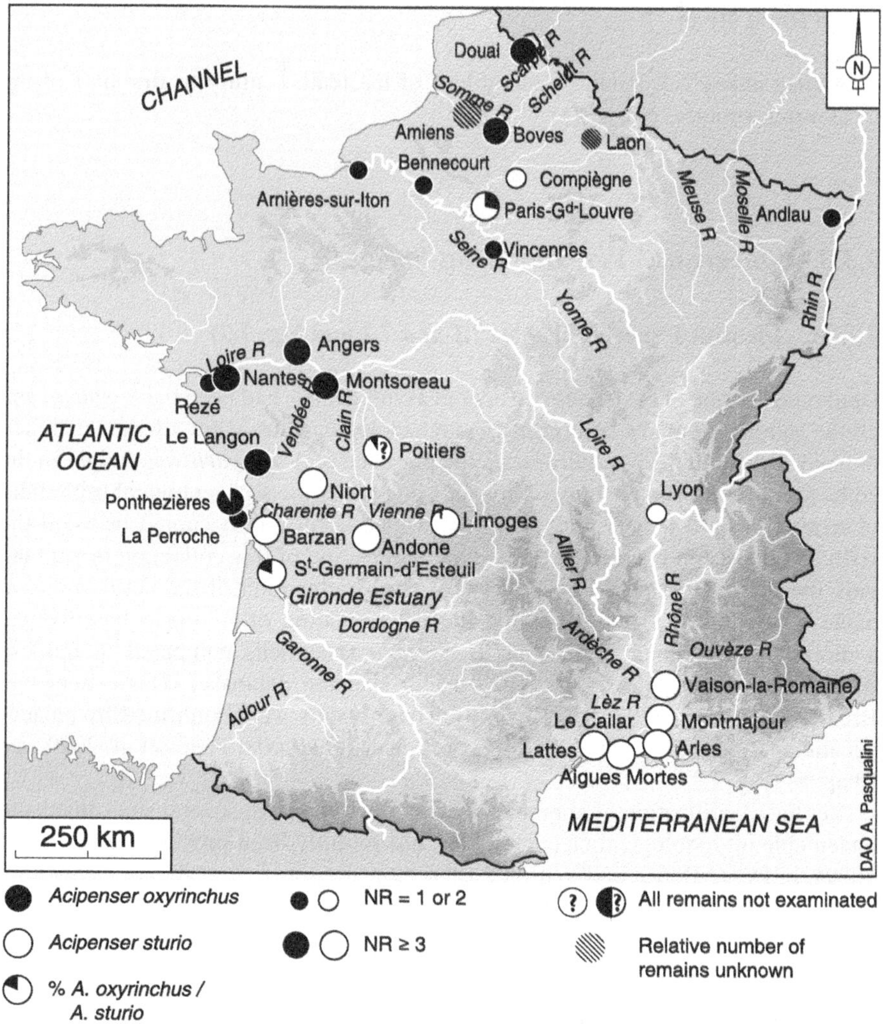

Fig. 7.7 Presence of archaeological sturgeon remains according to location and species

among 36, $z = 3.12$, $p = 0.002$). The only two remains coming from La Perroche are *A. oxyrinchus* (Desse-Berset 2009a, b).

In the Gironde estuary, St Germain d'Esteuil from the third century BC onwards supports the sympatry of the two species (Desse-Berset 2009a).

In the Loire basin, this species is also the only one in the most ancient levels (first century BC), then in sympatry with *A. sturio* from the fifth century AD.

In the north, *A. oxyrinchus* is predominant during long periods of time, in all the Channel–North Sea basins (Somme, Seine, Scheldt), starting from its oldest attestation (second century BC). Referring to our inventory, the *A. oxyrinchus* does not seem to be in sympatry with the *A. sturio* before recent times (sixteenth to seventeenth century AD). This means *A. sturio* would have colonized north of France much later.

However, at the moment, the respective number of sites from which remains of both species of sturgeon have been excavated in those two areas (Atlantic and Channel–North Sea coasts and river drainages) are not significantly different ($z = 0.739$; $p = 0.46$). Sixteen sites contained *A. oxyrinchus* remains, while ten sites contained *A. sturio* remains, and in four sites both species were present.

But on the Mediterranean front, only *A. sturio* has been determined to date. No other species (neither *A. oxyrinchus* nor *A. naccarii*) has been detected in archaeological remains until now (Desse-Berset et al. 2008; Pagès et al. 2009).

7.4 Conclusions

Except for the site of Arles, the number of sturgeon bone remains is very limited, especially if we consider that one sole specimen includes approximately 130 bony pieces. This means that sturgeons were seldom caught and thus awarded the status of "royal fish", which is incompatible with a plentiful species. The sturgeon is a much sought-after fish which appears on exceptional menus without playing a considerable economic role. The role of sturgeon transport seems consequently unimportant in France. Henceforth it is logical to bear in mind the geographic relationship of the sites with the ecology.

The present findings coming from different French areas, the Mediterranean, the Atlantic and the Channel–North Sea, allow us to point out:

- *A. oxyrinchus* inhabited the French shores, estuaries, and rivers of the Atlantic and Channel–North Sea for a long period ranging from the end of the fourth millennium BC up to seventeenth century AD, that is to say until very recent times (Desse-Berset 2009a; Desse-Berset and Williot 2011).
- The presence of remains far upstream and/or of large specimens in estuaries over a lengthy period supports the hypothesis that *A. oxyrinchus* spawned in these drainages (Desse-Berset and Williot 2011).
- *A. oxyrinchus* and *A. sturio* were often in sympatry on the Atlantic side during this long period and spawned in the same river drainages (Desse-Berset 2009a).
- *A. oxyrinchus* is totally absent from the Mediterranean side, where the only attested species is *A. sturio* (Desse-Berset et al. 2008; Pagès et al. 2009).

It has been shown that during the twentieth century very few *A. sturio* from the Gironde–Garonne–Dordogne drainage migrated regularly in summer up to the North Sea (Castelnaud et al. 1991; Rochard et al. 1997). Similarly to *A. sturio*, *A. oxyrinchus* spends most of its life span in estuaries and coastal waters, sometimes far away from its mother-river (Scott and Crossman 1973; Waldman et al. 2002). Therefore, the geographic extension of *A. oxyrinchus* to Western Europe was most probably similar to that shown for *A. sturio*. This strongly suggests the earlier presence of *A. oxyrinchus* in the North Sea and its main tributaries in Belgium, the Netherlands (supported by present observation in the Scheldt River drainage) and Germany. Indeed, in Belgium support to this statement is given by two pictures referenced as *A. sturio* instead of *A. oxyrinchus* (Van Neer and Ervynck 1993, p 38 and on the cover of the book).

The species was present in Western Europe much earlier (Neolithic) than in the Baltic Sea, contrary to what has been suggested in previous work (from eighth century to twelfth century AD, Ludwig et al. 2002). In order to explain this unexpected result, Ludwig et al. (2002) suggested a west–east migration from Northern America to Western Europe. As the present findings predate those of Ludwig et al. (2002), we can assume that the source of the Baltic Sea *A. oxyrinchus* was European (Desse-Berset and Williot 2011). Moreover, as the period during which the species inhabited France is similar to or even longer than that of *A. sturio*, extending up until a very recent period, the species might be considered as a native species, like *A. sturio*.

The selection of sturgeon species for re-introduction programs has become a new key issue in France. Restoration programs should consider the native species as much as possible (Birstein et al. 1998; Brosse et al. 2009; Kirschbaum et al. 2009; Williot et al. 1997, 2009; Waldman 2000). Obviously, great care should be taken before re-introducing a species. However, the question has to be addressed, especially in a context of climate change (see Chap. 45) that may favour the species exhibiting the best adaptation potential, as shown by *A. oxyrinchus*, which has colonized all river drainages and adjacent ocean shores from the Saint Lawrence River (Quebec) down to Florida (Birstein and Doukakis 2000; Grunwald et al. 2008).

All the sturgeon populations in Eurasia are either threatened, on the verge of extinction, or already considered as extinct, but sympatry has never been pointed out as a potential cause for such dramatic declines (Williot et al. 2002), though natural hybridizations were reported (Rochard et al. 1991; Birstein et al. 1997 for synthesis). As the morphology of the two species (*A. oxyrinchus* and *A. sturio*) is very similar, this may have led to incorrect species determination. Great attention should then be paid to archaeological remains, to look for the former presence of *A. oxyrinchus* along Western Europe coasts: North Sea, British Islands, other Channel and Atlantic locations. *A. oxyrinchus* was a "shadow sturgeon species" for a long time, as the present re-examination of French archaeological collections of sturgeons has already shown. All additional investigations on the remains, especially in the field of genetics, dealing with a potential relationship (including hybridizations) between French and Northern European *A. oxyrinchus* and between French and American *A. oxyrinchus* should be encouraged.

By way of summary, we can state that *A. sturio* has been well-implanted on the Mediterranean side for over 2,500 years, where it is the only identified sturgeon species. It has found in the Rhône River the favourable conditions to reproduce. *A. oxyrinchus* inhabited French Atlantic and Channel shores and river drainages, and was often in sympatry with *A. sturio* for a very long time (for the Atlantic) up until a very recent period (for the Channel). The current indications of the presence of

A. oxyrinchus in the Gironde–Garonne–Dordogne River, Loire River, Seine River, Somme River and in the Scheldt River drainages support the theory of its presence in other Northern European countries. It can be suggested that *A. oxyrinchus* is a native European species. New key issues are highlighted with regard to Northern Atlantic sturgeon and sturgeon restoration programmes in France.

Acknowledgements I would like to thank:

P. Arcelin, E. Bernard, L. Bourgeois, P. Demolon, P. Garmy, L. Laporte, F. Péone, the archaeologist directors of the sites who have trustingly lent me the sturgeon remains and provided me all the chronostratigraphical information concerning their sites

A. Borvon, C. Bémilli, B. Clavel, B. Ephrem, O. Putelat, M.-H. Santrot, C. Vallet and J.-H. Yvinec, archaeozoologists or museum curators, who have given to me some sturgeon remains or some pictures that they had in their collections; C. Callou, curator of the archaeozoological and archeobotanical inventory of France (National Inventory of the Natural Heritage), National Museum of Natural History, Paris, for her collaboration on the present inventory

The Cemagref; the Giovanini fish farming Azienda Agricola V.I.P.

The Ostéothèque of Montreal and the company "Acadian Sturgeon and Caviar Inc.", Saint John, NB, Canada for gifts of sturgeon specimens

In CEPAM-CNRS, J.-D. Strich for the photographs of bones, C. Perrot and A. Pasqualini for the illustrations infography

J.-M. Paillard ("Starboard Silent Side") for the English revision of the text

Finally, P. Williot for his sturgeon knowledge and experience which he was willing to share with me

References

Artyukhin E, Vecsei P (1999) On the status of Atlantic sturgeon: conspecificity of European *Acipenser sturio* and North American *Acipenser oxyrhinchus*. J Appl Ichthyol 15:35–37

Bémilli C (2008) La faune du site antique d'Arnières-sur-Iton. Rapport d'étude

Bemis WE, Kynard B (1997) Sturgeon rivers: an introduction to acipenseriform biogeography and life history. Environ Biol Fish 48:167–183

Bemis WE, Findeis E, Grande L (1997) An overview of Acipenseriformes. Environ Biol Fish 48:25–71

Birstein VJ, DeSalle R (1998) Molecular phylogeny of Acipenserinae. Mol Phylogenet Evol 9(1): 141–155

Birstein VJ, Hanner R, DeSalle R (1997) Phylogeny of the Acipenseriformes: cytogenetic and molecular approaches. Environ Biol Fish 48:127–155

Birstein VJ, Bretts J, De Salle R (1998) Molecular identification of *A. sturio* specimens: a warning note for recovery plans. Biol Conserv 84:97–101

Birstein VJ, Doukakis P (2000) Molecular analysis of *Acipenser sturio* L., 1758 and Acipenser oxyrinchus Mitchill, 1815: A review. Boletín Instituto Español de Oceanografía 16:61–73

Boessneck J, von den Driesch A (1980) Tierknochenfunde aus vier südspanischen Höhlen. Studien über frühe Tierknochenfunde von der Iberischen Halbinsel 7:1–83

Borvon A (2006) Les animaux en Anjou de la fin du Xe au début du XIIe siècle: étude archéozoologique du site de Montsoreau (Maine et Loire, France). Bull Soc Sc Nat Ouest Fr 28(2):100–103

Bourgeois L (ed) (2009) Une résidence des comtes d'Angoulême autour de l'an mil: le castrum d'Andone. CRAHM, Caen, 562 p. ISBN 978-2-902685-66-0

Brosse L, Desse-Berset N, Berrebi P, Lepage M, Menella JY (2005) Étude de la faisabilité de la réintroduction de l'esturgeon dans le bassin du Rhône. Rapport d'étape-Phase 1: février-septembre 2005, Association MRM, Arles

Brosse L, Berrebi P, Desse-Berset N, Lepage M (2009) Sturgeon recovery plan in the Rhône River (France). Preliminary results on species determination and habitat suitability. In: Carmona R et al (eds) Biology, conservation and sustainable development of sturgeons. Springer, New York, pp 403–421, Chapter 25

Callou C (ed) (2009) Inventaire archéozoologique et archéobotanique de France (Inventaire national du Patrimoine naturel). Muséum national d'Histoire naturel

Castelnaud G, Rochard E, Jatteau P, Lepage M (1991) Données actuelles sur la biologie d'*Acipenser sturio* dans l'estuaire de la Gironde. In: Williot P (ed) Acipenser. Cemagref, Antony, pp 251–275

Chassaing O (2010) Organisation génétique des populations d'esturgeon européen *Acipenser sturio*: passé, présent et futur. PhD dissertation, Université Montpellier 2

Choudhury A, Dick TA (1998) The historical biogeography of sturgeons (Osteichthyes: Acipenseridae): a synthesis of phylogenetics, palaeontology and palaeogeography. J Biogeogr 25:623–640

Citerne P (2004) Présence de l'esturgeon dans le bestiaire figuré paléolithique: conditions et limites de l'analyse morphologique. Bull Soc Préhist Ariège-Pyrénées 59:71–92

Clavel B (2001) L'animal dans l'alimentation médiévale et moderne en France du Nord (XIIe-XVIIe siècles). Revue Archéologique de Picardie N° Spécial 19

Cleyet-Merle JJ (1987) Les figurations de poissons dans l'art paléolithique. Bull Soc Préhist Fr 84(10–12):394–402

Cleyet-Merle JJ (1990) La préhistoire de la pêche. Errance, Paris

Cravinho S, Desse-Berset N (2005) Les poissons du Bois-Ragot (Gouex, Vienne). In: Chollet A, Dujardin V (Mémoire coordonné par) La grotte du Bois-Ragot à Gouex (Vienne). Magdalénien et Azilien. Essais sur les hommes et leur environnement. Mémoire XXXVIII de la Société préhistorique française. SPF, Paris, pp 355–371

Creuzieux A (2008) Étude archéozoologique de l'espace Pierre Mendès France à Poitiers: le groupe épiscopal au milieu du XIIe siècle (Mémoire de Master2, université de Paris I)

Debus L (1999) Meristic and morphological features of the Baltic sturgeon (*Acipenser* sturio L.). J Appl Ichthyol 15:38–45

De la Blanchère H (1868–1869) Les pêches curieuses de la France, l'esturgeon. La Chasse Illustrée

Desse G, Desse J (1976) La pêche. In: de Lumley H (ed) La Préhistoire Française. CNRS, Paris, pp 697–702

Desse J, Desse-Berset N (1992) Pêches locales, côtières ou lointaines: le poisson au menu des Parisiens du Grand Louvre, du 14e au 18e siècle. Anthropozoologica 16:119–126

Desse J, Desse-Berset N (1998) Sans queue ni tête? Reconstitution de la taille et du poids des poissons à partir des restes osseux issus de sites archéologiques. In: Camps G (ed) L'homme préhistorique et la mer, 120e Congrès national des Sociétés historiques et scientifiques, Aix-en-Provence, octobre 1995, pp 366–376

Desse J, Desse-Berset N (2002) Le cortège de Neptune: les poissons de la Méditerranée durant l'Holocène. In: Gardeisen A (ed) Mouvements ou déplacements de populations animales en Méditerranée au cours de l'Holocène. BAR International Series, Oxford, pp 83–96

Desse J, Desse G, Desse-Berset N (2002) L'Archéoichtyologie. In: Miskovsky JJ (ed) Géologie de la Préhistoire (Réédition). Presses Universitaires de Perpignan, Paris, pp 815–822, 1519 p

Desse-Berset N (1994a) Sturgeons of the Rhône during Protohistory in Arles (6th-2nd century BC). In: Van Neer W (ed) Fish exploitation in the past. Proceedings of the 7th meeting of the ICAZ Fish Remains Working Group. Annales du Musée Royal de l'Afrique centrale vol 274, Tervueren, pp 81–90

Desse-Berset N (1994b) Saint Thibaud de Couz: les restes osseux de poissons. In: Bintz P (ed) Les grottes Jean Pierre 1 et 2 à Saint Thibaud de Couz (Savoie) Première partie:

Paléoenvironnement et cultures du Tardiglaciaire à l'Holocène dans les Alpes du Nord, vol 36. Gallia-Préhistoire, pp 218–224

Desse-Berset N (1995) La pêche est au bout du jardin... Deux îles, hier et aujourd'hui. IVe Colloque International de l'Homme et l'Animal, Société de recherche interdisciplinaire. Genève, novembre 1994. Anthropozoologica 21: 7–20

Desse-Berset N (1999) Les restes osseux de poissons du sanctuaire celtique. In: Bourgeois L (dir) Le sanctuaire rural de Bennecourt, du temple celtique au temple gallo-romain, vol 77. Documents d'Archéologie Française, pp 163–168

Desse-Berset N (2009a) First archaeozoological identification of Atlantic sturgeon (*Acipenser oxyrinchus* Mitchill 1815) in France. CR Palevol 8:717–724. doi:10.1016/j.crpv.2009.06.001

Desse-Berset N (2009b) La pêche dans l'économie de subsistance des sites de Ponthezières et de la Perroche (île d'Oléron, Néolithique final). In: Laporte L (ed) Des premiers paysans aux premiers métallurgistes sur la façade atlantique de la France (3500–2000 av. J.-C.). Mémoire XXXIII. Éd. Association des Publications Chauvinoises, Chauvigny, pp 584–609

Desse-Berset N, Williot P (2011) Emerging questions from the discovery of the long term presence of *Acipenser oxyrinchus* in France. J Appl Ichthyol 27:263–268

Desse-Berset N, Pagès M, Brosse L, Tougard C, Chassaing O, Hänni C, Berrebi P (2008) Specific identification of the extinct population of sturgeon from the Rhône River by mtDNA analysis from bone remains (Jardin d'Hiver, Arles, France, 6th to 2nd century BC). In: Béarez P, Grouard S, Clavel B (eds) Archéologie du poisson. Trente ans d'archéo-ichtyologie au CNRS. Hommage aux travaux de Jean Desse et Nathalie Desse-Berset. XVIIIe rencontres Internationales d'archéologie et d'histoire. APDCA, Antibes, pp 195–200

Doukakis P, Birstein VJ, De Salle R, Ludwig A, Ludwig N, Machordom A, Almodovar A, Elvira B (2000) Failure to confirm previous identification of two putative museum specimens of the Atlantic sturgeon, *Acipenser sturio* as Adriatic sturgeon, *A. naccarii*. Mar Biol 136:373–377

Garrido-Ramos MA, Soriguer MC, de la Herran RJM, Ruiz Rejon C, Domezain A, Hernando JA, Ruiz Rejon M (1997) Morphometric and genetic analysis as proof for the existence of two sturgeon species in the Guadalquivir River. Mar Biol 129:1–7

Goodrich ES (1930) Studies on the structure and development of vertebrates. Macmillan, London

Grassé PP (dir) (1958) Traité de zoologie. Anatomie, systématique, biologie. Tome III, Agnathes et poissons. 3 fascicules. Masson, Paris, p 2758

Gregory WK (1933) Fish skulls: a study of the evolution of natural mechanisms. Am Philos Soc 23:75–481

Grunwald CL, Maceda J, Waldman J, Stabile et I Wirgin (2008) Conservation of Atlantic sturgeon Acipenser oxyrinchus oxyrinchus: delineation of stock structure and distinct population segments. Conservation Genetics 9(5):1111–1124

Holcik J (1989) The freshwater fishes of Europe. Vol I Part II. General introduction to fishes. Acipenseriformes. Aula-Verlag, Wiesbaden, 469 p

Kirschbaum F, Wuertz S, Williot P, Tiedemann R, Arndt GM, Anders E, Krüger A, Bartel R, Gessner J (2009) Prerequesites for the restoration of the European Atlantic sturgeon, *Acipenser sturio* and the Baltic sturgeon (*A. oxyrinchus x A. sturio*) in Germany. In: Carmona R et al (eds) Biology, conservation and sustainable development of sturgeons. Springer, New York, pp 385–401, Chapter 24

Le Gall O (1984) L'ichtyofaune d'eau douce dans les sites préhistoriques. Ostéologie, paléo-écologie-palethnologie Les Cahiers du Quaternaire VIII, 196 p

Lelek A (1987) The freshwater fishes of Europe. Vol. 9 Threatened fishes of Europe. Aula Verlag, Wiesbaden, chapitre 3.33 Acipenseridae: 42–57

Lentacker A (1994) Fish remains from Portugal: preliminary analysis of the Mesolithic shell-midden sites of Cabeço da Amoreira and Cabeço da Arruda. In: Van Neer W (ed) Fish exploitation in the past. Proceedings of the 7th meeting of the ICAZ Fish Remains Working Group. Annales du Musée Royal de l'Afrique centrale vol 274, Tervueren, pp 263–271

Ludwig A, Gessner J (2007) What makes the difference? Sea sturgeon on both sides of the Atlantic Ocean. Am Fish Soc Symp 56:285–300

Ludwig L, Debus L, Lieckfeldt D, Wirgin I, Benecke N, Jenneckens I, Williot P, Waldman JR, Pitra C (2002) When the American sea sturgeon swam east. Nature 419:447–448
Ludwig A, Arndt U, Lippold S, Benecke N, Debus L, King TL, Matsumura S (2008) Tracing the first steps of American sturgeon pioneers in Europe. BMC Evol Biol 8:214–221. doi:10.1186/1471-2148-8-221
Ludwig A, Arndt U, Debus L, Rosello E, Morales A (2009) Ancient mitochondrial DNA analyses of Iberian sturgeons. J Appl Ichthyol 25:5–9
Magnin E (1959) Répartition actuelle des Acipenséridés. Rev Trav Inst Pêches Marit 23(3):277–285
Magnin E (1962) Recherches sur la systématique et la biologie des Acipenséridés *Acipenser sturio* L, *Acipenser oxyrhynchus* Mitchill, et *Acipenser fulvescens* Raf. Ann Stat Cent Hydrobiol Appl 9:7–242
Magnin E (1964) Validité d'une distinction spécifique entre les deux Acipenséridés: *Acipenser sturio* L. d'Europe et *Acipenser oxyrhynchus* d'Amérique du Nord. Travaux sur les pêcheries du Québec N° 1, pp 5–20
Magnin E, Beaulieu G (1963) Etude morphométrique comparée de l'*Acipenser oxyrinchus* Mitchill du Saint Laurent et de l'*Acipenser sturio* Linné de la Gironde. Le Naturaliste Canadien XC(1):5–38
Morales A, Rosello E, Canas JM (1994) Cueva de Nerja (prov Malaga): a close look at a twelve thousand year ichthyofaunal sequence from southern Spain. Annales du Musée Royal de l'Afrique centrale. Zool Sci 274:253–262
Pagès M, Desse-Berset N, Tougard C, Brosse L, Hänni C, Berrebi P (2009) Historical presence of the sturgeon *Acipenser sturio* in the Rhône basin determined by the analysis of ancient DNA cytochrome *b* sequences. Conserv Genet 10:217–224. doi:10.1007/s10592-008-9549-6
Piquès G (2010) Consommation et modes d'approvisionnement du poisson à Lattara au Ve s. av. n.è. Lattara 21:387–418
Piquès G, Hänni C, Silvino T (2008) L'approvisionnement de Lugdunum en poisson au IIIè siècle: les données de la fouille du Parc Saint-Georges (Lyon, France), In: Béarez P, Grouard S, Clavel B (eds) Archéologie du poisson, 30 ans d'archéo-ichtyologie au CNRS, Hommage aux travaux de Jean Desse et Nathalie Desse-Berset. XXVIIIe Colloque International d'Archéologie et d'Histoire d'Antibes, XIVth ICAZ Fish Remains Working Group meeting, octobre 2007, pp 255–268
Rochard E, Williot P, Castelnaud G, Lepage M (1991) Eléments de systématique et de biologie des populations sauvages d'esturgeons. In: Williot P (ed) Acipenser. Cemagref, Antony, pp 475–507
Rochard E, Lepage M, Meauzé L (1997) Identification et caractérisation de l'aire de répartition marine de l'esturgeon européen *Acipenser sturio* à partir de déclarations de captures. Aquat Living Resour 10:101–109
Scott WB, Crossman EJ (1973) Poissons d'eau douce du Canada. Bulletin 184. Ministère de l'Environnement. Service des pêches et des sciences de la Mer, Ottawa, p 1026
Sternberg M (1995) La pêche à Lattes dans l'Antiquité à travers l'analyse de l'ichtyofaune. Lattara 8:1–152
Sternberg M (1996) Rapport sur l'ichtyofaune de Vaison la Romaine. INRAP
Stouff L (1984) Saint-Pierre de Montmajour. Les Moines noirs (XIII-XIV siècles), Les cahiers de Fanjeaux N° 19, 423 p
Tabardel M (1994) Le point sur la situation de l'esturgeon (*Acipenser sturio* L.) en Méditerranée occidentale et possibilités de réintroduction dans le Rhône. E.N.S.A. Rennes, Rennes/Arles, 57 p (rapport non publié)
Vallet C (2007) Poitiers (Les Hospitalières) Rapport Final Tome 3-INRAP Août 2007, 367 p
Van Neer W, Ervynck A (1993) Archeologie en Vis. Herlevend Verleden 1, Instituut voor het Archeologisch Patrimonium, Zellik
Waldman JR (2000) Restoring Acipenser sturio L., 1758 in Europe: lessons from the Acipenser oxyrinchus Mitchill, 1815 experience in North America. Bol Inst Esp Oceanogr 16:237–244

Waldman JR, Grumwald C, Stabile J, Wirgin I (2002) Impacts of life history and biogeography on the genetic stock structure of Atlantic sturgeon *Acipenser oxyrinchus oxyrinchus*, Gulf sturgeon *A. oxyrinchus desotoi*, and shortnose sturgeon *A. brevirostrum*. J Appl Ichthyol 18:509–518

Whitehead P, Bauchot ML, Bauchot JP, Hureau JC, Nielsen J, Tortonese E (1984) Fishes of the North-Eastern Atlantic and the Mediterranean, 1st edn. UNESCO, Paris

Williot P, Rochard E, Castelnaud G, Rouault T, Brun R, Lepage M, Elie P (1997) Biological characteristics of European Atlantic sturgeon, *Acipenser sturio*, as the basis for a restoration program in France. Environ Biol Fishes 48:359–370

Williot P, Arlati G, Chebanov M, Gulyas T, Kasimov R, Kirschbaum F, Patriche N, Pavlovskaya L, Poliakova L, Pourkazemi M, Kim Yu, Zhuang P, Zholdasova IM (2002) Status and management of Eurasian sturgeon: an overview. Int Rev Hydrobiol 87:483–506

Williot P, Rochard E, Kirschbaum F (2009) Acceptability and prerequisites for the successful introduction of sturgeon species. In: Carmona R, Domezain A, Garcia Gallego M, Hernando-Casal J, Rodriguez F, Ruiz Rejon M (eds) Biology, conservation and sustainable development of sturgeons, Fish & fisheries series. Springer, New York, pp 369–384, 467 p

Yvinec JH (1997) Etude archéozoologique du site de la place des Hallettes à Compiègne (Oise) du Haut Moyen Âge au XIIe siècle. Revue archéologique de Picardie N° spécial 13

Chapter 8
Palaeogeographic Patterns of *A. sturio*

Olivier Chassaing, Nathalie Desse-Berset, Marilyne Duffraisse, Gaël Piquès, Catherine Hänni, and Patrick Berrebi

Abstract Determination of the evolutionary history of the European sturgeon is severely handicapped by its recent decline. Studying ancient biological material such as museum specimens or archaeological remains represents the only opportunity to access the past diversity of the species. Extracting and analyzing DNA from ancient material provide the means to compare genetically sturgeons from the entire former geographical range of the species. Here we present a combination of paleogenetic data which gives new insights on phylogeographic patterns of *A. sturio*. A high genetic diversity was present on the Iberian Atlantic coast, as well as in the Mediterranean and Adriatic Seas, which is concluded to represent the origin of the species range. Contrastingly, the diversity was low on the European Atlantic and

O. Chassaing
Paléogénétique et Evolution Moléculaire, Institut de Génomique Fonctionnelle de Lyon, Université de Lyon, Université Lyon 1, CNRS, INRA, Ecole Normale Supérieure de Lyon, 46 allée d'Italie, 69364 Lyon Cedex 07, France

Institut des Sciences de l'Evolution, UMR 5554 CNRS/UM2/IRD, Université Montpellier 2, cc065, Place Bataillon, 34095 Montpellier Cedex 05, France

N. Desse-Berset
Université de Nice-Sophia Antipolis – CNRS-CEPAM-UMR 6130, SJA3, 24 avenue des Diables Bleus, 06357 NICE Cédex 4, France

M. Duffraisse • C. Hänni
Paléogénétique et Evolution Moléculaire, Institut de Génomique Fonctionnelle de Lyon, Université de Lyon, Université Lyon 1, CNRS, INRA, Ecole Normale Supérieure de Lyon, 46 allée d'Italie, 69364 Lyon Cedex 07, France

G. Piquès
Archéologie des sociétés Méditerranéennes, UMR 5140 CNRS, 390 avenue de Pérols, 34970 Lattes, France

P. Berrebi (✉)
Institut des Sciences de l'Evolution, UMR 5554 CNRS/UM2/IRD, Université Montpellier 2, cc065, Place Bataillon, 34095 Montpellier Cedex 05, France
e-mail: patrick.berrebi@univ-montp2.fr

P. Williot et al. (eds.), *Biology and Conservation of the European Sturgeon Acipenser sturio L. 1758*, DOI 10.1007/978-3-642-20611-5_8,

the North Sea coasts as well as in the Black Sea, suggesting that these areas were colonized only recently.

8.1 Introduction

Biogeographic patterns of species – i.e. their spatial distribution and abundance – can be interpreted in the light of geology, geography, climatology, ecology and genetics (e.g. Taberlet et al. 1998). A time component can be introduced by using palaeontological or archaeological data, in order to infer past biogeographic patterns (e.g., Hadly et al. 2004). The goal of this chapter is to review palaeogeographic patterns of *A. sturio* by analyzing genetic data obtained from samples covering its former range.

Biogeographic approaches have been applied at the level of the order Acipenseriformes (Bemis and Kynard 1997; Choudhury and Dick 1998), and phylogeographic studies have been carried out for some sturgeon species, particularly in North America (see for example Bowen and Avise 1990; Wirgin et al. 2000; Waldman et al. 2002; Grunwald et al. 2008 for *A. oxyrinchus*). With regard to the European sturgeon *A. sturio*, data are limited because of the rarefaction of the species and the unavailability of fresh samples. Two centuries ago, *A. sturio* was indeed common in Europe – spawning in all major rivers from the North Sea to the Black Sea (see Chap. 6 for details) – but nowadays only one population remains, in the Gironde-Garonne-Dordogne basin system in France (Williot et al. 1997; Gonthier 2009). At the genetic level, little is known about extinct populations because they disappeared before the extensive use of molecular markers, which began in the 1960s. The only way to study genetic characteristics of extinct populations is thus to analyze museum specimens (naturalized or preserved in ethanol) or archaeological remains (mainly dermal bony scutes). Paleogenetic methods allow the access to genetic information from ancient samples (Orlando et al. 2002; Calvignac et al. 2008; Pages et al. 2009); we will therefore give a central place to paleogenetic analyses in this review, which is mainly based on the phylogeography of *A. sturio* described by Chassaing (2010). By analyzing the genetic structure of *A. sturio* in the main geographic regions from its former range, we expect (1) to obtain insights into its evolutionary history (geographical origin of the species, recent demographic events, relationships with sympatric species), and (2) its life history (migration patterns), but also (3) to quantify the loss of genetic diversity due to human activities, and (4) to make some assumptions about conservation of the remaining Gironde population.

8.2 Which Molecular Marker?

Many molecular markers are used to study intraspecific genetic variability and identify population structure. Among them, mitochondrial DNA has been the most extensively used in animal phylogeography because: (1) it is present in all animals,

(2) this haploid locus is maternally inherited without recombination, and (3) it evolves rapidly and provides character states from which phylogenetic relationships can be readily deduced (Avise et al. 1987). Furthermore, mitochondrial DNA is present in numerous copies per cell – between 1,000 and 10,000 depending on the tissue and the species – therefore being easier to recover from ancient samples than nuclear DNA being present in only two copies per cell (Paabo et al. 2004).

The higher the variability of a genetic marker, the more information it will provide. Many studies on animals have therefore focused on the major non-coding region of mitochondrial DNA, i.e., the control region or D-Loop, because it evolves faster than the average mitochondrial sequence and offers a higher variability. This is the case in fish in general, and in sturgeons in particular (Brown et al. 1993). A ~210 base pair (bp) fragment of the D-Loop has been commonly used in *A. sturio* (see Ludwig et al. 2000 for PCR amplification details). Here we will focus on results obtained with this marker.

8.3 Paleogenetics Studies: A Necessity

A. sturio from the Gironde basin from wild catches or from hatchery-reared individuals constitutes the only source of fresh DNA samples available for the entire species, and have therefore been used as an *A. sturio* standard in genetic studies (Birstein and DeSalle 1998; Ludwig et al. 2000). To have access to the genetic variability of the species, samples from all over its former range have to be analyzed. These samples exist, but consist in: (1) archaeological remains, mainly excavated from human occupation sites (see Chap. 7 for French examples), and (2) naturalized or alcohol-preserved specimens, collected since the creation of museums of natural history starting from the eighteenth century. Until the 1980s, molecular techniques did not allow recovering DNA from ancient samples, but with the invention of the polymerase chain reaction (PCR) by Saiki et al. in 1985, the dream came true.

Specific methodologies have been developed to extract and analyze ancient DNA, giving birth to paleogenetics (Paabo 1989; Hänni et al. 1990). To avoid problems of contamination by fresh DNA and to ensure ancient DNA sequence authentication, strict criteria have been established (Cooper and Poinar 2000). They include: (1) the use of specially dedicated facilities (independent from labs where fresh DNA is handled, over-pressurized, decontaminated by UV-lights and cleaned with bleach or other sterilants), (2) the monitoring of contamination by performing several blanks in parallel to the experiments, (3) the reproduction of the results in independent experiments, and (4) the cloning and sequencing of independent PCR products to rule out artefactual substitutions, imitating mutations, linked to post-mortem DNA degradation. Following these criteria is essential to avoid errors.

8.4 *A. sturio* Samples Included

Samples used in this study are listed in Table 8.1, comprising 136 ancient *A. sturio* samples, spreading among seven geographical regions covering the former range of the species, and dating from 2500 BP to present. Seventy-seven fresh samples were also considered, but were not taken into account for phylogeographic reconstructions, as they consist of closely related individuals from the relict Gironde population, and all analyzed sturgeons shared the same mitochondrial haplotype (Ludwig et al. 2000, 2002).

The definition of geographical regions was performed based on barriers that potentially limit migrations between sturgeon populations from adjacent regions: (1) Oresund and Danish straits between Baltic and North seas, (2) Dover Strait between North Sea and Atlantic Ocean, (3) Gulf of Cap Breton between French and Iberian Atlantic Ocean, (4) Gibraltar Strait between Atlantic Ocean and Mediterranean Sea, (5) the Straits of Sicily and Messina between Western and Eastern Mediterranean, (6) Otranto strait limiting the Adriatic Sea and (7) the Dardanelles-Bosphorus straits between Mediterranean and Black Seas. Points 2 and 6 are quite hypothetical, as migrations through these straits have been recorded (Rochard et al. 1997; Economidis et al. 2000), but were conserved as they separate major seas.

Within a geographical region, several rivers are present, each inhabited by a specific sturgeon population. We considered that populations from a same region are more closely related between each other than with populations from other regions.

We also mixed samples of different historic periods to reduce sample size variations between geographical regions (see Table 8.1). Therefore, we will not consider here change of genetic diversity over time (see Chassaing 2010 for diachronic analysis of the genetic diversity of the Rhône River population).

8.5 *A. sturio* Phylogeography

Twenty different haplotypes were identified for *A. sturio* on the 210 bp fragment of the mitochondrial D-Loop (see Table 8.1 for accession numbers), among which 14 were newly described in Chassaing (2010). They can be assembled into three main groups – A, B1, and B2 – when plotted on a phylogenetic network (Fig. 8.1). Group B1 is the most diverse and includes ten haplotypes – among which the only haplotype found in the extant Gironde population – whereas groups A and B2 both contain five haplotypes. *A. sturio* haplotypes from groups A and B2 are the most closely related with those from its sister species *A. oxyrinchus*.

The genetic composition of the samples of *A. sturio* in each of the seven predefined geographical areas is given in Fig. 8.2. Haplogroup B1 was present in the samples from each region from the Baltic Sea to the Black Sea. It was dominant in all samples except in the Black Sea, and was the only group found in the samples

Table 8.1 Geographical origin, date and number of *A. sturio* samples used in this review

Geographical origin	Date	Sample type	Sample number	Reference/provider	Genbank accession nos.
Gironde	Extant	Fresh tissue	$n = 67$		AJ249673
North Sea	19th–20th cent. AD	Museum specimen	$n = 12$		AJ249673
Western Mediterranean	19th cent. AD	Museum specimen	$n = 1$		AJ428274
Adriatic Sea	19th cent. AD	Museum specimen	$n = 1$	Ludwig et al. (2002)	AJ428274
Baltic Sea	8th–13th cent. AD	Archaeological bone (Ralswiek, Germany)	$n = 1–7$	Ludwig et al. (2008)	AJ428274
Iberian Atlantic	4th cent. BC–15th cent. AD	Archaeological bone (Castro Marim, Portugal; La Cartuja, Spain)	$n = 5$	Ludwig et al. (2009)	EU420003-7
Gironde	Extant	Fresh tissue	$n = 10$	Williot P. (Cemagref, Bordeaux, France)	
French Atlantic	19th–20th cent. AD	Museum specimen	$n = 13$	Muséum National d'Histoire Naturelle (MNHN, Paris) and Muséum d'Histoire Naturelle de Nantes (MHNNantes)	
Iberian Atlantic	5th cent. AD	Archaeological bone (Troia, Portugal)	$n = 1$	Desse-Berset N. (Cepam, Nice, France)	
Western Mediterranean	19th–20th cent. AD	Museum specimen	$n = 20$	Musée Requien (MR, Avignon, France), Muséum National d'Histoire Naturelle (MNHN, Paris, France), Muséum d'Histoire Naturelle de la ville de Genève (MNHG, Genève, Switzerland). Museo di Storia Naturale dell'Università di Firenze, Zoological Section "La Specola" (MZUF, Florence, Italy) and Muséum d'Histoire Naturelle de Nîmes (MHNNîmes, Nîmes, France), de Lyon (MHNL, Lyon, France), de Grenoble (MHNG, Grenoble, France), de Nice (MHNNice, Nice, France) de Marseille (MHNM, Marseille, France)	HQ239243-HQ239358

(continued)

Table 8.1 (continued)

Geographical origin	Date	Sample type	Sample number	Reference/provider	Genbank accession nos.
Western Mediterranean	6th cent. BC–18th cent. AD	Archaeological bone (Jardin d'Hiver and Montmajour, France)	$n = 50$	Desse-Berset N. (Cepam, Nice, France)	
Western Mediterranean	5th cent. BC–3rd cent. AD	Archaeological bone (Lattes and Parc St-Georges of Lyon, France)	$n = 6$	Piquès G. (Archéologie des Sociétés Méditerranéennes Lattes, France)	
Adriatic Sea	19th–20th cent. AD	Museum specimen	$n = 14$	Muséum National d'Histoire Naturelle (MNHN, Paris, France) and Museo di Storia Naturale dell'Università di Firenze, Zoological Section "La Specola" (MZUF, Florence, Italy)	
Black Sea	19th–20th cent. AD	Museum specimen	$n = 4$	Muzeul National de Istorie Naturale Grigore Antipa (MGAB, Bucarest, Romania)	

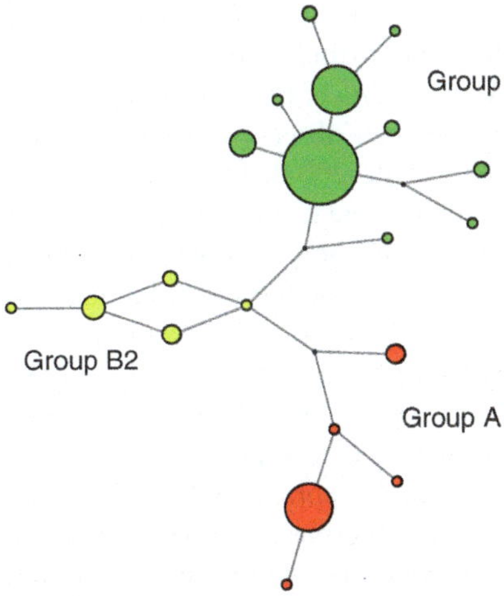

Fig. 8.1 Phylogenetic network of all known *A. sturio* D-Loop haplotypes, realized with Network Publisher 1.2 (Fluxus Technology Ltd). *Circles* represent distinct haplotypes, their width is proportional to sample number, and their *colour* refers to haplogroups

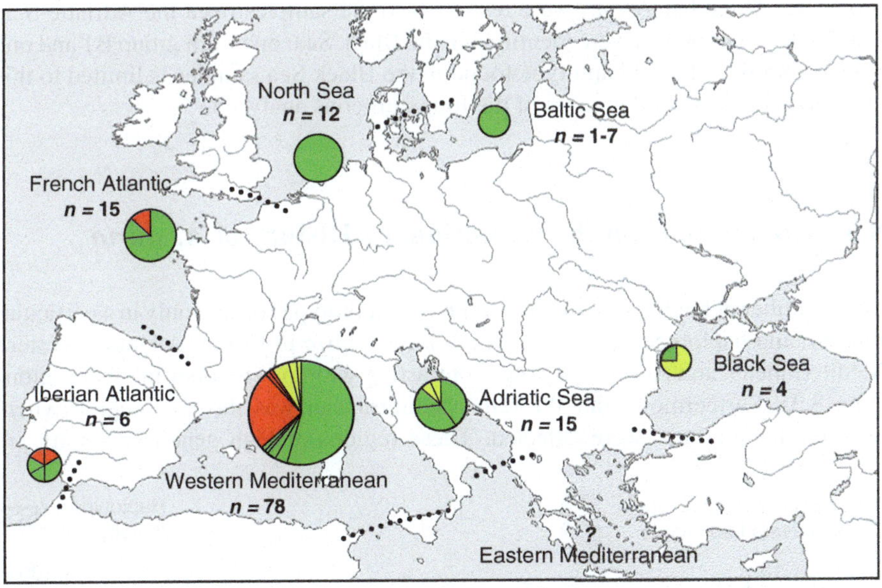

Fig. 8.2 Geographical repartition of *A. sturio* D-Loop haplogroups. *Circles* are proportional to sample size, *colours* refer to groups defined in Fig. 8.1, and *lines in the circles* separate distinct haplotypes. *Dashed lines* mark potential natural barriers delimiting geographical regions (see text)

from the North Sea and the Baltic Sea. In contrast, haplogroup A is present in both samples from the Atlantic Ocean and the Mediterranean Sea, but is absent from the samples of the two margins of the distribution, the North and Baltic Seas and the Adriatic and Black Seas. Haplogroup B2 was detected only in the Mediterranean, the Adriatic and the Black Seas.

Only one haplotype from haplogroup B1 was found in the seven scutes from the Baltic Sea attributed to *A. sturio* (out of 227, the majority of which were attributed to *A. oxyrinchus* ; Ludwig et al. 2008) suggesting a low genetic diversity. The same situation was observed among the 12 samples from the North Sea analyzed in Ludwig et al. (2002). Three haplotypes were detected on the French Atlantic coast in 15 samples, one from group A and two from group B1. The dominant haplotype was similar to the only one present in the extant Gironde population. Five haplotypes were found in the Iberian Atlantic region out of only six samples, three from group B1 and two from group A. Four of these haplotypes were restricted to this area. In Western Mediterranean, 11 haplotypes were observed in 77 samples. This was the only region where all three haplogroups were represented, with five, three and three haplotypes respectively for haplogroups B1, A and B2. Nevertheless, 80% of the samples correspond to only two haplotypes – which for this reason represent the most numerous haplotypes of groups A and B1 in this study (Fig. 8.1). Seven of the 11 haplotypes were restricted to this area. Five haplotypes were detected among 15 samples from the Adriatic Sea, three from haplogroup B1 and two from haplogroup B2. Three haplotypes were restricted to the samples from the Adriatic Sea. Finally, two haplotypes were identified in the Black Sea: one from group B1 and one from group B2. The B2 haplotype found in the Black Sea samples is limited to this area, and is represented in three of the four sturgeons analyzed.

8.6 Conclusions on the Evolutionary History of *A. sturio*

Total number of haplotypes, number of private haplotypes (found only in a particular region), and haplotype diversity are high for *A. sturio* in Iberian Atlantic, Western Mediterranean, and Adriatic regions, contrasting with lower values in other regions (Fig. 8.3). Furthermore, the Western Mediterranean was the only region where all three haplogroups were detected. These regions of high genetic diversity are

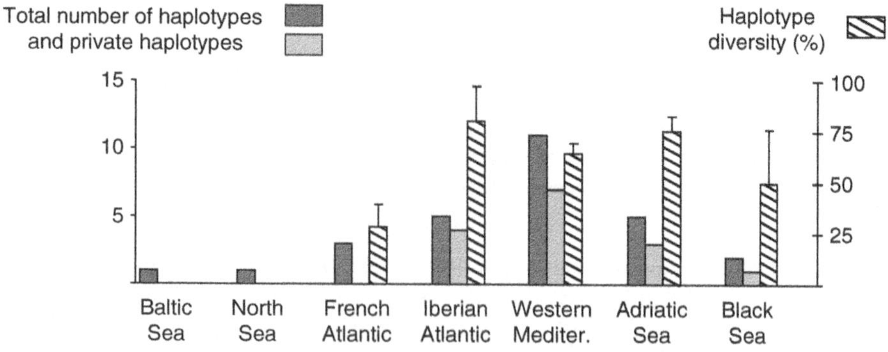

Fig. 8.3 Total number of haplotype, number of restricted haplotypes and haplotypic diversity for each geographical region defined in Fig. 8.2, computed with DnaSP (Librado and Rozas 2009)

considered as the ancient core area of the species' distribution prior to radiation. The extant sturgeons from the Gironde could be seen as a relict of a peripheral population.

A gradient of diversity is noticeable from the Baltic Sea to the Mediterranean Sea, northern populations being genetically homogeneous and southern populations being much more diverse. This pattern is similar to the one observed for *A. oxyrinchus* on the American East Coast (Wirgin et al. 2000; Grunwald et al. 2008), and could be the result of a recent colonization of northern areas after the Late Glacial Maximum [LGM, ca. 30–25 ka before present (BP) or the Younger Dryas cold interval (ca. 12 ka BP] (Chassaing 2010). Climatic conditions in Europe favoured a permafrost extension as far south as 45–50°N during the LGM, with most rivers glaciated during winter (Toucanne et al. 2009). It is likely that sturgeons were absent from glaciated rivers northward to 50°N and only survived in southern refugia – probably Iberian Peninsula, Mediterranean and Adriatic rivers, characterized by a high genetic diversity (Figs. 8.2 and 8.3). With climate warming, sturgeons from these refugia could have colonized deglaciated rivers. Only a few migrants are necessary to establish a self-sustainable population (Ludwig et al. 2008), and by founder effect, only a subset of source populations' genetic diversity was conserved in colonized areas. The most probable source for the colonization of previously glaciated rivers is the Iberian Peninsula or south-western French Atlantic rivers such as the Adour or the Gironde. Competition with *A. oxyrinchus* for river colonization could also have played a role in shaping the pattern of genetic diversity of *A. sturio* in the North of Europe (Chassaing 2010).

However, the diversity of *A. sturio* is still not known with the same precision all throughout its historic range. Additional samples from the Baltic and North seas should be analyzed to confirm the genetic homogeneity of the northern range. No samples from the Eastern Mediterranean were available, and only four samples from the Black Sea were analyzed, so that the eastern part of the former distribution of *A. sturio* remains poorly known.

There were probably few rivers in which sturgeons spawned on the east coast of the Adriatic Sea, in Eastern Mediterranean and in the Black Sea (see Chap. 6), and it could explain the rarity of available samples in these regions, along with poor sample preservation. With regard to the Black Sea, an aggravating factor could be that *A. sturio* was the least frequent sturgeon species among six that were present in the area (Bacalbasa-Dobrovici and Holcik 2000); *A. sturio* was present along with *A. ruthenus*, *A. stellatus*, *A. gueldenstaedtii*, *A. nudiventris* and *Huso huso*. Three phylogeographic hypotheses can be envisaged for the Black Sea population of *A. sturio*: (1) it originated from Mediterranean populations that entered the Black Sea after the connection between the two basins between 10 and 7 ka BP (Aksu et al. 2002), (2) it was already present at the time of the connection, and represents an ancient isolated population, or (3) it was an ancient isolated population, but also received migrants from Mediterranean populations after the Bosphorus-Dardanelles opening. Samples from the Eastern Mediterranean and more samples from the Black Sea are needed to solve this question.

8.7 Conclusions on the Life History of *A. sturio*

The fact that northern European rivers were glaciated, and that Baltic and North Seas emerged during the LGM, means that colonization of northern Europe by *A. sturio* could have occurred relatively quickly on the geological timescale (less than 20 ka). This capacity to colonize new available habitats was also observed for *A. oxyrinchus* on the American East coast (Wirgin et al. 2000), and means that mature adults can sporadically spawn in non-native rivers. This capacity should be linked with the migratory behaviour of several sturgeon species and the possibility to colonize new river systems (Wirgin et al. 2007).

However, private haplotypes were identified in each geographical region – except putatively those recently colonized – with a peak in the Western Mediterranean (Fig. 8.3). The presence in a region of haplotypes that are not shared with neighbouring regions is a sign of genetic differentiation, result of low gene flow between adjacent regions. Natural barriers and homing behaviour are the most plausible explanations for this observation. Migratory behaviour and homing are not exclusive. We know that any homing behaviour suffers a small proportion of exceptions, and that occurrence of straying fish does not imply reproduction.

8.8 Implications for Conservation

The location of the hotspot of *A. sturio* genetic diversity southward to the Gironde River, and especially in the Mediterranean, could enhance reintroduction programs in rivers of the Iberian Peninsula, Mediterranean Sea and Adriatic Sea, which constitute the historical core of the species (Chassaing 2010). Conditions for a reintroduction could become less favourable with global warming (Lassalle et al. 2010), but sturgeons survived in these area during warmer periods than today.

Nowadays, only one haplotype is found in the remaining Gironde population, whereas at least 20 existed throughout the historic range of the species (Fig. 8.1). At least 95% of the mitochondrial genetic diversity of *A. sturio* has been lost over the last 2 centuries. The loss of Iberian, Mediterranean and Adriatic genetic diversity is particularly worrying, because we may think that these populations were perhaps more prone to adapt to warm climate than the population from the Gironde, which is genetically homogeneous. With respect to global warming, potential of adaptation could have been beneficial for the species (Chassaing 2010). Some rare fish are still caught by chance in Mediterranean waters (Paschos et al. 2008), even if a functional population no longer exists. It could perhaps be worthwhile to use them for artificial reproduction, in order to save this potential of adaptation.

The low number of *A. sturio* samples from the Baltic Sea available in museums and archaeological sites is explained by the dominant presence of *A. oxyrinchus* until the recent sturgeon disappearance in this area (Ludwig et al. 2008). Little is known about biotic interactions between these species, but they can naturally hybridize (Ludwig et al. 2008; Chassaing 2010) and it is likely they may compete for food resources. It would therefore be important to protect the extant fragile Gironde population from contacts with *A. oxyrinchus* and to restrict *A. oxyrinchus* reintroduction to the Baltic region, despite its historical presence being attested by archaeological and molecular evidence on the French coast (Desse-Berset 2009; Desse-Berset and Williot 2011; Chassaing 2010).

Acknowledgements We would like to thank all museum curators (Paris, Lyon, Nice, Grenoble, Genève, Nîmes, Avignon, Marseille, Florence, Bucharest) for allowing us to sample sturgeons in their collections, Laurent Brosse and Valentin Radu for sampling some of the museum specimens, and all members of the teams Paleogenetics and Molecular Evolution (IGFL, UMR5242) and Métapopulations, Coévolution et Conservation (ISEM, UMR5554) for their support.

References

Aksu AE, Hiscott RN, Kaminski MA, Mudie PJ, Gillespie H, Abrajano T, Yasar D (2002) Last glacial-Holocene paleoceanography of the Black Sea and Marmara Sea: stable isotopic foraminiferal and coccolith evidence. Mar Geol 190:119–149

Avise JC, Arnold J, Ball RM, Bermingham E, Lamb T, Neigel JE, Reeb CA, Saunders NC (1987) Intraspecific phylogeography – the mitochondrial–DNA bridge between population-genetics and systematics. Ann Rev Ecol Syst 18:489–522

Bacalbasa-Dobrovici N, Holcik J (2000) Distribution of *Acipenser sturio* L., 1758 in the Black Sea and its watershed. Bol Instit Esp Oceanogr 16:37–41

Bemis WE, Kynard B (1997) Sturgeon rivers: an introduction to acipenseriform biogeography and life history. Environ Biol Fishes 48:167–184

Birstein VJ, DeSalle R (1998) Molecular phylogeny of Acipenserinae. Mol Phylogenet Evol 9:141–155

Bowen BW, Avise JC (1990) Genetic structure of Atlantic and Gulf of Mexico populations of Sea Bass, Menhaden, and Sturgeon – influence of zoogeographic factors and life-history patterns. Mar Biol 107:371–381

Brown JR, Beckenbach AT, Smith MJ (1993) Intraspecific DNA-sequence variation of the mitochondrial control region of white sturgeon (*Acipenser transmontanus*). Mol Biol Evol 10:326–341

Calvignac S, Hughe S, Tougard C, Michaux J, Thevenot M, Philippe M, Hamdine W, Hanni C (2008) Ancient DNA evidence for the loss of a highly divergent brown bear clade during historical times. Mol Ecol 17:1962–1970

Chassaing O (2010) Organisation génétique des populations d'esturgeon européen Acipenser sturio: passé, présent et futur. PhD dissertation, Université Montpellier 2

Choudhury A, Dick TA (1998) The historical biogeography of sturgeons (Osteichthyes: Acipenseridae): a synthesis of phylogenetics, palaeontology and palaeogeography. J Biogeogr 25:623–640

Cooper A, Poinar HN (2000) Ancient DNA: do it right or not at all. Science 289:1139

Desse-Berset N (2009) First archaeozoological identification of Atlantic sturgeon (*Acipenser oxyrinchus* Mitchill 1815) in France. CR Palevol 8:717–724, 10.1016/j.crpv.2009.06.001

Desse-Berset N, Williot P (2011) Emerging questions from the discovery of the long term presence of *Acipenser oxyrinchus* in France. J Appl Ichthyol 27(2):263–268, doi:10.1111/j.1439-0426.2010.01649.x

Economidis PS, Koutrakis ET, Bobori DC (2000) Distribution and conservation of *Acipenser sturio* L., 1758 and related species in Greek waters. Bol Inst Esp Oceanogr 16:81–88

Gonthier P (2009) European sturgeon . . . A come back? Cah Agric 18:195–198

Grunwald C, Maceda L, Waklman J, Stabile J, Wirgin I (2008) Conservation of Atlantic sturgeon *Acipenser oxyrinchus oxyrinchus*: delineation of stock structure and distinct population segments. Conserv Genet 9:1111–1124

Hadly EA, Ramakrishnan U, Chan YL, van Tuinen M, O'Keefe K, Spaeth P, Conroy CJ (2004) Genetic response to climatic change: Insights from ancient DNA and phylochronology. PLoS Biology 2(10):e290

Hänni C, Laudet V, Sakka M, Begue A, Stehelin D (1990) Amplification of mitochondrial DNA fragments from ancient human teeth and bones. C R Acad Sci III 310:365–370

Lassalle G, Crouzet P, Gessner J, Rochard E (2010) Global warming impacts and conservation responses for the critically endangered European Atlantic sturgeon. Biol Conserv. doi:10.1016

Librado P, Rozas J (2009) DnaSP v5: a software for comprehensive analysis of DNA polymorphism data. Bioinformatics 25:1451–1452

Ludwig A, May B, Debus L, Jenneckens I (2000) Heteroplasmy in the mtDNA control region of sturgeon (*Acipenser, Huso* and *Scaphirhynchus*). Genetics 156:1933–1947

Ludwig A, Debus L, Lieckfeldt D, Wirgin I, Benecke N, Jenneckens I, Williot P, Waldman JR, Pitra C (2002) When the American sea sturgeon swam east – a colder Baltic Sea greeted this fish from across the Atlantic Ocean in the Middle Ages. Nature 419:447–448

Ludwig A, Arndt U, Lippold S, Benecke N, Debus L, King TL, Matsumura S (2008) Tracing the first steps of American sturgeon pioneers in Europe. BMC Evol Biol 8:221

Ludwig A, Arndt U, Debus L, Rosello E, Morales A (2009) Ancient mitochondrial DNA analyses of Iberian sturgeons. J Appl Ichthyol 25:5–9

Orlando L, Bonjean D, Bocherens H, Thenot A, Argant A, Otte M, Hanni C (2002) Ancient DNA and the population genetics of cave bears (Ursus spelaeus) through space and time. Mol Biol Evol 19:1920–1933

Paabo S (1989) Ancient DNA – extraction, characterization, molecular-cloning, and enzymatic amplification. Proc Nat Acad Sci USA 86:1939–1943

Paabo S, Poinar H, Serre D, Jaenicke-Despres V, Hebler J, Rohland N, Kuch M, Krause J, Vigilant L, Hofreiter M (2004) Genetic analyses from ancient DNA. Annu Rev Genet 38:645–679

Pages M, Desse-Berset N, Tougard C, Brosse L, Hanni C, Berrebi P (2009) Historical presence of the sturgeon Acipenser sturio in the Rhone basin determined by the analysis of ancient DNA cytochrome b sequences. Conserv Genet 10:217–224

Paschos I, Perdikaris C, Gouva E, Nathanailides C (2008) Sturgeons in Greece: a review. J Appl Ichthyol 24:131–137

Rochard E, Lepage M, Meauze L (1997) Identification and characterisation of the marine distribution of the European sturgeon Acipenser sturio. Aquat Living Resour 10:101–109

Saiki RK, Scharf S, Faloona F, Mullis KB, Horn GT, Erlich HA, Arnheim N (1985) Enzymatic amplification of beta-globin genomic sequences and restriction site analysis for diagnosis of sickle-cell anemia. Science 230:1350–1354

Taberlet P, Fumagalli L, Wust-Saucy AG, Cosson JF (1998) Comparative phylogeography and postglacial colonization routes in Europe. Mol Ecol 7:453–464

Toucanne S, Zaragosi S, Bourillet JF, Cremer M, Eynaud F, Van Vliet-Lanoe B, Penaud A, Fontanier C, Turon JL, Cortijo E, Gibbard PL (2009) Timing of massive 'Fleuve Manche' discharges over the last 350 kyr: insights into the European ice-sheet oscillations and the European drainage network from MIS 10 to 2. Quat Sci Rev 28:1238–1256

Waldman JR, Grunwald C, Stabile J, Wirgin I (2002) Impacts of life history and biogeography on the genetic stock structure of Atlantic sturgeon *Acipenser oxyrinchus oxyrinchus*, Gulf sturgeon *A. oxyrinchus desotoi*, and shortnose sturgeon *A. brevirostrum*. J Appl Ichthyol 18: 509–518

Williot P, Rochard E, Castelnaud G, Rouault T, Brun R, Lepage M, Elie P (1997) Biological characteristics of European Atlantic sturgeon, Acipenser sturio, as the basis for a restoration program in France. Environmental Biology of Fishes 48:359–372

Wirgin I, Waldman JR, Rosko J, Gross R, Collins MR, Rogers SG, Stabile J (2000) Genetic structure of Atlantic sturgeon populations based on mitochondrial DNA control region sequences. Trans Am Fish Soc 129:476–486
Wirgin I, Grunwald C, Stabile J, Waldman J (2007) Genetic evidence for relict Atlantic sturgeon stocks along the mid-Atlantic coast of the USA. North Am J Fish Manag 27:1214–1229

Chapter 9
Sturgeon in Iberia from Past to Present

Arne Ludwig, Arturo Morales-Muñiz, and Eufrasia Roselló-Izquierdo

Abstract Archaeozoological data from the Upper Paleolithic onwards provide evidence of the presence of sturgeon in the Iberian Peninsula suggesting both a refuge during the last glaciations and their apparent scarcity during ancient times. The species recorded in all cases was apparently the European sturgeon *Acipenser sturio*. Genetic data demonstrate a sub-structuring in the Iberian Peninsula. Populations increased during the eighteenth and nineteenth century resulting in commercial exploitations in the Guadalquivir and Ebro rivers. Overfishing and river damming have been key factors causing extinction during the early twentieth century. In addition to the European sturgeon, Atlantic sturgeon (*A. oxyrinchus*), beluga (*Huso huso*) and Adriatic sturgeon (*A. naccarii*) are discussed as presumably former endemic species. Such status remains open given that the occurrence of isolated specimens is not taken as conclusive evidence for the upgrading of these taxa to a native species status.

9.1 Sturgeon in Ancient Times

9.1.1 From Late Glacial Times to Early Medieval Times

The earliest post-glacial European records of sturgeon in the Iberian Peninsula date back to the early Holocene (13000–9000 BC). Although rare, these findings point to the existence of an Iberian refuge during the last glacial episodes. Western and Northern Europe's cold climate at the time prevented the presence of sturgeon,

A. Ludwig (✉)
Leibniz Institute for Zoo and Wildlife Research, Alfred-Kowalke-Str. 17, 10315 Berlin, Germany
e-mail: ludwig@izw-berlin.de

A. Morales-Muñiz • E. Roselló-Izquierdo
Laboratorio Arqueozoologia, Universidad Autónoma de Madrid, Darwin, 2, 28049 Madrid, Spain

P. Williot et al. (eds.), *Biology and Conservation of the European Sturgeon Acipenser sturio L. 1758*, DOI 10.1007/978-3-642-20611-5_9,

but a few thousand years later [5500–3000 BC (Ludwig et al. 2009a)] sturgeon became widely distributed throughout the subcontinent. Numbers in Western and Central Europe increased rapidly thereafter, reaching a peak during medieval times. As shown in Tables 9.1–9.5, the archaeozoological record in Iberia does not reveal any such trend. In fact, it appears that sturgeon have remained a constant, though rare, item in the peninsula, as can be inferred from the barely 200 remains retrieved from a mere 14 sites, representing 27 samples documented since Palaeolithic times (Fig. 9.1; Table 9.2). Sturgeon populations have been strongly influenced by humans, with overfishing and river damming being two major reasons for population declines in the recent past (Ludwig 2008). However, it seems highly unlikely that humans were solely responsible for the apparently low numbers of sturgeon in Iberia previous to modern times.

The archaeozoological record of sturgeon in Iberia dates back to a period of the Upper Palaeolithic era, the Solutrean period, from which almost no evidence for fishing exists, either in Iberia or in Western Europe (Cleyet-Merle 1990; Cortés et al. 2008). Radiocarbon dates available for the level below which two dermal plates of European sturgeon were retrieved at the Cueva de Nerja, Málaga, Spain (18420 ± 530 BP (UBAR-158), 17940 ± 200 BP (UBAR-98) and 15990 ± 260 BP (UBAR-157)) are coincident with the Last Glacial Maximum (LGM), and evidence that previous to the Magdalenian period, when fishing appears to have been a well-developed practice throughout Europe, modern humans in southern Spain were catching sturgeon in coastal (i.e. marine) environments (Morales and Roselló 2008).

From a chrono-cultural perspective, there exists a rather equilibrated distribution of sturgeon finds from the Solutrean period to the Modern Age (Table 9.5). The exceptions are the Bronze Age (a period of Iberian prehistory when fishing essentially vanishes from sight) and the Iron Age [when fishing resumed in the shores of southern Iberia, probably triggered by the arrival of the Eastern Mediterranean colonists (Morales and Roselló 2008)]. The statistical reliability of these patterns is

Table 9.1 Evidence of sturgeon on European archaeological sites according to periods: I 15,000–9,000 BC, II 9000–7000 BC, III 7000–5500 BC, IV 5500–3000 BC, V 3000–1000 BC, VI 1000 BC–0, VII 0–600 AD, VIII 600–1500 AD, IX 1500–1900 AD

Geographic region	I	II	III	IV	V	VI	VII	VIII	IX
Sweden				2	3		1	6	
Denmark					1		1	1	1
Great Britain								10	
The Netherlands				1	10	8	5	5	
Belgium								8	5
Germany				2		2	7	19	2
Poland				2	1	4		50	3
France				1		1	1	2	
Spain	5		1	1	2	4	1	1	
Portugal			2	1				1	

Note that the table does not count the total amount of sturgeon remains; it counts only the number of locations with sturgeon remains (*source*: Ludwig et al. 2009a)

Table 9.2 Overview of sturgeon remains from Iberian sites arranged in chronological order

Sample	Site	Location	Km	Chronology	NISP (% fish)	Description	Reference
1	Cueva de Nerja	Málaga	5	Solutrean (V)	2 (0.1)	Dermal plate	Morales and Roselló (2008)
2	Cueva de Nerja	Málaga	3	Magdalenian (V)	31 (1)	–	Aura et al. (2002)
3	Cueva de Nerja	Málaga	3	Magdalenian (T)	1 (0.5)	–	Aura et al. (2002)
4	Cueva de Nerja	Málaga	1	Epipaleolithic (T, M)	5 (1)	Dermal plate NISP = 1; NISP = 2 skull plate (2)	Boessneck and Driesch et al. (1980)
5	Cueva de Nerja	Málaga	1	Epipaleolithic/ Neolithic (T)	1 (0.5)	Cleithrum	Boessneck et al. (1980)
6	Cabeço Amoreira		1	Mesolithic	2 (0.1)	Dermal plate	Lentacker (1990–1991)
7	Cabeço de Arruda		1	Mesolithic	2 (0.4)	Dermal plate	Lentacker (1990–1991)
8	Cueva de Nerja	Málaga	0.5	Early Neolithic (T, M)	2 (2)	Dermal plate, pectoral fin ray	Boessneck et al. (1980)
9	Cueva de Nerja	Málaga	–	Unspecified (T)	1 (20)	Parasphenoid	Boessneck et al. (1980)
10	Zambujal		1	Copper	2 (6)	Dermal plate, skull plate	Lepiksaar (1976)
11	Valencina	Sevilla	40/80	Copper	3 (50)	Dermal plate	Hain (1982)
12	La Fonteta	Alicante	1	Phase I (725–700 BC)	4 (6.5)	Dermal plate	Sternberg (2007)
13	La Fonteta	Alicante	1	Phase II (700–650 BC)	1 (2.2)	Dermal plate	Sternberg (2007)
14	C. Doña Blanca	Cádiz	0	675–625 BC (Fo. 18)	3 (2–3)	Dermal plate	Roselló and Morales (1994)
15	C. Doña Blanca	Cádiz	0	(Fo. 16)	3 (2–3)	Dermal plate	Roselló and Morales (1994)
16	C. Doña Blanca	Cádiz	0	650–600 BC (Fo. 15)	3 (2–3)	Dermal plate	Roselló and Morales (1994)
17	C. Doña Blanca	Cádiz	0	(Fo. 14)	1	Dermal plate	Roselló and Morales (1994)

(continued)

Table 9.2 (continued)

Sample	Site	Location	Km	Chronology	NISP (% fish)	Description	Reference
18	C. Doña Blanca	Cádiz	1?	600–575 BC (Fo. 10)	15 (7)	Dermal plate	Roselló and Morales (1994)
19	C. Doña Blanca	Cádiz	1?	575–550 BC (Fo. 9)	22 (11)	Dermal plate	Roselló and Morales (1994)
20	Puerto 29	Huelva	0	650–635/600 BC (N2)	1 (0.9)	Dermal plate	Roselló 1990
21	Puerto 10	Huelva	0	Iron (VII–IVth BC)	3 (4)	Dermal plate	Roselló (1989) and Roselló et al. (1994)
22	Castro Marim	Algarve	0	Iron	4 (1)	Dermal plate	Morales and Roselló (2008)
23	Cerro Macareno	Sevilla	40/80	Iron	1 (33)[1]	Dermal plate	Amberger (1985)
24	Munigua	Sevilla	80/100	Roman	1 (33)[1]	Pterotic	Boessneck and von den Driesch (1980)
25	Castro Marim	Algarve	0	Roman	3 (0.9)	Dermal plate	Morales and Roselló (2008)
26	Mertola	Baixo Alentejo	70	XIII century AD	1 (0.8)	Dermal plate	Roselló (1994)
27	La Cartuja	Sevilla	40/70	XV/XVIth century AD	39 (9)	Dermal plate	Roselló et al. (1994)

The “km” column refers to approximate distance of the site from the coastline. Data within brackets in the “chronology” column either specify levels within a sequence (e.g., Castillo de Doña Blanca, Puerto 29), archaeologically derived dates (e.g., La Fonteta, Puerto 10), cultural periods (e.g., Cerro Macareno) or excavated sectors within the cave of Nerja (V, Sala del Vestíbulo; T, Sala de la Torca; M, Sala de la Mina)

Table 9.3 Sturgeon records (NISP) distributed according to rivers (the site of Cueva de Nerja has been counted as strictly littoral because it is not connected to any river system; (see Table 9.2)

River	Sites	%	NISP	%
Sizandro	1	7	2	1
Tagus	2	14	4	2.5
Guadiana	2	14	8	5
Tinto/Odiel	2	14	4	2.5
Guadalquivir	4	28	44	28
Guadalete	1	7	47	30
Segura	1	7	5	3
Total	13	93	114	73

Table 9.4 Sturgeon records (see Table 9.2) arranged in terms of distance to the coast

Distance to the coast	Sites	%	NISP	%
0–1 km	6	24	18	11.5
1–5 km	11	44	84	54.5
5–10 km	–	–	–	–
+10 km	8	32	52	34
Total	25	100	154	100

Note that in some cases (Zambujal, Doña Blanca) these distances refer to the paleoshore when the site was in operation, and do not represent the present-day shoreline

Table 9.5 Sturgeon records (see Table 9.2) distributed in terms of the cultural periods from which they derived

Period	No.	%	NISP	%
Palaeolithic	3	12	34	21.5
Mesolithic	3	12	9	5.5
Neolithic	1	4	2	1
Copper	2	8	5	3
Bronze	–	–	–	–
Iron	12	48	65	41
Roman	2	8	4	2.5
Medieval	2	8	40	25
Total	25	100	158	100

The Iron Age is also taken into account for Phoenician and Punic sites

nevertheless questionable, due to the low overall number of sites and remains, yet the idea that sturgeon fishing has been a regular practice in Iberia since the Upper Palaeolithic period is one worth testing in the future.

Related to the latter issue, one should remark that sturgeon: (1) apparently remained a minor "catch" or only a by-catch throughout ancient times, (2) are mostly recorded on sites close to or on the coast (Table 9.4), and (3) seem restricted to the southern half of the peninsula (i.e. from the Sizandro River in the Portuguese Estremadura to the Segura River in the Province of Alicante; Table 9.3). Also, and except for a few critically small samples where the abundance of sturgeon may rise beyond 30% of the total, sturgeon NISPs (i.e. number of identified remains) have always oscillated around values of 1% (Table 9.2).

Fig. 9.1 Major rivers of the Iberian Peninsula with location of the archaeological sites mentioned in the text. (1) Zambujal, (2) Cabeço de Amoeira, (3) Cabeço de Arruda, (4) Mértola, (5) Castro Marim, (6) Puerto 29, (7) Puerto 10, (8) Cartuja, (9) Munigua, (10) Cerro Macareno, (11) Valencina de la Concepción, (12) Castillo de Doña Blanca, (13) Cueva de Nerja, (14) La Fonteta

Biases of various kinds may be involved in the generation of these patterns. In this way, low NISP numbers may relate to the fact that many elements of the sturgeon's skeleton are cartilaginous. Indeed, bony scutes of the ossified exoskeleton are the elements most frequently found when sturgeon remains are retrieved in Iberian sites (Table 9.2).

Likewise, although the distribution of finds around the southern half of the Peninsula seems to be indicative of a presence restricted to this area, coastal dynamics and the placement of sites upon highly vagile environments (e.g., dunes, deltas, etc.) may explain why no sturgeon finds appear on Mediterranean sites above the Segura River. This contrasts with the comparatively large number of sturgeon locations along the West European and Baltic coastlines during ancient times (Table 9.1).

Whenever sturgeon remains are found outside coastal sites, they are usually placed on the banks of large rivers [e.g., Munigua, Cerro Macareno and Valencina de la Concepción (Guadalquivir) and Mertola (Guadiana)]. These findings support the idea that in Iberia, although sturgeon fishing was probably not restricted to the marine environment, only large rivers harboured stable populations of the species.

In addition, and although these archaeological finds probably represent local catches, the transport of sturgeon products into terrestrial "environments" (i.e. markets) has been occasionally documented (Morales et al. 2009). For such reasons we believe that the near absence of historical records from towns located along the Tagus (the largest of the Iberian rivers) suggests that stable populations of sturgeon had disappeared from it by the Middle Ages (Sánchez Quiñones 2006).

This scarcity of archaeological records also prevents one from carrying out diachronic analyses of sturgeon abundances, be these local or regional. Only at Castillo de Doña Blanca is there a restricted sequence revealing a gradual increase of sturgeon remains from the oldest until the most recent periods (i.e. 700–550 BC; Table 9.1). Such an increase is probably a consequence of the filling of the present-day Bay of Cadiz, originally an open-sea environment. When the large Phoenician town was abandoned at the end of the fourth century BC, the environment around the site was a delta, and sturgeon represented 11% of the total catch (Roselló and Morales 1994). It is presumed that at those times sturgeon were fished when entering or leaving the Guadalete River during their upstream or downstream migrations.

To summarise, one can conclude that during post-glacial times, Iberian sturgeon were never an important element of the aquatic faunas, whether marine, fresh or brackish water. Their Pleistocene distribution was restricted to a geographically dispersed river network, a restriction that became all the more pressing with the onset of the Holocene period when the pluvial regime in many parts of Iberia became seasonal, and often torrential. Such non-favourable conditions prevented the establishment of large sturgeon populations in Iberian rivers, as was the case in other European regions (Ludwig and Gessner 2007; Desse-Berset 2009; Pages et al. 2009 and others).

9.1.2 From Medieval Times to the Twentieth Century

In February 1321, the King of Portugal and the Algarve, Diniz (1261–1325), released a document reporting the capture of a sturgeon at Valada, Santarém, some 100 km from the mouth of the river. This fish measured 3.75 m and weighed some 275 kg (Almaça and Elvira 2000). Today, it is not possible to decide what the reasons were for the release of the royal document, but it is likely that the size of the fish, or its discovery far from the river mouth, were not common feats. Three centuries later, André de Resende (1500–1573) mentions that sturgeon were rare in the Duero River, far more common in Minho, and very rare in the Tagus River (translated from Rosado-Fernandes 1986 and cited in Almaça and Elvira 2000). At that time, specimens of moderate size entered the Guadiana River from March until the summer.

There is no doubt that the Guadalquivir and Ebro rivers have been the most important of the Iberian sturgeon rivers. Presumably, the fishes found in both of them travelled along the Iberian coast, entering other rivers from time to time. Long-distance migrations and colonisation events are described for some sturgeon

Table 9.6 Sturgeon records in the Guadalquivir River basin from 1300 to 1992

	1300	1526	1624	1700	1840	1857	1893	1919	1931	1985	1987	1990	1992
Presence/absence	+	+	+	+	+	+	+	*	+	–	–	–	+

+presence – absence *no data

Note that Granado-Lorencio (1991) listed them never as abundant for this river. The table is modified after Granado-Lorencio (1991), including the last record of Guadalquivir sturgeon (Almaca and Elvira 2000)

species (e.g. for *A. sturio* see Ludwig et al. 2002); however, environmental conditions in Iberia seemed to be unsuitable for the establishment of self-sustaining populations outside of the Guadalquivir and the Ebro. Apparently, the Tagus River was never a preferred river for sturgeon (Baldaque da Silva 1891, cited in Almaça and Elvira 2000), and there are some reports of sturgeon catches outside the Guadalquivir and the Ebro (reviewed in Almaça and Elvira 2000). Sturgeon records from the Guadalquivir are available from the thirteenth century until 1992 (Table 9.6). From 1931 onwards, the spawning migration was interrupted by the building of the Alcala del Rio dam, causing a decline in all the migratory species (Granado-Lorencio 1991). Ironically, a caviar and smoked-flesh factory started to run that same year (Fernández-Pasquier 1999). The foundation of a factory processing sturgeon tells two things, namely: (1) that sturgeon were abundant in the Guadalquivir during this time (between 1932 and 1954, 3,186 specimens were caught, including 2,544 females), and (2) that the factory accelerated their decline through an over-exploitation of the stocks [from the 1970s onwards they became a rarity (Almaça and Elvira 2000)]. Additionally, human-induced aggressions such as the building of river dams and the establishment of gravel extraction factories (Granado-Lorencio 1991) combined with environmental phenomena such as dry springs, which restricted the spawning of sturgeon in the Guadalquivir (Fernández-Pasquier 1999), also had a negative impact on Iberian sturgeon. In retrospect, one can see that there were no chances of survival for the Guadalquivir sturgeon. The last specimen was caught in 1992 near the mouth of the river (Table 9.6). Nowadays, there are only a very few occasional catches, but it is highly unlikely that these fishes constitute the relicts from former native populations. More probably, they represent escaped individuals from aquaculture farms or releases from garden ponds and aquaria. For example, one specimen of Siberian sturgeon (*A. baerii*) archived at the Doñana Biological Station (no. EBD-8175) was reported one decade ago (Hernando et al. 1999). Today, nearly all European river basins have been contaminated by the release or escape of alien species, often being Siberian sturgeon specimens (Gessner et al. 1999; Maury-Brachet et al. 2008; Ludwig et al. 2009b; Reinartz et al. 2011).

The Ebro, a tributary of the Mediterranean Sea, is the second most important sturgeon river of the Iberian Peninsula. During historic times, sturgeon migrated upstream to the city of Tudela, some 450 km from the river mouth (Farnós and Porres 1999 cited in Almaça and Elvira 2000). This migration was interrupted when the weir at Xerta was built in the fifteenth century (Almaça and Elvira 2000). Although many records of the Ebro sturgeon are available in museum collections and scientific publications (reviewed in Almaça and Elvira 2000), the fish is extinct today. An artisanal sturgeon fishery persisted in the delta of the Ebro (province of Tarragona) well into the twentieth century, and the last of these specimens were caught in 1965, 1966 and 1970 (Farnós and Porres 1999, cited in Almaça and Elvira 2000). In addition to the Ebro and Guadalquivir, sturgeon were also present at the mouth of smaller Mediterranean rivers (e.g. Turia and Jŭcar), but there is no evidence for spawning there (Almaça and Elvira 2000). Finally, marine captures, although rare, are reported from time to time.

To summarise, one can conclude that (1) even though the Iberian Peninsula was probably a sturgeon refuge during the last glaciations, sturgeon were less abundant there than they were in later times in Western and Central Europe, (2) the numbers of Iberian sturgeon did not increase during the Middle Ages, (3) there are no sturgeons living in the wild today in Iberia and (4) river damming and overfishing have been the major reasons for the decline and extinction of sturgeon during recent times.

9.2 Phylogeny of Iberian Sturgeon

9.2.1 Endemic Species in Iberia

The supportive stocking of sturgeon is not only a worthwhile goal for nature conservation; it also became a very profitable business for hatchery owners during the last decade. For this reason, it is not surprising that nearly all countries that were inhabited by sturgeon in former times began restoration projects. Originally, these restoration projects focused on the European sturgeon (*A. sturio*) in west and southeast Europe, because Carl von Linnè (1707–1778) described it as the only valid species for the "European Sea" in his epoch-making opus *Systema Naturae* of 1758. Later, the Adriatic (*A. naccarii*) and beluga (*Huso huso*) sturgeon were discussed as additional Iberian species (Hernando et al. 1999; de la Herrán et al. 2004 and others). Recently, the North American Atlantic sturgeon (*A. oxyrinchus*) was added to the list of Iberian species (Garrido-Ramos et al. 2009). Considering all facts about the presence of sturgeon on the Iberian Peninsula, only populations of the European sturgeon have been proven to exist beyond doubt (Elvira et al. 1991; Almaça and Elvira 2000; Garrido-Ramos et al. 2009; Ludwig et al. 2009c). Ludwig et al. (2009c) used jointed ancient DNA techniques and morphological comparison to identify 12 archaeological remains from five Iberian sites [650 BC–1500 AD (one sample dated 10.1–11.8 KY)]. Mitochondrial (mt) DNA of five samples was successfully amplified. All samples had mtDNA haplotypes of the European sturgeon, but nuclear DNA analyses failed in all samples because of poor preservation of nuclear DNA (Ludwig et al. 2009c). These results support the theory that *Acipenser sturio* was the prevalent species in Spanish rivers, a hypothesis which is in concordance with the morphological identifications

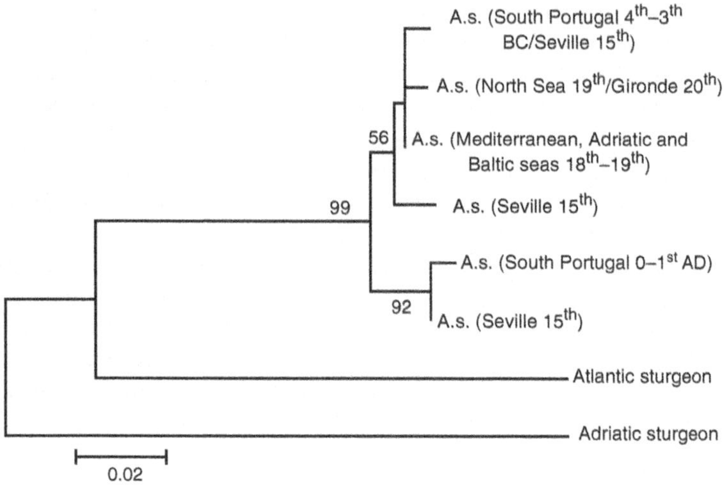

Fig. 9.2 Phylogenetic tree of Iberian sturgeons based on mitochondrial DNA sequences (modified after Ludwig et al. 2009c). Reference sequences of Atlantic sturgeon, Adriatic sturgeon and European sturgeon (*A.s.*, *Acipenser sturio*) were taken from Ludwig et al. (2002, 2003)

done for this study and others (Elvira and Almodóvar 2000a, b; Rincón 2000). In addition, two separated phylogenetic clades were found within Iberian *A. sturio* (Fig. 9.2), evidencing a sub-population, or subspecies, structuring of these fish. Both clades are separated by three diagnostic substitutions, and specimens of them were found in different locations. Almaça (1988) postulated different genotypes for Portuguese rivers, depending on reproductive isolation by distance, a theory which is supported by the outcome of this study. Interestingly, Iberian specimens of clade I are very similar to the Gironde (France) sturgeon and European sturgeon from the Adriatic, Mediterranean, North and Baltic Seas (Ludwig et al. 2002) indicating their post-glacial migration into western and northern Europe. Otherwise it could be ancestral polymorphism and an indication for a panmictic population in West and South Europe. However, the number of samples is too small to draw any definitive conclusions about the genetic structure or to classify Iberian sturgeon into subspecies.

Former populations of all the other species discussed previously (i.e. Adriatic sturgeon, beluga sturgeon and Atlantic sturgeon) are debatable, since none have been proven to exist beyond doubt. For example, only one specimen of beluga sturgeon [Museum of Oporto University, sample no. 1 (Hernando et al. 1999)], and of Atlantic sturgeon [Gabinete Salvador Collection of the Botanical Garden Barcelona, sample code GS (Garrido-Ramos et al. 2009)] have been reported thus far. But singular finds cannot be used as conclusive evidence for the native occurrence of a species, as sturgeon are migratory fish well-known for their long-distance migrations. For example, the Atlantic sturgeon crossed the North Atlantic from North America to Europe a few millennia ago (Ludwig et al. 2002, 2008). Consequently, only a viable population makes the difference between a native and a non-native species. The situation appears to be more confusing in the case of the

Adriatic sturgeon. A group of Spanish scientists who carried out intensive research on the subject recently concluded that some specimens originally classified as European sturgeon are, in fact, Adriatic sturgeon (for details see Table 2.1 in Garrido-Ramos et al. 2009). Such findings are all the more remarkable considering that, prior to these studies, only tributaries of the northern and southern Adriatic Sea and its coast have been considered as the native area of the Adriatic sturgeon (Tortonese 1989).

9.3 Adriatic Sturgeon in Spain?

Today, mitochondrial DNA polymorphisms and variations of nuclear genes or non-coding regions are used for sturgeon species identification (Ludwig 2008), because all sturgeon species comprise various degrees of intraspecific and interspecific genetic differentiation. The amount and pattern of the genetic divergence is determined by the formation of new genetic lineages through mutation, random lineage sorting through drift, the rate of genetic exchange among populations, and selective advantages or disadvantages of mutations. Traditionally, molecular phylogenetic analyses are based on genotypes (e.g. microsatellites, Restriction Fragment Lengths Polymorphisms – RFLP, Single Nucleotide Polymorphisms – SNP) or on haplotypes (when considering haploid genomes such as mtDNA or sex chromosome sequences), and, by including these phylogenetic relationships, information about the history of populations can be obtained. Recently, the history of the Adriatic sturgeon in Europe was analysed using both mitochondrial DNA and nuclear markers (Zane et al. 2002; Ludwig et al. 2003). The outcome of this study demonstrated that *A. naccarii* is a post-glacial newcomer to Europe, with an origin in the Caspian Sea region. It is also remarkable that *A. naccarii*-like mitochondrial haplotypes were observed in Russian sturgeon (*A. gueldenstaedtii*) from the northern Caspian Sea (Birstein et al. 2005), supporting their joint history and closely related phylogeny (Peng et al. 2007; Krieger et al. 2008). Based on post-glacial river captures between the Danube and Po rivers, the patterns of genetic divergence and the historical distribution along the North Adriatic shores, it is suggested that the colonisation of the Adriatic Sea probably occurred during the early Holocene from the Middle Danube towards the northern Adriatic Sea. This route of colonisation was also used by other fish such as the gobiid *Knipowitschia caucasica* or the beluga sturgeon (Economidis and Miller 1990). Between 13200 and 11000 BP, temperatures increased, resulting in the melting of glaciers, followed by a 120 m rise of seawater levels and a decrease in salinity in the northern Adriatic region (Zonneveld 1996). This timeframe of postglacial climate shift matches the molecular dating of divergence time of around 10600–12700 BP between the Adriatic sturgeon populations from the Buna River (Albania) and the Po River (Italy) (Ludwig et al. 2003). Since then, both populations evolved independently, resulting in significant molecular differences. Later, gene flow between both populations was interrupted by the restricted adaptation of Adriatic sturgeon to higher levels of salinity.

In view of these results, the description of a historic Spanish population of Adriatic sturgeon (Garrido-Ramos et al. 1997) comes as a surprise. Nevertheless, these were not the first scientists to claim a wider distribution for the Adriatic sturgeon. Several nineteenth century studies mention the species as endemic to the Iberian Peninsula (reviewed in Hernando et al. 2009). In contrast, most ichthyologists consider *A. naccarii* restricted to the Adriatic Sea (Holcik et al. 1989). In their work, Garrido-Ramos et al. (1997) used morphological and genetic tools to investigate museum specimens. Genetic analysis were focussed on two specimens from the Biological Station of Doñana, Sevilla, Spain (collection no. EBD 8173, EBD 8174). Originally, both individuals were classified as *A. sturio* (Hernando 1975), but Garrido-Ramos et al. (1997) identified them as *A. naccarii*, based on some morphological characteristics (e.g., snout shape, position of the barbels), and the presence of a DNA sequence that is species-specific for the Adriatic sturgeon. Soon afterwards, experts began to raise doubts about the scientific reliability of these investigations. For example, ontogenetic allometry of the morphological characteristics was not taken into consideration (Rincón 1999, 2000) and genetic results could not be reproduced by other researchers (Doukakis et al. 2000). The Spanish team considered these criticisms in extended analyses: firstly, they proofed the species-specificity, broadening their satellite DNA analyses (Ruiz-Rejón et al. 2000; Robles et al. 2004, 2005); secondly, they extended the number of genetic markers up to three nuclear marker systems and two mitochondrial regions (de la Herrán et al. 2004). Finally, in 2005 they organised a meeting of national and international sturgeon specialists in Granada discussing this issue (the outcome was published in Carmona et al. 2009). After all this, the team remained convinced that *A. naccarii* was a native species in Iberia to the point of recommending a release programme of Adriatic sturgeon in the Guadalquivir River (Domezain 2009). In our view, serious doubts persist over Adriatic sturgeons ever harbouring endemic viable populations in Iberia. First of all, as argued by Doukakis et al. (2000), the results of the aforementioned Spanish team were never reproduced outside of their own laboratories. A reproduction by a second, independent laboratory is an essential must for every DNA study of archived specimens or archaeological remains (Poinar and Cooper 2000; Hofreiter et al. 2001; Ludwig et al. 2009d), and several studies of ancient DNA have already been withdrawn because of artefacts or contamination. Secondly, recent genetic studies of archaeological remains evidence *A. sturio* as the dominating species inhabiting the Iberian Peninsula during ancient times (Ludwig et al. 2009c). No Adriatic sturgeon has been detected in ancient Iberian samples so far, and it is also significant that no Adriatic sturgeon have been ever found among the archaeological remains of sturgeon from French Mediterranean rivers, especially the Rhone River (Desse-Berset 1994; Desse-Berset et al. 2008). Assuming that *A. naccarii* colonised Iberia from the Adriatic Sea during post-glacial times, one should find signs of its historical presence on the French Mediterranean coast, because long-distance colonisation is highly unlikely for this species due to its poor tolerance of high levels of salinity (McKenzie et al. 2001a, b). For such reason, river hopping seems to be far more likely for its colonisation of new water basins.

When all these data are pooled, although it is true that one cannot definitively rule out the possibility that *A. naccarii* was endemic to rivers outside the Adriatic Sea, more and more compelling evidence will be required before claiming the species as a native Iberian taxon. The retrieval of isolated specimens does not prove that there ever existed a viable population of this species in the Peninsula. Given the ecological risks that the introduction of a non-native species brings to an ecosystem (Ludwig 2006; Gessner et al. 2007; Williot et al. 2009), and in particular to the fragile rivers of Iberia, this warning should be made loud and clear. Once a species is successfully introduced into a habitat, it is often near impossible to eliminate it, as over these past decades Iberian ichthyologists have so often learned the hard way (Maceda-Veiga et al. 2010).

References

Almaça C (1988) On the sturgeon, *Acipenser sturio* Linnaeus, 1758, in the Portuguese rivers and sea. Folia Zool 37:183–191

Almaça C, Elvira B (2000) Past and present distribution of Acipenser sturio L., 1758 on the Iberian Peninsula. Bol Inst Esp Oceanogr 16(1–4):11–16

Amberger G (1985) Tierknochenfunde vom Cerro Macareno/Sevilla. Studien über frühe Tierknochenfunde von der Iberischen Halbinsel 9:76–105

Aura T, Jordá Pardo F, Pérez Ripoll M, Rodrigo García MJ, Badal García E, Guillem Calatayud P (2002) The far south: the Pleistocene–Holocene transition in Nerja Cave (Andalucía, Spain). Quat Int 93–94:19–30

Birstein VJ, Ruban G, Ludwig A, Doukakis P, DeSalle R (2005) The enigmatic Caspian Sea Russian sturgeon: how many cryptic forms does it contain? Syst Biodivers 3:203–218

Boessneck J, von den Driesch A (1980) Tierknochenfunde aus vier südspanischen Höhlen. Studien über frühe Tierknochenfunde von der Iberischen Halbinsel 7:1–83

Carmona R, Domezain A, Garcia-Gallego M, Hernando JA, Rodriguez F, Ruiz-Rejon M (2009) Biology, conservation and sustainable development of sturgeons, vol 29, Fish & fisheries series. Springer, New York

Cleyet-Merle JJ (1990) La Préhistoire de la Pêche. Editions Errance, Paris

Cortés M, Morales A, Simón MD, Bergadà MM, Delgado A, López P, López-Sáez JA, Lozano MC, Riquelme JA, Roselló E, Sánchez A, Vera JL (2008) Palaeoenvironmental and cultural dynamics of the coast of Málaga (Andalusia, Spain) during the Upper Pleistocene and Holocene. Quat Sci Rev S1:2176–2193

de la Herrán R, Robles F, Martinez-Espin E, Lorente JA, Rejón CR, Garrido-Ramos MA, Rejón MR (2004) Genetic identification of western Mediterranean sturgeons and its implication for conservation. Conserv Genet 5:545–551

Desse-Berset N (1994) Sturgeon of the Rhône during Protohistory in Arles (6th–2nd) century BC. In: Van Neer W (ed) Proceedings of the 7th meeting of the ICAZ. Fish Remains Working Group. Annales du Musée Royal de l'Afrique Centrale. Sciences Zoologiques n° 274, Tervuren, pp 81–90

Desse-Berset N (2009) First archaeozoological identification of Atlantic sturgeon (*Acipenser oxyrinchus* Mitchill 1815) in France. C R Palevol 8:717–724

Desse-Berset N, Pages M, Brosse L, Tougard C, Chassaing O, Hänni C, Berrebi P (2008) Specific identification of the extinct population of sturgeon from the Rhône River by mtDNA analysis from bone remains (Jardin d'Hiver, Arles, France, 6th – 2nd century BC). In: Béarez P, Grouard S, Clavel B (eds) Archéologie du poisson. 30 ans d'archéo-ichtyologie au CNRS. Hommage aux travaux de Jean Desse et Nathalie Desse-Berset, XVIIIe rencontres internationales d'archéologie et d'histoire. Editions APDCA, Antibes, pp 195–200

Domezain A (2009) Main steps and proposals for a recovery plan of sturgeon in the Guadalquivir River (Spain). In: Carmona R, Domezain A, Garcia-Gallego M, Hernando JA, Rodriguez F, Ruiz-Rejón M (eds) Biology, conservation and sustainable development of sturgeons, vol 29, Fish & fisheries series. Springer, New York, pp 423–452

Doukakis P, Birstein VJ, DeSalle R, Ludwig A, Ludwig AN, Machordom A, Almodóvar A, Elvira B (2000) Failure to confirm previous identification of two putative museum specimens of the Atlantic sturgeon, *Acipenser sturio*, as the Adriatic sturgeon, *A. naccarii*. Mar Biol 136: 373–377

Economidis PS, Miller PJ (1990) Systematics of freshwater gobies from Greece (Teleostei: Gobiidae). J Zool 221:125–170

Elvira B, Almodóvar A (2000a) Further observations on the morphological characters of *Acipenser sturio* L., 1758 from the Iberian Peninsula: a comparison with North and Adriatic Sea populations. Bol Inst Esp Oceanogr 16(1–4):89–98

Elvira B, Almodóvar A (2000b) Morphology and taxonomy of the Atlantic sturgeon, *Acipenser sturio* from Spain. Folia Zool 49:221–230

Elvira B, Almodóvar A, Lobón-Cervià J (1991) Recorded distribution of sturgeon (*Acipenser sturio* L, 1758) in the Iberian Peninsula and actual status in Spanish waters. Arch Hydrobiol 121:253–258

Fernández-Pasquier V (1999) *Acipenser sturio* L. in the Guadalquivir River, Spain. Water regulation and fishery as factors in stock decline from 1932 to 1967. J Appl Ichthyol 15:133–135

Garrido-Ramos MA, Soriguer MC, de la Herrán R, Jamilena M, Ruiz-Rejón C, Domezain A, Hernando JA, Ruiz-Rejón M (1997) Morphometric and genetic analysis as proof of the existence of two sturgeon species in the Guadalquivir River. Mar Biol 129:33–39

Garrido-Ramos MA, Robles F, de la Herrán R, Martinez-Espin E, Lorente JA, Ruiz-Rejón C, Ruiz-Rejón M (2009) Analysis of mitochondrial and nuclear DNA. Markers in old museum sturgeons yield insights about the species existing in Western Europe: *A. sturio*, *A. naccarii* and *A. oxyrinchus*. In: Carmona R, Domezain A, Garcia-Gallego M, Hernando JA, Rodriguez F, Ruiz-Rejón M (eds) Biology, conservation and sustainable development of sturgeons, vol 29, Fish & fisheries series. Springer, New York, pp 25–49

Gessner J, Debus L, Filipiak J, Spratte S, Skora SK, Arndt GM (1999) Catches of sturgeons in German and adjacent waters since 1980. J Appl Ichthyol 15:136–142

Gessner J, Arndt GM, Ludwig A, Kirschbaum F (2007) Remediation of Atlantic sturgeon *Acipenser oxyrinchus* in the Baltic Sea – background, status, and perspectives. Am Fish Soc Symp 56:301–317

Granado-Lorencio C (1991) The effect of man on the fish fauna of the River Guadalquivir, Spain. Fish Res 12:91–100

Hain F (1982) Kupferzeitliche Tierknochenfunde aus Valencina de la Concepción/Sevilla. Studien über frühe Tierknochenfunde von der Iberischen Halbinsel 8:1–178

Hernando JA (1975) Notas sobre distribución de los peces fluviales en el Suroeste de España. Doñana Acta Vert 2:263–264

Hernando JA, Vasil'eva ED, Arlati J, Vasil'ev VP, Santiago JA, Belysheva-Polyakova L, Domezian A, Soriguer MC (1999) New evidence for a wider distribution area of two species of European sturgeons: *Acipenser naccarii* and *Huso huso* (Acipenseridae). J Ichthyol 39: 803–806

Hernando JA, Domezain A, Zabala C, Cabrera R, Domezain J, Soriguer MC (2009) The regression of sturgeons in southern Europe. In: Carmona R, Domezain A, Garcia-Gallego M, Hernando JA, Rodriguez F, Ruiz-Rejón M (eds) Biology, conservation and sustainable development of sturgeons, vol 29, Fish & fisheries series. Springer, New York, pp 3–23

Hofreiter M, Serre D, Poinar HN, Kuch M, Paabo S (2001) Ancient DNA. Nat Rev Genet 2: 353–359
Holcik J, Banarescu P, Evans D (1989) General introduction to fishes. In: Holcik J (ed) The freshwater fishes of Europe. AULA Verlag, Wiesbaden, pp 18–147
Krieger J, Hett AK, Fuerst PA, Artykhin E, Ludwig A (2008) The molecular phylogeny of the order Acipenseriformes revisited. J Appl Ichthyol 24(S1):36–45
Lentacker A (1990–1991) Archaeozoölogisch onderzoek van Laat-Prehistorische Vindplaatsen uit Portugal. Unpublished PhD, Rijksuniversiteit Gent, Gent
Lepiksaar J (1976) Die Fischknochen aus Castro do Zambujal. Studien über frühe Tierknochenfunde von der Iberischen Halbinsel 5:139–142
Ludwig A (2006) A sturgeon view on conservation genetics. Eur J Wildl Res 52:3–8
Ludwig A (2008) Identification of Acipenseriformes species in trade. J Appl Ichthyol 24(S1):2–19
Ludwig A, Gessner J (2007) What makes the difference? – sea sturgeon on both sides of the Atlantic Ocean. Am Fish Soc Symp 56:285–300
Ludwig A, Debus L, Lieckfeldt D, Wirgin I, Benecke N, Jenneckens I, Williot P, Waldman JR, Pitra C (2002) When the American sea sturgeon swam east. Nature 419:447–448
Ludwig A, Congiu L, Pitra C, Fickel J, Gessner J, Fontana F, Patarnello T, Zane L (2003) Nonconcordant evolutionary history of maternal and paternal lineages in Adriatic sturgeon. Mol Ecol 12:3253–3264
Ludwig A, Arndt U, Lippold S, Benecke N, Debus L, King TL, Matsumura S (2008) Tracing the first steps of American sturgeon pioneers in Europe. BMC Evol Biol 8:221
Ludwig A, Makowiecki D, Benecke N (2009a) Further evidence of trans-Atlantic colonization of Western Europe by American Atlantic sturgeons. Archaeofauna 18:183–190
Ludwig A, Lippold S, Debus L, Reinartz R (2009b) First evidence of hybridization between endangered sterlets (*Acipenser ruthenus*) and exotic Siberian sturgeons (*Acipenser baerii*) in the Danube River. Biol Invasions 11:753–760
Ludwig A, Arndt U, Debus L, Rosselló E, Morales A (2009c) Ancient mitochondrial DNA analyses of Iberian sturgeons. J Appl Ichthyol 25:5–9
Ludwig A, Pruvost M, Reissmann M, Benecke N, Brockmann GA, Castaños P, Cieslak M, Lippold S, Llorente L, Malaspinas AS, Slatkin M, Hofreiter M (2009d) Coat color variation at the beginning of horse domestication. Science 324:485
Maceda-Veiga A, Monleón-Getino A, Caiola N, Casals F, de Sostoa A (2010) Changes in fish assemblages in catchments in north-eastern Spain: biodiversity, conservation status and introduced species. Freshw Biol 55:1734–1746
Maury-Brachet R, Rochard E, Durrieu G, Boudou A (2008) The 'Storm of the Century' (December 1999) and the accidental escape of Siberian sturgeons (*Acipenser baerii*) into the Gironde estuary (Southwest France). Environ Sci Pollut Res Int 15:89–94
McKenzie DJ, Cataldi E, Romano P, Owen SF, Taylor EW, Bronzi P (2001a) Effects of acclimation to brackish water on the growth, respiratory metabolism, and swimming performance of young-of-the-year Adriatic sturgeon (*Acipenser naccarii*). Can J Fish Aquat Sci 58:1104–1112
McKenzie DJ, Cataldi E, Romano P, Taylor EW, Cataudella S, Bronzi P (2001b) Effects of acclimation to brackish water on tolerance of salinity challenge by young-of-the-year Adriatic sturgeon (*Acipenser naccarii*). Can J Fish Aquat Sci 58:1113–1121
Morales A, Roselló E (2008) 20,000 years of fishing in the strait: archaeological fish and shellfish assemblages from southern Iberia. In: Erlandson J, Torrey R (eds) Human impacts on ancient marine environments. University of California Press, Berkeley, pp 243–278
Morales DC, Roselló E, Morales A (2009) Pesquerías medievales hispanas: las evidencias arqueofaunísticas, vol 1. Monografias de la Sociedad de Estudios Medievales, Madrid, pp 145–166
Pages M, Desse-Berset N, Tougard C, Brosse L, Hanni C, Berrebi P (2009) Historical presence of the sturgeon Acipenser sturio in the Rhone basin determined by the analysis of ancient DNA cytochrome b sequences. Conserv Genet 10:217–224

Peng Z, Ludwig A, Wang D, Diogo R, Wei Q, He S (2007) Age and biogeography of major clades in sturgeons and paddlefishes (Pisces: Acipenseriformes). Mol Phylogenet Evol 42:854–862

Poinar HN, Cooper A (2000) Ancient DNA: do it right or not at all. Science 5482:1139

Reinartz R, Lippold S, Lieckfeldt D, Ludwig A (2011) Population genetic analyses of *Acipenser ruthenus* as a prerequisite for the conservation of the uppermost Danube population. J Appl Ichthyol 27:477–483

Rincón PA (1999) Putative morphometric evidence of the presence of *Acipenser naccarii* Bonaparte, 1836 in Iberian rivers, or why ontogenetic allometry needs adequate treatment. Bol Inst Esp Oceanogr 16(1–4):217–229

Rincón PA (2000) Big fish, small fish: still the same species. Lack of morphometric evidence of the existence of two sturgeon species in the Guadalquivir River. Mar Biol 4:715–723

Robles F, de la Herrán R, Ludwig A, Ruiz-Rejón C, Ruiz-Rejón M, Garrido Ramos M (2004) Evolution of ancient satellite DNAs in sturgeon genomes. Gene 338:133–142

Robles F, de la Herrán R, Ludwig A, Ruiz-Rejón C, Ruiz-Rejón M, Garrido Ramos M (2005) Genomic organization and evolution of the 5S ribosomal DNA in the ancient fish sturgeon. Genome 48:18–28

Roselló E (1989) *Arqueoictiofaunas Ibéricas. Aproximación Metodológica y Bio-Cultural.* Publicaciones de la Universidad Autónoma de Madrid, Madrid

Roselló E (1990) Puerto 29 (Huelva): La Ictiofauna. Huelva Arqueológica XII:143–166

Roselló E (1994) Análisis de los peces recuperados en la ciudad de Mértola (Portugal). Arqueología Medieval 2:277–283

Roselló E, Morales A (1994) The fishes. In: Roselló E, Morales A (eds) Castillo de Doña Blanca: archaeo-environmental investigations in the Bay of Cádiz, Spain (750–500 BC). BAR (International Series) vol 593. Archeopress, Oxford, pp 91–142

Roselló E, Morales A, Morales DC (1994) La Cartuja/Spain: anthropogenic Ichthyocenosis of culinary nature in a paleocultural contex. Offa 51:323–331

Ruiz-Rejón M, de la Herrán R, Ruiz-Rejón C, Garrido-Ramos MA (2000) Genetic characterization of Acipenser sturio L., 1758 in relation to other sturgeon species using satellite DNA. Bol Inst Esp Oceanogr 16(1–4):231–236

Sánchez Quiñones J (2006) La pesca en el reino de Toledo. La cuenca media y alta del Tajo en los siglos XII al XVI. British Archaeological Reports (International Series) 1489. Archeopress, Oxford

Sternberg M (2007) Pêche et exploitation des milieux aquatiques. In: Gailledrat E, Sala F, Rouillard P (eds) L'établissement protohistorique de La Fonteta (fin VIIIe – fin VIe siècle av. J.-C.).Fouilles de la Rábita de Guardamar II. Casa de Velázquez, Madrid, pp 372–398

Tortonese E (1989) Acipenser naccarii. In: Holcik J (ed) The freshwater fishes of Europe. AULA Verlag, Wiesbaden, pp 285–293

Williot P, Rochard E, Kirschbaum F (2009) Acceptability and prerequisites for the successful introduction of sturgeon species. In: Carmona R, Domezain A, Garcia-Gallego M, Hernando JA, Rodriguez F, Ruiz-Rejon M (eds) Biology, conservation and sustainable development of sturgeons, vol 29, Fish & fisheries series. Springer, New York, pp 369–384

Zane L, Patarnello T, Ludwig A, Fontana F, Congiu L (2002) Isolation and characterization of microsatellites in the Adriatic sturgeon (*Acipenser naccarii*). Mol Ecol Notes 2:586–588

Zonneveld KAF (1996) Palaeoclimatic reconstruction of the last deglaciation (18–8 ka B.P.) in the Adriatic Sea region; a land–sea correlation based on palynological evidence. Palaeogeogr Palaeoecol 122:89–106

Chapter 10
Biological Cycles and Migrations of *Acipenser sturio*

M.L. Acolas, G. Castelnaud, M. Lepage, and E. Rochard

Abstract Current knowledge of the biological cycle and migration behaviour is detailed in this chapter. Most of the data come from the Gironde population, and a diagram of the biological cycle is proposed. Spawning occurs in freshwater, and juveniles progressively migrate to the estuary and then to the sea to grow. Early life history in rivers, juvenile life in the estuary and at sea and finally adult life at sea and reproductive migration are described from the literature.

10.1 Introduction

The European sturgeon is an anadromous species. Reproduction generally occurs in the lower parts of rivers, and juvenile growth takes place mainly in estuaries and at sea. The main biological cycle data were collected on the population in the Gironde basin (Magnin 1962; Williot et al. 1997; Elie 1997; Brosse 2003; Castelnaud et al. 1991; Roule 1922; Rochard 1992; Trouvery et al. 1984). According to the terminology of fish development (Balon 1975), after the embryonic period sturgeon are called larvae, at which stage they start exogenous feeding until the formation of the axial skeleton (life in rivers at that stage). They then become juveniles, when they have their definitive shape with fully differentiated fins and definitive organs with their energy reserves for growth (life in rivers, estuary and at sea), until they start to mature, which corresponds to the adult phase. This adult phase is the period of life at sea and reproduction migration through the sea and the estuary towards the spawning grounds in the rivers (Castelnaud et al. 1991;

M.L. Acolas (✉) • G. Castelnaud • M. Lepage • E. Rochard
Cemagref, Estuarine Ecosystems and Diadromous Fish Research Unit, 50 avenue de Verdun, 33612 Cestas Cedex, France
e-mail: marie-laure.acolas@cemagref.fr

P. Williot et al. (eds.), *Biology and Conservation of the European Sturgeon Acipenser sturio L. 1758*, DOI 10.1007/978-3-642-20611-5_10,

Trouvery et al. 1984). Once growth has slowed and gametes are no longer produced, the senescence period can last for many years (Balon 1975). The reproductive cycle will be described in more detail in Chap. 12.

10.2 State of the Art

To explain the biological cycle, Fig. 10.1 presents a diagram with the migrations through different environments (rivers, estuary and sea) that are necessary to achieve growth and reproduction.

10.2.1 Early Life History in Rivers

The early life history of *A. sturio* occurs in freshwater. Adults migrate from the sea to the spawning grounds located in the lower part of rivers (Fig. 10.1). We know that these sites include the places where spawners gather and where eggs are laid, because only concentrations of spawners have been reported here; reproduction itself, however, has never been observed in the field (Elie 1997; Jego et al. 2002). The characteristics of spawning grounds are as follows (Jego et al. 2002): the site should include a sector with a minimum depth of 5 m, the substrate is heterogeneous (diameter, 3–250 mm), and water current should be at least 0.5 m/s and up to 1.5 m/s to allow eggs to gently disperse before sticking to the substrate. Areas below dams are often considered as forced spawning grounds. Once fish have found the relevant site, reproduction occurs between May and June (Magnin 1962). Reproduction is oviparous, and females are believed to reproduce with several males.

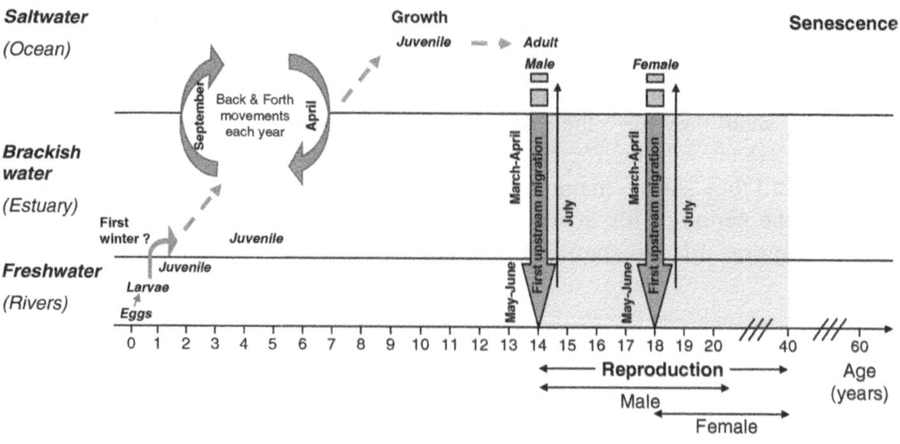

Fig. 10.1 Biological cycle of *Acipenser sturio*

Eggs are laid in the water column, and they are adhesive in order to stick rapidly to the substrate. Hatching occurs about 4 days after fertilization, at 18°C (Rouault et al. 2008; Williot et al. 2002, 2009) and embryos measure about 10 mm (Jatteau 1998). First feeding occurs about 6 days after hatching (Rouault et al. 2008), which corresponds to the beginning of the larval stage. The juvenile stage is believed to begin about 1 month after hatching, according to the literature on Acipenseridae (Doroshov 1985); however, the exact timing in *A. sturio* has still to be determined.

Knowledge of the biology of *A. sturio* larvae and young of the year juveniles is scarce, due to the difficulty of investigating in the wild. First data on activity and habitat utilization were obtained recently in experimental conditions on juveniles of 3 months old. Charles et al. 2009 have observed that fish are more active at night and colonize the water column; they also seem to prefer fine substrate (sand and gravel). A higher activity at night was also observed in 1-year-old fish (about 27 cm in length) in previous experiments (Staaks et al. 1999). With regard to feeding behaviour in freshwater, fish are believed to feed on worms (Oligochaeta), insect larvae, chironomids, and crustaceans (Gammaridae) (Ninua 1976; Magnin 1962; Acolas et al. 2009).

To enhance their growth and to start frequenting salt waters, juveniles will leave the river for the estuary. The timing for this is not well-documented, but at least some individuals leave the river for the estuary during their first winter (Magnin 1962; Rochard et al. 2001; Rochard 1992). Rochard et al. (2001) has documented the occurrence of *A. sturio* in the estuary from a 1994 cohort which was sampled and tagged. The youngest individuals captured by trawling were 1-year-old fish (about 27 cm); they were localized in the upper estuary by the end of their first winter. Moreover, the authors mentioned above have established that no sturgeon under 46 cm has been captured in salinity above 2.5‰. However, sampling occurs in only a few sectors of the estuary and not in the river; thus, juvenile downstream migration behaviour from the river to the estuary is largely unknown.

10.2.2 Juvenile Life in the Estuary and at Sea

The juvenile phase in the estuary is the best-known period in terms of habitat use (Brosse 2003; Taverny et al. 2002; Lepage et al. 2005), feeding regime (Brosse 2003) and population dynamic (Castelnaud et al. 1991; Rochard et al. 2001; Rochard 1992). Knowledge has been acquired mainly for fish between 2 and 7 years old, thanks to extensive tagging and sampling (Rochard 1992; Castelnaud et al. 1991) associated with an improved age-determination method (Rochard and Jatteau 1991). In the estuary, *A sturio* feed mainly on polychaeta and some crustaceans (Brosse 2003; Brosse et al. 2000; Ninua 1976; Magnin 1962), and their localization can be explained by prey distribution (Brosse 2003; Taverny et al. 2002). Movements of juveniles of 4 and 5 years old were studied using telemetry tools, and they are oriented in the same direction as the tidal current (Taverny et al. 2002; Lepage et al. 2005).

By the end of their second winter, marine incursion can be observed for a few individuals, with fish of over 15 months being highly tolerant to salinity variations (Rochard et al. 2001). A regular back and forth movement between the lower part of the estuary and the sea (called locally "Mouvée de la Saint Jean") has been observed with a seasonal rhythm. Fish entered the sea in the autumn and returned to the estuary in spring, but the reasons for this are not well-known (Castelnaud et al. 1991; Trouvery et al. 1984). Juveniles exhibit this behaviour up to about 7 years old before remaining at sea definitively to continue their growth.

10.2.3 Adult Life at Sea and Reproductive Migration

Life at sea has only been documented thanks to incidental capture reports between the Bay of Biscay and Scandinavia (Letaconnoux 1961; Lassalle et al. 2010; Rochard et al. 1997; Williot et al. 1997, 2002; Trouvery et al. 1984). Capture occurs at depths of over 100 m (Letaconnoux 1961; Castelnaud et al. 1991), and main distribution has been shown to be restricted to the continental shelf, with fish being mainly found (71%) between 10 and 40 m (Rochard et al. 1997). The size of fish captured at sea ranges between 35 and 244 cm (average 113 cm, data 1980–1994). According to Letaconnoux (1961), sturgeon were mainly captured between October and March, whereas Rochard et al. (1997) showed that about two-thirds of the captures occurred in spring and summer, with the last third being distributed almost equally between autumn and winter. This difference may be due to the area and period considered (sturgeons captured by vessels from La Rochelle in the 1950s for Letaconnoux 1961), and all the western part of Europe in 1980–1990 for Rochard et al. (1997). However, knowledge of life at sea is very limited (restricted to a few reports on incidental captures), and displacements between estuary and coastal areas have yet to be documented.

This species has a long life cycle, with a late maturity. In the wild, males and females are mature at 13–15 and 19–22 years old respectively (Magnin 1962). At that time, adults leave the sea to enter the estuary in March and April, and begin their upstream migration to the spawning grounds. After reproduction, adults return to the sea by the end of July. The species is semelparous, and males are believed to reproduce more frequently than females until senescence (see Chap. 12 for a detailed analysis of reproduction).

10.3 Conclusion

The most well-known phase of the biological cycle corresponds to the period of growth in the lower estuary. Further studies are still needed on migration tactics and also on young-of-year behaviour and habitat utilization in rivers and upstream estuary, as well as on reproduction activity and life at sea.

References

Acolas ML, Rouleau E, Roqueplo C, Le Barh R, Rochard E (2009) Action n°10 Localisation et caractérisation des habitats fluviaux fréquentés par les jeunes. In: Rochard E (ed) Programme de recherche et de conservation de l'esturgeon européen *Acipenser sturio*; bilan scientifique et technique 2008. Cemagref de Bordeaux, étude n°133, pp 64–78

Balon EK (1975) Terminology of intervals in fish development. J Fish Res Board Can 32(9): 1663–1975

Brosse L (2003) Caractérisation des habitats des juvéniles d'esturgeon européen, Acipenser sturio, dans l'estuaire de la Gironde. Doctorat, Université Paul Sabatié, Toulouse

Brosse L, Rochard E, Dumont P, Lepage M (2000) Premiers résultats sur l'alimentation de l'esturgeon européen, *Acipenser sturio* Linnaeus, 1758 dans l'estuaire de la Gironde et comparaison avec la macrofaune estuarienne présente. Cybium 24(3):49–61

Castelnaud G, Rochard E, Jatteau P, Lepage M (1991) Données actuelles sur la biologie d'*Acipenser sturio* dans l'estuaire de la Gironde. In: Williot P (ed) Acipenser. Cemagref, Antony, pp 251–275

Charles K, Roqueplo C, Jatteau P (2009) Action n°9 : Identification expérimentale des préférences d'habitat des jeunes stades. In: Rochard E (ed) Programme de recherche et de conservation de l'esturgeon européen Acipenser sturio; bilan scientifique et technique 2008. Cemagref de Bordeaux, étude, n°133, pp 54–64

Doroshov SI (1985) Biology and culture of sturgeon Acipenseriformes. In: Muir J, Roberts RJ (eds) Recent advances in aquaculture, vol 2. Westview, Boulder, pp 252–274

Elie P (1997) Restauration de l'esturgeon européen *Acipenser sturio*. Contrat Life, rapport final du programme d'exécution. Cemagref de Bordeaux, Bordeaux

Jatteau P (1998) Etude bibliographique des principales caractéristiques de l'écologie des larves d'Acipenséridés. Bull Fr Pêche Piscic 350–351:445–464

Jego S, Gazeau C, Jatteau P, Elie P, Rochard E (2002) Les frayères potentielles de l'esturgeon européen *Acipenser sturio* L. 1758 dans le bassin Garonne–Dordogne. Méthodes d'investigation, état actuel et perspectives. Bull Fr Pêche Piscic 365–366:487–505

Lassalle G, Crouzet P, Gessner J, Rochard E (2010) Global warming impacts and conservation responses for the critically endangered European Atlantic sturgeon. Biol Conserv 143(11): 2441–2452

Lepage M, Taverny C, Piefort S, Dumont P, Rochard E, Brosse L (2005) Juvenile sturgeon (*Acipenser sturio*) habitat utilization in the Gironde estuary as determined by acoustic telemetry. In: Spedicato MT, Lembo G, Marmulla G (eds) Aquatic telemetry: advances and applications. Proceedings of the Fifth Conference on Fish Telemetry held in Europe, Ustica, Italy, 9–13 June 2003, FAO/COISPA, Rome, pp 169–177

Letaconnoux R (1961) Note sur la fréquence de distribution des captures d'esturgeons (*Acipenser sturio* L.) dans le Golfe de Gascogne. Revue des Travaux de l'Institut des Pêches Maritimes 25:253–261

Magnin E (1962) Recherches sur la systématique de la biologie des Acipensérides. Annales de la station centrale d'hydrobiologie appliquée 9:7–242

Ninua NS (1976) Atlanticheskii osëtr reki Rioni (L'esturgeon atlantique de la rivière Rioni). Metsniereba, Tbilisi

Rochard E (1992) Mise au point d'une méthode de suivi de l'abondance des amphihalins dans le système fluvio-estuarien de la Gironde, application à l'étude écobiologique de l'esturgeon, Acipenser sturio. Doctorat, Université de Rennes I, Rennes

Rochard E, Jatteau P (1991) Amélioration de la méthode de détermination de l'âge de l'esturgeon commun acipenser sturio et premières applications. In: Williot P (ed) Acipenser. Cemagref, Antony

Rochard E, Lepage M, Meauzé L (1997) Identification et caractérisation de l'aire de répartition marine de l'esturgeon européen *Acipenser sturio* à partir de déclarations de captures. Aquat Living Resour 10(2):101–109

Rochard E, Lepage M, Dumont P, Tremblay S, Gazeau C (2001) Downstream migration of juvenile European sturgeon *Acipenser sturio* L. in the Gironde Estuary. Estuaries 24(1): 108–115

Rouault T, Chèvre P, Rochard E, Jatteau P, Jacobs L, Gonthier P (2008) Programme de recherche et de conservation de l'esturgeon européen *Acipenser sturio*; bilan scientifique et technique 2007. Cemagref de Bordeaux, étude n°127, 79p

Roule L (1922) Etude sur l'esturgeon du Golfe de Gascogne et du bassin girondin. Office scientifique et technique des pêches maritimes 20:12

Staaks G, Kirschbaum F, Williot P (1999) Experimental studies on thermal behaviour and diurnal activity rhythms of juvenile European sturgeon (*Acipenser sturio*). J Appl Ichthyol Z Angew Ichthyol 15(4–5):243–247

Taverny C, Lepage M, Piefort S, Dumont P, Rochard E (2002) Habitat selection by juvenile European sturgeon *Acipenser sturio* in the Gironde estuary (France). J Appl Ichthyol 18(4–6): 536–541

Trouvery M, Williot P, Castelnaud G (1984) Biologie et écologie d'Acipenser sturio, étude de la pêcherie. Etude Cemagref/AGEDRA

Williot P, Rochard E, Castelnaud G, Rouault T, Brun R, Lepage M, Elie P (1997) Biological characteristics of European Atlantic sturgeon, *Acipenser sturio*, as the basis for a restoration program in France. Environ Biol Fish 48(1–4):359–372

Williot P, Rouault T, Brun R, Pelard M, Mercier D (2002) Status of caught wild spawners and propagation of the endangered sturgeon Acipenser sturio in France: a synthesis. Int Rev Hydrobiol 87(5–6):515–524

Williot P, Rochard E, Kirschbaum F (2009) Acceptability and prerequisites for the successful introduction of sturgeon species. In: Carmona R, Domezain A, Garcia Gallego M, Hernando-Casal J, Rodriguez F, Ruiz Rejon M (eds) Biology, conservation and sustainable development of sturgeons, Fish & fisheries series. Springer, New York, pp 369–384, 467p

Chapter 11
Habitat, Movements and Feeding of Juvenile European Sturgeon (*Acipenser sturio*) in Gironde Estuary

Laurent Brosse, Catherine Taverny, and Mario Lepage

Abstract This work identifies and characterizes the zones of essential habitats for juveniles of *Acipenser sturio* in the Gironde estuary. It is also attempts to identify the reasons for sturgeons' preferential habitat utilization. For this purpose, results from monthly surveys in the Gironde estuary carried out between 1995 and 2000, as well as telemetric surveys of 16 sturgeons in 1999, are used. Additionally, results from investigations on the diet of juvenile sturgeons, as well as biotic and abiotic characterization of habitat, are utilized. Three zones where sturgeon concentrate are distributed along a salinity gradient. They show common denominators in the subtidal area as far as depth, muddy or sandy sediment are concerned, and similar current velocity. These habitats constitute an important feeding ground rich in tube-dwelling polychaetes (*Polydora* sp. and *Heteromastus filiformis*), which have been identified as the dominant prey items of European sturgeon juveniles.

11.1 Introduction

Initially, very little was known about the diet of the European sturgeon, *Acipenser sturio*. The diet of juveniles caught at the mouth of the Gironde estuary described by Magnin (1962) consists mainly of polychaetes, mysids, and shrimps. According to Ninua (1976), the Rioni juveniles mainly feed on small crustaceans.

This lack of knowledge and data for *A. sturio* largely contrasts with the situation for other sturgeon species, such as lake sturgeon, *Acipenser fulvescens* (Nilo 1996; Chiasson et al. 1997; Beamish et al. 1998), the shortnose sturgeon, *Acipenser*

L. Brosse (✉)
Aqua-Logiq, 527 rue Zac Petite Camargue, 34400 Lunel, France
e-mail: laurent.brosse@aqua-logiq.fr

C. Taverny • M. Lepage
Cemagref, Estuarine Ecosystems and Diadromous Fish Research Unit, 50 Avenue de Verdun, 33612 Cestas Cedex, France

P. Williot et al. (eds.), *Biology and Conservation of the European Sturgeon Acipenser sturio L. 1758*, DOI 10.1007/978-3-642-20611-5_11,

brevirostrum (Dadswell 1979), the Gulf sturgeon, *Acipenser oxyrinchus desotoi* (Mason and Clugston 1993), Russian sturgeon, *Acipenser gueldenstaedtii* (Levin 1989; Polyaninova and Molodsteva 1995; Zolotarev et al. 1996) or white sturgeon, *Acipenser transmontanus* (McCabe et al. 1993). All these studies reveal that sturgeon selectively feed on benthic invertebrates, while their diet varies according to species and their location (Zolotarev et al. 1996).

Moreover, little was known about characteristics of juvenile European sturgeon habitats (location, boundaries and role) and habitat use in estuary. Some authors have shown that juvenile European sturgeon enter the upper estuary at the end of their first winter and stay several years in the estuary, gradually descending towards the ocean and making an annual journey to the ocean before leaving their estuarine habitats for their oceanic habitats (Roule 1922; Letaconnoux 1961; Magnin 1962; Castelnaud et al. 1991; Rochard and Jatteau 1991; Rochard et al. 1991; Rochard 1992). The studies also show that the estuary is used unevenly, and that some areas are much more intensively populated than others (Magnin 1962; Castelnaud et al. 1991; Rochard et al. 2001), which indicates the existence of preferred habitat in the Gironde estuary for juvenile European sturgeon.

The scarcity of information about juvenile European sturgeon biology in the estuary is also related to the rarity of this species and its threatened status, which has created some technical difficulties in collecting data about diet without damaging the fish. Identification of the main aggregation areas for juvenile *A. sturio* in estuarine environment, and the characterization of these habitats in terms of biotic and abiotic environment variables, were carried out over several years in a series of studies concerning two cohorts (1994 and 1995) (Lochet et al. 2004).

11.2 Material and Methods

Between 1995 and 2000 a large capture–mark–recapture survey was realized in the Gironde estuary by trawling. Georeferenced catch data were collected on two sturgeon cohorts (see Part III, Chap. 29). Furthermore, an acoustic telemetry survey on 16 juvenile sturgeons over a period of 6 months was performed in 1999 (Taverny et al. 2002; Lepage et al. 2004). At the same time, a study of the European sturgeon diet was carried out, adapting a gastric lavage method for the study of stomach contents (Meehan and Miller 1978; Hyslop 1980; Gaudin et al. 1981; Haley 1998; Fig. 11.1). The device includes a hand pump with a reservoir capacity of 7 l, with an injection pipe fitted with a flexible tube of 6 mm external diameter inserted into a tube of 12 mm outer diameter for recovery of prey items. Some preliminary tests for safety and efficiency of this application were conducted on Siberian sturgeon as a biological model (*Acipenser baerii*). These tests concluded that the method is effective but not without risk, and that precautions are necessary in order to prevent, as much as possible, injuries caused by the stomach washing (Brosse et al. 2002).

For the characterization of the biotic (benthic invertebrate fauna) and abiotic (depth, sediments nature, salinity . . .) habitat parameters, a survey of the macrobenthic

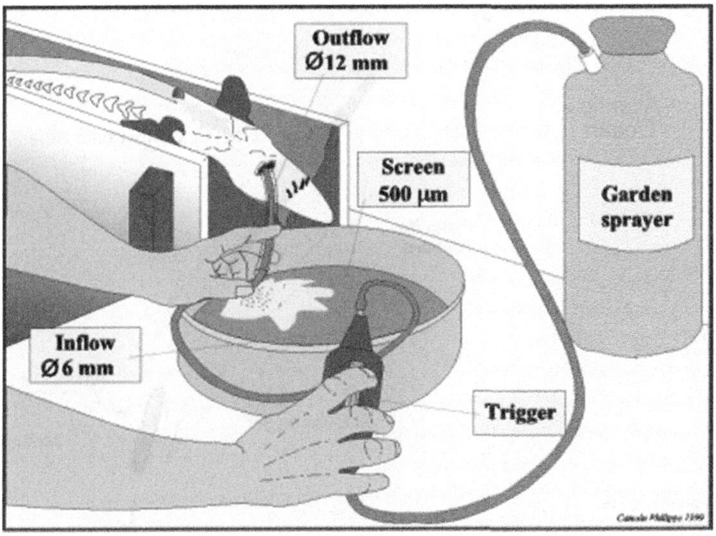

Fig. 11.1 The gastric lavage device once in place (*source*: Cemagref)

invertebrates was carried out. Data from benthic fauna were gathered during two sampling campaigns (June 2000 and May 2001) in the Gironde estuary using a Shipek grab. The data on depth came from surveys performed by the Port of Bordeaux. Sediment nature and distribution came from surveys performed by University of Bordeaux I, and current velocity was derived from a hydraulic model developed by the same University.

These elements made possible the characterization of typical juvenile *A. sturio* estuarine habitat, by the use of key environmental variables and key elements to differentiate these habitats from all other habitats found in the estuary.

11.3 Results

11.3.1 Habitat Identification

According to the catch data analysis based upon the trawl survey, the juvenile sturgeon habitat in the Gironde estuary (Brosse 2003; Lepage et al. 2004) showed three main aggregation areas where the level of capture was elevated compared to other areas (Fig. 11.2).

These relatively small aggregation zones were located along a salinity gradient (from 1 PSU in the upper estuary to 25 PSU in the lower estuary) which represents less than 15% of the estuary's surface, and were observed to be utilized for a period of 6 years by the 1994 and 1995 cohort. The utilization of discrete areas over extended periods of time has been described for other species (Buckley and Kynard

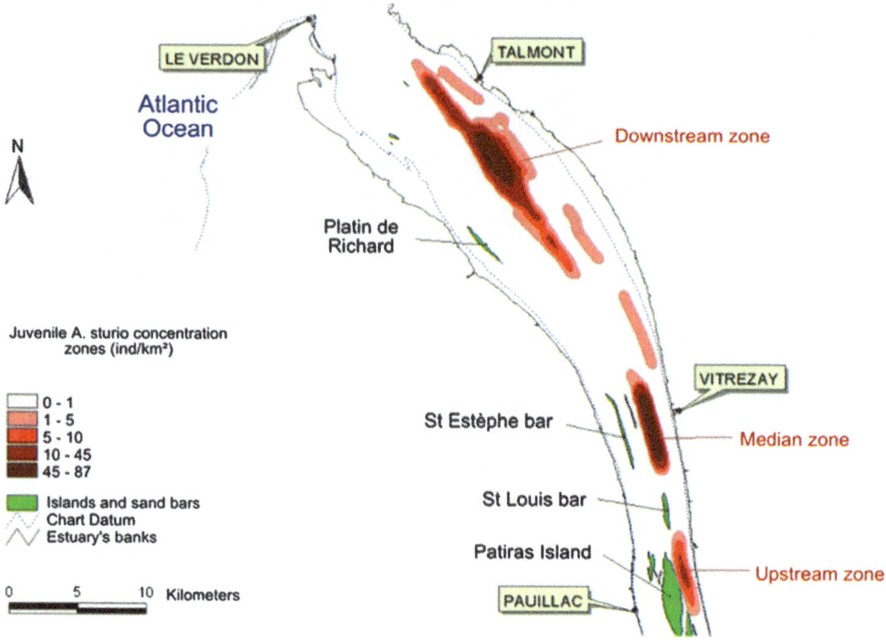

Fig. 11.2 Juvenile sturgeon habitat zones in the Gironde estuary determined from trawling survey (*source*: Cemagref)

1985; Hall et al. 1991; Kieffer and Kynards 1993; Bain 1997; Foster and Clugston 1997; Collins et al. 2000). Seasonal distribution patterns (Fig. 11.3) between these zones were observed. Preferential use of some areas was hypothesized to be related to sturgeon age and osmoregulation ability (Brosse 2003; Lepage et al. 2004), but could not be confirmed. The smallest individual captured in the upper estuary was 25 cm (TL) and was not weighed. In the lower estuary, the smallest one was 46 cm (TL) for 350 g. This observation strongly suggests that juveniles, having the ability to move easily, have the ability in osmoregulation as already stated by Magnin (1962). Some seasonal movements between these zones may also be linked to water temperature or food availability.

Results from the acoustic telemetry survey on 4 to 5-years-old sturgeon (88–122 cm) confirm the most downstream concentration zone, and make it possible to locate it more precisely (Fig. 11.4). This experiment showed that sturgeon exhibited movements mainly oriented following the direction of the tidal current: downstream movements during ebb and upstream movements during flood. Nevertheless, swimming speeds were lower than current speeds. No significant difference was observed in the swimming speed between flood and ebb tide, and no significant differences were observed in the swimming speed in day- and night-time (Lepage et al. 2004). During the continuous 24-h surveys, juvenile European sturgeon never used intertidal areas, not even during high-tide periods. Unfortunately, no data were available for upstream habitats, because these large juveniles were too old and didn't remain in this part of the estuary.

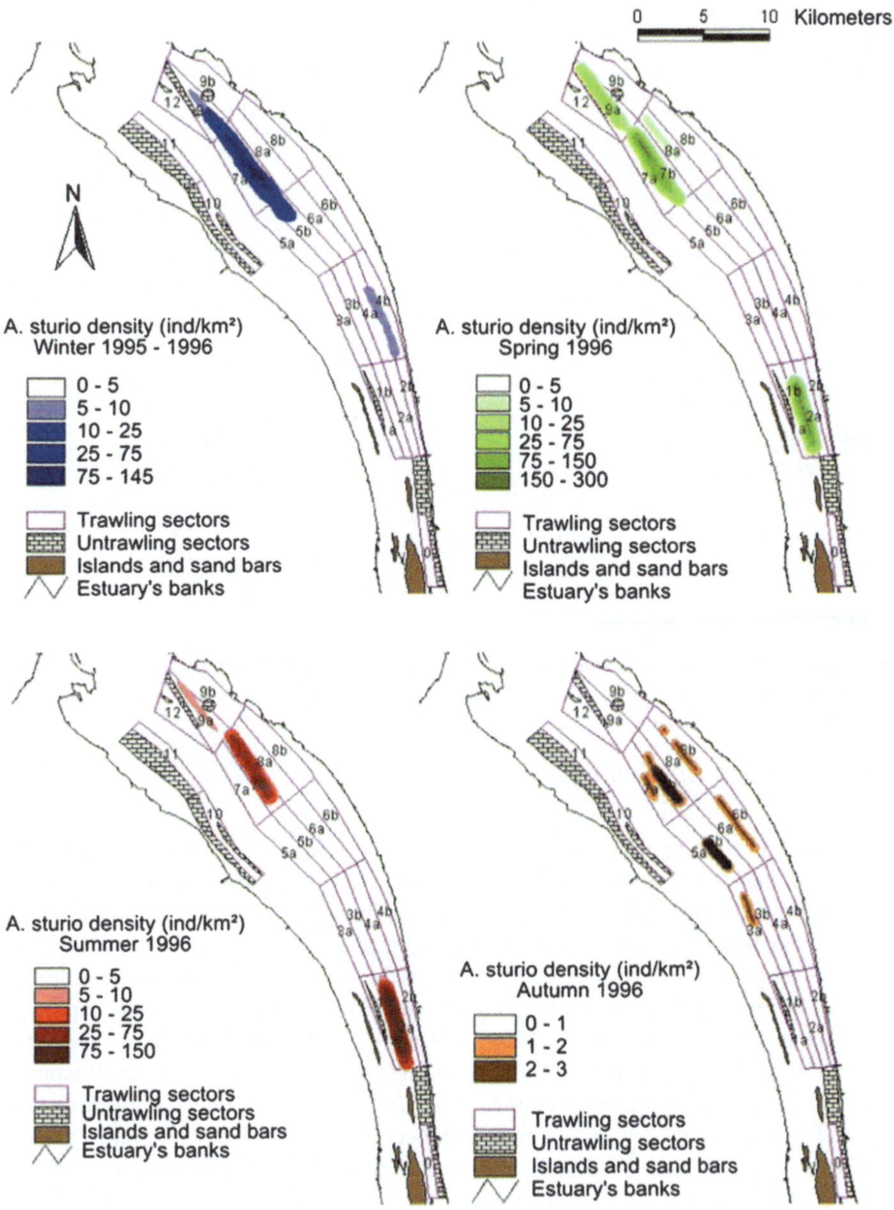

Fig. 11.3 Seasonal distribution pattern for juvenile sturgeon in the Gironde estuary as observed from December 1995 to December 1996 (*source*: Cemagref)

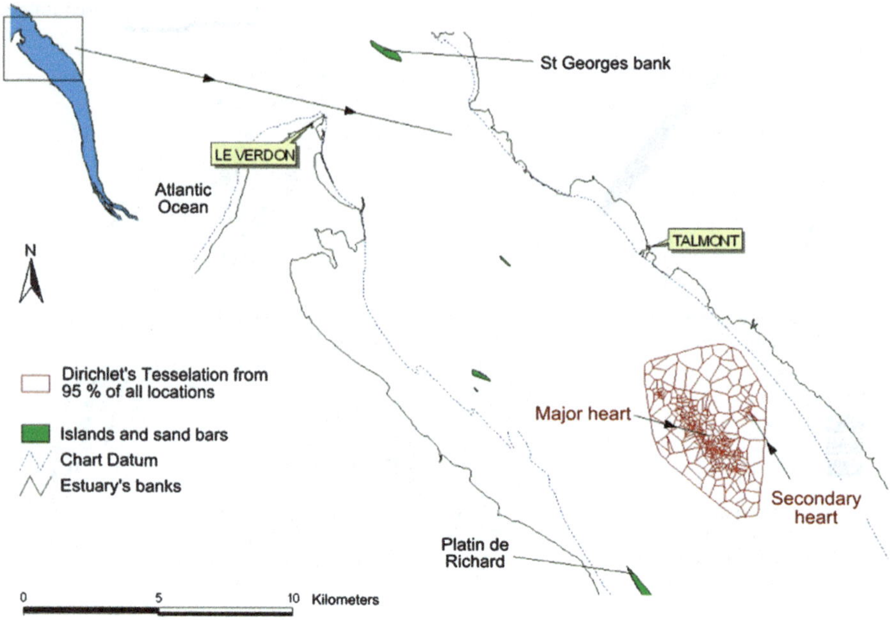

Fig. 11.4 Sturgeon main concentration area observed with acoustic telemetry survey during 1999 (*source*: Cemagref)

11.3.2 Feeding Behaviour

From May 1998 to October 2000, the stomach contents of 94 juveniles from 3 to 7 years old (63–141 cm total length) were collected by gastric lavage. Specimens were mainly caught during spring and summer in the lower two aggregation areas.

Fourteen taxa of prey were found. The highest proportions of prey in number and in occurrence for both middle and downstream congregation areas consisted of tube-dwelling polychaetes, mainly *capitellidae* (*Heteromastus filiformis* Claparede, 1864) and *spionidae* (*Polydora ligni* Bosc, 1802). Small Crustaceans were the second most abundant group of prey, with high occurrence for some small isopods (Fig. 11.5). Free polychaetes (mainly *Nereidae* never more than 10 cm long) are found frequently, but only a few individuals each time. In sturgeon caught outside of the aggregation areas, stomachs contained a majority of small crustaceans, while other prey items were negligible.

Both *capitellidae* an *spionidae* are small tube-dwelling annelid polychaetes (5–20 mm long) living in large colonies covering several square metres in muddy sediments (Fig. 11.6). During the survey in the Gironde a very high density was found, with up to 95,000 individuals per square metre in the upper estuary (Brosse 2003). In the Gironde, the young sturgeon were shown to have a preference for small soft-bodied prey organisms, but this could be linked to the availability of this type of prey rather than a deliberate choice.

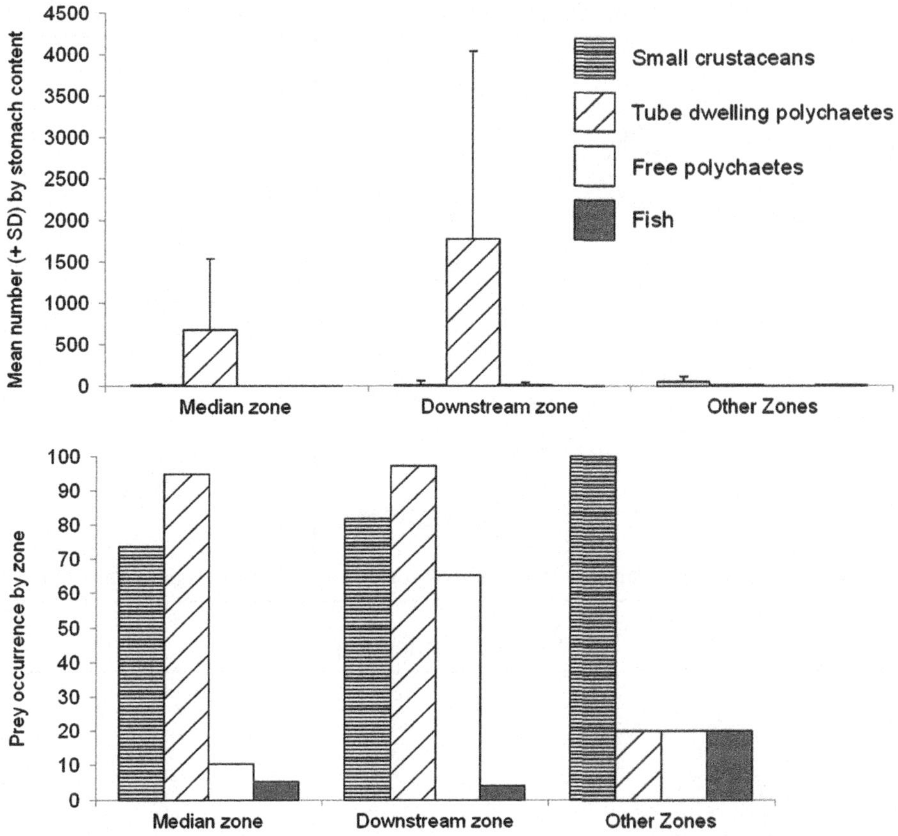

Fig. 11.5 Mean number (±SD) of major prey categories by stomach content by habitat zone (*upper graphic*) and major prey categories occurrence by habitat zone (*lower graphic*) determined on the 96 stomach contents studied between 1998 and 2000

Finally the diet of *A. sturio* is mainly based on arthropods and annelids as *A. oxyrinchus* (Mason and Clugston 1993; Johnson et al. 1997) and *A. stellatus* (Zolotarev et al. 1996).

11.3.3 Characteristics of Nursery Grounds

The surveys revealed a clear dominance in surface area of the tube-dwelling polychaetes compared to other taxa (Brosse 2003). These polychaetes inhabit two large areas or the estuary, one in the middle and the other in the lower estuary Fig. 11.7.

Sediment nature and distribution for the entire Gironde estuary were determined by an intensive ($N = 644$) sampling (Kapsimalis et al. 2000), and results during

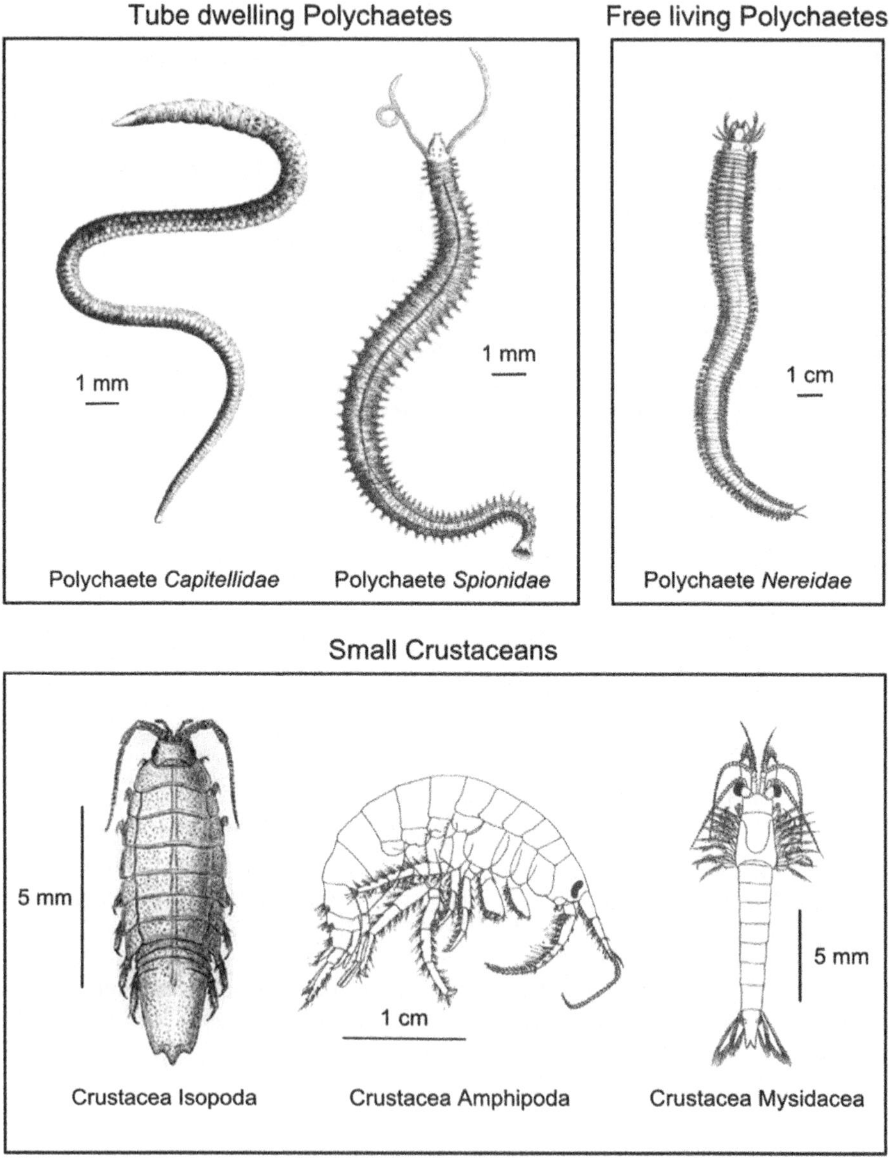

Fig. 11.6 Major prey items found in *A. sturio* bolus [from Hayward and Ryland (1994) except Isopoda from Sars (1899)]

1998 and 1999 were provided by the Department of Oceanographic Geology – Bordeaux I. Bottom substrate was dominated by muddy sediment (49% of the total surface) compared to sand (27%) and muddy sand (8%).

Depth determination in the Gironde estuary used for characterisation of sturgeon habitat was provided by the Port Authority of Bordeaux. The depth used was a

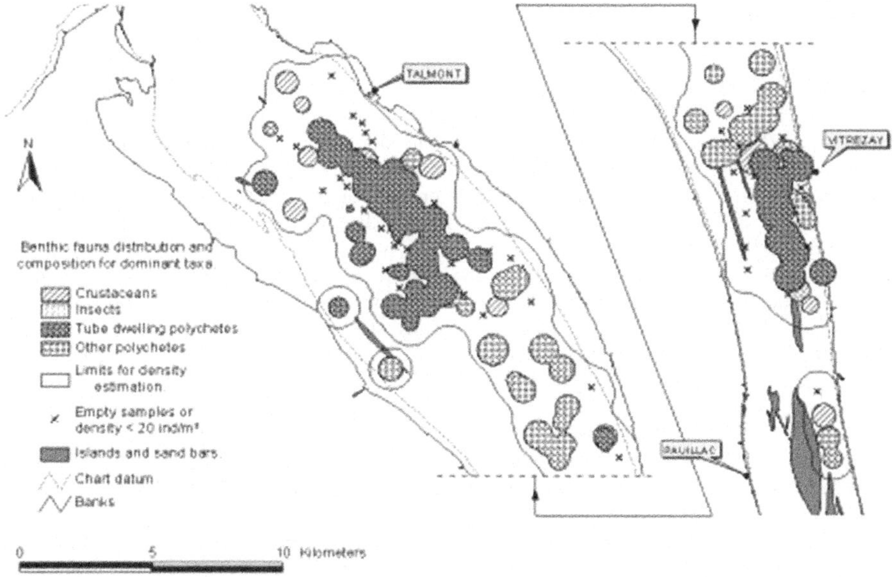

Fig. 11.7 Benthic fauna distribution and composition in the Gironde estuary determined from year 2000 and 2001 samplings using Shipek grab (*source*: Cemagref)

normalized depth, and did not take into account the variation linked to tidal range in the estuary (maximum 5 m during a tidal cycle).

Strong correspondence between areas with high densities of the favourite prey of the sturgeon (tube-dwelling polychaetes) and the aggregation of sturgeon in this area were observed (Brosse 2003; Lepage et al. 2004). Thus, even if the location of rich feeding grounds seems predominant for habitat choice by juvenile sturgeons, some other factors like bathymetry, sediment size and current velocity appear to play an important role (Brosse 2003). The intertidal mudflats of the Gironde estuary are known to have several taxa of oligochaetes, polychaetes and small crustaceans in a higher proportion than in the subtidal muddy substratum (Quintin 2006); however, no sturgeon were found in the shallow waters (0–2 m) of the intertidal area. During the study period, the distribution of all captures according to depth shows that the maximum is observed between 4 and 8 m (Brosse 2003). Sediment size and current intensity are linked together, and to the habitat requirements of the sturgeon prey (small tube-dwelling polychaetes). Taverny et al. (2002) showed that the sturgeon swim partly against the current (maximum swimming speed calculated 110 cm s^{-1}), and tend to stay in a restricted area. According to Trump and Leggett (1980), Brodersen et al. (2008) and Ohlberger et al. (2006), swimming against the current velocity should affect the energetic cost of the foraging activity, but we could reasonably think that this over-cost is balanced by the benefits represented by being able to feed easily and in high quantity.

References

Bain MB (1997) Atlantic and shortnose sturgeons of the Hudson River: common and divergent life history. Environ Biol Fish 48:347–358

Beamish FWH, David LGN, Rossiter A (1998) Feeding ecology of juvenile Lake Sturgeon, *Acipenser fulvescens,* in northern Ontario. Can Field Nat 112(3):459–468

Brodersen J, Nilsson PA, Ammitzboll J, Hansson LA, Skov C, Brönmark C (2008) Optimal swimming speed in head currents and effects on distance movement of winter-migrating fish. PLoS One 3:e2156

Brosse L (2003) Caractérisation des habitats des juvéniles d'esturgeon européen, *Acipenser sturio*, dans l'estuaire de la Gironde: Relations trophiques, hiérarchisation et vulnérabilité des habitats. PhD Thesis, Toulouse, Toulouse III University

Brosse L, Dumont P, Lepage M, Rochard E (2002) Evaluation of a gastric lavage method for sturgeons. N Am J Fish Manag 22:955–960

Buckley J, Kynard B (1985) Yearly movements of shortnose sturgeons in Connecticut River. Trans Am Fish Soc 114:813–820

Castelnaud G, Rochard E, Jatteau P, Lepage M (1991) Données actuelles sur la biologie d'*Acipenser sturio* dans l'estuaire de la Gironde. In: Williot P (ed) Acipenser. Cemagref, Antony, pp 251–275

Chiasson WB, David LGN, Beamish FWH (1997) Habitat, benthic prey and distribution of juvenile lake sturgeon (*Acipenser fulvescens*) in northern Ontario rivers. Can J Fish Aquat Sci 54:2866–2871

Collins MR, Smith TIJ, Post WC, Pashuk O (2000) Habitat utilization and biological characteristics of adult Atlantic sturgeon in two South Carolina rivers. Trans Am Fish Soc 129:982–988

Dadswell MJ (1979) Biology and population characteristic of the shortnose sturgeon, *Acipenser brevirostrum* LeSueur 1818 (Osteichthyes: Acipenseridae), in the Saint John River estuary, New Brunswick. Can J Zool 57:2186–2210

Foster AM, Clugston JP (1997) Seasonal migration of Gulf sturgeon in the Suwannee River. Fla Trans Am Fish Soc 126:302–308

Gaudin P, Martin E, Caillère L (1981) Le tubage gastrique chez les poissons: Mise au point d'un équipement et test chez le chabot *Cottus gobio* L. Bull Fr Piscic 282:8–15

Haley N (1998) A gastric lavage technique for characterizing diet of sturgeons. N Am J Fish Manag 18:978–981

Hall JW, Smith TIJ, Lamprecht SD (1991) Movements and habitats of shortnose sturgeons, *Acipenser brevirostrum* in the Savannah River. Copeia 1991(3):695–702

Hayward PJ, Ryland JS (1994) The marine fauna of the British Isles and north-west Europe. Volume 1: Introduction and protozoans to arthropods. Oxford Science, Oxford, 627 pp

Hyslop EJ (1980) Stomach content analysis – a review of methods and their application. J Fish Biol 17:411–429

Johnson JH, Dropkin D, Warkentine S, Barbara E, Rachlin JW, Andrews WD (1997) Food habitat of Atlantic Sturgeon off the Central New Jersey coast. Trans Am Fish Soc 126:166–170

Kapsimalis V, Tastet JP, Massé L (2000) Modern sedimentology of macrotidal estuaries: the example of the Gironde (SW France). 6th Hellenic Symposium of Oceanography and Fishery, Chios

Kieffer MC, Kynard B (1993) Annual movements of shortnose and Atlantic sturgeons in the Merrimack River. Mass Trans Am Fish Soc 122:1088–1103

Lepage M, Taverny C, Piefort S, Dumont P, Rochard E, Brosse L (2004) Juvenile sturgeon (*Acipenser sturio*) habitat utilization in the Gironde estuary as determined by acoustic telemetry. In: FAO (ed) Aquatic telemetry: advances and applications. FAO, Rome, pp 169–177

Letaconnoux R (1961) Note sur la fréquence de la distribution des captures d'esturgeons (*Acipenser sturio* L.) dans le Golfe de Gascogne. Revue des Travaux de l'Institut des Pêches Maritimes 25:253–261

Levin AV (1989) Characteristics of feeding behaviour of juvenile Russian sturgeon, *Acipenser gueldenstaedti*, in relation to food availability. J Ichtyol 27(3):41–47

Lochet A, Lambert P, Lepage M, Rochard E (2004) Croissance de juvéniles d'esturgeons européens *Acipenser sturio* (Acipenseridae) sauvages et issus d'alevinage, durant leur séjour dans l'estuaire de la Gironde (France). Cybium 28:91–98

Magnin E (1962) Recherches sur la systématique et la biologie des Acipenséridés. Extraits des Annales de la Station Centrale d'Hydrobiologie Appliquée

Mason WT Jr, Clugston JP (1993) Foods of the Gulf sturgeon in the Suwannee River. Fla Trans Am Fish Soc 122:378–385

McCabe GT Jr, Emmett RL, Hinton SA (1993) Feeding ecology of juvenile white sturgeon (*Acipenser transmontanus*) in the lower Columbia River. Northwest Sci 67:170–180

Meehan WR, Miller RA (1978) Stomach flushing: effectiveness and influence on survival and conditions of juvenile salmonids. J Fish Res Board Can 35:1359–1363

Nilo P (1996) Force des classes d'âge, habitats et alimentation des esturgeons jaunes (*Acipenser fulvescens*) juvéniles du système Saint Laurent. Masters Thesis in Biology, University of Quebec, Montreal

Ninua NS (1976) The Atlantic sturgeon from the Rioni River (L'esturgeon atlantique de la rivière Rioni). Tbilisi, 121 pp (in Russian)

Ohlberger J, Staaks G, Hölker F (2006) Swimming efficiency and the influence of morphology on swimming costs in fishes. J Comp Physiol Biochem Syst Environ Physiol 176:17–25

Polyaninova AA, Molodsteva AI (1995) The benthos–sturgeon feeding relationship for the Caspian Sea. In: Proceedings of International Sturgeon Symposium, Vniro

Quintin JY (coord) (2006) Surveillance écologique du site du Blayais, année 2005. Rapport IFREMER RST DYNECO/VIGIES/06-07, mars 2006, 230 pp

Rochard E (1992) Mise au point d'une méthode de suivi de l'abondance des amphihalins dans le système fluvio-estuarien de la Gironde, application à l'étude écobiologique de l'esturgeon, Acipenser sturio. Thèse de Doctorat, Rennes, Université de Rennes I

Rochard E, Jatteau P (1991) Amélioration de la méthode de détermination de l'âge de l'esturgeon commun *Acipenser sturio* et premières applications. In: Williot P (ed) Acipenser. Cemagref, Antony, pp 193–208

Rochard E, Williot P, Castelnaud G, Lepage M (1991) Eléments de systématique et de biologie des populations sauvages d'esturgeons. In: Williot P (ed) Premier Colloque International sur l'Esturgeon, Bordeaux. Cemagref, Antony, pp 475–507

Rochard E, Lepage M, Dumont P, Tremblay S, Gazeau C (2001) Downstream migration of juvenile European sturgeon *Acipenser sturio* L. in the Gironde estuary. Estuaries 24(1):108–115

Roule L (1922) Etude sur l'esturgeon du Golfe de Gascogne et du bassin Girondin. Office Scientifique et Technique des Pêches Maritimes. Notes et Mémoires 20:12

Sars GO (1899) An account of the Crustacea of Norway with short descriptions and figures of all the species. Vol. II. Isopoda. Bergen Museum, Bergen, pp I–X

Taverny C, Lepage M, Piefort S, Dumont P, Rochard E (2002) Habitat selection by juvenile European sturgeon *Acipenser sturio* in the Gironde estuary (France). J App Ichthyol 18(4–6):536–541

Trump CL, Leggett WC (1980) Optimum swimming speeds in fish: the problem of currents. Can J Fish Aquat Sci 37:1086–1092

Zolotarev PN, Shlyakhov VA, Akselev OI (1996) The food supply and feeding of the Russian Sturgeon *Acipenser gueldenstadtii* and the Starred Sturgeon *Acipenser stellatus* of the North-western part of the Black Sea under modern ecological conditions. J Ichthyol 36(4):317–322

Chapter 12
Characteristics of the Reproductive Cycle of Wild *Acipenser sturio*

Patrick Williot, Thierry Rouault, Rémi Brun, and Jörn Gessner

Abstract This chapter provides data on age at puberty, duration of sexual activity, frequency of spawning, period of spawning, and zones of spawning grounds. Previous data as well as recent information are reported, mainly for the Atlantic population in France. When available, other data have also been mentioned, particularly from the River Guadalquivir (Sp) and the Elbe/Eider Rivers (D). Puberty was reached between 13 and 15 years for males and 19 and 22 years for females. Corresponding length (TL) was 140–145 cm for males and 185 cm for females. Duration of sexual activity was probably in the range of 6–10 years for males and 17–22 years for females. Frequency of spawning is mostly under-documented. The main spawning season is May–June in the Garonne basin. Some characteristics varied with latitude, e.g., age at puberty, onset of reproduction migration. Characteristics of puberty and sexual cycles in farmed sturgeon are given, including the few available data on *A. sturio*.

12.1 Introduction

Knowledge of reproductive cycles is a key issue in fisheries science and species conservation, since these are primary biological data used in population dynamics studies. Some specific issues should be addressed: What is the age at puberty? Over

P. Williot (✉)
Sturgeon Consultant, 4 Rue du Pas de madame, 33980 Audenge, France
e-mail: williot.patrick@neuf.fr

T. Rouault • R. Brun
Cemagref, Estuarine Ecosystems and Diadromous Fish Research Unit, 50 Avenue de Verdun, 33612 Cestas Cedex, France

J. Gessner
Leibniz-Institute of Freshwater Ecology and Inland Fisheries, Müggelseedamm 310, 12587 Berlin, Germany

P. Williot et al. (eds.), *Biology and Conservation of the European Sturgeon Acipenser sturio L. 1758*, DOI 10.1007/978-3-642-20611-5_12,

what life span does sexual activity last? What is the spawning frequency? When is the spawning period? Where are the spawning grounds? Are there climatic factors that may influence spawning?

Currently, some of these issues are poorly documented because of difficulties in successfully obtaining data, especially as sturgeons are long-lived fish with late puberty and mostly non-yearly oogenesis. In this respect, *Acipenser sturio* is no exception. The great majority of data presented here originate from the French populations, with particular focus on the Atlantic population which spawns in the Garonne basin.

This short section summarizes the fragmentary data available in this field, covering a specific aspect of the more general subject of biological cycles (see Chap. 10). The relative numbers of both genders of adult brood fish, i.e., the sex ratio, will be analyzed in Chap. 27, as its impact on reproduction is crucial.

12.2 Age at Puberty

Magnin (1962) determined age at puberty and length (TL) by comparing the distribution of mature–immature fish according to age and length (Figs. 12.1 and 12.2). From these results, the age at puberty was 13–15 years for males and 19–22 years for females. Corresponding length was 140–145 for males and 185 cm for females. Therefore, as for most other sturgeon species, *A. sturio* is a late-puberty fish. It is worth noting that the transition from pre-adult stage to adulthood in females appears to be more abrupt than that in males, given the figures.

The characteristics of puberty depend on the geographical location, as synthesized in Table 12.1. Magnin (1962) outlined a South–North gradient, with

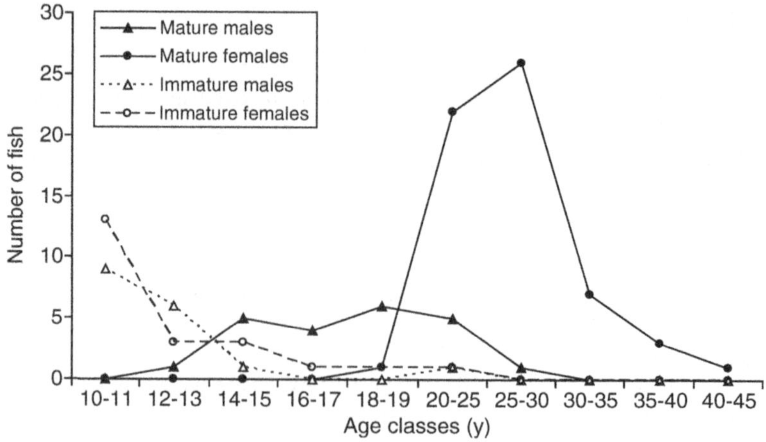

Fig. 12.1 Distribution of mature and immature *A. sturio* (Garonne basin population) according to sex and age (data drawn from Magnin 1962)

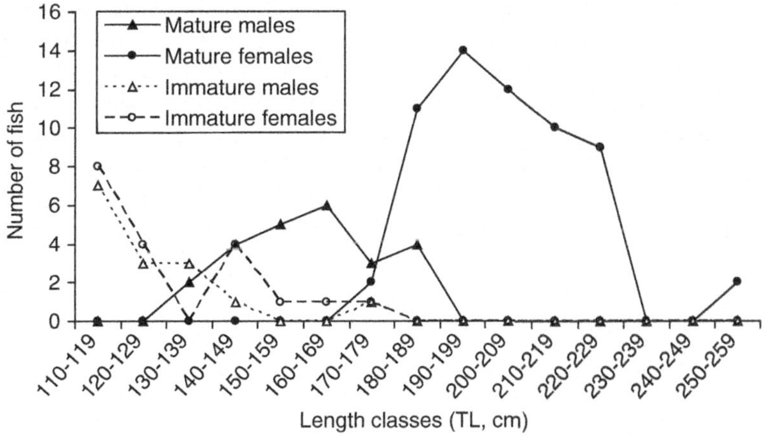

Fig. 12.2 Distribution of mature and immature *A. sturio* (Garonne basin population) according to sex and length (data drawn from Magnin 1962)

Table 12.1 Age and size of wild *Acipenser sturio* at puberty according to geographical area and sex (completed and updated from Magnin 1962 and Holčik et al. 1989)

	Age (y)		Size (TL, cm)		
Catching area	Males	Females	Males	Females	Sources
Northern Sea (Elbe)			120	160	Ehrenbaum (1927)[a]
Atlantic coast	13–15	19–22	145	185	Magnin (1962)
(Garonne, Dordogne)	~10	~15	145	185	Extrapolated from Rochard and Jatteau (1991)[b]
(Guadalquivir)	10–11	14–15	120	150	Classen (1944)
Adriatic Sea (Skadar lake)	7–9	8–14	89–135	95–187	Poljakov et al. (1958)[c]
Adriatic Sea (Pô)	9–10	11–12	120	155	Paccagnella (1948)[a]
Black Sea					
(Rioni)	7–9	8–14			Kozhin (1964)
	9–16	17–18			Berg (1932)[a]
	7–9	8–14	110	137	Marti (1939)[a]
(Azov Sea)	8	9			Chugunov (1927)[a]

[a]Cited in Magnin (1962)
[b]Ages are extrapolated from growth curve established for 1988 year class
[c]Cited in Holčik et al. (1989)

the more southern populations experiencing earlier puberty than the northern ones. Therefore, it might be concluded that puberty is related to growth, and the faster the growth, the earlier the puberty.

A rather wide range in age at puberty is observed, with males being more precocious than females. Recently, an updated age determination method (Rochard and Jatteau 1991; see Chap. 23) revealed a more rapid growth in the 1988 cohort when compared with previous data by Magnin (1962). The most likely explanation would seem to be a density-dependent growth effect due to the dramatic decline of the species recorded in the 1970s and the resulting population size in the estuary (Williot et al. 1997). Therefore, puberty for the most recent wild brood fish would be expected to occur at even younger ages compared with previous records. The dramatic decline in the population meant that this hypothesis was not verified.

Preliminary results obtained from farmed specimens reveal precocious puberty at 13 years for females (Williot et al. 2009a; see Chap. 32). With the improvement in farming conditions, it is likely that puberty will occur in even younger fish in the future.

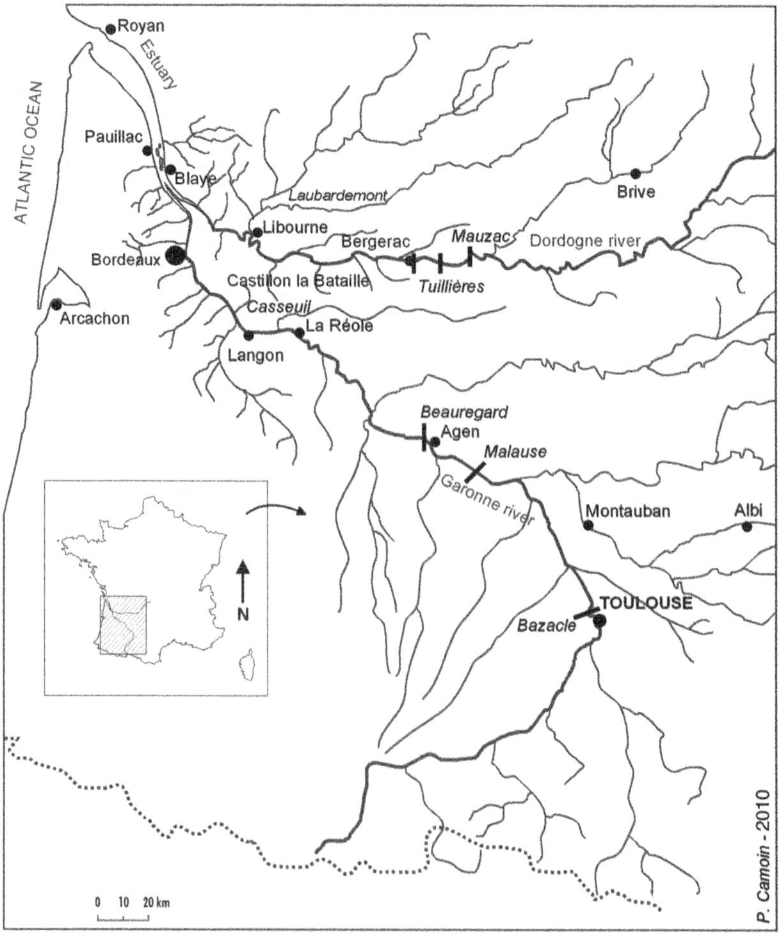

Fig. 12.3 Garonne River basin (French Atlantic coast) with river section in the Garonne (La Réole–Agen) and in the Dordogne (Libourne–Bergerac) where potential spawning areas still exist

12.3 Location of Spawning Grounds

Spawning grounds in the River Garonne (Fig. 12.3) are located in the river section between La Réole (rkm[1] ~170) and the Beauregard weir downstream from Agen at rkm ~270, with well-known spawning grounds at around rkm 180–190 (Vibert 1945). In the River Dordogne, spawning grounds are located between Libourne (rkm ~100) and the Bergerac dam (rkm ~210). The corresponding river sections have been mapped recently, and the main morphological and hydrological characteristics have been summarized by Jego et al. (2002). Most of the studied sites correspond to the required criteria for sturgeon spawning grounds.

Spawning grounds in the River Rhône are restricted to the portion of river upstream from Arles and downstream from the Vallabrègues dam (erected in 1970), i.e., between rkm 50 and rkm 60 (Tabardel 1994; Brosse et al. 2009) (Fig. 12.4).

In Spain, in the River Guadalquivir, spawning grounds were located near Cordoba, i.e., 230 rkm from the mouth. To our knowledge, no observations of

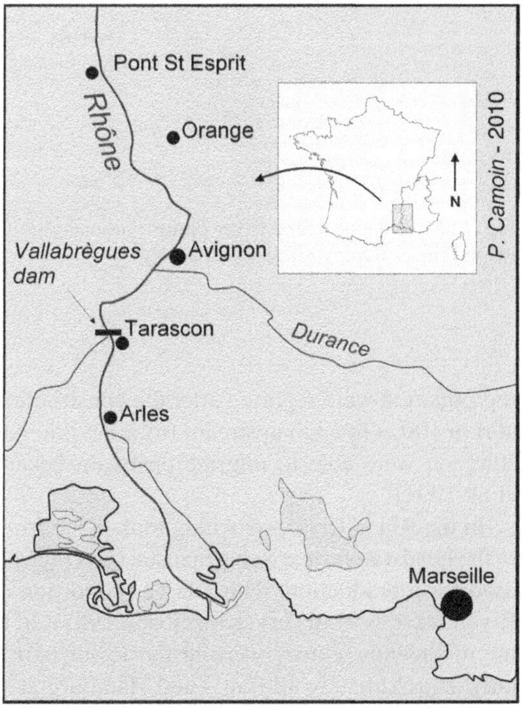

Fig. 12.4 Rhône River basin (French Mediterranean coast) with river section (Arles–Vallabrègues dam)

[1]rkm = number of kilometers from the river mouth to the point in question.

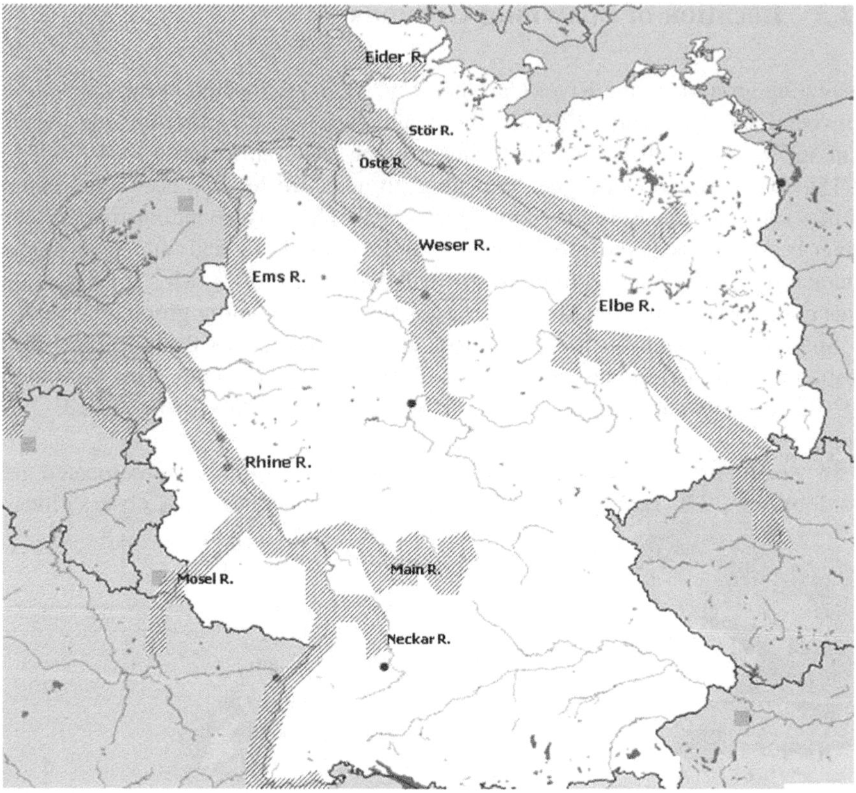

Fig. 12.5 Rhine and Elbe River basins (Germany) with river sections where spawning grounds existed (map: http://www.wikipedia.org)

reproduction were reported after the construction of the Alcala del Rio dam in 1930 at rkm 100, a few km upstream from Sevilla. Although a fish pass was installed, no sturgeon were able to migrate upstream because of its unsuitable design (Elvira et al. 1991).

In the Rhine River, spawning took place from the lower river in the Netherlands to the headwaters near Schaffhausen (D) (Fig. 12.5). The larger tributaries were also used for reproduction, with a large proportion of the fish migrating into the Mosel River as far as Toul (Fr) (Kinzelbach 1987). In the Elbe, the main spawning sites in the nineteenth century were in the lower part of the river between Brunsbüttel – only approximately 40 rkm – and Hamburg at 110 rkm (Quantz 1903). Also in the vicinity of Magdeburg (400 rkm), large sturgeon abundance and substrate availability indicated the presence of historic spawning grounds (Kluge 1904). Dredging for navigation purposes removed gravel banks from the river and increased saltwater inflow into the upriver section, destroying the spawning sites up to rkm 80.

12.4 Spawning Period

Based on a 5-year period (1957–1961), the seasonal occurrence of mature *A. sturio* in the fishery was described by Magnin (1962) (Fig. 12.6). The abundance increased in early May to late June. The more recent data (1981–2001) are rather similar, all the more so since the very limited number of specimens means that excessive generalization is avoided (Fig. 12.7). However, the most recent results suggest that males start to migrate earlier than the females. A similar observation has been reported in the Guadalquivir River, where the peak male migration occurs 2–3 weeks before that of the females (Classen 1944; Gutierrez-Rodriguez 1962).

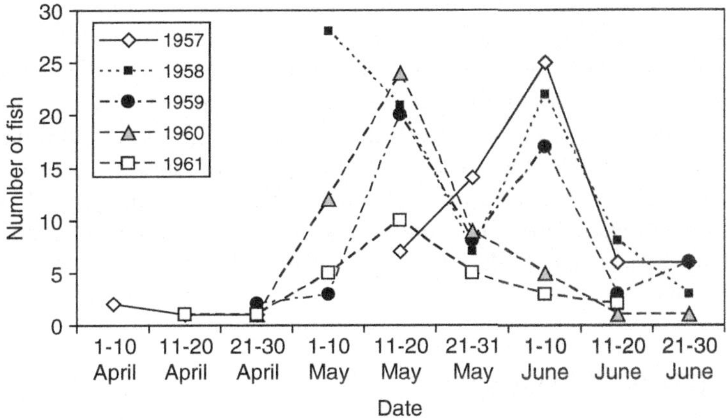

Fig. 12.6 Distribution of mature *A. sturio* in catches in the Garonne and Dordogne Rivers according to date of upstream migration within a year (data drawn from Magnin 1962; $N = 290$ fish; 1957–1961)

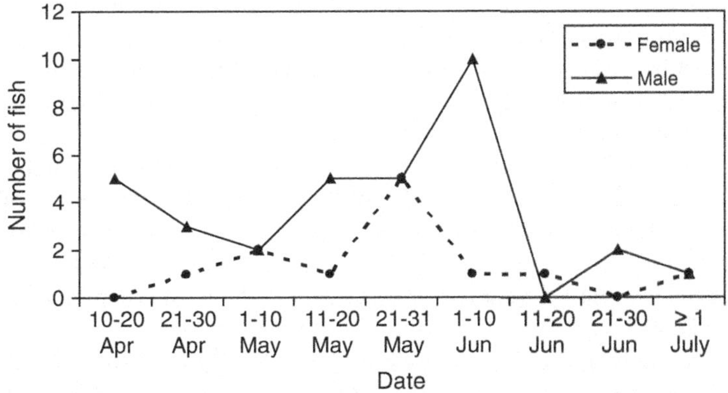

Fig. 12.7 Period of upstream migration through the by-catch of wild brood *A. sturio* (Garonne basin population) (1981–2001) (after Williot et al. 2002)

The spawning migration occurred earlier in the Southern part of its geographical distribution [Black Sea – April and May in River Rioni (Ninua 1976), in Spain February–April in River Guadalquivir (Classen 1944)] and later in the Northern part [North Sea – April to August with peak migration in June and July (Ehrenbaum 1936; Blankenburg 1912; Quantz 1903)], with first fish entering the river by April and the last in August–September. Also, catches from the fourteenth to sixteenth centuries in the middle Rhine between Cologne and Speyer revealed similar catch levels in May, June and July (Kinzelbach 1987). In the Atlantic population in France, the timing of migration is intermediate (Magnin 1962; Holčik et al. 1989).

Magnin (1962) described how as soon as *A. sturio* started migrating upstream into the Gironde, they ceased feeding.

The factors that influence the onset of migration are somewhat speculative, although water temperature is most often evoked. The tide has been cited as a key factor in the River Guadalquivir (Classen 1944) and by Vehrey (1949) in Magnin (1962) in the River Meuse. Also, Sulak and Clugston (1999) have described a close relationship between the moon phase and the timing of spawning in Gulf sturgeon, *A. oxyrinchus desotoi*. Recruitment has been positively correlated with river discharge during the last 2 months prior to spawning in *Acipenser transmontanus* in central California (Kohlorst et al. 1991) and in Gulf sturgeon *Acipenser oxyrinchus desotoi* (Randall and Sulak 2007). Similar observations were reported by Kinzelbach (1987) for the numbers of fish migrating upstream in different years.

12.5 Spawning Frequency

Data regarding the reproductive frequency of *A. sturio* are very scarce. No directed studies have been carried out either in the Gironde or in the Guadalquivir, since the data obtained originated from fisheries (Magnin 1962). The author reported that most of the scientists who used to work in the field thought that the males might be able to breed each year, while the females would spawn every 2 to 3 or even 4 years. These assumptions were based on: (1) females exhibiting different levels of sexual advancement being present at the same time, and (2) the low probability that such a large amount of eggs would be produced annually. A similar uncertainty is reported by Holčik et al. (1989).

From tagging wild brood fish, it has been possible to assume that a male *A. sturio* matures three times at 2-year intervals (Williot et al. 1997). Iteroparity of females has been tentatively evaluated by the presence of atretic ovarian follicles showing black pigment in the wild population of *Acipenser oxyrinchus* (Van Eenenaam et al. 1996), suggesting that a 2-year recurring cycle was the most common pattern in the oldest female fish.

Thanks to tagging, the figures observed for wild-originated males reared under controlled conditions were extremely variable in *A. sturio* (Williot and Rouault 2008; see Chap. 32). This illustrates the great inter-individual variability for this species.

In farmed sturgeon brood fish, both tagging and staging were used to identify recurrent cycles in *Acipenser transmontanus* (Doroshov et al. 1997) and in *Acipenser baerii* (Williot and Brun 1998). For both species, a 2-year recurring cycle in females was most common. However, the latter species exhibited a rather large spectrum of recurring reproductive cycles of 1 and 2 years in the same individual. Annual and triennial cycles were also observed. The preliminary observations on farmed *A. sturio* suggest a 2-year recurring cycle in females (see Chap. 32).

12.6 Duration of Sexual Activity

The question of the period of sexual activity has been approached using both the age distributions (Fig. 12.1) and size distribution of mature fish at their upstream migration in the Garonne basin (Fig. 12.2) by Magnin (1962). The main sexual period was in the range of 14–15 to 20–25 years for males, and from 18–19 to 35–40 years for females. The corresponding lengths (TL, cm) ranged from 140–149 to 180–189 for males, and 180–189 to 220–229 cm for females. This means that the males remained sexually active for 6–10 years, while the females were sexually active for 17–22 years. It should be noted that the author observed a very small number of mature females at 40–45 years measuring 250–259 cm TL. Even larger spawners were mentioned in the past (Holčik et al. 1989). French illustrations of a few fish of about 2.40 m were reported during the first half of the twentieth century (see Chap. 19). Gutierrez-Rodriguez (1962) reported that sexual activity lasted around 8 years for the males and 30 years for females in the Guadalquivir River. These last records show that it is most likely that the duration of female sexual activity is longer than that determined by Magnin. And this comment is supported by German catches which report that until 1900, fish of more than 3 m TL were caught annually. In the 1840s, maximum-sized fish of 4.5 m were still being caught, but several reports mention the decline of large fish of more than 4 m in catches from the seventeenth to the nineteenth century. It is known that senescence in fish is delayed compared with other vertebrates (Reznik et al. 2002). However, this does not mean that fish, especially sturgeon, keep their reproductive potential intact until they die, and this was supported by a recent work on farmed sterlet (*Acipenser ruthenus*), which showed age-related deterioration in reproductive capacities (Williot et al. 2005).

However, the ranges mentioned above, for French populations at least, corresponded to the given population at the time Magnin (1962) carried out his studies in the late 1950s. It is very likely that the population was already negatively impacted by fisheries, which had taken out the larger fish first, so that the reported ranges represent a minimum.

In farmed *A. sturio* males of wild origin, maturation has been observed over a period of 9 years (see Chap. 32). Therefore, at present, the results for farmed males are in a similar range to that reported earlier for wild fish.

12.7 Conclusions

The difficulties encountered historically in assessing age at puberty have been outlined here (see Chap. 24). These observations highlight the necessity for a long-term survey, which must be emphasized (Williot et al. 2009b), as well as an awareness campaign to encourage fishermen not to keep sturgeon by-catch at sea (see Chap. 37). It is encouraging that a large proportion of French rivers contain potential spawning grounds (Jego et al. 2002). Thanks to the pioneering work by Magnin (1962), the ecological characteristics of the species such as age at maturity, growth, and intervals between mating have been described, and have been used as a guideline for ex situ brood stock management (e.g., spawning period , starvation period, see Chap. 32) as well as for managing releases. More recent data on the characteristics of the reproductive cycle of forthcoming brood fish will be needed, so that changes can be assessed.

References

Blankenburg A (1912) Die Fischerei in der Unterelbe. 35 Jahresbericht des Central-Fischerei-Vereins e.V. für Schleswig-Holstein 1911–1912, pp 125–140

Brosse L, Berrebi P, Desse-Berset N, Lepage M (2009) Sturgeon recovery plan in the Rhône River (France); Preliminary results on species determination and habitat suitability. In: Carmona R, Domezain A, Garcia Gallego M, Hernando-Casal J, Rodriguez F, Ruiz Rejon M (eds) Biology, conservation and sustainable development of sturgeons, Fish & fisheries series. Springer, New York, pp 403–421

Classen TEA (1944) Estudio bio-estadistico del esturion o sollo del Guadalquivir. Instituto Espanol de Oceanografia, Ministerio de Marina, Trabajos, N° 19, 112p + XVII planches

Doroshov SI, Moberg GP, Van EZenennaam JP (1997) Observations on the reproductive cycle of cultured white sturgeon, *Acipenser transmontanus*. Environ Biol Fishes 48:265–278

Ehrenbaum E (1936) Naturgeschichtliche und wirtschaftliche Bedeutung der Seefische Nordeuropas. In: Lübbert H, Ehrenbaum E (eds) Handbuch der Seefischerei Nordeuropas vol 2. E. Schweizerbart, Stuttgart, pp 3–5

Elvira B, Almodovar A, Lobon-Cervia J (1991) Sturgeon (*Acipenser sturio* L., 1758) in Spain. The population of the River Guadalquivir: a case history and claim for a restoration programme. In: Williot P (ed) Acipenser. Cemagref, Antony, pp 337–347

Gutierrez-Rodriguez F (1962) El esturion del rio Guadalquivir. Ministerio de Agricultura, Servicio Nacional de pesca fluvial y caza. 5 Folleto informativo (temas piscicolas), 58p + photographs

Holčik J, Kinzelbach R, Sokolov LI, Vassilev VP (1989) *Acipenser sturio* Linnaeus, 1758. In: Holčik J (ed) The freshwater fishes of Europe. Aula Verlag, Wiesbaden, pp 367–394

Jego S, Gazeau C, Jatteau P, Elie P, Rochard E (2002) Les frayères potentielles de l'esturgeon Européen *Acipenser sturio* L. 1758 dans le bassin Garonne-Dordogne. Méthodes d'investigation, état actuel et perspectives. Bull Fr Pêche Piscic 365(366):487–505

Kinzelbach R (1987) Das ehemalige Vorkommen des Störs, *Acipenser sturio* (Linnaeus 1758) im Einzugsgebiet des Rheins (Chondrostei: Acipenseridae). Z Angew Zool 74(2):167–200

Kluge M (1904) Zum Störfang am Cracauer Wehr in Magdeburg. Fischereizeitung 7:153–155/ 187–188

Kohlorst DW, Botsford LW, Brennan JS, Caillet GM (1991) Aspects of the structure and dynamics of an explored central California population of white sturgeon (*Acipenser transmontanus*). In: Williot P (ed) Acipenser. Cemagref, Antony, pp 277–293

Kozhin NI (1964) Sturgeon from USSR and their reproduction. Trudy VNIRO LII:21–58 (in Russian)

Magnin E (1962) Recherche sur la systématique et la biologie des Acipenséridés *Acipenser sturio* L., *Acipenser oxyrinchus* Mitchill et *Acipenser fulvescens* Raf. Ann Stat Cent Hydrobiol Appl 9:7–242

Ninua NSh (1976) Atlantic sturgeon of the Rioni River. Editions Metsniereba, Tbilissi, p 122 (in Russian)

Quantz H (1903) Störfischerei und Störzucht im Gebiete der deutschen Nordseeküste. Mitteilungen des Deutschen Seefischerei-Vereins XIX:176–204

Randall MT, Sulak KJ (2007) Relationship between recruitment of gulf sturgeon and water flow in the Suwannee River, Florida. In: Munro J (ed) Anadromous sturgeons: habitats, threats, and management. Symposium, vol. 56. American Fisheries Society, Bethesda, MD, pp 69–83

Reznik D, Ghalambor C, Nunney L (2002) The evolution of senescence in fish. Mech Ageing Dev 123:773–789

Rochard E, Jatteau Ph (1991) Amélioration de la méthode de détermination de l'âge de l'esturgeon commun, *Acipenser sturio* et premières applications. In: Williot P (ed) Acipenser. Cemagref, Antony, pp 193–208

Sulak KJ, Clugston JP (1999) Recent advances in life history of Gulf of Mexico sturgeon, *Acipenser oxyrinchus* desotoi, in the Suwannee River, Florida, USA: a synopsis. J Appl Ichthyol 15(4–5):116–128

Tabardel M (1994) Le point sur la situation de l'esturgeon (*Acipenser sturio* L.) en Méditerranée occidentale et possibilités de réintroduction dans le Rhône. Mémoire de fin d'études ENSA, Rennes, Rennes/Arles, 57p

Van Eenenaam JP, Doroshov SI, Moberg GP, Watson JG, Moore DS, Linares J (1996) Reproductive conditions of the Atlantic sturgeon (Acipenser oxyrinchus) in the Hudson River. Estuaries 19(4):769–777

Vibert R (1945) Les poissons migrateurs dans l'économie du sud-ouest. Bull Franc Piscic 136:121–135

Williot P, Brun R (1998) Ovarian development and cycles in cultured Siberian sturgeon, *Acipenser baeri*. Aquat Living Resour 11(2):111–118

Williot P, Rouault T (2008) Basic management for spawning the cultured sturgeon, *Acipenser sturio* L., 1758, a critically endangered species. Cybium 32((2 suppl):334–335

Williot P, Rochard E, Castelnaud G, Rouault T, Brun R, Lepage M, Elie P (1997) Biological characteristics of European Atlantic sturgeon, *Acipenser sturio*, as the basis for a restoration program in France. Environ Biol Fish 48:359–370

Williot P, Rouault T, Brun R, Pelard M, Mercier D (2002) Status of caught wild spawners and propagation of the endangered sturgeon *Acipenser sturio* in France: a synthesis. Inter Rev Hydrobiol 87:515–524

Williot P, Brun R, Rouault T, Pelard M, Mercier D, Ludwig A (2005) Artificial spawning in cultured sterlet sturgeon, *Acipenser ruthenus* L., with special emphasis on hermaphrodites. Aquaculture 246:263–273

Williot P, Rouault T, Pelard M, Mercier D, Jacobs L (2009a) Artificial reproduction of captive endangered European Atlantic sturgeon, *Acipenser sturio*. Endanger Species Res 6:251–257. doi:10.3354/esr00174

Williot P, Rochard E, Kirschbaum F (2009b) Acceptability and prerequisites for the successful introduction of sturgeon species. In: Carmona R, Domezain A, Garcia Gallego M, Hernando-Casal J, Rodriguez F, Ruiz Rejon M (eds) Biology, conservation and sustainable development of sturgeons, Fish & fisheries series. Springer, New York, pp 369–384, 467p

Chapter 13
Sturgeon Fishing, Landings, and Caviar Production During the Twentieth Century in the Garonne Basin and the Coastal Sea

Gérard Castelnaud

Abstract Fishing techniques are described. These generally concern large-sized sturgeons, which could be adults or pre-adults, while small-sized juveniles were also fished at sea and in the estuary. The main fishing net used in the Garonne basin was the drift trammel net. Captures at sea by trawling were also reported. Around 100 fishermen were targeting sturgeon in the Garonne basin in the 1950s. In 1945, there were ten active caviar manufacturing sites. Data on landings and caviar production, which really only appeared after World War I in the Garonne basin, are presented and analysed, in order to assess the evolution of landings and caviar production. According to our estimates, with the intensive fishing of juveniles, landings reached 120 t in 1920, with a production of 3,000 kg of caviar. The sharp decrease after World War II up to 1980 is described.

13.1 Introduction

The sturgeon *Acipenser sturio* is believed to have been abundant in the Middle Ages (Roule 1922), and present until World War II in the main French rivers such as the Seine, the Loire, the Adour, and the Rhône (Magnin 1962). However, the fact that it has been fished as a target species is demonstrated and documented only for the Garonne basin and the coastal sea. Usable data on sturgeon and caviar capture began to be available in the Garonne basin with the launching of the manufacturing of caviar at the very beginning of the 1920s. This paper will focus on sturgeon fishing, fishermen population, sturgeon landings, and caviar production in the Garonne basin (Fig. 13.1) and coastal sea on either side of the estuary mouth during the twentieth century.

G. Castelnaud (✉)
Cemagref, Estuarine Ecosystems and Diadromous Fish Research Unit, 50 Avenue de Verdun, 33612 Cestas Cedex, France
e-mail: gerard.castelnaud@cemagref.fr

P. Williot et al. (eds.), *Biology and Conservation of the European Sturgeon Acipenser sturio L. 1758*, DOI 10.1007/978-3-642-20611-5_13,

13.2 Material and Methods

The area covered in this historical overview of sturgeon fishing is the Garonne basin with the adjoining coastal sea of the Bay of Biscay. The coastal sea extends from the mouth of the Loire River to the Spanish coast and 80 miles out, from the coast to the open sea. The Garonne basin (Fig. 13.1) is made up of the Garonne River and the Dordogne River with their tributaries and their common Gironde estuary up to the sea limit (Fig. 13.2). The Garonne River and the Dordogne River end at the salt limit, and the Gironde estuary comprises the area between the salt limit and the sea limit. The estuary has two banks, the left Médoc bank and the right Saintonge bank.

The Beauregard dam near Agen on the Garonne River and the Bergerac dam on the Dordogne River determine respectively the low part of the Garonne River and the low part of the Dordogne River, which, with the estuary, constitute the low part of the Garonne basin (Fig. 13.1). The tidal limit at Casseuil on the Garonne River and the tidal limit at Castillon-La-Bataille on the Dordogne River determine the tidal part of the Garonne basin, extending to the sea limit (Fig. 13.2).

From these definitions, we will use the following terminology: Garonne basin, Garonne River, Dordogne River, estuary, Médoc bank, Saintonge bank; low Garonne basin, low Garonne River, low Dordogne River; tidal Garonne basin. We will refer to the Garonne basin when the literature does not specify the location of data on the basin.

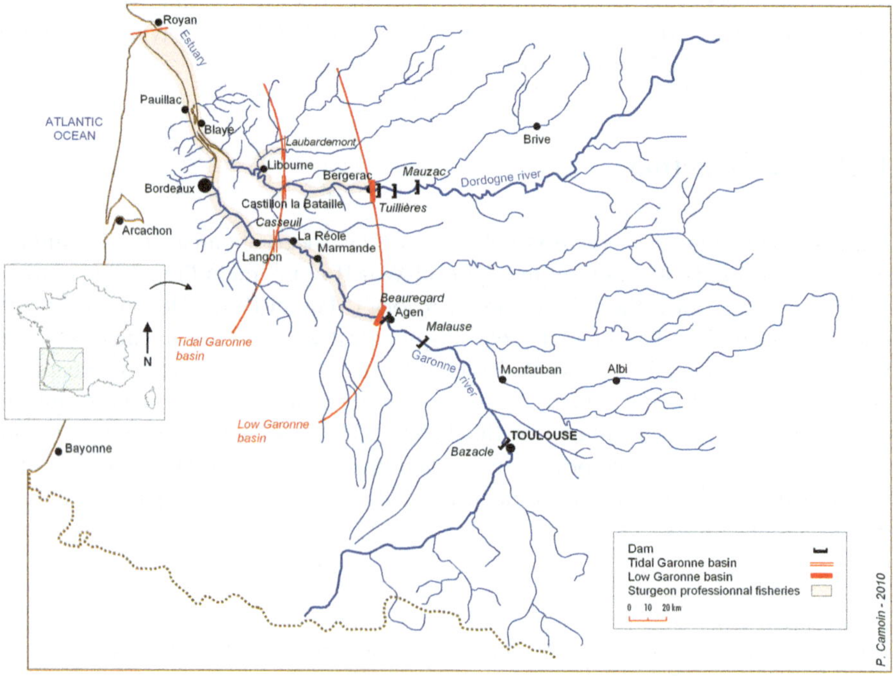

Fig. 13.1 Map of the Garonne basin with dams, limits of the Tidal Garonne basin and of the Low Garonne basin, with sturgeon professional fisheries area

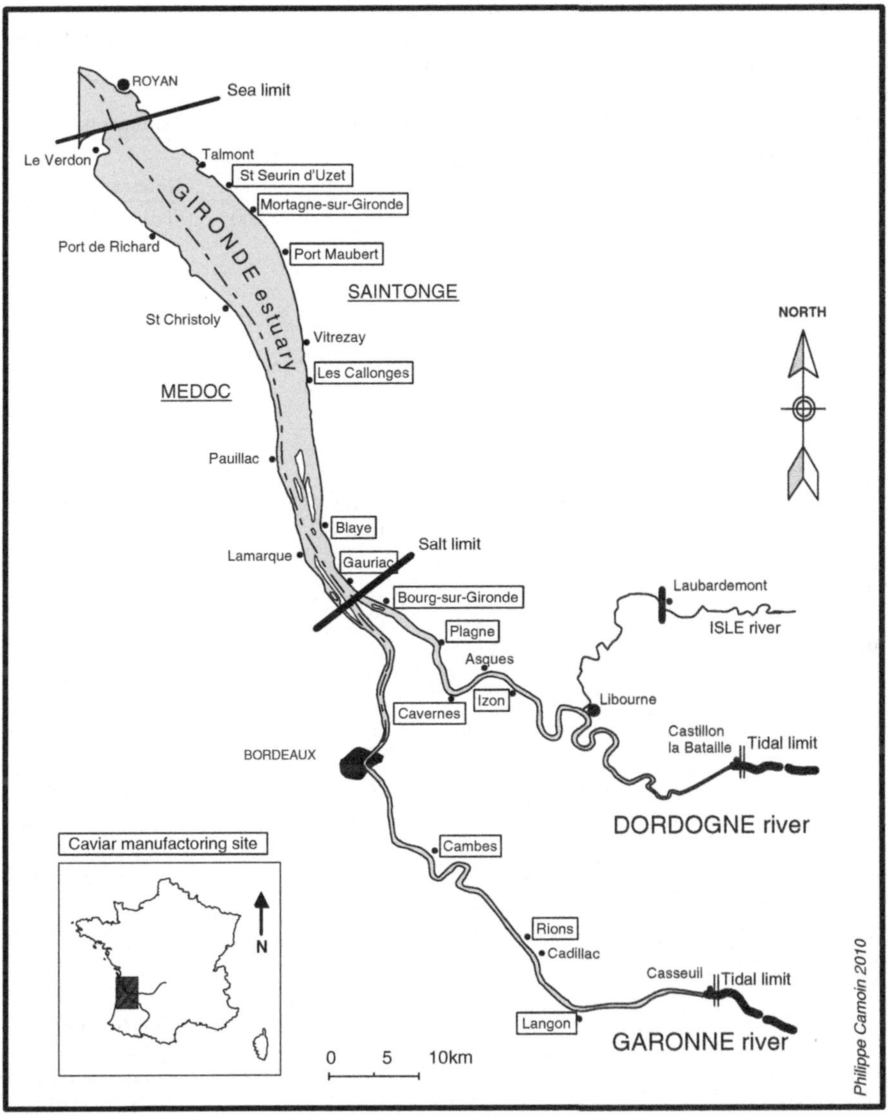

Fig. 13.2 Map of the Tidal Garonne basin with its limits and the caviar manufacturing sites

This overview of sturgeon fishing covers the compilation and analysis of the available literature, which comprises technical and scientific papers, historical surveys and stories, and fisheries management reports. The global figures for landings and caviar production for the period 1920–1980 in the Garonne basin are given (1) in the official statistical records from the Marine Marchande Administration, used by Roule (1922), CRDP (1977) and Fournet (1986), and (2) in personal reports and letters from caviar manufacturers. These global figures with the references are reported in Table 13.1.

Table 13.1 Landings (t) from 1920 to 1980 from the different origins: Sea Census, Basin Census, Basin Fishermen, Garonne basin with caviar production (kg) from Basin fishermen and Garonne basin (references associated to Garonne basin landings and caviar)

Fishing area	Sea Census	Basin Census	Basin Fishermen	Basin Fishermen	Garonne basin	Garonne basin	Garonne basin
Year	Total landings	Total landings	Total landings	Total caviar	Total captures	Total caviar	References
1920					50	3,000	Tribondeau (1983)
1921							
1922							
1923					32–37/ 40–45	1,200	Lorin de Reure (1924)/Scott (1948)
1936	8.25						
1937							
1938					20	1,918	Scott (1948)
1939							
1940					12–15		Scott (1945)
1945					12–15		Scott (1945)
1946							
1947		3.75					
1948	3.6	3.66	4.35	110	12	1,000	Scott (1948)
1949	2.37	4.74	12	1,015.5			
1950	1.17	5.7	4.5	388	50/9/19	3,000	Prioux (1957)/ CRDP (1977)/ Fournet (1986)
1951	0.66	8.1	13.95	1,055.5	23		Fournet (1986)
1952	0.24	5.88	12.45	258.5	20		Fournet (1986)
1953	0.21	9.48	6.3	166.5	23.5		Fournet (1986)
1954	0.3		8.4	715.5	7/20		CRDP (1977)/ Fournet (1986)
1955	0.48		6.6	922			
1956	0.42		3.75	450			
1957			2.85	252			
1958	0.06		13.8	1,155	12		CRDP (1977)
1959	0.06		6.15	307	8	1,375	Clément (1960, personal communication)
1960	0.06		13.8	902.5	8		CRDP (1977)
1961			10.35	654			
1962			6	363.5	8		CRDP (1977)
1963		5.85	1.95	345	6	250	Castelnaud et al. (1985)
1964		5.49	3.15	253	3.7		CRDP (1977)
1965		5.13	1.65	200			
1966		3.48	1.05	48.5	0.73		CRDP (1977)
1967		5.01	2.7	290.5			
1968	0.03	3.21	2.55	50	1		CRDP (1977)
1969	0.06	1.08	3	228.5			
1970	0.03	3.18	2.7	142.5	0.5		CRDP (1977)

(continued)

Table 13.1 (continued)

Fishing area	Sea Census	Basin Census	Basin Fishermen	Basin Fishermen	Garonne basin	Garonne basin	Garonne basin
Year	Total landings	Total landings	Total landings	Total caviar	Total captures	Total caviar	References
1971	0.12	1.98	1.05	101.5			
1972	0.03	1.32	1.8	159.5	1.2		CRDP (1977)
1973	0.06	1.41	1.65	95	1–2		CTGREF (1973)
1974	0.27	1.02	3.15	208.5	0.65		CRDP (1977)
1975	0.15	0.6	0.45	15			
1976	0.09	0.18	0.3				
1977		0.48					
1978		0.48					
1979	0.03	0.09					
1980		0.36			0.4	25	Castelnaud et al. (1985)

Table 13.2 Captures (number) of sturgeon juveniles, males, pre-adult females and adults for 1936 and from 1948 to 1960 in the port of La Rochelle (France) by Letaconnoux (1961)

Captures	Juveniles	Male or pre-adult female	Adults	Total captures
Year	1–1.45 m	1.45–1.85 m	>1.85 m	La Rochelle
1936				275
1948	101	19	0	120
1949	66	10	3	79
1950	30	7	2	39
1951	13	6	3	22
1952	2	5	1	8
1953	1	4	2	7
1954	0	7	3	10
1955	3	10	3	16
1956	1	5	8	14
1957				
1958				2
1959				2
1960				2

The only captures at sea for which data have been collected were for 1936 and for the subsequent period 1948–1960 in the port of La Rochelle (Table 13.2) (Letaconnoux 1961). This data series has been completed (Autissier and Décamps 1980 personal communication) by the captures at sea (not dissociated) recorded at La Rochelle from 1968 to 1979 and at Arcachon from 1973 to 1976. The weight of these captures per year has been calculated at a mean weight of 0.03 t per individual, using the data available from Scott (1945) and Lorin de Reure (1924) and from fish caught by Cemagref. This result is called "Sea Census landings" in the text, and reported in Table 13.1 and Fig. 13.3.

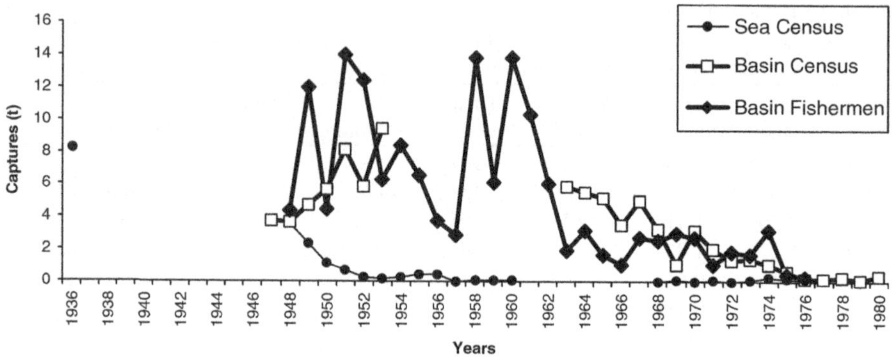

Fig. 13.3 Superimposing 1936 to 1980 total landings (t) from the different origins: Sea Census, Basin Census and Basin Fishermen

A detailed but incomplete data series on captures of males and females in the low Garonne basin has been established for the most part by Vignaud (1979) as water bailiff for the period 1947–1978, and by Castelnaud et al. (1985) for the years 1979 and 1980 (Fig. 13.5). Additional information has also been reported by Pustelnik (1978, personal communication) concerning captures in the low Dordogne River close to the Bergerac dam (Fig. 13.1) for 1940 and for the period 1946–1973. The weight of total captures for the low Garonne basin from Vignaud (1979) and from Castelnaud et al. (1985) has been calculated, with a mean weight of 0.03 t per individual. This result is called "Basin Census landings" in the text, and reported in Table 13.1 and Fig. 13.3.

Another detailed data series on sturgeon captures and caviar production covering the period 1948–1975 was recorded from eight personal fishing books belonging to old fishermen on the low Garonne basin. Each book covers a sample from one or several fishermen in different fishing ports. They are represented in the text and in Table 13.3 by the name of the port.

The sample of fishermen whose captures are recorded in seven of the fishing books is composed, depending on the year, of a minimum of eight fishermen and a maximum of 18, and represents 1/5 of the evaluated sturgeon fishermen population in the (low) Garonne basin in the 1950s (100 fishermen, see Sect. 13.3.2). To obtain an estimate (certainly minimized) of captures in weight per year for all fishermen in the (low) Garonne basin, we have to multiply the number of captures in the sample by 5, giving a mean weight of 0.03 t per individual. The result is called "Basin Fishermen landings" in the text, and reported in Table 13.1 and Fig. 13.3.

The sample of fishermen concerned with caviar production corresponds to six fishing books, and is composed, depending on the year, of a minimum of seven fishermen and a maximum of 17. The same operation was carried out to obtain an estimated global production of caviar for the Garonne basin per year (also

Table 13.3 Production for the period 1948 to 1975 of caviar (kg) with the number of associated females (*F*), per fishing port representing fishermen and their fishing books, in the low Garonne basin

Fishing Port	Vitrezay		Les Callonges	Blaye		Gauriac		Izon		Libourne		Low Garonne basin
Year	Caviar	*F*	Caviar	Caviar	*F*	Caviar	*F*	Caviar	*F*	Caviar	*F*	Total caviar
1948				22	4							22
1949				115	20			88.1	9			203.1
1950				28	5			49.6	7			77.6
1951				150	30			61.1	9			211.1
1952				38	8			13.7	2			51.7
1953								33.3	6			33.3
1954						120	16	23.1	4			143.1
1955						100	15	84.4	9			184.4
1956						90	13					90
1957						26.5	5	23.9	5			50.4
1958				107	23	71.5	15	52.5	8			231
1959				28	7	15	7	18.4	4		1	61.4
1960				110	26	16	3	32.6	5	21.9	3	180.5
1961			43			74	12	13.8	3			130.8
1962			14.6			37.5	7	20.6	4			72.7
1963			20			49	8					69
1964			19.6			31	6					50.6
1965			29			11	3					40
1966			9.7									9.7
1967			20.6			37.5	7					58.1
1968	10	2										10
1969	14.4	2	8.5	16.5	4			6.3	1			45.7
1970			16.5	8	1	4	1					28.5
1971			9.1	8	2			3.2	1			20.3
1972			21.9	10	2							31.9
1973			6	13	3							19
1974				34	8					7.7	1	41.7
1975				3	1							3

certainly minimized). The result is called "Basin Fishermen caviar" in the text, and is reported in Table 13.1 and Fig. 13.3.

The data gathered in Table 13.1 has been analysed and compared in order to build up the most representative and realistic time series of sturgeon landings and caviar production for the period 1920–1980, based on some key years for which data are available and usable (Fig. 13.4). For the landings, we have combined sea sturgeon landings, Garonne basin sturgeon landings and juvenile sturgeon landings. The juveniles are separated as a specific fishing target in the estuary, even though some are included in captures at sea and in captures in the Garonne basin.

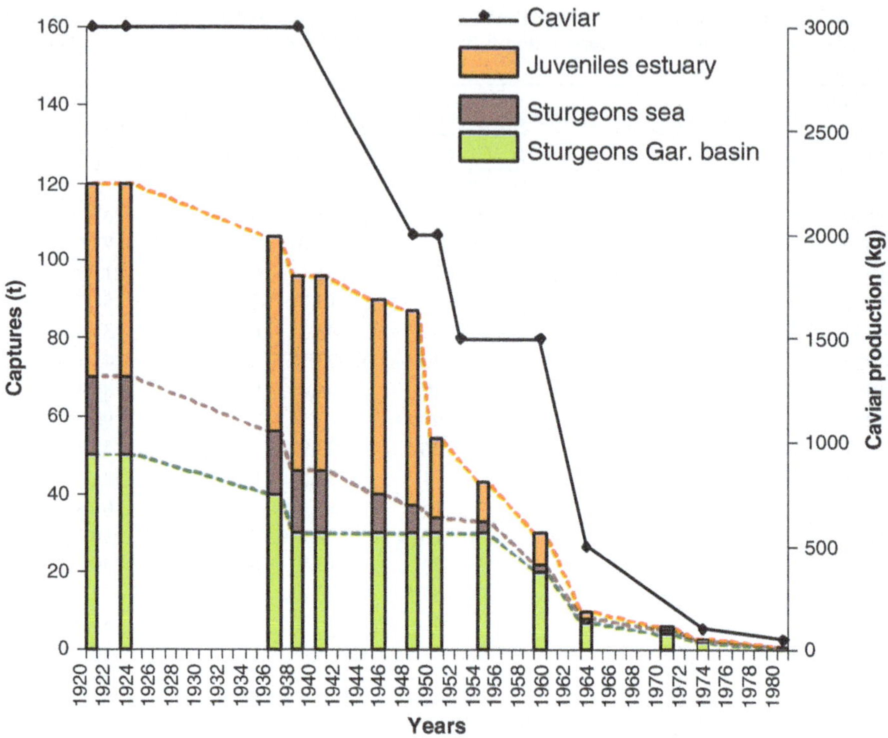

Fig. 13.4 Reference estimates for the period 1920–1980 of (1) landings of sturgeons (t) separated into sturgeons from sea, sturgeons from the Garonne basin and capture of juveniles in the estuary, (2) caviar production (kg)

13.3 Results

13.3.1 Fishing Techniques, Fish Stages, Periods, and Associated Fishing Areas

Although the legal size of sturgeon has varied over time, the fishing techniques described below generally concern sturgeons that are large in size, which may be adults or pre-adults. Apparently, the fishing of small juveniles has occurred at sea and in the estuary at least since the beginning of the twentieth century, and was described by Roule (1922) and Scott (1948). According to Scott (1948), the juveniles were caught in the estuary in spring and summer with drift trammel nets of appropriate mesh size (nets for *Platichthys flesus*, *Liza ramada*) and with fixed gill nets called "courtines". According to the data he gathered (Table 13.2), Letaconnoux (1961) indicated that captures at sea by trawling occurred from October to March, within a territory of 40–80 miles around the mouth of the

Garonne basin estuary, including the ports of La Rochelle and Arcachon. The largest fish caught was 3.25 m long, and the majority of the captures were between 1 and 3.25 m in length. Earlier, the fishing of young sturgeons by sailing trawlers was reported by Roule (1922) and Lorin de Reure (1924), who located this fishing between the mouth of the Loire River and the Spanish coast.

According to Le Masson du Parc (1727), sturgeon fishing in the Garonne basin began in February and ended in August or September. During the twentieth century, after World War I, there was sturgeon fishing in the Garonne basin between February and June (Scott 1945; Tribondeau 1952, 1983). More precisely, Scott (1945) indicates that during the full season, which extended from 15 April to 30 June, 90% of the females were caught. Nevertheless, it has to be noted that it was possible to continue sturgeon fishing legally in summer, according to the regulations until 1939 in the Garonne and Dordogne rivers and until 1950 in the estuary (Trouvery et al. 1984; Castelnaud et al. 1985). Some of the techniques used for sturgeon fishing have been described by Le Masson du Parc (1727), and remained valid until the end of sturgeon-targeted fishing in the 1970s. The main fishing net used in the Garonne basin (Fig. 13.1) was the drift trammel net, which runs with the current and entangles the moving fish. This net is called a "créaquière" after the local name for the sturgeon which is "créac". The stretched mesh size of the central panel was 0.24 and 0.62 m for the two outside panels; the total length of the net varied from 100 to 160 m, and the height was 4 m (Scott 1936; Tribondeau 1983; Castelnaud et al. 1985). Sturgeons were also captured as a by-catch during shad fishing (*Alosa alosa*) with the specific drift trammel net called the "bichareyre", with a stretched mesh size of the central panel of 0.11 m (Lorin de Reure 1924).

The drift trammel net was operated by the fishermen from a typical small boat of 6 m long, the "filadière", equipped with sails and oars (Tribondeau 1952, 1983). At the beginning of the twentieth century, after World War I, a new boat, the "yole" appears (CRDP 1988; Tribondeau and Sylvius 1999; Calero 2006), of similar size, and as it was gradually equipped with an engine (Tribondeau 1983), it became dominant. This increased the fishing power by allowing the fishermen to follow the fish as they migrated, into the tidal Garonne basin (Castelnaud et al. 1985).

According to Le Masson du Parc (1727), the trammel net was also operated fixed to the bank of the Garonne and Dordogne rivers. Besides the trammel net, another important net was widely used, the seine, called the "garolle" on the Médoc bank, the "escave" on the Dordogne River and the "tresson" on the Garonne River (Castelnaud et al. 1985; Calero 2006). The stretched mesh size was 0.06–0.08 m, and the length varied from 50 to 190 m according to the location and the fishing method. On the Médoc bank, the seine was hauled by two boats or from the bank by five to six fishermen. This second method was used in the rivers, also by five to six fishermen on the low Dordogne River and as many as ten on the low Garonne River. At the end of the nineteenth century, the castan appeared, which simplified the operation (Castelnaud et al. 1985).

The cast net was also used in the lower Garonne basin upstream of the tidal limits of the rivers (Cassou-Leins 1981; Vital 1984; Calero 2006), and anyone with a skilful technique was able to catch males and females on the spawning grounds.

The general structure of the different nets, weighted with lead and with cork for floatation, did not change throughout the period of sturgeon commercial fishing. Scott (1945) indicates that from 1939 until 1945 at least, some caviar manufacturers provided hemp thread, leads and corks for their associated fishermen. Hemp and flax threads were gradually replaced after Word War II by nylon threads, first monofilament and later multifilament (CRDP 1988; Tribondeau and Sylvius 1999; Calero 2006). Because in the past nets had to be hung out to dry in order to prevent rotting, fishing possibilities were limited (Castelnaud et al. 1985). This innovation made it possible to increase the number of hauls, and to really take advantage of the motorisation of the boat to follow the fish along its migration course.

After 1920, when the roes of the females began to be processed as caviar, the fishing of females was certainly intensified, and the number of fishermen targeting sturgeons and the number of fishing zones being searched increased. Thus, the fishing strategies and tactics changed, but not the fishing gear and techniques.

13.3.2 Fishermen Population and Caviar Manufacturing Sites

In his survey, Laborie in Le Masson du Parc (1727) counted 285 boats in 1727 on the tidal Garonne basin, and on this basis Castelnaud et al. (1985) proposed the possible number of 600–800 fishermen in all on the Garonne basin. Castelnaud et al. (1985) also counted a total of 1,000 fishermen in 1830 for the Garonne basin but indicated that at this period, fishing was a minor activity which developed at the end of the nineteenth century until World War I. This number during the period between World Wars I and II is not confirmed by the data gathered by Castelnaud et al. (1985), which remain incomplete. In any case, only some of the fishermen (although numbers certainly increased during this period as explained above in Sect. 13.3.1) were fishing sturgeon, as is demonstrated next. After World War II, in 1949, the Inscription Maritime of Bordeaux counted around 200 boats in the estuary, while in 1952 Beaulaton (2008) estimated a total of 421 professional fishermen operating in the tidal Garonne basin. However, at this period, apparently no more than 40 fishermen were strongly involved in sturgeon fishing, according to the detailed information given by Barnagaud (1952, personal communication) in connection with the caviar production sites in the same fishing area. This number is slightly increased according to the report from Vignaud (1979), which for 1952 listed 41 sturgeon fishermen on the tidal Dordogne river and 17 for the estuary. As the Garonne River is not considered in this report, and the number of fishermen in the estuary is under-evaluated, it can be estimated that in the 1950s, around 100 fishermen were targeting sturgeon in the Garonne basin, which means in the low Garonne basin (Fig. 13.1), with the majority fishing in the tidal Garonne basin (Fig. 13.2). Numbers of specialised fishermen and also occasional fishermen gradually decreased as the sturgeon stocks declined, until the end of the targeted sturgeon fishery in the early 1970s.

The manufacturing of caviar began to be really organised at the very beginning of the 1920s (Scott 1948) with the Prunier Caviar House, which remained the largest on the Garonne basin. In 1923, ten caviar manufacturing sites were created in the tidal Garonne basin in association with fishermen, and sometimes with the setting-up of a small factory (Scott 1945; Lorin de Reure 1924): St Seurin d'Uzet, Port Maubert, Les Callonges, and Blaye on the Saintonge bank, Plagne and Cavernes on the Dordogne River, and Cambes and Rions on the Garonne River (Fig. 13.2). Some fishermen had created their own artisanal factory, as in Saint-Seurin d'Uzet (Val 2006).

Three other caviar houses were set up progressively, while other fishermen created their own artisanal factory, at least two of which were registered in Table 13.3 as located in Blaye and Izon between 1948 and 1962. We can infer from the detailed information given by Scott (1945) that none or very few of the other caviar houses had a factory. In 1945, ten caviar manufacturing sites were active for the Prunier Caviar House (Scott 1945): Saint-Seurin d'Uzet, Mortagne-sur-Gironde, Port Maubert, Les Callonges, and Gauriac on the Saintonge bank, Bourg-sur-Gironde and Plagne on the Dordogne River, and Cambes, Rions, and Langon on the Garonne River (Fig. 13.2). It should be noted that no manufacturing site is indicated on the Médoc bank, whereas we can confirm from the personal fishing books collected that there were specialised sturgeon fishermen on this side. It is likely that at least some fishermen were making caviar.

13.3.3 Analysis of Data on Landings of Sturgeons and Caviar Production

Some historical sources cited by Scott (1936), Tribondeau (1952), Tribondeau (1983), Vital (1984), and Fournet (1986) have demonstrated the existence of sturgeon fishing in the past since the Middle Ages and earlier, but no figures are given for production; at the very most, some individual captures with weights are available. Le Masson du Parc (1727), describing sturgeon fishing in the Garonne basin in his time, did not report any data on landings. Roule (1922) protested about the destruction of juveniles at sea in 1905, and Fournet (1986) talked about an increase in sturgeon captures which led in 1910 to more than 20 t for the Saintonge bank. Some attempts at processing caviar with the roes from the females were made around 1890 and continued during this period, but the result was of poor quality (Lorin de Reure 1924; Scott 1945; Val 2006). The data on fish and caviar production appears really after World War I, as indicated above in Sect. 13.3.2, in conjunction with the beginning of the manufacturing of true caviar with appropriate methods and organisation.

As specified by Tribondeau (1983), the first data collected after World War I on landings of fish (50 t) and caviar production (3,000 kg) concerns 1920 (Table 13.1). Roule (1922) reported the official production for one of the more productive ports

Table 13.4 Detailed captures of males and females in number and weight (kg) and caviar production (kg) in 1923 for several fishing ports in the low Garonne Basin reported by Lorin de Reure (1924)

Product	Male		Female		Caviar
Fishing harbour	Number	Weight	Number	Weight	Weight
Garonne river					
Rions	29	930	16	913	145
Cambes	20	599	14	799	85
Dordogne river					
Plagne	3	83	5	250	31
Cavernes	41	753	12	568	83
Estuary					
Mortagne-sur-G.	85	1,350	41	2,650	352
Total	178	3,715	88	5,180	696

on the Garonne river: 4.16 t in 1920, 4 t in 1921, and 6.8 t in 1922. Because of the numerous ports specialising in sturgeon fishing (see above in Sect. 13.3.2), these figures at least support the global production figures given for 1920.

For 1923, Lorin de Reure (1924), on the basis of his collection of detailed data from fishing ports reported in Table 13.4 for captures of males and females and roes converted into caviar, suggested a total landing of 12 t and a production of 1,200 kg of caviar. Lorin de Reure (1924) added the landings of small-sized sturgeons from coastal sea fishing, estimated at 20–25 t, giving a total of 32–37 t (Table 13.1). For the same year, Scott (1948) indicated a total of 40–45 t of fish. The other data from Scott (1945, 1948) cover several years from 1938 until 1948, and show a decrease for captures and caviar. Scott (1948) underlined the fact that around 50–60 t of small juveniles were destroyed at this time in the lower part of the estuary (see Sect. 13.3.1). The official data reported by CRDP (1977), and the three evaluations in 1963 and 1980 by Castelnaud et al. (1985) and in 1973 by CTGREF (1973), show a continued decrease from 1950 to 1974 (Table 13.1). However, the other official data reported by Fournet (1986), of the same origin as CRDP (1977) but from a special survey, gave a total capture for 1950–1954 of more than 20 t, at the same level as 1938. Moreover, Fournet (1986) supposed that these figures were underestimated, and that the true production was around 40 t. This can be related to the estimates of Prioux (1957) for 1950: 50 t of landings and 3,000 kg of caviar, going back to the level of 1920!

The captures at sea collected (Letaconnoux 1961) were divided into three groups: juveniles, immature males or females, and adult males and females (Table 13.2). Letaconnoux (1961) specified that sturgeon was not a fishing target at sea, and that the landings were composed of a majority of juveniles, but also of large adult fishes. This data series for 1936 and for the following period 1948–1960 indicates a strong decline in the number of captures. This is clearly confirmed by the data series for 1968–1979 (Autissier and Décamps 1980 personal communication), and is shown over a period of 40 years by the "Sea Census landings", reported in Table 13.1 and Fig. 13.3.

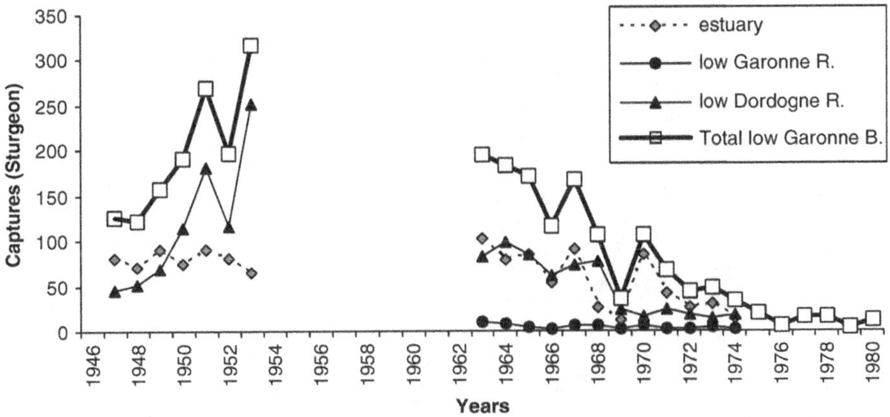

Fig. 13.5 Captures (number) from 1947 to 1980 for the estuary, the low Garonne River, the low Dordogne River and for the low Garonne basin by Vignaud (1979) and Castelnaud et al. (1985)

The captures recorded by Vignaud (1979) and by Castelnaud et al. (1985) for the period 1947–1978 are reported in Fig. 13.5, separated into estuary, low Garonne River, and low Dordogne River, with a total for the low Garonne basin. In the first period, 1947–1953, the trend for the total captures for the lower Garonne basin is determined by the captures in the low Dordogne River, while in the second period 1963–1974, it is determined by captures both in the estuary and in the low Dordogne River. The total captures for the low Garonne basin decreased at the end of the 1960s, as shown in Fig. 13.5 and by the "Basin Census landings" in Table 13.1 and Fig. 13.3.

The data from Letaconnoux (1961) in Table 13.2, and the data from Vignaud (1979) and Castelnaud et al. (1985) in Fig. 13.5, are more precise than the global data from various authors listed in Table 13.1, but their reliability and representativeness is no more guaranteed, due to the recognised difficulty in investigating fishermen and their captures (Castelnaud and Beaulaton 2008). The data for the sample of fishermen recorded in the eight personal fishing books, according to method of collection, although discontinuous, can clearly be supposed to be of good quality and reliable.

The captures obtained from seven fishing books corresponding to seven fishing ports concern both sexes, the number of females being sometimes equal to or higher than the number of males. The total captures for the two sexes grouped by year for the estuary, for the low Dordogne River, and for the low Garonne basin are reported in Fig. 13.6. The captures in the estuary predominate, and determine the trend of the total captures for the low Garonne basin. It is of interest to note that this trend is consistent with the trend for total captures from Vignaud (1979) for the low Garonne basin (Fig. 13.5) during the period 1947–1953, which is driven by the capture of the low Dordogne River as described above. According to the trend of the curve for total captures for the Garonne basin ports during the period 1954–1962, we can suppose that for Vignaud (1979) the curve for total captures in the Garonne basin remains stable, at least at the 1953 level, during this period

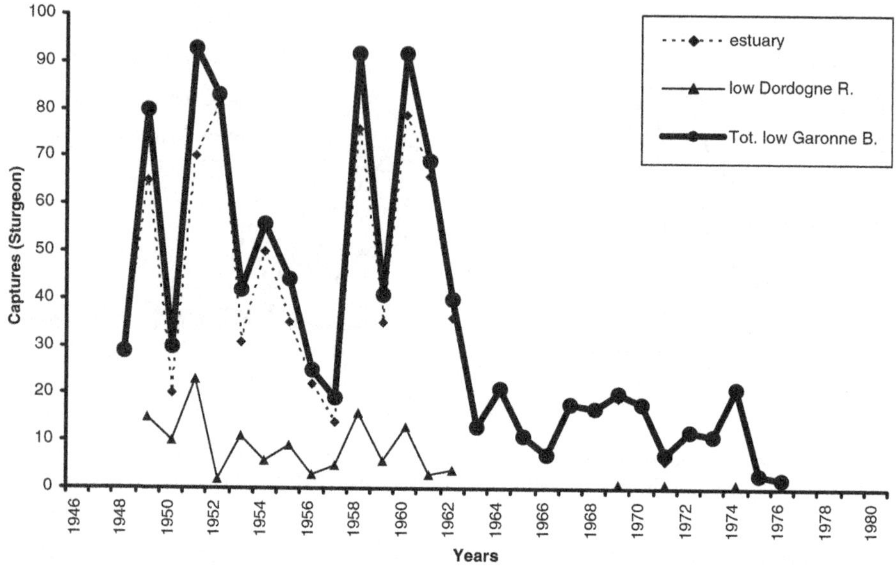

Fig. 13.6 Captures (number) from 1948 to 1976 per fishing port representing fishermen and their fishing books, for the estuary, the low Dordogne River and total for the low Garonne basin

when the data are missing. This is shown in Fig. 13.3 by superimposing "Basin Fishermen landings" (Table 13.1) and "Basin Census landings". The result is more a bell curve than a steady curve between 1954 and 1962. The trend between 1948 and 1975 for total landings in the low Garonne basin from the two different origins is not reflected by the strongly decreasing trend of the "Sea Census".

Caviar production in kg, from the six fishing books that correspond to six fishing ports, is given per year for the period 1948–1975 in Table 13.3. The number of females associated with the caviar production enables us to calculate that in this sample 5.6 kg of caviar was obtained per female, and that caviar represented 11% of the female's weight (9.5% according to Scott 1948). From the estimate for total caviar production in the Garonne basin per year obtained with "Basin Fishermen caviar" (Table 13.1), with a result of 1,015.5 kg in 1949 and 1,055.5 kg in 1951, it is possible to confirm the figure of 1,000 kg indicated in 1948 by Scott (1948) (Table 13.1).

13.3.4 Attempt to Estimate Total Landings and Caviar Production 1920–1980

In this section we present reference estimates of sturgeon landings and caviar production based on some key years, for the period 1920–1980 (Fig. 13.4).

For the years 1920 and 1923, according to the "Sea Census landings" (Table 13.1) and the data detailed in Sect. 13.3.3 from the different authors cited in Table 13.1, we add 50 t of sturgeon landings from the Garonne basin, 50 t of juveniles from the estuary and 20 t of sturgeon from sea, which gives 120 t (Fig. 13.4). In 1936, 1938, and 1940, we maintain that juvenile captures remained at the same level, while captures at sea decreased and captures in the Garonne basin decreased also, according to Scott (1948), so from 120 t in 1920, totals decreased to 106 t in 1936, and then to 96 t in the period 1936 to 1940.

In 1945, after World War II, and until 1950, the Garonne basin sturgeon landings remained at 30 t, considering that part of the global evaluation of 50 t by Prioux (1957) is included in the capture of juveniles taken into account, and it stayed at 50 t until 1948, then decreased to 20 t in 1950. From 1948, the data from "Sea Census landings" at sea, from "Basin Census landings", and from "Basin Fishermen landings" in the Garonne basin all become available (Table 13.1). The estimate of sea sturgeon landings from 1945 to 1950 decreased sharply. Total landings ranged from 90 t in 1945 to 87 t in 1948, and then to 54 t in 1950. Following this year, the majority of data are official, and we have based the estimates mainly on data from "Sea Census landings", from "Basin Census landings" and from "Basin Fishermen landings" in the Garonne basin.

Between 1954 and 1963, landings at sea and the fishing of juveniles continued to decrease, while the Garonne basin landings dropped from 30 to 7 t. Total landings decreased from 43 t in 1954 to 30 t in 1959 and then to 10 t in 1963. In 1970, total landings were evaluated at 6 t and in 1973 at 3 t, of which 2 t came from the Garonne basin (1–2 t for CTGREF 1973); 0.5 t corresponds to juveniles, and 0.5 t to captures at sea. In 1980, we counted total landings of 1 t (0.4 t for Castelnaud et al. 1985).

For caviar production in the Garonne basin, we refer to the global figure of 3,000 kg from Tribondeau (1983) in 1920. This production was maintained from 1923 until 1938 (Fig. 13.4), according to the partial productions announced by Scott (1945, 1948) with details and comparison with production by other manufacturers. During World War II, sturgeon fishing continued, but the production of caviar decreased and reached 2,000 kg in 1948. Because of the clear statement by Scott (1948) concerning the situation of the sturgeon fishery and the manufacturing of caviar, we maintain this figure in 1950, even if Prioux (1957) and Fournet (1986) give an estimate of 3,000 kg. In the period 1952 to 1959, according to the quantities indicated confidentially by Barnagaud (1952 personal communication) and officially by Clément (1960 personal communication), from the "Revue Economique de la Gironde", we estimate production to be 1,500 kg. In combination with the evaluations from CTGREF (1973) and Castelnaud et al. (1985), our estimate of caviar production is 500 kg in 1963, 100 kg in 1973, and possibly 50 kg in 1980.

13.4 Conclusion

It becomes clear, when analysing historical documents, even from twentieth century authors, that some reports on the fishing situation, on the status of the stocks, are far removed from reality, and are very different from the results of investigations carried out in the field (along the river banks, in contact with the fishermen, with their practices, their captures). Nevertheless, concerning the most active and consequently the most documented period of sturgeon fishing in the Garonne basin, i.e., 1920–1980, it has been estimated that around 100 fishermen were specialised in this fishing, which concerned adult males and females for flesh and caviar, and juveniles for flesh. Sturgeons were fished at sea at the same stages, but not as a systematic target. With the intensive fishing of juveniles, landings reached 120 t in 1920, with a production of 3,000 kg of caviar until 1938 which then dropped to 2,000 kg after the end of World War II. During the same period, landings decreased to 87 t in 1948, then fell rapidly to 54 t in 1950, 30 t in 1959, 10 t in 1963, and to no more than 1 t in 1980. Logically, caviar production followed a similar trend, but remained apparently at 1,500 kg between 1952 and 1959, decreased to 500 kg in 1963 and fell to no more than 50 kg in 1980. Then, quickly, the species became protected in the Garonne basin and fishery was forbidden. That was the end of a glorious story of fishing for this European species of sturgeon that laid "golden eggs", with its myths and secrets, and which for some fishermen was life-changing.

Acknowledgements To Géraldine Lassalle and Christian Rigaud for their help with the figures, to Patrick Williot for his advice on the manuscript

References

Beaulaton L (2008) Analyse comparative et fiabilité des indicateurs halieutiques obtenus sur les espèces amphihalines et les crustacés dans le bassin de la Gironde. Ph.D., University Paul Sabatier, Toulouse

Calero C (2006) La fièvre de l'or noir. Série illustres ancêtres. Le Pêcheur Professionnel 57:8–11

Cassou-Leins F (1981) Recherches sur la biologie et l'halieutique des migrateurs de la Garonne et principalement de l'alose: Alosa alosa L. Doctoral dissertation, Institut National Polytechnique de Toulouse, Toulouse

Castelnaud G, Beaulaton L (2008) Indicateurs d'abondance et de pression issus des pêcheries. In: Adam G, Feunteun E, Prouzet P, Rigaud C (eds) L'anguille Européenne: indicateurs d'abondance et de colonisation. Savoir faire. QUAE, Versailles, pp 189–220

Castelnaud G, Coutancier B, Cerezuelle D, Guchan A (1985) La pêche des migrateurs en Gironde: part.1 Analyse historique du dix-huitième siècle à nos jours: bilan et perspectives. Cemagref, Bordeaux

CRDP (1977) La pêche à l'esturgeon et la fabrication du caviar dans l'estuaire de la Gironde, vol 21.10. Documents d'Aquitaine. Centre Régional de Documentation Pédagogique, Bordeaux

CRDP (1988) L'esturgeon espèce protégée. Centre Régional de Documentation Pédagogique, Bordeaux

CTGREF (1973) Rapport sur la pêche en Gironde. CTGREF, Bordeaux

Fournet P (1986) La pêche du caviar et la fabrication du caviar en Gironde 1920–1980. Pêche Maritime 1294:45–48

Le Masson du Parc F (1727) Procès verbaux des visites faites par ordre du Roy concernant la pesche en mer (1727). Pêches et Pêcheurs du domaine maritime aquitain au XVIIIe siècle. Amirautés de Bayonne & de Bordeaux. Les Editions de l'Entre-deux-Mers, Camiac et Saint-Denis

Letaconnoux R (1961) Note sur la fréquence de la distribution des captures d'esturgeons (*Acipenser sturio* L.) dans le Golfe de Gascogne. Revue des Travaux de l' Institut des Pêches Maritimes 25:253–261
Lorin de Reure H (1924) Note sur la pêche de l'esturgeon. Bulletin Trimestriel de l'Enseignement professionnel Technique des Pêches Maritimes 29(1):9–16
Magnin E (1962) Recherches sur la systématique et la biologie des Acipenséridés. Annales de la Station Centrale d'Hydrobiologie Appliquée 9:7–242
Prioux MG (1957) L'esturgeon en France. Rivières et Fôrets 6:39–44
Roule L (1922) Etude sur l'esturgeon du Golfe de Gascogne et du bassin girondin
Scott A (1936) Etude sur les esturgeons et le caviar français. Maison Prunier, Paris
Scott A (1945) Pêche à l'esturgeon et préparation du caviar en 1945. Maison Prunier, Paris
Scott A (1948) La pêche à l'esturgeon en Gironde. Maison Prunier, Paris
Tribondeau J (1952) Caviar en Gironde. Pétrole et Progrés 4:45–49
Tribondeau J (1983) Le caviar des esturgeons de Gironde. Sciences et Avenir, pp 497–501
Tribondeau J, Sylvius A (1999) Pêches traditionnelles des rives Saintongeaises de la Gironde. Confluences edn., Société des amis de Talmont, Talmont
Trouvery M, Williot P, Castelnaud G (1984) Biologie et écologie d'*Acipenser sturio*: étude la pêcherie. Cemagref, Bordeaux
Val R (2006) La véritable histoire du caviar de Gironde. Bonne Anse edn, Société des amis de Talmont, Talmont
Vignaud E (1979) Liste des frayères; captures d'esturgeons en Gironde. DDA de Gironde, Bordeaux
Vital P (1984) Requiem pour une Garonne défunte. Wallada edn, Bordeaux

Chapter 14
Historic Overview on the Status of the European Sturgeon (*Acipenser sturio*) and Its Fishery in the North Sea and Its Tributaries with a Focus on German Waters

J. Gessner, S. Spratte, and F. Kirschbaum

Abstract The historic range and status of *A. sturio* in the tributaries to the North Sea, its past utilization and its population development is presented to the extent that it can be derived from available fisheries data. The alteration in fishing techniques and fishing sites, from a strictly riverine fishery to coastal fishery average in the 1880s and subsequently, the increased marine catch is illustrated. Its effects on fish size as well as on the population are documented. While the fishery initially targeted the fish for meat over several hundred years, caviar became an additional driver to the fishery after 1840 in Germany. The endpoint of a targeted fishery occurred in the early twentieth century for the majority of rivers. River constructions to improve navigation, pollution, as well as overharvest interlinking with insufficient protection led to the extirpation of the species in the North Sea and its tributaries.

14.1 Historic Status of *A. sturio* in the North Sea Drainage

The North Sea represents the north-eastern border of *A. sturio*'s historic range, where the species developed large populations of high economic impact in the tributaries and the coastal waters.

The presence of *A. sturio* in the North Sea and its tributaries is documented since approximately 4,500 BP (Ludwig and Gessner 2007). Its utilization by local

J. Gessner (✉)
Leibniz-Institute of Freshwater Ecology and Inland Fisheries, Müggelseedamm 310, 12587 Berlin, Germany
e-mail: sturgeon@igb-berlin.de

S. Spratte
Landesamt für Landwirtschaft, Umwelt und ländliche Räume des Landes Schleswig-Holstein (LLUR), Abt. 3 Fischerei, Dezernat Binnenfischerei und Aquakultur, Hamburger Chaussee 25, 24220 Flintbek, Germany

F. Kirschbaum
Faculty of Agriculture and Horticulture, Biology and Ecology of Fishes, Humboldt University, Philippstr. 13, Haus 16, 10115 Berlin, Germany

P. Williot et al. (eds.), *Biology and Conservation of the European Sturgeon Acipenser sturio L. 1758*, DOI 10.1007/978-3-642-20611-5_14,

inhabitants is confirmed by excavations revealing a strong focus on the lower reaches of the rivers Rhine, Ems, Weser, Elbe (with Oste and Stör) and Eider since 100 BC (Kinzelbach 1997; Zawatka and Reichstein 1997). Excavations indicate a widespread utilization of sturgeons in Roman settlements along the Rhine (Kinzelbach 1987) and also during medieval times in settlements along the North Sea coast, although the rise in sea level in these areas has caused a large-scale loss of evidence (Lepiksaar and Heinrich 1977; Prummel 1983). The European sturgeon was widely distributed throughout Germany (Fig. 14.1).

All major rivers of the North Sea drainage supported populations of the species at least until the middle of the twentieth century. In the Netherlands the Meuse and in Belgium the Schelde rivers were inhabited by European sturgeon. On the North Sea

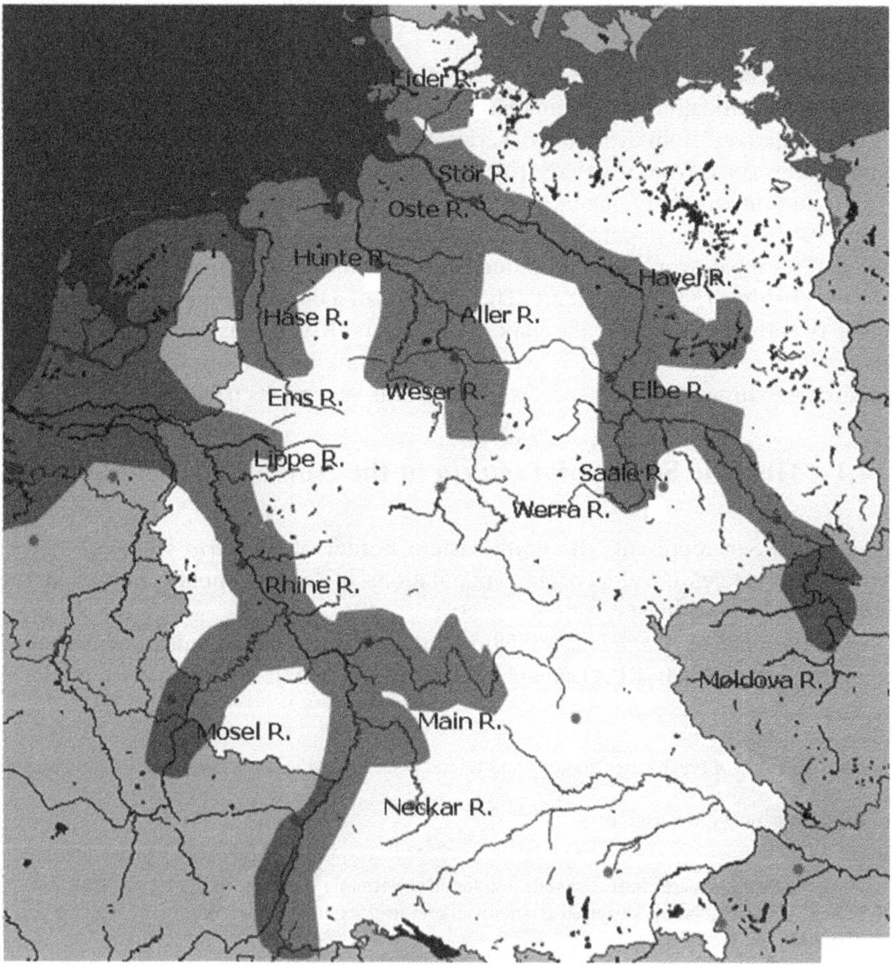

Fig. 14.1 Range of *A. sturio* (*shaded*) in the German North Sea tributaries, (map: http://www.Wikipedia.org)

coast of Great Britain, sturgeon historically entered the rivers Thames, Trent and Don; the species status is unclear, since Parnell (1838) described the presence of two species. A fish caught in 2006 near the Severn was identified as *A. oxyrinchus* (Ludwig pers. comm.). In Denmark and Norway, the species was not noted to have maintained self-sustaining populations. According to Kottelat and Freyhoff (2007) as well as Holčik et al. (1989), in Germany the European sturgeon ascended the River Eider to Rendsburg, the Elbe Mulde and River with its Czech tributary (the Moldau) to Prague, the Ohre to Kadan, the lower Saale, the Havel and Spree, the Geeste, Oste and Stör Rivers. In the Weser, the fish were observed in the Werra and Fulda Rivers, as well as in the smaller tributaries Aller, Leine and Hunte; in the Ems, the fish went upstream to Rheine and into the Hase River; in the Rhine, to Schaffhausen and into its tributaries the Neckar to Heilbronn, and the Mosel to Toul (France), as well as the Saar, Sauer, and Nahe Rivers. The Mosel had a strong sturgeon run compared to the upper Rhine River (Kinzelbach 1987). Whether this was associated with the historic role of the Mosel as the main upriver section of the Rhine prior to the connection of the Upper Rhine Valley, or a result of increased productivity, is not known. The species also occurred in the Main River to Würzburg and Schweinfurt, the Lippe to Lünen and in the Stever to Senden. In the lower River Rhine, the species was more abundant than in the middle and upper reaches. In the Netherlands it was found throughout the Rhine Delta, including tributaries such as the Ijssel Rivers.

14.2 Historic Fisheries

Sturgeon, like other migratory fishes with distinct migration peaks, were a preferred target for fisheries because of seasonal abundance (Albrecht 1960; Rehbein 2003). Due to their value, sturgeons have been considered a target worthwhile the effort of spending time for their catch, as long as the populations were sufficiently large (Breckwoldt 1894; Rübcke 1910). Documents on sturgeon and its catch can be found in a multitude of records since the Mesolithic (Kinzelbach 1997).

The sturgeon fishery was an integral part of the income for fishermen along the large rivers, especially in areas where aggregations persisted due to the utilization of habitat for spawning or pre-spawning residence. Sturgeon meat has widely been used for human consumption (Von Bingen 1159; Breckwoldt 1894; Mohr 1952). Additionally, juvenile sturgeons have been caught to serve as food, fertilizer or as pig feed in several areas of high abundance, mainly in the lower sections of rivers (Hessel 1675).

14.2.1 Historic Catching Methods

How have sturgeons been targeted in riverine fisheries over the centuries? Directed sturgeon fisheries were – with only few exceptions – focusing on adult fish during their spawning migrations. Juveniles were mostly caught in unspecific fishing gear locally. One main prerequisite to catch large fish is the quality of the material used. The effort required to produce specialized nets makes its feasibility dependant upon the amount of fish available as well as on the time to be spent for directed fisheries – e.g., the presence of alternative targets.

The first documents concerning sturgeon catches reveal that due to the size of the fish and most probably the insufficient durability of the nets made available during the Mesolithic and Neolithic times, spearing was widely employed throughout Europe as for other large fish (Brandt 1964; Amacher 2006). The universal gear that also caught sturgeons was the beach seine (Fig. 14.2), which is still frequently used in the Volga and Ural Rivers for this purpose. These nets were employed in many river fisheries as well as in the salmon and shad fishery. Suitable sites were rather popular, and were fished by several crews in alternating shifts (Kisker 1926). Drift nets also targeted typical river fish including salmon, while sturgeon mainly occurred as bycatch due to mesh size limitations. Seines and drift nets required river sections with clear even bottom structure, minimizing the risk of hooking the net on obstacles.

Nets in the early medieval period were made from natural fibre such as stinging nettle (Uticacea) and flax: only after the seventeenth century, hemp and still later

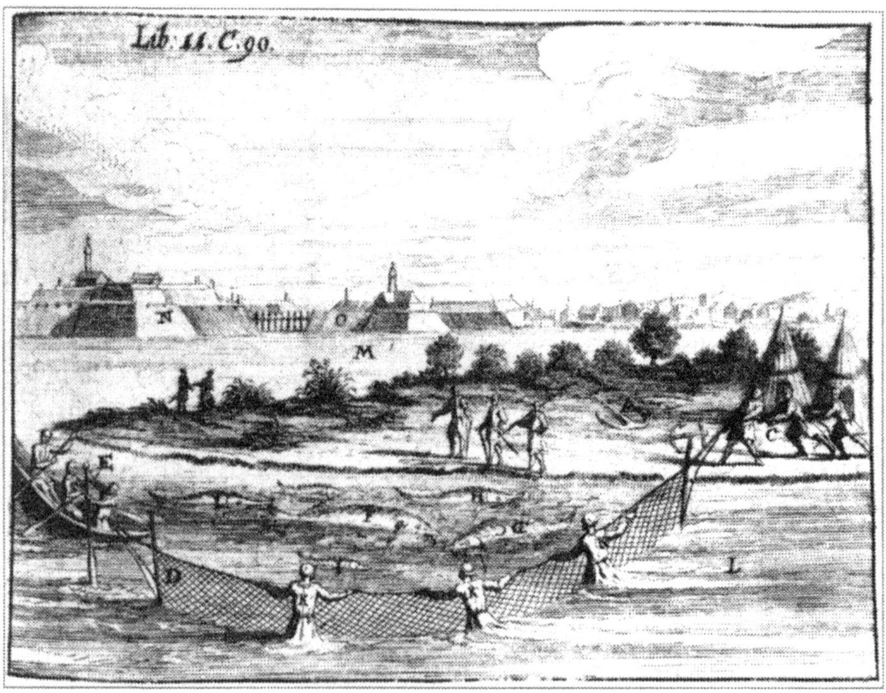

Fig. 14.2 Traditional beach seine fishery from a fishing camp harvesting sturgeons; fishery probably located on the Danube River (Nürnberg 1716 after Haase 2000)

cotton became widely distributed. Due to the low durability of the natural fibre, the nets had to be thoroughly maintained. In Pillau (Baltic Sea coast), for preservation, the nets were treated in smoke houses to prevent fouling. Tar and pitch were frequently used to impregnate the nets, although they made the material stiff and less flexible. To limit decay, the nets were hung up to dry on the river banks on posts during high tide, contributing to the characteristic view of the fishing harbours (Mahn 1979).

The specialized drift net fishery for migrating sturgeon predominantly utilized single-wall, large-mesh nets. The most refined specialization of a drift net is the "Pümpelgarn" (Fig. 14.3), a drift net which in the Elbe was 4–6 m high and 60–100 m long, with a mesh size of 16–17 cm (knot to knot). The dimensions of the nets varied between regions. The upper part of the net is suspended by 20–24 bottle shaped floats (Pümpel), which were attached to the top line by ropes to allow an adjustment to the water depth (Benecke et al. 1886). No lead line was used, but the lower 25–30% of the net dragged on the bottom during the drift.

The net was fished from a boat that drifted alongside the net to correct its drift and to maintain it. A barrel was used as a float to control the drift at the far end. The net was only fished at or close to slack low water while drifting over deep holes in

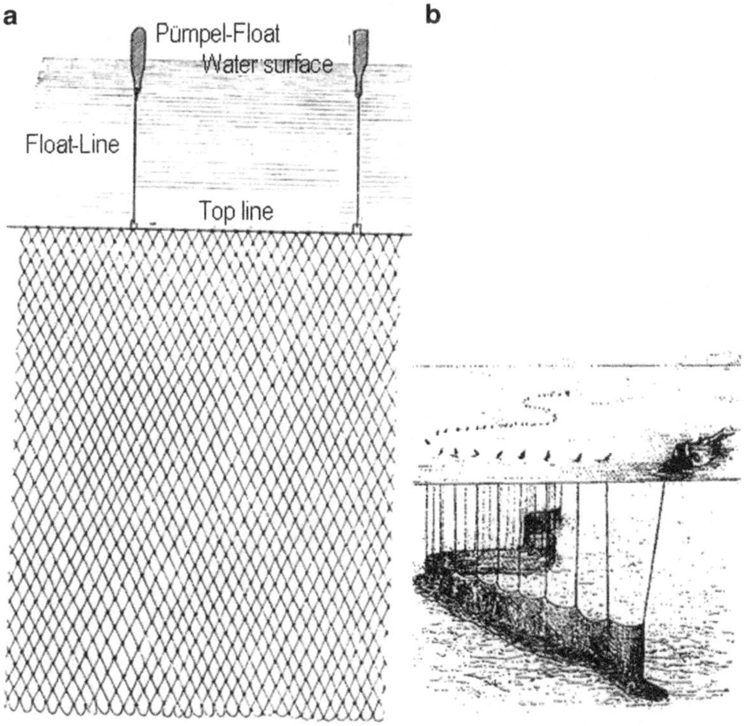

Fig. 14.3 **a** Setup of a "Pümpel" Drift-Net with mesh, floatlines and floats (Pümpel) modified after Benecke et al. (1886) and **b** illustration of utilization during low tide (after Mohr 1952)

the river with clear sand or mud bottom. In suitable river sections, several nets were fished drifting one after another, and it is described that only the third or fourth net caught a fish while the first nets were drifting over top of it immediately before (Quantz 1903). During the drift the floats signalled where a fish hit the net, and due to the rope attachments of the floats, the net could be lifted where one or several of the floats were submerged.

14.2.2 Fishing Sites

Prominent fishing sites (Fig. 14.4) were described for the lower Elbe River in or close to the main flow, for instance at Brunsbüttel (1), between Brockdorf and Hollerwettern (2), at the Krautsand island, near Kollmar (3), in the "Hannöver Nebenelbe", a diversion of the mainstem river (4), and in the southern branch of the river (5) in the Hamburg Harbour region. In the Stör River, sites in the city of Itzehoe and predominantly near Beidenfleth (6) were prominent fishing locations, while in the Oste River, Bentwisch to Oberndorf, (7) Sethlerhemm, Großenwörden,

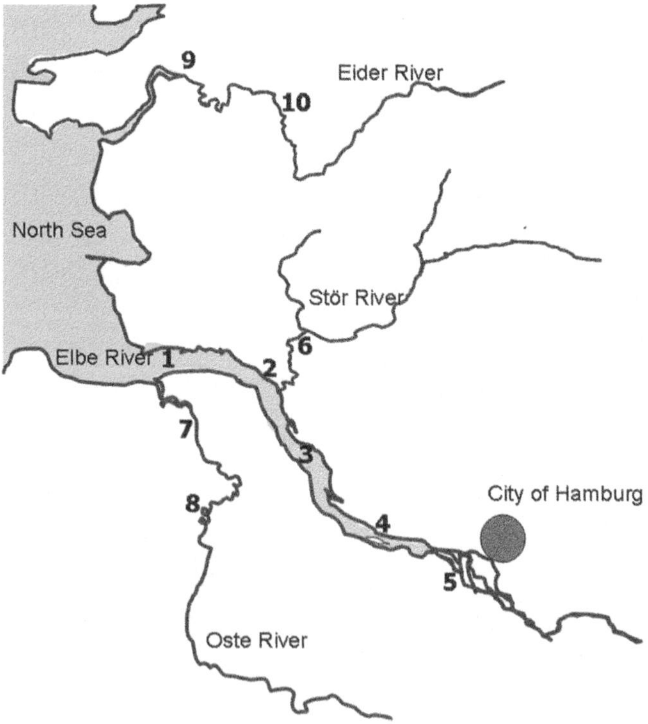

Fig. 14.4 Location of main fishing sites in the Lower Elbe River and tributaries, the numbers are referred to in the text

Breitenwisch, Hechthausen and Klimt (8), and in the Eider River Süderstapel, Friedrichstadt (9), Pahlhude and Horst (10) were productive sites (Breckwoldt 1894; Quantz 1903; Rübcke 1910). These sites were characterized by deep holes or channels. All of the mentioned sites prior to the onset of the major dredging operations were located in tidal freshwater areas (Kausch 1996).

14.3 Fisheries Landings and Population Trends

14.3.1 Fisheries Landings

Despite their importance for fisheries and trade, sturgeon population assessments are missing, and even catch data are rarely available until the fifteenth century. The catch data for Magdeburg were by no means reliable (Kisker 1926), since even the fishermen did not know what their colleagues caught, a fact that was assumed to be intentional. The size of populations, therefore, can only be guessed in relation to the catch and from anecdotal evidence. Based on these sources (Kinzelbach 1997; Blankenburg 1910), it seems that the populations of the North Sea tributaries did reveal large-scale fluctuations in abundance during reproductive migrations over time (Fig. 14.5).

When comparing the Late Medieval catch data from the Elbe River at Magdeburg (Fig. 14.5) it is evident that the catches in the Upper Rhine are an order of magnitude lower than the Elbe catches. It has to be noted that the accuracy of the catch data is widely unknown, while the trends would be correct due to the consistency in data acquisition. The fluctuations in catch are based on 5-year harmonic means – the Elbe population reveals fluctuations in 14–20 year cycles

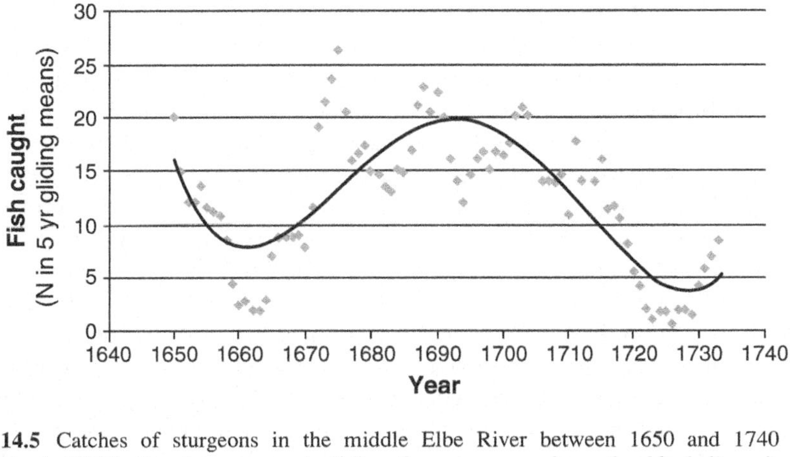

Fig. 14.5 Catches of sturgeons in the middle Elbe River between 1650 and 1740 (after Kinzelbach 1997); the *dots* represent gliding 5-year mean values, the *black line* gives the polynomic approximation

(Fig. 14.5). In comparison, the catches from the Upper Rhine show a 12–17 year cycle (Kinzelbach 1997). The timing of minima and maxima in the catches during the period 1650–1695 reveals an offset by 2–3 years.

Cycles in sturgeon recruitment over the years have also been described in *A. oxyrinchus* in the St. Lawrence River, representing the northernmost population in the range of the species and are considered typical for the extremes of the range of a species (Antonovics et al. 2006). Hydrological, morphological and climatic impacts are considered to be among the main abiotic drivers for reproductive success in diadromous fishes (Béguer et al. 2007). In addition, the effects of population density might trigger intraspecific effects such as concurrence and overpopulation of spawning sites that have profound effects on the recruitment (Schmitt Kline et al. 2010).

Unfortunately, the catch data for German Rivers, as well as those on anthropogenic impacts like fishing, are too scarce to carry out an in-depth analysis over longer periods. For instance, the fluctuation in landings in the mid nineteenth century do reflect both increasing exploitation and the general decline of the populations of the Rhine, Weser and middle Elbe Rivers due to environmental alterations and overharvest (Bauch 1958).

14.3.2 Case Study Lower Elbe River

The average annual sturgeon catch in the first half of the nineteenth century was assessed to vary between 8,000 and 10,000 fish per year (Anonymus 1876; Brunn et al. 1894; Linde 1908). Prior to 1830, the market of the city of Hamburg was supplied by catches of local fishermen (Fig. 14.4 site 4 and mainly site 5). Breckwoldt (1894) describes the abundance of in the Elbe River in the Hamburg harbour area until the end of the 1880s. "During the first half of the nineteenth century, the sturgeons were so numerous in the southern Elbe River branch at the Köhlbrand stretch that it was impossible to make use of the catch. My grandfather and his employee caught 1,100 fish, and my father an additional 1,000 fish between April and August". Fishermen from the lower River Elbe in the area of Glückstadt (Fig. 14.4 site 1 and 2) entered the fishery only in 1830 (Kluge 1926). The high level of catches with 4,000–7,000 mature fish annually was maintained until the late 1880s, with increasing effort. Maximum catches on the Elbe River between Hamburg and the North Sea in the 1870s were assessed to have comprised at least 10,000 fish in 1875, based on the fact that a single wholesaler in Hamburg bought 8,051 sturgeon swim-bladders (Anonymus 1876). These data are supported by Lindemann (1888), who assessed the catch for 1883 to have comprised 8,000 fish and for 1884 6,000 fish. A shift in effort became noticeable already in the late 1880s. Breckwoldt (1894) mentions the sharp decline in the Hamburg Harbour catches after 1888: "In 1893 on the Köhlbrandt only two fish were caught, so the fishery was rendered unprofitable at that time". The decline of the population of mature fish became visible from 1889 on (see Fig. 14.6), when approximately 50% reductions in catch were recorded annually.

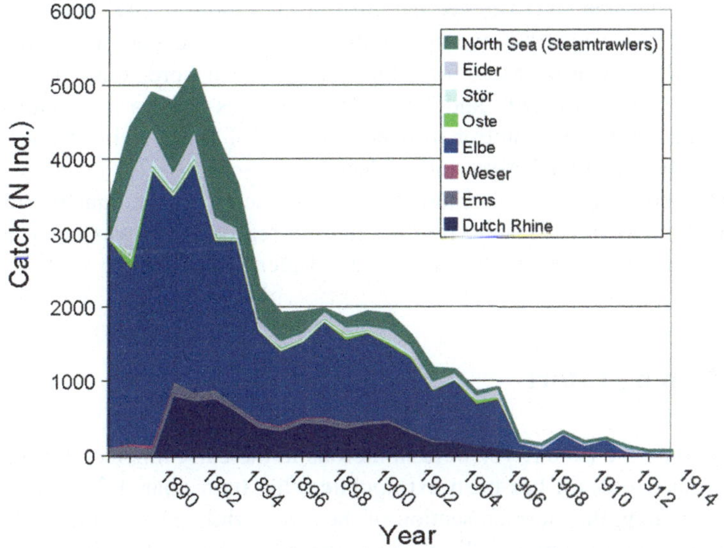

Fig. 14.6 Sturgeon catch in individuals between 1888 and 1915 for the German North Sea tributaries and the Dutch Rhine (after Demoll and Maier 1940)

The fishery was relying solely on the catch of mature fish to this point (Quantz 1903). According to a survey between 1887 and 1889, only three fish of below 1.2 m (the size at which male fish were considered to become mature) were caught (Kiel Commission 1894). It was concluded that "The population ... seemed to be rather healthy, despite the fact that the fish caught did not reach the sizes that were frequently noted in the 17th century" (ibid). This becomes evident when considering the reports for the seventeenth century stating that fish of 18 ft (5.4 m) were rare but not exceptional (Schonevelde 1624; Hessel 1675). In contrast, in the eighteenth century, fish of 3–3.6 m were reported to be rather common (Bloch 1785), while in the late nineteenth century the average size of females mainly comprised fish between 2.0 and 2.4 m (Quantz 1903), although Oesau (1962) reports a fisherman in Glückstadt during the second half of the nineteenth century who caught one fish of 14.5 ft (4.35 m).

Not only the sizes but also the numbers of fish caught were declining. While after the 1890s fishing focused on the lower Elbe River and its tributaries as well as on the Eider River and the Wadden Sea, other German rivers lost economic relevance with regard to their sturgeon fishery. By 1910 the Elbe River catch had also declined to approximately 1% of the 1880s catch (see Fig. 14.6). In 1911, only two mature fish were caught in the southern Elbe River branch in Hamburg, and in 1912 the fishery achieved a total of only 45 sturgeon throughout the lower Elbe River comprising one percent of historic average catches (Anon 1913).

Subsequently, sturgeon became rare already after 1930. The catches from the lower Elbe River were limited to single reports of large (i.e., old) individuals (Bauch 1958; Wilkens and Köhler 1977). The last juveniles measuring 25–105 cm were observed between 1962 and 1968 in the Oste and the Elbe river mouths (Fig. 14.4 site 1). The last catch from the mouth of the Oste River was a dead fish of 2.37 m in 1985, which was transferred to the Zoological Museum in Hamburg where it was prepared for exhibition. Two fish were reported to be caught from the North Sea in 1993 and 1995. The first was sold to the canteen of the Federal Ministry for the Interior in Bon,n the city of Bonn while the second still survives in the Helgoland Aquarium to date.

The sturgeon fishery in the Eider River continued after the construction of the Nordfeld Weir in 1936, despite the fact that no reproduction was noted after this event. The last fish to be caught in the lower Eider River dates to 1969.

14.4 Fishing Effort

Fisheries along the Elbe River were numerous in the early nineteenth century, and from various sources it has been extrapolated that more than 1,200 fishing rights were issued along the German section of the river, while every 2 km of the river a professional fishermen made a living from the river. Directed sturgeon fisheries on the Elbe River were carried out by a fragment of these fishermen, mainly in reproduction and aggregation areas. Nevertheless, sturgeon were caught occasionally (Rehbein 2003) in standard gear such as driftnets and beach seines (Schiemenz 1967). The decline of this traditional fishery was caused by the loss of suitable fishing sites, due to continuous regulation and improvement of the navigation conditions on the river after 1815. Therefore, records of sturgeon catches after 1860 were restricted to a few areas of the upriver section where the traditional beach seines and drift nets continued to be used.

Kluge (1924) mentions that the onset of the directed sturgeon fishery on the Lower River Elbe in 1830 was by no means successful for those fishermen. Only eight fish were caught in the first season. Nevertheless, the number of fishermen participating in the fishery drastically increased with their success. The increased intensity of the fishery was also associated to the fact that fishermen from Hamburg started to fish in the lower reaches of the Elbe when the sites located closer to the city became less productive (Rönnau 1987). The increasing activity and number of fishermen on the downstream sites led to an increased expansion of the fishery, with more fishermen moving further out of the lower river reach beyond (Fig. 14.4 site 1) and into the coastal waters.

In 1884 Lindemann (1888) counted a total of 231 sturgeon nets from the coastal waters of the Lower Elbe, Elbe River mouth and west coast of Schleswig-Holstein. He mentions that more than 300 nets are employed during the main catch season in May. In Kollmar (Fig. 14.4 site 3), 130 fishermen were fishing for sturgeon in 1870, while only 13 were left in 1902 (Quantz 1903).

Based upon the data summarized above, the drastic increase of the fishing effort between 1830 and 1880 becomes evident. During the same time the catches remained relatively constant, at least over a period of 40 years until the end of the 1880s.

A further increase in fishing pressure resulted from the onset of drift netting in coastal waters, which was unknown to the local inhabitants at that time (Hansen 1877; Müller 1917). Cornils et al. (2008) date the onset of directed sturgeon fishing along the coast of the Wadden Sea back to 1877. The nets were provided by the Sea Fisheries Association as a loan during the experimental fishery, to increase regional income (Müller 1917). Furthermore, the state made available cheap advances or credits for fishermen in the area to purchase their own gear (ibid). The fishing season for the coastal driftnet fishery was limited to late spring time, when the juvenile fish utilized the Wadden Sea and the mature fish gathered for their upriver migration. In this fishery, due to the reduced mesh sizes applied (10–12.5 cm knot to knot), mainly juveniles were landed.

In addition to the expansion of the traditional fishing methods, trawlers began to operate in coastal waters as a result of technical improvements such as the increased distribution of steam engines for boats by 1887. This technique increased the range of the fishery, and made it possible to develop more intensive fishing techniques such as beam trawls and bottom trawls. Despite the rapid increase in the steam trawler fleet after 1890 (for the development of the German steam trawler fleet see Fig. 14.7), the contribution to the reported sturgeon harvest did not exceed 20–30% of the Elbe River catch (Fig. 14.6). In the trawl fishery, some boats targeted sturgeon effectively in their wintering sites in the Wadden Sea between December and March at water depths of 40–50 m. The marine catch by the trawler fleet ceased after 1895 because it became increasingly infeasible to target the fish directly (Backhaus 1935).

Also, at the onset of the 1920s the catch with drift and set nets in coastal waters commenced with motored boats from mid April to end of June, resulting in a total catch of 50–150 fish per year which were of extremely different sizes. Schnakenbeck (1928) reports that at the onset of the twentieth century the catch sites in the Wadden Sea were mainly located at the westcoast of Sylt Island. By 1927, the fishery had been abandoned due to poor results in the area. The reported

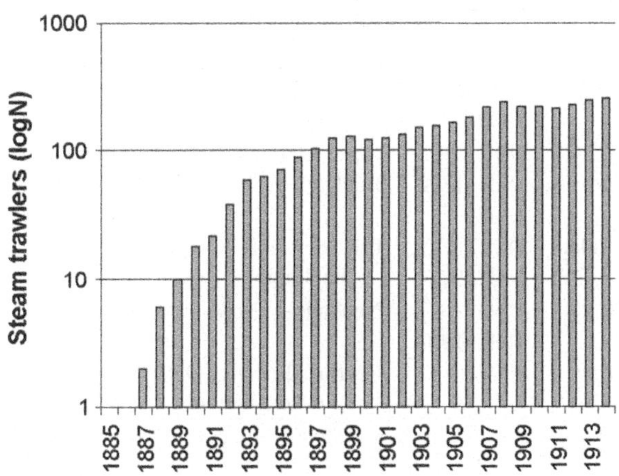

Fig. 14.7 Development of the German steam trawler fleet between 1885 and 1914 (after Schnakenbeck 1928)

decline of the catch to insignificant numbers is supported by Dröscher (1926), who claimed that in the middle of the 1920s the sturgeon catch in the North Sea coastal fisheries had become completely irrelevant.

14.5 Processing and Products

Sturgeon meat was sought after for consumption (Rönnau 1987), despite the fact that they were considered a heavy meal that was not recommended for people with a weak constitution (Von Bingen 1159), while in the first half of the eighteenth century sturgeon roe was sold cheaply to fishermen who used it as bait for the eel-pots very effectively (Rübcke 1910).

The caviar produced in the North German fishery was considered of inferior quality (Goedecken 1969). Descriptions of the production process given by Kienau (1924) for the Elbe River reveal the drawbacks in production. When the fish became accessible for the fishery during spawning season in May until August, water and air temperatures tended to reach or exceed 20°C. The fish were butchered in the open at the landing sites or in the fishing harbours.

Only around 1840 did the production of caviar, applying Russian processing techniques, start in the lower Elbe at Glückstadt (Fig. 14.4 site 3), while in Hamburg some caviar had already been processed since the end of the seventeenth century (ibid). Also, hot smoking of sturgeons "increased the utilization of the meat".

The specificy of the fishery up to 1896 is demonstrated by the dominance of females in the catch, expressed in the sex ratio of male/female sturgeons in the catch reaching 0.36 (Table 14.1). Decreasing mesh sizes altered the sex ratio in the catch by 1902, when the ratio males/females from all of the catches analyzed in the North

Table 14.1 Proportion of male and female sturgeon in selected catches and sex ratio over time as well as at different catch sites

River	Year	Catch site	Males (*N*)	Females (*N*)	Sex Ratio M/F
Elbe	1886	Schleswig-Holstein total	813	1371	0.59
	1884	Glückstadt	37	101	0.36
	1885–1891		289	718	0.40
	1902		43	29	1.48
	1902	Outer Elbe River below Brockdorf	28	27	1.10
Stör	1885–1896	Beidenfleth	195	61	3.2
	1902		13	21	0.62
Eider	1876	Friedrichstadt, Süderstapel, Horst, Pahlhude	251	236	1.06
	1901/2		44	9	4.89
Oste	1893	Neuhaus – Großenwörden	97[a]	3	32.2
	1897–1902		528[a]	62	8.52
Ems	1902	Leerort, Terborg, Oldersum, Ditzum, Weener, Soltburg	13	26	0.5

[a]Fish below 200cm TL were generally considered males in the Oste River, no determination was carried out (Quantz 1903)

Table 14.2 Proportion of the females in the catch (in %) and stage of maturity determined (developmental stages were transformed according to Chapter 25 this volume)

Source	Stage II–III	Stage IV	Stage V	Overmature or spent
Investigation Kiel Commission (1890) $N = 1{,}923$ females	30.4%	23.2%	36.7%	9.7%

Sea tributaries ($N = 253$) reached 1.2). It was noted, though, that only 36% of the 113 females were sufficiently mature to provide good quality caviar (Table 14.2).

Hygienic conditions were poor, with river water being used to wash the fish. Also, the roe after salting was stored in cans and glasses that rarely saw hot water before being used. The amount of salt used varied largely (a "handful" or more per kilo roe), with produced variable results with regard to taste and preservation ability.

14.6 Sturgeon Trade

Sturgeon were landed alive by leading a rope through the mouth and gills and tying the tail to tow the fish alongside the boat. In the lower River Elbe, the fish were maintained alive for a day or two but sometimes for as long as a week, by tying them to a pole or chain in the harbour, to overcome periods of low prices.

Several centres for sturgeon trade were established in close vicinity to the main catch sites. Although even during Roman times the long-distance transport of live fish was well-established (Kinzelbach 1987), the trade of fish began to suffer from decreasing infrastructure during the medieval period (Amacher 2006). Therefore, only a limited trade of fresh fish was seen with regard to distance and quantity. The main markets utilized pickled sturgeon, salted eggs and dried meat in trade (Benecke et al. 1886). Since fishing rights had been transferred to the church in many places (Benecke 1881), clerical settlements (Magdeburg, Xanten, Cologne, Speyer) also played an important role for fish trade (Kinzelbach 1987).

Marketing of sturgeons was not easy when large quantities of fish were available. Rübcke (1910) recalls the historic abundance of the sturgeon of the Elbe River from the first half of the nineteenth century and the problems associated to it: "The catch was plenty, but the utilization (of the fish) was difficult. The meat of sturgeons was offered for 0.05 Mark per pound, and often did not find a buyer even at this price." An alternative indication for the low prices is given by Köhn and Wilkes (1979): "On average, a sturgeon fisherman caught 50–60 sturgeons per season which weighed up to 500 pounds. Between 1860 and 1890 the sturgeon meat was so cheap that even the farmers in the area (widely known not to waste money) served sturgeon meat at least once or twice a week". Since cold storage was unavailable on a large scale and the fish were numerous, the markets even of

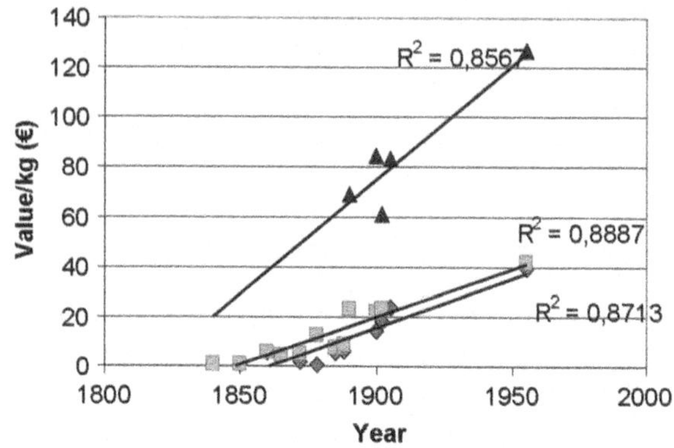

Fig. 14.8 Wholesale price for sturgeon meat (minimum values *light grey*, maximum values *dark grey*) and caviar (*black*) in the Lower River Elbe region between 1840 and 1955; the prices have been standardized to € at a value for 2008

large cities like Hamburg with 75,000 inhabitants in 1750 and 160,000 inhabitants in 1850 quickly reached market saturation.

On the Elbe River, the dominant markets for sturgeon landings in the eighteenth and nineteenth century were the city of Hamburg, the city of Altona, and Magdeburg. Glückstadt on the Lower Elbe was a local fisheries centre, and some smaller cities along the tributaries, such as Oberndorf on the Oste River, Friedrichstadt on the Eider River, and Bremen on the Weser, became regional centres for sturgeon trade. From here in the second half of the nineteenth century the fish were shipped either on ice by train or alive in ships to the larger markets (Oesau 1962). The long-distance trade uncoupled the production from the limited capacity of the local markets, increased the income for the fishermen (Fig. 14.8), and put the resource at risk through increased effort and resulting overexploitation. Therefore, sturgeon as well as the North Cape whale (Oesau 1980) are first examples for the Tragedy of the Commons (Hardin 1968).

14.7 Legal Background of the Sturgeon Fishery and Attempts to Manage It

Fishing rights emerged with the hierarchical structures of state-like organizations, in order to regulate access to the resource in the fishery. Sturgeons, for example, were often excluded from common fishery rights (Benecke 1881). During the medieval period, the state representatives owned the fishing rights too. These rights were given to the cities, to monasteries or to individuals, and were utilized either by renting them or by employing fishermen to work for the landlord. In medieval times, guilds as first professional organizations represented the main commercial harvesting entities (Nadler 2010).

Apart from compliance issues, active fisheries management, i.e., activities that respond to the changes in availability of the resource in a dynamic way, were first noted in the sixteenth century and became widespread thereafter. Two main targets can be discriminated in these regulations; the limitation of access to the resource, and the attempt to ensure continuous benefit from any resource for the fishermen. In general, three different management tools were utilized: catch regulations by (a) size limits or mesh size, (b) closed areas, or (c) closed seasons.

Effective fisheries management measures such as the prohibition of gear or the limitation of mesh sizes were only attempted in the late nineteenth century. To improve the management of the sturgeon fishery and increase its sustainability, a variety of measures were suggested by fisheries inspectors, scientists, fisheries associations, and fishermen themselves. Interestingly, despite support by the fishermen, the state authorities did not accept the changes suggested for long periods of time (Ehrenbaum 1923). The full protection of the species by inhibition of catches was mainly implemented after it no longer affected the commercial fisheries, e.g., after the collapse of populations and their extirpation. Even at this stage, protection was not effectively enforced.

14.7.1 Gear Restrictions

Gear restrictions to prohibit adverse fisheries techniques had already been enacted during Roman times, and violations resulted in punishment in the medieval period (Nadler 2010). Direct measures to improve protection of sturgeon juveniles and of spawners were discussed in Germany from the 1880s onwards. Mesh sizes exceeding 15 cm knot to knot, in conjunction with selection of fishing sites, provided effective means of protection. It is reported that until 1890, the minimum size of fish landed commonly exceeded 1.6 m (Kiel Comission 1890; Mohr 1952). Only following the introduction of smaller mesh sizes and the utilization of other fishing sites did the proportion of juvenile sturgeons increase drastically (Ehrenbaum 1894).

With increasing fishing pressure and increased effort on the spring and summer feeding grounds at sea, the mesh size of the nets decreased. After 1888, the mesh size was reduced from 17 cm to 12 and 10 cm (knot to knot), which increased the proportion of sublegal-sized sturgeons in the catch from 3% in 1888 up to 88% in subsequent years (Ehrenbaum 1916).

In order to avoid mass mortalities of juvenile sturgeons in riverine or estuarine habitat due to baited hooklines, additional regulations of gear were effected. Baited hooklines commonly used in eel fishery were prohibited in the Eider River in 1915 and in the Oste River in 1918.

14.7.2 Size Limits

Catch regulations in Germany varied to a large extent. In the 1890s, a main concern in the commercial fishery and in the fisheries administration was the trend of increased exploitation of juvenile fish even below the size limit of 1 m, which already was considered insufficient (Ehrenbaum 1913, 1916). This phenomenon of juvenile exploitation was already known in the seventeenth century (Hessel 1675). Therefore, legal size limits were discussed, especially after the onset of the fishery on the Wadden Sea where large numbers of juveniles were targeted.

Attempts to increase size limits to 1.5–2.0 m failed in 1886 because the number of juveniles taken in the fishery was considered too low by the earlier judgement of the Kiel Commission (1890) to justify the attempt. Only a deliberate size limit of 1.25 m was effected in 1904 by the fisheries co-operatives along the lower Elbe River (Blankenburg 1910). In 1923 the size limit of 1.5 m was finally adopted, but was abolished again a year later (Ehrenbaum 1926).

14.7.3 Closed Areas and Closed Seasons

Spawning refuges were closed for the fishery in the Oste River (a tributary of the lower Elbe River) in 1898 until 1918. This measure was applied to 7 km of the river in an area where fertilized eggs had been detected (Quantz 1903). This refuge was extended in 1914 to a river section of 19 km (Ehrenbaum 1916). Further increases were prohibited by owners of fisheries rights, who asked for large-scale compensation (Tiedemann-Wingst 2001). In the River Elbe annually varying fishing leases were defined after 1898 to limit the effort applicable.

Closed seasons in both the Elbe River and tributaries as well as in the Eider, beginning in 1890 from July 26th to August 26th, were effected both by regional authorities in Lower Saxony and Schleswig-Holstein. This closure was increased to a fisheries ban lasting from July 15th to August 26th in 1898, and in 1904 was modified to a fisheries ban from August 1st until the end of the year. Nevertheless, exemptions of the closed season could be applied for, and generally were provided once a poor season or a late run were observed (Quantz 1903).

The combination of various measures was also applied in later years. On the Eider River in 1915, a series of legal protection measures were effectuated to protect the sturgeon population and to support a sustainable harvest (Mohr 1915). These included: (a) the prohibition of baited hooklines using worms as bait, (b) increased intensity of surveys to prevent marketing of sublegal (<100 cm) sturgeons, (c) improving enforcement by increasing the number of fisheries officers for the Eider River, (d) increase of the legal size limit to 150 cm TL, and e) the attempt to buy fishing rights along the river to install protection areas. Point (d) was abolished again in 1916 (Ehrenbaum 1916), and the protected areas were not established until the dam at Nordfeld was completed in 1936, which blocked the migration of the adults to the spawning sites.

14.7.4 Stocking

From the 1870s onwards, controlled reproduction and stocking of fish was a common measure to enhance fisheries. In 1871, the first successful reproduction of sturgeon was carried out by J. Knoch with sterlet (*A. ruthenus*) in Russia (Von dem Borne 1877). In 1875 Seth Green succeeded in reproducing *A. oxyrinchus* on the Hudson. The first German attempts to promote the controlled fertilization and incubation for wild-caught sturgeons were carried out starting in 1874 on the Elbe through the Fisheries Societies which are presented in detail in Table 14.3. After the first successful attempts in 1877, the German Sea-Fisheries Association and the German Fisheries Association provided collection sites for spawners, sites for reproduction, and rewards for the successful reproduction and subsequent release of sturgeon larvae to the open waters (Quantz 1903).

The main drawback already limiting the applicability of the release strategy in the 1890s was the decreasing numbers of running ripe females (Elsner and Stemann 1886; Blankenburg 1910; Mohr 1952). Due to a lack of appropriate biotechnology, it was impossible to induce ovulation in females. The only solution was to fish on or near the presumed spawning sites for fish that were ovulating or had ovulated already.

Males were held on site, being tied to a pole in the river by a rope tied through the mouth and gills. This technique was limited with regard to the duration of holding, since the fish started to develop fungus after approximately 1 week (Quantz 1903). Also the milt production ceased within a few days (Frauen 1877). If milt was unavailable after prolonged holding, the testes were minced and mixed with the eggs with some success (Frauen 1877; Elsner 1886).

Eggs were thoroughly washed in river water for de-adhesion before being transferred into wooden incubation boxes with fine mesh screens placed in the rivers to facilitate water exchange (Fig. 14.9). The incubation boxes initially were 60 cm long, 40 cm wide, and 20 cm high (later modified to 40 cm height). They were made of wood, and were equipped with a fine metal mesh on the bottom. The mesh was painted with tar to prevent adhesion of the eggs. To maintain the boxes afloat, wooden boards were mounted along the sides. They were anchored in moderate current, but had to be moved at slack tide to maintain the eggs in motion (Anonymus 1877).

The results were largely dependant upon the condition of the spawners (Frauen 1877; Anonymus 1877; Elsner 1886; Elsner and Stemann 1886). In ovulated females being maintained over prolonged periods, the eggs resulted in poor survival following fertilization. In these cases, according to the description, the eggs showed signs of resorbtion. Also, it was noted that eggs that did not flow freely (e.g., were fully ovulated) but were obtained by Cesarian section revealed low fertilization and increased mortality during incubation (Anonymous 1877). Incubation mainly suffered from high egg densities that resulted in elevated fungal infestation and subsequent mortality. Cleaning and maintenance were required on a regular basis.

While the releases of larvae exceeded 1 million individuals between 1886 and 1896, feeding and ongrowing was carried out at an experimental scale only (see Table 14.3, 1891 on the River Oste). Therefore, no releases took place with fingerling size or even larger fish (Quantz 1903). Additionally, due to a lack of appropriate tagging techniques (Blankenburg 1910), no verification of the impact of the releases was possible.

Table 14.3 Fertilization and incubation attempts of sturgeon in German rivers

Date	Location	Results	Reference
1874	Lower Elbe River and tributaries	Instructions to prepare the incubation boxes were provided	Anonymous (1874)
4.7.1877	Beidenfleth, Stör River	Initial attempts to fertilize sturgeon eggs using different de-adhesion techniques, incubation without success, eggs died after neurulation	Frauen (1877)
15.7.1877	Beidenfleth, Stör River	M. Frauen fertilizes 25 kg of roe with sperm of two males, five incubation boxes are maintained with river water at 17.5–21°C, first hatch observed after 6 days; yolk sac is absorbed after 6–7 days	Frauen (1877)
1878/1879	Beidenfleth, Stör River	Frauen und Elsner successfully incubate sturgeon larvae for the release in the Stör and Elbe Rivers	Quantz (1903)
July 1881	Beidenfleth, Stör River	Frauen, Beidenfleth, fertilizes and incubates eggs at 20–20.6°C for 70 h until hatch	Quantz (1903)
1883	Tielenhemme, Eider River	Installation of the first reproduction site	Kroezus (1953a, b)
17.6.1885	Tielenhemme, Eider River	H. Söth, incubation of sturgeon eggs is unsuccessful	Quantz (1903)
July 1885	Glückstadt, Elbe River	Elsner and Mohr fertilize and incubate eggs from 8 fish 4 males and 4 females with/-out success	Elsner (1887)
2.7.1886	Glückstadt, Elbe River	J. Mohr fertilizes and incubates eggs, hatch occurs on July 6th	Elsner and Stemann (1886) and Quantz (1903)
5.7.1886	Glückstadt, Elbe River	J. Mohr fertilizes 2.5 kg of eggs, hatch occurs after 84 h, after 91 h hatch completed, 150–200,000 released into the Elbe River	Elsner and Stemann (1886) and Quantz (1903)
15.7.1886	Glückstadt, Elbe River	J. Mohr fertilizes and incubates eggs with mashed sturgeon testis, hatch occurs after 82 h, 50,000 larvae. released	Elsner and Stemann (1886) and Quantz (1903)
16.7.1886	Glückstadt, Elbe River	J. Mohr 400,000 larvae hatched and were released into the Elbe River	Elsner and Stemann (1886) and Quantz (1903)
26.7.1886	Glückstadt, Elbe River	J. Mohr 400,000 sturgeon eggs incubated, 9,000 larvae released, two incubation boxes were washed into the river by wave	Elsner and Stemann (1886) and Quantz (1903)
31.7.1886	Glückstadt, Elbe River	J. Mohr fertilizes eggs, 9,000 hatchlings released	Elsner and Stemann (1886) and Quantz (1903)
23.7.1886	Kollmar, Elbe River	J. Lau incubates 25–30,000 eggs to hatch, fish released	Elsner and Stemann (1886) and Quantz (1903)
20.7.1891	Großenwörden, Oste	H. Schwolert fertilizes 5–6,000 eggs of a spent female, after 65 h at 21.5°C larvae hatch, 700 fish are stocked into a pond near Cuxhaven; out of eight larvae in water barrel, five survive for 14 days	Quantz (1903)

(continued)

Table 14.3 (continued)

Date	Location	Results	Reference
1891	Glückstadt, Elbe River	1.5 Mio larvae are released	Koos (1924)
1.7.1896	Osten, Oste River	Brinkmann inseminates 0.25 kg eggs that are overripe, no success	Quantz (1903)
10.7.1896	Großenwörden, Oste	Grote and Torborg fertilize 18,000 eggs (females was almost spent), hatch occurs after 70–78 h resulting in 28 (!) larvae	Quantz (1903)
1896	Glückstadt, Elbe River	Fishermen incubate 200,000 eggs, larvae released	Cordes (1958)
1913	Glückstadt, Elbe River	Closure of the hatchery after 17 years without reproduction	Koos (1924)

Fig. 14.9 Incubators for sturgeon eggs in operation in the Stör River (after Koos 1935)

14.7.5 The Eider River Example

The last attempts to support and maintain a wild population of the species were carried out in the Eider River, where the fish were still caught in small amounts (see

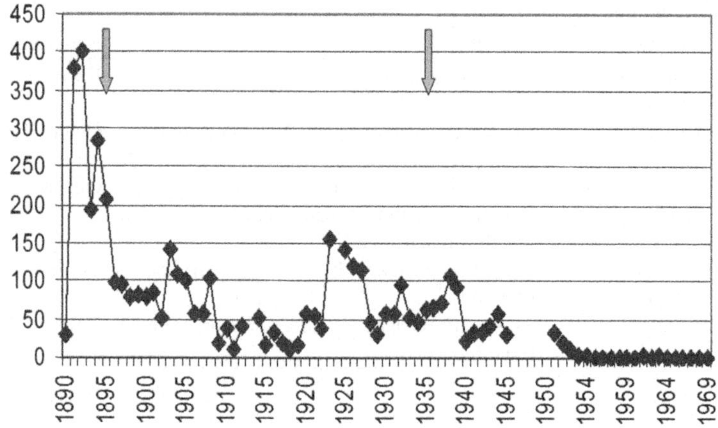

Fig. 14.10 Catch data for *A. sturio* in the Eider River between 1890 and 1970, *arrows* indicate the loss of 30% of the catchment area due to construction of the Kiel Canal in 1896 and damming of Eider River at Nordfeld 1936

Fig. 14.10) in the 1950s. In 1883, a first reproduction facility was opened near Tielenhemme on the lower Eider River.

To protect and enhance the population, controlled reproduction and subsequent stocking measures were initiated in 1939. For this purpose, a collection facility was installed in Friedrichstadt after the weir and sluice at Nordfeld interrupted the migrations to the spawning sites (Anonymous 1939). In 1953, a comprehensive plan of actions was finally effectuated by the cooperative "Lower Eider River" to improve the status of sturgeons in the river (Kroezus 1967), comprising several prerequisites: male sturgeon had to be delivered to the rearing facility, females had to registered alive, the location of fertilization and incubation of the eggs was designated, release sites in the middle Eider River were identified, refunding for provision of fish was guaranteed and the level determined, and the fish after fertilization also had to be returned to the fisherman. Finally, the penalty for non-compliance was outlined, which comprised deprivation of the fishing licence.

Despite the fact that 71 fish were still caught between 1950 and 1960, no successful controlled reproduction was carried out. The reasons were manifold. Fish caught predominantly comprised mature females, and no mature males became accessible synchronously. Furthermore, the fishermen were dissatisfied with the attitude of the authorities, who had permitted the construction of the Nordfeld dam and sluice which successively impacted the fish stocks and thereby the income of the fishermen. Thirdly, the aims and perspective of the planned support measures, as well as the role of the fishermen, were poorly communicated, which did not stimulate the fishermen to cooperate. And finally, no fisherman lost his licence for not complying with the regulation.

14.8 Indirect Impacts Facilitating the Decline of the Sturgeon

In addition to the described impact of fisheries on the species, several other alterations of the environment resulted from human activities (Gessner 2000). These impacts had their onset when river corridors became the main areas for settlement. As a consequence, deforestation increased sediment transport (Hoffmann 1996). Mill weirs and rafting activities changed the river flow and bed-shape, but the impacts upon structure and quality peaked during the industrial revolution in the eighteenth to twentieth century (Schirmer 1994). Especially following the Congress of Vienna in 1815, alterations to river structure became vigorous (Kausch 1996; Schiemenz 1967), since demand for mass transportation of goods and raw material made the rivers a prime objective for logistics (Albrecht and Kirchhoff 1987). Removal of by-products and wastes from settlements, industrial production and agriculture was facilitated by the same waterways. Although the various environmental alterations have been discussed as reasons for the decline of the anadromous fish species (Benecke 1881; Schimenz 1908; Volk 1910; Kluge 1904; Schiemenz 1913; Schiemenz 1921; Seligo 1931; Bauch 1958), practical actions have not been taken. Laws to protect water resources, and thereby aquatic communities, have been neglected by the respective administrations (Bonne 1905).

Two major non-fisheries-related factors that contributed to this development for the Elbe River can be identified. The cities of Hamburg and Altona began to remove their communal wastes (sewage) by a newly installed canalization into the Elbe River at 1862 (above the main spawning grounds at Köhlbrandt) and 1868 respectively (Bonne 1905), resulting in oxygen-devoid areas in the Hamburg harbour area. Development of industrial settlements in the middle reaches of the Elbe River (Kisker 1923) as well as in the vicinity of Hamburg (Bonne 1905) utilized the river to remove production wastes (Kirschbaum and Gessner 2002).

Furthermore, in 1870–1895 the navigation requirements led to a variety of measures to alter the river from the middle reaches of the Elbe to the river mouth, thereby destroying the spawning habitats (comp. Kausch 1996). The river corrections in the lower sections of the river removed shallow banks and deep holes that predominantly held sturgeons. The historic spawning sites were still located upstream from Brunsbüttel (Fig. 14.4 site 1) in 1891 (Ehrenbaum 1894). These sites (Fig. 14.4 site 1 up to site 3) subsequently became influenced by brackish water due to the continuous deepening of the navigation channel. In any case, the anthropogenic impact on the species was not assessed in quantitative or qualitative means (Riedel-Lorje and Gaumert 1982).

For the Eider River, the decline of the population started with the construction of the Kaiser Wilhelm Canal (now the Kiel Canal) in 1896. One third of the drainage of the Eider River was used for maintaining the headwaters of the canal. The decreased river discharge enhanced the transport of sediments from the Wadden Sea into the river, also increasing the frequency of floods. As a prevention measure, in 1934 the construction of the Nordfeld weir and sluice were started. Its completion in 1936 blocked the migration of the fish to its spawning sites in the middle part of the river from Pahlhude and Lexfähre to Rendsburg (Fig. 14.4 sites 9 and 10). Since the fish accumulated below the sluice during migration, they became an easy target for the fishery. In 1969 the last fish was obtained from the Eider River, being sold on the Kiel fish market.

Summarizing the attempts to protect *A. sturio*, it must be concluded that the measures taken were not sufficient to prevent the species from extirpation (Gessner 2000). The fisheries regulations, in particular the size limits applied, have been ineffective or came too late. This was already known at the time the size limits were discussed, as was outlined by Ehrenbaum (1916). Sufficient size limits, that would have protected the spawners for at least one reproduction, were not applicable from a political point of view.

Closed seasons protected fish migrations only at the end of the season or protected the fish after the migration and spawning season. Additionally, the restrictions could be circumvented by dispensation upon request. Furthermore, it did not include marine fisheries on the wintering sites, and therefore it can be classified as completely ineffective.

Closed areas were too small in size, or were not accompanied by measures protecting the fish in associated river stretches. Furthermore, greed of the owners of fishery rights prevented a timely expansion of these measures (Tiedemann-Wingst 2001). Compliance with restrictions was not enhanced or even monitored.

Due to the lack of biotechnology to reproduce sturgeons that were not fully mature at time of catch, the attempt to support the population by stocking was dependant upon chance. In addition, the lack of methods for ongrowing of larvae prevented alternative strategies for release.

In general, protection was considered only after the stocks were reduced to an economically insignificant factor. The subsequent prohibition of catches throughout the range, their background and intentions were not communicated to the fisheries effectively. Therefore, despite the protection status, sturgeons were caught and sold without any consequences for the fishermen. Overall population sizes continued to decrease, and finally the species became extirpated in the 1970s to 1990s.

The disappearance was accompanied by a listing in the fisheries legislations of the German federal states. Following the inclusion of *A. sturio* in the Washington Convention Annex I in 1973, the species was protected on a national level, listed in the Red Book on species protection. The international protection measures are listed by Rochard et al. (Chapter 18 this volume).

References

Albrecht M-L (1960) Die Elbe als Fischgewässer. Wasserwirtschaft – Wassertechnik 10(10): 461–465

Albrecht J, Kirchhoff N (1987) Ökologie der Weser. In: Bachmann J, Hartmann H (eds) Der Fluß als Lebensraum im Wandel der Zeit. Schiffahrt, Handel, Häfen. J.C.C. Bruns, Minden, pp 295–325

Amacher U (2006) Geschichte der Fischer und Fischerei im Mittelalter. In: Hüster Plogmann H (ed) Fisch und Fischer aus zwei Jahrtausenden. Eine fischereiwirtschaftliche Zeitreise durch die Nordwestschweiz. Forschungen in Augst, vol 39. Römermuseum Augst, Augst, Switzerland, pp 102–103

Anonymus (1876) Bericht der Königlichen Regierung in Schleswig an den Herrn Landwirtschaftsminister über die Fischerei des Jahres 1875. Circulare des Deutschen Fischerei-Vereins im Jahre 1876, pp 89–95

Anonymus (1877) Aus einem Schreiben des Störfischereivereins zu Itzehoe 26. Juli 1877. Circulare des Deutschen Fischerei-Vereins im Jahre 1877, p 171

Anonymus (1939) Maßnahmen zur Vermehrung des Störes in der Eider. Fischerei-Zeitung Neudamm 27(42):322–323

Antonovics J, McKane AJ, Newman TJ (2006) Spatiotemporal dynamics in marginal populations. Am Nat 167(1):16–27

Backhaus J (1935) Wo ist der Stör geblieben? Der Fischerbote 3:115–116

Bauch G (1958) Untersuchungen über die Gründe für den Ertragsrückgang der Elbfischerei zwischen Elbsandsteingebirge und Boizenburg. Z Fisch VII(3–6):161–437

Béguer M, Beaulaton L, Rochard E (2007) Distribution and richness of diadromous fish assemblages in Western Europe: large-scale explanatory factors. Ecol Freshwat Fish 16:221–237

Benecke B (1881) Fische, Fischerei und Fischzucht in Ost- und Westpreußen. Hartungsche Verlagsdruckerei, Königsberg, 331 pp

Benecke B, Dallmer E, von dem Borne M (1886) Handbuch der Fischzucht und Fischerei. Paul Parey, Berlin, 705 pp

Blankenburg A (1910) Von der Störfischerei in der Elbe. Der Fischerbote 2(1):7–12

Bloch ME (1785) Oekonomische Naturgeschichte der Fische Deutschlands. Buchhandlung der Realschule, Berlin, 375pp

Bonne G (1905) Die Vernichtung der deutschen Flußfischerei durch die Verunreinigung unserer Gewässer mit besonderer Berücksichtigung der Verhältnisse auf der Unterelbe. Zeitschrift für Fischkunde 12:1–28

Brandt AV (1964) Fish catching methods of the world. Wiley, London, 348 pp

Breckwoldt JJ (1894) Historische Nachrichten über die Elbinsel Altenwerder. Danckwert, Harburg, 66 pp

Brunn v, Hatzfeld-Trachenberg, Weigelt, Adickes (1894) Zum Störfang und dessen künstlicher Erbrütung; in: Die Verhandlungen des II. Deutschen Fischereirathes. Zeitschrift für Fischerei 2:218–305

Cordes JJ (1958) Die Störe sterben aus. Ndt Heimatblatt 98

Cornils E, Cornils NP, Rust J (2008) Von Ording an den Südpol. Johannes Cornils – ein Fischerleben. Cobra Verlag, Langenhorn, 95 S

Demoll R, Maier HN (1940) Handbuch der Binnenfischerei Mitteleuropas, Band III B. Schweizerbart'sche Verlagsbuchhandlung, Stuttgart, 336 pp

Dröscher W (1926) Die Küstenfischerei. In: Jahresbericht über die deutsche Fischerei 1925. Reichsministerium für Ernährung und Landwirtschaft, Berlin, pp 238–286

Ehrenbaum E (1894) Beiträge zur Naturgeschichte einiger Elbfische (*Osmerus eperlanus* L., *Clupea finta* Cuv., *Acerina cernua* L., *Acipenser sturio* L.). Beilage zu den Mitt Deutsch Seefisch Ver 10, 49 pp

Ehrenbaum E (1913) Vorschläge zur Hebung der Störfischerei. Bericht vom 7. Oktober 1913 an den Regierungs-Präsidenten zu Schleswig. 15 S

Ehrenbaum E (1916) Zum Antrag auf Erhöhung des gesetzlichen Mindestmaßes für den Stör. Der Fischerbote 8(1/2):31–33

Ehrenbaum E (1923) Die Eider als Störfluss und die Schonung des Störs. Fischerbote 15(5):77–83

Ehrenbaum E (1926) Die Eider als Störfluss und die Schonung des Störs. Der Naturforscher 2(10):10–15

Elsner B (1886) Bericht über die Arbeiten in 1885. 9. Jahresbericht des zentralen Fischereiverbandes Schleswig-Holstein, Rendsburg

Elsner B (1887) Die künstliche Vermehrung des Störs in Schleswig-Holstein: Experimente im Fluß die Stör im Jahre 1877. Itzehoher Nachrichten 87
Elsner B, Stemann v (1886) Vorläufiger genereller Bericht über die Störerbrütung im Sommer 1886. Circulare des Deutschen Fischerei-Vereins 4:212
Frauen M (1877) Störzucht in Holstein. Circulare des Deutschen Fischerei-Vereins 6:166–170
Gessner J (2000) Reasons for the decline of *Acipenser sturio* L. 1758 in Europe and attempts for its restoration. Boletín Instituto Espanol de Oceanografía 16:117–126
Goedecken H (1969) Der Königliche Kaviar. Praxis der Fischwirtschaft, vol 6. Heinr. Siepmann Verlag, Hamburg, 112 pp
Haase H (2000) Faszination Fisch – Geschichtliches zum Fisch und seinem Fange. Findling, Neuenhagen
Hansen CP (1877) Chronik der friesischen Uthlande. Lühr and Dirks, Garding, 320 pp
Hardin G (1968) The Tragedy of the Commons. Science 162:1243–1248
Hessel P (1675) Hertzfließende Betrachtungen vom Elbe-Strom. DeLeeu; Faksimile, Schäfer Hannover, Biblii rari, 220 pp
Hoffmann RC (1996) Economic development and aquatic ecosystems in mediaeval Europe. Am Hist Rev 101:631–669
Holčik J, Kinzelbach R, Sokolov LI, Vasil'ev V (1989) *Acipenser sturio* Linnaeus, 1758. General introduction of fishes. Acipenseriformes. In: Holčik J (ed) The freshwater fishes of Europe, vol 1. AULA Verlag, Wiesbaden, pp 367–394
Kausch H (1996) Die Elbe – ein immer wieder veränderter Fluss. In: Lozan JL, Kausch H (eds) Warnsingnale aus Flüssen und Ästuaren. Paul Parey, Berlin, pp 42–51
Kienau R (1924) Jann Rogenstör. Uns' Modersprak, vol 6. Quickborn, Hamburg, 46 pp
Kinzelbach R (1987) Das ehemalige Vorkommen des Störs, Aciopenser sturio (Linnaeus 1758) im Einzugsgebiet des Rheins (Chondrostei: Acipenseridae). Z Angew Zool 74(2):167–200
Kinzelbach R (1997) The sturgeon (Acipenser sturio L. 1758) in Europe. Zeitschrift für Ökologie und Naturschutz 6:129–135
Kirschbaum F, Gessner J (2002) Perspectives for the re-introduction of the European sturgeon, Acipenser sturio L., in the river Elbe system. Zeitschr Fischkunde suppl 1:217–232
Kisker A (1926) Die Fischerei in der mittleren Elbe. Zeitschrift für Fischkunde 24:9–17
Kluge H (1904) Zum Stöorfang am Cracauer Elbwehr bei Magdeburg. Fisch Ztg Neudamm 7 (10):153–155
Köhn G, Wilkes W (1979) Alt-Glückstadt in Bildern. G. Köhn und W. Wilkes, Hamburg, 304 pp
Koos H (1924) Die Elbstörfischerei. Fisch. Ztg. Neudamm 27(47):563–565
Koos H (1935) Die Störfischerei in der Elbe. Der Fischerbote 6:209–211
Kottelat M, Freyhoff J (2007) Handbook of European freshwater fishes. Kottelat, Cornol, pp 1–646
Kroezus E (1953a) Das Ende des Störfanges in der Eider? Der Fischwirt 3(11):371–372
Kroezus E (1953b) Letzter Rettungsversuch für den heimischen Stör. Der Fischwirt 3(5):152–154
Kroezus E (1967) Der gemeine Stör stirbt aus. Der Fischwirt 17:193–199
Lepiksaar J, Heinrich D (1977) Untersuchungen an Fischresten aus der früh-mittelalterlichen Siedlung Haithabu. In: Schietzel K (ed) Berichte über die Ausgrabungen in Haithabu vol 10, p 17
Linde R (1908) Die Niederelbe. Monographien zur Erdkunde, 28th edn. Velhagen and Klasing, Bielefeld, 200 pp
Lindemann M (1888) Beiträge zur Statistik der Deutschen Seefischerei. Im Auftrage der Sektion für Küsten- und Hochseefischerei. W. Moeser, Berlin, 247pp
Ludwig AN, Gessner J (2007) What makes the difference? Sea sturgeon on both sides of the Atlantic Ocean. Am Fish Soc Symp 56:285–300
Mahn H (1979) In Glückstadt um 1890. In: Köhn G, Wilkes W (eds) Alt-Glückstadt in Bildern. Köhn und Wilkes, Hamburg, pp 82–89, 304 pp
Mohr J (1915) Jahresbericht des Zentral-Fischerei-Vereins für Schleswig-Holstein. Der Fischerbote 3/4:102–103
Mohr E (1952) Der Stör. Brehmsche Verlagsbuchhandlung, Leipzig
Müller F (1917) VIII. Fischereiwesen. In: Das Wasserwesen an der schleswig-holsteinischen Nordseeküste, I. Teil – Die Halligen-, Band 1, pp 369–377

Nadler J (2010) http://www.schleifischer.de/html/mittelalter5.htm
Oesau W (1962) Glückstadts Heringslogger unter Segel. Rautenberg, Glückstadt, 26 pp
Oesau W (1980) Schleswig-Holsteins Grönlandfahrt auf Walfischfang und Robbenschlag vom 17. bis 19. Jahrhundert. Augustin Verlag, Glückstadt
Parnell R (1838) Prize essay on the fishes of the district of the Firth of Forth Wernerian Natural History Society, Edinburgh, 440 pp
Prummel W (1983) Excavations at Dorestad 2. Early medieval Dorestad. An archaeozoological study. Amersfoort
Quantz H (1903) Störfischerei und Störzucht im Gebiete der deutschen Nordseeküste. Mitteilungen des Deutschen Seefischerei-Vereins 19:176–204
Rehbein KH (2003) Unsere drei fahren jeden Tag nach den Fischen. Books on Demand, Norderstedt, 450 pp. ISBN 3-8330-1061-4
Riedel-Lorje JC, Gaumert T (1982) 100 Jahre Elbe–Forschung. Hydrobiologische Situation und Fischbestand 1842–1943 unter dem Einfluß von Stromverbau und Sieleinleitungen. Arch Hydrobiol suppl 61:317–376
Rönnau K (1987) Historische Entwicklung und gegenwärtiger Stand der Fischerei auf der Unterelbe. Diplom Arbeit, Universität Kiel, 97 pp
Rübcke M (1910) Die Finkenwärder Elbfischerei in früherer Zeit. Der Fischerbote 6:148–149
Schiemenz P (1913) Über den Rückgang der Fischerei in den Flüssen und Strömen. Zeitschrift für Fischerei 17:3–15
Schiemenz P (1921) Die Fischerei in unseren Strömen einst und jetzt. D Fisch Ztg 44:276, ff
Schiemenz F (1967) Wurde der Stromausbau durch die Opferung der Fischerei rentabel? Denkschrift, Hannover, 36 pp
Schimenz P (1908) Weitere fischereiliche Studien über anorganische Abwässer. Zeitschrift für Fischerei 13:49–81
Schirmer M (1994) Ökologische Konsequenzen des Ausbaus der Ästuare von Elbe und Weser. In: Lozan JL, Rachor E, Relse K, Westernhagen HV, Lenz W (eds) Warnsignale aus dem Wattenmeer. Blackwell, Berlin, pp 164–171, 368 pp
Schmitt Kline K, Bruch RM, Binkowski FP, Rashid B (2010) People of the Sturgeon. Wisconsin Historical Society Press, Madison, 292 pp
Schnakenbeck W (1928) Die Nordseefischerei. Handbuch der Seefischerei Nordeuropas, vol 5, issue 1. Schweitzerbart, Stuttgart, 229 pp
Schonevelde S (1624) Ichthyologia et nomenclatvræ animalium marinarum, fluviatilium, lacustrium quæ in Florentissimis ducatibus Slesvici et Holsatiæ and celeberrimo Emporio Hamburgo occurrent triviales: ac plerorumâq bactenus desideratorum imagines, breves descriptiones, & explicationes auctore. Ex Bibliopolio Heringiano in Hamburgi, 87 pp
Seligo A (1931) Die Seefischerei von Danzig. Handbuch der Seefischerei Nordeuropas, vol 8, issue 7. Schweitzerbart, Stuttgart, pp 25–28
Tiedemann-Wingst G (2001) Hamburger Caviar. Jahrbuch der Männer vom Morgenstern 79:71–97
Volk R (1910) Die Bedeutung der Sielabwässer von Hamburg Altona für die Ernährung der Elbfische. Der Fischerbote 2(3):57, ff
Von Bingen H (1159) Das Buch von den Fischen. In: Riethe P (ed) Hildegard von Bingen – Das Buch von den Fischen. Otto Müller, Salzburg, pp 36–111, 150 pp
Von dem Borne M (1877) Dr. Knoch's Reise zur Wolga zwecks Sterlettbefruchtung, im Bulletin de la Socièté Imperiale des Natualistes du Moscou, TOME XLIV 1871. Circulare des Deutschen Fischerei-Vereins im Jahre 1877. 185–186 S
Wilkens H, Köhler A (1977) Die Fischfauna der unteren und mittleren Elbe: die genutzten Arten, 1950–1975. Abh Verh Naturwiss Ver Hamburg 20:185–222
Zawatka D, Reichstein H (1997) Untersuchungen an Tierknochenfunden von den römischen Siedlungsplätzen Bentumersiel und Jegumkloster an der unteren Ems (Ostfriesland). PKN 12:85–128

Chapter 15
History of the Sturgeon in the Baltic Sea and Lake Ladoga

Ryszard Kolman, Andrzej Kapusta, and Jacek Morzuch

Abstract This paper presents a short review of the information on the history regarding Atlantic sturgeon *Acipenser oxyrinchus* Mitchill in the Baltic Sea and Lake Ladoga. Due to overfishing and habitat alternation, including damming and pollution, the Atlantic sturgeon was extirpated from this area. The history of the fisheries exploitation of the Atlantic sturgeon population dates back to the Neolithic period.

15.1 Introduction

The zoogeographic phylogeny of the *Acipenser sturio* group, which includes *Acipenser sturio* L. and *Acipenser oxyrinchus* Mitchill, has one of the longest histories of all the sturgeons. It dates back to the Late Cretaceous period, which means that it has lived on the earth for about 95 million years (Berg 1948; Nesov and Kaznyshkin 1983; Birstein et al. 1998). Throughout its long and experience-rich history, the sturgeon has witnessed drastic environmental changes that were tragic for many groups of animals, which either lost their significance or became extinct forever. This paper aims to provide a general characteristic of the past status of Atlantic sturgeon *A. oxyrinchus* on the Baltic Sea and the Ladoga Lake.

15.2 Sturgeon in the Prehistory Period

Following the Pleistocene glacier period, which ended about 11,500 years ago, sturgeon, similarly to other fish, began to increase their areas of occurrence and began to gradually settle in newly-formed seas and river systems. The post-glacial,

R. Kolman (✉) • A. Kapusta • J. Morzuch
Department of Ichthyology, Inland Fisheries Institute, Oczapowskiego 10, 10-719 Olsztyn-Kortowo, Poland
e-mail: kolrys@infish.com.pl

P. Williot et al. (eds.), *Biology and Conservation of the European Sturgeon Acipenser sturio L. 1758*, DOI 10.1007/978-3-642-20611-5_15,

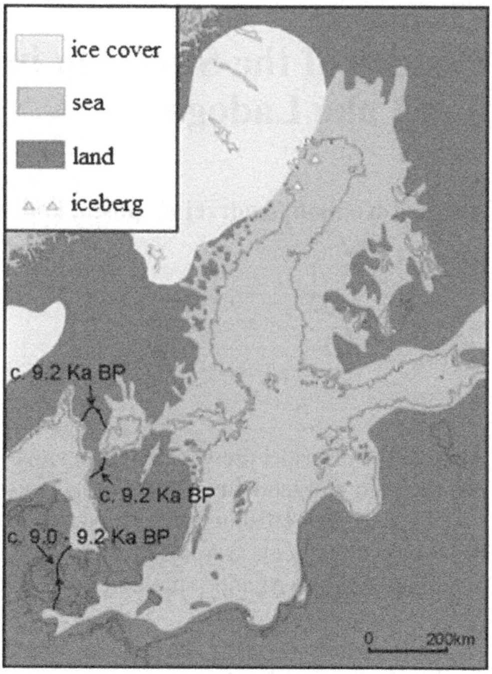

Fig. 15.1 Ancylus Lake, about 9,200 years ago (according to Uścinowicz 2003)

freshwater Lake Ancylus (Fig. 15.1), which was located in part where today's Baltic Sea lies, was joined definitively with the North Sea in approximately 8500 BCE. Saline oceanic waters began to flow gradually into the lake, since it was at a lower elevation. During the Boreal period of the Holocene, which was about 7500–6900 BCE, this process intensified, and resulted in a significant increase in salinity. The resulting environmental conditions were suitable for marine animals to inhabit this newly-formed sea, and this included sturgeon, which probably appeared in the Littorina Sea (Fig. 15.2) towards the end of this period (Makowiecki 2003).

During the Atlantic period (6900–3700 BCE), the climate warmed even further and the sea assumed a shape that resembles today's Baltic; it also attained a wide, permanent connection with the North Sea. This was the optimal period for the development of the Baltic sturgeon population. Its abundance was so high during this period that this fish became a fishing target. This is indicated by finds at archeological excavations along the Baltic coast. The oldest sturgeon bone fragments were found in Dąbki (near Darłowo, central coast) at a proto-Neolithic site (first half of the Atlantic Period). Other sturgeon finds come from the Neolithic and late Neolithic periods in Rzucewo near Puck (Gdańsk Bay) and from Szczecin (Makowiecki 2003).

The history of the fisheries exploitation of the Atlantic sturgeon population dates back to the Neolithic period (Makowiecki 2003). Sturgeon ascending rivers for spawning migrations were relatively easy catch for these ancient fishermen, who,

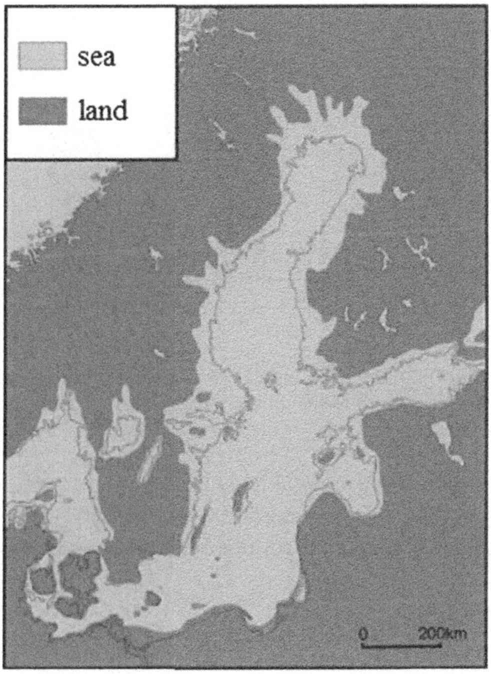

Fig. 15.2 Littorina Sea about 7,500 years ago (according to Uścinowicz 2003)

even by the Neolithic period, deployed fishing methods that were sufficiently effective. The intensity of sturgeon catches increased along with the growing population inhabiting the Baltic coastal regions. This is confirmed by the results of archeological excavations conducted on Wolin Island and nearby Old Gdańsk, Staraya Ladoga, and Pstkow, where, in former human settlements, the share of sturgeon scutes among the fishbones discarded after catch and consumption was very high (Lebedev 1960; Urbanowicz 1965; Filipiak and Chełchowski 2000).

15.3 Sturgeon in the Middle Ages

In the Middle Ages there was a royal monopoly on the catching of sturgeon (Mamcarz 2000). Unfortunately, negative impacts of human fishing efforts were noted in the declining abundance of the Baltic sturgeon population as soon as in the early Middle Ages. During the period from the seventh to the tenth centuries, the relative number of sturgeon caught near Staraya Ladoga decreased by 150%, while catches declined nearly sevenfold in the vicinity of Old Gdańsk from the tenth to the thirteenth centuries (Fig. 15.3) (Lebedev 1960; Urbanowicz 1965; Kolman 2003). Population numbers of Atlantic sturgeon from Baltic Sea continued to decline, as is evidenced in the late Middle Ages by fewer exports from Piława,

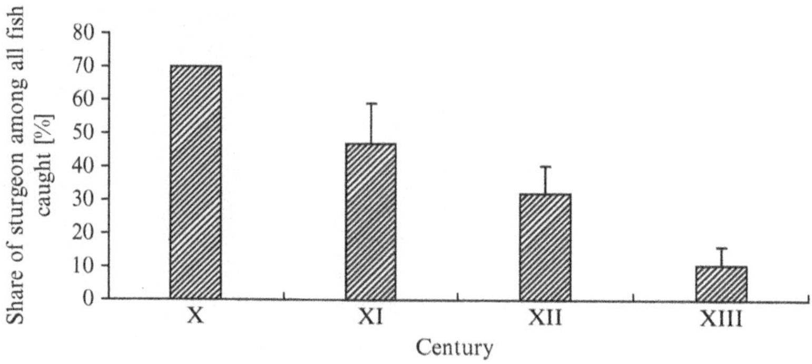

Fig. 15.3 Decrease in relative catch size of Baltic sturgeon in the Vistula Mouth during the period from the tenth to the thirteenth centuries (according to Urbanowicz 1965, Kolman 2003)

which was a notable sturgeon trade center in the seventeenth and eighteenth centuries (Dovydenko 2004; Cios 2007). Similar data from late-eighteenth century Gdańsk also confirm this phenomenon (Cios 2007).

15.4 Sturgeon in the Modern Era

By the end of the nineteenth century, the fate of the sturgeon was sealed by swift industrial development and parallel developments in water transport, constructions on rivers, and increasing pollution (Kolman 1996). The synergistic impact of these factors, combined with the intense catches mentioned previously, led to the extinction of the Baltic sturgeon population. At the beginning of the twentieth century, the Baltic sturgeon was still a significant commercial fish, and catches of it exceeded 200 tons annually, with half of this catch originating from the Gdańsk Bay and the Vistula River (Fig. 15.4).

In the first decades of the twentieth century, the decrease in the numbers of sturgeon in this region was so sudden that by the end of the 1920s catches of this species were counted not in tons, but by the number of fish that were caught (Kulmatycki 1932; Grabda 1968). In the second half of the twentieth century when the sturgeon was under strict protection, which was introduced in 1938, a total of 28 individuals were caught in Poland (Kolman 1996; Mamcarz 2000).

The Atlantic sturgeon's population survived longer in eastern part of Baltic Sea. In the 1930s, annual Atlantic sturgeon catches were about 6 tons (Chalikov 1949). A unique fresh-water population from Lake Ladoga existed until the mid-1980s. The spawning grounds of this population were found in the Volkhov (Kuderskii 1983). The last sturgeon in the Ladoga Lake was caught in 1984 (Podushka 1999).

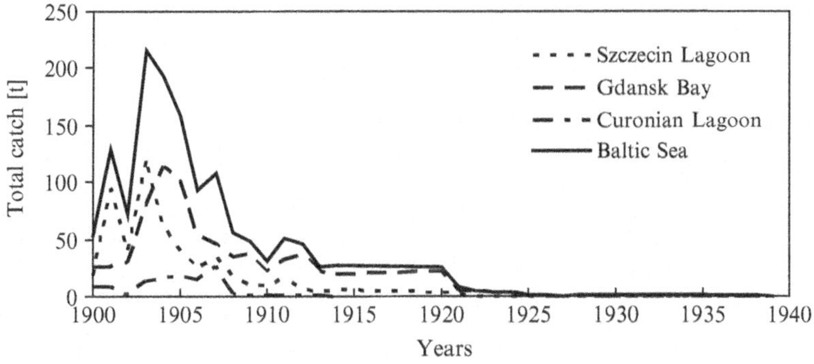

Fig. 15.4 Decreases in the size of Baltic sturgeon catches in the Baltic Sea in the twentieth century (according to Kolman 2003)

15.5 Conclusion

Probably the last catches of Baltic sturgeon from the Vistula population were made in the Vistula River upstream from Toruń in 1965 and 1967 and in the mouth of the Vistula in 1972 (Anon. 1965; Kolman et al. 2008). In 1996, a female sturgeon with a weight of 135 kg and a body length of 2.7 m was caught in fishing gear in Estonian territorial waters near the island of Saarema (Paaver 1996). These were the last reports of catches, which is evidence of the probable total extinction of the population of this species.

References

Anon. (1965) Sturgeon caught. Gosp Ryb 12:21 (in Polish)

Berg LS (1948) Ryby presnykh vod SSSR i sopredelnykh stran, vol 1. AN SSSR, Moskwa

Birstein VI, Betts J, DeSalle R (1998) Molecular identification of Acipenser sturio specimens: a warning note for recovery plans. Biol Conserv 84:97–101

Chalikov BG (1949) Atlanticheskij osetr -Acipenser sturio L. Promyslovye Ryby SSSR.Rass T.S. (red.), Pishchepromizdat. Moskva.:69–71

Cios S (2007) Fish in the lives of Poles from the tenth to nineteenth centuries. Wyd. IRS, Olsztyn, p 251 (in Polish)

Dovydenko LB (2004) Tayny Pillau. Ocherki iz istorii Goroda. Izd. Terra Baltika, Kaliningrad

Filipiak J, Chełchowski Z (2000) Osteological characteristic of fish remains from early medieval sedimentary layers of the port in the town of Wolin. Acta Ichth Piscat 30:135–150

Grabda E (1968) Endangered sturgeon. Ochrona Przyrody 33:177–191 (in Polish)

Kolman R (1996) The past and the future of the sturgeons in Poland. Zool Pol 41(suppl):171–178, in Polish

Kolman R (2003) Which sturgeon became extinct in the Baltic? Komun Ryb 1:1–3 (in Polish)

Kolman R, Kapusta A, Szczepkowski M, Duda A, Bogacka-Kapusta E (2008) Baltic sturgeon *Acipenser oxyrhynchus oxyrhynchus* Mitchill. Wyd. IRS, Olsztyn (in Polish)

Kuderskii LA (1983) Osetrovye ryby v basseinakh Onezhskogo i Ladozhskogo ozer. Sbornik nauchnykh trudov GosNIORKH 205:128–149

Kulmatycki W (1932) About protection of the sturgeon in Polish rivers. Ochrona Przyrody 12:8–28 (in Polish)

Lebedev VD (1960) Presnovodna chetvertichnaya ikhtiofauna evropeyskoy chasti SSSR. Izd. Moskovskogo Universiteta, Moskva, pp 39–44

Makowiecki D (2003) History of fish from the Holocene period in Polish lowlands as seen through archeological studies. Wyd. Instytut Archeologii i Etnologii PAN, Poznań (in Polish)

Mamcarz A (2000) Decline of the Atlantic sturgeon Acipenser sturio L., 1758 in Poland: an outline of problems and prospects. Bol Inst Esp Oceanogr 16:191–202

Nesov LA, Kaznyshkin MN (1983) New sturgeons form the Cretaceous and Paleogene of the USSR. Contemporary problems of paleoichthyology. Nauka, Moscow, pp 68–76

Paaver T (1996) A common or Atlantic sturgeon, Acipenser sturio, was caught in the Estonian waters of the Baltic Sea. Sturgeon Q 4:3–7

Podushka SB (1999) Poimka atlanticheskogo osetra Acipenser sturio v Ladozhskom ozere. Izd INENKO, St. Petersburg, 5–10

Urbanowicz K (1965) Catches of European sturgeon *Acipenser sturio* L. in the early medieval Gdansk in the light of archaeological findings. Przeglad Zologiczny 9:372–375 (in Polish)

Uścinowicz S (2003) How the Baltic changed. Publication in service Polish Geological Institute. http://www.pgi.gov.pl/geologia-morza-kopalnia-lewe/387-jak-zmienias-siastyk (in Polish)

Chapter 16
The Historical and Contemporary Status of the European Sturgeon, *Acipenser sturio* L., in Italy

Paolo Bronzi, Giuseppe Castaldelli, Stefano Cataudella, and Remigio Rossi

Abstract The European sturgeon, *Acipenser sturio* L., was present in rivers of the north-western Adriatic Coast together with the Adriatic sturgeon, *Acipenser naccarii* Bonaparte 1836, and the great sturgeon, *Huso huso* L., while it was the only species on the Tyrrhenian Coast.

It was the most common sturgeon species in the Po River, where the first signal of decline, at the end of the nineteenth century, became a progressive trend in the first half of the following century and accelerated in the 1950s, when industrial development took place. In a different way, the most important Tyrrhenian population in the Tiber River underwent a rapid decline and disappearance in the 1920s, due to increasing fishing and pollution.

In Italy, sturgeon fishing was declared illegal in the 1980s, but in a few years the species disappeared. Nowadays, a reintroduction plan could take advantage of the experience gained with the recovery plan carried out for *A. naccarii*.

16.1 Historical Distribution

In Italy, the European sturgeon, *Acipenser sturio* L., called "storione comune," lived sympatrically with two other species of sturgeons, the "cobice," *Acipenser naccarii,* and the "ladano" *Huso huso,* distributed exclusively in the Po, in its main

---

P. Bronzi (✉)
World Sturgeon Conservation Society, via Piave, 8, 20854 Vedano al Lambro, Italy
e-mail: aerespe@mac.com

G. Castaldelli • R. Rossi
Department of Biology and Evolution, University of Ferrara, via L. Borsari, 46, 44121 Ferrara, Italy

S. Cataudella
Department of Biology, University of Tor Vergata, via della Ricerca Scientifica, 00173 Rome, Italy

P. Williot et al. (eds.), *Biology and Conservation of the European Sturgeon Acipenser sturio L. 1758*, DOI 10.1007/978-3-642-20611-5_16,

tributaries and in the other major rivers of the Padano, Atesino and Veneto catchments, flowing into the northern part of the Adriatic Sea, while it was the only sturgeon present along all the other coasts of Italy and in the Tiber River in the latter reported as not abundant also during the Roman Empire (Bonaparte 1832-41; Canestrini 1872; Vinciguerra 1889; Parona 1898; Tuttolomondo 1899; Pavesi 1907; D'Ancona 1924b). This picture was confirmed later also by Tortonese (1970) and Holcik (1989).

In the Po, the largest Italian river and the most important one for sturgeons, Scotti (1898) and Pavesi (1907) confirmed the presence of *A. sturio* from Ferrara to Casale Monferrato, and Festa (1892) indicated the first decrease in the distribution of this species, which originally extended upstream to Turin, after the construction of the Casale Monferrato Dam in 1874. Based on fishery reports, D'Ancona (1924b) gave some quantitative indications, describing *A. sturio* as rare in the Ligurian Sea and Tyrrhenian Sea, including Sicily. Also in the Ionian Sea, and in the lower and middle Adriatic Sea as well as in the Adige, Piave, Arno, Liri, Volturno and Garigliano rivers, he considers the species rare. In contrast, it was frequent and commonly fished in the upper Adriatic Sea, the Po River and its tributaries, and in the Tiber River (Fig. 16.1). The presence of *A. sturio* in the Po River and in the other rivers flowing into the northwestern Adriatic Sea, as well as in the coastal lagoons from Venice to the Slovenian border, was confirmed by Paccagnella (1948), while Bini (1971) confirmed its presence in the rivers Tiber, Arno, and Volturno flowing into the Tyrrhenian Sea.

Until at least the 1980s, the species was present in the Po River and its main Alpine tributaries, the Mincio, Adda, and Ticino rivers, as well as in the Adige, Brenta, Piave, Livenza, and Tagliamento rivers flowing into the Adriatic Sea (Tortonese 1970; Gandolfi and Zerunian 1987; Gandolfi et al. 1991).

Although we can not exclude the possibility that some additional information on the distribution of *A. sturio* could be still added, due to the difficulty of recovering records scattered among various locations (museum, hydrobiological stations, universities, etc.) and not readily available, in this research we summarize the most important legislative, commercial and scientific documents, collected in private and public archives and libraries. In Table 16.1 and in Figs. 16.1 and 16.2, the reported fisheries and accidental captures, including area, site, year, and authorities are given. Further, more recent records are not verified, since sturgeon fishing became illegal after the Decree of the Ministry of Mercantile Marine (G.U. no. 158, 1989) and subsequent regional regulations came into force.

16.2 Biological Data

Knowledge of the morphological characteristics of the Italian *A. sturio* is limited to descriptions by Tortonese (1970) and Gandolfi et al. (1991). The most informative reports of biology of *A. sturio* inhabiting Italian waters are those of D'Ancona (1923, 1924a, b) on the Tiber population and Paccagnella (1948) on the Po River population. They give details of the reproductive periods of the two populations,

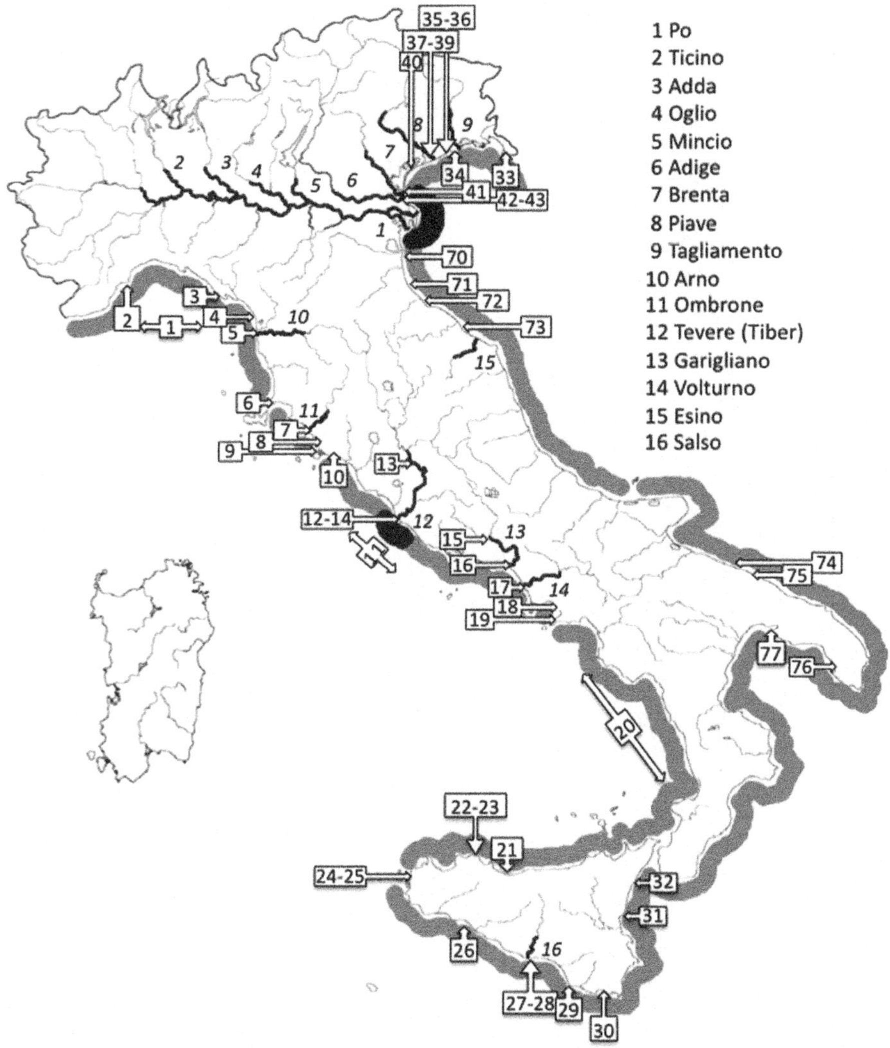

Fig. 16.1 Map showing sites where *A. sturio* was recorded: rivers are in *black* (with names listed beside) and coastal areas in *dark gray*, if records were frequent, and *light gray*, if rare. Numbers in frame with *arrow/s* refer to the details given in Table 16.1

which were similar, starting with migration in March and April, with a reproductive period in May and June in the Po, extending into the early summer. Spawning grounds were shallow, running water areas with sandy, pebble bottoms, originally located upstream to Casale Monferrato and Turin on the Po River (Pavesi 1907) and upstream to Todi on the Tiber (Silvestri 1892). Paolucci (1901) reports also the importance for reproduction of river mouths and brackish waters.

D'Ancona (1924a) and Paccagnella (1948) reported different growth patterns for juveniles, growing more rapidly in the Tiber than in the northern part of the Adriatic

Table 16.1 *A. sturio* captures in Italian waters listed according to area, catch site, year of catch, and source

Area	No.	Site	Desc.	Place	Year	Reported by	Cited in	Note
Ligurian Sea								
	1	Mare Ligusticum	Ligurian Sea			Sassi	Carus	
	2	Castelfranco	Town		1880	Giglioli	Carus	
	3	Monterosso	Town		1924	D'Ancona a		Last catch March 1917
	4	Livorno	Town		1880	Giglioli	Carus	Rare at the mouth
Tyrrenian Sea								
	5	Arno	River		1898	Pesca in Italia, Pavesi	Scotti, D'Ancona a	
	6	Portovecchio di Piombino	Town		1924	D'Ancona a		Last catch 1904
	7	Ombrone	River	Pescaia Poggio Cavallo	1905	Del Rosso	D'Ancona b	1892-at Grosseto
	8	Albegna	River		1900	Del Rosso	D'Ancona b	
	9	Porto S. Stefano	Town		1924	D'Ancona a		Very rare
	10	Fiora	River	Mouth	1924	D'Ancona b		1–2/year
	11	Roman coasts	Coastal area		1890	Bonaparte, Vinciguerra	Carus	
	12	Tiber	River		1832 1841	Bonaparte		
	13	Tiber	River	–>Todi, Orte	1892	Vinciguerra, Silvestri	Pavesi, D'Ancona b	
	14	Tiber	River		1924	D'Ancona	D'Ancona a	
	15	Liri	River		1871 1874	Pesca in Italia	Pavesi, Scotti, D'Ancona b	
	16	Garigliano	River	Mouth	1869	Costa, Pavesi	D'Ancona a	
	17	Volturno	River	Capua, mouth	1898	Bindi, Pavesi	Pavesi	Rare at the mouth
	18	Napoli	Town		1869	Costa, Staz. Zool.	Carus	

	19	Sarno	River	Mouth	1924	D'Ancona b		Last catch 1904
	20	Calabria	Region		1869	Costa	Carus	
Sicily	21	Termini Imerese	Town		1924	D'Ancona b		One (30 kg) in 1904; one (15 kg) July 1922
	22	Oreto	River	Mouth	1881	Doderlain	Carus, Pavesi	
	23	Palermo	Town			Targioni	Carus	
	24	Trapani	Town		1881	Doderlain	D'Ancona b	Rare
	25	Trapani	Town		1896	Vinciguerra	D'Ancona a, b	1 catch every 5–6 years
	26	Sciacca	Town		1881	Doderlain	Carus	
	27	Licata	Town			Gemmellaro	Carus	
	28	Salso	River	Mouth	1881	Doderlain	Carus, Pavesi	
	29	Scoglitti	Town		1924	D'Ancona b		2–3/year (1923)
	30	Pozzallo	Town		1924	D'Ancona b		2–3/year (1923)
	31	Catania	Town	Maritime Dep.		D'Ancona b		Accidental, one every 3–4 years
	32	Riposto	Town			Aradas	Carus	
Adriatic Sea								
	33	Trieste	Town			Grube, Graeffe	Carus	
Veneto basin	34	Noncello	River		1895	Tellini	Pavesi, Scotti, D'Ancona b	
	35	Livenza	River		1895	Tellini	Pavesi, Scotti, D'Ancona b	
	36	Livenza	River	Quinto di Treviso	1991	Arlati p.c.	MSN Venezia	
	37	Piave	River		1898	Ninni	Pavesi, Scotti, D'Ancona b	
	38	Piave	River	Cortellazzo	1976	Arlati p.c.		12 kg
	39	Piave	River	Passerella	1991	Arlati p.c.		6–7 kg

(continued)

Table 16.1 (continued)

Area	No.	Site	Desc.	Place	Year	Reported by	Cited in	Note
	40	Venezia	Town		1912	Ninni	Carus	
	41	Chioggia	Town			Renier	Carus	Average b.w. 2–8 kg
Atesino basin	42	Adige	River	–>Zevio	1862	De Betta	Pavesi, D'Ancona	
	43	Adige	River	Ronco<>Alberedo	1902	Bettoni	Pavesi	Big fish
	44	Tartaro	River			De Betta	Scotti, D'Ancona b	
Padano basin	45	Po	River	Alessandria	1871–1874	Pesca in Italia	Scotti, D'Ancona b	
					1871			
	46	Po	River	Casale	1874	Pesca in Italia	Scotti, D'Ancona b	
					1871			
	47	Po	River	Pavia	1874	Pesca in Italia	Scotti, D'Ancona b	
	48	Po	River	Portalbera		Prada, Pavesi	Scotti	Big fish
	49	Po	River	Cervesina		Pavesi		Big fish
					1871			
	50	Po	River	Piacenza	1874	Pesca in Italia	Scotti, D'Ancona b	
					1871			
	51	Po	River	Lodi	1874	Pesca in Italia	Scotti, D'Ancona b	
					1871			
	52	Po	River	Cremona	1874	Pesca in Italia	Scotti, D'Ancona b	
	53	Po	River	Ferrara		Giglioli	Scotti, D'Ancona b	
	54	Po	River	Ferrara<>Casale	1892	Festa	Pavesi, Plehn	
	55	Po	River	mouth	1924	D'Ancona b		Relatively abundant
	56	Po	River	Ficarolo	1942	Paccagnella		
	57	Po	River	Calto	1942	Paccagnella		
	58	Po	River	Carbonara Po	1942	Paccagnella		
	59	Po	River	Po di Pila	1942	Paccagnella		
	60	Ticino	River	> Pavia	1907	Pavesi		
	61	Lambro	River		1907	Pavesi		

	62	Olona	River		1907	Pavesi		
	63	Adda	River	–> Serio	1907	Pavesi		
					1871			
	64	Taro	River		1874	Pesca in Italia	Scotti, D'Ancona b	
	65	Serio	River	When flooding			Scotti, D'Ancona a	
	66	Sarca	River		1907	Pavesi	Scotti, D'Ancona a	
	67	Mincio	River	Lower strech	1907	Pavesi		
	68	Secchia	River		1869	Bonizzi	Scotti, D'Ancona b	
	69	Gravallone	River		1877	Pavesi		
	70	Ravenna	Town	Capitaneria di Porto	1924	D'Ancona b		3,000 fish/year
	71	Cesenatico	Town	off shore	1977	Il Pesce, 4/1986		73 kg, off shore
	72	Rimini	Town		1880	Giglioli	Carus	
	73	Fano	Town	Lab. Biol. Pesca Fano	1970	Bronzi p.c.		30 cm into the sea
	74	Molfetta	Town		1869	Costa	Carus	
	75	Bari	Town		1924	D'Ancona b		Accidental
Jonian Sea								
	76	Gallipoli	Town		1924	D'Ancona b		1–2/year
	77	Taranto	Town		1924	D'Ancona b		Accidental

-> : up to; <> : from. . .to; > : upstream

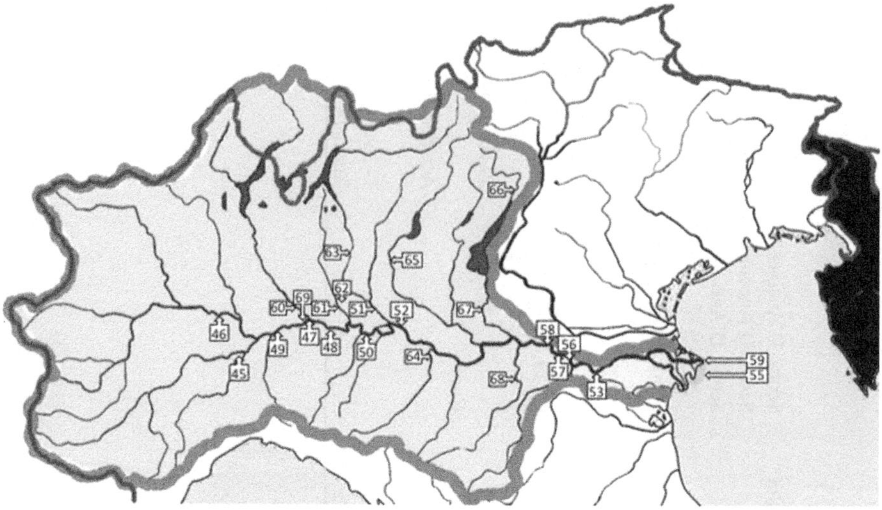

Fig. 16.2 Map showing rivers and sites in the Po basin (*gray area*) where *A. sturio* was recorded; numbers refer to Table 16.1, for details

Sea. These observations agree with the expected trend due to a more favorable latitudinal gradient from the Po to the Tiber, but unfortunately were based on too few specimens. By pooling the data, however, it is possible to give the following length intervals: for 0+ 24–27 cm, for 1+ 31–48 cm, and for 2+ 55–67cm.

In both sites, juveniles 0+ and 1+ generally remained in the lower reaches of the rivers, near the mouth, in fresh or brackish water. In contrast, the adults returned to the sea in autumn, as demonstrated by catches at depths ranging from 10 to 40 m, not far from the mouths of the rivers. According to reports of mature specimens caught in front of the mouths of the Po River Delta, onset of migration was from the end of February to the end of March. In the same study, attempts to verify the age of fish at first maturity were carried out by Paccagnella (1948) by examination of clear and dark layers of dorsal scutes, since sectioning pectoral fin first ray was considered unreliable with specimens over the age of 14. Age at first maturation was determined on a sample of 20 specimens, captured in the 50 km reach of the lower Po River upstream from Ferrara. Maturation did not occur before 9–10 years in males, at a minimum length of 120 cm and weight of 12–13 kg, was reached. In females, first maturation was observed at 11–12 years, at an average minimum length of 155 cm and minimum weight of 42 kg. The same methodology for age determination was carried out by D'Ancona (1924b), who found the sections of dorsal scutes more reliable than those of otholiths.

Unfortunately, no definitive research, other than D'Ancona's (1924a, b) and Paccagnella's (1948) and descriptions of biology and distribution, have been conducted on this species. Genetics studies on the karyotype of Italian population have been carried out by Fontana and Colombo (1974, also see this volume Chapter 2).

16.3 Role in Italian Fishery

The European sturgeon was an important species to Italian fisheries since Roman times, when it was described as "nobilis" by Ovid in his *Halieuticon*, and cited also by Pliny the Elder, Eliano, and Marziale. Its importance was also demonstrated by a baş-relief, originally in the "Forum piscium" (fish market) in the Portico of Ottavia in Rome, and since 1500 located in the Capitoline Museum in Rome in the *Sala dei Conservatori*, showing a sturgeon 115 cm long, that was used as "regulum" i.e., the reference measure to be used by authorized fishermen as minimum legal size for catches (Fig. 16.3). It is documented that in the Middle Ages fishermen of the Tiber were organized in a corporation operating on most of the river course, from the upper reach of the town of Orte to the mouth. From 1400 to 1798, in the case of capture of a sturgeon longer than the "regulum" they were obliged to consign the head, sectioned at the pectoral fins, as a tribute to the authorities (Cataudella 1991). We do not have references proving or disproving the presence of other sturgeon species in the Tiber at that time. However, it is very likely that the use of that "regulum" concerned *A. sturio*, reported to be the only sturgeon species present in the Tiber River (please note in the previous section 16.1) and which in an official document dated 1447 is reported to be the most expensive fish in Roman markets (Cataudella 1991). The quality of sturgeons of the Tiber River was celebrated also by Giovio (1524).

Fig. 16.3 Sturgeon carved in stone used as "regulum" indicating the minimum legal size for fishing (Sala dei Conservatori, Museo Capitolino, Rome) (photo by S. Cataudella)

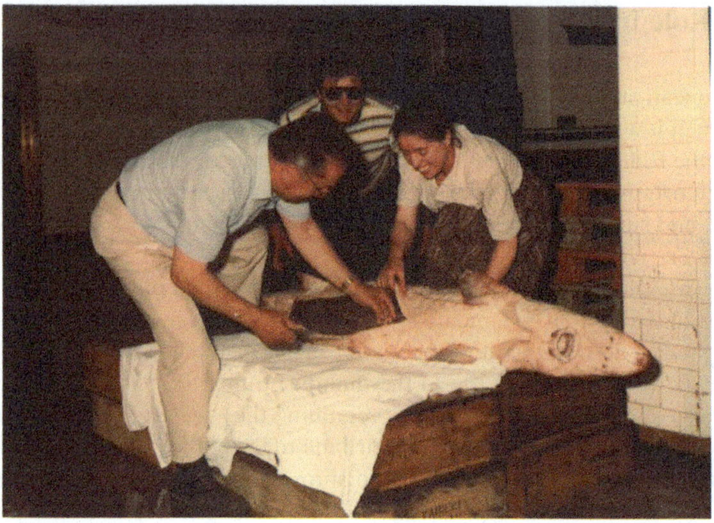

Fig. 16.4 Collecting caviar from *A. sturio* in the 1970s in Ferrara (by courtesy of C. Maresi, Le Occare, Ferrara)

Until the World War II, sturgeon was an important source of income for fishermen of the Po River and its delta. A single large catch could provide sufficient revenue to maintain a family for a year (Zaccaria and Masini 1991). The meat was sold in the markets at Milan and other major cities, while the eggs were used to produce caviar. In Ferrara, a historically important town for fisheries and trade, located on the Po River just upstream the delta, a shop called "la Nuta" sold a special type of cooked caviar until 1972, when it closed because sturgeon became very rare (Fig. 16.4). Through interviews conducted in the most important sites for sturgeon fishing and in the last market receiving wild sturgeons caught in the Po River Delta, in Donada, Rovigo Province, as well as through documents from public archives and libraries in Ferrara (Messisbugo 1549; Archivio di Stato di Ferrara 1722–1797, 1879; Archivio Storico Comunale di Ferrara 1700–1900), it is apparent that the caviar in northern Italy was made preferably from *A. sturio*. This was the reason for a great difference in price between specimens of the two more similar species present along the Adriatic coast, *A. sturio* and *A. naccarii*, the former being much more expensive. In turn, this is the reason why fishermen and dealers were perfectly able to recognize species and sex by sight.

16.4 Decline of the Species

Records show that until WWII *A. sturio* was the most abundant sturgeon in Italian waters, accounting for about 80% of the total catch in 1934 (Brunelli 1934), followed by *A. naccarii*, while *H. huso* was by far the least abundant.

The first note on the decline in sturgeon population is given by Festa in 1892, along with that of other species, due to the barrage constructed on the Po River at Casale Monferrato in 1874 to permit water diversion for irrigation. Festa reports that after the construction of the dam, sturgeons disappeared from the upstream reach, together with other anadromous species, such as twaite shad *Alosa fallax* and sea lamprey *Petromyzon marinus*. Since then, a progressive reduction was reported by Scotti (1898) and D'ancona (1924a). Nevertheless, at the beginning of the twentieth century, sturgeons were still common on Italian fish markets, caught mainly in the Po River basin. At the Trieste fish market, 1,235 kg of *A. sturio* and 764 kg of *A. naccarii* were sold in 1914, while only 445 and 250 kg, respectively, were sold in 1920. Other data concern the Venice fish market, where 51,810 kg of all the three species of sturgeon were sold in the second half of 1919, 69,230 kg in 1920, 69,700 kg in 1921 and 36,680 in the first 10 months of 1922. In 1924, the total annual sturgeon catch in the Po River basin was estimated at about 3,000 fish. In the same year, *A. sturio* essentially disappeared from the Tiber River, and then was found only sporadically along the Tyrrhenian coast (D'Ancona 1924b).

From the 1920s onward, there is a general lack of information, and we do not know when the decline of the Adriatic population accelerated. It is assumed that the major decline took place by the end of the 1950s and throughout the 1960s. Only later, the results of a monitoring program specifically designed for sturgeon species in the Po River low reaches and delta, conducted systematically from September 1972 to February 1975 as well as from March 1987 to September 1989, documented the disappearance of this species over a 15-year period (Rossi et al. 1991). From 1972 to 1975, the overall catch only comprised 20 *A. sturio* (26.1–273.0 cm and 0.052–83 kg) out of a total of 95 sturgeons. This demonstrates a noticeable contraction of *A. sturio* compared to data reported for the area by Brunelli (1934), indicating *A. sturio* accounted for 80% of the total catch. However, both adults and juveniles were present, indicating that recruitment occurred until the beginning of the 1970s (Rossi et al. 1991). From March 1987 to September 1989, all of the 142 sturgeon were caught in the low Po River and delta, and they were all *A. naccarii* (Rossi et al. 1991). Through interviews with local professional fishermen during the same time period, the catch of only two *A. sturio* of weight above 50 kg in this reach of the river was reported. Results of a specific monitoring program, conducted from 1996 to 2006 in the same stretches of the Po for the compilation of the Ferrara Province and Emilia-Romagna Region Fish Inventories (Castaldelli et al. 2008; Castaldelli and Rossi 2008; Lanzoni et al. 2010), confirmed the previously demonstrated trend, with no *A. sturio* caught. The last captures were reported in 1991 in the Piave and Livenza rivers (Arlati, G. personal communication; one specimen at Museo di Storia Naturale di Venezia, no. 14038) (Fig. 16.5).

In the last century, several disturbances may have contributed to its disappearance, but the general scarcity of information together with the complexity of its lifecycle do not allow to make univocal hypothesis. However, it is very likely that sturgeon suffered from different actions in different areas. The sudden disappearance in the 1920–1930 period in the river Tiber could probably be attributed to overfishing during that period of economic development. Moreover, the Tiber is a

Fig. 16.5 Director of the fish market in Cesenatico measuring the last large specimen (2 m, 73 kg) of *A. sturio* legally caught in 1977 (by courtesy of "Il Pesce," 1986 no. 4, p. 80)

relatively small river, and where it was not operatively difficult to intensify fishing effort to an unsustainable level. D'Ancona (1924a, b) reported the importance of increasing pollution as one of the main factors for their decline besides the catches of individuals before reproductive age.

For the rivers along the Adriatic coast, and particularly for the Po River Basin, the decline was more progressive in the first half of the twentieth century, and accelerated in the 1950s. We try here to hypothesize some possible synergic causes.

Fishing (legal and illegal) – the increase of fishing impacts on sturgeon in the Po River Basin may be dated to the 1960s. Professional fishing in the rivers, lagoons, and coastal waters benefited from the introduction of plastic nets and motorized boats, and sturgeon were heavily affected as target (and also, both adults and juveniles, as non-target) species by an increasing fleet of small trawlers fishing more intensively at the mouths of the main rivers flowing into the north-western Adriatic Sea. Moreover, economic development also increased the number of sport fishermen and gear efficiency, in the 1965–1985 period, without the introduction of any regulation for sturgeon protection. Although it is difficult to attribute the relative responsibility of licensed sport fishing, it may have played a significant role, as appeared to be the case through oral interviews with fishermen and their associations, confirming a significant number of specimens recently caught (probably *A. naccarii*, but without the possibility of verification because released or eaten). Last but not least poaching may have had a serious impact of an unknown extent.

Damming – two dams were constructed on the Po main course; one barrage constructed at Casale Monferrato in 1874 to divert part of hydraulic load for irrigation, and the hydroelectric dam of Isola Serafini, built in 1962 for electricity production. Both dams impacted the range of *A. sturio* in the Po River, and still are a barrier to sturgeon migrations (Bronzi 2008) but this does not necessarily exclude spawning, since some suitable areas are located downstream of it, at the inlets of Adda, Oglio, and Mincio rivers. Another fact is the construction, in the period 1950–1970, of several barriers on all the main Po left tributaries (the Ticino, Adda, Oglio, and Mincio) and on all the main rivers flowing into the north-western Adriatic (the Adige, Brenta, and Piave), leading to the loss of longitudinal connectivity in all the most important basins for sturgeon migration.

River canalization – since the 1950s, all the above-cited rivers underwent heavy flow alteration and canalization, resulting in cutting off meanders and bends and eliminating sand bars, and leading to a dramatic loss of habitat heterogeneity and transversal connectivity.

Pollution – the rapid industrial development that took place in the north of Italy since the late 1950s had a serious impact on all large rivers of the Pianura Padana, due to the uncontrolled discharge of synthetic and natural pollutants, potentially interfering at various levels with sturgeon reproductive and growth cycles.

Exotic fish species – since the 1970s, the introduction of several exotic fish species, and particularly of the European catfish, *Silurus glanis* L., (Rossi et al. 1992), may have exerted a negative effect on sturgeon, through predation on smaller sizes and competition for resources.

16.5 Conclusions

Today, there may be some uncertainty as to whether the species is truly extinct, or still present in Italian waters. The absence of capture of juveniles or sub-adults of *A. sturio* demonstrates the lack of recruitment in the past 15 years. Reported rare captures in drop nets (10 × 10 m), which are commonly used in the Po River Delta branches, have all been of *A. naccarii*. However, this does not preclude the possibility that a few *A. sturio* remain, for which the likelihood of being caught by non-targeted fishing is minimal. Furthermore, due to preservation laws, accidental captures may have gone unreported, and poachers are unlikely to reveal illegal catches. Thus, we cannot be certain that no capture has occurred. Currently, it should therefore be considered "missing" as well as *H. huso*, which disappeared in the 1970s, whereas *A. naccarii* continues to persist, due to controlled reproduction initiated in 1988 using broodstock caught in the wild at the beginning of the 1970s and maintained in captivity (Arlati et al. 1988).

Therefore, the perspective of *A. sturio* restoration in Italy should first take accurately into account almost all the above-cited terms of disappearance: damming, remained stable, and pollution, noticeably decreased with respect to the 1970s and 1980s, but still affecting river health. Professional and sport fishermen have acquired an higher consciousness with regard to the necessity for wildlife protection but, mainly in the Po River, poaching for the European catfish is common, and this might become a serious risk for introduced sturgeon if no additional measures are taken. Lastly, among several recently introduced exotic fish species, the European catfish has become an ulterior constraint to sturgeon restoration. The capture of specimens of more than 100 kg is not an exception in the Po River, and measures of limitation have to be included in a plan for *A. sturio* reintroduction.

Although the cited constraints need to be addressed at an interregional level, a plan for recovering *A. sturio* in Italy may take advantage of the experience gained during the last 20 years of the plans for recovering *A. naccarii*, carried out starting from a wild broodstock. Scientific and technical knowledge, together with the availability of suitable structures, may be easily adapted and become strengths for a recovery plan for the European Atlantic sturgeon.

Acknowledgements The authors wish to thank Dr Giovanni Arlati, Dr Emilia Cataldi, Prof Francesco Fontana, Dr Mattia Lanzoni and Dr Mauro Mariani for their help in data and references collection and Mr Sante Asferri, Mrs Elena Benedetti and Mrs Cristina Maresi for kindly providing photographic and documentary material.

References

Archivio di Stato di Ferrara (1722–1797) Arte degli sprocani e degli speziali

Archivio di Stato di Ferrara (1879) Statistiche sulla pesca

Archivio Storico Comunale di Ferrara (1700–1900) Decreti legislativi regolamentativi

Arlati G, Bronzi P, Colombo G, Giovannini G (1988) Induzione della riproduzione nello storione italiano (*Acipenser naccarii*) allevato in cattività. Riv Ital Acquacol 23:94–96

Bini G (1971) Atlante dei pesci delle coste italiane, vol II. Ed. Mondo Sommerso, Milano, 311 pp

Bonaparte CL (1832–1841) Iconografia della fauna italica per le quattro classi degli animali vertebrati. Salviucci, Roma

Bronzi P (2008) The sturgeon's status in Italy with special emphasis on *Acipenser naccarii*. In: Rosenthal H, Bronzi P, Spezia M, Poggioli C (eds) Passages for fish. World Sturgeon Conservation Society, Neu Wulmstort, Special Publication n°2

Brunelli G (1934) Rapporto sulla biologia dello storione. Rapp DU Comm Int Expl Sci Medit 8:77–79

Canestrini G (1872) Pesci. Fauna d'Italia. pt. 3. Vallardi, Milano, 208 pp

Carus JV (1889–1893) Prodromus faunae mediterraneae. Schweizerbart, Stuttgart

Castaldelli G, Rossi R (2008) Carta ittica dell'Emilia-Romagna Zone B e A Regione Emilia-Romagna Ed. Greentime, Bologna, p 324 pp

Castaldelli G, Lanzoni M, Rossi R (2008) La fauna ittica del tratto terminale del fiume Po ieri e oggi. Il Pesce 6:99–109

Cataudella S (1991) La pesca fiumarola e il mercato ittico a Roma. Amm. Prov. Di Roma, Ufficio Pesca, Rome, p 96

D'Ancona U (1923) Dati per la determinazione dell'età e per lo studio dell'accrescimento negli storioni. Rend R Accad Lincei Ser5° Vol XXXII, 1° sem

D'Ancona U (1924a) Contributo alla biologia degli storioni nelle acque italiane. Libreria dello stato, Roma

D'Ancona U (1924b) Dati per la biologia degli storioni nelle acque italiane. Monitore Zoologico Italiano. Anno XXXV n°6–7:126–133

Festa E (1892) I pesci del Piemonte. Boll Musei Zool Anat Comp R Univ Torino 7(129):6

Fontana F, Colombo G (1974) The chromosomes of Italian sturgeons. Experientia 30:739–742
Gandolfi G, Zerunian S (1987) I pesci delle acque interne italiane: aggiornamenti e considerazioni critiche sulla sistematica e la distribuzione. Atti Soc It St nat Milano 128(1/2):3–56
Gandolfi G, Zerunian S, Torricelli P, Marconato A (1991) I Pesci delle acque interne italiane. Ist. Poligr. e Zecca dello Stato, Rome, p 617
Giovio P (1524) De romanis piscibus libellus. Franciscus Minutius Calvus, Roma
Holcik J (ed) (1989) The freshwater fishes of Europe. Vol. 1, Part II. General introduction to fishes Acipenseriformes. AULA-Verlag, Wiesbaden, 469 pp
Lanzoni M, Castaldelli G, Caramori G, Turolla E, Fano EA, Rossi R (2010) Popolamenti ittici del Delta del Po. Biologia Ambientale 24(1):157–166
La pesca in Italia (1871–1874) Documenti raccolti per cura del Ministero di Agricoltura, Industria e Commercio del Regno d'Italia, ordinati da Ad. Targioni Tozzetti. Tip. del R. Istituto sordo-muti, Genova
Messisbugo C (1549) Libro novo nel qual s'insegna a far d'ogni sorte di vivande et apparecchio generale. Biblioteca Comunale Ariostea, Ferrara
Paccagnella B (1948) Osservazioni sulla biologia degli storioni del bacino Padano. Arch Oceanogr Limnol 5(1/3):141–154
Paolucci L (1901) Le pescagioni nella zona italiana del medio Adriatico. Regia Camera di Commercio, p 59
Parona C (1898) La pesca marittima in Liguria. Att. Soc. Lig. Sci. Nat, Genova
Pavesi P (1907) Gli Acipenseri nostrali. Rend R Ist Lombardo Serie II XL:332 (in Italian)
Plehn M (1909) I pesci del mare e delle acque interne. Hoepli, Milano
Rossi R, Grandi G, Trisolini R, Franzoi P, Carrieri A, Dezfuli BS, Vecchietti E (1991) Osservazioni sulla biologia e la pesca dello storione cobice *Acipenser naccari* Bonaparte nella parte terminale del fiume Po. Atti Soc Ital Sci Nat Museo Civ Storia Nat Milano 132: 121–142
Rossi R, Trisolini R, Rizzo MG, Dezfuli MS, Franzoi P, Grandi G (1992) Biologia ed ecologia di una specie alloctona, il siluro (*Silurus glanis* L.) (Osteichthyes, Siluridae), nella parte terminale del fiume Po. Atti Soc Ital Sci Nat Museo Civ Storia Nat Milano 132(7):69–87
Scotti L (1898) La distribuzione dei pesci d'acqua dolce in Italia. Stab. Tip. Crivelli, Roma. 47 pp (In Italian). Giorn. Ital. pesca e acquicolt, nn 1–6. Roma
Silvestri F (1892) I pesci dell'Umbria. Tipografia Boncompagni, Perugia
Tortonese E (1970) Fauna d'Italia. Vol. XI. Osteichthyes, part 2. Calderini, Bologna, p 545
Tuttolomondo A (1899) Fauna ittiologica del Compartimento marittimo di Catania – Girgenti
Vinciguerra D (1889) Guida del Museo di Zoologia della R. Università di Roma. Fauna locale. Specie animali della Provincia di Roma esistenti nella nuova collezione. Parte 3°. Pesci – Lo Spallanzani, anno XXVII, p 340
Vinciguerra D (1896) Relazione intorno alla pesca di acqua dolce e di mare in Sicilia e ai modi di aumentarne il prodotto. Min Agricolt Bollett Notizie agrarie Anno XVIII, p 105
Zaccaria M, Masini G (1991) Il commercio dello storione. Quaderni del Centro Etnografico Ferrarese 31:39–42

Chapter 17
European Sturgeon, *Acipenser sturio* in Georgia

Ryszard Kolman

Abstract The purpose of this chapter is to present the current status of European sturgeon, *Acipenser sturio* L. population in Georgia. The anadromous European sturgeon occurs in Rioni River and in the Black Sea of Georgia. The size of this population has never been estimated. Available information suggests the population in Georgia is disappearing. At present, there are occasional catches of individual sturgeon run to the Rioni River for spawning. The coastal waters of Georgia are predominant feeding grounds for sturgeons, since they offer an unusually rich food base that provides feeding grounds for adults and sub-adults alike.

Keywords Sturgeon • Black Sea • Rioni • Biology • Conservation

17.1 Introduction

The European sturgeon, *Acipenser sturio* L. was historically widely distributed in Europe, but is extirpated in its historic range, apart from a relict population in the Gironde–Garonne–Dordogne basin, France (Lepage and Rochard 1995; Williot et al. 2009), and Rioni, Georgia (Zarkua et al. 2006). The European sturgeon inhabiting the Black Sea was first described by the Rumanian ichthyologist Antipa (1905). The occurrence of the European sturgeon in the Rioni River was confirmed unequivocally by Tikhii (1929). In this river, sturgeon was a common species exhibiting migrations reaching far from the river mouth (Tikhii 1929; Zarkua et al. 2006). Population decline was related to overfishing, mostly for their meat and caviar, pollution, habitat destruction. At present, there are occasional catches of

R. Kolman (✉)
Department of Ichthyology, Inland Fisheries Institute, Oczapowskiego 10, 10-719 Olsztyn-Kortowo, Poland
e-mail: kolrys@infish.com.pl

P. Williot et al. (eds.), *Biology and Conservation of the European Sturgeon Acipenser sturio L. 1758*, DOI 10.1007/978-3-642-20611-5_17,

individual sturgeon run to the Rioni River for spawning. The purpose of this chapter is to present the status and population characteristics of European sturgeon in Georgia.

17.2 Species Description and Differences from the Other Populations Described

The sturgeon from the Rioni River differed significantly from those from the Gironde River in the numbers of dorsal and lateral scutes and gill rakers, but both populations have identical tubercular structures. Tikhii (1929) compiled a description of the meristic characters of individuals from this sturgeon population, and compared it with European sturgeon (Table 17.1).

Male European sturgeon from the Black Sea population attain sexual maturity between the ages of 7 and 9 years at a mean body length of 95–110 cm, while females do so from 2 to 6 years later at longer mean lengths (110–135 cm). Absolute fecundity in the Black Sea population varies depending on fish size from 789 to 1,815 thousand eggs, which corresponds to 22–25% of female body weight (Ninua 1976). Sturgeon fry begin to descend to the sea about 2–2.5 months after hatching. As early as late July, fry are observed in the Rioni Delta with a mean weight of 4.5–6 g, and by November, they measure from 19 to 23 cm in length at a weight of 30–40 g. From late fall, the majority of the fry inhabit the fecund waters of the river mouth where both marine (amphipods, shrimps, small fish fry) and freshwater food organisms (larval *Chironomidae* and other insects) are found. When the fry reach an age of between 2 and 3 years, they move along the coasts of the Black Sea to the north. The European sturgeon is exceptionally mobile, and this permits it to penetrate the entire sea littoral zone in search of food. Its primary food is fish, especially the European anchovy, which winters in large shoals in the littoral zones of the Caucasus region (Kolman and Zarkua 1999).

Table 17.1 Comparisons of selected morphometric characters of European sturgeon, *Acipenser sturio* L., populations. Sd dorsal fin, Sl lateral scutes, Sv lateral scutes, S.br. gill rakers

Character	Gironde River (Magnin 1962; Artyukhin and Vecsei 1999)	Rioni River (Marti 1939; Ninua 1976)
Scute surface	Tubercular	Tubercular
Number Sd	12.74 ± 0.13	14.27 ± 0.09
Number Sl	35.13 ± 0.20	32.36 ± 0.12
Number Sv	11.3 ± 0.10	10.68 ± 0.08
Number S.br.	20.19	24.89

17.3 Range Including Predominant Feeding Grounds, Sympatric Species

In the first half of the twentieth century, mature European sturgeon occurred in most of the rivers in the Black Sea basin from Turkish Anatolia to the coasts of Bulgaria, Rumania, Ukraine, Russia, and Georgia (Tikhii 1929; Marti 1939). It has also been recorded occasionally in the Danube River, where it occurred mostly in its delta (Bacalbaşa-Dobrovici and Holčík 2000). Along with the European sturgeon, five other sturgeon species occurred in the area (*Huso huso*, *A. gueldenstaedtii*, *A. colchicus*, *A. stellatus*, and *A. nudiventris*).

In the second half of the twentieth century (after 1970), the area of occurrence of this species was limited to the region of Poti–Ochamchira, Georgia, while commercially interesting species were maintained through hatchery programs in the Azov Sea and former Soviet countries. The coastal waters of Georgia are predominant feeding grounds for sturgeons since they offer an unusually rich food base that provides feeding grounds for adults and sub-adults alike.

17.4 Reproduction Sites

Until 1970 and 1993 respectively, the sturgeon of this area ascended two rivers on spawning migrations, the Inguri and the Rioni rivers. The artificial barrier that spans the Inguri River about half-way up its course (30 km from the river mouth) renders the spawning grounds farther upstream inaccessible, and now the only spawning grounds in the area for this species are located in the Rioni River (Fig. 17.1).

The source of the river is in the glaciated, southern slopes of the Caucasus Mountains (Fig. 17.2). It is 327 km in length, and it flows into the southeastern part of the Black Sea near the port city of Poti. The river mouth has two arms which extend around the city to the north and the south. Most of the river water flows through the northern channel, and this is the main spawning migratory route for sturgeon. The amount of water that flows through the channels is regulated by a system of locks (Fig. 17.3), which are not a barrier for migrating fish. The Rioni River is characterized by a variable flow rate; the long-term mean measured in its lower reaches increases from about 244 $m^3 s^{-1}$ in September to about 581 $m^3 s^{-1}$ in June.

In the 1960s, access to spawning grounds was limited by the construction of a hydroelectric power plant. Access to spawning grounds in the Inguri River was eliminated, while that in the Rioni River was limited by about one third (Kolman and Zarkua 2002). The crude oil facility that was opened in the vicinity of the town of Supsa has undoubtedly had a negative impact on the quality of the environment, and consequently on the feeding and growth of sturgeon.

Increased flow rates also mean increased turbidity, which is a signal to mature sturgeon that it is time to begin spawning migrations. Currently, the sturgeon

Fig. 17.1 Rioni River, where potential spawning grounds are located

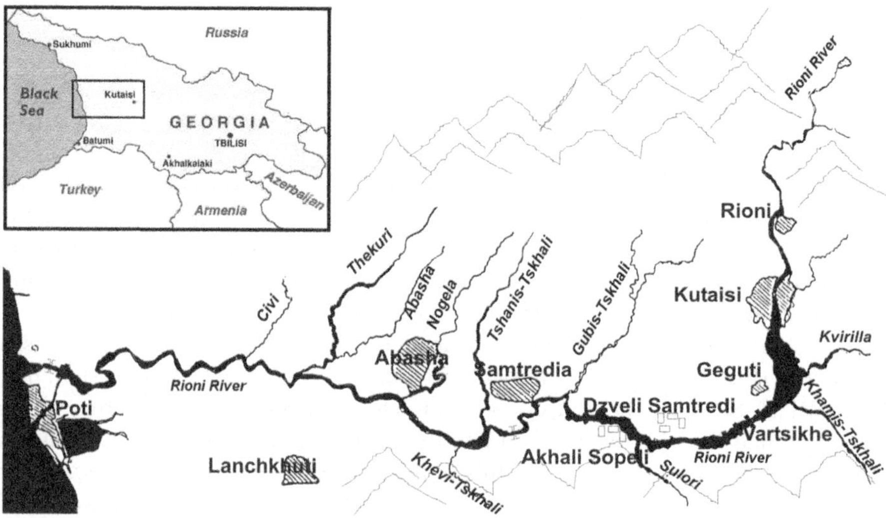

Fig. 17.2 Rioni River basin

spawning grounds are located about 120–130 km from the river mouth, near the locality of Samtredia (Kolman and Zarkua 1999). The width of the river in this stretch is about 150–250 m, while the maximum depth ranges from 2 to 4 m. During the spawning period, the linear water flow rate has been reported to fluctuate from 1.5 to 2.1 m s^{-1} at water temperatures of 13–15°C (Kolman and Zarkua 2002). The river bed splits into two channels, forming islands, which form places with different hydrology. This gives the spawning fish the opportunity to choose a spot with optimal spawning conditions (Fig. 17.1).

Fig. 17.3 Locks regulating water flow in the northern and southern canals of Rioni Delta

17.5 Catch and Decrease of the Populations

In the early twentieth century, catches of the European sturgeon in coastal Georgian waters were estimated to have comprised approximately 100 tons annually (Ninua 1976). However, the population size of the European sturgeon decreased sharply because of the intensity of catches, including poaching, and limited access to spawning grounds. In 1967, this species was designated as endangered, and listed in the Georgian Red Book of Endangered Species. Unfortunately, this did not limit either fishing or poaching; this species was caught as by-catch by commercial fishers during intense fishing for European anchovy *Engraulis encrasicolus* (L.) within the 5-mile zone of Poti–Ochamchira (Zarkua et al. 2006). Adult individuals were especially susceptible to poachers during their spawning migrations. Until the early twenty-first century, sturgeons were still fished with dangerous set hook gear (Fig. 17.4) (Kolman and Zarkua 2000).

The process of the disappearance of this sturgeon population is illustrated perfectly by the results of experimental sturgeon catches conducted in the littoral zone of the Poti–Ochamchira area in the 1975–1991 period by scientists from the Marine Ecology and Fisheries Institute in Batumi (Table 17.2).

17.5.1 Current Status of the Population

In recent years (2002–2008) single specimens of *A. sturio* weighing from 15 to 60 kg and measuring from 1.20 to 1.75 m in length have been caught in the Rioni River (Kolman, unpublished data). Small numbers of sturgeon from various age groups have also been caught in fishing gear deployed in the Poti–Ochamchira region.

The European sturgeon of the Rioni is currently the only remaining population in the Georgian zone of the Black Sea, and it is probably the only population left spawning in the entire area of the Black Sea (Zarkua et al. 2006).

Fig. 17.4 Poaching in the Rioni River in 2001

Table 17.2 Decreases in the numbers of adult European sturgeon individuals in the Poti–Ochamchira region (Zarkua 1998; Zarkua et al. 2006)

Year	Total catch (individuals)
1975	1,820
1978	1,500
1980	1,200
1984	720
1988	500
1991	400

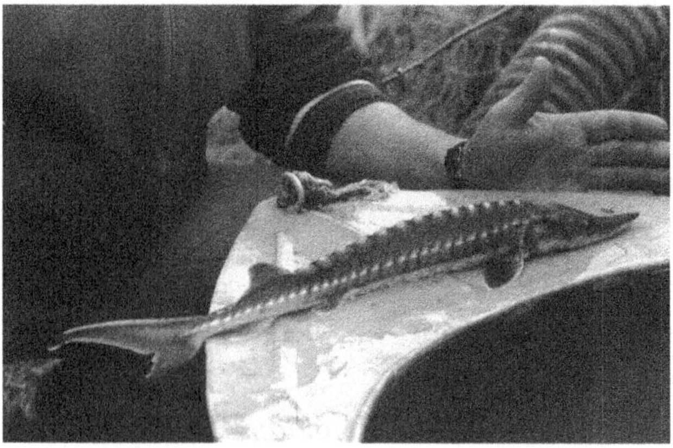

Fig. 17.5 European sturgeon fry caught in fall 1993 in the lower reaches of the Rioni River

Experimental catches were confirmed for the last time in 1993; less than 20 fry individuals under 20 cm (Fig. 17.5) and three adults were caught. This was evidence that the sturgeon were, at the time, spawning effectively and naturally in the Rioni River.

17.6 Protection Measures

The European sturgeon has been designated an endangered species in Georgia since 1967, and commercial catches of it are banned. Immediately prior to the fishing ban, approximately 500–600 fish migrating to spawn were caught annually in the Rioni River (Ninua 1976).

17.7 Perspective

According to information from the Georgian Ministry of the Environment, nearly every year there are accidental catches of individual sturgeon ascending the river to spawn (Archil Guchmanidze, pers. comm.). It is, therefore, possible to conserve this sturgeon population by establishing an ex situ stock, conducting controlled reproduction, and rearing stocking material. This is all the more possible since work in these areas has already been done. For the first time in 1968, European sturgeon was bred in the field, and a small number of hatches were obtained (Milshteyn et al. 1968). Further work on artificial breeding was performed at the Sturgeon Stocking Center that opened on the Rioni River in the early 1980s. This work was usually limited to issues of artificial spawning and excluded work on rearing fry, which meant that stocking was not especially effective. In the current situation, the only chance for rebuilding the disappearing population of European sturgeon is to reactivate and expand these endeavors. Additionally, it is extremely important to minimize the bycatch of sturgeons in the coastal waters as well as on the spawning migration. Bycatch by the commercial fleets harvesting *Engraulis* in the territorial waters of Georgia with purse seines, as well as frequent catches of artisanal fisheries, are reported to collect numerous sturgeons annually (Zarkua pers. comm.).

Acknowledgments I thank A Kapusta, J Morzuch, and J Gessner for their comments on an earlier version of this manuscript. This research was sponsored by Inland Fisheries Institute research project no. S-027/2010-2012.

References

Antipa G (1905) Die Störe und ihre Wanderungen in den europäischen Gewässern mit besonderer Berücksichtigung der Störe der Donau und des Schwarzen Meeres. Wien, Sitzungberichte der international Fischerei Kongress

Artyukhin E, Vecsei P (1999) On the status of European sturgeon: conspecificity of European Acipenser sturio and North American *Acipenser oxyrinchus*. J Appl Ichthyol 15:35–37

Bacalbaşa-Dobrovici N, Holčík J (2000) Distribution of *Acipenser sturio* L., 1758 in the Black Sea and its watershed. Bol Inst Esp Oceanogr 16:37–41

Kolman R, Zarkua Z (1999) Atlantic sturgeon (*A. sturio* L.) in Georgia. Komun Ryb 5:24–27, in Polish
Kolman R, Zarkua Z (2000) Sturgeon of Georgia. Komun Ryb 4:26–28 (in Polish)
Kolman R, Zarkua Z (2002) Environmental condition of Common sturgeon (*Acipenser sturio* L.) spawning in river Rioni (Georgia). EJPAU, Fisheries
Lepage M, Rochard E (1995) Threatened fishes of the world: Acipenser sturio Linnaeus, 1758 (Acipenseridae). Environ Biol Fish 43:28
Magnin E (1962) Recherches sur la systématique et la biologie des Acipensérides: *Acipenser sturio* L., *Acipenser oxyrhynchus* Mitchill et *Acipenser fulvescens* Raf. Annal St Cent Hydrobiol Appl 9:1–242
Marti VY (1939) Biology and harvest of *Acipenser sturio* in the Black Sea. Zool Zh 18:435–442 (in Russian)
Milshteyn VV, Popova AA, Ninua NS, Shverdashvili RS (1968) Reproduction of Atlantic sturgeon. Rybn Khoz 12:15 (in Russian)
Ninua NS (1976) Atlantic sturgeon of Rioni River. Metsniereba, Tbilisi (in Russian)
Tikhii MI (1929) The common sturgeon in Rioni. Priroda 4:369 (in Russian)
Williot P, Rochard E, Rouault T, Kirschbaum F (2009) *Acipenser sturio* recovery research actions in France. In: Carmona R, Domezain A, García-Gallego M, Hernando JA, Rodríguez F, Ruiz-Rejón M (eds) Biology, conservation and sustainable development of sturgeons, vol 29, Fish & fisheries series. Springer, New York, pp 247–263
Zarkua Z (1998) Sturgeon/Acipenseridae. In: Komakhidze A (ed) Black Sea biological diversity. Georgian National Report. United Nations Publications, New York, pp 29–31
Zarkua ZG, Bolkvadze LG, Khintibidze MS, Gogotishvili MA, Variadis LD, Kolman R (2006) Current state of sturgeon populations in Georgian Black Sea coastal waters. Komun Ryb 4:20–24 (in Polish)

Chapter 18
Present Legal Status of the European Sturgeon *Acipenser sturio*

E. Rochard

Abstract The present legal status of the European sturgeon depends mainly on its conservation status. As a critically endangered species and an emblematic fish, the European sturgeon is listed in many conventions, and is nowadays strictly protected. Official recovery plans are now beginning in France and Germany.

18.1 Introduction

Conventions, directives and other legal texts dedicated to the conservation of *A. sturio* refer to the assessments of the conservation status of the species.

Assessment processes vary, but nowadays IUCN status constitutes a key criterion for inclusion in red lists and conservation regulations (Lamoreux et al. 2003). A revision of the status of *A. sturio* was carried out by the IUCN sturgeon specialist group in 2009, which confirmed its status of "critically endangered species." This means that the species is facing an extremely high risk of extinction in the wild in the immediate future. This ranking is mainly related to the small size of the only remaining population and its drastically decreasing numbers. (http://www.iucnredlist.org/apps/redlist/details/230/0).

18.2 Legal Status

As a critically endangered species, the European sturgeon is listed in numerous national laws and international conventions and directives (Table 18.1). The species is strictly protected in all the countries of the present distribution area,

E. Rochard (✉)
Cemagref, Estuarine Ecosystems and Diadromous Fish Research Unit, 50 Avenue de Verdun, 33612 Cestas Cedex, France
e-mail: eric.rochard@cemagref.fr

P. Williot et al. (eds.), *Biology and Conservation of the European Sturgeon Acipenser sturio L. 1758*, DOI 10.1007/978-3-642-20611-5_18,

Table 18.1 Conventions and directives listing *A. sturio*

Name	Location and date of ratification	Listing of *A. sturio* in the text
Convention on International Trade in Endangered Species of Wild Fauna and Flora. (CITES)	Washington, 1975	Appendix I means that CITES prohibits international trade in specimens of these species except for scientific research (http://www.cites.org/eng/resources/pub/checklist08/Checklist.pdf)
Convention on Migratory Species (CMS)	Bonn, 1979	Appendix I lists endangered migratory species (http://www.cms.int/documents/appendix/cms_app1_2.htm#appendix_I)
Bern Convention on the Conservation of European Wildlife and Natural Habitats	Bern, 1979	Appendix II lists animal species requiring special protection (http://www.coe.int/t/dg4/cultureheritage/nature/bern/default_en.asp), action plans go with this listing
Council Directive on the conservation of natural habitats and of wild fauna and flora (CEE 1992)	Brussels, 1992	Appendix II lists animal and plant species of community interest whose conservation requires the designation of special areas of conservation (Natura 2000 network) (http://www.central2013.eu/fileadmin/user_upload/Downloads/Document_Centre/OP_Resources/HABITAT_DIRECTIVE_92-43-EEC.pdf)
Convention protecting and conserving the North-East Atlantic and its resources (OSPAR)	Oslo–Paris, 1992	The list of threatened and/or declining species (http://www.snh.gov.uk/docs/B469310.pdf) is to guide the setting of protection priorities by the parties

either directly by national laws (e.g., France, Ministère de l'Écologie et du Développement durable 2004) or under the application of International Conventions and European Directives (e.g., Ministère des Affaires étrangères 1990a, b) (Table 18.1).

CITES is concerned with the international trade of species, and can be a very strong tool; however, nowadays there is no trade in *A. sturio*. The Bern Convention (for European species), CMS (for migratory species) and OSPAR (for the North East Atlantic region) aim to conserve wild flora and fauna and their natural habitats, and to promote co-operation in that field. The Bern Convention helped to establish an international action plan (Rosenthal et al. 2007) which has now to be enforced by the parties, if necessary by establishing national action plans (see Sect. 18.4).

As a result, the capture, transport, sale and handling in captivity of *A. sturio* specimens is forbidden in all cases, and the use of specimens for research purposes is possible but strictly controlled. For an exhaustive list of national laws concerning *A. sturio*, see Rosenthal et al. (2007); for an administrative and legal analysis, see Léthier (2005).

18.3 Enforcement of These Regulations

The implementation of the above mentioned regulations is facing several difficulties in order to achieve effective protection for individuals in the wild, and two in particular: (1) short-sighted behavior on the part of some fishermen who still catch the species, and try to sell it with no respect for the regulations (see Chap. 22), and (2) a lack of knowledge and willingness on the part of the administration in charge of fisheries (see Williot et al. 2002 and Chap. 20 for a wider analysis).

The more severe problems that have to be addressed in the mid-term are associated with the importance of conservation in society. Enforcement of strict individual protection might be possible by education and control. However, the protection of the species when addressing the ban on fisheries practices or habitat protection interferes with larger economic pressures, and therefore priorities need to be clearly set on a national or regional scale. This lack of priorities has led to the extirpation of the species from some countries in the past, since the role of ecosystems and biodiversity has not been considered as a value as such, and therefore has been set aside in the conflicts over economic development. In order to cope with this persistent perspective, alternative models for assessing the value of biodiversity are being developed.

However, a few positive signals reflect a change, for example: (1) following media coverage of an illegal capture and sale of an *A. sturio* in Les Sables d'Olonne (France) in June 2004 (Fig. 18.1, top) the French Ministry of the Environment launched an administrative inquiry (Guth and Laurent 2004) which concluded that responsibility lay with the fisheries administration which had misinformed the fisherman, and (2) when a fisherman tried to sell a specimen illegally in Concarneau (France) in January 2007, he clashed with the director of the auction hall, who returned the fish to the water (and received official congratulations from the Fisheries Ministry, which can be considered as a positive consequence of the awareness campaign among marine fishermen – see Chap. 37). Similarly, positive trends should be listed on a wider scale (see the gravel extraction in the Gironde in 2000 and after), problems for instance with the development of navigation routes and increased support for conservation as a result of the WFD.

Un esturgeon de 1,38 m et 14 kg à la criée

Frédéric Benoist (à gauche), patron de « L'Hexagone », et Marc Penard, matelot, encore amusés d'avoir pris un esturgeon dans leurs filets.

Fait rarissime de mémoire de mareyeur. Hier vendredi matin, un esturgeon de 14 kg a été vendu sous la criée des Sables. La « bête » a été achetée par un poissonnier de La Rochelle autour de 10 € le kilo. L'histoire ne dit pas si elle était pleine d'œufs, auquel cas l'acheteur pourrait avoir fait une excellente affaire au goût de caviar. L'esturgeon, mesurant 1,38 m, s'est pris dans les filets de *L'Hexagone* alors que le côtier était en pêche devant Les Sables. Une première pour le patron Frédéric Benoist et son matelot Marc Penard, plutôt amusés par cette prise.

PAGINA 11

Vangst GO 26 voor IJmuiden

Forse steur voor de consumptie

IJMUIDEN – Vorige week vrijdag werd op de afslag van IJmuiden een forse steur ter verkoop aangeboden. De vis werd gekocht door de firma Gebr. Bakker, in commissie voor horecagroothandel ISPC.

Steuren kunnen tot zes meter lang worden. Zo lang was het door de GO 26 van Lokker gevangen exemplaar zeker niet, maar met zeker een meter lengte en een gewicht van 33 kilo mocht de vis er zijn. Bijzonder aan de primitief ogende vis is met name de uitstulpbare bek omringd door zachte 'tanden', waarmee de vis wormen en andere bodemorganismen uit de grond graaft. De vis kwam vorige week donderdag in het net. Volgens schipper Gert Lokker was het een unieke vangst voor de lokatie: 30 mijl ten westen van IJmuiden.

Gebr. Bakker betaalde 17,84 de kilo. ISPC (met twee vestigingen in Nederland en twee in België) heeft de vis tentoongesteld in Gent, waarna de vis opgegeten wordt. Of er kaviaar in de vis zat is nog niet bekend. De vis was dicht aangevoerd en mocht van de koper niet gestript worden.

Fig. 18.1 Two newspaper articles (Ouest France 26–27 June 2004, *top*; Het Visserijnieuws 20 July 2007, the Netherlands, *low*) reporting capture and sale of *A. sturio* without mentioning the protection status of the species and the illegality of the capture and sale

18.4 Work in Progress

The species is mentioned in the French and German national strategies for biodiversity as a flagship for the recovery of aquatic biodiversity (Ministère de l' Écologie et du Développement durable 2005). France and Germany are finalizing their national Action Plans, and welcome mutual participation of experts from both countries at the coming meetings.

References

CEE (1992) Directive 92/43/CEE du conseil du 21 mai 1992 concernant la conservation des habitats naturels ainsi que de la faune et de la flore sauvages. Official Journal L 206, 22/07/1992 P. 0007–0050

Guth MO, Laurent JL (2004) Retour d'expérience sur la capture et la vente illicite d'un esturgeon en criée aux Sables d'Olonne (Vendée). Rapport de l'inspection générale de l'environnement. Ministère de l'Ecologie et du Développement durable, Paris

Lamoreux J, Resit Akcakaya H, Bennun L, Collar NJ, Boitani L, Brackett D, Brautigam A, Brooks TM, da Fonseca GAB, Mittermeier RA (2003) Value of the IUCN Red List. Trends Ecol Evol 18(5):214–215

Léthier H (2005) Vers un plan international de restauration de l'esturgeon (*Acipenser sturio*), document préliminaire d'orientation. EMC pour WWF France, Paris

Ministère de l'Écologie et du Développement durable (2004) Arrêté du 20 décembre 2004 relatif à la protection de l'espèce *Acipenser sturio* (esturgeon). Journal officiel de la république française

Ministère de l'Écologie et du Développement durable (2005) Stratégie nationale pour la biodiversité, plans d'action patrimoine naturel. Ministère de l'Écologie et du Développement durable, Paris

Ministère des Affaires étrangères (1990a) Décret n° 90/962 portant publication de la convention sur la conservation des espèces migratrices appartenant à la faune sauvage, faite à Bonn le 23 juin 1979, telle qu'amendée à Bonn le 16 octobre 1985. Journal officiel de la République française

Ministère des Affaires étrangères (1990b) Décret portant publication de la convention relative à la conservation de la vie sauvage et du milieu naturel de l'Europe, ouverte à la signature à Berne le 19 septembre 1979, vol 90/56. Journal officiel de la République française

Rosenthal H, Bronzi P, Gessner J, Moreau D, Rochard E, Lasen C (2007) Draft action plan for the conservation and restoration of the European sturgeon (*Acipenser sturio*). Council of Europe, Convention on the Conservation of European Wildlife and Natural Habitats, Strasbourg

Williot P, Arlati G, Chebanov M, Gulyas T, Kasimov R, Kirschbaum F, Patriche N, Pavlovskaya LP, Poliakova L, Pourkazemi M, Kim Y, Zhuang P, Zholdasova IM (2002) Status and management of Eurasian sturgeon: an overview. Int Rev Hydrobiol 87(5–6):483–506

Part II
Iconography of the European Sturgeon

Chapter 19
Iconography of the European Sturgeon in France

Nathalie Desse-Berset and Patrick Williot

19.1 Introduction (Plate 19.1 and 19.2)

Although not numerous, sturgeon representations can be found in sculpture, painting, engraving and numismatics. Several paintings, especially those of the Flemish school, show sturgeons, among other fish species, in still life or in marketplace scenes (among which those of Frans Snijders or Willem Ormea, seventeenth century painters). The species and the size of the specimens are generally imprecise, even though the sturgeon sizes are often evoked in textual data. Sauvage in 1883 writes that the common sturgeon can reach a size of 5–6 m long, but that this event is becoming rarer and rarer. He says that during the previous century its presence on the Picardie, Normandie and Bretagne coasts was accidental. It sometimes swam up the Seine River; accidentally some individuals ended up in the Yonne River (Moreau 1897). Some sturgeons were captured beyond Sens between Laroche and Auxerre.

The sturgeons seldom swam all the way up to Paris. However, in 1800 a sturgeon was once found in Neuilly-sur-Seine near Paris. The fish weighed 200 french pounds (=100 kilos) and measured 6½ ft (= about 2 m long).

Some sturgeons swam up the Loire River all the way to Les Ponts-de-Cé (near Angers) and even to Saumur, but very rarely. In the sixteenth century the French naturalist Pierre Belon reports that during King François 1er's stay in Montargis, he was shown a 5.40 m-long sturgeon that had been captured in the Loire River.

N. Desse-Berset (✉)
Université de Nice-Sophia Antipolis – CNRS-CEPAM-UMR 6130, SJA3, 24 avenue des Diables Bleus, 06357 NICE Cédex 4, France
e-mail: berset.desse@hotmail.com; berset@cepam.cnrs.fr

P. Williot
Sturgeon Consultant, 4 Rue du Pas de madame, 33980 Audenge, France

P. Williot et al. (eds.), *Biology and Conservation of the European Sturgeon Acipenser sturio L. 1758*, DOI 10.1007/978-3-642-20611-5_19,

Plate 19.1 "Coming back from sturgeon fishing, Les Callonges" by the painter Mathurin Méheut (around 1920). This site was an important fishing harbour on the Gironde estuary (N. Desse-Berset collection)

Several authors put forward the same information (Larousse 1870; Moreau 1898; Sonnini in Sauvage 1883). One of the last captures in this river is represented on a postcard: a sturgeon measuring 256 cm long and weighing 88 kg was fished in Saint-Firmin-sur-Loire on 26 May 1904 (Pouillot 2001).

In 1806 a 2.40 m-long sturgeon was captured in the Doubs River.

All these mentions represent precious evidence, but sometimes some of those sizes, apparently gigantic, should be reconsidered.

Thus, in the Great Dictionary of Cuisine, Alexandre Dumas reports that in 1769 a 20-m-long sturgeon had been captured, and that the fish "weighed 1,155 kilograms and provided 3,030 kilograms eggs (sic). Often the sturgeons that are caught weigh around 1,400 kilograms and measure around 13 meters" (Dumas 1873). This information seems, to say the least, exaggerated!

Therefore, ancient photographs, despite their sometime poor quality, represent a remarkable source of information, especially when they still bear the origin and the date of the events.

These pages are of course far from being exhaustive concerning this vast subject. Their goal is principally to bring selective information about sturgeon captures belonging to a recent past but which minds have already forgotten: who today remembers that these fish were frequently caught in the Rhône River, the Loire River, the Seine River, or the Somme River?

These images can be backed up with some anecdotes.

Thus Grimod de la Reynière in his "Almanach des Gourmands" (Grimod de la Reynière 1802–1812, pp 17–20) evokes "sturgeon pâtés" prepared by a pastry cook from Abbeville (Somme River) in May 1810: "It is hard to understand how a sturgeon that is a firm, often a little tough and hardly digested fish becomes a tender, smooth and very healthy fish in the hands of M. Richard. . .This variant sturgeon preparation gives it a different taste than the fresh sturgeon that one usually eats and that is still so sought-after. The sturgeons used by M. Richard are caught in the Somme River. One knows that this fish, which just like the salmon swims up the rivers – sometimes even a long distance – tastes much better when caught in a river.

The sturgeons captured in the Somme River are known for their extreme fineness and their delightful delicacy. . . Therefore the "sturgeon pâtés" from Abbeville are famous all over Europe."

With regard to the Rhône River, "although this magnificent fish can be captured in most of French rivers, it is never as common and as big as in the Rhône River, where it reigns alone. In other fresh waters, the sturgeon has to fight for royalty with the salmon." (De La Blanchère 1868–1869, p 396).

The documents presented hereinafter date mainly back to the first half of the twentieth century, and are mostly photographs (or postcards) taken in France by fishermen families who were gracious enough to give them for this book

Plate 19.2 "Filadière", traditional small boat of the Gironde River, in Talmont, around 1930 (Yvan Robert collection)

(some of these photographs have already been published in two works about the Gironde estuary: one about fishing, the other one about caviar (Bouquin et al. 1999; Val 2006).

These documents inform us about the places and locations where the fish was caught and about its size. When the size was not mentioned, the authors have tried to estimate the total length of the fish. These are probably the most beautiful catches that have ever been documented, as we were reminded of by written sources already two centuries ago: "Because of its prize and its rarity, the sturgeon only appears on sovereign tables" (Grimod de la Reynière 1802).

The available illustrations have been gathered into four sections: (A) Ancient captures, (B) sturgeon exploitation, (C) population study, and (D) philately.

19.2 Ancient Captures (Plates 19.3–19.15)

These ancient captures are presented according to their geographical origin: the Gironde estuary (Les Callonges; Port-Maubert/Mortagne; Saint-Seurin-d'Uzet; Talmont); the Loire River (La Bernerie); the Somme River (Montigny-lez-Nampont); Fort-Mahon in the Baie d'Authie; the Rhône River (Aramon; Avignon; Pont-Saint-Esprit).

The sturgeon size estimation (total length): when the size was not mentioned in the source document, it was calculated as follows. The measure of the fish's total length was compared to the size estimated by a fisherman on enlarged photographs. When the fish was in a curved position, the length of the midline was taken as reference. For the human size, an average of 1.65 m was accepted, knowing that the average size of humans has risen during the last decades (the majority of these pictures are prior to 1950).

What is the margin of error?

For a fish measuring15 cm and a man measuring 12.5 cm on a photograph, the estimated length of the fish varies from 1.92 to 2.10 m according to whether the size of the fisherman is estimated at 1.60 or 1.75 m.

The absolute gap is of the order of 10%, and of the order of 5% if we accept an average value (in our example, we obtain the most probable value of ~2.0 ± 0.1 m (minimum–maximum)).

Plate 19.3 **a**) Saint-Seurin-d'Uzet. Photograph taken in 1927 by Edouard Val. Estimated length: 240 cm. Given weight: 120 kg. It is very likely that this fish actually weighed 120 french pounds (= 60 kg) (René Val collection). **b**) Port Maubert. trammel net: this rectangular net, measuring around 100 m long and measuring 3–6 m high, is used as a drift net. It is made of three layers attached together with 15–20 cm- to 60-cm-wide meshes (N. Desse-Berset collection)

Plate 19.4 390-cm-long and 300-kg-heavy male sturgeon, captured between Mortagne-sur-Gironde and Port Maubert in June 1944 (René Val collection)

Plate 19.5 Talmont, Saint-Fort-sur-Gironde. Estimated length: 243 cm (Yvan Robert collection)

Plate 19.6 Talmont, Saint-Fort-sur-Gironde. Estimated length: 215 cm (René Val collection)

Plate 19.7 Saint-Seurin-d'Uzet. 1950. Estimated length: 230 cm (René Val collection)

Plate 19.8 Saint-Seurin-d'Uzet. 1950. Estimated length: 230 cm (René Val collection)

Plate 19.9 Saint-Seurin-d'Uzet, March 1941. The weighing of a sturgeon (René Val collection)

Plate 19.10 Talmont, Saint-Fort-sur-Gironde. Estimated length: 192 cm (Yvan Robert collection)

Plate 19.11 A 65-kg-heavy sturgeon captured in Baie d'Authie, in Fort-Mahon (Somme). Estimated length: 230 cm (postcard, web site http://www.Delcampe.net)

Plate 19.12 **a**) La Bernerie (south of the Loire River estuary), 18 May 1924. Legend: Germain Durand, fisherman: one of his beautiful captures. Total length: 287 cm, 146 kg. Illustration taken from the work "La Bernerie, son histoire à travers les âges" by Louis Lacroix (published in 1953, republished by Le Livre d'histoire – Lorisse in 2004). **b**) A 68-kg-heavy sturgeon, measuring 236 cm, captured in Montigny-lez-Nampont (Somme) on 29 May 1928 (postcard, web site http://www.Delcampe.net, seller: Tonton_78)

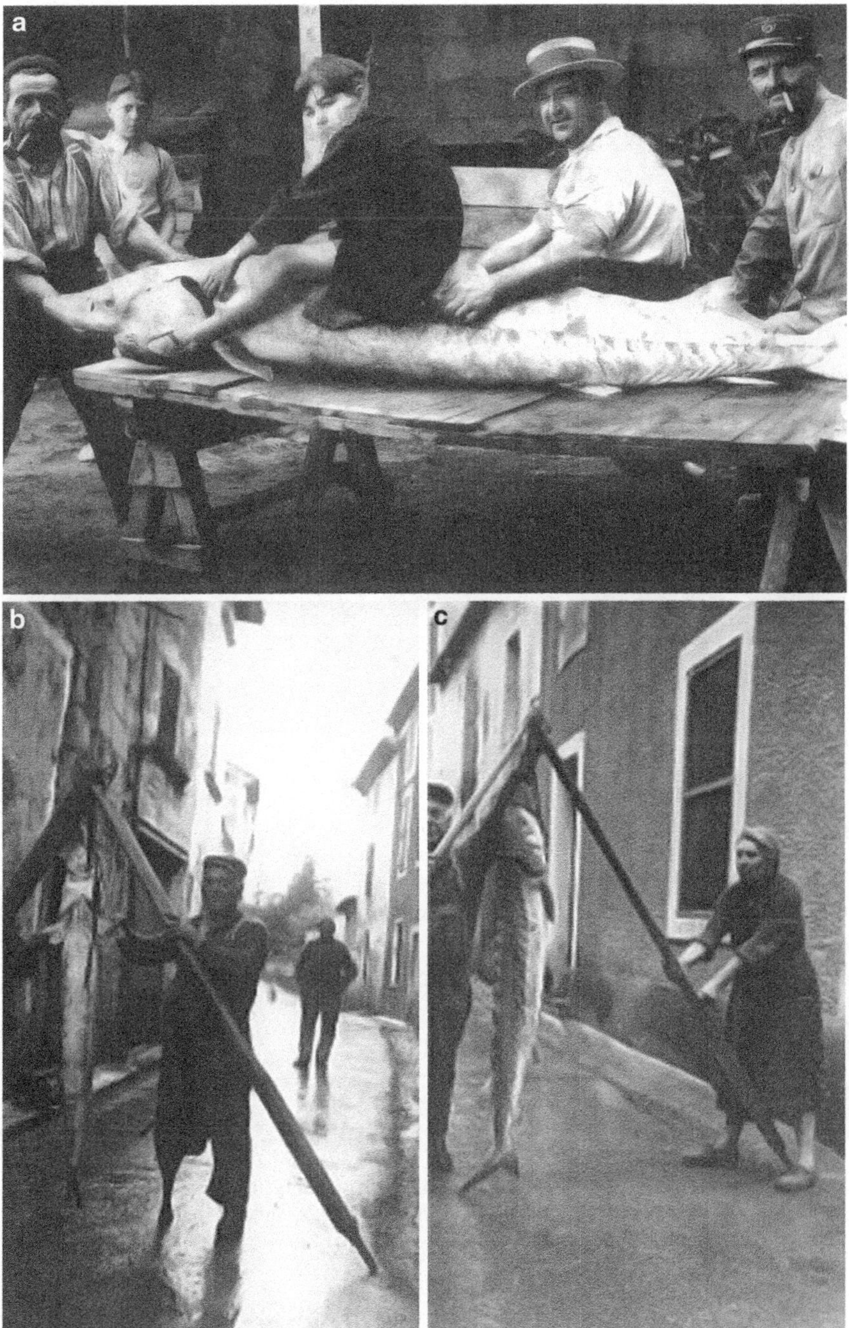

Plate 19.13 **a**) Rhône River, Aramon, 1950–60. Estimated length: 245 cm (probably the same as the one on board Plate 19.14a) (original photograph by Mr Claude Bonjean, Bonjean/Toucheboeuf property). **b**) Aramon, 1950–60. Estimated length: 170 cm (Gabrilot collection, Aramon). **c**) Aramon, 1950–60. Estimated length: 170 cm (Gabrilot collection, Aramon)

Plate 19.14 **a**) Avignon. Estimated length: 245 cm (probably the same as the one on board Plate 19.13a) (original photograph, Mrs Soucheyre property, Avignon). **b**) Pont-Saint-Esprit 1932 (web site http://www.Delcampe.net, seller Saint-Victor)

Plate 19.15 Pont-Saint-Esprit 1932 Estimated length: 197–203 cm, the same as board Plate 19.14b (web site http://www.Delcampe.net, seller Saint-Victor)

19.3 Sturgeon Exploitation: Flesh and Caviar (Plates 19.16–19.19b)

These photographs refer to the transport of these big fish, to the first preparation phases, and to the several caviar preparation steps.

The flesh was usually marketed locally: the fish were transported whole on small handcarts after the extraction of the ovaries of the females. In order to allow cutting-up, the scutes were pulled off from tail to head before slicing the sturgeon.

Even though sturgeon exploitation has been attested for a long time, the value of caviar suddenly increased from the beginning of the 1920s because of a strong demand by experts, which triggered a stricter preparation methodology. Some photographs show the extraction of the ovaries, the sieving and the tin packaging of the caviar.

Caviar processing was essentially performed in Saint-Seurin-d'Uzet in the Gironde estuary. Most of the females ready for caviar processing were captured in the estuary nearby the village. It is located about 80 and 150 km downstream from the nearest spawning grounds in the River Dordogne and the River Garonne respectively. The pictures show well-developed ovaries and the colour, already darkish, is typical of the last growing stage. However, these brood females were not yet ready for spawning. This would have been occurred later (2–4 weeks) under the stimulation of endocrine signals mediated by the surrounding cells of the oocytes. The oocyte imbedded in the cells is ovarian follicle. It is these surrounding cells that enable the ovarian follicle to be resistant to salting. Thus, the screening consists in separating the ovarian follicles one by one, i.e., in taking off the vascularised tissues which keep the ovarian follicles connected to the tissues. After this, the next step of caviar processing can be undertaken.

Plate 19.16 Ancient caviar tin from Kaspia Company, Paris (original photograph by Claude Businelli)

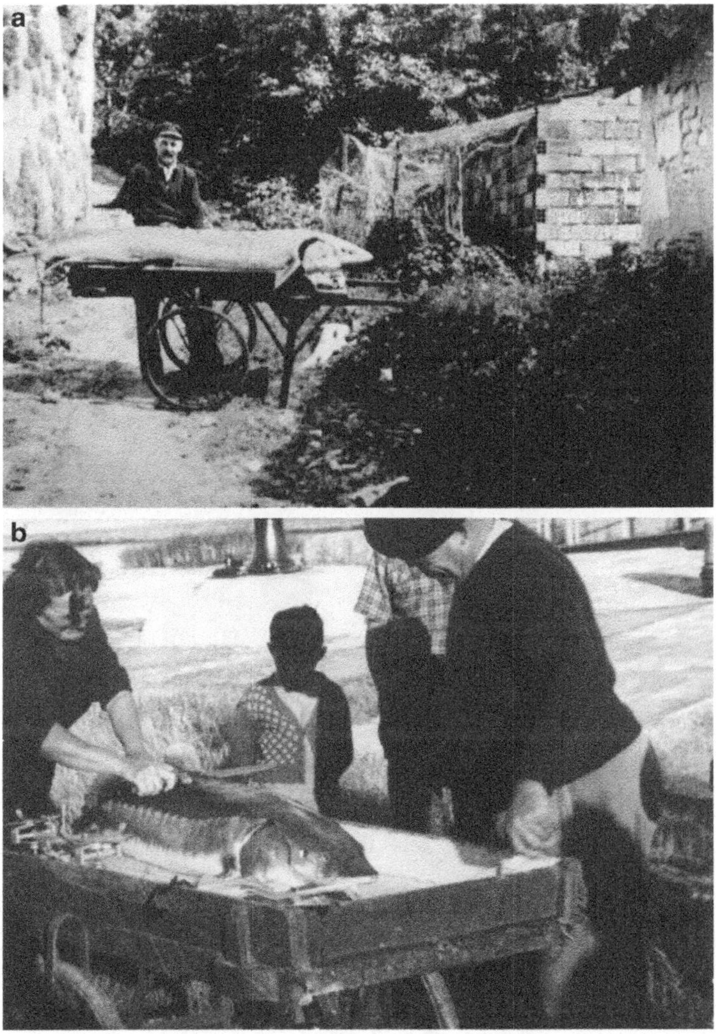

Plate 19.17 a) Cart used for transporting ready-to-be-sliced sturgeon in Mortagne-sur-Gironde and the surrounding area (Yvan Robert collection). **b**) Cutting sturgeon up for its flesh: before the slicing, one must first remove the scutes (hereafter the dorsal scutes row) (Yvan Robert collection)

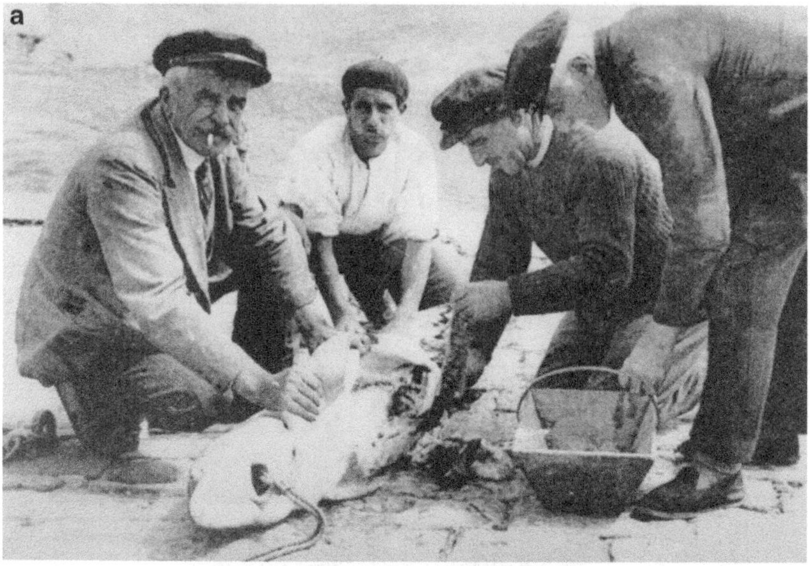

Plate 19.18 **a**) Saint-Seurin-d'Uzet. Extraction of sturgeon ovaries for caviar preparation (René Val collection). **b**) Extraction of ovaries, detail (René Val collection)

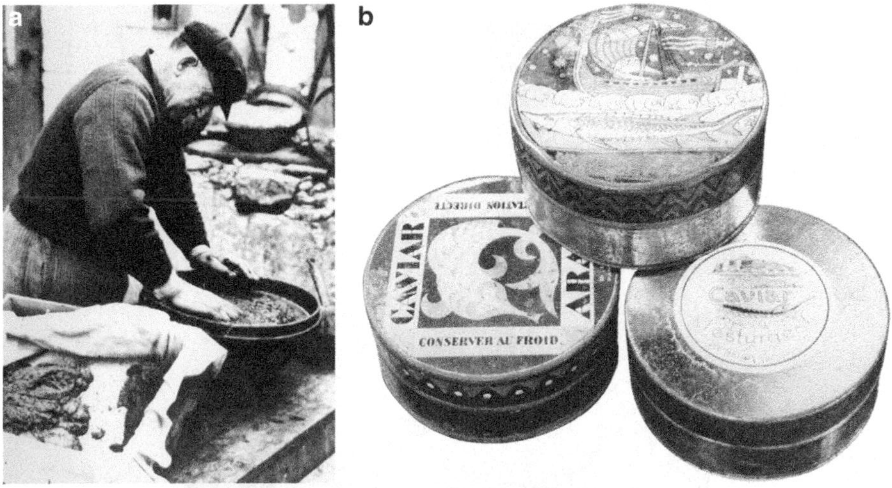

Plate 19.19 **a**) Saint-Seurin-d'Uzet. Caviar sieving, around 1950 (René Val collection). **b**) Caviar tin packaging (René Val collection)

19.4 Population Study (Plates 19.20a–19.21b)

In 1982, sturgeon fishing was prohibited in France. From that point, the main objective was to study the dynamic of this population, which was until then totally unknown. Campaigns were led by the Cemagref from the beginning of the 1980s with the active participation of professional fishermen of the estuary (see Chap. 20).

The pictures illustrate the final phases of the juvenile tagging campaigns in the downstream areas of the Gironde estuary. These captures were led during the summertime, a period when one can observe a particular behaviour of this population: the juveniles between 2 and 7 years old gather for several months in the lower part of the estuary around benthic micro-fauna rich zones (see Chap. 10 about the migration cycles, and Chap. 11 about the food habits of juveniles in estuaries).

The chances of catching sturgeons were thus higher. Captures were realised thanks to trammel nets drifting at the bottom of the river during brief periods in order not to damage the fish. These specimens were then measured, tagged and put back into the water. The re-capture of some of these tagged fish made it possible to provide new information about some aspects of their migration, of their demography and of their growth, and made it possible to give a first estimation of illegal captures (Castelnaud et al. 1991; Rochard and Jatteau 1991; Rochard et al. 1997).

Plate 19.20 a) Around 1985. The fisherman Yvan Robert is holding a 145-cm-long sturgeon (estimated length), that has been tagged (Petersen disc visible at the root of the dorsal fin) before putting it back into the water (Yvan Robert collection). **b**) Around 1985. Yvan Robert is putting back a tagged sturgeon into the water from his skiff. Skiffs are motorboats that have followed the sailing boats ("filadières") (Yvan Robert collection)

Plate 19.21 **a**) Around 1985. Sturgeon tagging (Petersen disc visible at the root of the dorsal fin) (Yvan Robert collection). **b**) The same sturgeon as above returning to its natural habitat (Yvan Robert collection)

19.5 Philately (Plates 19.22a–19.24b)

The retained stamps are the few which illustrate the European sturgeon (*Acipenser sturio*), the protection of sturgeon, and the Atlantic sturgeon (*Acipenser oxyrinchus*), a species that was in sympatry with the European sturgeon in different European countries (Ludwig et al. 2002; Tiedemann et al. 2007; Desse-Berset 2009; Desse-Berset and Williot 2011; see Chap. 7).

a

b

Plate 19.22 **a**) Former USSR, 1975. Translation of the inscription: "Save the 'ocietre' (sturgeon), the most precious fish of our patrimony. USSR post". The term "ocietre" is at the same time the generic name for sturgeons, and especially the name of the *Acipenser gueldenstaedtii* species. **b**) Label of the Natural History Museum of Bordeaux about big European migratory fish, among which *A. sturio*. Not a single stamp representing sturgeons has been produced in France

a

b

c

Plate 19.23 **a**) *A. sturio*, Romania, 1994. Is part of a WWF logo series. **b**) *A. sturio*, Albania. **c**) *A. sturio*, Lithuania, 2006

a

b

Plate 19.24 **a)** *A. oxyrinchus*, Canada (Atlantic sturgeon, black sturgeon). **b)** *A. naccarii*, Croatia. 2009

Since the possible presence of the Adriatic sturgeon (*Acipenser naccarii*) had been evoked in the Mediterranean Sea, a stamp representing this species was added, although today only the *A. sturio* (Garrido-Ramos et al. 1997) has been highlighted (Desse-Berset 1994; Desse-Berset et al. 2008; Pagès et al. 2009).

19.6 Some Conclusions

The majority of these ancient sturgeon fishing pictures represent large specimens (specimens equal to or even longer than 2 m). None of these pictures was taken after 1960. The few pictures of individuals taken near the mouth of the Loire River

(La Bernerie) or the Somme River (Baie d'Authie and Montigny-lez-Namponts) show fish with an established or estimated length of about 2.40 m. These documents are isolated, which can be interpreted as showing a declining density of the population in these zones. As a matter of fact, the species is no longer present in the Loire River and the Somme River. The photographed animals are among the last specimens native to these rivers, or probably stray fish roaming through the continental shelf (Letaconnoux 1961; Castelnaud et al. 1991; Rochard et al. 1997).

These last photographs were taken during the 1920s and the 1930s. The age of the pictures on which these large specimens are represented shows the global collapse of the populations. In France, the last animal of this size was captured in the Garonne River in 1999 (it was a female measuring 2.20 long and weighing 60 kg, and which abruptly died for an unknown reason)

In all these pictures, it is impossible to determine with certainty if we are dealing with *A. sturio* or *A. oxyrinchus*. More detailed photographs would certainly have allowed this determination.

The pictures concerning the study of the population reproducing in the Garonne basin show the determined participation of some estuarine professional fishermen in the estuary in this part of the restoration–conservation program launched in the beginning of the 1980s.

Finally, it is striking to notice that the rare countries which have produced stamps representing the *A. sturio* species are outside of its settlement area. These countries have apparently not hosted any numerous populations which justified its exploitation. France and Germany have never produced any stamps representing sturgeons, despite the presence of the fish in these countries' rivers and seas.

Acknowledgements We would like to warmly thank all the people who have allowed us not only to get a hold of and work on their own collections, but also to reproduce them, and the people who helped us to obtain them.

These people are: René Val in Saint-Seurin-d'Uzet; Yvan Robert, also known as the "Capitaine", and his wife Monique in Saint-Fort-sur-Gironde; Claude Businelli in Blanquefort; the Bonjean, Toucheboeuf, and Gabrilot families in Aramon; Mrs Soucheyre in Avignon; Laurent Brosse from Aqua-Logiq in Lunel-Viel; the postcard-selling web site http://www.Delcampe.net; the Publishers of "Le Livre d'Histoire" in Lorisse.

A. Pasqualini for the illustration infography and J.-D. Strich for the photographs (CEPAM/ CNRS-University of Nice–Sophia Antipolis); and finally J.-M. Paillard ("Starboard Silent Side") for the English revision of the text

References

Bouquin A et al. (1999) Pêches traditionnelles des rives saintongeaises de la Gironde. Éditions Confluence, Société des amis de Talmont

Castelnaud G, Rochard E, Jatteau P, Lepage M (1991) Données actuelles sur la biologie d'*Acipenser sturio* dans l'estuaire de la Gironde. In: Williot P (ed) Acipenser. Cemagref, Antony, pp 251–275

de La Blanchère H (1868–1869) Les pêches curieuses de la France. L'esturgeon. La Chasse Illustrée, pp 396–398, 404–406

Desse-Berset N (1994) Sturgeons of the Rhône during Protohistory in Arles (6th-2nd century BC). In: Van Neer W (ed) Fish exploitation in the past. Proceedings of the 7th meeting of the ICAZ Fish Remains Working Group. Annales du Musée royal de l'Afrique centrale vol 274, Tervueren, pp 81–90

Desse-Berset N (2009) First archaeozoological identification of Atlantic sturgeon (*Acipenser oxyrinchus* Mitchill 1815) in France. CR Palevol 8:717–724. doi:10.1016/j.crpv.2009.06.001

Desse-Berset N, Williot P (2011) Emerging questions from the discovery of the long term presence of *Acipenser oxyrinchus* in France. J Appl Ichthyol 27:263–268, Blackwell Verlag, Berlin. doi:10.1111/j.1439-0426.2010.01649.x

Desse-Berset N, Pagès M, Brosse L, Tougard C, Chassaing O, Hänni, C, Berrebi P (2008) Specific identification of the extinct population of sturgeon from the Rhône River by mtDNA analysis from bone remains (Jardin d'Hiver, Arles, France, 6th to 2nd century BC). In: Béarez P, Grouard S, Clavel B (eds) Archéologie du poisson. Trente ans d'archéo-ichtyologie au CNRS. Hommage aux travaux de Jean Desse et Nathalie Desse-Berset. XVIIIe rencontres internationales d'archéologie et d'histoire. Éditions APDCA, Antibes, pp 195–200

Dumas A (1873) Le Grand Dictionnaire de cuisine. Editions Pierre Grobel, Paris (republished byTchou ed. Paris in 1965)

Garrido-Ramos MA, Soriguer MC, de la Herran RJM, Ruiz Rejon C, Domezain A, Hernando JA, Ruiz Rejon M (1997) Morphometric and genetic analysis as proof for the existence of two sturgeon species in the Guadalquivir River. Mar Biol 129:1–7

Grimod de la Reynière (1802–1812) Almanach des Gourmands. 8 volumes (Réédition: Mercure de France, coll. Le Petit Mercure, 2003)

Lacroix L (1953) La Bernerie, son histoire à travers les âges (published in 1953, republished by Le Livre d'Histoire – Lorisse in 2004)

Larousse P (1870) Grand Dictionnaire universel du XIXe siècle, vol 16. Pierre Larousse, Paris

Letaconnoux R (1961) Note sur la fréquence de la distribution des captures d'esturgeons (*Acipenser sturio* L.) dans le Golfe de Gascogne. Revue des Travaux de l' Institut des Pêches Maritimes 25:253–261

Ludwig L, Debus L, Lieckfeldt D, Wirgin I, Benecke N, Jenneckens I, Williot P, Waldman JR, Pitra C (2002) When the American sea sturgeon swam east. Nature 419:447–448

Moreau E (1898) Les poissons du département de l'Yonne. Bulletin de la Société des Sciences historiques et naturelles de l'Yonne. 51e volume, Paris

Pagès M, Desse-Berset N, Tougard C, Brosse L, Hänni C, Berrebi P (2009) Historical presence of the sturgeon *Acipenser sturio* in the Rhône basin determined by the analysis of ancient DNA cytochrome *b* sequences. Conserv Genet 10:217–224. doi:10.1007/s10592-008-9549-6

Pouillot G (2001) Loire en Loiret. Le patrimoine du Val de Loire en images. Edition du Cercle des Cartophiles du Loiret

Rochard E, Williot P, Castelnaud G, Lepage M (1991) Eléments de systématique et de biologie des populations sauvages d'esturgeons. In: Williot P (ed) Acipenser. Cemagref, Antony, pp 475–507

Rochard E, Lepage M, Meauzé L (1997) Identification et caractérisation de l'aire de répartition marine de l'esturgeon européen *Acipenser sturio* à partir de déclarations de captures. Aquat Living Resour 10:101–109

Rochard E, Jatteau P (1991) Amélioration de la méthode de détermination de l'âge de l'esturgeon commun Acipenser sturio et premières applications. In: Williot P (ed) Acipenser. Cemagref, Antony, pp 193–208

Sauvage HE (1883) La grande pêche. Jouvet, Paris

Tiedemann R, Moll K, Paulusk B, Scheer M, Williot P, Bartel R, Gessner J, Kirschbaum F (2007) Atlantic sturgeons (*Acipenser sturio, Acipenser oxyrinchus*): American females successful in Europe. Naturwissenschaften 94(3):213–217. Category short communication. doi:10.1007/s00114-006-0175-1

Val R (2006) René Val ou la véritable histoire du caviar de la Gironde. Société des amis de Talmont

Part III
Restoration: Conservation of *Acipenser sturio*, Scientific and Political Management

Chapter 20
Historic Overview of the European Sturgeon *Acipenser sturio* in France: Surveys, Regulations, Reasons for the Decline, Conservation, and Analysis

Patrick Williot and Gérard Castelnaud

Abstract This chapter deals with an historical overview of sturgeon fishery in France and with preservation and conservation issues. It includes conclusions and recommendations of the many scientific surveys, which have almost always been neglected. The cause of the decline in the population, fishery regulations, and the deterioration of the environment are described in detail. Proposals for the preservation and conservation of the species from the late 1970s on are briefly presented. The subsequent three-leg strategy is described: (1) the necessity to improve our knowledge of the biology of the remaining population, (2) the inevitable restocking programme to sustain the population, and (3) the need to use another sturgeon species as a biological model to acquire experience on life-cycle characteristics and to develop methods of artificial reproduction for subsequent application to *Acipenser sturio*. Conservation measures, both in situ and ex situ, are summarised. The analysis highlights the absence of (1) a resource-oriented fishery policy, and (2) measures to preserve biodiversity.

20.1 Introduction

Acipenser sturio was one of the most widespread sturgeon species, occurring from the Baltic Sea to the Black Sea via the North Sea, the European Atlantic coast and the northern Mediterranean coast (Magnin 1962; Holčik et al. 1989) (Fig. 20.1 left side), inhabiting river drainage systems and adjacent continental shelves.

P. Williot (✉)
Sturgeon Consultant, 4 Rue du Pas de madame, 33980 Audenge, France
e-mail: williot.patrick@neuf.fr

G. Castelnaud
Cemagref, Estuarine Ecosystems and Diadromous Fish Research Unit, 50 Avenue de Verdun, 33612 Cestas Cedex, France

P. Williot et al. (eds.), *Biology and Conservation of the European Sturgeon Acipenser sturio L. 1758*, DOI 10.1007/978-3-642-20611-5_20,

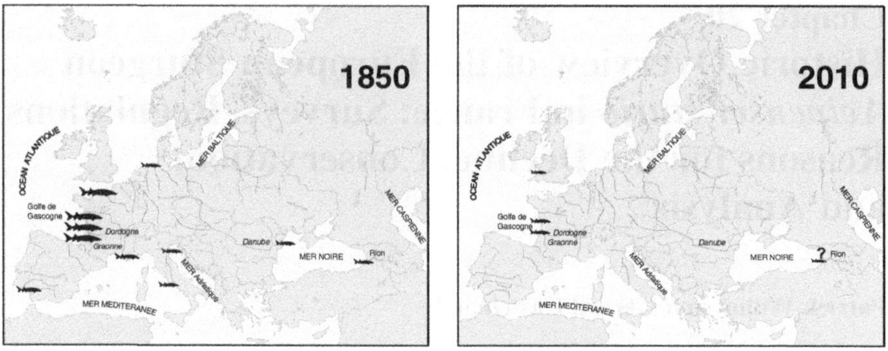

Fig. 20.1 History of the geographic distribution of the European sturgeon *Acipenser sturio*. The 1850 distribution is based on Magnin (1962) and that of 2009 takes into account a recent synthesis (Rochard et al. 1990) and the most recent captures after 2000 reported in Table 20.3

The species was present and apparently abundant in France in the Middle Ages, as we know from regulations and taxes that were applied (Roule 1922).

In the early 1970s, the last remaining population of *Acipenser sturio* was no longer an economic resource, due to the dramatic decline in captures (Ctgref 1973; Trouvery et al. 1984).

Though protected in France since 1982, the population decreased to such an extent that it was classified as in danger of extinction (Appendix 1 of the Convention on Migratory Species) in 2005 (see Chap. 18).

In this chapter, we attempt to summarize the changes in fishery practices and in the environment as well as conservation actions, in order to understand the mechanisms behind the decline in what is today the last population of the species worldwide.

20.2 Material and Method

The history of the sturgeon fishery is presented, using all the available literature, comprising papers from scientific or technical reviews, reports, notes, and also grey literature. This includes for the first time the proceedings of primary inquiries carried out by the first general fisheries inspector in the early eighteenth century, as well as the available archaeological data in the field (see Chap. 7). The opportunity was taken to include some publications dealing with the species in other European waters.

French regulations with regard to sturgeon fishery are described. The latest illegal records of the species in Western Europe are included. Conservation measures are reported with regard to in situ and ex situ activities.

From the data obtained, we were able to identify some milestones which we used to suggest a preliminary analysis of the causes for the species decline. It should be mentioned that the present synthesis does not include data on the Atlantic sturgeon

Acipenser oxyrinchus; its earlier presence in France has been indicated recently by Desse-Berset (2009) (see Chap. 7).

In this paper, the term Garonne basin is used instead of the currently used expression Gironde–Garonne–Dordogne, because the Garonne is the main river and the Gironde is the estuary.

20.3 Results

20.3.1 Geographical Distribution

At the edges of its large extension area, *A. sturio* was rare, e.g., in Ireland from time immemorial (Went 1948), in Russia (Barannikova and Holcik 2000), in Portugal (Almaça and Elvira 2000), and in the Black Sea (Antipa 1934; Antoniu-Murgoci 1937; Marti 1939). Marti (1939) reported that *A. sturio* was thought to be a migrant from the Mediterranean Sea until the late 1920s, when spawning grounds were established in the River Rioni (Georgia) and some rivers in Anatolia (Turkey). According to Antoniu-Murgoci (1937), the presence of the species in the Black Sea was first reported by Antipa in 1909. Indeed, the species was of little economic importance in Romanian waters of the Black Sea (Antipa 1934). It was present along southern Mediterranean coasts with erratic vagrant specimens (Furnestin et al. 1958; Magnin 1962).

In contrast, its significant presence was confirmed until World War II in Greece, the Aegean Sea in the vicinity of the Evros River (Belloc in Economidis et al. 2000), the Northern Adriatic, namely the Po River basin (It), the Rhône River (Fr), the Iberian Peninsula (Spain and Portugal), Rivers Guadalquivir (Sp) and Guadiana (P & Sp), the French Atlantic coast [Adour, Garonne, Dordogne & Loire Rivers (Fr)], the Channel [Seine River (Fr)], the North Sea [Rhine River (NL & D)], Ems, Weser, Eider and Elbe Rivers (D) and the Baltic Sea [River Oder (D & Pol), Vistula River (Pol)] (Magnin 1962; Kinzelbach 1997; Mamcarz 2000, see Chap. 6).

Since that time, the geographical extension of *A. sturio* has considerably decreased, and the species now exists as a critically endangered population in France in the Garonne basin and its northern adjacent continental shelves; some vagrants are occasionally caught as far away as the English Channel and the North Sea (Trouvery et al. 1984; Rochard et al. 1990, 1997a) (Fig. 20.1 right side). A much more detailed analysis of the geographical distribution is given in Chap. 6.

20.3.2 The History of Sturgeon Fishery

Sturgeons have long been used as a food resource by humans. Bone remains found close to the River Rhône indicate the presence of the species from the sixth to the second century BC (Desse-Berset 1994), and correspondingly in the Gironde

Estuary and its northern adjacent area around 3000 years BC (Desse-Berset 2009; see Chap. 7). It is worth noting that the author provided confirmed species identification only for *Acipenser sturio* in the Rhône basin by paleo-molecular analysis (Desse-Berset et al. 2008; Pagès et al. 2009), and for both *Acipenser sturio* and *Acipenser oxyrinchus* on the French Atlantic coast (Desse-Berset 2009). With regard to the French Mediterranean coast, especially the Rhône River and its adjacent coastlines, there are signs of the presence of a stable sturgeon population during the aforementioned period, because of the occurrence of a large range of year classes (Desse and Desse-Berset 1993; Desse-Berset 1994).

Scott (1936 in Val 2006) reported that the species was so common in the sixteenth century that it was used to fertilise the land. However, large stranded fish have long been considered as "Royal fish", like salmon (Le Masson du Parc 1727–1728). The species was so highly valued at that time that archbishops paid a high price for live specimens as they knew how to keep them alive (Bernard 2004).

However, to our knowledge, the first mention of the State's interest in sturgeons as an aquatic resource dates from 1727 by Le Masson du Parc, the primary General Fisheries officer. The author mentioned that the species was caught incidentally in the Adour River as well as in all other French rivers, with the exception of the Garonne River with its estuary and its tributaries where there was an organised fishery with specific fishing gear.

From the beginning of the twentieth century (Table 20.1), various surveys were either ordered by French fisheries institutes or conducted independently by scientists or people educated in fish exploitation.

The decline in the population started at the latest in the nineteenth century, and in the early 1920s the Garonne basin was the last French river basin system where the species still occurred with a significant population (Roule 1922, 1925; Prioux 1957). This population was thought to be disappearing at the beginning of the 1960s (Castaing 1963), predominantly in the Garonne then in the Dordogne River. The authors highlighted the main causes for this decline as: (1) the building of dams in the mid-nineteenth century on both the Garonne and Dordogne Rivers, (2) the deterioration in spawning grounds due to diverse changes in the rivers, and (3) the capture of juveniles both in the rivers and on the continental shelf (Table 20.1). Scott (1936 in Val 2006) reported huge quantities of juveniles being captured at a time when they were migrating downstream in autumn. It is interesting that it was a forestry commissioner, who was nevertheless not unfamiliar with sturgeon fishery and its biology, who claimed that everything was done in France to protect the species, and that the species would have to cope with the changing environment (Charpy 1951).

Further reports pointed out that the effectiveness of the regulations was not monitored, and drew attention to the fact that every human activity should be ecologically sustainable (Vibert 1953), while indicating the lack of biological data and the lack of education provided for fishermen (Prioux 1957). The authors proposed to ensure recruitment and introduce a size limit of 1.5 m, as in the Dutch legislation, as no biological data were available on the French sturgeon population. The authors of most of the surveys emphasized the extremely negative impact of legal captures of juveniles.

Table 20.1 Surveys on the *Acipenser sturio* population in France and its fishery in the Garonne basin

Date	Outcomes, results, proposals	Source
1905	Juveniles are captured at sea	Roule (1922)
1921	Onset of caviar exploitation	Prioux (1957)
1922	First scientific investigation. Main points highlighted are: – The species has been eradicated from the River Seine – Juveniles are captured at sea and in the estuary as well Roule suggests – Protect spawning and recruitment – Set minimum size limit at 1.50 m	Roule (1922)
1936	Second survey – Juvenile fish are caught in excess – Spawning should be protected – Restocking should be promoted	Scott (1936) in Val (2006)
1950	Third survey pointed out: – Species is quite unknown to the public – Absence of scientific data Authors suggest – Apply the regulations – Train the fishermen	Darlet and Prioux (1950)
1951	Study of the situation in the Spanish River Guadalquivir; it was proposed/stated that: – Everything should be done in France to protect the species from 1950 onwards – The species is incapable of adapting to new conditions – Knowledge of reproductive biology is very poor (forestry commissioner)	Charpy (1951)
1953	Fourth scientific survey which highlighted the value of sturgeon fishery and strongly suggests – Apply the fishery regulations – Make sure that every human activity respects water quality	Vibert (1953)
1957	Fifth survey stressed: – The population has been facing damming, captures of juveniles and non-respected regulations – Fishermen are probably not trained – Absence of scientific data on the species	Prioux (1957)
1962	Publication of the first biological–ecological compendium on the species (with the exception of population dynamics)	Magnin (1962)
1968	Erection of a dam on River Garonne at Malause	
1973	Sixth scientific investigation highlighted: – Two different regulations from upstream to downstream are applied (see Table 20.2 for details) – Absence of a state policy for fishery – Professional organisations are diverse – Assessment of lost income	Ctgref (1973)
1981–1986	– Setting up of a marine professional organization in the estuary – Establishment of statutes for professional fishermen in rivers, with a separate professional organisation in each river	
2004	Three specimens were auctioned in the UK, NL, and Fr, the last one with the authorisation of French regional administration	Guth and Laurent (2004)

(continued)

Table 20.1 (continued)

Date	Outcomes, results, proposals	Source
2005	The species is now included in Appendix 1 as a species on the verge of extinction (Bonn Convention)	
2007	Editing of the European action plan	Rosenthal et al. (2007)

The first two important papers that focused on the biology and the ecology of the species were published in the early 1960s. They dealt with the marine geographical distribution of captures close to the mouth of the estuary and adjacent areas, and on basic biological and ecological issues (Letaconnoux 1961; Magnin 1962). Ninety-one per cent of captures were concentrated in a circlular area with a radius of about 111 km centred on the mouth of the Gironde estuary. Magnin (1962) provided a huge compendium on sturgeon biology, focusing on migrations, growth, age determination, feeding, and hydro-mineral balance. The population decline was illustrated by the difficulties he experienced in getting enough specimens in 1958 (Magnin 1962) when he started, on his own, his investigation into sturgeons, and he was unable to perform a tagging campaign to assess the dynamics of the population.

The last scientific survey, carried out in the early 1970s, aimed at providing an overview of the fishing of migratory fish. The following points were highlighted: (1) there were three different fishery regulations from upstream to downstream, (2) there was no state fishery policy, (3) there were multiple professional bodies, often with conflicting interests, and (4) fishermen suffered a loss in income due to the dramatic decline in sturgeon landings. In the absence of an official system of recording landings, the loss of income post World War II was confirmed by only very few data obtained from professional fishermen and river bailiffs (Ctgref 1973).

It is worth noting that the fishing activity of fishermen harvesting in the river was recognised and codified as late as 1984. Before this date, these fishermen did not exist, in administrative terms (labour, health insurance, and fisheries): in contrast, 1,500 professional and 4,000 recreational fishermen were registered along the French Atlantic estuaries in the late 1970s (Castelnaud and de Verdilhac 1982).

20.3.3 Fishery Regulation

The first regulations on aquatic animals were imposed in August 1681 through a royal statute, and they contained two main ideas: protection of (1) the fishermen themselves against their own greed, and (2) the fish fry (Le Masson du Parc 1727 in Lieppe 2004).

The first regulation on sturgeon fishery is the decree of 1853, stating that a minimum square mesh of 8 cm (Scott 1936 in Val 2006) and a minimum sturgeon length of 27 cm from eye to fork of the caudal fin had to be respected (Lorin de Reure 1924).

However, the authors remarked that this did not provide protection for the juveniles. Indeed, this mesh size and the minimum stipulated length allowed only young juveniles, less than 1 or up to 2 years old, to escape.

The fishery regulations varied depending on the area in which they applied, and there were wide variations in size limit over time (Table 20.2). It is noteworthy that not one regulation was applied in rivers until the late 1930s. Most of the time, juveniles, including very young specimens, were allowed to be captured in the whole extension area. There were, however, two exceptions: (1) In 1923, the minimum legal size was raised to 1.5 m. This threshold level was quickly changed back to the lower size after pressure from the fishermen. It has to be mentioned that even 1.5 m was insufficient for the species, as females do not mature below a length of 1.85 m (mean value) (Magnin 1962); (2) From 1950 onwards (Table 20.2), any captures in the marine and estuarine areas were forbidden during the second part of the year.

Finally, the species was totally protected in France from 1982 as a result of a three-step procedure: protection in rivers from 1980, in marine and estuarine areas from 1981 on, and in all French waters from 1982.

Apart from the regulation per se, and whatever its content may be, enforcement is also important. From the earliest times, there are indications that the law has not been applied, or in only a very limited way. In this respect, Le Masson du Parc (1727, 1727–1728) pointed out that neither admiralty officers nor forestry commissioners carried out their duty according to the law.

In the 1980s and 1990s, many people (including the authors and colleagues) working in contact with fishermen were informed about captures of juvenile

Table 20.2 Changes over time in fishery regulations (minimum size) of *A. sturio* in France depending on time and area (adapted and updated from Trouvery et al. 1984)

Marine and Estuarine areas		River	
Date	Minimum size (fish or mesh)	Date	Minimum size
1853	0.27 m[a] Minimum square mesh size of nets 8 cm[b]		
1890	0.14 m[c]		
1923	1.50 m[c]		
1924–1927	1 m[c]		
1928	Mesh surface 10 cm^2		
1935	1 m[c]		
		1939	1.50 m
1950	1.30 m (every size forbidden between 1/07 and 31/12)	1950	1.30 m
1952	1.45 m[d] (every size forbidden between 1/07 and 31/12)		
		1958	1.45 m
		1980	Every size forbidden
1981	Every size forbidden		
≥1982	Protected species (no capture permitted[d])		

[a]From eye to fork of the caudal fin (Lorin de Reure 1924)
[b]Scott (1936) in Val (2006)
[c]Total length
[d]Except for scientific purposes (under supervision)

Table 20.3 Incidental captures of *Acipenser sturio* in Western Europe auctioned after capture

Date	Location	Biometry TL (*W*)	Observation
04/2003	Leiden (NL)[a]	~1.40 m	Dead-auctioned-confiscation
06/2004	Swansea (Wales, UK)	~2.70 m (120 kg)	♀Dead-auctioned-confiscation
06/2004	Sables d'Olonne (Fr)	1.38 m (~14 kg)	Dead-auctioned-confiscation[b]

[a]The area where the specimen was caught was not ascertained. There are suspicions that it was in Belgium
[b]Guth and Laurent (2004)

sturgeon in the mouth of the Gironde estuary and the adjacent fishing port. This was confirmed later on (Lepage and Rochard 1997; see Chap. 22).

Though the species was totally protected in Europe from 1998 onwards, after 2000 some captures were confiscated at auctions in three European countries, France, UK, and the Netherlands (Table 20.3). Moreover, the fishermen in France, amazed by regional fisheries administrations which ignored the complete protected status of the species. This illustrtaes the limited concern by the authorities for the protected status of the species and the need to enforce it.

20.3.4 Deterioration of the Environment

Alterations to the aquatic environment have had negative impacts on the sturgeon population. Dams reduced the access to spawning grounds, modified the hydrology, and stopped the long-term renewal of the river bed. Mining of the river bed by gravel extraction destroyed the existing or potential spawning grounds, and also altered those located downstream by transporting sediment and thus filling the interstices between gravel and blocks. These alterations are chronologically listed for both drainage systems where they have been documented, the River Rhône on the Mediterranean coast and the Rivers Garonne and Dordogne on the Atlantic coast (Table 20.4). These actions began as early as the mid-twelfth century by damming the River Garonne, and this was later expanded in the nineteenth century in both the Garonne and Dordogne Rivers. Roule (1922) indicated that the migration of adults had been stopped by the dam at Beauregard near Agen on the River Garonne in the mid-nineteenth century, whereas before this date sturgeons migrated as far as Toulouse (Fig. 20.2). Similarly, the upper migratory limit in the River Dordogne was Bergerac. Mining of the river bed started after World War II in the River Garonne and remained very active up to the mid-1990s, though some limitations were put in place in the 1980s in both the Garonne and Dordogne. Castaing (1963) reported that the main area of the spawning grounds above Bordeaux in the River Garonne had been intensively mined since the end of World War II for about 15 years. At the end of 1985, under pressure from the research institute Cemagref, a well-recognised sturgeon spawning ground located at Meilhan, some km upstream from La Réole (Fig. 20.2) in the River Garonne, was partially

Table 20.4 Deleterious physical causes (damming and mining) acting on the *Acipenser sturio* population and their effects in France

Date	Designation	Sources
1177	Bazacle dam (weir) on the Garonne river at Toulouse	
1843	Dam on the Dordogne River at Mauzac	
1846	Damming (weir) the River Garonne at Beauregard (close to Agen)	Trouvery et al. (1984)
1851–1852	Damming the River Dordogne at Bergerac, downstream from Mauzac	Trouvery et al. (1984)
1908	Dam on the Dordogne River at Tuilères (between Mauzac and Bergerac)	
>1940	Mining of the river bed	
1968	Erection of a dam on the River Garonne at Malause Golfech (upstream from Agen)	Trouvery et al. (1984)
1970	Erection of a dam at Vallabrègues on the River Rhône (upstream from Arles)	Tabardel (1994)
1973	Dam on the Garonne River at Malause Golfech (upstream from Agen)	Trouvery et al. (1984)
1973–1975	Erection of dams at Avignon and Caderousse, on River Rhône upstream from that at Vallabrègues	Tabardel (1994)
1980	Five out of 15 previous spawning grounds are destroyed in the River Dordogne Fifteen out of 17 previous spawning grounds are destroyed in the River Garonne	Ctgref (1980), Castelnaud et al. (1981), and Castelnaud et al. (1985)
1981–1991	Some limitations in mining the beds in both Rivers Garonne (La Réole–Agen) and Dordogne	Castelnaud et al. (1985)
1985	Protection of the Meilhan spawning ground in the River Garonne (La Réole–Agen)	Jego et al. (2002)
1989–2008	Project to extract gravel in the middle part of the estuary	Lepage et al. (2000)

protected. As a result of all these alterations, the number of potentially usable spawning grounds decreased dramatically, by 33 and 88% in the Dordogne and Garonne respectively (Table 20.4).

More recently, and only after a long fight, gravel extraction was finally stopped in the Gironde estuary (Lepage et al. 2000) adjacent to the sturgeon nursery ground.

In the River Rhône, the dam located at Vallabrègues restricted the spawning grounds up to an upper limit located a few km upstream of Arles (Fig. 20.3).

Finally, in the 1980s and 1990s a significant decrease in the motility of spermatozoa (Fig. 20.4) was observed, indicating a reduced fertility of *A. sturio* males. A possible cause (Williot et al. 1997) could be a negative effect of molecules accumulated in the sediment as a result of human activities. Indeed, juveniles (up to 7–8 years old) spent several months in summer time each year foraging in the lower estuary, and at this age differentiation has occurred and

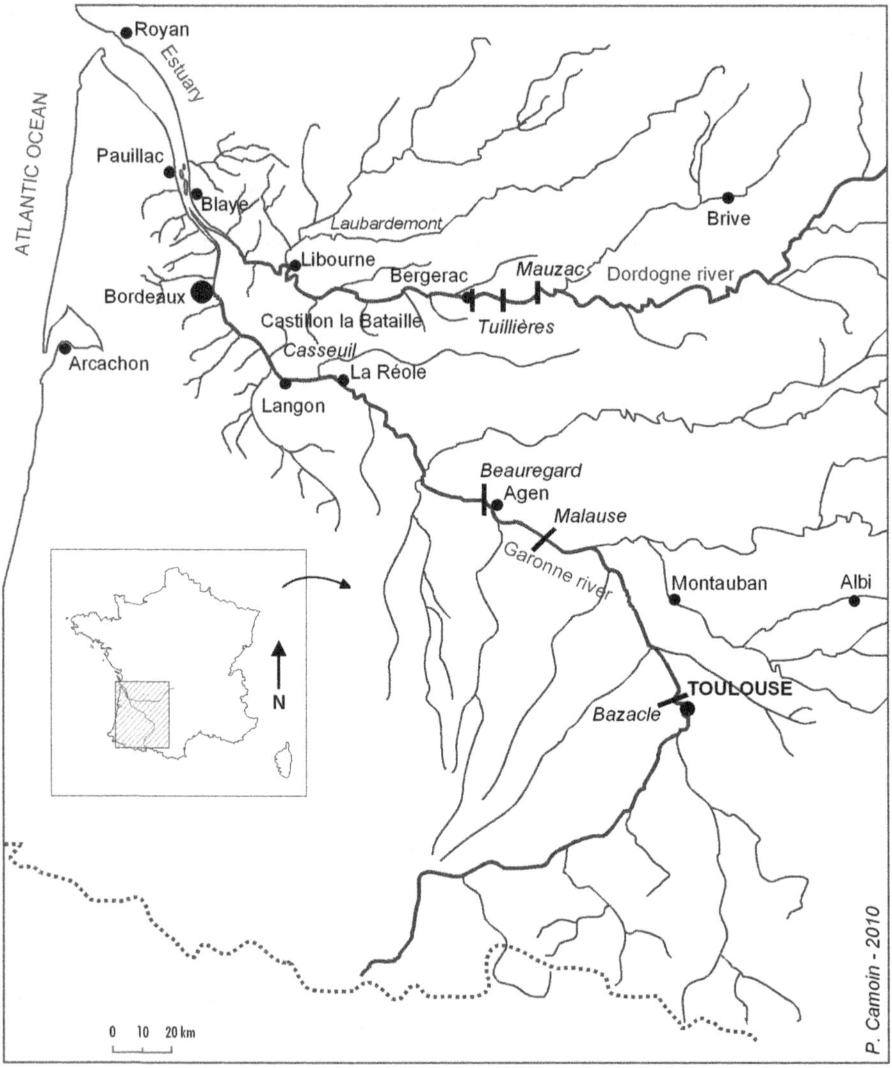

Fig. 20.2 Garonne basin (Fr) showing rivers, dams and weirs locations

gonads have started their low growth (see Chaps. 10 and 12). A recent work on the contamination of fish by heavy metals showed that benthic carnivorous species provided some support for this hypothesis (Durrieu et al. 2005). Moreover, Pierron et al. (2008) demonstrated how cadmium can affect the eel's entire reproductive system, including its reproductive migration. Cadmium is known to be present in the Garonne–Gironde drainage system at a very high level due to old mining in the River Lot, a right tributary of the River Garonne (IFREMER 2009).

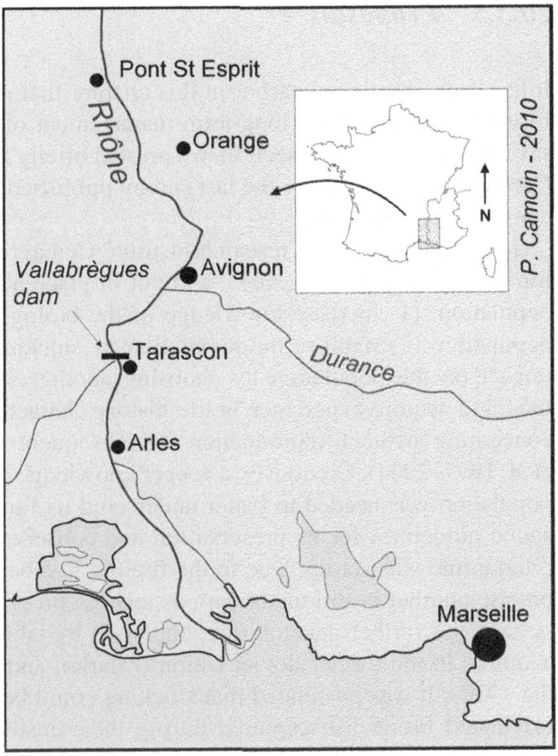

Fig. 20.3 Rhône basin (Fr) showing rivers, dam and locations

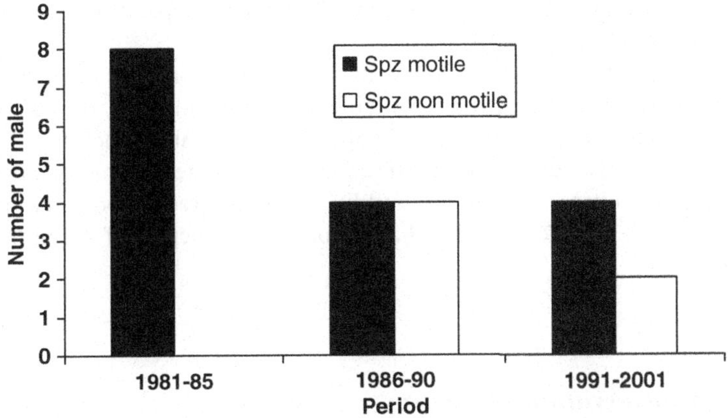

Fig. 20.4 Quality of reproduction potential of incidentally captured wild males in the Gironde–Garonne–Dordogne basin during their upstream migration deduced from the motility of the spermatozoa (redrawn from Williot et al. 2002b)

20.3.5 Proposals

It has been mentioned earlier in this chapter that many suggestions have been put forward to improve the long-term management of the sturgeon population and/or save the species. In this section, we present briefly the strategy that was set up in the Garonne basin soon after the last survey published in 1973, and give a synthesis of conservation measures.

In the late 1970s the research institute Cemagref decided to take responsibility for the protection of *A. sturio*, and put in place a three-leg strategy to restore the population: (1) increase knowledge of the biology and ecology of the remaining population, (2) stabilise the population by stocking, and (3) avoid any additional impact on the population by choosing another sturgeon species as a biological model to acquire experience in life history characteristics, and to develop methods concerning artificial reproduction for subsequent application to *A. sturio* (Williot et al. 1997, 2004). Obviously, a deeper knowledge of the biology and ecology of the population was needed to better understand its functioning and thus provide more sound guidelines for its preservation and conservation. For this reason, a tagging programme was set up. Due to the limited number of fish in the wild (to date the precise number is still unknown), a stocking programme appeared the only way to sustain and further develop the population by taking advantage of the knowledge acquired in the former Soviet Union (Charlon and Williot 1978; Williot 1984). In the 1970s, it was postulated that stocking could be achieved by reproducing wild-originated brood fish captured during their upstream migration, and subsequent rearing of fingerlings to be released into the wild. As indicated above, the use of another sturgeon species as a biological model appeared inevitable if we were to avoid any further negative effects on the remaining population. As the main model species, the Siberian sturgeon (*Acipenser baerii* Brandt) was selected, and to a lesser extent the Sterlet (*Acipenser ruthenus*). Specimens of *A. baerii* were obtained twice, in 1975 and 1982, under the French–Soviet cooperation programme in oceanology (Williot et al. 2004).

The assumption that there was an unlimited number of good quality wild-originated brood fish of *A. sturio* available was not confirmed during the 1980s or later on (Williot et al. 2002a; see Chap. 27). The strategy therefore had to be changed in the early 1990s, and the research programme became one of conservation rather than preservation; the building of a cultured brood stock became a key point in order to increase the chance for the production of young fish to be released.

20.3.6 Conservation Measures

All conservation measures were initiated and/or promoted by the research institute Cemagref in the late 1970s and early 1980s (Table 20.5).

Table 20.5 Main in situ and ex situ conservation actions performed in France/Garonne basin

Date	Actions/results	Source
1981	– First artificial reproduction of Siberian sturgeon and subsequently of *Acipenser sturio* (wild broodfish) – Juvenile fish are caught in the estuary and tagged – Building of first experimental rearing facility	Williot and Rouault (1982) and Williot et al. (1997)
1981–1992	Survey of juveniles (≥2,000 fish tagged in the period 1984–1989, some fish were ≥130 cm)	Castelnaud et al. (1991)
1985	– Complete ovulation of two wild females and high rates of fertilisation with two different males	Williot et al. (1997)
1989	Capture of YOY, revealing spawning in 1988	
1991–1993	– New research facility for Cemagref – Start of building an ex situ brood stock	Williot et al. (1997)
1993	New research vessel enabling experimental trawling in the Gironde estuary	
1993–2000	~800 juveniles tagged	
1994–1997	First Life programme (1994–1997)	Elie (1997)
1995	– First larval rearing and further stocking with 7,000 marked fingerlings – Capture of YOY, revealing spawning in 1994	Williot et al. (2005)
≥1996	Setting up of cooperation between Cemagref (Fr) and IGB (D)	
1996–1997	Recapture of some marked specimens released before	Lochet et al. (2004)
2001	First spermiation of reconditioned wild males Feeding habit and movements of juveniles in estuary	Williot et al. (2007), Brosse et al. (2000), and Taverny et al. (2002)
2002	Second Life Programme (1998–2002)	Rochard (2002)
2005	*Acipenser sturio* is now included in Appendix 1 as a species on the verge of extinction (Bonn Convention)	
2006	Publication by the French Research Institute Cemagref of the proposed research actions to contribute to the restoration of the species	Rochard and Williot (2006)
2007	First reproduction of farmed specimen of *A. sturio*	Williot et al. (2009a)

YOY young of the year

20.3.6.1 In Situ Measures in the Garonne Basin

The first comprehensive study on population dynamics was published thanks to a large tagging programme throughout the 1980s (Castelnaud et al. 1991). Two campaigns were set up, the first between 1981 and 1992 with 2,000 fish tagged, and the second between 1993 and 2000 with about 800 tagged fish. In addition, the relative number of juveniles by year-class, their relative strength, an estimate of the total number of fish, and their growth were investigated (Castelnaud et al. 1991). Age determination methods were updated (Rochard and Jatteau 1991), as well as data on the marine distribution area (Rochard et al. 1997a). These works provided evidence for a possible regeneration of the juvenile population with its rapid development between 1981 and 1988 under protection and a scientific presence in the field in cooperation with local fishermen. However, this population, composed progressively of pre-mature individuals of 1.30–1.60 m in length, suddenly decreased (Castelnaud et al. 1991; Williot et al. 1997). As a consequence, very few adults were recruited and migrated into the rivers for reproduction. As mentioned before, the main cause for the decrease in this population of pre-mature individuals was trawl fishing at sea (see Chap. 22).

Later on, potential spawning grounds and downstream migrations of juveniles were characterised (Jego et al. 2002; Rochard et al. 2001), as well as habitat and trophic relationships of juveniles in the estuary (Taverny et al. 2002; Brosse et al. 2000, see Chap. 11). In addition to these scientific approaches, efforts were made to support the changes in the regulations (protective status) and in the protection of some key biotopes such as spawning and nursery grounds.

20.3.6.2 Ex Situ Measures

Soon after the first successful artificial reproduction of *A. baerii* (Williot and Rouault 1982), the first successful artificial reproduction from a pair of *A. sturio* wild brood fish was achieved in 1981 (Williot et al. 1997) (Table 20.5).

After fighting for several years with its own management headquarters, the leading laboratory Cemagref was finally authorised to build a research facility designed specifically for and dedicated primarily to the conservation of *A. sturio*, because simultaneously a private hatchery was established for the reproduction of the Siberian sturgeon. The facility started operations in 1991 (Table 20.5), mainly with financial support from the Ministry of Research provided that the know-how of brood fish management, breeding and larval rearing of *A. baerii* was transferred to the private company. No financial support was obtained from the Ministry of Environment.

Due to constraints in obtaining wild brood fish on a regular basis, ex situ conservation was initiated in the early 1990s (Williot et al. 1997; see Chap. 27). This objective rapidly became the most promising way to ensure future potential stocking, and since the status of wild brood fish was deteriorating (Williot et al. 2002a), the building of further brood stock was started (Williot et al. 2005; see Chap. 31).

Throughout the whole programme, i.e., from the early 1980s, only four artificial reproductions of wild-originated brood fish have been successful, with the last occurring in 1995 (Williot et al. 2000; see Chap. 27). This success ensured the first successful larval rearing and the release of marked juveniles (Williot et al. 2005). Some of the stocked fish were recaptured, and survival and growth could be analysed (Lochet et al. 2004). The authors demonstrated that the growth of these fish was similar to that of wild fish, and records from recaptured specimens and further computational analysis (Rochard et al. 1997b) enabled us to estimate the survival rate as being in the range of 3–4%. Both outcomes provided key support for the releasing programme.

Most of the aforementioned work was performed in the context of two LIFE programmes (Table 20.5), Life I from 1994 to 1997 (Elie 1997) and Life II from 1998 to 2002 (Rochard 2002).

A significant result was obtained in 2007 with the first artificial reproduction of fish reared in the brood stock facility from the juvenile stage on (Williot et al. 2009a; see Chap. 32).

20.4 Analysis of the French European Sturgeon Fishery and Species Conservation Histories

In the past, regulations concerning the protection of the sturgeon stocks were never based on a knowledge of life-history characteristics. Officially, there was no rule against catching very young juveniles, as there was for catching large juveniles. It is worth noting that regulations (concerning both fishery and species protection) were not consistent along the basin from the sea to the upstream river. This reflects a total absence of a biological approach for the resource, which had to match the artificial administrative organisation of territories. The only exception was the complete fishing ban in the second part of the year from 1950 onwards in marine and estuarine zones.

The two criteria that were used to support the regulations were size (of fish or of the net mesh size) and fishing period. As there was no obligation to declare the captures, there are neither records of landings (volume and size) nor abundance indicators to give a long-term trend for the state of the population. The only possible records were auction records which registered all species. This allowed Letaconnoux (1961) to analyse official landings at the fishing port of La Rochelle. The absence of a fishing survey and a research programme based on the biology of the exploited resource meant that three types of change could not be assessed: (1) CPUE[1] of juveniles as an indirect indicator of recruitment, illustrated by Peterson et al. (2000) for the Atlantic sturgeon (*Acipenser oxyrinchus*) in the River Hudson (USA) could not be applied to *A. sturio*, (2) computational analysis of the mean length of fish illustrating the deterioration of the Stellate sturgeon (*Acipenser stellatus*) population in the Iranian part of the Caspian Sea (Moghim and Nielson 1999) was not applicable to *A. sturio*, and (3) due to the absence of any survey there was no information on sturgeon migration patterns, e.g., summer migrations, possibly trophic movements (called locally "mouvée de la Saint-Jean") from the marine continental shelf into the lower part of the Gironde estuary. This was not documented in Letaconnoux (1961) nor in Magnin (1962), and could be described for the first time in the late 1970s. In the 1980s, because of the potential

[1]Captures Per Unit of Effort.

concentration of fish in the estuary, a tagging programme was set up with the participation of teams of marine fishermen from the estuary, in order to manage the operation and the protection of the species (Castelnaud et al. 1991). In addition to the expected biological information, this operation was used to document the increase in the juvenile population until the stage of pre-maturity, in the hope of rebuilding the population.

In contrast to this absence of knowledge, emphasized as early as 1950 (Darlet and Prioux 1950, Table 20.1), all the authors of surveys, mostly scientists, proposed sound measures to protect the species. Scientists' conclusions were never taken into account, however, or they were rapidly swept away under pressure from fishermen, as exemplified by the regulation for the minimum size of 1.5 m for fish put in place in 1923 and changed soon afterwards in 1924.

Another aspect has been stressed thanks to these surveys, i.e., the absence of training for fishery administration officers and fishermen in fish biology and fisheries management. Unfortunately, this remains predominantly the case today.

It has been shown that the protective status was put in place very late, and only after pressure from scientific bodies. As for fishing periods, other preservation regulations were not enforced, and both fishermen and marine fisheries officers were either unable or unwilling to organise their supervision. As a result, the status of the species deteriorated in spite of its complete protection, with ineffective application of the regulations.

All the conservation actions were initiated by a public research institute, Cemagref, which was nevertheless not commissioned for such action. The initiators were very committed to getting the appropriate funds and facilities. It is remarkable that the Cemagref research station in St-Seurin-sur-l'Isle was achieved, funded mainly due to the development of the Siberian sturgeon to farming conditions. This made France one of the leaders in sturgeon farming (Williot et al. 2001). This is another unexpected justification for our primary long-term three-leg strategy, which has often been wrongly presented and understood, especially by environmentalists including the Ministry of the Environment. There is no doubt that the *A. sturio* conservation programme could not have been developed without the experience and without applying methods developed with the help of the model species. This conservation programme, which included a model species which rapidly supported farming activities, is thus a good example of the complementary effects between fisheries management (including species-population preservation) and aquaculture (Williot 2000, 2004).

At present, there is no certainty that the last population of the species will be safeguarded, even though awareness campaigns targeted towards professional marine fishermen were organised at sea (though too late) (see Chap. 37). However, one bottleneck has disappeared with the successes in reproducing farmed specimens for a period of 3 years now (2007–2009) (Williot et al. 2009a, see Chap. 32) and the recent progress in larval and juvenile rearing (Williot et al. 2005; see Chaps. 28 and 33) Two opportunities were missed in the course of the programme. The first was in 1985 when, after artificial reproduction, we were unable to feed the larvae (Williot et al. 1997) successfully. There are two possible reasons for this failure: (1) people were not as experienced as they were later on, and (2) people had to put more effort into other programmes as the *A. sturio* conservation programme was not funded at that time. The second missed opportunity was the fact that no blood samples were taken from wild fish at the time of tagging campaigns, though this was proposed. As a result, no data were collected on endocrinology, osmoregulation, haematology, etc. Those data might have been very useful as standard reference for the species as well as for setting up the ex situ programme, because of the absence of any references in applied physiology and husbandry (see Chap. 31). The absence of training in biology applied to conservation of some of those involved, and the decision-making process, are at least partly responsible for these missed opportunities.

Taken altogether, the present history of *A. sturio* with its focus on fisheries management, if any, and on species conservation, is a good example of the extinction vortex (Gilpin and Soulé 1986). At present, we are unable to rank the main reasons for the decline in the species, as already mentioned (Williot et al. 2002b). However, the present work strongly suggests the following causes which appeared chronologically: (1) damming and fishing of juveniles, (2) overfishing of all year classes (greed for caviar), and (3) degradation of habitats and pollution.

From the beginning of the conservation programme, i.e., from the late 1970s, the restoration–conservation programme was partly based on restocking. This measure was questioned by Roule (1922) and later by Scott (1936). Roule (1922) preferred to protect natural spawning, and Scott (1936) regretted the absence of development in sturgeon farming. Sturgeon restocking was mainly carried out in the former USSR in the Caspian Sea (Williot 1984). Despite the huge releasing programme, sturgeon fishery collapsed considerably (Birstein 1993; Williot et al. 2002b). There is therefore a great temptation to consider restocking as ineffective if not counterproductive. Indeed, Secor et al. (2000) estimated that restocking contributed less than 30% of the adult stock of Russian (*A. gueldenstaedtii*) and Stellate sturgeon (*A. stellatus*) in the Caspian Sea. Some of the reasons for this situation were proposed by Carré (1978). He underlined the great changes in specific landings in the last few decades as a consequence of all anthropogenic activities that had greatly impacted the environment (damming, oil pollution, etc.). There are some prerequisites for restocking to be successful: a safe environment and a monitoring programme of released fish are two key points which unfortunately are very rarely controlled and applied (Williot et al. 2009b). However, the Beluga (*Huso huso*) is still present in the Ponto-Caspian basin thanks to releasing programmes, as brood fish are unable to access their natural spawning grounds which are located far upstream from the dams (Barannikova 1987; Secor et al. 2000). However, restocking, and thus ex situ conservation in the case of a critically endangered species like *A. sturio*, might be the only solution to potentially reinitiate the installation of a viable population. Stocking has been recognised as a potentially useful measure for sturgeon (Beamesderfer and Farr 1997), on condition that the quality of the environment is good.

What lesson can be drawn from switching to another sturgeon species as a biological model? In our case this proved to be a key point in terms of reproduction. This is because a collaborative programme oriented towards fundamental and applied science was launched in reproductive biology, the applied part of which is given in Chaps. 27 and 32. Apart from highly unexpected scientific outcomes, which had considerable impact, e.g., the influence of food components on endocrinology (Pelissero and Le Menn 1991; Le Menn et al. 1999), the easy availability of live sturgeon allowing testing techniques for marking (Brun et al. 1998), gastric washing (Brosse et al. 2002), as well as the use of the model species for testing the potential toxicity of the farming environment (Williot et al. 2007). The biological tools in endocrinology (Pelissero and Le Menn 1991; Cuisset et al. 1991) are either still in use or were the forerunners of a new generation of tools to apply to *A. sturio* (Davail-Cuisset et al. 2008). *A. baerii* was not a good model for larval rearing, as the Siberian sturgeon and the Sterlet could be easily fed on compound diets (Gisbert and Williot 1997 and Williot unpublished results, respectively) from the beginning of exogenous feeding, whereas this technique could not be applied to *A. sturio*. Preliminary attempts at larval rearing of *A. sturio* based on compound diets provided poor results (Williot et al. 2005; see Chap. 28), this however was recently overcome (see Chap. 33). There is a last unexpected outcome of using another sturgeon as a biological model, which proved to be a decisive one. The only way to get permission and to raise funds to build the research facility at St-Seurin-sur-l'Isle where we could start to build a brood stock was to engage a private company to build a commercial hatchery dedicated to the production of fingerlings of the model species, the Siberian sturgeon. Finally, conducting both a conservation (*A. sturio*) and a farming (*A. baerii*) programme enabled us to hold ACIPENSER, the first International Symposium on Sturgeon (ISS1) in Bordeaux in 1989, with the peer-reviewed proceedings being published soon after (Williot 1991). ACIPENSER provided the opportunity to launch for the first time the well-accepted idea of a non-profit international sturgeon society. Through lack of time the project could not be put into action, but the idea was taken up later, with the World Sturgeon Conservation Society established in 2003. This is the early conviction of the great importance of European cooperation especially for the conservation of the European species (i.e., *A. sturio*), and this led to the establishment of a cooperation programme between France and Germany through the respective Institutes Cemagref and Leibniz-IGB (see Chap. 38) which proved to be a key point for continuing this conservation action.

20.5 Conclusions

A. sturio was exploited due to regulations which were not based on biology and without any knowledge of the resource, with no awareness of the change in fishing, i.e., fishing power (new trammel net threads, boat motorization), fishing tactics (fishermen following the adults along their migration course) or of the economic context (value of the processed caviar) (see Chap.13). The multiple scientific alarms raised about fishing practices were not respected. This led to the dramatic decline in the population, and later to the late protective status which proved to be ineffective. These are typical statements for an unmanaged resource, which led the species to the verge of extinction. The French version of the "Action plan for the conservation and restoration of the European sturgeon" (Rosenthal et al. 2007) has still to be established. This is a final example of the lack of consideration given by administrations to species conservation. As a result, there is no resource-oriented policy for either fisheries or for species conservation. For *A. sturio*, now that all farming steps have been overcome, the last chance to avoid extinction depends on the long-term responsible management of the brood stock for restocking. Obviously, these actions will only be successful in the future when there is effective protection for the species and its habitats.

Acknowledgements We would like to thank Yamina Larabi for kindly providing copies of ancient references.

References

Almaça C, Elvira B (2000) Past and present distribution of *Acipenser sturio* L., 1758 on the Iberian Peninsula. Bol Inst Esp Oceanogr 16(1–4):11–16

Antipa G (1934) Les esturgeons de la Mer Noire. Leur biologie et les mesures nécessaires pour leur protection. Rapports et procès verbaux des réunions de la Commission Internationale pour l'Exploration Scientifique de la Mer Méditerranée, vol 8. pp 61–75

Antoniu-Murgoci A (1937) Observations concernant Acipenser sturio L., de la Mer Noire. Annales Scientifiques de l'Université de Jassy, vol 23. pp 109–116

Barannikova IA (1987) Review of sturgeon farming in the Soviet Union. J Ichthyol 27:17–27

Barannikova IA, Holcik J (2000) Past and present distribution of *Acipenser sturio* L., 1758 in Russia, and problems involving its restoration. Bol Inst Esp Oceanogr 6(1–4):55–59

Beamesderfer RCP, Farr RA (1997) Alternatives for the protection and restoration of sturgeons and their habitat. Environ Biol Fishes 48:407–417

Bernard J (2004) D'Ausone à Le Masson du Parc. In: Procès verbaux des visites faites par ordre du Roy concernant la pesche en mer (1727). Pêches & Pêcheurs du domaine maritime aquitain au XVIIIe siècle. Amirautés de Bayonne & de Bordeaux, Les Editions de l'Entre–deux-Mers, 33420, Camiac et Saint Denis, France, pp III-IX

Birstein VJ (1993) Sturgeon and paddlefishes: threatened fishes in need of conservation. Conserv Biol 7:773–787

Brosse L, Lepage M, Dumont P (2000) First results on the diet of the young Atlantic sturgeon *Acipenser sturio* L., 1758 in the Gironde estuary. Bol Inst Esp Oceanogr 16:75–80

Brosse L, Dumont P, Lepage M, Rochard E (2002) Validation of a gastric lavage method for the study of the European sturgeon diet, *Acipenser sturio*, Linnaeus. North Am Fish Manag 22:955–960

Brun R, Pelard M, Williot P (1998) Utilisation de deux méthodes pour marquer les cohortes d'esturgeons. In: Rauta M, Bacalbasa-Dobrovici N, Vasilescu G, Oprea L (eds) Fisheries management in the Danube River. University of Dunarea de Jos, Galati, pp 56–57

Carré F (1978) Les pêches en mer Caspienne. Annales de Géographie 479:1–39

Castaing G (1963) Pêche à l'esturgeon. Mesures de protection dans le département de la Gironde. Rapport DDA 33, Bordeaux, 8 pp

Castelnaud G, de Verdilhac P (1982) Etat de la production des pêches d'estuaires: importance de cette évaluation et méthodologie. Colloque sur la production et la commercialisation du poisson d'eau douce, 30 Mars-1 Avril 1982. Association Internationale des Entretiens Ecologiques, Dijon, cahier n° 12, pp 98–107

Castelnaud G, Gadrat S, Trouvery M (1981) La protection des migrateurs contre les extractions de granulats, réflexions, méthodes et résultats pour la Dordogne en aval de Bergerac. Actes du XXVIe Congrès de l'A.F.L., Orléans, juin 1981, SRAE éd, pp 242–246

Castelnaud G, Coutancier B, Cerezuelle D, Guchan A (1985) La pêche des migrateurs en Gironde : 1ère partie : Analyse historique : du 18e siècle à nos jours : bilan et perspectives. Cemagref de Bordeaux, Division ALA/MSHA/Ministère de l'Environnement/Département de la Gironde, p 177, annexes

Castelnaud G, Rochard E, Jatteau P, Lepage M (1991) Données actuelles sur la biologie d'*Acipenser sturio* dans l'estuaire de la Gironde. In: Williot P (ed) Acipenser. Cemagref, Antony, France, pp 251–275

Charlon N, Williot P (1978) Elevage d'esturgeons de repeuplement et de consommation en URSS. Bul Centr Etud Rech Scient Biarritz 12(1):7–156

Charpy R (1951) Observations sur la pêche de l'esturgeon (*Acipenser sturio* L.) dans le Guadalquivir durant l'année 1950. Bull Franç Piscic 163:49–56

Ctgref (1973) Rapport sur la pêche en Gironde. Ctgref, Bordeaux-Cestas, France, p 24

Ctgref (1980) Situation de la pêche aux filets et aux engins en Gironde. Etude générale de l'impact des extractions de granulats sur les espèces migratrices et la pêche en Dordogne. CTGREF Bordeaux, Division ALA, p 19

Cuisset B, Pelissero C, Le Menn F, Nunez-Rodriguez J (1991) Elisa for Siberian sturgeon (*Acipenser baerii* Brandt) vitellogenin. In: Williot P (ed) Acipenser. Cemagref, Antony, pp 107–111

Darlet M, Prioux G (1950) L'esturgeon et le caviar Français. Bull Français Piscicult 158:5–13

Davail-Cuisset B, Lacomme S, Viaene E, Williot P, Lepage M, Gonthier P, Davail S, Rouault T (2008) Hormonal profile in adult European sturgeon Acipenser sturio, adapted to hatchery conditions in France. Cybium 32:169–170

Desse J, Desse-Berset N (1993) Pêche et surpêche en Méditerranée: le témoignage des os. XIIIèmes Rencontres Internationales d'Archéologie et d'Histoire d'Antibes. Editions APDCA, Juan les Pins, pp 327–340

Desse-Berset N (1994) Sturgeon of the Rhône during Prohistory in Arles (6th–2nd) century BC. In: Van Neer W (ed) Proceedings of the 7th meeting of the ICAZ. Fish Remains Working Group. Annales du Musée Royal de l'Afrique Centrale. Sciences Zoologiques n° 274, Tervuren, pp 81–90

Desse-Berset N (2009) First archaeozoological identification of Atlantic sturgeon (*Acipenser oxyrinchus* Mitchill 1815) in France. CR Palevol 8:717–724

Desse-Berset N, Pages M, Brosse L, Tougard C, Chassaing O, Hänni C, Berrebi P (2008) Specific identification of the extinct population of sturgeon from the Rhône River by mtDNA analysis from bone remains (Jardin d'Hiver, Arles, France, 6th–2nd century BC. In: Béarez P, Grouard S, Clavel B (eds) Archéologie du poisson. 30 ans d'archéo-ichtyologie au CNRS. Hommage aux travaux de Jean Desse et Nathalie Desse-Berset, XVIIIèmes Rencontres Internationales d'archéologie et d'histoire. Editions APDCA, Antibes

Durrieu G, Maury-Brachet R, Girardin M, Rochard E, Boudou A (2005) Contamination by heavy metals (Cd, Zn, Cu, Hg) of eight fish species in the Gironde Estuary (France). Estuaries 28:581–591

Economidis PS, Koutrakis ET, Bobori DC (2000) Distribution and conservation of *Acipenser sturio* L., 1758 and related species in Greek waters. Bol Inst Esp Oceanogr 16:81–88

Elie P (coord) (1997) Restauration de l'esturgeon européen *Acipenser sturio*. Contrat Life B4-3200/94/754. Rapport final du programme d'exécution, Etude Cemagref (Bordeaux, GMA, RAC) n°24, Etude Gemagref n° 24, p 381

Furnestin J, Dardignac J, Maurin C, Vincent A, Coupe R (1958) Données nouvelles sur les poissons du Maroc Atlantique. Rev Trav Instit Pêches Marit 22:382–493

Gilpin ME, Soulé ME (1986) Minimum viable population: processes of species extinction. In: Soulé ME (ed) Conservation biology. The science of scarcity and diversity. Sinauer Associates, Sunderland, MA, pp 19–34

Gisbert E, Williot P (1997) Larval behaviour and effect of the timing of initial feeding on growth and survival of Siberian sturgeon larvae under small scale hatchery production. Aquaculture 156:63–76

Guth MO, Laurent JL (2004) Retour d'expérience sur la capture et la vente illicite d'un esturgeon en criée aux Sables–d'Olonne (Vendée). Ministère de l'Ecologie et du Développement Durable, Rapport de l'Inspection Générale de l'Environnement, n° IGE/04/2004, p 57

Holčik J, Kinzelbach R, Sokolov LI, Vassilev VP (1989) *Acipenser sturio* Linnaeus, 1758. In: Holčik J (ed) The freshwater fishes of Europe. Aula Verlag, Wiesbaden, pp 367–394

IFREMER (2009) Bulletin de la surveillance de la qualité du milieu marin littoral Résultats acquis jusqu'en 2008. Ifremer/LST.RER/AR/09.007 Laboratoire Environnement Ressources d'Arcachon, p 100

Jego S, Gazeau C, Jatteau P, Elie P, Rochard E (2002) Les frayères potentielles de l'esturgeon Européen *Acipenser sturio* L. 1758 dans le bassin Garonne-Dordogne. Méthodes d'investigation, état actuel et perspectives. Bull Fr Pêche Piscic 365/366:487–505

Kinzelbach RK (1997) The sturgeon (*Acipenser sturio* L. 1758) in Europe. Z Okol u Natur 6:129–135

Le Masson du Parc F (1727) (2004) Procès verbaux des visites faites par ordre du Roy concernant la pesche en mer (1727). Pêches & Pêcheurs du domaine maritime aquitain au XVIIIe siècle. Amirautés de Bayonne & de Bordeaux, Les Editions de l'Entre–deux-Mers, 33420, Camiac et Saint Denis, France

Le Masson du Parc F (1727–1728) (2009) Procès verbaux des visites faites par ordre du Roy concernant la pesche en mer en 1727 et en 1728. Pêches & Pêcheurs du domaine maritime et des Iles adjacentes de Saintonge, d'Aunis et du Poitou au XVIIIe siècle. Amirautés de Marennes, de La Rochelle & des Sables d'Olonne, Les Editions de l'Entre–deux-Mers, 33420, Camiac et Saint Denis, France

Le Menn F, Pelissero-Benetteau C, Williot P, Cuisset-Davail B (1999) Female Siberian sturgeon, Acipenser baerii: an interesting model for reproduction physiology studies in fish. J Appl Ichthyol 15:315–316

Lepage M, Rochard E (1997) Estimation des captures accidentelles d'*Acipenser sturio* réalisées en mer. In: Elie P (ed) Restauration de l'esturgeon européen *Acipenser sturio*. Contrat Life rapport final du programme d'exécution, n° 24. Cemagref étude, Bordeaux, pp 377–381

Lepage M, Rochard E, Castelnaud G (2000) Atlantic sturgeon *Acipenser sturio* L., 1758 restoration and gravel extraction in the Gironde estuary. Bol Inst Esp Oceanogr 16(1–4):175–179

Letaconnoux R (1961) Note sur la fréquence de la distribution des captures d'esturgeons (*Acipenser sturio* L) dans le golfe de Gascogne. Rev Trav Inst Pêches Marit 25:253–261

Lieppe D (2004) Prémices d'une politique gouvernementale de la pêche, François Le Masson du Parc (1671–741): Intinéraire d'un «commis» des classes. In: Procès verbaux des visites faites par ordre du Roy concernant la pesche en mer (1727). Pêches & Pêcheurs du domaine maritime aquitain au XVIIIe siècle. Amirautés de Bayonne & de Bordeaux, Les Editions de l'Entre–deux-Mers, 33420, Camiac et Saint Denis, France, pp III-XLVII

Lochet A, Lambert P, Lepage M, Rochard E (2004) Croissance de juvéniles d'esturgeons européens *Acipenser sturio* (Acipenseridae) sauvages et issus d'alevinage, durant leur séjour dans l'estuaire de la Gironde (France). Cybium 28:91–98

Lorin de Reure H (1924) Note sur la pêche à l'esturgeon. Bul. Trim. de l'Enseignement professionnel et technique des Pêches Maritimes. XXIXème année, n° 1. Siège Social: Hôtel des Sociétés Savantes, Rue Serpente, Paris, pp 9–16

Magnin E (1962) Recherches sur la systématique et la biologie des acipenséridés *Acipenser sturio* L., *Acipenser oxyrinchus* Mitchill et *Acipenser fulvescens* Raf. Ann Stat Centr Hydrobiol Appl 9:7–242

Mamcarz A (2000) Decline of the Atlantic sturgeon *Acipenser sturio* L., 1758 in Poland: An outline of problems and propects. Bol Inst Esp Oceaogr 16:191–202

Marti VY (1939) Biology and small fishery of *Acipenser sturio* in Black Sea. Zool J 18(3):435–442 (in Russian)

Moghim M, Nielson JD (1999) Imminent collapse of the Caspian sea stellate sturgeon (*Acipenser stellatus*): Evidence from the Iranian fishery. Royal Swed Acad Sci, Ambio 28(4):372–373

Pagès M, Desse-Berset N, Tougard C, Brosse L, Hänni C, Berrebi P (2009) Historical presence of the sturgeon *Acipenser sturio* in the Rhône basin determined by the analysis of ancient DNA cytochrome b sequences. Conserv Genet 10:217–224. doi:10.1016/j.crpv.2009.06.001

Pelissero C, Le Menn F (1991) Evolution of sex steroid levels in males and first time maturing females of the Siberian sturgeon (Acipenser baerii) reared in a French fish farm. In: Williot P (ed) Acipenser. Cemagref, Antony, pp 87–97

Peterson DL, Bain MB, Haley N (2000) Evidence of declining recruitment of Atlantic sturgeon in the Hudson River. North Am J Fish Manag 20:231–238

Pierron F, Baudrimont M, Dufour S, Elie P, Bossy A, Baloche S, Mesmer-Dudons N, Gonzalez P, Bourdineaud JP, Massabiau JC (2008) How cadmium could compromise the completion of the European eel's reproductive migration. Environ Sci Technol 42:4607–4612

Prioux MG (1957) L'esturgeon en France. Rivières et Forêts, cahier n6:39–44

Rochard E (coord) (2002) Restauration de l'esturgeon européen *Acipenser sturio*. Rapport scientifique Contrat LIFE nB-3200/98/460, Cemagref

Rochard E, Jatteau P (1991) Amélioration de la méthode de détermination de l'âge de l'esturgeon commun *Acipenser sturio* et premières applications. In: Williot P (ed) Acipenser. Cemagref, Antony, pp 193–208

Rochard E, Williot P (coord) (2006) Actions de recherches proposées pour contribuer au plan international de restauration de l'esturgeon européen *Acipenser sturio*. Etude Cemagref groupement de Bordeaux n 103, p 51

Rochard E, Castelnaud G, Lepage M (1990) Sturgeons (Pisces: Acipenseridae); threats and prospects. J Fish Biol 37A:123–132

Rochard E, Lepage M, Gazeau C, Lambert P (1997b) Tableau de bord de la population. Estimation de l'abondance des différentes classes d'âge. In: Elie P (coord) Restauration de l'esturgeon européen *Acipenser sturio*. Contrat Life B4-3200/94/754. Rapport final du programme d'exécution, Etude Cemagref (Bordeaux, GMA, RAC), n 24, pp 349–374

Rochard E, Lepage M, Meauze L (1997b) Identification et caractérisation de l'aire de répartition marine de l'esturgeon européen *Acipenser sturio* à partir de déclarations de captures. Aquat Living Resour 10:101–109

Rochard E, Lepage M, Dumont P, Tremblay S, Gazeau C (2001) Downstream migration of juvenile European sturgeon *Acipenser sturio* L. in the Gironde estuary. Estuaries 24:108–115

Rosenthal H, Bronzi P, Gessner J, Moreau D, Rochard E (2007) Action plan for the conservation and restoration of the European sturgeon. Council of Europe Publishing, Nature and Environment, n 152

Roule L (1922) Etude sur l'esturgeon du golfe de Gascogne et du basin girondin. Office Scientifique et Technique des Pêches Maritimes, Notes et Mémoires N20, p 12

Roule L (1925) Famille des Acipenséridés. In: Les poissons des eaux douces de la France. Presses Universitaires de France, Paris, pp 9–13

Scott A (1936) Etude sur les esturgeons et le caviar français. Rapport manuscrit repris in extenso dans Val (2006)

Secor DH, Arefjev V, Nikolaev A, Sharov A (2000) Restoration of sturgeons: lessons from the Caspian Sea sturgeon ranching programme. Fish Fish 1:215–230

Tabardel M (1994) Le point sur la situation de l'esturgeon (*Acipenser sturio* L.) en Méditerranée occidentale et possibilités de réintroduction dans le Rhône. Mémoire de fin d'études ENSA, Rennes, Rennes/Arles, p 57

Taverny C, Lepage M, Piefort S, Dumont P, Rochard E (2002) Habitat selection by juvenile European sturgeon *Acipenser sturio* in the Gironde Estuary (France). J Appl Ichthyol 18:536–541

Trouvery M, Catelnaud G, Williot P (1984) Biologie et Ecologie d'*Acipenser sturio*. Etude de la pêcherie, Etude n° 17, Cemagref-Agedra, p 79

Val R (2006) La véritable histoire du caviar de la Gironde. Editions Bonne Anse, Société des Amis de Talmont, France, p 93
Vibert R (1953) Les poissons migrateurs dans l'économie piscicole du Sud Ouest. Bull Franç Piscicult 136:121–135 (écrit en 1944)
Went AEJ (1948) The status of the sturgeon, *Acipenser sturio* L., in Irish waters now and in former days. Ir Nat J IX:172–174
Williot P (1984) L'expérience soviétique en matière d'exploitation des stocks d'esturgeons en mer d'Azov et mer Caspienne. Cemagref, Etude n 20 Série Esturgeon n 3, p 50
Williot P (1991) Acipenser. Actes du premier colloque international sur l'esturgeon. Cemagref, Antony, 519 p
Williot P (2000) Relationship between fishing and restocking aquaculture: example of an endangered species, the sturgeon *Acipenser sturio*. In: Guastavino JM, Holzman L, Johansson L, Pelouin A, Saunier L, Viatte L (eds) Man and coastal areas towards a sustainable aquaculture. Service technique et technologique, Ambassade de France en Suède, pp 31–33
Williot P (2004) Rôle de l'aquaculture dans la conservation des espèces: exemple des esturgeons. In: Boujard T (ed) Aquaculture et environnement, vol 26. Les cahiers de l'environnement de l'INRA, pp 59–66
Williot P, Rouault T (1982) Compte rendu d'une première reproduction en France de l'esturgeon sibérien *Acipenser baeri*. Bull Franç Piscic 286:255–261
Williot P, Rochard E, Castelnaud G, Rouault T, Brun R, Lepage M, Elie P (1997) Biological characteristics of European Atlantic sturgeon, *Acipenser sturio*, as the basis for a restoration program in France. Environ Biol Fishes 48:359–370
Williot P, Brun R, Pelard M, Mercier D (2000) Unusual induced maturation and spawning in an incidentally caught pair of adults of the critically endangered European sturgeon *Acipenser sturio* L. J Appl Ichthyol 16(6):279–281
Williot P, Sabeau L, Gessner J, Arlati G, Bronzi P, Gulyas T, Berni P (2001) Sturgeon farming in Western Europe: recent developments and perspectives. Aquat Living Resour 14:367–374
Williot P, Arlati G, Chebanov M, Gulyas T, Kasimov R, Kirschbaum F, Patriche N, Pavlovskaya L, Poliakova L, Pourkazemi M, Kim Yu, Zhuang P, Zholdasova IM (2002a) Status and management of Eurasian sturgeon: an overview. Int Rev Hydrobiol 87:483–506
Williot P, Rouault T, Brun R, Pelard M, Mercier D (2002b) Status of caught wild spawners and propagation of the endangered sturgeon *Acipenser sturio* in France: a synthesis. Intern Rev Hydrobiol 87:515–524
Williot P, Rouault T, Rochard E, Castelnaud G, Lepage M, Gonthier P, Elie P (2004) French attempts to protect and restore *Acipenser sturio* in the Gironde: Status and prospective, the research point of view. In: Gessner J, Ritterhoff J (eds) Species differentiation and population identification in the sturgeons *Acipenser sturio* L. and *Acipenser oxyrinchus*. Bundesamt für Naturschutz, Bonn, pp 83–99
Williot P, Brun R, Rouault T, Pelard M, Mercier D (2005) Attempts at larval rearing of the endangered western European sturgeon, *Acipenser sturio* L. (ACIPENSERIDAE), in France. Cybium 29:381–387
Williot P, Rouault T, Pelard M, Mercier D, Lepage M, Davail-Cuisset B, Kirschbaum F, Ludwig A (2007) Building a broodstock of the critically endangered sturgeon *Acipenser sturio* L.: problems and observations associated with adaptation of wild-caught fish to hatchery conditions. Cybium 31:3–11
Williot P, Rouault T, Pelard M, Mercier D, Jacobs L (2009a) Artificial reproduction of captive endangered European Atlantic sturgeon, *Acipenser sturio*. Endanger Species Res 6:251–257. doi:10.3354/esr00174
Williot P, Rochard E, Kirschbaum F (2009b) Acceptability and prerequisites for the successful introduction of sturgeon species. In: Carmona R, Domezain A, Garcia Gallego M, Hernando-Casal J, Rodriguez F, Ruiz Rejon M (eds) Biology, conservation and sustainable development of sturgeons, Fish & fisheries series. Springer, New York, pp 369–384, 467 p

Chapter 21
Restoration of the European Sturgeon *Acipenser sturio* in Germany

Frank Kirschbaum, Patrick Williot, Frank Fredrich, Ralph Tiedemann, and Jörn Gessner

Abstract The onset of practical restoration measures on *Acipenser sturio* in Germany is represented by the transfer of 40 one-year-old juveniles from the French Cemagref to the German Leibniz-Institute of Freshwater Ecology and Inland Fisheries in 1996 as a result of a bilateral cooperation. Molecular genetic studies revealed that these fish from the Garonne river system represented suitable material for the re-introduction into the tributaries of the North Sea. The fish were used for behavioural, feeding, and rearing experiments to improve the husbandry practice in fresh water. In 2005, a first female developed ripe gonad for the first time but artificial reproduction had not been accomplished to date. Subsequent transfers of juveniles in 2007, 2009, and 2010 allowed the expansion of the German brood stock, as well as first experimental releases both with tagged and telemetry transmitters to verify habitat utilization, migration patterns, and – through catch data – growth of the fish in the rivers Elbe, Oste, and Stör.

F. Kirschbaum (✉)
Faculty of Agriculture and Horticulture, Biology and Ecology of Fishes, Humboldt University, Philippstr. 13, Haus 16, 10115 Berlin, Germany
e-mail: Frank.kirschbaum@staff.hu-berlin.de; frank.kirschbaum@staff.hu-berlin.de

P. Williot
Sturgeon Consultant, 4 Rue du Pas de madame, 33980 Audenge, France

F. Fredrich • J. Gessner
Leibniz-Institut für Gewässerökologie und Binnenfischerei, Müggelseedamm 310, 12587 Berlin, Germany

R. Tiedemann
Unit of Evolutionary Biology/Systematic Zoology, Institute of Biochemistry and Biology, University of Potsdam, Potsdam, Germany

P. Williot et al. (eds.), *Biology and Conservation of the European Sturgeon Acipenser sturio L. 1758*, DOI 10.1007/978-3-642-20611-5_21,

21.1 Introduction

21.1.1 Decrease of the Populations

Historically, *Acipenser sturio* ranged from the Black Sea via the Mediterranean and the Eastern North Atlantic to the North, Baltic and White Seas (Holčik et al. 1989). During the nineteenth and particularly the twentieth century, the stocks decreased drastically. Today only one relict population in the Gironde–Garonne–Dordogne basin in France is documented (see Chaps. 1 and 20).

In German waters, *A. sturio* played a major role in the fisheries in former times (Benecke 1881; Quantz 1903; Blankenburg 1910; Seligo 1931; Ehrenbaum 1936; Mohr 1952; Kinzelbach 1987, 1997; for details see Chap. 14).

In the River Elbe, which represented the river with the highest productivity for sturgeon, between 4,000 and 7,000 specimens were caught annually between 1880 and 1890. However, after 1890, stocks dramatically decreased, leading to the economic insignificance of the species, indicated by insignificantly low catches within 25 years. The main reasons were overfishing, hydroconstruction, and pollution.

If fishing pressure, however, especially by the coastal marine fishing, had been the major reason for the extirpation of *A. sturio* in all large German rivers, it should have been obvious in the Eider River, too. The difference between the Eider and the other German rivers (Fig. 21.1) was that only small settlements and little industry were located along the river in the nineteenth century. Apart from damming the tributaries Sorge and Treene in the sixteenth and seventeenth century, no major hydroconstruction took place until 1890, when the construction of the Kiel Canal had a detrimental impact on the river habitat, eliminating approximately 35% of the catchment area from the main river (Fock and Ricklefs 1996). The tidal zone and the sediment transport from the Wadden Sea into the river severely increased due to flow reduction, thus resulting in more frequent floods. To prevent these floods, a second, even more detrimental measure for the sturgeon population of the Eider was taken in 1934 with the erection of the dam at Nordfeld, which blocked the migration route to the spawning sites (Ehrenbaum 1923). As a consequence of the recruitment failure, the sturgeon catches declined in the 1950s to incidental catches of single individuals (Spratte 2001). The last sturgeon in the Eider River was caught in 1969.

The historic description of the dynamics of the populations in the North Sea tributaries has to be reconsidered due to the recent discovery of the historic presence of *A. oxyrinchus* in the North Sea (see Chap. 7).

The development of the sturgeon in the Eider demonstrates that each river system has to be analyzed separately to identify the underlying reasons for the decline of the population.

The last natural reproduction in German waters was documented in 1957 in the Oste River (Gaumert and Kämmereit 1993), a tributary of the lower Elbe River, while juveniles up to 100 cm were observed in the Elbe estuary until 1968

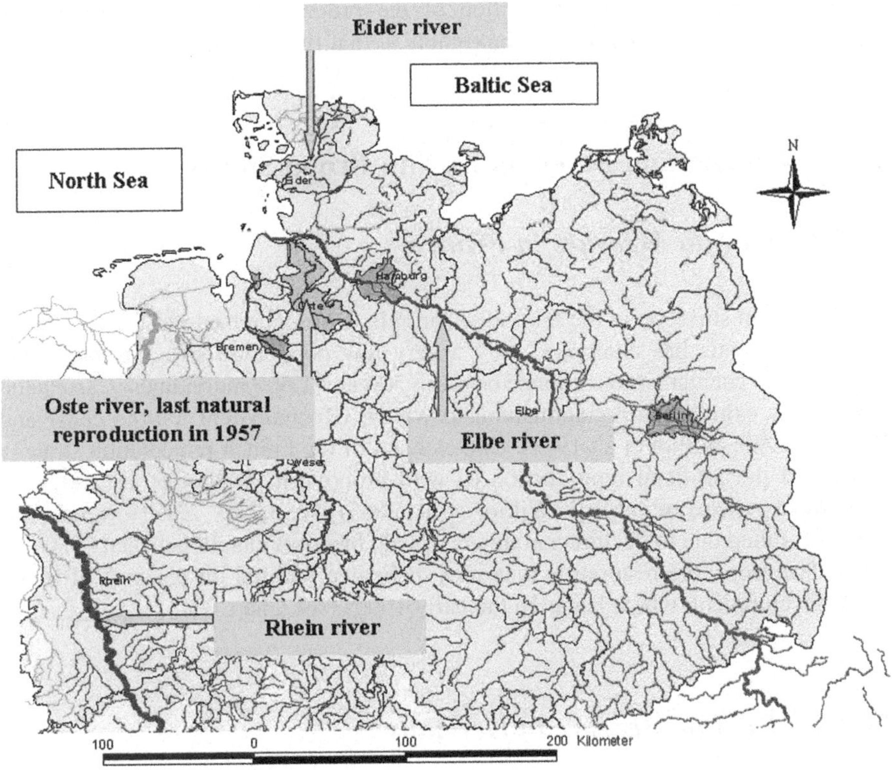

Fig. 21.1 A map of Northern Germany showing the most important sturgeon rivers

(Diercking and Wehrmann 1991). In the last two decades of the twentieth century, only occasional catches of *A. sturio* were reported in Northern Germany. The last large sturgeons were caught in the lower River Elbe in 1987 (Anon 1985), in the North Sea near Heligoland in 1993 (Arndt et al. 2000).

21.1.2 First German Remediation Attempts

With the intensive harvesting of sturgeons at the second half of the nineteenth century, concerns grew that management should be applied to increase the number of juveniles. Stocking was considered a simple and effective means to support recruitment. Because hormone injections inducing ovulation and spermiation had not yet been applied at that time (these biotechniques were established in the 1930s), fully mature fish had to be obtained to perform artificial reproduction. Incubation took place in the river itself. Juveniles were released prior to the onset of exogenous feeding. These attempts dated back to 1874 on the Elbe River (see Chap. 14). With decreasing catches, the number of mature fish available ceased, and this led to the inhibition of the programme. A validation of the effectiveness of the methods was impossible at that time (Blankenburg 1910).

21.2 Remediation Based on Ex Situ Brood Stocks

21.2.1 Ex Situ Measure in France

The European sturgeon in the Gironde–Garonne–Dordogne basin in France during the past 30 years has comprised only a very low population size. Currently, the population is calculated to comprise between 500 and 1,500 individuals (Castelnaud et al. 1991), exhibiting irregular natural spawnings (Rochard et al. 1990; Lepage and Rochard 1995; Williot et al. 1997). One element of the French remediation strategy comprised the artificial reproduction of wild brood fish (Williot et al. 2002) for subsequent release of the fry (Williot et al. 2005) (see Chap. 27). However, this approach failed due to declining numbers of wild breeders caught between 1980 and 1995. Therefore, the restoration measures in France focused on research of brood stock development, which began in the early 1990s (see Chap. 31)

21.2.2 Ex Situ Measure in Germany

21.2.2.1 Juveniles Transferred in 1996

The interest in the restoration of *Acipenser sturio* in Germany arose in the early 1990s (see e.g. Arndt and Anders 1997), and is documented in the establishment of the "German Society to Save the Sturgeon (*Acipenser sturio*) e.V." in 1994, aimed at the national coordination of the remediation measures. Subsequently between 1996 and 1998, in the framework of a project financed by the "Federal Agency of Nature Conservation" (BfN), attempts were made to localize remaining specimens of the species in its original large distribution area, e.g. in Germany, Georgia, and Albania. However, no remaining specimens were detected. Thus, brood stock development from these sources failed.

In parallel, a research proposal for joint investigation of the biology of the species was developed after the International Sturgeon Symposium in Moscow 1993 between the IGB and Cemagref researchers (see Chap. 38). In 1995 the French Cemagref at Bordeaux and the Leibniz Institute of Freshwater Ecology and Inland Fisheries (IGB) in Berlin started a cooperation on the restoration issue (see Chap. 38). In the framework of this cooperation, 40 juveniles were transferred to the IGB in 1996. This was the actual start of the restoration measures in Germany (Kirschbaum and Gessner 2000). Behavioural studies (Stoaks et al. 1996) showed that there juveniles are most active during dawn and that they do not show a clear thermal preference behaviour. There juveniles constituted the basis for the establishment of a brood stock, making it necessary to focus the research on adapting rearing and feeding strategies for the species.

21.2.2.2 Rearing Strategy

The first experience with this species indicated that this fish is difficult to maintain under captive conditions (Williot et al. 1997), in contrast to many other sturgeon species.

The Cemagref and the IGB in their research cooperation therefore decided to apply different strategies during the establishment of the brood stock, in order to gain as much knowledge as possible. In France, fresh, brackish, and salt water were used for rearing and maturing the fish (Williot et al. 2007; see Chap. 31); at the IGB, only freshwater was applied. In France, for years very low light intensities, a natural photoperiod and an annual temperature variation between 25 and 10°C were applied, whereas at the IGB the fish were reared for about 10 years at constant temperatures around 20°C, at high light intensities, and with a natural photoperiod. The weighing intervals were several months in France, but 2–4 weeks at the IGB. In France, the fish were fed on different species of shrimp only (Williot et al. 1997, 2007), whereas at the IGB a variety of natural food items and composed feeds were tested.

Feeding Experiments

The aim of the feeding experiments was to reach the minimum size for first sexual maturity as early as possible, and to learn more about the feeding strategy of *A. sturio*. The minimum size for reproduction in males reported for German waters was 1.20 m and approx. 1.50 m in females (Mohr 1952).

The 40 juveniles fed with small frozen chironomids had reached a mean total length of 58 cm and a mean weight of 838 g at an age of 29 months (Kirschbaum et al. 1998); this growth performance was slightly slower than that of wild fish of the same age.

In 1998, at the end of the third year, the juveniles had reached a median length of 68 cm (56–76 cm) and a median weight of 1,281 g (512–2,097 g). The range was quite large, and the largest fish revealed sizes similar to those of wild fish of the same age (Elie 1997). Small frozen chironomids were fed for more than 2 years, and to this food were then added frozen large chironomids, krill and small marine fish (Kirschbaum et al. 2000).

During the following 2½ years, the remaining 27 fish (13 fish had died accidentally due to a technical defect) were used for feeding experiments with biweekly weighing intervals. The four different food components which had been fed simultaneously before were subsequently tested as single food items. The large chironomids resulted in the highest growth rates; in addition, the fish showed

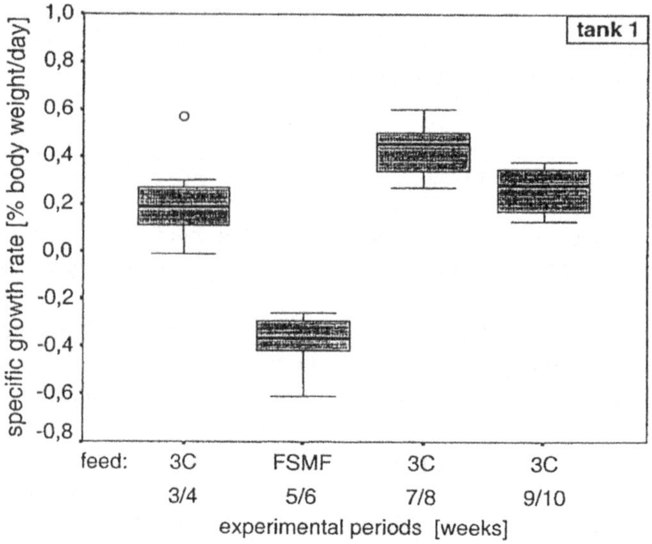

Fig. 21.2 Specific growth rates (SGR) of 12 large *Acipenser sturio* juveniles in tank I during an 8-week feeding trial. During weeks 3 and 4, fish were fed a three-component diet [3C: Frozen small marine fish (FSMF), frozen small chironomids, and large chironomids]. During weeks 5 and 6, fish were only fed FSMF, resulting in a significant decrease in SGR. In the following 2 weeks (7 and 8), again the 3C feed was offered, leading to a significant increase in SGR (compensatory growth). During weeks 9 and 10, SGR values dropped to the normal values. Feed was administered ad libitum (after Hensel et al. 2002)

pronounced compensatory growth (Fig. 21.2) and large individual differences concerning food preference (Fig. 21.3). Marine fish (smelt, sprat and herring) were used as whole fish if no longer than 3 cm or minced. Smelt were readily accepted, while sprat and herring were refused. Krill was consumed in small amounts only, both latter feed items making frequent cleaning of the tanks a necessity. Weaning to dry food was not successful (Hensel et al. 2002).

In these feeding experiments, we found out that it takes a long time to accustom the fish to different food items. Large tanks (11.6 m^3) and a low stocking density (3.8 kg m^{-3}) led to a better growth than small tanks (6.8 m^3) and high stocking densities (9.6 kg m^{-3}) (Kirschbaum et al. 2006a).

Mortalities

In 2002, several fish reduced feed uptake and decreasing growth was observed (Fig. 21.4) (Kirschbaum et al. 2006b). Some fish lost buoyancy, and also malformations occurred. For several months these fish were force-fed. However, weight gain initially observed did not last for a very long time. Subsequently, more than 50% of the 27 fish still alive in 2002 were lost: only 11 survived up to 2010. It was determined that the large chironomids contained high levels

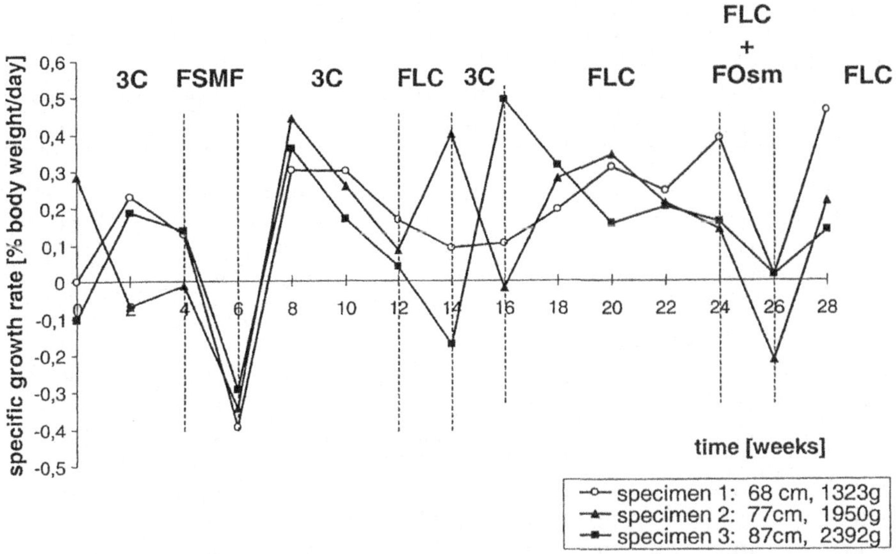

Fig. 21.3 Large differences in individual specific growth rates of three large juvenile *Acipenser sturio* in tank 1 (total number of specimens comprised 12) over an experimental period of 28 weeks. The fish were fed various feed items: three component diet (3C: frozen small marine fish (FSMD), frozen small chironomids, and frozen large chironomids (FLC)), and pieces of frozen smelt, *Osmerus eperlanus* (FOsm). Feed was administered ad libitum (after Hensel et al. 2002)

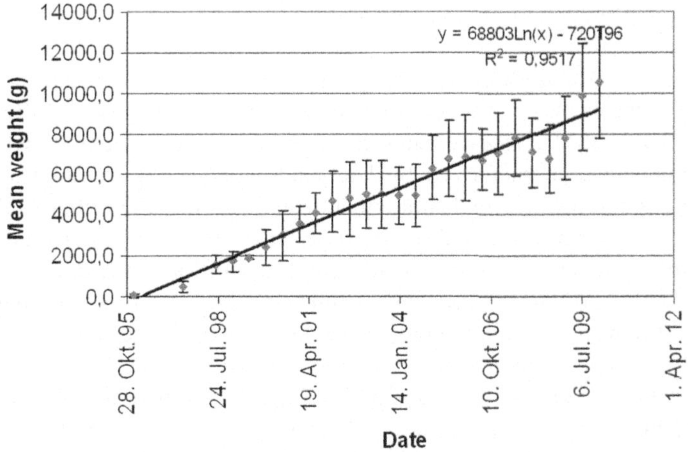

Fig. 21.4 Growth of the German brood stock (1995-year class) over the last 15 years. $N = 40$ in the beginning; $N = 27$ from 1999 on; $N = 15$ from 2005 on, $N = 11$ from 2008 on

of pesticides, which had accumulated in the fish (Kirschbaum et al. 2009b). From 2004 on, we therefore accustomed our fish over a period of several months to two kinds of shrimps; the same food that was successfully used for feeding the French

Table 21.1 Comparison of weight, total length and sexual maturity in ex situ stock of *Acipenser sturio* (year class 1995) in 2005 and 2010 (gonad maturation according to Le Menn and Pelissero 1991; see also Chap. 26, Table 26.1)

	Total length (cm)		Weight (g)		Reproductive stage	
Tag no.	2005	2010	2005	2010	2005	2010
Females						
81EF	136	Dead	12,595	Dead	F5	Dead
1906	127	147	9,133	15,560	F1	F4
DDAE	112	130	7,211	12,450	F1	F4
F322	101	126	5,235	11,260	F1	F3
2A50	108	122	5,970	9,640	F1	F2
Males						
8138	118	137	6,926	11,360	M2	M5
6EC8	113	118	7,100	8,710	M2	M3
8557	122	136	7,550	12,088	M2	M3
A3CA	123	141	7,378	12,170	M1	M3
7A59	103	114	6,153	7,330	M1	M3
221C	122	126	8,214	7,980	M2	M2
FACO	104	108	4,937	6,970	M1	M2

brood stock in Saint-Seurin-sur-l'Isle for many years. Indeed, our fish finally regained weight and showed length increase (Table 21.1; Fig. 21.4).

Monitoring and Induction of Gametogenesis

The ten largest fish biopsied in 2005 comprised five males in advanced stages of spermatogenesis, stage 3 (according to Le Menn and Pelissero 1991; see also Chap. 26, Table 26.1), four females representing stage 2 females, whereas the largest female was characterized by late vitellogenesis representing stage 5, as subsequently demonstrated by histological investigation of the follicles. Controlled reproduction was not attempted at the time, since the males were not yet fully mature resulting in a lack of sperm availability. In addition, we had to move our fish from one facility to another one at that time, which caused drastic alterations in rearing conditions associated with increased temperatures resulting in resorbtion of the eggs. This stage 5 female died in subsequent years.

From 2009 on, vernalization was applied to synchronize gonad growth in the brood fish. In 2009, temperature was decreased from 22 to 14°C for 3 weeks; in 2010, temperature was decreased from 20 to 14°C within 30 days, then kept at this temperature for 6 weeks, and afterwards increased to 20°C over 20 days.

Despite the fact that somatic growth of the 11 fish remaining by 2010 was pronounced since 2005 (Fig. 21.4), only small reserves were accumulated. The four remaining females only revealed early stages of gonad development. Ultrasonic assessment of the status in gonad development confirmed the slow gonad development in the individuals by 2010. The echograms showed homogenous

Table 21.2 Captive *Acipenser sturio* year classes 2007–2009, reproduced artificially in France; wet weight (mean, minimum, maximum) total length and number in ex situ stock at the Leibniz-Institute of Freshwater Ecology and Inland Fisheries in Berlin in October 2010

	Wet weight (g)			Total length (cm)	*N*
Year class	Mean	Min.	Max.	Mean	
2007	612	190	3,208	50.7	164
2008	319	158	1,542	38.9	248
2009	147	109	760	33.6	281

gonad tissues. A biopsy carried out in July 2010 verified early gonad development (Table 21.1).

21.2.2.3 Expansion of the Brood Stock by Integration of Juveniles Transferred 2007–2010

The largest juveniles transferred between 2007 and 2010 have in the meantime gained maximum weights of 3,208 g (2007-year class (YC), 1,542 g (2008 YC), and 760 g (2009 YC) (Table 21.2). These large growing fish had all accepted being fed on formulated diets (see next section); they represent the first candidates for the integration into the German brood stock.

21.2.2.4 Growth of the Juveniles Transferred in 2007, 2009, and 2010

In addition to the 40 juveniles transferred from France to Germany in 1996, further juveniles were transferred to Berlin after successful artificial reproduction in France between 2007 and 2009: (1) 300 specimens (11 cm; 7.1 g) in autumn 2007 after reproduction in 2007, (2) 1,000 specimens (25 cm; 62 g) in spring 2009 originating from the 2008 reproduction, and (3) 300 specimens (16.6 cm; 19.8 g) in spring 2010 based on the reproductions in 2009 (see Chap. 32).

These fish, being intended to increase the future German brood stock and to serve for experimental adaptation of rearing conditions, were reared under controlled indoor conditions. The temperatures fluctuated between 17 and 25°C over the year. Weaning experiments were conducted to identify means to adapt the fish to formulated diets at an early stage of development.

Several weaning trials were conducted, including the utilization of mixed formulated and natural diets including dry, semi-moist, and natural diets. One weaning trial was carried out consisting of a series of 14-day feeding experiments in batches of 18 fish each. The fish were fed chironomids (5% body weight per day (bw day^{-1})), a mixture of 2/3 chironomids and 1/3 formulated diet mixed to a soft semi-moist consistency fed at 3% bw day^{-1}, as well as two formulated commercial starter diets (P1 and P2) containing 57–64% digestible protein and 9–15% fat (Table 21.3). The digestible energy content of the formulated diet is given as 15–19 Kcal/MJ. The feed was administered at 2% bw day^{-1}.

The results consistently revealed a high variability in growth response as a result of the food administered (Fig. 21.5). Pronounced differences in acceptance of the mixed and formulated diets were observed. Attractiveness of the mixed diet was consistently higher, resulting in active search behaviour and feed bites. Differences between different commercial diets were readily to be recognized. The response towards formulated diet P1 (see Table 21.3) increased search behaviour, while administration of the food particles of the formulated diet P2 revealed avoidance reactions. This also resulted in differences in body mass over the 14-day test period, with P1 maintaining body weight, while the group being fed P2 revealed decreasing body weight. Growth of fish being solely fed with 2% rations of dry diets decreased,

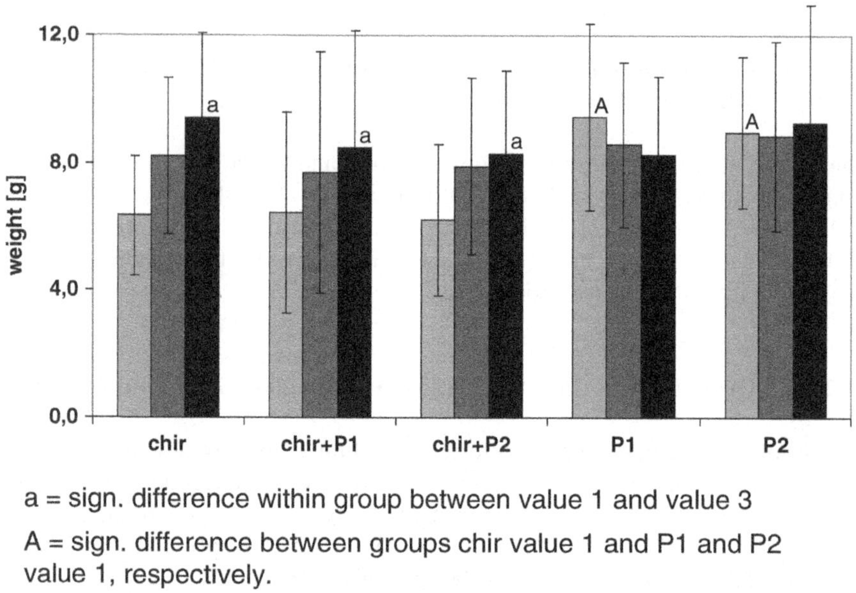

Fig. 21.5 Growth response (in wet weight) in 14-day feeding trials of juvenile *Acipenser sturio* utilizing chironomids (chir), dry diet 1 (P1) and dry diet 2 (P2) mixed with chironomids, as well as on P1 and P2 exclusively; both dry diets contain 52% protein, 14% fat content (see also Table 21.3). First column always represents value at the beginning of the experiment; second column value after 7 days, and last column after 14 days respectively. Significant values: a = Chir 0, 7, 14 days; b = Chir + P1 0, 7, 14 days; c = P1 0, 7, 14 days; A = Chir 1(1)-P1(1) or P2(1)

Table 21.3 Characteristics of the two commercial starter diets used during the 14-day feeding trials (see Fig. 21.5)

	P1	P2
Size (mm)	0.2–0.6	0.4–0.6
Crude protein (%)	64	57
Fat (%)	9	15
NFE (%)	6	8.5
Ash (%)	13	11
Crude fiber (%)	0.5	0.6
Digestable energy (Kcal/MJ)	15.4	19.1

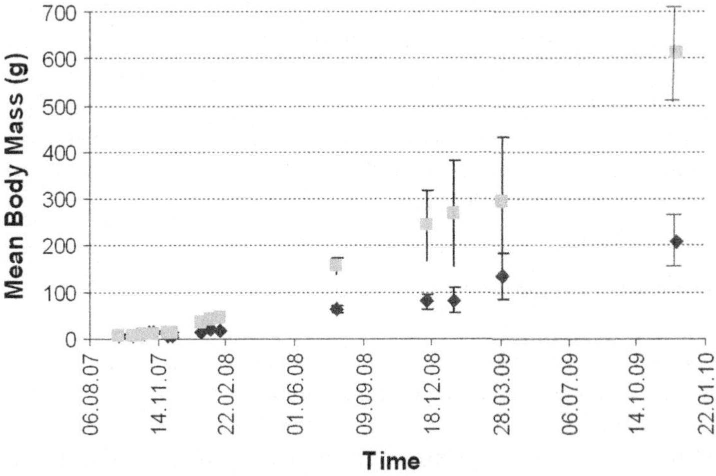

Fig. 21.6 Growth of the 2007-year class being weaned to 2% bw commercial dry diet (*grey squares*) and being fed ad libitum on chironomids (*black squares*)

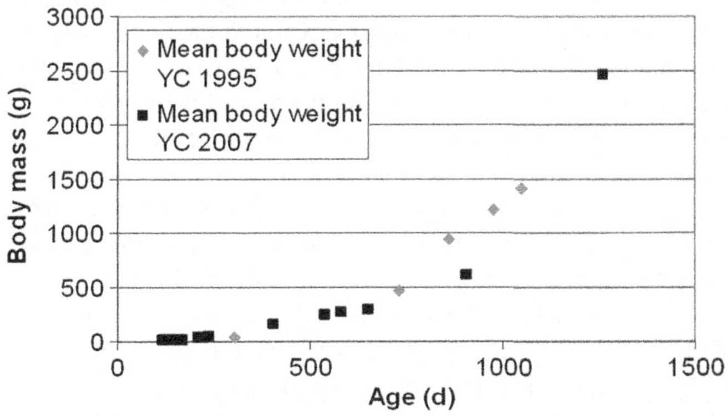

Fig. 21.7 Comparison of the 1995-year class (YC) and the 2007 YC of *Acipenser sturio* during the first 1,200 days of rearing, indicating that: (a) the weaning growth does not differ significantly, (b) if feeding and rearing was suboptimal for the 1995 YC, it was still so for the YC 2007, and (c) increased efforts have to be undertaken to accustom the fish to formulated diets sooner (see Chap. 33)

revealing significant differences in comparison to fish being fed with 5% chironomids, which is contrasted by the energy contents of the food. Mixed diets (50% chironomids and 50% dry diets) revealed similar growth to that of the group fed chironomids alone, indicating that the acceptance or palatability of the diet was increased by addition of the natural diet. The short duration of the experiments was necessary to avoid losses of the fish due to starvation.

Those specimens of the 2007-year class being successfully weaned to 2% bw commercial dry diet finally grew better than the specimens being fed ad libitum on chironomids (Fig. 21.6)

The growth of the 2007-year class was compared to the growth of the 1995-year class (transferred in 1996) over 1,200 days for the same ages. No significant difference in growth was observed between the two groups (Fig. 21.7), indicating that growth of the 2007-year class still has to be considered suboptimal as was the case for the 1995-year class. One reason for this result is the extended period of time that it took to adapt the fish to formulated diets. The process was completed for 60% of the population after about 700 days.

21.3 Habitat Assessment

The assessment of available habitat for reproduction and early life phases is an important part of the preparation for rehabilitation of sturgeon in the North Sea tributaries. In contrast to the Oder river assessments (see Chap. 41), the verification of the current status of potential spawning sites in the North Sea drainages was carried out based mainly upon published data. For the Elbe River, data acquisition for the management plans, in accordance with the Water Framework Directive and through the navigation authorities as well as specific surveys, was carried out by the regional authorities to assess potential spawning habitat for anadromous fish species (Fladung et al. 2004; Scholle et al. 2008); they provide a sound data base for this work.

In the non-tidal middle part of the Elbe River upstream of Geesthacht, suitable substrate quality is encountered only locally because this river stretch is dominated by sand substrate. The presence of suitable gravel banks is compromised by continuous transport of fines on the regulated river. This symptom is typical for all large German rivers, which predominantly serve as waterways. Therefore, large-scale habitat is available between the Czech border and the city of Torgau on 260 km of the river, as well as locally in the vicinity of Magdeburg (rkm 320), near Tangermünde (rkm 380), Havelberg (rkm 420), and the larger tributaries like the Saale and Mulde rivers. Several spawning sites have been destroyed by alterations to the river course, e.g., near Lenzen, or by reduced erosion and sedimentation processes. All of the sites in the middle reach of the river have only become accessible with the construction of the most recent fish passage at the weir in Geesthacht (rkm 580), being 120 km from the river mouth. This weir was constructed in 1957 to ensure navigation by blocking the tide from the upstream reach of the Elbe River. All previous fish passage facilities were not suited for adult sturgeon due to restrictions in size and water level.

Below the weir, in the lower river, the situation is even more complex. Historic spawning sites in this river stretch have supported the majority of the Elbe River population (Blankenburg 1910; Quantz 1903; Mohr 1952). The sites have been subjected to continuous hydroconstructions and dredging. Through removal of coarse substrate as well as morphological changes, the habitat structure was considerably altered. Due to the deepening of the channel, the salt water reaches some 70 km further upstream than it did in the 1870s (Kausch 1996). Therefore, the lower Elbe River and its estuary have to be considered as habitat for juveniles rather than as potential spawning habitat. Only the major tributaries in the lower reach, the Oste and Stör rivers, have the potential to still provide suitable spawning habitat, as can be witnessed from the substrate mapping by the navigation authorities (WSA 1997). The verification of substrate quality was carried out on the Stör River in its tidal reach from the river mouth to Itzehoe in cooperation with the navigation authorities on the 2nd of June in 2010. The video analysis applied in the Oder River (see Chap. 41) was not applicable due to the high contents of suspended solids in the tidal water. Mechanical substrate analysis revealed the dominance of fine substrate in the river bed, with only a few local aggregations of hard substrate.

21.4 First Experimental Releases

One hundred and sixty-eight fish (see below) have been released since 2008 in the rivers Elbe, Oste, and Stör. Individual fish were used for telemetric studies (see next paragraph), and the remaining fish were tagged with external markers for determination of migration patterns, habitat suitability, and fisheries impact through bycatch data.

The reported recaptures of these 168 releases comprise five fish. Three fish were caught within 2 days post release in the middle Elbe River, and two fish were reported from the coastal fishery in the Wadden Sea in April 2010, approx. 10 months after release in the Lower Elbe tributaries. While the first group only indicated that telemetry results concerning migration speed were in accordance to the migration of the fish caught, the second group revealed that migration into coastal marine waters occurred at an age of 2 years. Between release and catch, the fish revealed 100% increase in length over the 10 months following release.

In order to increase awareness in the fishery, the provision of information leaflets designed by the French Association of Commercial Marine Fisheries (CNPMEM) for European fishermen were distributed both directly and through the fisheries associations. Additionally, media coverage was used to promote the release of bycatch and to stimulate the reporting of incidental captures. However, direct contacts or the use of middlemen was considered more effective in triggering reports from fishermen. This measure involves the coastal fisheries both through local representatives and the fisheries inspections.

21.5 Telemetry

21.5.1 Elbe River

On 4 September 2008, the first *A. sturio* were released into the Elbe river near Lenzen (river km 458), at least 50 years after the disappearance of this species from its natural habitat (Kirschbaum et al. 2009a). Fifty specimens (62 g, 26 cm) tagged with grey Floy T-Bar anchor tags attached to the base of the dorsal fin were released into the river (see Chap. 24). In addition, one specimen (35 cm long, 158 g of

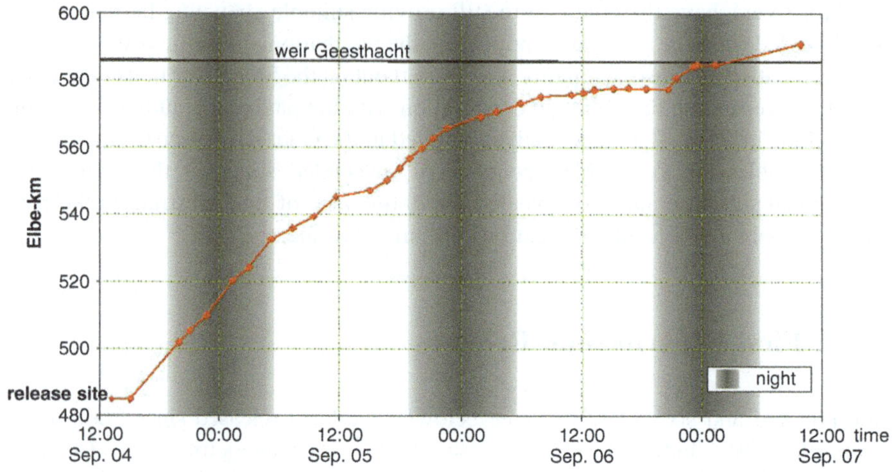

Fig. 21.8 Migration pattern (*red line*) of the juvenile *Acipenser sturio* (35 cm, 158 g; equipped with an ultrasonic transmitter) released on September 4th 2008 into the Elbe River at Lenzen (river km 485). Up to river km 586 this stretch of the Elbe River is not tide influenced (after Kirschbaum et al. 2009a)

weight) was released equipped in addition with an ultrasonic transmitter (Veco V9-1H). Tracking was carried out by boat with a hydrophone receiver combination.

The specimen equipped with the ultrasonic transmitter did not move very much for about 2 h after release (Fig. 21.8). Then, downstream migration was observed with an average speed of 3.5 km h^{-1} until the next morning. During the next days, migration speed slowed down from 1.5 km h^{-1} to below 0.5 km h^{-1} while approaching the weir at Geesthacht. For 2 h, the fish was tracked in the headwater until it crossed the weir during the night.

On reaching the tidal part of the river, the fish altered its migration pattern (Fig. 21.9). In this river stretch the fish mainly moved with the tide, especially during the night. Over 5 days, these movements took place in the same area, not resulting in net distances covered. The water depth utilized during migration resembled those during stationary phases, with the fish preferring the deepest channels and associated structures. As a result of small-scale movements with the tide, the fish reached the harbour area through the southern branch of the Elbe River, but returned upstream in the mornings for 4 days.

After 20 September, the fish was not recorded again upstream or downstream of the harbour during the following weeks. This might indicate that the transmitters run out of battery before the fish left the harbour again.

21.5.2 Stör and Oste Rivers

In 2009 and 2010 from April till September, telemetric studies were performed to investigate the behaviour of juvenile sturgeon in the tidal influenced tributaries of

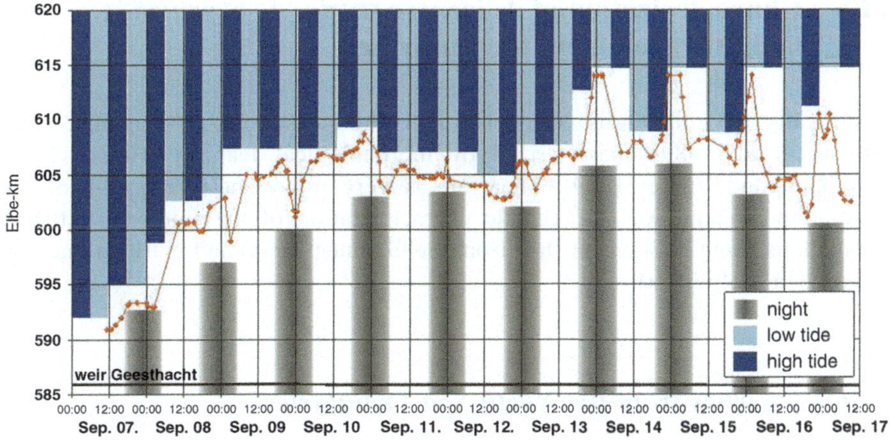

Fig. 21.9 Migration pattern (*red line*) of the juvenile *Acipenser sturio* (35 cm, 158 g; equipped with an ultrasonic transmitter) released on September 4th 2008 into the Elbe River at Lenzen (river km 485). Here, the migration pattern in the Elbe River below the weir in Geesthacht, influenced by tide, is shown (after Kirschbaum et al. 2009a)

the Elbe River catchments. In two tributaries of the Lower Elbe, the Stör River on the right bank and the Oste River on the left bank, 114 juvenile sturgeon, born in 2008 and 2009, respectively, were released, 14 of which were equipped with ultrasonic transmitters. Six of these 14 juveniles were released into the Stör River and eight into the Oste River.

The Stör River has a length of 87 km, of which more than 40 km are influenced by the tide. The whole river is hydrologically modified. It has been straightened, and the surrounding pasture land has been drained. The banks are fixed with old, deteriorated fascines. The lower 30 km of the river are navigable during high tide, and this part of the river is maintained as a Federal waterway. Here, the banks of the river are protected with stones. Because of the strong and permanent tide, water turbidity is high in the lower part and submersed macrophytes are not encountered. The six juveniles released left the river downstream in a few days. The present state of the river is not very favourable for sturgeon (Gessner et al. 2009), and this might explain the migration pattern of the juveniles.

The Oste River is 153 km long, with 78 km upstream of the confluence with the Elbe River being limited to the upstream reach by a weir, dividing the tidal lower part of the river from the non-tidal upper part. The Oste River since 2008 has no longer been maintained as a Federal waterway. The water is clear, and on the sandy bottom submersed macrophytes occur. In this stretch of the river two tagged sturgeon were released, but they were lost after some days. as in the Stör River; however, this part has much higher structural diversity. The variability of depth and width and current velocity is higher. In this stretch of the river, six sturgeons with telemetry transmitters were released. According to preliminary data two of the six sturgeons left this lower part of the Oste after 2 months, into the Elbe River. Analysis of migration patterns is currently being carried out.

21.6 Characterization of *Acipenser sturio*, *A. oxyrinchus*, and Sturgeon from the Baltic Sea

Historically, *A. sturio* was considered to comprise several geographical races (Holčik et al. 1989) characterized by morphological features (Holčik 2000; Artyukhin and Vecsei 1999). Therefore, a very important prerequisite of the restoration strategy was the genetic characterization of the European Atlantic populations using both recent (fish from the Garonne basin) and museum material (North Sea, Baltic Sea).

21.6.1 Fishes from the North Sea Tributaries

Ludwig et al. (2000) performed mtDNA studies that demonstrated a great genetic similarity between the *A. sturio* populations in the Gironde River and the North Sea. The Gironde fish, which are the basis of the German brood stock, therefore constitute suitable material for the re-introduction of *A. sturio* into the North Sea tributaries, in particular the rivers Elbe and Rhine (Kirschbaum and Gessner 2002).

However, the analysis of five microsatellites showed a decrease in allele numbers between 1823 and 1992. Similarly, the recent information on decreased heterogeneity in microsatellites between the 1988-, 1994-, and 1995-year classes demonstrate a rapid loss of diversity in the ex situ stock (see Chap. 48). This is especially true for the 1995-year class of *A. sturio* serving as the nucleus of the German brood stock, which comprises full siblings only.

In addition, a set of six microsatellites and a sequence fragment of the highly variable mitochondrial control region (D-loop) were used to characterize the natural population and the brood stock at the nuclear and the mitochondrial genome level. All specimens shared the same mitochondrial haplotype, and allelic richness at the microsatellite loci was low compared to various populations of the North American sister species, *A. oxyrinchus* (Ludwig et al. 2004).

21.6.2 Fishes from the Baltic

Notably, the mtDNA studies also revealed that ten archived specimens from the Baltic and one from the Oste River (North Sea) carried the *A. oxyrinchus* haplotype A, typical for the most northern populations of *A. oxyrinchus* inhabiting the east coast of North America (Ludwig et al. 2002). This finding further supported the investigation of the differential morphology of the scutes of both *A. sturio* and *A. oxyrinchus*, indicating the colonization of the Baltic by *A. sturio* about 3,000 years ago, the presence of *A. oxyrinchus* in the Baltic about 1,200 years ago, and (after the sympatric occurrence of both species for several hundred years) a dominance of the *A. oxyrinchus* mitochondrial haplotype A. The investigation of one microsatellite did not reveal any sign of hybridization in the sympatric populations of the two Atlantic sturgeons. Based on these results, an international workshop in 2002 proposed (though there were some concerns from the French side) to change the restoration strategy: re-introduction of *A. oxyrinchus* into the Baltic Sea, (Gessner et al. 2006) instead of *A. sturio* (Kirschbaum et al. 2004), and re-introduction of *A. sturio* into the North Sea tributaries, in particular the rivers Elbe and Rhine (Kirschbaum and Gessner 2002).

However, recently Tiedemann et al. (2007) reported extensive hybridization between the two Atlantic sturgeons in the Baltic Sea, based on MHCII genes. In addition, Ludwig et al. (2008, 2009) reported different data, e.g. hybridization between the two species at a lower level than observed by Tiedemann et al. (2007).

Hybridization would not be surprising, as many sturgeon species do hybridize (Rochard et al. 1991; Billard and Lecointre 2001) and *A.sturio* and *A. oxyrinchus* are genetically very similar (Birstein and DeSalle 1998), morphologically (Magnin 1962; Magnin and Beaulieu 1963; Artyukhin and Vecsei 1999), and at the cytogenetic level (Fontana et al. 2008).

Interesting in this respect is the finding that *A. oxyrinchus* apparently colonized the French Atlantic coast for thousands of years BP up to the seventeenth century. There are even data indicating that this species inhabited tributaries of the North Sea (Desse-Berset 2009; Desse-Berset and Williot 2011; see Chap. 7).

21.7 Perspective

The future of the European sturgeon in Germany depends upon three main factors: availability of quality stocking material, regional support for habitat improvement, and cooperation of fisheries in coastal and marine waters to avoid excess mortality through bycatch.

21.7.1 Development of the Brood Stock

Remediation of the European sturgeon in Germany strongly suffers from the lack of stocking material both for the establishment of a self-sustaining population in one of the North Sea tributaries and for the commitment and support of the relevant agencies and stakeholders at a regional level. With the number of fish currently available in the ex situ stock, the improvement of this bottleneck will be at hand only in 6–8 years; therefore, the programme heavily relies on the success of the French attempts to reproduce the ex situ stock.

The fish of the German brood stock were kept for about 10 years at constant temperatures around 20°C and at the natural photoperiod of Berlin. Under these conditions, a first gonad development of a female and of several males occurred in

2005 (Table 21.1). In the years 2009 and 2010, lowering of the temperature down to 14°C was performed (see Sect. 21.2.2.1). However, up to the year 2010 no complete gonad development of the remaining four females was observed. Only one male showed advanced maturation of gonads. The reason for the incomplete growth of the fish of the brood stock could be the incomplete and erratic application of vernalization, also not in synchrony with the natural photoperiodic change.

A further prerequisite of successful gonad maturation is the availability of appropriate food sources. From 2009 on, the brood stock was in addition to the shrimp feeding supplied with formulated diets. The increase in growth in 2009 and 2010 might indicate an influence of this new feeding strategy.

21.7.2 Breeding Plan and Cryoconservation

The genetic diversity of the French and German brood stock is rather low (see Sect. 21.6.1). Ludwig et al. (2004) used a set of six microsatellites to explore the specimens with a high number of rare alleles; these fish should be used for crosses to preserve the genetic heterogeneity of the brood stock and to avoid in- and outbreeding. In Ludwig's paper (2004), however, the specimens investigated were not explicitely characterized with regard to the gender.

Tiedemann et al. (Chap. 34) analyzed 12 autosomal polymorphic microsatellite loci with proven diploid inheritance to detect genetic heterogeneity in 23 specimens of the French brood stock and 11 specimens of the German brood stock still alive in 2011. On the basis of these results, they proposed a breeding plan. It became apparent that the genetic diversity present in the French brood stock is only partially represented in the German brood stock.

One major obstacle related to the effective application of a breeding plan is the long maturation phase and time lapse between subsequent reproductions. This limits the applicability, due to the potential for mismatch of intended mating pairs. The only means of circumventing this problem is to apply cryoconservation. Cryoconservation of sperm (see Chaps. 35 and 48) is one method for achieving independence of maturation in males for the controlled reproduction. In *A. sturio*, Kopeika et al. (2000) carried out the first experiments using sperm from the reproduction of the wild-caught *A. sturio* used for reproduction in France. Adaptation of the methods established by Horváth et al. (2008) was carried out for *A. sturio* in 2007 and 2008 (see Chap. 35). Development of a cryopreserved sperm bank has been carried out since 2008 in France, and is currently being prepared for the German brood stock. The main advantage of the method is the availability of male genotypes for reproduction, independent of the physical presence of the individual. For this reason, exchange of gametes is a viable and necessary means to ensure genetic diversity in small brood stocks. In this context, some French males have been genetically identified which appear particularly suitable as mating partners in order to maximize the genetic diversity in the offspring (see Chap 34).

21.7.3 Size at Release and Fitness for Survival

Size at release reveals a typical conflict of interest. On one side, the release should take place as soon as possible to avoid adverse impacts of artificial rearing on the behaviour of the fish, and to facilitate the adaptation to conditions of the river to be released into. On the other side, size reduces predation. Unfortunately, there is practically nothing known about predation on *A. sturio*. The release scenario largely depends upon the availability of fish, and the chances that juveniles are suffering substantially from predation. *A. sturio* actually faces a limitation in numbers of fish to be released. For this reason, it is currently considered more promising to increase the chance for survival by rearing the fish over winter and carrying out releases prior to the vegetation period.

21.7.4 Imprinting of Juveniles to the River of Release

A further prerequisite for successful remediation comprises the imprinting of the fry. No data are available on this issue with regard to *A. sturio*. Boyko (1999) concludes that in Russian sturgeon the imprinting starts at the onset of exogenous feeding. In *Acipenser oxyrinchus desotoi*, genetic data (Waldman et al. 2002) indicate that return rates of more than 94% occur per generation, and straying takes place between neighbouring basins. Homing has also been reported (Bain, pers com) in *A. oxyrinchus oxyrinchus* stocked as juveniles with 2 kg in the Hudson River, although homing rate has not been assessed due to the low absolute numbers of fish observed. Based on the preliminary information about imprinting in sturgeons, the strategy for release currently should provide the potential for high return rates in stocked fish. This would include early releases or (temporal) rearing in water originating from the future natal river. This strategy, however, also depends on the number of fry available (see preceding paragraph).

21.7.5 Active National Protection Approach and Remediation Strategy

The remediation of the European sturgeon has received strong support from the Federal Agencies over the last 10 years. In 2009, the remediation programme was rewarded by its inclusion as a lighthouse project for the national biodiversity strategy. The future strategy for the rehabilitation of the European sturgeon is comprised in the National Action Plan that transformed the Bern Convention Action Plan (Rosenthal et al. 2007) and adopted it to the federal structure of Germany, taking into consideration the specific requirements of the system (Gessner et al. 2010). As such, the Action Plan is considered the baseline for the management plans that are to be implemented for the respective river management plans. With regard to limitations on long-term management, see Chap. 49.

21.7.6 Habitat Remediation

In 2010, the Federal Ministry for traffic, construction and regional planning (BMVBS) took the lead in mitigation of migration obstacles and implementation of the corridor functions of large rivers as waterways for migratory fish species.

21.7.6.1 Fish Passage Facility at the Weir at Geesthacht on the Elbe River

For the Elbe River, sturgeon today plays a major role for the development of the criteria of migration-assisting structures. The river has one weir 123 km upstream from the river mouth that since 1961 blocked the migration of migratory fishes. A first amendment of state-of-the-art techniques was carried out in the 1990s by building a side channel. Hydrological conditions did not allow large-scale migrations through the system. In 2010, the new fish passage facility was opened that utilizes sturgeon of 3.5 m as the reference for the layout and hydrology. The target is to allow all migratory species to utilize the 500-m-long and 11-m-wide double-slit pass. The performance of the facility is being monitored for 4 years to verify its function and to allow improvements if necessary.

21.7.6.2 Habitat Remediation in Stör and Oste Rivers

In some of the tributaries, namely the Stör and Oste Rivers, habitat improvement is under construction to reduce the sediment load and to improve structural diversity. The main target is the provision of clean gravel beds in the middle and upper reaches, together with the creation of floodplains that would reduce the flood risk at high water.

21.7.7 Increasing Awareness and Cooperation in the Fishery

To avoid the loss of *A. sturio* through bycatch, a Europe-wide information campaign was initiated by the French Association of Commercial Marine Fisheries (CNPMEM), comprising flyers and information material to be distributed through fisheries associations, fisheries administrations, and research institutions to increase their awareness and to result in a higher proportion of released fish (see Chap. 37).

21.7.8 Stakeholder Support

The results of the experimental releases will help to develop strategies for the improvement of habitat and reduction of anthropogenic stressors. Therefore, the involvement of stakeholders in the river catchment is of vital importance for the longterm success of the remediation. A local network of practitioners, administration and management bodies will be important to solve the open questions arising. A baseline document in this respect comprises the German Action Plan (Bern Convention), established according to the European Action Plan.

21.8 Conclusion

The success of remediation measures depends to a large degree on the development of the brood stocks on a long-term basis. Due to the low number of specimens of the German brood stock, the artificial reproduction of these fish in upcoming years can only occur on an irregular basis. Therefore, the continuous supply of fish from the French brood stock will be in the future, as was the case in the past, the prerequisite for the restoration measures in Germany. In addition, the restoration measures will need long-term federal and regional support to guarantee its success. Proper identification of drawbacks is an important objective of the monitoring programme, thus leading to active development of alternative approaches to circumvent the main causes of disturbances. For this approach, a management unit is necessary that includes all relevant stakeholders. A first attempt for this involvement was applied for the negotiations on the national AP, which can serve as a nucleus for the subsequent stages of decision-making.

Acknowledgements The authors thank the Federal Agency of Nature Conservation (BfN) and the Federal Ministry for Education and Science (BMBF) for their strong and long-term support and funding (BfN: Grants Az.: Z1.3-892 11-4/96; Z 1.3-892 11-7/99; Z1.3-892 11 8/05; Z1.3-892 11 8/09 and BMBF: Grants FKZ 0330532, FKZ 0330718), which were essential for the work performed.

References

Anon. (1985) Krabbenfischer zog riesigen Stör an Deck. Niederelbezeitung 26.6.1985, p 13

Arndt G-M, Anders E (1997) Der Stör (*Acipenser sturio*) und seine Wiedereinbürgerung in Mecklenburg-Vorpommern und Deutschland. Fischerei in Mecklenburg-Vorpommern 4/97:14–20

Arndt GM, Gessner J, Anders E, Spratte S, Filipiak J, Debus L, Skora K (2000) Predominance of exotic and introduced species among sturgeons captured from the Baltic and North Seas and their watersheds, 1981–1999. Symp. on Conservation of the Atlantic sturgeon *Acipenser sturio* L. 1758 in Europe. Madrid Bol Inst Oceanogr 16:29–36

Artyukhin EN, Vecsei P (1999) On the status of Atlantic sturgeon: conspecificity of European *Acipenser sturio* and North American *Acipenser oxyrinchus*. J Appl Ichthyol 15:35–37
Benecke B (1881) Fische, Fischerei und Fischzucht in Ost- und Westpreussen. Hartungsche Verlagsdruckerei, Koenigsberg
Billard R, Lecointre G (2001) Biology and conservation of sturgeon and paddlefish. Rev Fish Biol Fish 10:355–392
Birstein VJ, DeSalle R (1998) Molecular phylogeny of *Acipenserinae*. Mol Phylogenet Evol 9:141–155
Blankenburg A (1910) Von der Störfischerei in der Elbe. Der Fischerbote 11:7–11
Boyko NE (1999) Formation of olfactory imprinting and thyroid hormone metabolism in early ontogenesis of Russian sturgeon. J Appl Ichthyol 15(4–5):287–288
Castelnaud G, Rochard E, Jatteau P, Lepage M (1991) Données actuelles sur la biologie d'*Acipenser sturio* dans l'estuaire de la Gironde. In: Williot P (ed) Acipenser. Cemagref, Antony, pp 251–275
Desse-Berset N (2009) First archaeozoological identification of Atlantic sturgeon (*Acipenser oxyrinchus* Mitchill 1815) in France. CR Palevol 8:717–724
Desse-Berset N, Williot P (2011) Emerging questions from the discovery of the long term presence of *Acipenser oxyrinchus* in France. J Appl Ichthyol 27:263–268
Diercking R, Wehrmann I (1991) Artenschutzprogramm Fische und Rundmäuler in Hamburg, 38th edn. Umweltbehörde Hamburg-Naturschutzamt, Hamburg, p 126
Ehrenbaum E (1923) Die Eider als Störfluß und die Schonung des Störs. Fischerbote norddeutscher Fischer 5:77–83
Ehrenbaum E (1936) Naturgeschichtliche und wirtschaftliche Bedeutung der Seefische Nordeuropas. In: Lubbert H, Ehrenbaum E (eds) Handbuch der Seefischerei Nordeuropas, vol 2. E. Schweizerbarth'sche Verlagsbuchhandlung, Stuttgart, pp 3–5
Elie P (coord) (1997) Restauration de l'esturgeon européen *Acipenser sturio*. Contrat Life B4-3200/94/754. Rapport final du programme d'exécution, Etude Cemagref (Bordeaux, GMA, RAC) n°24, Etude Gemagref, EPBX, Bordeaux
Fladung E, Scholten M, Wirtz C (2004) Verfügbarkeit und Nutzung von Sand- und Kiesbänken im Hauptstrom der unteren Mittelelbe als Laich- und Aufwuchshabitate für Fische. Verh Ges Ichthyol Vol 4(2004):25–47
Fock H, Ricklefs K (1996) Die Eider – Veränderungen seit dem Mittelalter. In: Lozan JL, Kausch H (eds) Warnsignale aus Flüssen und Ästuaren. Paul Parey, Berlin, pp 39–42
Fontana F, Lanfredi M, Kirschbaum F, Garrido-Ramos MA, Robles F, Forlani A, Congiu L (2008) Comparison of karyotypes of *Acipenser oxyrinchus* and *A. sturio* by chromosome banding and fluorescent *in situ* hybridisation. Genetica 132:281–286
Gaumert D, Kämmereit M (1993) Süßwasserfische in Niedersachsen. Niedersächsisches Landesamt für Ökologie, Hildesheim, pp 1–162
Gessner J, Arndt GM, Tiedemann R, Bartel R, Kirschbaum F (2006) Remediation measures for the Baltic sturgeon: status review and perspectives. J Appl Ichthyol 22(suppl 1):23–31
Gessner J, Spratte S, Kirschbaum F (2009) Störe für die Stör – Wem hilft ein lebendes Fossil? Steinburger Jahrbuch 2010, 54. Jahrg, pp 247–273
Gessner J, Tautenhahn M, von Nordheim H, Borchers T (2010) Nationaler Aktionsplan zum Schutz und zur Erhaltung des Europäischen Störs (Acipenser sturio). Bundesministerium für Umwelt, Naturschutz und Reaktorsicherheit (BMU), Bundesamt für Naturschutz (BfN). Silber Druck, Niestetal
Hensel E, Kirschbaum F, Williot P, Gessner J (2002) Restoration of the European Sturgeon *Acipenser sturio* L. 1758 in Germany: effect of different food items on specific growth rates of large juvenile fish. Int Rev Hydrobiol 87:539–551
Holcik J (2000) Major problems concerning the conservation and recovery of the Atlantic sturgeon *Acipenser sturio* L. 1758. Symposium on Conservation of the Atlantic Sturgeon *Acipenser sturio* L. 1758 in Europe, Madrid. Bol Inst Esp Oceanogr 16:139–148

Holcik J, Kinzelbach R, Sokolov LI, Vasil'ev VP (1989) *Acipenser sturio* Linnaeus, 1758. In: Holcik J (ed) The freshwater fishes of Europe, 2. Acipenseriformes. AULA Verlag, Wiesbaden, pp 367–394

Horváth Á, Wayman WR, Dean JC, Urbányi B, Tiersch TR, Mims SD, Johnson D, Jenkins JA (2008) Viability and fertilizing capacity of cryopreserved sperm from three North American acipenseriform species: a retrospective study. J Appl Ichthyol 24:443–449

Kausch H (1996) Fahrwasservertiefungen ohne Grenzen? In: Lozan JL, Kausch H (eds) Warnsignale aus Flüssen und Ästuaren. Paul Parey, Berlin, pp 162–176

Kinzelbach RK (1987) Das ehemalige Vorkommen des Störs, *Acipenser sturio* (Linnaeus 1758) im Einzugsgebiet des Rheins (Chondrostei: Acipenseridae). Z Angew Zool 74:67–200

Kinzelbach RK (1997) The Sturgeon (*Acipenser sturio* L. 1758) in Europe. Z Ökologie Naturschutz 6:29–135

Kirschbaum F (2002) La dimension européenne de la stratégie de sauvegarde de l'esturgeon. In: Etats généraux de la Dordogne, Actes du séminaire, Quel avenir pour l'esturgeon européen? Libourne, France, 04–05.10.2001, EPIDOR, Castelnaud-la-Chapelle, pp 118–132

Kirschbaum F, Gessner J (2000) Re-establishment programmeme for *Acipenser sturio* L. 1758: the German approach. Symposium on Conservation of the Atlantic Sturgeon *Acipenser sturio* L. 1758 in Europe, Madrid. Bol Inst Esp Oceanogr 16:149–156

Kirschbaum F, Gessner J (2002) Perspectives for the re-introduction of the European sturgeon, Acipenser sturio L. in the Elbe River. Z Fischk Suppl 1:217–232

Kirschbaum F, Gessner J, Williot P (1998) Growth performance of *Acipenser sturio* reared under experimental indoor conditions. AQUAROM '98 – Fisheries management in the Danube river basin. Galati, Romania, 18–22.5.98, extended abstracts, pp 227–228

Kirschbaum F, Gessner J, Williot P (2000) Restoration of *Acipenser sturio* L. 1758 in Germany, I: growth characteristics of juvenile fish reared under experimental indoor conditions. Symposium on Conservation of the Atlantic Sturgeon *Acipenser sturio* L. 1758 in Europe, Madrid. Bol Inst Esp Oceanogr 16:157–165

Kirschbaum F, Ludwig A, Hensel E, Wuertz S, Kloas W, Williot P, Tiedemann R, Gessner J (2004) Status of the protection and restoration of Atlantic sturgeon in Germany: background, actual situation, and perspectives. In: Gessner J, Ritterhoff J (eds) BfN-Skripten, vol 101. BfN Federal Agency for Nature Conservation, Bonn, pp 36–53

Kirschbaum F, Hensel ECK, Williot P (2006a) Feeding experiments with the European Atlantic sturgeon, *Acipenser sturio* L. 1758 to accustom large juveniles to a new feed item and the influence of tank size and stocking density on growth. J Appl Ichthyol 22(suppl 1):307–315

Kirschbaum F, Würtz S, Williot P, Tiedemann R, Arndt GM, Anders E, Bartel R, Gessner J (2006b) Prerequisites for the restoration of Atlantic sturgeons, *Acipenser sturio* and *A. oxyrinchus*, in Germany – Report on the twelve-year preparatory period. Verh Ges Ichthyol 5:79–93

Kirschbaum F, Fredrich F, Williot P, Gessner J (2009a) Wiedereinbürgerung des Europäischen Störs, *Acipenser sturio*, in Deutschland – Vorbereitende Maßnahmen und erster Besatz. Naturschutz und Landschaftspflege in Brandenburg 18(3):76–82

Kirschbaum F, Wuertz S, Williot P, Tiedemann R, Arndt G-M, Anders E, Krüger A, Bartel R, Gessner J (2009b) Prerequisites for the restoration of the European Atlantic sturgeon, *Acipenser sturio* and the Baltic sturgeon (*A. oxyrinchus* ♀ x *A. sturio* ♂) in Germany. In: Carmona R, Domezain A, Garcia Gallego M, Hernando-Casal J, Rodriguez F, Ruiz Rejon M (eds) Biology, conservation and sustainable development of sturgeons, vol 29, Fish & fisheries series. Springer, New York

Kopeika E, Williot P, Goncharov B (2000) Cryopreservation of Atlantic sturgeon *Acipenser sturio* L., 1758 sperm: first results and associated problems. Bol Inst Esp Oceanogr 16:167–173

Le Menn F, Pelissero C (1991) Histologtical and ultrastructural studies of oogenesis of the Siberian sturgeon *Acipenser baerii*. In: Williot P (ed) Acipenser. Cemagref, Antony, pp 113–127

Lepage M, Rochard E (1995) Threatened fishes of the world: *Acipenser sturio* Linnaeus, 1758 (Acipenseridae). Environ Biol Fishes 43:28

Ludwig AN, Jenneckens I, Debus L, Ludwig A, Becker J, Kirschbaum F (2000) Genetic analyses of archival specimens of the Atlantic sturgeon *Acipenser sturio* L. 1758. Symposium on Conservation of the Atlantic sturgeon *Acipenser sturio* L. 1758 in Europe, Madrid. Bol Inst Esp Oceanogr 16:221–230

Ludwig A, Debus L, Lieckfeld D, Wirgin I, Benecke N, Jenneckens I, Williot P, Waldman JR, Pitra C (2002) When the American sea sturgeon swam east. Nature 419:447–448

Ludwig A, Williot P, Kirschbaum F, Liekfeld D (2004) Genetic variability of the Gironde sturgeon population. In: Gessner J, Ritterhoff J (eds) BfN-Skripten, vol 101. BfN Federal Agency for Nature Conservation, Bonn, pp 54–72

Ludwig AN, Arndt U, Lippold S, Benecke N, Debus L, King TL, Matsamura S (2008) Tracing the first steps of American sturgeon pioneers in Europe. BMC Evol Biol 8:221–236

Ludwig A, Makowiecki DU, Benecke N (2009) Further evidence of trans-Atlantic colonization of Western Europe by American Atlantic sturgeons. Archaeofauna 18:185–192

Magnin E (1962) Recherche sur la systématique et la biologie des Acipensérìdés *Acipenser sturio* L., *Acipenser oxyrinchus* Mitchill et *Acipenser fulvescens* Raf. Annales de la Station Centrale d'Hydrobiologie Appliquée 9:7–242

Magnin E, Beaulieu G (1963) Etude morphométrique comparée de *l'Acipenser oxyrhynchus* Mitchill du Saint-Laurent et *l'Acipenser sturio* Linné de la Gironde. Nat Can 90:5–38

Mohr E (1952) Der Stör. Die neue Brehm-Bücherei, Geest und Portig, Leipzig

Quantz H (1903) Störfischerei und Störzucht im Gebiet der deutschen Nordseeküste. Mitt des Deutschen Seefischerei-Vereins 19:176–204

Rochard E, Castelnaud G, Lepage M (1990) Sturgeons (Pisces: *Acipenseridae*): threats and prospects. J Fish Biol 37(suppl A):123–132

Rochard E, Williot P, Castelnaud G, Lepage M (1991) Eléments de systématique et de biologie des populations sauvages d'esturgeons. In: Williot P (ed) Acipenser. Cemagref, Antony, pp 475–507

Rosenthal H, Bronzi P, Gessner J, Moreau D, Rochard E (2007) Action plan for the conservation and restoration of the European sturgeon. Convention on the Conservation of European wildlife and natural habitats (Bern Convention). Council of Europe Publishing, Nature and Environment n° 152

Scholle J, Schuchardt B, Rückert P, Bildstein T (2008) Konzeption zur Umsetzung der ökologischen Durchgängigkeit in den Fließgewässern in Sachsen-Anhalt. Landesbetrieb für Hochwasserschutz und Wasserwirtschaft (LHW), Sachsen-Anhalt, Gewässerkundlicher Landesdienst, p 97

Seligo A (1931) Die Seefischerei von Danzig. In: Lubbert H, Ehrenbaum E (eds) Handbuch der Seefischerei Nordeuropas, VIII (7)th edn. E. Schweizerbart'sche Verlagsbuchhandlung, Stuttgart, pp 25–28

Spratte S (2001) Aussterben des Störs (*Acipenser sturio* L.) in der Eider. In: Verband Deutscher Sportfischer (ed) Der Stör – Fisch des Jahres 2001. Verlag M. Faste, Kassel, pp 66–86

Tiedemann R, Moll K, Paulus KB, Scheer M, Williot P, Bartel R, Gessner J, Kirschbaum F (2007) Atlantic sturgeons (*Acipenser sturio, A. oxyrinchus*): American females successful in Europe. Naturwissenschaften 94:213–217

Waldman JR, Grunwald C, Stabile J, Wirgin II (2002) Impacts of life history and biogeography on the genetic stock structure of Atlantic sturgeon *Acipenser oxyrinchus oxyrinchus*, Gulf sturgeon *A. o. desotoi*, and shortnose sturgeon *A. brevirostrum*. J Appl Ichthyol 18:509–518

Williot P, Rochard E, Castelnaud G, Rouault T, Brun R, Lepage M, Elie P (1997) Biological and ecological characteristics of European Atlantic sturgeon, *Acipenser sturio*, as foundations for a restoration programme in France. Environ Biol Fish 48:359–370

Williot P, Brun R, Pelard M, Mercier D (2000) Unusual induced maturation and spawning in an incidentally caught pair of adult of the critically endangered European sturgeon *Acipenser sturio* L. J Appl Ichthyol 16:279–281

Williot P, Brun R, Rouault T, Pelard M, Mercier D (2005) Attempts at larval rearing of the endangered western European sturgeon, *Acipenser sturio* L. (Acipenseridae), in France. Cybium 29(4):381–387

Williot P, Rouault T, Pelard M, Mercier D, Lepage M, Davail-Cuisset B, Kirschbaum F, Ludwig A (2007) Building a broodstock of the critically endangered sturgeon, *Acipenser sturio* L: problems associated with the adaptation of wild-caught fish to hatchery conditions. Cybium 31:3–11

WSA (1997) http://www.wsa-hamburg.wsv.de/wasserstrassen/Neubau/Fahrrinnenanpassung/beweissicherung/index.html

Chapter 22
Fishery By-Catch of European Sturgeon in the Bay of Biscay

M. Lepage and E. Rochard

Abstract Almost 3,000 sturgeons have been tagged since 1984, and incidental captures of sturgeons tagged ($N = 154$) and untagged ($N = 86$) have been reported by marine fishermen. We suspected that these data did not reflect the total captures at sea. Therefore, we carried out a large survey which provided information about captures that were not previously declared. The high fishing pressure on sturgeon by incidental captures and poaching at sea was demonstrated. It confirmed that several hundreds of sturgeons were captured in near the Gironde estuary in 1997–1998. The stock assessment made on juvenile in estuary and the absence of breeders returning to their spawning sites lead us to think that most probably capture at sea was the first cause of mortality for these cohorts. While the stocking is in progress, the communication effort must continue, and effective enforcement is urgently required to eliminate poaching.

22.1 Introduction

The present situation of the European sturgeon *Acipenser sturio* L. raises much concern. The only population proven to persist originates from the Gironde basin in France. Despite the fact that the European sturgeon is one of the most protected fish species (see Chap. 18), no other population of the species seems to exist in its former range (see Chap. 6). The Gironde population started to decline dramatically in the 1960s (see Chap. 13). As a consequence, its reproduction has become very sporadic, and only three wild born age classes have been detected since 1985 (1985, 1988, 1994). These cohorts represent the backbone of the wild stock, with probably today less than 1,000 individuals.

M. Lepage (✉) • E. Rochard
Cemagref, Estuarine Ecosystems and Diadromous Fish Research Unit, 50 Avenue de Verdun, 33612 Cestas Cedex, France
e-mail: mario.lepage@cemagref.fr

P. Williot et al. (eds.), *Biology and Conservation of the European Sturgeon Acipenser sturio L. 1758*, DOI 10.1007/978-3-642-20611-5_22,

Since 1995, only two mature females and nine mature males have been captured in rivers (Fig. 22.1), while several thousand juveniles left the Gironde estuary for their marine phase during the early 1980s (Castelnaud et al. 1991). One can wonder about what happened to these fish. Some hypotheses have been developed and verified (see also Chap. 30). Mortality due to predation could have been a potential impact, but there is no known predator for sturgeon exceeding 80 cm in length in the north east Atlantic, unlike the reports of shark attacks in the Gulf of Mexico (Sulak and Randall 2002). Diseases or parasites could increase mortality, but no obvious sign of disease or malicious external or internal parasites have been reported since the beginning of the sturgeon survey in 1980. Straying to other river basins could also have been a possible cause of local disappearance as a result of deceasing water quality, but the number of reported captures in or near other European estuaries has not supported this hypothesis (Rochard et al. 1997; Lassalle et al. 2010). Therefore, the only feasible reason for the absence of expected spawning runs to the Gironde system could be marine by-catch. We have evidence supporting the last point. Several incidental captures have occurred every year, and even poaching has been reported at certain periods.

In 1999, a new awareness campaign called "Operation Atlantic Sturio" was initiated. In contrast to previous initiatives mainly involving scientists (e.g., Castelnaud 1988) this operation was a joint venture between politicians, scientists, and environmental NGOs (Pustelnik and Guerri 2000). The campaign aimed at the creation and support of an information network. Main targets were the dissemination

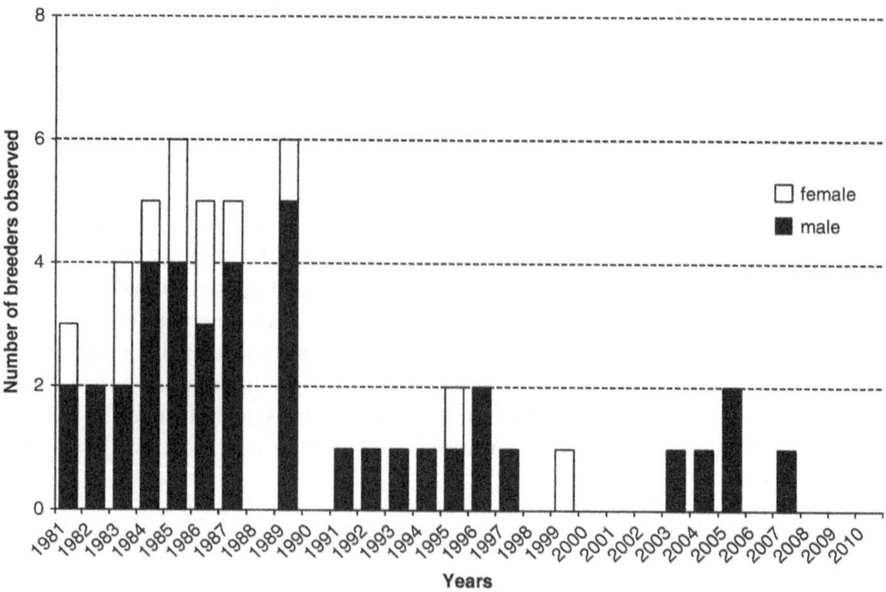

Fig. 22.1 Number of European sturgeon breeders recorded in the Gironde–Garonne–Dordogne system from 1981 to 2010. (Updated from Williot et al. 2002)

of information about the status and threats concerning the European sturgeon, and the retrieval of reliable data concerning sturgeon captures at sea (Mayer and Lepage 2001).

22.2 Methods

22.2.1 Communication (See Also Chaps. 36 and 37)

In order to spread information, we met all the representatives of fishermen's organizations, fishery control services, fish auction sites, coastal scientific stations, public aquariums and specialized journalists.

Several communication tools (sticker, writing pad ...) were developed, produced and distributed to inform about the situation of the species and to present what to do in case of a sturgeon capture. Among other information, attention was focused on tag-presence, and on explaining the procedure for reporting the tag number if present, as well as on the importance of releasing the fish and contacting an emergency phone number.

A former booklet, "L'esturgeon ne doit pas disparaître", edited in 1987 was replaced by a new one describing the restoration plan for the sturgeon, and a voluntary declaration form was provided. These two documents have been given to almost every commercial fisherman active on the French Atlantic coast in 1999–2000. As a reward, fishermen declaring a capture could choose between a reward of 25 € or a book describing traditional fishing activities.

In order to create social pressure and to educate people, starting with the young, some communication actions were directed towards the general public. As a communication tool, a special comic strip dedicated to youth, "Les aventures de sturio 007, sauvetage en haute mer" was edited, and an itinerant exposition with real-size sturgeon models was developed. A large communication effort was made towards radio and television interviews, news papers and specialized magazine articles.

22.2.2 Network

The network resulting from this work spreads over 17 coastal French Departments, 50 ports on the Atlantic coast and in the English Channel. The collaboration of 180 persons in charge of organizations dealing with the fisheries was obtained. These organizations represent 4,020 fishing boats from 6 to 25 m, and 12,000 commercial fishermen.

The network is based on institutional partners such as the French Ministry for Agriculture and Fisheries, the fish auction sites, the Regional and Local

Committees for Fisheries and Marine Aquaculture, several coastal research stations, sport fishermen, together with some temporary facilities that can be used to take over fish when fishermen bring their captures back to the port alive.

To analyze the behaviour of fishermen, we considered the number of sturgeons voluntarily declared, the proportion of tagged and non-tagged fish, and the proportion of declared fish compared to the estimation of the number of sturgeon at risk at sea (we postulate that this corresponds to fish older than 3 years; see Chap. 30). To detect any changes, we use cumulative sum (Cusum) charts (Woodward and Goldsmith 1970) which make it possible to identify rupture points in time series. On Cusum charts, we present the cumulative sum of differences between the values and the average of the observations of the series. A segment of the chart with an upward slope indicates a period where the values tend to be above the average. Likewise a segment with a downward slope indicates a period of time where the values tend to be below the average. Inflexion points correspond to changes in the trend of the series.

22.3 Results and Discussion

Results from 1980 to 1994 were analyzed by Rochard et al. (1997). Additionally, we here update the information for the period from 1995 to 2010.

The voluntarily declared data of capture vary considerably in both detail and quality. Very often, fishermen do not want to give the precise site of their capture, or the targeted species or even the fishing gear. It was noticed that among all captures, 57% of the sturgeon were declared dead, but if we only consider fish longer than 1 m, almost 100% were declared dead. It is unrealistic to expect that the bigger fish are more subject to increased fishing or handling mortality than the smaller ones. So we hypothesize that the behaviour of the fishermen has a significant effect on the survival of the fish. We knew from 15 years of trawling campaigns on sturgeon that these fish are resistant, and that most of them should be still alive when they are captured.

The number of sturgeon by-catch declared (Fig. 22.2) is a combined response of three factors: (1) the abundance of sturgeon at sea, which mainly depends on the strength of the cohorts, (2) the areas frequented and the gear used to target fish (we assume it has not changed during the study period), and (3) the interest in cooperation of the fishermen.

Analyzing the voluntarily declared sturgeon captures, it was found that fish were caught in three types of fishing gears: the bottom trawl (31%), the trammel net (32%), and the gill net (35%). Most of the captures occurred at depths between 20 and 30 m, and no capture was reported at depths exceeding 80 m.

Since 1980, most of the captures (66%) have occurred from January to June (Fig. 22.3). That may coincide with an increasing activity of fish, and for adults the beginning of the migration towards the Gironde estuary.

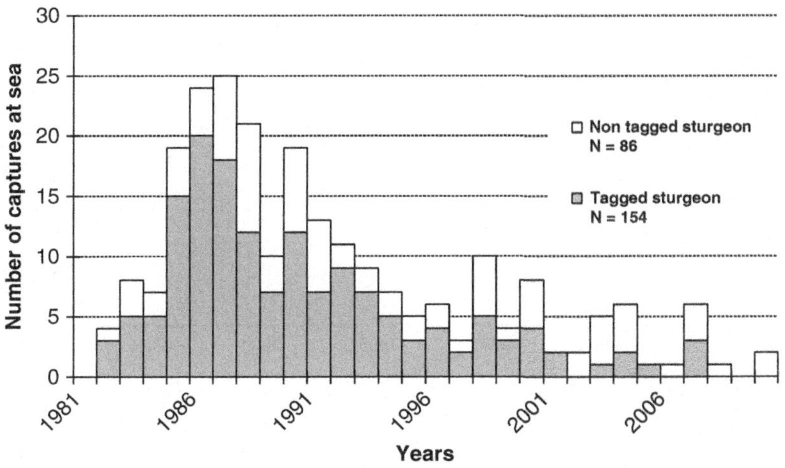

Fig. 22.2 Distribution of the voluntarily declared sturgeon captures at sea with and without tags from 1981 to 2010 ($N = 241$)

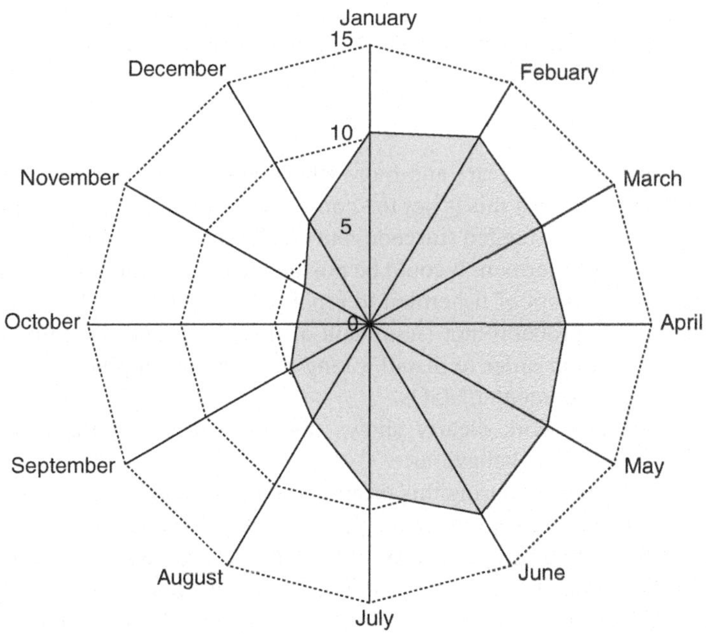

Fig. 22.3 Monthly distribution of the declared captures of sturgeon at sea from 1981 to 2010 ($N = 209$)

From 2001 onwards, the number of non-tagged fish reported exceeded the number of tagged fish declared (Fig. 22.4), even for large fish. At the same time, the number of sturgeon by catch declared, compared to the estimation of the

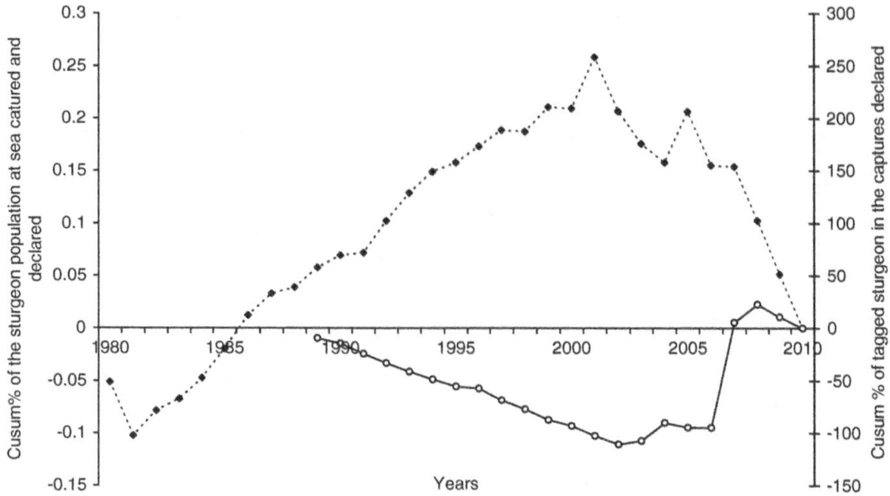

Fig. 22.4 Cumulative sums (Cusum) of the differences between the number of by-catch declared compared to the number of sturgeons estimated at sea and the average along the period 1989–2010 (*white circles*, *y*-axis *left*) and cumulative sums (Cusum) of the differences between the percentages of tagged European sturgeon in the captures declared 1 year and the average along the period 1980–2010 (*black diamonds*, *y*-axis *right*)

number of sturgeons of 3 years and more (those which can be found at sea), also increased. We assume that this is not the consequence of an increased fishing effort or more captures of non-tagged sturgeon, but an effect of the information campaign on the behaviour of fishermen. It could be postulated that the more frequent the fish, the lower the willingness of fishermen to report catches. The level of commitment of the fishermen is probably not significant at this point, and the communication effort must continue in order to install a long-term partnership between scientists, fishermen, and environmental NGOs.

Nevertheless, this work clearly shows that the success of the restoration is strongly linked to human behaviour.

The fieldwork realized during this programme has provided the information that several sturgeons are still captured each year and are sold in the official commercial circuit, even though there is a fishery officer in each fish auction site (Guth and Laurent 2004). Fortunately, the large communication effort carried out for more than 20 years has resulted in increased awareness steadily limiting such behaviour.

The study also highlights the high fishing pressure which is effected upon sturgeon by incidental capture and by poaching. The investigations in each main port on the French Atlantic coast suggest that several hundred fish (probably mainly from cohort 1994) were poached from 1996 to 1998 near the Gironde estuary. Some fishermen had discovered a congregation area between the Oleron Island and the Gironde estuary, and targeted sturgeons until the fish left. Previously, it was known only that congregation areas existed in the Gironde estuary (Brosse 2003).

Captures at sea are an important factor for the future of the population. Enforcement is urgently needed to eliminate the few poaching acts which could turn detrimental for the conservation efforts, especially as they may intercept large number of juveniles. This aspect must be solved by local authorities. The second major impact to address is diffuse incidental captures. Here, measures are much more difficult to apply, since common management tools such as protected marine areas or restriction of gears are not adapted. The topic of by-catch affects a large number of fishermen; most of them have never encountered any sturgeon, and they might not imagine the importance of these few fish. Here, a commitment would be needed form every fisherman to release accidental sturgeon captures. However, this concerns more than 4,000 fishing vessels in France alone, while European sturgeon is also captured by other EU fleets. Among the 241 records of by-catches, 84% occurred in France, 7% in UK, 3% in Germany, 2% in The Netherlands, 2% in Denmark, 1% in Belgium, and 1% in Norway. Other initiatives such as communication actions in order to minimize the mortality due to fishing pressure were carried out 2010 in Belgium and The Netherlands. They are urgently needed to expand awareness to the European level.

While reproduction and stocking are becoming an established technique, it is essential that we optimize the survival of individuals at sea, as well as leading the releases to a sustainable effect.

References

Brosse L (2003) Caractérisation des habitats des juvéniles d'esturgeon européen, Acipenser sturio, dans l'estuaire de la Gironde: relations trophiques, hiérarchisation et vulnérabilité des habitats. Ph.D., Université Toulouse III, Toulouse

Castelnaud G (1988) The sturgeon tagging programme in the Gironde estuary (France): a European dimension. Occasional paper. ICES, Bergen

Castelnaud G, Rochard E, Jatteau P, Lepage M (1991) Données actuelles sur la biologie d'*Acipenser sturio* dans l'estuaire de la Gironde. In: Williot P (ed) Acipenser. Cemagref, Antony, pp 251–275

Guth MO, Laurent JL (2004) Retour d'expérience sur la capture et la vente illicite d'un esturgeon en criée aux Sables d'Olonne (Vendée). Rapport de l'inspection générale de l'environnement. Ministère de l'Ecologie et du Développement durable, Paris

Lassalle G, Crouzet P, Gessner J, Rochard E (2010) Global warming impacts and conservation responses for the critically endangered European Atlantic sturgeon. Biol Conserv 143: 2441–2452

Mayer N, Lepage M (2001) Sauvegarde et restauration de l'esturgeon européen – action de communication et de sensibilisation, Opération Atlantique sturio. Agedra report for Life nature programme

Pustelnik G, Guerri O (2000) Analysis of partnership and conservation requirements for a theatened species, *Acipenser sturio* L., 1758: Towards the implementation of a recovery plan. Boletin Instituto Espanol de Oceanografia 16(1–4):209–216

Rochard E, Lepage M, Meauze L (1997) Identification et caractérisation de l'aire de répartition marine de l'esturgeon européen *Acipenser sturio* à partir de déclarations de captures. Aquat Living Resour 10(2):101–109

Sulak KJ, Randall M (2002) Understanding sturgeon life history: Enigmas, myths, and insights from scientific studies. J Appl Ichthyol 18(4–6):519–528

Williot P, Rouault T, Brun R, Pelard M, Mercier D (2002) Status of caught wild spawners and propagation of the endangered sturgeon Acipenser sturio in France: a synthesis. International Review of Hydrobiology 87(5–6):515–524

Woodward RH, Goldsmith PL (1970) Les sommes cumulées. In: Verhulst M (ed) Mathématiques et statistiques pour l'industrie. ICI, Paris, Monographies 3–1 et 3–8

Chapter 23
Age Assessment in European Sturgeon

Philippe Jatteau, E. Rochard, M. Lepage, and Christine Gazeau

Abstract Age determination is a key tool in population dynamics for acquiring data on age structure and cohort strength. In the case of a protected species, the method is required to have no adverse effects on the fish. A method using a thin section of the first pectoral fin ray has been set up on *Acipenser sturio* according to these prerequisites, and has been validated. The main limitations concerning old fish or fish held in captivity are discussed.

23.1 Introduction

Population management requires a good understanding of population dynamics, particularly in threatened populations such as the European sturgeon (*Acipenser sturio*) population.

To acquire sufficiently precise data on age structure and growth, a distinction between cohorts is essential. In long-living species such as sturgeon, statistical length–age relationships are not accurate enough to provide a reliable age assessment (Rochard 1992), and therefore an individual age assessment method is necessary. Numerous methods and various calcified structures have been used for sturgeons (see Brennan and Cailliet 1991 for a review), but in the case of an endangered species like the European sturgeon, the method was required not to be lethal and to cause minimal effects. Considering these prerequisites, pectoral fin rays were the most widely used structures for age assessment.

Historically, the entire leading ray of the pectoral fin was removed (Brennan and Cailliet 1989; Priegel and Wirth 1971; Wilson 1987; Xin et al. 1991). This sampling obviously had no repercussions when sampled fish came from commercial captures

P. Jatteau (✉) • E. Rochard • M. Lepage • C. Gazeau
Cemagref, Estuarine Ecosystems and Diadromous Fish Research Unit, 33612 Cestas Cedex, France
e-mail: philippe.jatteau@cemagref.fr

P. Williot et al. (eds.), *Biology and Conservation of the European Sturgeon Acipenser sturio L. 1758*, DOI 10.1007/978-3-642-20611-5_23,

(Guenette et al. 1992; Stevenson and Secor 2000). However, for animal safety reasons, the effect of this total removal had to be addressed. Kohlhorst (1979) demonstrated that survival was negatively impacted in white sturgeon (*A. transmontanus*). Collins and Smith (1996), on the contrary, recorded no effect on growth and survival in shortnose sturgeon (*A. brevirostrum*) or Atlantic sturgeon (*A. oxyrinchus*), and Parsons et al. (2003) detected no modification in station-holding ability in shovelnose sturgeon (*Scaphirhynchus platorynchus*). Nevertheless, in a very threatened species, it was necessary to minimize any risk for the individuals, which in turn led to the exclusion of any method based on the removal of the entire fin ray for the European sturgeon.

23.2 Establishing an Appropriate Method of Age Determination for the European Sturgeon

The method applied and adopted to the species was first described by Rochard and Jatteau (1991). It consists of the removal of only a thin section of the first pectoral fin ray (Cochnauer et al. 1985; Peterson et al. 2002), thereby limiting any possible adverse impact on survival, growth or swimming ability. The method was then modified, mainly in the preparation of fin ray sections, which is illustrated in Fig. 23.1.

23.2.1 Sampling

A thin transverse section (2–3 mm width) was removed from the first pectoral fin ray at 1 cm from the point of articulation of the pectoral fin ray, using a fine-toothed jeweller's saw and a scalpel. After being cleaned in distilled water, the sample was dried and stored in a labelled paper envelope.

23.2.2 Preparation

The fin ray section was embedded in silicon mould with araldite resin. The mould was dried at 40°C for 3 h and subsequently for 24 h at room temperature.

The resin block was removed from the mould and mounted on a slide with thermolabile Crystalbond glue. The section was ground with an automatic polisher with 1,200 and 4,000 waterproof silicon carbide papers successively. The readability of a section (contrast between translucent and opaque rings) varied between fish, so appropriate thickness varied too. Regular checking of the reading quality was required so that grinding could be stopped when necessary. Nevertheless, section thickness must be between 0.5 and 0.25 mm (Rochard and Jatteau 1991).

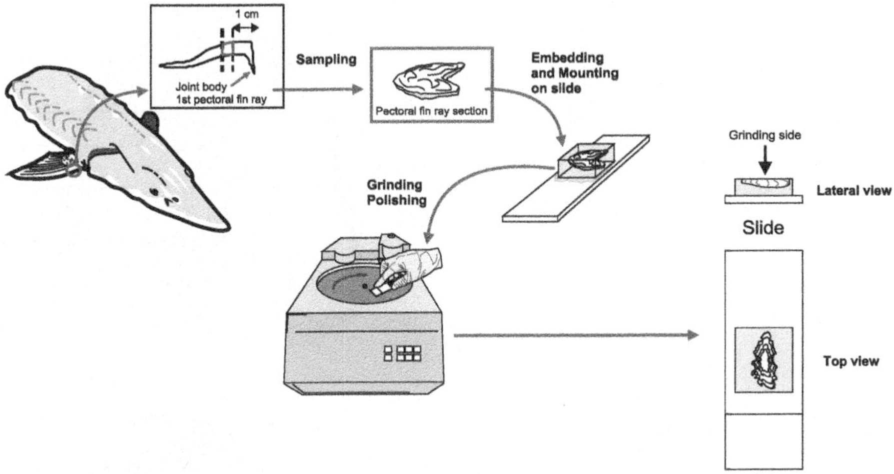

Fig. 23.1 Steps of fin ray section preparation from sampling to observation

The section was then polished with a 1 μm diamond and colloidal silica suspension.

23.2.3 *Observation*

Observations were performed with a compound microscope (Nikon Eclipse 90i) under transmitted light. The section was observed after adding a drop of ethanol, and pictures were recorded. Translucent rings (corresponding to low growth periods) or opaque rings were enumerated. Each fin ray section was observed by three different readers. In case of disagreement over aging, a second observation was performed. If there was still disagreement, the section was not aged.

Figures 23.2 and 23.3 show fin ray sections of two European sturgeons of 6 and 1 years old respectively.

23.3 Validation of Age Determination

The age determination method was based on the assumption of one thin translucent annulus deposit during the winter period. This hypothesis was tested by sampling wild fish several times at intervals of at least 9 months and including a winter period. Results from fish between 3 and 8 years old demonstrated the periodicity of deposits, with a yearly sequence of one translucent and one opaque ring, a translucent ring being formed during the winter period (Rochard 1992).

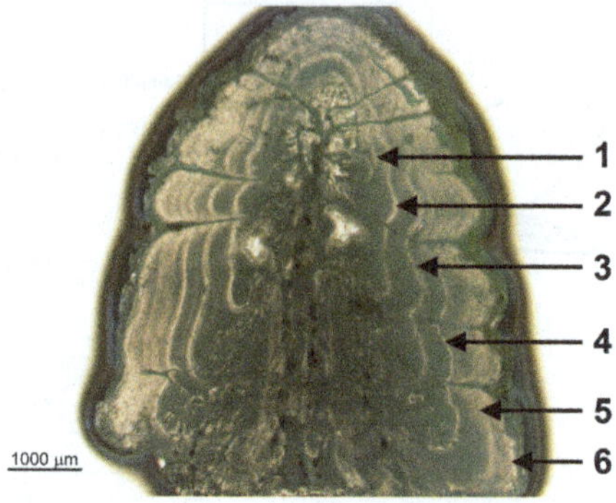

Fig. 23.2 Fin ray section of a European sturgeon caught on 26 June 1989 (length 120 cm, weight 10.3 kg) showing six translucent annuli – age 6+ (cohort 1983) (photo Cemagref)

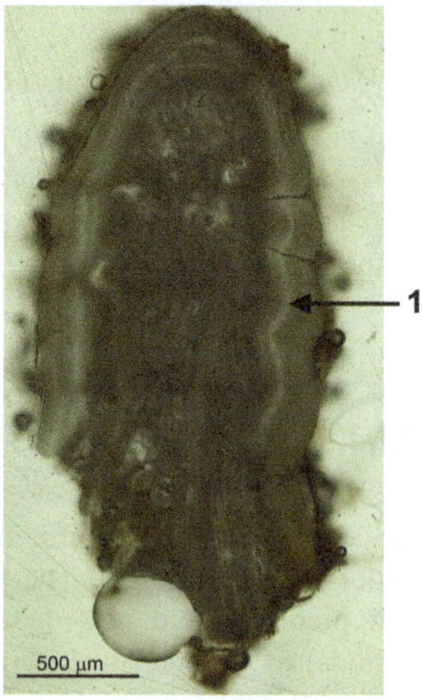

Fig. 23.3 Fin ray section of a European sturgeon caught on 9 December 2009 (length 77 cm, weight 2.8 kg) showing one translucent annulus – Age 1+ (cohort 2008) (photo Cemagref)

23.4 Critical Analysis of the Method

Sampling a section of the pectoral fin ray can be more difficult on a moving boat than removing the entire ray, but sampling took less than 5 min. This method caused no bleeding phenomena, and healing was effective in less than 3 months (Rochard 1992).

Deposition of yearly annuli was also verified in other species, Atlantic sturgeon (Stevenson and Secor 2000) or lake sturgeon (Rossiter et al. 1995) for instance, showing that pectoral fin ray was an appropriate structure for age assessment in sturgeons. However, some authors have pointed out difficulties in age estimates from pectoral fin rays (Hurley et al. 2004; Paragamian and Beamesderfer 2003; Rien and Beamesderfer 1994). The contrast between opaque and translucent rings could be altered if fish did not experience sufficient variations in water temperature and food resources (Cochnauer et al. 1985). Moreover, age assessment quality decreases when fish are very old: growth rate is lower and the interannuli space narrows, or does not form any detectable growth zone every year (Rien and Beamesderfer 1994). Bruch et al. (2009) specified that age estimation by pectoral fin rays was correct in lake sturgeon until year 11, but the method underestimated the real age afterwards, and error increased with age. This information shows that age determination data from pectoral fin ray must be used with appropriate care in large fish.

Age determination in fish caught in the wild and held in captivity for ex situ stock conservation purposes (see Chap. 31) is difficult. The contrast between translucent and opaque annuli is diminished, probably as a result of too little difference in thermal regimes and of constant food availability. We therefore recommend that in this situation fish are sampled as soon as they arrive in captivity.

All fin ray samples were stored to constitute a "sturgeon fin ray bank", with now more than 400 samples from age 1 to 33.

References

Brennan JS, Cailliet GM (1989) Comparative age-determination techniques for white sturgeon in California. Trans Am Fish Soc 118:296–310

Brennan JS, Cailliet GM (1991) Age determination and validation studies of white sturgeon, *Acipenser transmontanus*, in California. In: Williot P (ed) Acipenser, Actes du premier colloque international sur l'esturgeon, Bordeaux, 1989. Cemagref, Antony, pp 209–234

Bruch R, Campana S, Davis-Foust SL, Hansen MJ, Janssen J (2009) Lake sturgeon age validation using bomb radiocarbon and known-age fish. Trans Am Fish Soc 138:361–372. doi:10.1577/T08-098.1

Cochnauer TG, Lukens JR, Partridge FE (1985) Status of white sturgeon, *Acipenser transmontanus*, in Idaho. In: Binkowski F, Doroshov S (eds) North American sturgeons: biology and aquaculture potential. Dr W. Junk, Dordrecht, pp 127–133

Collins MR, Smith TIJ (1996) Sturgeon fin ray removal is nondeleterious. N Am J Fish Manag 16:939–941

Guenette S, Goyette D, Fortin R, Leclerc J, Fournier N, Roy G, Dumont P (1992) La périodicité de la croissance chez la femelle de l'esturgeon jaune (*Acipenser fulvescens*) du fleuve Saint-Laurent est-elle reliée à la périodicité de la reproduction ? Can J Fish Aquat Sci 49:1336–1342

Hurley KL, Sheehan RJ, Heidinger RC (2004) Accuracy and precision of age estimates for pallid sturgeon from pectoral fin rays. N Am J Fish Manag 24:715–718

Kohlhorst DW (1979) Effect of first pectoral fin removal on survival and estimated harvest rate of white sturgeon in the Sacramento–San Joaquin estuary. Calif Fish Game 65:173–177

Paragamian VL, Beamesderfer RC (2003) Growth estimates from tagged white sturgeon suggest that ages from fin rays underestimate true age in the Kootenai River, USA and Canada. Trans Am Fish Soc 132:895–903

Parsons GR, Hoover JJ, Killgore KJ (2003) Effect of pectoral fin ray removal on station-holding ability of shovelnose sturgeon. N Am J Fish Manag 23:742–747

Peterson DL, Gunderman B, Vecsei P (2002) Lake sturgeon of the Manistee River: a current assessment of spawning stock size, age and growth. Am Fish Soc Symp 28:175–182

Priegel GR, Wirth TL (1971) The lake sturgeon, its life history, ecology and management. Wisconsin Department of Natural Resources, Madison

Rien TA, Beamesderfer RC (1994) Accuracy and precision of white sturgeon age estimates from pectoral fin rays. Trans Am Fish Soc 123(2):255–265

Rochard E (1992) Mise au point d'une méthode de suivi de l'abondance des amphihalins dans le système fluvio-estuarien de la Gironde, application à l'étude écobiologique de l'esturgeon, *Acipenser sturio*, Université de Rennes I, Rennes

Rochard E, Jatteau J (1991) Amélioration de la méthode de détermination de l'âge de l'esturgeon commun *Acipenser sturio* et premières applications. In: Williot P (ed) Acipenser, Actes du premier colloque international sur l'esturgeon, Bordeaux, 1989. Cemagref, Antony, pp 193–208

Rossiter A, Noakes DLG, Beamish FWH (1995) Validation of age estimation for the Lake Sturgeon. Trans Am Fish Soc 124:777–781

Stevenson JT, Secor DH (2000) Age determination and growth of Hudson River Atlantic sturgeon, *Acipenser oxyrinchus*. Fish Bull 98:153–166

Wilson NC (1987) Age determination of lake sturgeon (*Acipenser fulvescens*) by use of the marginal pectoral fin ray. In: Olver CH (ed) Workshop on the lake sturgeon, Ontario Fisheries Technical Report Series, Toronto, ON, pp 77–83

Xin D, Zonglin D, Mingyan C (1991) Spawning population characteristics of *Acipenser sinensis* in Yangtze River just below Gezhouba Dam. In: Williot P (ed) Acipenser, Actes du premier colloque international sur l'esturgeon, Bordeaux, 1989. Cemagref, Antony, pp 235–242

Chapter 24
Tagging European and Atlantic Sturgeons in Europe

Philippe Jatteau, G. Castelnaud, E. Rochard, J. Gessner, and M. Lepage

Abstract Tagging is a widespread method used to acquire data on fish populations. Successive steps to obtain a practical tagging method for European sturgeon are detailed, and a list of different tags used in *Acipenser sturio* and *Acipenser oxyrinchus* in Europe is drawn up. The need to bring together tagging information at the European level is pointed out.

24.1 Introduction

Fish tagging has been widely used for more than a century for scientific purposes (Thorsteinsson 2002). Using this methodology, it has been possible to obtain data on fish migration and movements, biological traits (e.g., growth), and population dynamics, and to assess the efficiency of stocking programmes (see Williot et al. 2009 for a recent synthesis on sturgeon stocking programmes). The need to identify individuals or groups of fish has led to the development of a huge variety of tags (see McFarlane et al. 1990 for an extensive review).

Thorsteinsson (2002) defined tags as "man-made objects attached to the fish", which this author differentiated from marks, which were defined as "identifiable characteristics either natural or applied to a fish". Therefore, all physical objects, inserted or attached externally or internally to a fish are considered as tags.

Sturgeon tagging studies have mainly been carried out in North American species (Bordner et al. 1990; Smith and Jenkins 1991; Smith et al. 1990), but there

P. Jatteau (✉) • G. Castelnaud • E. Rochard • M. Lepage
Cemagref, Estuarine Ecosystems and Diadromous Fish Research Unit, 33612 Cestas Cedex, France
e-mail: philippe.jatteau@cemagref.fr

J. Gessner
Leibniz-Institute of Freshwater Ecology and Inland Fisheries, Müggelseedamm 310, 12587 Berlin, Germany

P. Williot et al. (eds.), *Biology and Conservation of the European Sturgeon Acipenser sturio L. 1758*, DOI 10.1007/978-3-642-20611-5_24,

were no trends to identify a satisfactory method. During controlled studies, Collins et al. (1994) found that a T-anchor tag inserted in the abdominal cavity gave the best tag retention 1 year after tagging, while a spaghetti tag inserted in the posterior part of the dorsal fin gave best results after 2 years in *Acipenser oxyrinchus* juveniles tagged in the Columbia River (Rien et al. 1990). Tag retention in the field could be markedly lower than controlled studies tended to suggest (Smith et al. 2002a), as external tag shedding (Carlin and T-anchor tags) could reach 72% (Smith et al. 2002b). In more recent studies, internal tags were used in sturgeon juveniles. Coded wire tags (CWT) have been used in Atlantic sturgeon (Smith and Clugston 1997; St-Pierre 1999), but the fish must be sacrificed if individual identification is needed. Passive integrated transponder tags (PIT tags) have been implemented to evaluate stocking success in juvenile white sturgeon (Ireland et al. 2002; Justice et al. 2009).

Since 1980, Cemagref has initiated a restoration programme on the *Acipenser sturio* population inhabiting the Gironde estuary and the Garonne and Dordogne Rivers (see Chap. 20). In order to get information mainly on migration, growth and population dynamics, tagging operations have been carried out on juveniles in estuarine habitats during their first years of life.

24.2 History of Tagging Operations in the Gironde Estuary (Castelnaud et al. 1991)

As professional fishermen were involved in scientific surveys, tags had to be easily spotted and readable. Moreover, individual identification was needed to reach the study targets. Therefore, internal tagging did not match our requirements and was not considered as a suitable technique in these conditions.

Tagging operations were begun in June 1981. From 1981 to 1983, 348 fish were caught and tagged with Carlin tags. The tag was attached with a stainless steel wire at the base of the dorsal fin (see Dietrich and Cunjak 2006 for insertion method). Inserting the tag was rather difficult on a small unsteady boat. Also, the wire was not strong enough, and it broke frequently when the knot was being tied. Of the fish tagged, 38 had been caught again by the end of 1983. Twenty percent of them presented scars or injuries at the base of the dorsal fin, clear evidence of a lost tag.

Drawbacks of using the Carlin tag have already been reported in salmonids. Hansen (1988) concluded that marked smolts had a higher mortality rate, which could be due to severe wounds observed around tag insertion (Dietrich and Cunjak 2006; Strand et al. 2002). The reported rate of tag loss was about 10% in Atlantic salmon, and increased with time from tagging to recapture (McAllister et al. 1992).

Obviously, Carlin tags were not suitable for sturgeon under our conditions, and this led us to change to a different type of tag in 1984.

From 1984 to 1987, plastic and metallic Petersen discs were used with two different insertion methods: on the pectoral fin (Fig. 24.1) and at the base of the dorsal fin. Six hundred and ninety three juveniles were tagged using the first type of

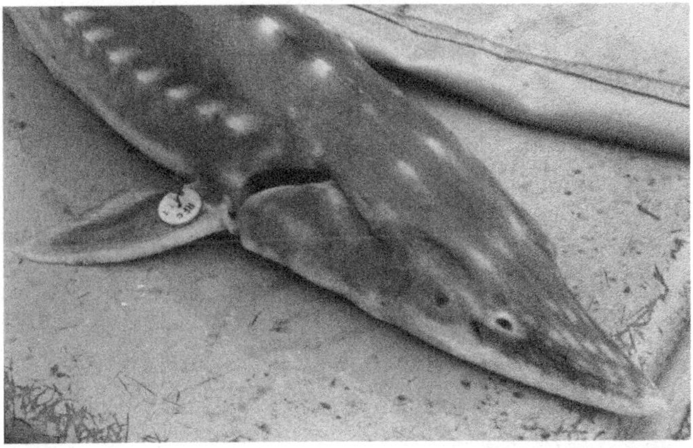

Fig. 24.1 European sturgeon with a Petersen disc tag attached to the pectoral fin ray (photo Cemagref)

attachment. This tag proved to be stronger, but when fish were recaptured, a lot of the tags were entangled in the net and pulled out by the net because of the attachment system. Thus, from 1986 the attachment system was modified. Preparation of the tag consisted of fixing one Petersen disc at the end of a stainless steel wire about 15 cm long. The other end of the wire was cut to obtain a pointed tip, and was inserted at the anterior base of the dorsal fin, then attached near the knot fixing the Petersen disc, to form a loop. The loop was about 3 cm in diameter to allow for growth. Up to the end of 1987, 697 juveniles had been tagged using this method. The rate of entangled tags was reduced to 13%, but entangling remained around the disc.

A special system was thus set up, with the following constraints:

- Use of a wire which allowed insertion without a tool
- Possibility of forming a loop
- Limited risk of entanglement

This special tag was manufactured by Hallprint (Victor Harbor, Australia) from 1988. A yellow 3 cm long plastic tube was threaded onto a 0.7 mm diameter stainless steel wire (AISI 316L). The printing "Cemagref France No. *(of the tag)*" on the plastic tube was protected by transparent film. The tag was attached at the anterior base of the dorsal fin (Fig. 24.2) following the second Petersen attachment method. A total of 198 fish were tagged in 1988, 1989 and 1990, and fish caught with this tag were not entangled in the net. Tag insertion was rapid and easy; the tag was ready to use (no preparation), and allowed growth with a loop of 3 centimetres in diameter.

When fish with a Petersen tag were recaptured, the tag was replaced with this new system.

From 2008, a modified tag was used for fish less than 40 cm long (around 300 g). The stainless steel wire was replaced by a steel wire with a twisted Kevlar sheath.

Fig. 24.2 European sturgeon with a Hallprint loop tag attached to the anterior base of the dorsal fin ray (photo Cemagref)

The wire was thus thinner (0.4 mm diameter) and more flexible. The plastic tube with the same printing was held in place by two stop-floats placed both sides of the plastic tube.

Around 560 fish were tagged from 2008 with the traditional or modified Hallprint loop tag. All these fish were also PIT-tagged in the dorsal musculature.

24.3 Tagging Techniques Applied in Germany and in the Baltic Sea Tributaries

Two sturgeon species were tagged in Germany: European sturgeon *A. sturio* and Atlantic sturgeon *A. oxyrinchus*.

One hundred and fifty six juvenile European sturgeons were released into the Elbe, Oste and Stör Rivers in 2008 and 2009 as part of an experimental study to determine habitat utilization and migration patterns. They were tagged with grey Floy T-bar anchor tags inserted in the dorsal musculature at the left base of the dorsal fin. Each tag had the following printed on it: "GRS 07/XXXX", where XXXX was the identification number. Additionally, all these fish received a PIT tag inserted in the dorsal musculature. Ten fish were also equipped with ultrasonic transmitters implanted in the body cavity (scar on the ventral part of the body).

In the Baltic Sea tributaries, a total of 70,000 juvenile Atlantic sturgeons of different age classes from feeding larvae to subadults were released into the Oder and Vistula Rivers between 2006 and 2009. Of these fish, approximately:

(a) 3,000 1+ fish were tagged with yellow Floy T-bar anchor tags inserted in the musculature at the dorsal fin base (left side of the fish) (Fig. 24.3). The tag print

Fig. 24.3 Atlantic sturgeon with a Floy T-anchor tag (*red circle*) inserted at the anterior base of the dorsal fin (photo IGB)

Fig. 24.4 Attachment of a Carlin tag to Atlantic sturgeon (photo IGB)

was A xxxx where xxxx was the identification number (maximum identification number 4480)

(b) 160 subadults were released in 2009 tagged with yellow Floy T-Bar Anchor tags, and 20 of these fish were equipped with external (white) Data Storage tags to record temperature, salinity, and depth. T-bar tag print was M 050 to M 099, F 050 to F 060 and between A 4652 and A 4875

(c) 900 0+ fish were released in 2009 with Floy Carlin dangler tags (yellow, numbers from 001 to 899)

(d) 2,500 1+ fish were released between 2007 and 2010 tagged with Polish Carlin tags (red) with stainless steel attachment at the anterior base of the dorsal fin (Fig. 24.4)

Except for the subadults released in 2009, no PIT tags were applied.

According to current information, tag loss varies dramatically, even with the same tag types, indicating the importance of handling procedure. It is suggested that all sturgeon catches be reported regardless of tag presence, and the report should preferably be accompanied by a picture.

24.4 Conclusion and Perspectives

The setting-up of an appropriate tagging method in European sturgeon was rather difficult because of skin thickness and toughness, and the working conditions on small unsteady boats. The Hallprint loop tag, specifically adapted for smaller fish, gives quite good retention results. Nevertheless, permanent external tagging in sturgeon does not exist, due to the lifespan and the growth of this species. The PIT tag, now usually inserted in fish that are sufficiently large, allows individual identification without retention problems, but external tagging remains necessary for collecting information on migration or growth from by-catch fishing and fisherman awareness. Paradoxically, this external tagging also constituted a protection for the fish. Fishermen considered tagged fish differently; they reported their capture and put them back in the water more frequently than non-tagged fish (Rochard et al. 1997).

Facing recent developments in stocking operations in France with the European sturgeon, in Germany with the European and Atlantic sturgeons, and in Poland with Atlantic sturgeon, it is essential to enable non-specialized people to correctly identify the species of a caught sturgeon. To this end, knowledge of the tags used is very useful. A summary of information on tagging operations, including data on tags used (type, colour, printing), species, stocking sites, and number of fish concerned, will be published on a specific page of the Diadfish website which is dedicated to sharing diadromous fish information http://www.diadfish.org.

References

Bordner CE, Doroshov SI, Hinton DE, Pipkin RE, Fridley RB, Haw F (1990) Evaluation of marking techniques for juvenile and adult white sturgeons reared in captivity. Am Fish Soc Symp 7:293–303

Castelnaud G, Rochard E, Jatteau P, Lepage M (1991) Données actuelles sur la biologie d'*Acipenser sturio* dans l'estuaire de la Gironde. In: Williot P (ed) Acipenser, Actes du premier colloque international sur l'esturgeon, Bordeaux, 1989. Cemagref, Antony, pp 251–275

Collins MR, Smith TIJ, Heyward LD (1994) Effectiveness of six methods for marking juvenile shortnose sturgeons. Prog Fish Cult 56:250–254

Dietrich JP, Cunjak RA (2006) Evaluation of the impacts of Carlin tags, fin clips, and panjet tattoos on juvenile Atlantic salmon. N Am J Fish Manag 26:163–169. doi:10.1577/M05-032.1

Hansen LP (1988) Effects of Carlin tagging and fin clipping on survival of Atlantic salmon (*Salmo salar* L.) released as smolts. Aquaculture 70:391–394. doi:10.1016/0044-8486(88)90122-6

Ireland SC, Beamesderfer RCP, Paragamian VL, Wakkinen VD, Siple JT (2002) Success of hatchery-reared juvenile white sturgeon (*Acipenser transmontanus*) following release in the Kootenai River, Idaho, USA. J Appl Ichthyol 18(4–6):642–650

Justice C, Pyper BJ, Beamesderfer RC, Paragamian VL, Rust PJ, Neufeld MD, Ireland SC (2009) Evidence of density- and size-dependent mortality in hatchery-reared juvenile white sturgeon (*Acipenser transmontanus*) in the Kootenai River. Can J Fish Aquat Sci 66:802–815. doi:10.1139/F09-034

McAllister KW, McAllister PE, Simon RC, Werner JK (1992) Performance of nine external tags on hatchery-reared rainbow trout. Trans Am Fish Soc 121:192–198

McFarlane GA, Wydoski RS, Prince ED (1990) Historical review of the development of external tags and marks. Am Fish Soc Symp 7:9–29

Rien TA, Beamesderfer RC, Foster CA (1990) Retention, recognition, and effects on survival of several tags and marks on white sturgeon. Calif Fish Game 76:174–180

Rochard E, Lepage M, Meauze L (1997) Identification et caractérisation de l'aire de répartition marine de l'esturgeon européen *Acipenser sturio* à partir de déclarations de captures. Aquat Living Resour 10(2):101–109

Smith TIJ, Clugston JP (1997) Status and management of Atlantic sturgeon, *Acipenser oxyrinchus*, in North America. Environ Biol Fish 48(1–4):335–346

Smith TIJ, Collins MC, Post WC, McCord JW (2002a) Stock enhancement of shortnose sturgeon: a case study. Am Fish Soc Symp 28:31–44

Smith TIJ, Jenkins WE (1991) Development of a shortnose sturgeon, *Acipenser brevirostrum*, stock enhancement program in North America. In: Williot P (ed) Acipenser, Premier Colloque International sur l'esturgeon, Bordeaux, 1989. Cemagref, Antony, pp 329–336

Smith TIJ, Lamprecht SD, Hall JW (1990) Evaluation of tagging techniques for shortnose sturgeon and Atlantic sturgeon. Am Fish Soc Symp 7:134–141

Smith TIJ, McCord JW, Collins MC, Post WC (2002b) Occurrence of stocked shortnose sturgeons *Acipenser brevirostrum* in non-target rivers. J Appl Ichthyol 18:470–474

St-Pierre RA (1999) Restoration of Atlantic sturgeon in the northeastern USA with special emphasis on culture and restocking. J Appl Ichthyol 15:180–182

Strand R, Finstad B, Lamberg A, Heggberget TG (2002) The effect of Carlin tags on survival and growth of anadromous Arctic charr, *Salvelinus alpinus*. Environ Biol Fish 64:275–280. doi:10.1023/A:1016091619937

Thorsteinsson V (2002) Tagging methods for stock assessment and research in fisheries. Report of Concerted Action FAIR CT.93.1394. Marine Research Institute, Reykjavik

Williot P, Rochard E, Kirschbaum F (2009) Acceptability and prerequisites for the successful introduction of sturgeon species. In: Carmona R, Domezain A, Garcia Gallego M, Hernando-Casal J, Rodriguez F, Ruiz Rejon M (eds) Biology, conservation and sustainable development of sturgeons, vol 29, Fish & fisheries series. Springer, New York, pp 369–384

Chapter 25
Mass Marking in European Sturgeon

Philippe Jatteau and Aude Lochet

Abstract When assessing stocking programmes, there needs to be discrimination between stocked fish and wild fish, and between different groups (i.e., cohorts, release sites) within stocked fish. Few methods are available for marking a large number of fish at an early stage with minimum handling. Chemical mass marking matches these prerequisites, and this method has been retained for the European sturgeon. Because the sampling has no adverse effects on individuals, marking efficiency has been checked on the pectoral fin ray. After qualified results were recorded in 2007 and 2008, this led to the testing of the possible effect of chemical or size/age on marking efficiency and mark quality. A tetracycline bath at 600 ppm for 8 h has given satisfactory results in fish not younger than 4 months.

25.1 Introduction

The last century has seen dramatic declines in numerous fish species, including diadromous species (Limburg and Waldman 2009). Stocking programmes are commonly used to rehabilitate endangered fish species (Brown and Day 2002; St-Pierre 2003). Their success, although rarely assessed except in cases of salmonids, can be evaluated in terms of survival rates at different stages (Cote and Pomerleau 1985), return rates (Aprahamian et al. 2003; Salminen et al. 2007), growth or ecological interactions (Cowx 1994).

To evaluate the success of such programmes, and following the recommendations of the IUCN Re-introduction Specialist Group (1998), hatchery-reared individuals have to be discriminated from wild ones (Baer and Rösch 2008; Champigneulle and Cachera 2003). At a finer scale, discriminating groups of

P. Jatteau (✉) • A. Lochet
Cemagref, Estuarine Ecosystems and Diadromous Fish Research Unit, 33612 Cestas Cedex, France
e-mail: philippe.jatteau@cemagref.fr

P. Williot et al. (eds.), *Biology and Conservation of the European Sturgeon Acipenser sturio L. 1758*, DOI 10.1007/978-3-642-20611-5_25,

stocked individuals can be used to improve stocking practises, by assessing the relative survival of specimens released at different ages and sites, for example (Caudron and Champigneulle 2009).

The European sturgeon is a greatly threatened species and the Gironde–Garonne–Dordogne watershed is the last to be inhabited by this species. In spite of its protected status in France since 1982, the population was declining and stocking was considered necessary to support the population (Williot et al. 2009b). The assessment of stocking efficiency was judged to be of primary importance for the management of the *Acipenser sturio* restoration programme in Europe. We therefore had to develop a mass marking procedure for this species without any reference to sturgeon species.

Following artificial reproduction, the very first attempts at chemical mass marking were carried out on European sturgeon larvae or juveniles stocked in 1995 (Brun et al. 1998; Williot et al. 2000, 2005). As marking efficiency proved to be poor, experiments and tests were carried out to define a marking protocol and mark checking method with larvae produced and released in 2007, 2008 and 2009.

Here we present a rapid overview of marking practises in sturgeon species, then describe the development of an appropriate mass marking protocol for sturgeons and the marking efficiency results obtained.

25.2 Overview of Marking Practises in Sturgeon Species

Stocking of Eurasian sturgeon species has been widely used to support fisheries or to restore populations (Williot et al. 2002). As no mass marking operations were in place to identify wild and hatchery-produced fish, the success of these programmes was never correctly assessed (Secor et al. 2000a; Williot et al. 2009a). The absence of marking operations within stocking programmes has been recorded worldwide, in Europe with *Acipenser ruthenus* in the Danube basin (Holcik et al. 2006), and in China with *Acipenser sinensis* (Zhu et al. 2006) for instance. Procedures to identify released fish in the wild have been implemented in the USA using coded wire tags (CWT) (Secor et al. 2000b; St-Pierre 1999). This technique required individual handling, so too few fish were able to be processed at once compared with the yearly number of fish involved in restoration or enhancement programmes. Moreover, fish must reach a minimum size to receive CWT, and smaller fish could be stocked unmarked (Smith et al. 2002). Therefore, these operations could not be considered as mass marking operations. Nevertheless, CWT or passive transponder tags are very useful for studying the ecology of released fish (Ireland et al. 2002; Justice et al. 2009 – see also Chaps. 24 and 29).

In the absence of any reference in sturgeon mass marking, chemical marking appeared to be a possible and original way to implement mass marking in very young stages of European sturgeons.

25.3 Setting Up an Appropriate Mass Marking Method

When determining an appropriate marking method, there are several criteria and constraints that must be met. The first concern is the number of fish that have to be marked.

Basically, in restoration programmes, the goal is to tag several tens of thousands of fish. External and internal tags involve individual handling, which does not match mass marking requirements. Chemical mass marking is a good way to mark fish at a very young stage, without individual handling (Secor et al. 1991; Taylor et al. 2005; Tsukamoto 1988). With this technique, fish are immersed in a bath containing chemicals that fluoresce under UV light and bind to hard tissues like otoliths, fin rays, and scales. In this way, a large number of fish can be marked simultaneously. Marks have been successfully produced in young stages of fish using chemicals such as oxytetracycline (Hendricks et al. 1991; Reinert et al. 1998), tetracycline (Dabrowski and Tsukamoto 1986), alizarine complexone (Van der Walt and Faragher 2003), alizarine RedS (Lagardère et al. 2000), and calcein (Mohler 1997).

The second concern is related to the assessment of marking efficiency. Survival and behaviour are the first criteria which can be immediately assessed. Survival indicates which chemical and which concentration can be withstood by the fish, and results can vary according to species and/or age (Beckman and Schulz 1996; Brooks et al. 1994; Rojas-Beltran et al. 1995).

Mark quality and persistence over time (Jenkins et al. 2002; Reinert et al. 1998) are the main criteria when selecting the marking protocol. Mark quality is generally observed in otoliths, which are considered as a reference structure as they appear early in life, grow throughout the entire fish life, and do not resorb (Campana 1999). However, using the otolith as a checking structure for mark quality involves sacrifice of the fish, which is not conceivable for a protected and endangered species like European sturgeon. Two other calcified structures can be sampled in sturgeons without fish sacrifice: fin rays and scutes. Age determination methods in sturgeon species have been developed on the first pectoral fin ray (see Chap. 23), and are now widely used (Brennan and Cailliet 1989, 1991; Peterson et al. 2002; Rien and Beamesderfer 1994; Rochard and Jatteau 1991). The sampling of a small cross-section is rapid, easy, non-lethal (Collins and Smith 1996), and has no effect on swimming behaviour (Parsons et al. 2003). Moreover, oxytetracycline marks can be used to validate age determination (Miller and Beckman 1993; Rien and North 2002). Thus, the pectoral fin ray was used to check marking efficiency.

The weak occurrence and the low number of European sturgeons available to set up a marking method (only four times in 14 years), made it impossible to carry out all the tests that were planned and required, and a biological model (*Acipenser baerii*) had to be used. Results obtained with this species must be considered carefully because of a possible difference in hardiness between species.

After the successful artificial reproduction in 1995 (Williot et al. 2000), batches of juveniles were marked before stocking. First attempts were mainly assessed according to survival, behaviour, and mark quality, with mark persistence not monitored experimentally.

Table 25.1 Summary of mass-marking treatments performed on European sturgeon from 1995

Cohort	Marker	Number of fish	Age at marking (days)	Weight at marking (g)	Concentration mg l^{-1}	Duration	Mortality %
1995	Oxytetracycline	6,000	51	1	7,500	6 min	2.5
1995	Oxytetracycline	2,000	131	6.5	8,000	7 min	0
2007	Tetracycline hydrochloride	6,150	86	3	300	8 h	0
2008	Tetracycline hydrochloride	84,500	56–113	1–7	300	8 h	0.1
2009	Tetracycline hydrochloride	33,100	60–96	2–11	600	8 h	0.05

According to results recorded on the biological model, it was decided to use oxytetracycline hydrochloride ($C_{22}\,H_{24}\,N_2\,O_9 - HCl$) to mark *Acipenser sturio* late larvae and juveniles that were to be stocked. Two different protocols were applied according to age/size (Table 25.1)

To evaluate mark persistence, all sturgeons from the 1995 cohort caught in the Gironde estuary were checked for chemical mark. A total of 35 fish from the 1995 cohort were caught in the Gironde estuary during sampling campaigns between winter 1995 and spring 2001. Only 14 of these fish exhibited a mark on the pectoral fin ray (Lochet et al. 2004).

The success of these first marking trials on European sturgeons was not convincing. The marking protocol was modified in 2007 by using tetracycline hydrochloride ($C_{22}\,H_{24}\,N_2\,O_8 - HCl$), and in 2009 by increasing chemical concentration. Table 25.1 summarises marking treatments carried out on European sturgeon from 1995.

The mass marking procedure applied from 2007 was as follows:

- Feeding was stopped 18 h before treatment
- Concentrated marker solution was prepared in order to obtain 300 ppm tetracycline hydrochloride in the rearing tank
- Concentrated solution was buffered with a mix of sodium phosphate (Na_2HPO_4, $7H_2O$) and potassium phosphate (K_2HPO_4, $3H_2O$) in a ratio of 2:1 (Hendricks et al. 1991)
- Concentrated solution was gradually added to the rearing tank
- Inflow was stopped and pure oxygen added. A pump was installed to provide a good distribution of the marker and oxygen in the tank
- Temperature, pH, dissolved oxygen, and fish behaviour were recorded every hour until the end of the bath
- At the end of the bath, water inflow was turned on, and the pump removed

25.4 Mass Marking Efficiency

To monitor mark quality with time, a batch of 50 fish from the 2007 and 2008 cohorts was kept in the Cemagref field station. These fish were marked at age 101 days (3.9 g) in 2007, and double-marked at age 56 days (1.2 g) and 127 days (17.17 g) in 2008 (Table 25.2).

Table 25.2 Marking success (expressed in % of fish marked and mark morphology) in European sturgeons kept at the Cemagref field station

Cohort	Age at marking (days)	Weight at marking (g)	Number of detected marks per sample	% Marked	Mark morphology (continuous + discontinuous)
2007 (single marking)	101	3.9	1	51%	16% + 35%
2008 (double marking)	56–127	1.22–17	1	100%	91% + 9%

Marking treatment was 300 ppm tetracycline hydrochloride for 8 h. Marking success was assessed 6 months after marking on pectoral fin ray sections

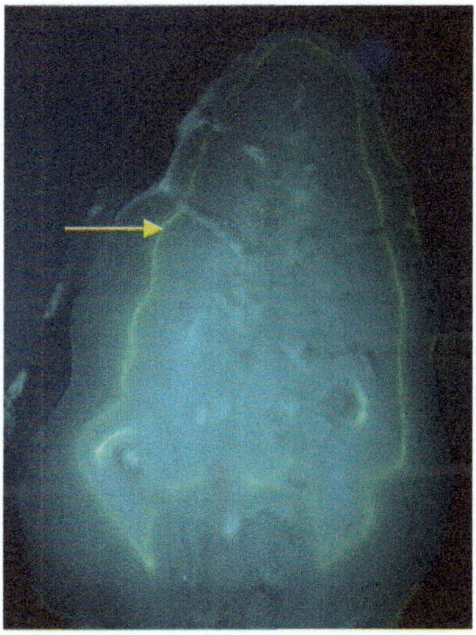

Fig. 25.1 Fin ray section of a European sturgeon showing a continuous mark (*arrow*) – marked at 4 months old (tetracycline hydrochloride 300 ppm, 8 h) – observed 6 months after marking (photo Cemagref)

Mark quality was checked 6 months after treatments performed in 2007 and 2008, on a section of the first pectoral fin ray. Fifty-one percent of fish marked in 2007 presented a distinguishable mark, but some marks were like continuous lines of fluorescence (Fig. 25.1) while others were discontinuous and appeared as spots of fluorescence (Fig. 25.2). Discontinuous marks were not considered to be of suitable quality. Discontinuous mark detection was fairly difficult in 6-month-old fish, and will probably become more difficult in older fish with material deposition in the pectoral fin ray during growth and fading of the mark intensity with time (Frenkel et al. 2002). Only 16% of the fish marked in 2007 presented a continuous mark, which was described as a "good mark". In 2008, 100% of fish observed presented a single mark, although the fish had been marked twice, at 56 and 127 days old. In many cases, the mark was located at the edge of the fin ray, suggesting that it had been produced by the second treatment. In other cases, it was more difficult to attribute the mark to the first or the second treatment.

These results show the limited success of the method used in 2007 and 2008. The low rate of marked fish and the high proportion of discontinuous marks are

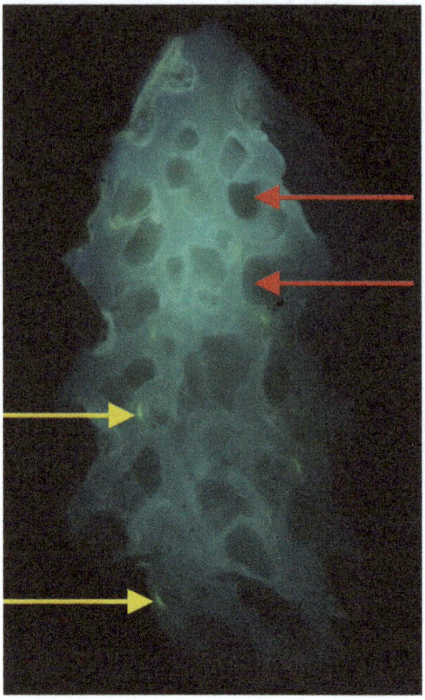

Fig. 25.2 Fin ray section of a 2-month-old Siberian sturgeon observed under UV light – marked at 1 month old (tetracycline hydrochloride 600 ppm, 8 h) and observed 1 month after marking. *Yellow arrows* show discontinuous mark appearing as spots of fluorescence, *red arrows* show vascular cavities (photo Cemagref)

questionable. Two hypotheses could be considered to explain the difference in mark morphology. First, the tetracycline hydrochloride may have a limited ability to be incorporated into the pectoral fin ray. Second, with regard to age/size at marking, there may be a possible threshold effect in relation to the development of calcified structures during ontogenesis.

25.4.1 Effect of Chemical on Mark Quality

Experiments were carried out with 1-month-old Siberian sturgeons *Acipenser baerii* (mean weight 0.6 g). Batches of 30 fish were exposed to two different treatments using Alizarine Red s ($C_{14}H_7NaO_7S$) at 200 ppm for 12 h and tetracycline hydrochloride at 300 ppm for 8 h. Operating method was as described above with respect to chemicals. One month after treatment, 15 fish per batch were sacrificed in a concentrated bath of eugenol, and a section of the first pectoral fin ray was sampled to check mark quality.

Alizarine bath enhanced marking efficiency (Table 25.3) but the rate of marked fish remained low (60%), and results for the two chemicals were not significantly different in terms of the proportion of fish marked (Pearson χ^2, $p > 0.05$). The proportion of continuous marks among marked fish was the same, and remained at an unsatisfactory level (66%).

Table 25.3 Effect of chemical, alizarine RedS and tetracycline hydrochloride, on marking efficiency in Siberian sturgeon, 1 month old (0.6 g)

	Alizarine	Tetracycline
% of marked fish	60	40
% of continuous mark	66.6	66.6

Table 25.4 Effects of age/size and tetracycline hydrochloride concentration on marking efficiency in Siberian sturgeon (small fish, 1 month old, 0.6 g – large fish, 14 months old, 15 g)

	300 ppm		600 ppm	
	Small	Large	Small	Large
% of marked fish	40	100	93	100
% of continuous mark	66.6	100	71.4	100

25.4.2 Effect of Age/Size on Mark Quality

Influence of age/size was addressed using Siberian sturgeons. Two batches of large fish (14 months old, mean weight 15 g) and two batches of small fish (1 month old, 0.6 g) were exposed to tetracycline hydrochloride for 8 h at two concentration levels, high (600 ppm) and low (300 ppm). Operating method was as described above, and mark quality was checked 1 month after treatment.

Increase in chemical concentration enhanced marking efficiency in small fish (Pearson χ^2, $p = 0.02$). Results showed a clear difference according to age at low chemical concentration (Pearson $\chi^2, p = 0$), but not at high concentration (Table 25.4). All large fish were marked with a clear continuous mark. Although 93% of small fish were marked with the 600 ppm treatment, 29% of these marked fish still presented a discontinuous mark (Fig. 25.2).

25.5 Conclusion and Perspectives

Results from the first experiments showed that whatever the chemical used, alizarine or tetracycline hydrochloride, the proportion of discontinuous marks remained high, and mark morphology was not linked to the chemical nature.

We highlighted a possible effect of age/size on the mark quality with the second experiments. Our first investigations on double-marking (2 and 4 months) in European sturgeons led us to hypothesise that better mark quality (continuous mark) could be obtained in older fish. These preliminary data were confirmed with experiments carried out on Siberian sturgeon of different ages. These results could be explained by the pectoral fin ray structure in young stages of sturgeons. In fish with a high metabolism, which is the case in larvae and juveniles, vascular cavities are numerous (Meunier 2002). An illustration of this phenomenon is shown in Fig. 25.2. The presence of vascular cavities would prevent a good impregnation of marker in the calcified tissue, thus producing discontinuous marks, with a detrimental effect on mark quality and detection. Holding fish in the hatchery until they reached an appropriate size to be marked would create management difficulties. The main consequences are on costs, handlings and space required (number of tanks), which increase with the number of fish to be released. For example, a fourth month rearing 10,000 European sturgeon juveniles means a 75% increase in costs compared to the 3-month rearing, with the use of tanks representing a rearing surface of at least 55 m^2, and at least two additional handlings for grading (Patrick Chèvre personal communication, see also Chap. 33). Sequential releases from the same cohort (i.e., at different sizes) according to the method proposed by Burtsev (2009) could reduce management pressures and costs, which still must be considered when defining a recovery plan for the European sturgeon.

Although chemical mass marking on very young stages of European sturgeons was possible, mark checking with pectoral fin ray did not reach a satisfactory level of efficiency.

The use of the pectoral fin ray to detect chemical marks in European sturgeon marked and stocked under our conditions is therefore not recommended.

Other structures for mark checking, like dorsal scutes, have been identified. Further studies are needed to verify marker impregnation in these calcified structures, and to test the effects of dorsal scute removal on survival and behaviour. Dorsal scute removal was used in *Acipenser oxyrinchus* to distinguish wild fish from stocked fish, but no information concerning potential effects was available (Ireland et al. 2002; Justice et al. 2009).

Finally, other mass marking approaches could be addressed. Genetic identification by means of microsatellite markers on stocked fish, born of parents of known genetic characteristics, is a possibility currently being investigated, and will be implemented in 2010 in European sturgeon. Another possibility uses rare isotopes to identify stocked fish. Different methods are used to introduce rare isotopes into calcified tissue. They could be mixed with food (Yamada et al. 1979), added to the rearing tank (Munro et al. 2008), or injected in female broodstock to obtain a transgenerational marking (Munro et al. 2009). Once again, however, the choice of a reliable checking structure to detect isotopes incorporated in the calcified structures in European sturgeon will need to be determined.

References

Aprahamian MW, Martin Smith K, McGinnity P, McKelvey S, Taylor J (2003) Restocking of salmonids – opportunities and limitations. Fish Res 62(2):211–227

Baer J, Rösch R (2008) Mass-marking of brown trout (*Salmo trutta* L.) larvae by alizarine: method and evaluation stocking. J Appl Ichthyol 24:44–49, 10.1111/j.1439-0426.2007.01038.x

Beckman DW, Schulz RG (1996) A simple method for marking fish otoliths with alizarin compounds. Trans Am Fish Soc 125:146–149

Brennan JS, Cailliet GM (1989) Comparative age-determination techniques for white sturgeon in California. Trans Am Fish Soc 118:296–310

Brennan JS, Cailliet GM (1991) Age determination and validation studies of white sturgeon, *Acipenser transmontanus*, in California. In: Williot P (ed) Acipenser, Actes du premier colloque international sur l'esturgeon, Bordeaux, 1989. Cemagref, Antony, pp 209–234

Brooks RC, Heidinger RC, Kohler CC (1994) Mass-marking otoliths of larval and juvenile walleyes by immersion in oxytetracycline, calcein, or calcein blue. N Am J Fish Manag 14:143–150

Brown C, Day RL (2002) The future of stock enhancements: lessons for hatchery practice from conservation biology. Fish Fish 3:79–94

Brun R, Pelard M, Williot P (1998) Utilisation de deux méthodes pour marquer les cohortes d'esturgeons. In: Rauta M, Bacabasa-Dobrovoci N, Vasilescu G, Oprea L (eds) Fisheries management in the Danube River Basin – Aquarom '98. Galati, Romania, pp 56–57

Burtsev IA (2009) Towards the definition of optimal size–weight standards of hatchery-reared sturgeon fry for restoration. In: Carmona R, Domezain A, Garcia Gallego M, Hernando-Casal J, Rodriguez F, Ruiz Rejon M (eds) Biology, conservation and sustainable development of sturgeons, vol 29, Fish & fisheries series. Springer, New York, pp 359–368

Campana SE (1999) Chemistry and composition of fish otoliths: pathways, mechanisms and applications. Mar Ecol Prog Ser 188:263–297

Caudron A, Champigneulle A (2009) Multiple marking of otoliths of brown trout, *Salmo trutta* L., with alizarin redS to compare efficiency of stocking of three early life stages. Fish Manag Ecol 16:219–224, 10.1111/j.1365-2400.2009.00661.x

Champigneulle A, Cachera S (2003) Evaluation of large-scale stocking of early stages of brown trout, *Salmo trutta*, to angler catches in the French-Swiss part of the River Doubs. Fish Manag Ecol 10:79–85. doi:10.1046/j.1365-2400.2003.00325.x

Collins MR, Smith TIJ (1996) Sturgeon fin ray removal is nondeleterious. N Am J Fish Manag 16:939–941

Cote Y, Pomerleau C (1985) Survie et dispersion d'alevins de saumon Atlantique (*Salmo salar*) ensemencés en milieu naturel. Nat Can 112:548–557

Cowx IG (1994) Stocking strategies. Fish Manag Ecol 1:15–30

Dabrowski K, Tsukamoto K (1986) Tetracycline tagging in coregonid embryos and larvae. J Fish Biol 29:691–698

Frenkel V, Kindschi GA, Zohar Y (2002) Noninvasive, mass marking of fish by immersion in calcein: evaluation of fish size and ultrasound exposure on mark endurance. Aquaculture 214:169–183

Hendricks ML, Bender TR, Mudrak VA (1991) Multiple marking of American shad otoliths with tetracycline antibiotics. N Am J Fish Manag 11:212–219

Holcik J, Klindova A, Masar J, Meszaros J (2006) Sturgeons in the Slovakian rivers of the Danube River basin: an overview of their current status and proposal for their conservation and restoration. J Appl Ichthyol 22(suppl 1):17–22

Ireland SC, Beamesderfer RCP, Paragamian VL, Wakkinen VD, Siple JT (2002) Success of hatchery-reared juvenile white sturgeon (*Acipenser transmontanus*) following release in the Kootenai River, Idaho, USA. J Appl Ichthyol 18(4–6):642–650

IUCN (1998) Guidelines for re-introduction. IUCN/SSC Re-introduction Specialist Group, Gland

Jenkins WE, Denson MR, Bridgham BC, Collins MR, Smith TIJ (2002) Retention of oxytetracycline-induced marks on sagittae of red drum. N Am J Fish Manag 22:590–594

Justice C, Pyper BJ, Beamesderfer RC, Paragamian VL, Rust PJ, Neufeld MD, Ireland SC (2009) Evidence of density- and size-dependent mortality in hatchery-reared juvenile white sturgeon (*Acipenser transmontanus*) in the Kootenai River. Can J Fish Aquat Sci 66:802–815. doi:10.1139/F09-034

Lagardère F, Thibaudeau K, Bégout Anras ML (2000) Feasibility of otolith markings in large turbot, *Scophthalmus maximus*, using immersion in alizarin-red S solutions. ICES J Mar Sci 57:1175–1181. doi:10.1006/jmsc.2000.0804

Limburg KE, Waldman JR (2009) Dramatic declines in North Atlantic diadromous fishes. Bioscience 59:955–965. doi:10.1525/bio.2009.59.11.7

Lochet A, Lambert P, Lepage M, Rochard E (2004) Croissance de juvéniles d'esturgeons européens *Acipenser sturio* (Acipenseridae) sauvages et issus d'alevinage, durant leur séjour dans l'estuaire de la Gironde (France). Cybium 28(suppl 1):91–98

Meunier F (2002) Différents types de pièce calcifiée: le squelette. In: Panfili J, De Pontual H, Troadec H, Wright PJ (eds) Manuel de sclérochronologie des poissons. Ifremer-IRD, Brest, pp 65–88

Miller AI, Beckman LG (1993) Age and growth of juvenile white sturgeon in the Columbia River downstream from Bonneville Dam. In: Beamesderfer RC, Nigro AA (eds) Status and habitat requirements of the white sturgeon populations in the Columbia River downstream from McNary Dam, vol 2. US Fish and Wildlife Service, Portland, pp 201–216

Mohler JW (1997) Immersion of larval Atlantic salmon in calcein solutions to induce a non-lethally detectable mark. N Am J Fish Manag 17:751–756

Munro AR, Gillanders BM, Elsdon TS, Crook DA, Sanger AC (2008) Enriched stable isotope marking of juvenile golden perch (*Macquaria ambigua*) otoliths. Can J Fish Aquat Sci 65:276–285. doi:10.1139/F08-010

Munro A, Gillanders BM, Thurstan S, Crook DA, Sanger AC (2009) Transgenerational marking of freshwater fishes with enriched stable isotopes: a tool for fisheries management and research. J Fish Biol 75:668–684. doi:10.1111/j.1095-8649.2009.02352.x

Parsons GR, Hoover JJ, Killgore KJ (2003) Effect of pectoral fin ray removal on station-holding ability of shovelnose sturgeon. N Am J Fish Manag 23:742–747

Peterson DL, Gunderman B, Vecsei P (2002) Lake sturgeon of the Manistee River: a current assessment of spawning stock size, age and growth. Am Fish Soc Symp 28:175–182

Reinert TR, Wallin J, Griffin MC, Conroy MJ, Van Den Avyle MJ (1998) Long-term retention and detection of oxytetracycline marks applied to hatchery-reared larval striped bass, *Morone saxatilis*. Can J Fish Aquat Sci 55:539–543

Rien TA, Beamesderfer RC (1994) Accuracy and precision of white sturgeon age estimates from pectoral fin rays. Trans Am Fish Soc 123(2):255–265

Rien TA, North JA (2002) White sturgeon transplants within the Columbia River. Am Fish Soc Symp 28:223–236

Rochard E, Jatteau J (1991) Amélioration de la méthode de détermination de l'âge de l'esturgeon commun *Acipenser sturio* et premières applications. In: Williot P (ed) Acipenser, Actes du premier colloque international sur l'esturgeon, Bordeaux, 1989. Cemagref, Antony, pp 193–208

Rojas-Beltran R, Gillet C, Champigneulle A (1995) Immersion mass-marking of otoliths and bone tissue of embryos, yolk-sac larvae and fingerlings of Artic charr *Salvelinus alpinus* (L.). Nord J Freshw Res 71:411–418

Salminen M, Alapassi T, Ikonen E (2007) The importance of stocking age in the enhancement of River Kymijoki salmon (*Salmo salar*). J Appl Ichthyol 23:46–52. doi:10.1111/j.1439-0426.2006.00794.x

Secor DH, White MG, Dean JM (1991) Immersion marking of larval and juvenile hatchery-produced striped bass with oxytetracycline. Trans Am Fish Soc 120:261–266

Secor DH, Arefjev V, Nikolaev A, Sharov A (2000a) Restoration of sturgeons: lessons from the Caspian Sea Sturgeon Ranching Programme. Fish Fish 1(3):215–230

Secor DH, Niklitschek EJ, Stevenson JT, Gunderson TE, Minkkinen SP, Richardson B, Florence B, Mangold M, Skjeveland J, Henderson Arzapalo A (2000b) Dispersal and growth of yearling Atlantic sturgeon, *Acipenser oxyrinchus*, released into Chesapeake Bay. Fish Bull 98:800–810

Smith TIJ, Collins MC, Post WC, McCord JW (2002) Stock enhancement of shortnose sturgeon: a case study. Am Fish Soc Symp 28:31–44

St-Pierre RA (1999) Restoration of Atlantic sturgeon in the northeastern USA with special emphasis on culture and restocking. J Appl Ichthyol 15:180–182

St-Pierre RA (2003) A case history: American shad restoration on the Susquehanna River. In: Limburg KE, Waldman JR (eds) Biodiversity, status, and conservation of the world's shads, AFS Symposium. American Fisheries Society, Bethesda, MD, pp 315–321

Taylor MD, Fielder DS, Suthers IM (2005) Batch marking of otoliths and fin spines to assess the stock enhancement of *Argyrosomus japonicus*. J Fish Biol 66:1149–1162. doi:doi:10.1111/j.1095-8649.2005.00678.x

Tsukamoto K (1988) Otolith tagging of Ayu embryo with fluorescent substances. Nippon Suisan Gakkaishi 54(8):1289–1295

Van der Walt B, Faragher RA (2003) Otolith marking of rainbow trout fry by immersion in low concentrations of alizarine complexone. N Am J Fish Manag 23:141–148

Williot P, Brun R, Pelard M, Mercier D (2000) Induced maturation and spawning in an incidentally caught adult pair of critically endangered European sturgeon, *Acipenser sturio* L. J Appl Ichthyol 16(6):279–281. doi:10.1046/j.1439-0426.2000.00238

Williot P, Arlati G, Chebanov M, Gulyas T, Kasimov R, Kirschbaum F, Patriche N, Pavlovskaya LP, Poliakova L, Pourkazemi M, Kim Y, Zhuang P, Zholdasova IM (2002) Status and management of Eurasian sturgeon: an overview. Int Rev Hydrobiol 87(5–6):483–506

Williot P, Brun R, Rouault T, Pelard M, Mercier D (2005) Attempts at larval rearing of the endangered western European sturgeon, *Acipenser sturio* (Acipenseridae), in France. Cybium 29:381–387

Williot P, Rochard E, Kirschbaum F (2009a) Acceptability and prerequisites for the successful introduction of sturgeon species. In: Carmona R, Domezain A, Garcia Gallego M, Hernando-Casal J, Rodriguez F, Ruiz Rejon M (eds) Biology, conservation and sustainable development of sturgeons, vol 29, Fish & fisheries series. Springer, New York, pp 369–384

Williot P, Rochard E, Rouault T, Kirschbaum F (2009b) *Acipenser sturio* recovery research actions in France. In: Carmona R, Domezain A, Garcia Gallego M, Hernando-Casal J, Rodriguez F, Ruiz Rejon M (eds) Biology, conservation and sustainable development of sturgeons, 29th edn, Fish & fisheries series. Springer, New York, pp 247–263

Yamada SB, Mulligan TJ, Fairchild SJ (1979) Strontium marking of hatchery-reared coho salmon (*Oncorhynchus kisutch*, Walbaum). J Fish Biol 14:267–275

Zhu Y, Wei Q, Yang D, Wang K, Chen X, Liu J, Li L (2006) Large-scale cultivation of fingerlings of the Chinese sturgeon *Acipenser sinensis* for restocking: a description of current technology. J Appl Ichthyol 22(suppl 1):238–243

Chapter 26
Sex Determination and Staging of Gonads

Patrick Williot

Abstract This outline of the methods developed to discriminate the gender and to stage the maturity of sturgeon includes: (1) the state of the art on genetic sex determination, (2) the peculiarities of the sturgeon reproductive system, (3) the main plasmatic molecules that have been developed, (4) the recent outcomes from plasmatic indicators, (5) an analysis of gametogenesis, and (6) a brief description of the different methods. Species-specific differences in gametogenesis, impacts of the origin of the fish (wild or farmed), and the methods are summarized, and their applicability is evaluated.

26.1 Introduction

Sturgeons have no reliable sexual dimorphism. A variety of characteristics have long been used to determine the gender of brood fish, such as size (total length), condition coefficient, and abdomen appearance (Charlon and Williot 1978). Even in wild sturgeon, the discrimination of gender is delicate, and in cultured specimens the use of external characteristics is totally ineffective. Moreover, the "external" methods, if any, are unable to document an assessment of the stage of development of the gametogenesis. Both sex discrimination and sexual development staging have become crucial prerequisites for species propagation, and methods have been developed in both conservation and farming. *Acipenser sturio* is no exception.

Additionally, in farming that is mainly caviar-oriented, it is preferable to culture females, and so many efforts have dealt with attempts at gynogenesis. This presupposes that one knows or can hypothesize on the mechanisms for sex determinism in sturgeon. Very little is known at present. It has been suggested that white

P. Williot (✉)
Sturgeon Consultant, 4 Rue du Pas de madame, 33980 Audenge, France
e-mail: williot.patrick@neuf.fr

P. Williot et al. (eds.), *Biology and Conservation of the European Sturgeon Acipenser sturio L. 1758*, DOI 10.1007/978-3-642-20611-5_26,

sturgeon, *Acipenser transmontanus* females (Van Eenennaam et al. 1999), and bester, *Huso huso* female × *Acipenser ruthenus* male (Omoto et al. 2005) are heterogametic. Females of the shortnose, *Acipenser brevirostrum* are considered non-homogametic (Flynn et al. 2006). More recent approaches using extensive genome screening in four sturgeon species, *Acipenser baerii*, *Acipenser naccarii*, *Acipenser gueldenstadetii*, and *Acipenser ruthenus*, failed to identify sex-specific markers (Wuertz et al. 2006a). A search for a Sox9 gene in *Acipenser sturio* was also unsuccessful (Hett et al. 2005). In an attempt to overcome the problem, recent proteomic approaches of sex marker determination also failed in *Acipenser persicus* (Keyvannshokooh et al. 2009). Confronted with these failures in proteomics, Keyvannshokooh and Gharaei (2010) suggested focusing on gene expression, as well as looking at proteomics of cell and tissues. A very recent study illustrates this suggestion, in showing that the expression of foxl2 gene was greater in females and was useful for identifying gender (Arnberg et al. 2010). Although most of these investigations are unfruitful to date, the fact that the sex ratio of all documented progenies was not significantly different from 1:1 strongly suggests a genetic sex determinism. The fact still remains that a genetic identification of the sexes in the early stages by PCR techniques has no basis, in spite of its potential advantage in that it is not necessary to wait for gonad differentiation.

So for a long time, different means have been explored to try and address these questions. Some of these methods are precise and some are less so; some are invasive and some less so, some might be better adapted to a certain situation, e.g., farming or wild, some may be dependent on the biological cycle, and some are species-dependent. Indeed, all of the methods based on blood-mediated indicators suppose that sexual differentiation has been completed, and this period of undifferentiation depends on the species and on environmental conditions such as temperature and food availability.

Among the blood-mediated indicators are those which are supposed to be sex-specific, such as steroid hormones, estrogens, i.e., estradiol (E2), and androgens, testosterone (T) and 11 ketotestosterone (11-KT). A steroid hormone (E2)-induced molecule is vitellogenin (VTG), which will be incorporated into the ovarian follicles to constitute the reserves of future eggs. Both estrogens and androgens are produced by the gonads. However, these general endocrine pathways proved to be unable to account for some situations.

The aim of this chapter is to summarize the available methods that were (are) used in sturgeon by describing the range, the advantages and drawbacks, and the limitations of each of them, focusing particularly on the preliminary results for *A. sturio*. In order to clarify some of the aforementioned statements, it seemed necessary to recall some of the peculiarities of the sturgeon reproductive system, and to propose a summary of the typical plasmatic profiles of the main plasmatic indicators.

The methods applied to stage and choose the brood fish (females and males) prior to hormonal stimulation have not been taken into account in this chapter. This requires different methods that are much more effective, and are described elsewhere (Williot 2002; Williot et al. 2002, 2009; see Chaps. 27 and 32).

26.2 Materials and Methods

26.2.1 Reproductive System

Not only do sturgeons not exhibit external sexual dimorphism, but the anatomy is similar for both genders. This was described a long time ago (Magnin 1962), and described in more recent publications (Conte et al. 1988; Kynard and Kieffer 2002). Both genders have Müllerian ducts. The oviducts open into the abdominal cavity by the funnel-shaped ostium (Fig. 26.1) which end in a one-way valve in the paramesonephric duct (Müllerian ducts), here called the genital duct. The ostium is situated approximately one third of the way along the length of the abdominal cavity from the anterior part. At the time of ovulation, eggs drop out from the follicular lamellae into the abdominal cavity and are expelled through the ostium of the funnel-shaped duct as far as the genital pore. This peculiarity explains why the eggs cannot be easily handstripped, as they can for salmonids or cyprinids. This means that when an ovarian follicle sample is needed in order to check the state of maturation, surgery is required. This peculiar structure of the sturgeon's reproductive system explains, at least in part, why simultaneous hermaphrodites occur with already fertilized eggs in the genital duct, as described, for example, in sterlet, *Acipenser ruthenus* (Williot et al. 2005).

Fig. 26.1 Abdominal cavity of a female Siberian sturgeon at stage III (onset of vitellogenesis). The *red arrow* indicates the ostium of the funnel-shaped oviduct ending in a one-way valve into the genital canal. The ovaries are *yellowish* (*dark arrow*) with the lipid part more translucent at the border of follicular lamellae (*white arrow*) (photo P. Williot)

26.2.2 Description of Gametogenesis Stages

The complete gametogenetic cycle has been described using a five-stage classification, in *Acipenser sturio* by Magnin (1962) and in *Acipenser baerii* by Le Menn and Pelissero (1991). Badenko et al. (1981) distinguished four stages, but subdivided the fourth one into IV1 and IV2 in *Acipenser stellatus*, with no clear boundary between the last two. Recently, Hurvitz et al. (2007) also distinguished five stages in *Acipenser gueldenstaedtii*, with a displacement of stage boundaries. Five stages have been distinguished in *Acipenser transmontanus* (Doroshov et al. 1991). Six stages characterized ovarian maturity in *A. transmontanus* (Linares-Cazenave et al. 2003), and seven stages in the same species by Webb et al. (2002). Six stages were also defined for *Acipenser stellatus* (Moghim et al. 2002). Surprisingly, in bester, Mojazi Amiri et al. (1996a, b) distinguished eight stages. From this brief overview, it is clear that no standard reference is being used, and thus comparisons are somewhat delicate. The studies mentioned above which did not use a five-stage scale to separate the gametogenesis took into account the different appearance in the ovarian follicles, but never referred to physiological processes (supported by key blood mediated indicators), which form the basis of the work of Le Menn and Pelissero (1991). We therefore retained this five-stage grid in the present work, and other results have been tentatively rescaled accordingly. Moreover, the advantage of such a classification grid is that although the duration of each stage may vary according to species and environmental conditions, the boundaries are consistent and have reliable reference points.

26.2.3 Recent Outcomes with Regard to Some Plasmatic Indicators

In the early 1980s, with the initiation of *A. sturio* restoration and *A. baerii* farming programmes in France, early sex determination became a relevant topic (Williot et al. 2004). The initial idea was to use vitellogenin (VTG), which was recognized as sex-specific, because of its role in ovarian follicle reserves. This complex protein is synthesized by the liver under the influence of estradiol (E2). High levels of VTG have rapidly been recorded in *A. baerii* males fed compound diets, and these are not significantly different from the VTG levels recorded in stage III females (Pelissero and Le Menn 1988, 1991; Pelissero et al. 1989; Cuisset et al. 1991). The authors hypothesized that diet was responsible for the high levels of VTG because of the high steroid content, and later by the high content of estrogeno-mimetic molecules. As a result, E2 and VTG contents are not reliable indicators to determine sex in stage III farmed sturgeon fed a compound diet. Moreover, great care should be taken with similar measurements in wild fish where there are suspicions of estrogen-like pollution. Previously, VTG was assessed through alkali-labile phosphoprotein (Pelissero et al. 1989; Doroshov et al. 1991). Enzyme-linked immunosorbent assay (ELISA) (Cuisset et al. 1991) and more recently calcium dosage (Doroshov et al. 1997) have been used to assess, either directly or indirectly, the plasmatic VTG content. However, calcium content has to be interpreted with care (Allen et al. 2009). In the late 1970s and early 1980s, protein dosage was used as an indicator of maturation, as the protein content declined when fish were in the last stage of maturation (e.g., Badenko et al. 1981). Indeed, this might be considered as an early approach to future VTG determination.

After the failure of the primary objective of using VTG to discriminate both genders and the corresponding difficulties of using E2, the only potential solution was to investigate steroid androgens, testosterone (T) and 11-ketotestosterone (11-KT), which are presumed to be male-specific. Initial investigations have proved this hypothesis by revealing a reliable way to discriminate male and female Siberian sturgeon aged between 27 and 39 months (Cuisset et al. 1994). However, further investigations on older fish revealed that 11-KT was also present in maturing females, including those in the very last stages of reproduction (Cuisset et al. 1995; Williot 1997). Moreover, it has been shown that 11-KT may be synthesized not only in the ovary but also by blood cells and the interrenal through cortisol (Cuisset et al. 1995). This means that a plasmatic interpretation of 11-KT content is probably more complex than previously expected, and consequently the physiological role of 11-KT still remains to be determined; it has even been suggested that it could be involved in migration (Ceapa et al. 2002).

26.2.4 Methodologies

There are two categories of sex determination, invasive methods and non-invasive methods [i.e., ultrasonography (also called echography)] (Chebanov and Chmyr 2005).

Among the first category, the biopsy is the reference method, since it enables the gonads to be sampled for further staging (morphologically) and serves as a reference for other methods. There are two alternatives to the biopsy. For young fish with low underdeveloped gonads, the operation may take a quarter of an hour, with the opening (~4 cm long) being cross-stitched. Fish recover perfectly well as long as precautions are taken prior to and during the surgery, including 1 day of starvation, a constant water supply in the mouth (Doroshov et al. 1983). Anaesthetizing (e.g., clover oil) is another precaution that is currently taken before any handling of *A. sturio*. The second alternative to biopsy is the current use of a trocar, which is made up of two pieces; one is very pointed and is inserted first into the abdomen at about a 45° angle, the second piece is hollowed out so that tissue can be sampled from the gonad. The latter method is recommended for gonads that are already developed, i.e., from stages II to III onwards, and does not need cross-stitches. Weekly sampling has been performed without any deleterious effect. A very recent new approach in sturgeon showed that the IGF-1 content of ovaries increases in maturing sterlet, *Acipenser ruthenus*, thus being a new indicator of maturing female vs non-maturing female (Wuertz et al. 2006b).

Apart from biopsy, sturgeons are currently blood-sampled in the caudal vasculature. A 2 ml blood sample can be taken fairly quickly with a heparinised syringe. This allows further analysis on plasma components (steroid hormones, VTG, proteins, calcium, etc.).

Finally, two other methods are somewhat invasive: the endoscope is inserted through an opening into the abdominal cavity (Hurvitz et al. 2007) and the borescope is inserted through the genital opening (Kynard and Kieffer 2002).

26.3 Results and Discussion

An overview of the different methods for discriminating gender and stage of sturgeon gonads is provided in Table 26.1. The first two lines of the table recall Magnin's description (1962) of the five gametogenetic stages in *A. sturio* and the five stages of oogenesis in *A. baerii* (Le Menn and Pelissero 1991). Biopsy is used as the method of reference to "standardize and scale" the other methods.

In addition to the different scaling of gametogenesis mentioned above, there are two other reasons why workers may have been confused, which thus led to imprecise observations and staging. First, ovarian development is not synchronous during the first gametogenesis, i.e., the oocytes do not grow all together. This means that at any given time, very different sizes might be observed, as shown in *A. baerii* (Williot and Brun 1998). By the end of ovarian development there are only two sizes of oocytes, the mature ones, which are coloured (dark, grey, green, . . .), and the next generation made up of white pre-vitellogenic oocytes, as shown in *A. baerii* and *A. transmontanus* by Williot and Brun (1998) and Doroshov et al. (1997) respectively. Second, within a year class, the fish did not exhibit a similar gonad cycle, which means that even in adult fish of the same age, very different stages may be observed from fish to fish (Williot and Brun 1998; Doroshov et al. 1997).

Much work has been done on blood-mediated indicators that have been obtained, and surgical biopsy makes staging of the gametogenesis possible. The androgen content (T and 11-KT) discriminates the genders fairly well as soon as differentiation has occurred, i.e., stage II, the only exception being *A. gueldenstaedtii*, which did not exhibit any sex-related difference (Barannikova et al. 2000). Though the sample is small, there is no difference in androgens between wild and farmed fish. At the end of gametogenesis (including from stage IV), the estradiol E2 content is higher in females than in males. However, E2 content in *A. sturio* should be interpreted carefully at present. Davail-Cuisset et al. (2008) reported a correlation between a high plasmatic content of E2 and the maturity of the female. Indeed, one female exhibited a very partial development of its ovaries and very few ovarian follicles matured (Williot and Rouault 2008).

Calcium content might be a good indicator for vitellogenesis, as it is easier to determine. However, some care should be taken when interpreting Ca^{++} content, as it could be influenced by the salinity level of the rearing water, as noted by Linares-Cazenave et al. 2003, and by the calcium content in the environment (Allen et al.

Table 26.1 Synthesis of indicators and/or methods used to sex and stage sturgeon gametogenesis depending on advancement in sexual development

Species – origin	Stages/indicators/method					Sources
	I	II	III	IV	V	
A. sturio (wild)	Undifferentiated gonads	Onset of differentiation. Oocytes are still not visible to naked eye Testes are thin. Spermatogonia are arranging	Onset of multiplication, Ovaries are white-yellowish and white spots of oocytes are visible Testes are milky-white and testis blisters are appearing	Ovaries are yellowish. Oocytes are about 1 mm in diameter and yolk granules are appearing in the cytoplasm	Oocytes are pigmented and ovaries are greyish. Large yolk granules are present in the cytoplasm mainly close to vegetal pole Testes blisters are distended	Magnin (1962)
A. baerii (farmed) ♀	High value of nucleoplasmic index, one nucleolus in the nucleus (20–80 μm). 1-year-old fish, ~40 cm, ~800 g	Nucleoli multiplication migrates towards the periphery The number of lipid globules increases in the cytoplasm Cortical alveoli start to be synthesized, 80–120 μm, 3 years old, 2 kg	First features of vitellogenin incorporation. Elaboration of zona radiata externa The zona radiata interna 1 is built, 120–600 μm, ≥3.5 years old, 4 kg	First appearance of pigment granules Elaboration of zona radiata interna 2 Yolk accumulation increases, 900–2,800 μm, 5 years old, 5 kg	Beginning of nucleus migration towards the periphery Zona radiata is completely synthesized Jelly coat (synthesized by follicular cells) is deposited at the periphery of the zona radiata externa Yolk accumulation ends at this stage, 2,800–3,400 μm, 6 years, 5.5 kg	Le Menn and Pelissero (1991)
Bester farmed ♀			[VTG] ↑ [T] ↑ [E2] ↑	[VTG] ↑ [T] ↑ [E2] ↑	[VTG] ↓ [T] ↑ [E2] ↓	Mojazi Amiri et al. (1996b)
Bester farmed ♂			[T] and [11KT] ↑	[T] and [11-KT] ↑	[T] and [11KT] ↑[a]	Mojazi Amiri et al. (1996a)
Bester farmed ♀			[VTG] ♀ ↑		[VTG] ♀ ↓	Fuji et al. (1991)

(continued)

Table 26.1 (continued)

Species – origin	Stages/indicators/method					Sources
	I	II	III	IV	V	
A. baerii (farmed)			[E2] ♂ ≈ [E2] ♀	[E2] ♀ > [E2] ♂	[E2] ♀	Pelissero et al. (1989)
A. baerii (farmed)			{VTG} ♂ ≈ {VTG} ♀			Cuisset et al. (1991)
A. baerii (farmed)		[11KT] ♂ > [11KT] ♀				Cuisset et al. (1994)
A. baerii (farmed)				[11KT] ♀ ↑		Cuisset et al. (1995)
H. huso (farmed)		[T] ♀ > [T] ♂? (no statistical test)				Yousefian (2006)
A. transmontanus (farmed)				[E2] ♀ ↑ Indications for ALPP ↑		Doroshov et al. (1991)
A. transmontanus (farmed)		[T] ♂ and [11KT] ♂ > 2 ng ml^{-1} vs. ♀ [E2] ♂ ≈ [E2] ♀				Feist et al. (2004)
A. sturio (farmed and fed natural food)					[T] and [11KT] ♂ ↑ [E2] ♀ ↑	Davail-Cuisset et al. (2008)
A. sturio (farmed and fed natural food)				E2 and ultrasonography are used to preselect maturing females		Chèvre pers. com.
A. stellqtus (wild)		[T] ♂ >> [T] ♀				Barannikova et al. (2000)
H. huso (wild)						
A. gueldenstaedtii (wild)		[T] ♂ ≈ [T] ♀				
A. oxyrinchus (wild)					[E2] ♀ > [E2] ♂ [Ca] ♀ > [Ca] ♂	Van Eenennaam et al. (1996)
A. stellatus (wild)					[VTG] ♀ > [VTG] ♂ [E2]/[T] = 0.002 discriminates genders [E2]/[11KT] = 0.02 discriminates genders	Ceapa et al (2002)
A. persicus (wild)					[T] ♂ > [T] ♀	

(continued)

Table 26.1 (continued)

Species – origin	Stages/indicators/method					Sources
	I	II	III	IV	V	
					[E2] ♀ > [E2] ♂	Viayeh et al. (2006)
A. brevirostrum (wild)			Borescope allowed to discriminate ♀ with ovarian follicle diameter ≥ 0.5 1.0 mm			Kynard and Kieffer (2002)
A. gueldenstaedtii (farmed)				Endoscope Discriminates genders and stages ♀		Hurvitz et al. (2007)
A. stellatus (wild)			Ultrasonography (echography) allowed to discriminate genders and maturing/non-maturing brood fish with 97% accuracy			Moghim et al. (2002)
A. ruthenus (farmed) *A. baerii* (farmed) *A. gueldenstaedtii* (farmed) *H. huso* (farmed)				Ultrasonography allowed to discriminate genders (see Table 26.2 for details on age and weight)		Chebanov and Chmyr (2005)

Plasmatic indicators have been determined after blood sample, and scaled using surgical biopsy

[a]At degenerative stage, T remained at a high level while 11-KT declined abruptly

A = Acipenser; H = Huso

Table 26.2 Minimum weight and age requirements for early sex determination by ultrasonography (echography) in sturgeon species (after Chebanov and Chmyr 2005)

	Warm-water fish farm		Natural conditions hatcheries	
Species	Weight (kg)	Age (year)	Weight (kg)	Age (year)
A. ruthenus	0.3–0.6	1–1+	0.3–0.6	2–2+
Huso huso	8–12	4–5	8–12	6–7
A. baerii	2–2.5	2–2+	2–2.5	3–4
A. gueldenstaedtii	1.5–3	1 ± 2	1.5–3	2–3

2009). A promising tool is the ratio between E2 and androgens for gender discrimination (Ceapa et al. 2002). Staging using blood-mediated indicators is an essential approach for a better understanding of the physiology (e.g., growth, osmoregulation, and reproduction) and also to improve rearing conditions, especially for newly investigated sturgeon species such as *A. sturio*, all the more so since most brood fish of the species are kept in brackish water (Williot et al. 2007, 2009; see Chap. 31).

Using a borescope inserted into the genital duct, it was possible to sort maturing females (probable stages III–IV) rapidly, with no negative impact on the duct in spite of the existence of the one-way valve (Kynard and Kieffer 2002). Males were not identified, and nor were immature females.

The endoscope inserted into the abdominal cavity was used to discriminate gender and to stage farmed *A. gueldenstadetii* with a fairly good level of effectiveness in sex discrimination for 3-year-old fish onwards, which corresponded to stage III in terms of oocyte size (Hurvitz et al. 2007).

Finally, ultrasonography has been mentioned as being very effective, even with fish in stage II (Chebanov and Chmyr 2005). The authors indicate that the best period is winter time after 2 months holding at low temperature. The fish should be food-deprived 10–12 days before the expertise. This method can also be useful to detect internal abnormalities. Several hundreds of fish a day can be checked, and this is the only method that is not invasive (Chebanov and Chmyr 2005).

Biopsy with surgery is currently accused of being stressful and leaving an opening for pathogens, which is indeed potentially true. However, no one has published any evidence or examples to support such statements. The author has repeatedly biopsied adult females with no apparent deleterious consequences. Moreover, some positive impacts of so-called stress might be suggested (Williot et al. 2011).

Overall, the degree of invasiveness of the techniques, in increasing order, could be ranked as follows: (0) ultrasonography, (1) blood sample in the caudal vasculature, (2) borescope, endoscope, and biopsy with trocar, and (3) biopsy with surgery, i.e., with cross-stitched opening.

Where all the current methods of sex discrimination prove disappointing is in gonad differentiation. This is species-characteristic, and is dependent on rearing conditions (food availability and temperature). Very few data are available in the field, and *A. sturio* is no exception to date. Magnin (1962) reported that 4-year-old wild *A. sturio* females have oocytes with a diameter ranging from 46 to 141 μm. Unfortunately, younger fish were not examined, and no more recent data exist.

26.4 Conclusions

The selection of a proper method of sex discrimination greatly depends on the objective and working conditions. On the one hand, biopsy with surgery is considered as the reference method, which allows scaling for all other methods. In contrast, it is the most time-consuming, with about 50 fish being dealt with daily for fish at stage III or over. On the other hand, with echography, or ultrasonography, several hundred fish a day can be dealt with. Very effective devices are now available, and have become much less costly than some years ago (Chebanov, personal commmunication). In between these two methods, the endoscope and borescope, though less accurate than the preceding methods, are well-adapted field methods. As for echography, several hundreds of fish can be checked daily. The biopsy with trocar is no more invasive than the last two methods, and offers the great advantage of providing gonad-tissue that can be observed and eventually used for different tests. The plasmatic indicators (through blood sampling) make a lab a necessity. Results are available only after a time lapse, but they provide very useful additional physiological information that can be correlated with data from the biopsy with surgery. The limitations with regard to interpretation have been given regarding E2 and VTG. In contrast to these limitations, the indicators may reveal unusual situations.

References

Allen PJ, Webb MAH, Cureton E, Bruch RM, Barth CC, Peake SJ, Anderson WG (2009) Calcium regulation in wild populations of a freshwater cartilaginous fish, the lake sturgeon *Acipenser fulvescens*. Comp Biochem Physiol C 154:437–450

Arnberg JJ, Goforth R, Stefanavage T, Sepálveda MS (2010) Sexually dimorphic gene expression in the gonad and liver of shovelnose sturgeon (*Scaphyrhinchus platorynchus*). Fish Physiol Biochem 36:923–932

Badenko LV, Korniyenko GG, Chikhacheva VP, Altukhova LA (1981) Methods for evaluating the quality of sturgeon spawners (as exemplified in the sevryuga, Acipenser stellatus) from the Kuban River. J Ichthyol 21(1):96–103

Barannikova IA, Bayunova LV, Geraskin PP, Semenkova TB (2000) Content of sex steroid hormones in blood serum of the Acipenseridae in the marine life period at different gonad states. J Ichthyol 40:197–202

Ceapa C, Williot P, Le Menn F, Davail-Cuisset B (2002) Plasma sex steroids and vitellogenin levels in stellate sturgeon (*Acipenser stellatus* Pallas) during spawning migration in the Danube River. J Appl Ichthyol 18:391–396

Charlon N, Williot P (1978) Elevage d'esturgeons de repeuplement et de consommation en URSS. Bull Cent Etud Rech Sci Biarritz 12(1):7–156

Chebanov MS, Chmyr YN (2005) Early sexing in live sturgeons by using ultrasound technique. A field guide. Federal Centre of Selection and Genetics for Aquaculture, Krasnodar, Russian Federation. Ministry of Agriculture of Russia, Krasnodar. 46 p

Conte F, Doroshov SI, Lutes PB, Strange EM (1988) Hatchery manual for the white sturgeon *Acipenser transmontanus* Richardson with application to other North American Acipenseridae. Cooperative Extension University of California, Division of Agriculture and Natural Resources, Oakland, CA, publication 3322, 104 p

Cuisset B, Pelissero C, Le Menn F, Nunez-Rodriguez J (1991) Elisa for Siberian sturgeon (*Acipenser baerii* Brandt) vitellogenin. In: Williot P (ed) Acipenser. Cemagref, Antony, pp 193–208

Cuisset B, Pradelles P, Kime DE, Kühn ER, Babin P, Davail S, Le Menn F (1994) Enzyme immunoassay for 11-ketotestosterone using acetylcholinesterase as label: application to the measurement of 11-ketotestosterone in plasma Siberian sturgeon. Comp Biochem Physiol 108C:229–241

Cuisset B, Fostier A, Williot P, Benneteau-Pelissero C, Le Menn F (1995) Occurrence and in vitro biosynthesis of 11-ketotestosterone in Siberian sturgeon, *Acipenser baeri* Brandt maturing females. Fish Physiol Biochem 14(4):313–322

Davail-Cuisset B, Lacomme S, Viaene E, Williot P, Lepage M, Gonthier P, Davail S, Rouault T (2008) Hormonal profile in adults of Atlantic European sturgeon, *Acipenser sturio*, adapted to hatchery in France. Cybium 32(2 suppl):169–170

Doroshov SI, Clark WH, Lutes PB, Swallow RL, Beer KE, McGuirre AB, Cochran MD (1983) Artificial propagation of the white sturgeon, *Acipenser transmontanus* Richardson. Aquaculture 32:93–104

Doroshov JN, Van Eenennaam JP, Chapman FA, Doroshov SI (1991) Histological study of the ovarian development in wild sturgeon, *Acipenser transmontanus*. In: Williot P (ed) Acipenser. Cemagref, Antony, pp 129–135

Doroshov SI, Moberg GP, Van Eenennaam JP (1997) Observations on the reproductive cycle of cultured white sturgeon, *Acipenser transmontanus*. Environ Biol Fish 48:265–278

Feist G, Van Eenennaam JP, Doroshov SI, Schreck CCB, Schneider RP, Fitzpatrick MS (2004) Early identification of sex in cultured white sturgeon, *Acipenser transmontanus*, using plasma steroid levels. Aquaculture 232:581–590

Flynn SR, Matsuoka M, Reith M, Martin–Robichaud DJ (2006) Gynogenesis and sex determination in shortnose sturgeon, *Acipenser brevirostrum* Lesueur. Aquaculture 253:721–727

Fuji K, Hirose K, Hara A, Shirashi M, Mayurama T (1991) Use of vitellogenin level as maturational indicator for artificial spawning of cultured hybrid sturgeon, *Huso huso* x *Acipenser ruthenus*. In: Williot P (ed) Acipenser. Cemagref, Antony, pp 381–388

Hett AK, Pitra C, Jenneckens I, Ludwig A (2005) Characterization of Sox9 in European Atlantic sturgeon (*Acipenser sturio*). J Hered 96(2):150–154

Hurvitz A, Jackson K, Degani G, Levavi-Sivan B (2007) Use of endoscopy for gender and ovarian stage determinations in Russian sturgeon (*Acipenser gueldenstaedtii*) grown in aquaculture. Aquaculture 270:158–166

Keyvannshokooh S, Gharaei A (2010) A review of sex determination and searches for sex-specific markers in sturgeon. Aquacult Res 41:e1–e7. doi:10.1111/j.1365-2109.2009.02463.x

Keyvannshokooh S, Kalbassi MR, Hosseinkhani S, Vaziri B (2009) Comparative proteomics analysis of male and female Persian sturgeon (*Acipenser persicus*) gonads. Anim Reprod Sci 111:361–368

Kynard B, Kieffer M (2002) Use of borescope to determine the sex and egg maturity stage of sturgeons and the effect of borescope use on reproductive structures. J Appl Ichthyol 18:505–508

Le Menn F, Pelissero C (1991) Histological and ultrastructural studies of oogenesis of the Siberian sturgeon *Acipenser baerii*. In: Williot P (ed) Acipenser. Cemagref, Antony, pp 113–127

Linares-Cazenave J, Kroll KJ, Van Eenennaam JP, Doroshov SI (2003) Effect of ovarian stage on plasma vitellogenin and calcium in cultured white sturgeon. Aquaculture 221:645–656

Magnin E (1962) Recherche sur la systématique et la biologie des Acipenséridés *Acipenser sturio* L., *Acipenser oxyrinchus* Mitchill et *Acipenser fulvescens* Raf. Annales de la Station Centrale d'Hydrobiologie Appliquée 9:7–242

Moghim M, Vajhi AR, Veshkini A, Masoudifard M (2002) Determination of sex maturity in *Acipenser stellatus* by using ultrasonography. J Appl Ichthyol 18:325–328

Mojazi Amiri B, Maebayashi M, Adachi S, Yamauchi K (1996a) Testicular development and serum sex steroid profiles during the annual sexual cycle of the male sturgeon hybrid, the bester. J Fish Biol 48:1039–1050

Mojazi Amiri B, Maebayashi M, Hara A, Adachi S, Yamauchi K (1996b) Ovarian development and serum sex steroid and vitellogenin profiles in the female cultured sturgeon hybrid, the bester. J Fish Biol 48:1164–1178

Omoto N, Maebayashi M, Adachi S, Arai K, Yamauchi K (2005) Sex ratios of triploids and gynogenetic diploids induced in the hybrid sturgeon, the bester (*Huso huso* female x *Acipenser ruthenus* male). Aquaculture 245:39–47

Pelissero C, Le Menn F (1988) Détermination des taux plasmatiques de stéroïdes sexuels et de la vitellogénine chez l'esturgeon sibérien *Acipenser baeri* élevé en pisciculture. CR Acad Sci Paris 307(Série III):749–754

Pelissero C, Le Menn F (1991) Evolution of sex steroid levels in males and first time maturing females of the Siberian sturgeon (*Acipenser baerii*) reared in a French fish farm. In: Williot P (ed) Acipenser. Cemagref, Antony, pp 87–97

Pelissero C, Cuisset B, Le Menn F (1989) The influence of sex steroids in commercial fish meals and fish diets on plasma concentration of estrogens and vitellogenin in cultured Siberian sturgeon *Acipenser baerii*. Aquat Living Resour 2:161–168

Van Eenennaam AL, Van Eenennaam JP, Medrano JF, Doroshov SI (1996) Evidence of female heterogametic genetic sex determination in white sturgeon. J Hered 90:231–233

Viayeh RM, Webb MAH, Hallajian A, Kazemi R, Yali MP (2006) Biochemical and morphometric parameters as indicators of sex and gonadal stages in maturing Persian sturgeon, *Acipenser persicus*. J Appl Ichthyol 22(suppl 1):364–368

Webb MAH, Feist G, Foster E, Shreck CB, Fitzpatrick MS (2002) Potential classification of sex and stage of gonadal maturity of wild white sturgeon using blood plasma indicators. Trans Am Fish Soc 131:132–142

Williot P (1997) Reproduction de l'esturgeon sibérien (*Acipenser baeri* Brandt) en élevage: gestion des génitrices, compétence à la maturation in vitro de follicules ovariens et caractéristiques plasmatiques durant l'induction de la ponte. PhD thesis n° 1822, Université Bordeaux I, France, 227 p

Williot P (2002) Reproduction des esturgeons. In: Billard R (ed) Esturgeons et caviar. Lavoisier Tech et Doc, Paris, pp 63–90

Williot P, Brun R (1998) Ovarian development and cycles in cultured Siberian sturgeon, *Acipenser baeri*. Aquat Living Resour 11(2):111–118

Williot P, Comte S, Le Menn F (2011) Stress indicators throughout the reproduction of farmed Siberian sturgeon *Acipenser baerii* (Brandt) females. Int Aquat Res 3:31–43

Williot P, Rouault T (2008) Basic management for spawning the cultured sturgeon, *Acipenser sturio* L., 1758, a critically endangered species. Cybium 32(2 suppl):334–335

Williot P, Rouault T, Brun R, Pelard M, Mercier D (2002) Status of caught wild spawners and propagation of the endangered sturgeon *Acipenser sturio* in France: a synthesis. Int Rev Hydrobiol 87:515–524

Williot P, Rouault T, Rochard E, Castelnaud G, Lepage M, Gonthier P, Elie P (2004) French attempts to protect and restore *Acipenser sturio* in the Gironde: status and perspectives, the research point of view. In: Gessner J, Ritterhoff J (eds) Species differentiation and population identification in the sturgeons *Acipenser sturio* L. and *Acipenser oxyrinchus*, vol 101. Bundesamt für Naturschutz, Bonn, pp 83–99

Williot P, Brun R, Rouault T, Pelard M, Mercier D, Ludwig A (2005) Artificial spawning in cultured sterlet sturgeon, *Acipenser ruthenus* L., with special emphasis on hermaphrodites. Aquaculture 246:263–273

Williot P, Rouault T, Pelard M, Mercier D, Lepage M, Davail-Cuisset B, Kirschbaum F, Ludwig A (2007) Building a broodstock of the critically endangered sturgeon *Acipenser sturio* L.: problems associated with the adaptation of wild-caught fish to hatchery conditions. Cybium 31:3–11

Williot P, Rouault T, Pelard M, Mercier D, Jacobs L (2009) Artificial reproduction of captive endangered European Atlantic sturgeon, *Acipenser sturio*. Endanger Species Res 6:251–257. doi:10.3354/esr00174

Wuertz S, Gaillard S, Barbisan F, Carle S, Congiu L, Forlani A, Aubert J, Kirschbaum F, Tosi E, Zane L, Grillasca JP (2006a) Extensive screening of sturgeon genomes by random screening techniques revealed no sex-specific marker. Aquaculture 258:685–688

Wuertz S, Nitsche A, Gessner J, Kirschbaum F, Kloas W (2006b) IGF-1 and its role in maturing gonads of female sterlet, *Acipenser ruthenus* Linnaeus, 1758. J Appl Ichthyol 22(suppl 1): 346–352

Yousefian M (2006) Sex differentiation by gonadogenesis and sex steroid hormones in cultured great sturgeon (*Huso huso*). J Appl Ichthyol 22(suppl 1):369–372

Chapter 27
Reproduction of Wild Brood Fish from Garonne Basin

Patrick Williot, Thierry Rouault, Rémi Brun, Marcel Pelard, and Daniel Mercier

Abstract The start of the restocking programme is described with reference to available knowledge and the difficulties faced (facilities, availability of wild-origin brood fish). Next, there is a summary of the number and quality of brood fish with respect to river (Garonne or Dordogne), gender (sex ratio), and the decline in by-catch over time following the total protection of the species. Changes in biometry over time are reported. The reproductive status of brood fish is analyzed with regard to the stage of the ovarian follicles and the sperm quality. The oosomatic index data are re-analysed. Hormonal stimulation treatment details are provided over time, with special reference to those used repeatedly to overcome the current non-simultaneity of both genders. Limited data on mating, embryogenesis, and larvae are provided.

27.1 Introduction

As outlined earlier (see Chap. 20), one of the three main aspects of the conservation strategy focuses on the release of juveniles to enhance the population, since its status was considered critical with regard to its ability to recover through natural reproduction. The initial attempt relied exclusively on incidental catches and subsequent reproduction of mature fish of wild origin, but in order to achieve this, holding facilities for the fish were required and the status of the fish had to be checked. These data, along with the results of the reproduction trials are provided and, where possible, compared with historic data.

P. Williot (✉)
Sturgeon Consultant, 4 Rue du Pas de madame, 33980 Audenge, France
e-mail: williot.patrick@neuf.fr

T. Rouault • R. Brun • M. Pelard • D. Mercier
Cemagref, Estuarine Ecosystems and Diadromous Fish Research Unit, Moulin de logerie, 33660 Saint-Seurin-sur-l'Isle, France

P. Williot et al. (eds.), *Biology and Conservation of the European Sturgeon Acipenser sturio L. 1758*, DOI 10.1007/978-3-642-20611-5_27,

27.2 The Context

To better understand the situation in which the work was carried out, three points should be recalled, especially with regard to the difficulties that were encountered. In the late 1970s and early 1980s, knowledge of sturgeon farming was limited almost exclusively to the former USSR. The author was fortunate to be able to spend 10 weeks in the USSR working with scientists and operators to acquire a minimum of experience and knowledge in the field of sturgeon reproduction (Charlon and Williot 1978). This time was spent almost entirely in sturgeon farms devoted to the production of fingerlings for release (Williot 1984). These farms relied exclusively on the capture of wild brood fish for reproduction and these fish were obtained from the river or estuary or delta during the onset of their reproductive migration upstream. The final maturation of the brood fish was stimulated with sturgeon pituitary suspension. It was claimed that pituitary extract from other fish species could not be used. In addition, no permission was provided to supply markets outside the former USSR with sturgeon pituitary. It is also worth remembering that all sturgeon-caviar related issues were considered a state secret until the 1990s in that country.

The second point is a consequence of the protection status of *A. sturio* in France since the early 1980s (see Chap. 20). Any capture and further marketing of the species were therefore forbidden. Despite this status, the request to fishermen to hand over incidental captures of brood fish would not have been successful without financial compensation. This was done via a non-profit society (Agedra[1]) which obtained yearly grants from the regional council to support the *A. sturio* restoration programme. The call for brood fish was renewed every year via the local newspapers.

Thirdly, at the very onset of the programme, no experimental hatchery facility was available to carry out the maintenance of the brood stock fish of the model species (*Acipenser baerii*). Thanks to volunteer support from different quarters, the *A. baerii* brood fish were maintained in earthen ponds, and for the first 2 years (1981 and 1982) were transported at the time of reproduction to a private hatchery several hundred km north of Bordeaux. A similar procedure was applied to *A. sturio* brood fish with regard to the hatchery process. During the following years, a small hatchery was built alongside an experimental trout farm (managed by another research institute), and some ponds were made available to us. This facility did have some severe limitations, however; in water flow, in the number of available ponds, and in the opportunity to manage our sturgeon-related activities separately. A new independent facility (owned by Cemagref) was built in the early 1990s, thanks to a know-how transfer of our knowledge of breeding the model species, the Siberian sturgeon, *A. baerii* Brandt (see Chap. 20).

[1]Agedra = Association Girondine pour le Développement des Ressources Aquatiques.

27.3 Material and Methods

At the beginning of the period of upstream migration for reproduction, a truck equipped with a tank (2 × 0.7 × 0.7; $L \times l \times h$ m) with an oxygen diffuser was held on standby in case of a by-catch of a mature sturgeon in either the Garonne or Dordogne rivers. Usually, the fishermen kept the fish in the water by a rope through the mouth and gills. A certificate was handed over to the fisherman for supplying the fish and he was paid based on a price per kg for the meat plus a 10% surcharge as an equivalent for caviar in females.

Fish were stripped on site; this allowed gender determination to be carried out in case of spermiating males. Upon arrival in the hatchery, the fish were weighed and checked for sexual maturation status by gonad biopsy. Brood fish were placed upside down on a V-shaped table, and received a continuous water flow in the mouth post-anaesthesia. Testes were graded according to texture, the softest being the most advanced. Ovarian follicles were checked for polarization index (PI) (Kazanskij et al. 1978). Later, a biotest was performed to assess the in vitro maturation competence (IVMC) of the ovarian follicles. The principle is to incubate ovarian follicles in standardized conditions in a medium to which progesterone has been added (or not added in the case of a control), and to see whether the envelope surrounding the nucleus, called germinal vesicle (GV), is present, or not present in cases of germinal vesicle breakdown (GVBD). The closer the GVBD to 100%, the better the ability of the ovarian follicles to mature, and the more likely the females are to react positively to a hormonal stimulation by producing good quality gametes. The details of the PI and IVMC methods are given in Chap. 32.

Hormonal stimulation was soon changed from sturgeon pituitary to carp pituitary due to its availability in western countries at that time. The dosage applied was 2 and 5 mg/kg body weight for males and females respectively. In cases of a two-injection procedure, priming was 10% of the total dose, and the remaining dose was applied a minimum of 12 h later (Williot et al. 2000).

Eggs were obtained by placing the female on the V-shape table with the continuous water supply. First, ovulated eggs were collected in bowls via the genital pore and later through a small abdominal incision 4 cm anterior of the genital opening. The incision was closed by cross stitches.

In males, the genital opening was carefully dried to eliminate any potential contact of semen with water or urine. Semen was collected by gentle abdominal stripping through a small polypropylene tube carefully inserted into the genital opening and running into a beaker.

Fertilization was performed according to current Russian procedure (10 ml of semen per kg eggs and dilution of semen in water in a 1:200 ratio (Dettlaff et al. 1993). De-adhesive treatment was performed for 1 h in a watery clay suspension at a rate of 280 g clay l^{-1} (Williot 2002).

27.4 Results and Discussion

27.4.1 Captures: Number, Sex Ratio, Biometry

To carry out the restocking strategy using wild-originated brood fish for controlled reproduction and rearing fingerlings for release (see Chap. 20), sufficient fish of both sexes were required. With the support of professional fishermen and water bailiffs, we were able to determine the number of brood fish captured in both the River Garonne and the River Dordogne according to sex in a 6-year period 1975–1980, i.e., prior to the fisheries ban (Table 27.1). The average number of captures was around nine fish per year, with four females, while no significant differences were determined in the occurrence of male and females.

Between 1981 and 2001, the number of captures declined dramatically, becoming severely reduced from the early 1990s onwards (Fig. 27.1). As a result, the

Table 27.1 Declared captures in the Garonne and Dordogne Rivers throughout a 6-year period according to sex (after Trouvery et al. 1984)

Year	Female	Male	Total
1975	8	12	20
1976	2	6	11
1977	3	6	9
1978	4	9	13
1979	1	3	4
1980	5	12	17
Total	23[a]	33[a]	56 (mean ~ 9)

Where letters in a given column or row are the same, this denotes the absence of significant difference ($P < 0.05$)
[a]z-test ($z = 1.05$; $P = 0.29$)

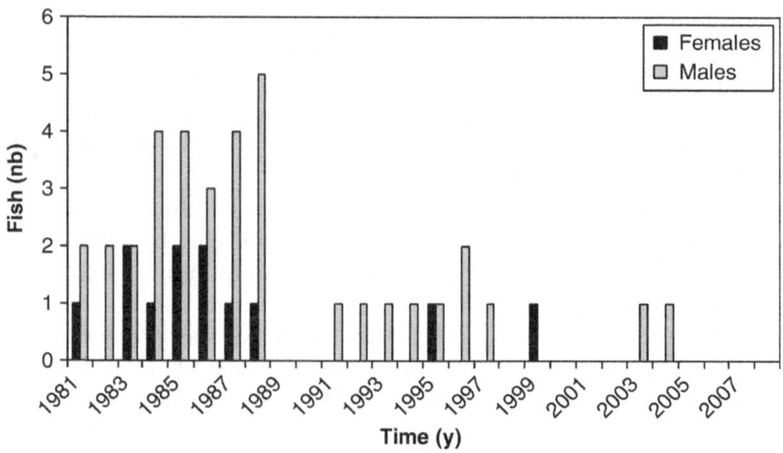

Fig. 27.1 Distribution of by-catch of wild-originated *A. sturio* brood fish over time in the period (1981–2007) (updated from Williot et al. 2002)

Table 27.2 Distribution of captures of wild *A. sturio* spawners according to river basin (1981–2001)

River	Female	Male	Total
Gironde (estuary)	2	6	8
Garonne	5	16	21[a]
Dordogne	4	8	12[a]
Total	11[b]	30[b]	41 (yearly mean ~ 2)

Where letters in a given column or row are the same, this denotes the absence of significant difference ($P < 0.05$) (completed from Williot et al. 2002)
[a]z-test ($z = 1.19$; $P = 0.23$)
[b]z-test ($z = 2.31$; $P = 0.02$)

yearly number of incidental wild-originated brood fish declined to a mean close to 2 (vs 9 for the preceding period) (Tables 27.1 and 27.2). There were no significant differences in the number of captures between the two rivers, Garonne and Dordogne (Table 27.2). However, captures of females became significantly lower than those of males. This is the opposite of the situation observed by Magnin (1962) who reported almost three times more females (76%) than males (24%) over a 3-year period, 1958, 1959, and 1961; and the same was the case in the River Guadalquivir in Spain during the period 1934–1942 (Classen 1944; Elvira et al. 1991). This means that the sex ratio may change greatly from period to period. Is this a phenomenon resulting from a more selective use of fishing gear, as suggested by Classen (1944)? A more convincing explanation is the shorter sexual period of males which could explain the relatively lower number of males (Magnin 1962). Preliminary results for the frequency of maturation in farmed males give some support to this explanation, as they show high maturation frequency over 2 successive years followed by a decline (see Chap. 32).

Throughout the experimental period (1981–2001), the females weighed on average 20 kg more than males (mean weight 43.5 and 23.8 kg respectively). Also, females were 30 cm longer than males (195.5 and 163.3 cm respectively) (Table 27.3). The mean length of mature females has decreased compared with that previously determined (204 cm), but remains similar for males (162 cm) compared to Magnin's data (1962). It is worth noting that for a given characteristic, variation coefficients were similar for both sexes, which means that the relative variability in weight and length are independent of gender (Table 27.3). With the exception of the female which spawned in 1995 and was most likely born in 1971, other large fish were probably born between 1981 and 1988 (Williot et al. 2007).

Surprisingly, there were no changes in weight throughout the experimental period 1981–2001 either for females (Fig. 27.2; $R^2 = 0.04$ for linear regression) or for males (Fig. 27.3; $R^2 = 0.22$ for linear regression). This might suggest that the population, namely the brood fish, belonged to a very limited number of year-classes, probably only one.

Weight and length (TL) of the fish caught during this period are plotted in Fig. 27.4. Log allometric relationships between weight (W) and length (TL) are: $\log W = -2.82 + 1.954 \log \mathrm{TL}$ ($R^2 = 0.74$), and $\log W = -5.73 + 3.21 \log \mathrm{TL}$ ($R^2 = 0.78$) for females and males respectively. The slopes are significantly

Table 27.3 Weight (*W* in kg) and length (TL in cm) of wild-caught *A. sturio* brood fish between 1981 and 2001 (after Williot et al. 2002)

		Female	Male
Weight (*W*, kg)	*n*	12	34
	W (mean ± sd)	43.4 ± 9.7	23.8 ± 6.7
	cv[a]	0.22	0.28
Length (TL, cm)	*n*	10	30
	TL (mean ± sd)	195.5 ± 18	163.3 ± 11.7
	cv	0.09	0.07

[a]cv = SD/mean

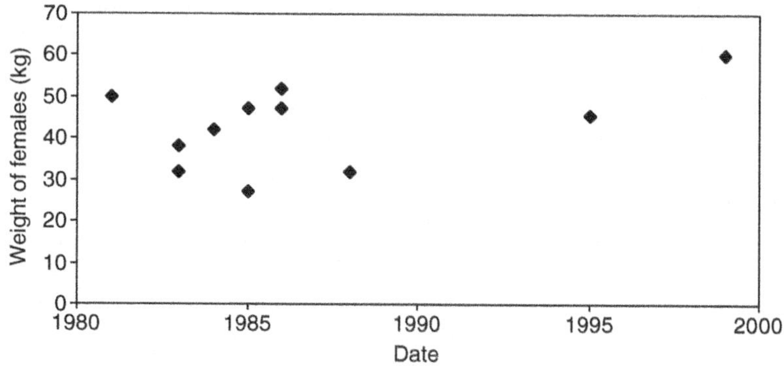

Fig. 27.2 Changes over time in the weight of wild-originated *A. sturio* females by-caught in the period 1981–2000

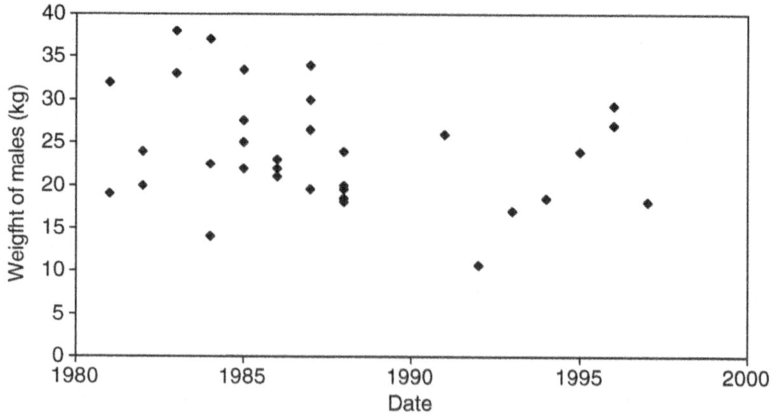

Fig. 27.3 Changes over time in the weight of wild-originated *A. sturio* males by-caught in the period 1981–2000

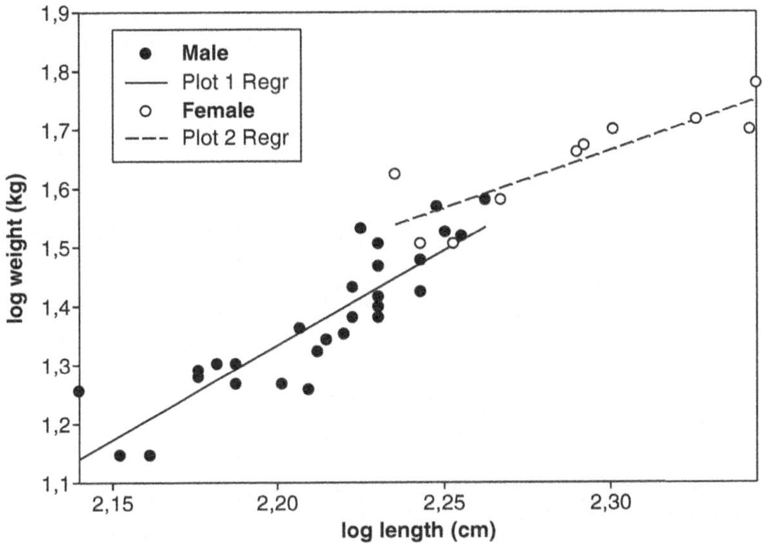

Fig. 27.4 Log allometric relationship between weight (kg) and total length (cm) according to sex in wild-originated *A. sturio* brood fish (period 1981–2001) (after Williot et al. 2007)

different ($P = 0.024$). This means that weight increased slightly more rapidly than length for females, and less rapidly for males; thus, the males were proportionally more elongated than the females.

27.4.2 Reproduction

An indirect indicator of the reproductive potential of the females is the oosomatic index (OSI). First records on OSI were obtained by Magnin (1962) and plotted (Fig. 27.5). The absence of data in the 12–32 kg range is associated with the minimum size for mature females, weighing 32–35 kg and measuring 180–189 cm (TL) (Magnin 1962). A large variability is observed in OSI, with no relationship to weight being revealed. This is supported by the absence of linear correlation ($R^2 = 0.12$) between the two variables when the few juvenile fish with an OSI in the range 0–0.01 are taken out of the sample. The mean value of OSI is about 0.15. Similar values have been reported in the Spanish *A. sturio* population from the River Guadalquivir with 0.19 in 1932 and 0.15 in 1954 (Classen 1944; Elvira et al. 1991).

The characteristics of the females' ovarian follicles upon arrival are given in Table 27.4. The diameter of the ovarian follicles ranged from 2.4 to 3.4 mm, and there was no relationship with weight ($P < 0.14$). Oocyte maturation, determined by polarization index (PI), was fairly constant over the years. PI was assessed at between 6 and 10%, but its capacity to show in vitro maturation competence

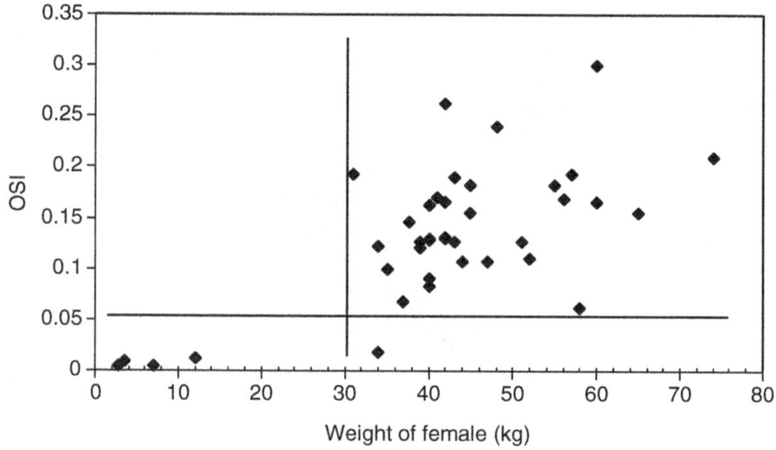

Fig. 27.5 Variation in the oosomatic index (OSI) according to the weight of *A. sturio* females (drawn from data in Magnin 1962)

Table 27.4 Morpho-physiological characteristics of the ovarian follicles in wild *A. sturio* spawners (1981–2001) on arrival in the hatchery (after Williot et al. 2002)

Year	Diameter (mm)	Polarization index (%)	GVBD (%)
1981	Small size, apparent germinal vesicle, brittle		
1984	2.6	nr	nr
1985	2.4	8	100
	2.5	7	100
1986	2.6	8	0
1987	3.1	7	9
1995	2.9	10	0
1999	3.4	6	100

nr not recorded

(GVBD) varied greatly (from 0 to 100%) without any intermediate response. Three out of the six females tested exhibited a maximum rate of GVBD upon arrival. Of these, the 1999-fish died before recovery, which means that some function remained unaffected (in vitro maturation competence) probably 6–12 h post mortem. Only the two females caught in 1985 exhibited complete GVBD upon arrival. In the female caught in 1995, there was an absence of any maturation potential. Despite the irregular darkish pigmentation of the nucleus – possibly correlated with the first steps of over-maturation as observed in Siberian sturgeon (Williot 1998) – sequential hormonal treatment was applied to the females caught in 1986, 1987, and 1995 to induce maturation. Only the 1995-fish responded to this treatment, and so the regular priming–resolving injections were applied (Table 27.5).

As a result of the above mentioned experience, it was concluded that after 7–10 days of maintaining fish in the hatchery, the reproductive status of the wild females would not improve. This means that only very few females could be used for induced reproduction.

Table 27.5 Protocols for inducing final maturation in wild *A. sturio* males and females under hatchery conditions in 1981–2001

Year	Sex	Holding time from arrival (days)	Hormonal treatment and dose (mg kg^{-1})	Result[a]
1981	F1	4	sph (0.4) + 12 h cph (6)	Poor
	M1	9	sph (2)	Average
	M2	4	cph (6)	Average
1984	F2	7	cph (6)	Poor
1985	F3	4	cph (5)	Fair
	F4	3	cph (5)	Fair
		6	GnRHa (0.05)	Average
	M3	26	cph (0.5) + 48 h cph (2)	Poor
		11	cph (2)	Zero
		16	cph (0.5) + 48 h cph (2)	Average
	M4	29	cph (0.5) + 48 h cph (2)	Median
1995	F5	13	cph (0.5) + 24 h cph (5)	Poor
	M5	6	cph (2)	Average

sph and cph are sturgeon and common carp pituitary homogenates respectively (after Williot et al. 2009)

[a]Global appreciation of gamete quality evaluated by quantity and quality using four-rank scale: zero, poor, average and fair

In males, the low quality of semen (based on spermatozoa motility) was the main drawback for utilization. The proportion of low-quality males started from 60 to 66% at the beginning of the fishing period, and decreased to 20–40%. In addition to this seasonal pattern, a decreasing trend in the quality of males was observed throughout the period (see Chap. 20).

Apart from the very small number of brood fish, the small proportion of appropriate reproductive status in females, and the increasingly poor-quality semen, the main obstacle to controlled reproduction was the lack of simultaneous availability of both sexes. Thus the available males had to be used repeatedly, and so different hormonal treatment sequences were applied during the captive rearing period (Table 27.5). A two-injection procedure with a 2-day interval was applied, leading to repeated maturation and spermiation over 16–29 days of rearing. Even without the possibility of controlling water temperature, average-quality semen (male no. 4 in 1985) was obtained. In 1995, a successful two-injection sequence was applied (Williot et al. 2000).

As a result of the constraints described above, very few attempts at reproduction have been successful during the period (Table 27.6), one in 1981, two in 1985, and one in 1995. Nor was it the use of carp pituitary extract that might be responsible for these rather poor results, as other sturgeon species have been bred successfully with this hormonal treatment, e.g., the Siberian sturgeon, *Acipenser baerii* (Williot and Rouault 1982), and the white sturgeon, *Acipenser transmontanus* (Doroshov et al. 1983).

The number of larvae produced depended largely on the quality of the eggs. The large number of larvae obtained in 1985 from two females was the result of the optimal reproductive status of these two fish upon arrival.

Table 27.6 Successful spawning attempts in *A. sturio* between 1981 and 2001, with the mating and number of larvae produced (after Williot et al. 2002)

Year	Mating	Larvae produced (10^3)
1981	♀1 × 2♂[a]	0.7
1985	♀1 × ♂1	220
	♀2 × ♂2	70
1995	♀1 × ♂1	23

[a]Fertilized by pooled sperm from two males

Table 27.7 Embryogenesis of fertilized eggs in 1995, incubated at 17 ± 0.5°C (after Williot et al. 2002)

Stage name (stage number)[a]	4th cleavage to early blastula (7–11)	Fusion of lateral plates (25–27)	C-shaped heart and elongated tail (29–30)	Onset of hatching (35)
Time post-fertilization (h)	7–9.30	57	71	98

[a]According to Dettlaff et al. (1993)

Until the early 1990s, wild brood fish were tagged and released into the upper part of the Gironde estuary. Later, fish were kept for brood stock building (see Chap. 31).

The incubation time at 17 ± 0.5°C for fertilized eggs until hatch varied from 100 to 120 h (beginning to end of hatching). Daily malachite green treatments (5 ppm for 20 min) were applied in 1995 (though this treatment is no longer officially allowed in France) to limit the development of *Saprolegnia* sp. For this temperature, the following embryonic stages were observed: stages 7–11 from 07 to 09.30 h post-fertilization, stages 25–27 57 h post-fertilization, stages 29–30 71 h post-fertilization and onset of hatching 98 h post-fertilization (Table 27.7).

On two occasions, 1985 and 1995, some abnormal larvae were observed. Despite the fact that hatchlings were not photosensitive from hatching to the age of 2 days, it was impossible to use this behaviour to separate normal larvae from abnormal ones, as they were unable to move. As a result, the sorting of larvae was difficult.

27.5 Conclusions

We must remember that when this activity was begun, the context was very different from the present day. The state of knowledge was very limited and sometimes obsolete, no experimental facility was available to carry out reproduction, and finally, despite the protected status of the species, the only way to obtain brood fish captured as by-catch was by financial compensation. The number of brood fish captures prior to and after the ban is analyzed according to river basin (Garonne and Dordogne), and gender. The sex ratio changed dramatically according to the period. The log allometric relationship between weight and total length for both genders during the last period (1981–2001) showed a significant difference between genders, the males being more elongated than the females. During the last period (1981–2001), females were slightly shorter (195 vs 204 cm) than during the preceding period; no changes were observed in the males with data of 163 and 162 cm. The number of brood fish in the by-catch further decreased after the species became protected. After 1990, this decline continued up to the present time. As a result, very few brood fish have been available for reproduction. Moreover, their reproductive status upon arrival at the experimental facility was rarely optimum, as judged from observations of the ovarian follicles (migration of the nucleus and in vitro maturation competence test) and semen (density, motility). Carp pituitary extracts enabled us to obtain gametes fairly quickly, and even repeatedly from the males, and this in turn enabled us to overcome the absence of simultaneity in the maturing of both genders. Few data on embryogenesis and larvae are given. Analysis of past data on the oosomatic index shows that it has no relationship with the weight of females.

Acknowledgements Thanks are due to Igor Burtsev and Tina Umanskaya for initiating PW into work with sturgeon in the former USSR, to the former owners of Le Blanc (Indre, France) hatchery and its manager, Peter Szabo, to Jean Duret, former President of both Bordeaux County Fish Anglers Association and Agedra society (also a non-profit society), to P. Luquet and B. Breton from INRA, for their constant support and help, and to O. Rooryck for his earlier assistance.

References

Charlon N, Williot P (1978) Elevage d'esturgeons de repeuplement et de consommation en URSS. Bull Cent Etud Rech Sci Biarritz 12(1):7–156

Classen TEA (1944) Estudio bio-estadistico del esturion o sollo del Guadalquivir. Instituto Espanol de Oceanografia, Ministerio de Marina, Trabajos, N° 19, 112 p + XVII planches

Dettlaff TA, Ginsburg AS, Schmalhausen OI (1993) Sturgeon fishes. Developmental biology and aquaculture. Springer, Berlin, 300 p

Doroshov SI, Clark WH, Lutes PB, Swallow RL, Beer KE, McGuire AB, Cochran MD (1983) Artificial propagation of the white sturgeon, *Acipenser transmontanus* Richardson. Aquacult 32:93–104

Elvira B, Almodovar A, Lobon-Cervia J (1991) Sturgeon (*Acipenser sturio* L., 1758) in Spain. The population of the River Guadalquivir: a case history and a claim for a restoration programme. In: Williot P (ed) Acipenser. Cemagref, Antony, pp 337–347

Kazanskij BN, Feklov YA, Poduska SB, Moloddtsov AN (1978) Express method for assessment of sexual maturity in sturgeon close to spawning. Rybnoe Khozaistvo 2:24–27 (in Russian)

Magnin E (1962) Recherche sur la systématique et la biologie des Acipensérides *Acipenser sturio* L., *Acipenser oxyrinchus* Mitchill et *Acipenser fulvescens* Raf. Ann Stat Cent Hydrobiol Appl 9:7–242

Trouvery M, Williot P, Castelnaud G (1984) Biologie et Ecologie d'*Acipenser sturio*. Etude de la pêcherie. Cemagref, Etude n 17, Série Esturgeon n 1, 79 p

Williot P (1984) L'expérience soviétique en matière d'exploitation des stocks d'esturgeons en mer d'Azov et mer Caspienne. Cemagref, Etude n 20, Série Esturgeon n 3, 50 p

Williot P (1998) Influence of yolk-blackish pigmentation of Siberian sturgeon on reproductive performance and larval survival. Aquacult Int 6:403–410

Williot P (2002) Reproduction des esturgeons. In: Billard R (ed) Esturgeons et caviar. Lavoisier Tech et Doc, Paris, pp 63–90

Williot P, Rouault T (1982) Compte rendu d'une première reproduction en France de l'esturgeon sibérien *Acipenser baeri*. Bul Franç Piscic 286:255–261

Williot P, Brun R, Pelard M, Mercier D (2000) Unusual induced maturation and spawning in an incidentally caught pair of adults of the critically endangered European sturgeon *Acipenser sturio* L. J Appl Ichthyol 16(6):279–281

Williot P, Rouault T, Brun R, Pelard M, Mercier D (2002) Status of caught wild spawners and propagation of the endangered sturgeon *Acipenser sturio* in France: a synthesis. Int Rev Hydrobiol 87:515–524

Williot P, Rouault T, Pelard M, Mercier D, Lepage M, Davail-Cuisset B, Kirschbaum F, Ludwig A (2007) Building a broodstock of the critically endangered sturgeon *Acipenser sturio* L.: problems associated with the adaptation of wild-caught fish to hatchery conditions. Cybium 31:3–11

Williot P, Rochard E, Rouault T, Kirschbaum F (2009) *Acipenser sturio* recovery research actions in France. In: Carmona R, Domezain A, Garcia Gallego M, Hernando-Casal J, Rodriguez F, Ruiz Rejon M (eds) Biology, conservation and sustainable development of sturgeons, Fish & fisheries series. Springer, New York, pp 247–263, 467 p

Chapter 28
Preliminary Results on Larval Rearing the European Sturgeon, *Acipenser sturio*

P. Williot, T. Rouault, R. Brun, M. Pelard, and D. Mercier

Abstract This chapter reports on the first successful larval rearing of *A. sturio* carried out in 1995. Four food items (two live prey and two compound diets) and water temperature were tested as first feeding, but with the constraint of saving as many fish as possible. As *Artemia salina* nauplii were first observed in the digestive tract, nauplii were added to all other batches. Nauplii alone provided the best growth until 23 days-post-hatch (dph). Water temperature of 21°C versus 17.5°C increased growth and mortality until 37 dph. Best survival post short weaning (3 days) at 17.5°C, depending on age at weaning initiation, was for 55 dph fish. For 50 dph fish, long weaning (3 weeks) provided better growth and survival than short weaning (3 days). Frozen chironomids proved to be impact-free on fish between 0.6 g and 2.9 g. Some conclusions are drawn for further improvements in larval rearing experiments.

28.1 Introduction

Very few successful reproductions were carried out during the 1980s and 1990s; thus, there were very few attempts at larval rearing, and indeed the first two, in 1981 and 1985, failed (Williot et al. 1997, 2004; see Chap. 27). Moreover, there was no published study on *A. sturio*'s sister-species *Acipenser oxyrinchus*. This means that at the time when some thousands of larvae hatched in 1995 (Williot et al. 2000), the experiment was limited to this species, with no published data on closely related species being available. However, it was known that live prey were most probably a

P. Williot (✉)
Sturgeon Consultant, 4 Rue du pas de madame, 33980 Audenge, France
e-mail: williot.patrick@neuf.fr

T. Rouault • R. Brun • M. Pelard • D. Mercier
Cemagref, Estuarine Ecosystems and Diadromous Fish Research Unit, Moulin de Logerie, 33660 Saint-Seurin-sur-l'Isle, France

P. Williot et al. (eds.), *Biology and Conservation of the European Sturgeon Acipenser sturio L. 1758*, DOI 10.1007/978-3-642-20611-5_28,

mandatory feeding practice, and that weaning on compound diets was also mandatory in the long term. Indeed, using live prey has a lot of drawbacks: supplying on a regular basis in terms of both quantity and quality, constraints in storage, production in the case of self-supplying.

Some apparently contradictory objectives have therefore been taken into account: (1) to determine the best rearing conditions with regard to growth and survival, (2) to determine the best feeding sequence including weaning, and (3) to save as many fish as possible for further releasing and brood stock building (Williot et al. 2005).

Most of the present section is a synthesis of already published data, reorganised and/or redrawn from Williot et al. (2005). Despite recent improvements in weaning and survival (see Chap. 33), some of the preliminary findings are worth noting, as they provided either some foundations or originality in the field.

28.2 Material and Methods

28.2.1 Fish and Rearing Conditions

Twenty-three thousand larvae were obtained from the reproduction of a pair of wild brood fish. The low number of progeny was due to the poor physiological reproductive status of the dam. As a precaution, first feeding was initiated at 7 dph with larvae weighing 23 mg. The larvae were randomly distributed in a grey fibreglass trough (1.00 × 0.5 × 0.15 m, L × W × H) supplied with running, degassed and re-aerated well water at 17.5 ± 0.2°C. Some trials were carried out in parallel at 21 ± 0.4°C during experiment 1 (Table 28.1). Flow rates were adjusted to maintain a minimum oxygen outlet level at around 7 mg l^{-1}.

Four different food items were tested at first feeding. Live prey consisted of either 1-day old *Artemia salina* nauplii or chironomids. Two compound diets were offered that had already proved their effectiveness on *A.baerii* and *A.ruthenus*. These were lanzi R1 (80–200 μm) (INVE™) with the following composition: 58% protein, 14% lipid, 12% ash, 7% moisture with vitamins A, D3, E and C added, and an INRA formula (80–120 μm) made by NACIP Inc., Montpellier, France with the following composition: 50% protein, 10% lipid, 6% ash, 4% moisture and with vitamins A, D3, E and C added. Refusals and faeces were removed twice a day. Dead larvae were removed and counted daily. 30 larvae were taken weekly for average live weight determination. A monthly bacterial survey was carried out.

28.2.2 Design

Four experiments were organised, a synthesis of which is given in Table 28.1.

Table 28.1 Main lines of the four experiments on larval feeding of *A. sturio* larvae in 1995

		Rearing conditions				
Experiment	Objective	Temperature (°C)	Duration (days)	Fish numbers per trough (mean weight at start)	Replicate	Age of fish at start (dph)
Experiment 1	First food feeding	17.5 and 21	30	1,500 (23 mg)	2	7
Experiment 2	Weaning on compound diets:					
	– Age at weaning	17.5	8 weeks (new trial each week)	200 (first 6) and 100 (last 2) (51 to 1,520 mg)	No	22–71
	– Length of weaning (short and long)	17.5	56	200 (460 and 639 mg)	No	50
Experiment 3	Influence of enrichment of *A. salina* nauplii	17.5	20	100 (135 and 148 mg)	No	49
Experiment 4	Fresh and frozen chironomids	17.5	40	100 (660 mg)	No	51

See text for more details on rearing conditions and management (after Williot et al. 2005)

Experiment 1 compared the four food items at two temperatures, with the exception of nauplii which were used only at 17.5°C. Nauplii were distributed four times a day at a daily rate of 40 g/1,500 larvae. Chironomids were distributed six times a day at a daily rate of 30 g/1,500 larvae. Compound diets were distributed every 2 h during working hours (08.00–18.00) and offered ad libitum (~10–15% body weight per day). The melanin plug corresponds "to the accumulation of granules of embryonic development in the digestive tube lumen" (Dettlaff et al. 1993). Depending on the sturgeon species, expulsion of the melanin plug either precedes first feeding [*Acipenser transmontanus* (Monaco et al. 1981); *Acipenser gueldenstaedti* (Dettlaff et al. 1993)] or follows it (*Acipenser baerii*, Gisbert and Williot 1997). Fish used in the other experiments came from experiment 1 – batches fed live food.

Experiment 2 tested the influence of age at weaning and the length of weaning. Each week (for 8 weeks) a new batch entered the experiment. After completion of weaning, larvae were offered a mix of equal quantities of the compound diets. Short weaning was performed in 3 days and long weaning in 3 weeks using 50 dph larvae. In both weaning procedures, there was a progressive switch to compound diet.

Experiment 3 was a 3-week experiment to test the effect of enriching nauplii for 1 day in the presence of a Super Selco (INVE™), with the enriched nauplii distributed at a daily rate of 80 g/trough. Mean weight of fish aged 49 dph at start was not significantly different.

Experiment 4 compared the growth and survival of 51 dph larvae weighing 650 mg on average, in a 40-day experiment with either fresh or frozen chironomids.

28.2.3 Statistics

Two-way ANOVA was used to analyse the effects of food items and water temperature on weight in experiment 1, and one-way ANOVA analysed the effects of food items on weight for a given temperature, with the Holm–Sidak method to estimate all pair-wise comparisons when needed. One-way ANOVA (Kruskal–Wallis, KW) and a further Dunn's test compared weight at 20 days and 34 days post initiation of weaning with that at initiation of weaning. Chi-square and *z*-test compared series of percentages and two proportions respectively. Retained significance level was $p < 0.05$. Normality was tested by computing the Kolmogorov–Smirnov test. Depending on normality, either the *t*-test or the U-test was used to compare two sets of weights. Results were either mean ± SD (normally-distributed data) or median (25–75%) percentiles.

28.3 Results

28.3.1 Experiment 1

The first excretion of the melanin plug occurred at 12 dph. *Artemia salina* nauplii were observed in the digestive tract for the first time at 16 dph, at a mean weight of 31 mg. In order to save as many fish as possible of that age, all four batches received nauplii in addition to their corresponding food item. Therefore, experiment 1 became a co-feeding experiment, *Artemia salina* nauplii being the standard. First faeces were observed at 19 dph. At 16 and 23 dph, weights recorded in batches reared at 17.5°C and fed nauplii were significantly higher than any other batch (not visible Fig. 28.1). Apart from the above-mentioned difference, there were no other differences between food items at 17.5°C (Fig. 28.1). At 30 dph, growth was better for batches reared at 21°C compared with 17.5°C (Fig. 28.2a, b, c). The absence of further differences is probably due to a wide range of variation in weight in all batches, as the variation coefficient (sd/mean) was sometimes close to 50%.

Cumulative mortality, at around 50% at 21°C, was rapidly higher than at 17.5°C, whatever the food item (Fig. 28.3a, b, c). At the end of the experiment, i.e., 37 dph, cumulated mortalities reached about 30% at 17.5°C and about 48% at 21°C. The worst result was recorded in the lanzi compound diet-fed batch (Fig. 28.4). Thus, despite the co-feeding that was applied, some differences did emerge.

28.3.2 Experiment 2

Growth of short-weaned fish of different ages (22–71 dph, Table 28.1) at the initiation of weaning was recorded at 20 and 34 days post initiation of weaning at

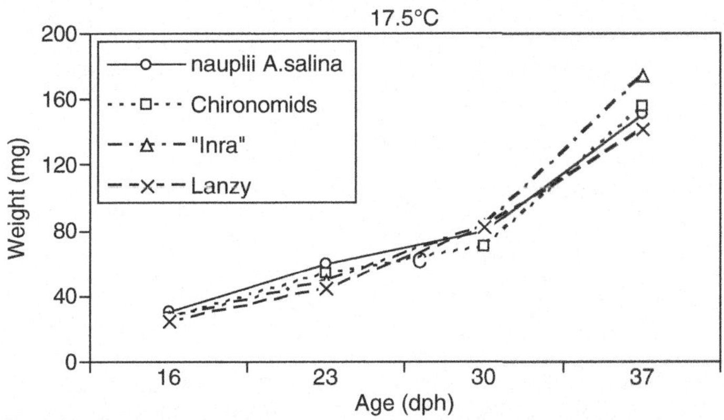

Fig. 28.1 Weight of *A. sturio* first-fed different food items at 17.5°C. Due to species status, from 16 dph, all batches were partially nauplii-fed. Batches fed only nauplii exhibited significantly better growth at 16 and 23 dph, not visible here (see text for more details) (drawn from Williot et al. 2005)

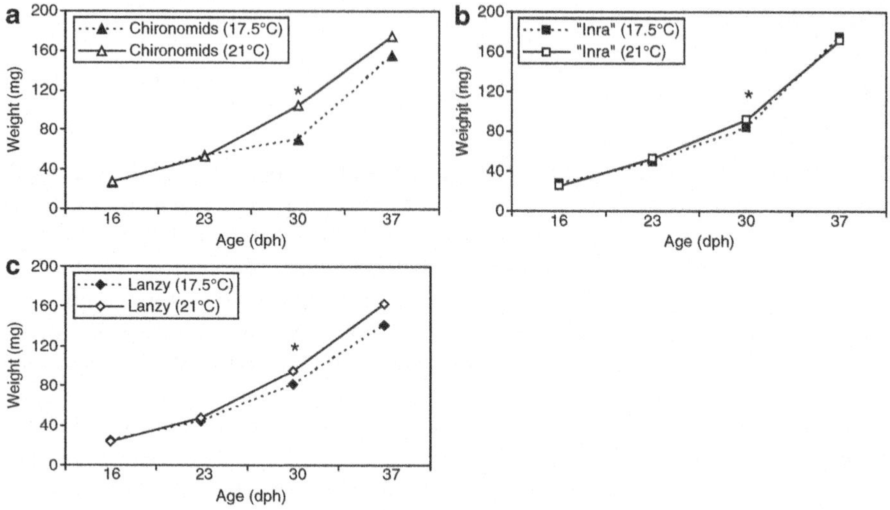

Fig. 28.2 Growth of first-fed *A. sturio* larvae depending on water temperature and food items: **a** live chironomids, **b** "INRA" compound diet, and **c** lanzi compound diet. *Asterisk* denotes a significant difference at $p = 0.05$. See text for more details (drawn from Williot et al. 2005)

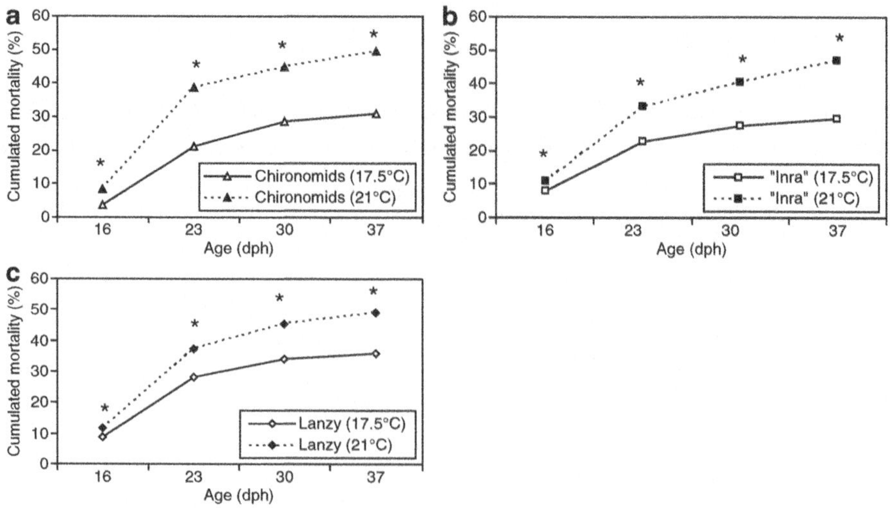

Fig. 28.3 Cumulative mortality (%) of *A. sturio* fed different food items at two water temperatures, 17.5°C and 21°C. Due to species status, from 16 dph all batches were partially nauplii-fed. *Asterisk* denotes a significant difference at $p = 0.05$. See text for more details (drawn from Williot et al. 2005)

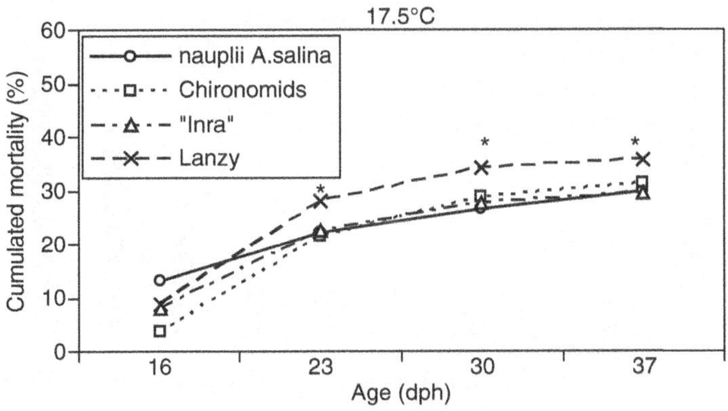

Fig. 28.4 Cumulative mortality (%) of *A. sturio* larvae fed different food items at 17.5°C. Due to species status, from 16 dph all batches were partially nauplii-fed. *Asterisk* denotes a significant difference of lanzi compound diet versus other food items. See text for more details (drawn from Williot et al. 2005)

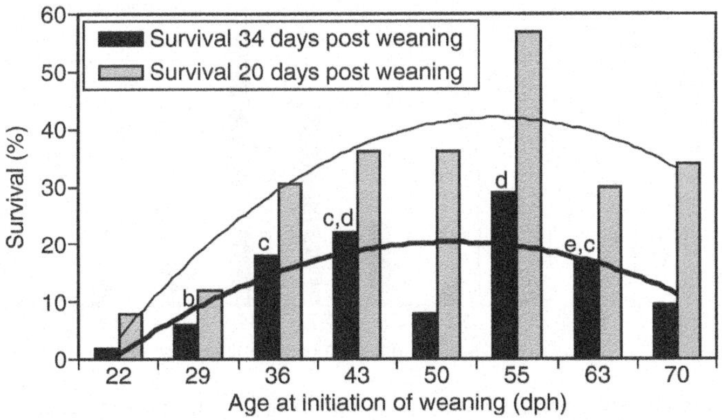

Fig. 28.5 Cumulative survival of short-weaned (3 days) *A. sturio* larvae according to age at initiation of weaning on a mix of compound diet (survival determined 20 and 34 days post initiation of weaning) at 17.5°C. Different letters denote significant difference at $p = 0.05$ (redrawn from Williot et al. 2005)

17.5°C. Median weight of control, 20 days post initiation of weaning, and 34 days post initiation of weaning were 1.3 g, 1.8 g, and 4.5 g respectively (not shown). Survival patterns at 20 and 34 days post initiation of weaning for the eight weaning trials are shown in Fig. 28.5. The best survival rate 34 days post initiation of short weaning was about 30% when fish were weaned at 55 dph (700 mg mean weight at initiation of weaning). Both younger and older fish had poorer survival rates.

Short (3 days) and long (3 weeks) weaning were compared for fish weaned at 50 dph. The long weaning showed better growth from 48 days post weaning onwards at 17.5°C, with final weights of around 2 g median weight and 3.5 g median weight for short and long weaning respectively (Fig. 28.6). Mortality

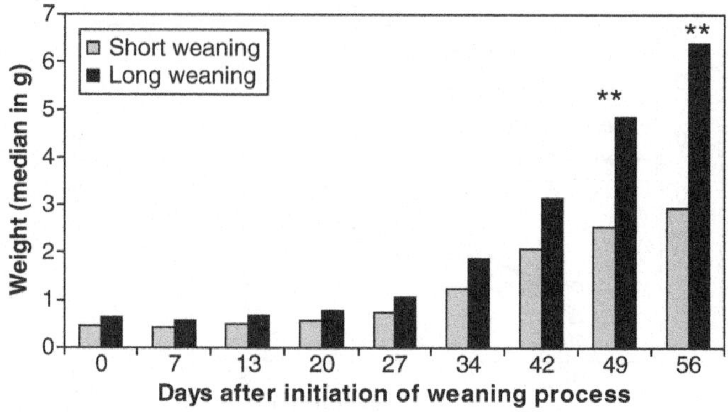

Fig. 28.6 Growth (median weight) of *A. sturio* larvae according to length of weaning [short (3 days) and long (3 weeks)] at 17.5°C. At initiation of weaning, larvae were 50 days post-hatch. *Two asterisks* denote a significant difference at $p = 0.01$ (after Williot et al. 2005)

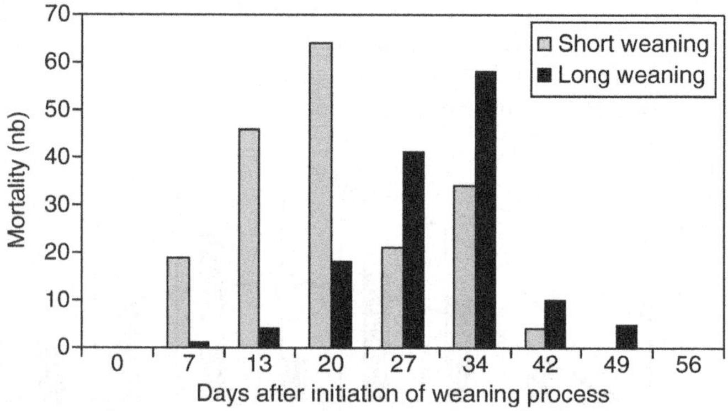

Fig. 28.7 Mortality distribution (number of individuals) in *A. sturio* according to length of weaning at 17.5°C [short (3 days) and long (3 weeks)]. At initiation of weaning, larvae were 50 days post-hatch (after Williot et al. 2005)

patterns were similar, but the peak of mortality for short weaning occurred 2 weeks earlier than for long weaning (Fig. 28.7). Mortality rate was greater (94%) for short weaning than for long weaning (68.5%).

28.3.3 Experiment 3

By the end of the experiment (3 weeks), the enriched *A. salina* nauplii-fed larvae weighed 700 mg, not significantly higher than the control (640 mg). A longer experiment should have been performed.

28.3.4 Experiment 4

The two batches fed either fresh or frozen chironomids exhibited similar final weights (2.88 and 2.87 g respectively) and mortality rates (8.5 and 6%). As a result, if necessary, these small fish at least might be fed frozen chironomids without any deleterious effect on growth and survival.

28.3.5 Growth Curve

Using the average weights results of nauplii-fed fish, a preliminary growth curve has been computed for *A. sturio* for a rearing water temperature of 17.5°C (Fig. 28.8). The best-fit curve ($R^2 = 0.997$) is a third degree polynomial equation.

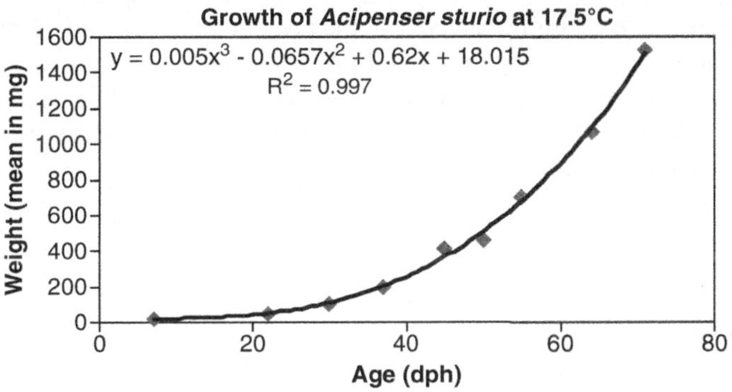

Fig. 28.8 Growth of *A. sturio* larvae fed *Artemia salina* nauplii then chironomids at 17.5°C (redrawn from Williot et al. 2005)

28.4 Conclusions

The first presence of *Artemia salina* nauplii was recorded at 16 dph, i.e., 4 days post expulsion of the melanin plug. This is later than the 10 dph reported for a very similar water temperature (17.1°C) with *Acipenser oxyrinchus* by Mohler (2000). This also means that first feeding of *A. sturio* is later than for *Acipenser baerii* [9 dph at the same temperature when fed compound diet (Gisbert and Williot, 1997)]. Most of the larvae were then fed a mix of natural and compound diet. However, at 16 and 23 dph, batches fed nauplii exhibited higher growth, which means that *Artemia salina* nauplii was the best accepted and the most effective food item tested. This practice of delivering multi-food items, called co-feeding, is usually applied as a progressive larval weaning in compound diets (Kolkovski 2001). Under this condition, survival at 38 dph ranged from 50 to 70% for fish weighing from 140 to 175 mg after 22 days feeding maximum. These results are lower than those recorded in the sister species, *A. oxyrinchus* (Mohler 2000). Mohler also demonstrated that growth was inversely proportional to the initial density when this was lower than 7.4 fish l^{-1} and this led to higher growth than for the 16–17 fish l^{-1} in the present experiment 1. Hypothesizing the same behaviour for *A. sturio* as that of *A. oxyrinchus* means that initial density in the small troughs should not be over about 650 fish, instead of the 1,500 in the present experiment 1.

Higher temperature (21 vs 17.5°C) led to better growth at 30 dph whatever the food items, and rapidly to higher mortality rates, of around 50%.

The survival of short-weaned fish (3 days) at 17.5°C, depending on their age at initiation of weaning, exhibited an unusual bell-shaped graph, with better survival 34 days post initiation of weaning of around 30% for fish weaned at 55 dph. Even though the present protocol may have exacerbated the results, this figure is still unusual and probably reveals species-specificity. Mohler (2000) showed that the best window for weaning *A. oxyrinchus* from *Artemia salina* nauplii to a compound diet was 20–26 dph when the fish were reared at an initial density of 7.4 fish l^{-1}.

Long progressive weaning (3 weeks in the present study) considerably improved both survival and growth, and thus should be favoured in the future.

This first successful larval rearing of *A. sturio* provided preliminary useful outcomes for further improvements, especially with regard to density, food items, and feeding management.

References

Dettlaff TA, Ginsburg AS, Schmalhausen OI (1993) Sturgeon fishes. Developmental biology and aquaculture. Springer, Berlin, p 300

Gisbert E, Williot P (1997) Larval behaviour and effect of the timing of initial feeding on growth and survival of Siberian sturgeon larvae under small scale hatchery production. Aquaculture 156:63–76

Kolkovski S (2001) Digestive enzymes in fish larvae and juveniles – implications and application to formulated diets. Aquaculture 200:181–201

Mohler JW (2000) Early culture of the American Atlantic sturgeon *Acipenser oxyrinchus oxyrinchus* Mitchill, 1815 and preliminary stocking trials. Bol Inst Esp Oceanogr 16 (1–4):203–208

Monaco G, Buddington RK, Doroshov SI (1981) Growth of white sturgeon (*Acipenser transmontanus*) under hatchery conditions. J World Maricul Soc 12:113–121

Williot P, Brun R, Pelard M, Mercier D (2000) Unusual induced maturation and spawning in an incidentally caught pair of adults of the critically endangered European sturgeon *Acipenser sturio* L. J Appl Ichthyol 16(6):279–281

Williot P, Brun R, Rouault T, Pelard M, Mercier D (2005) Attempts at larval rearing of the endangered western European sturgeon, *Acipenser sturio* L. (Acipenseridae), in France. Cybium 29:381–387

Williot P, Rochard E, Castelnaud G, Rouault T, Brun R, Lepage M, Elie P (1997) Biological characteristics of European Atlantic sturgeon, *Acipenser sturio*, as the basis for a restoration program in France. Environ Biol Fish 48:359–370

Williot P, Rouault T, Rochard E, Castelnaud G, Lepage M, Gonthier P, Elie P (2004) French attempts to protect and restore *Acipenser sturio* in the Gironde: status and perspectives, the research point of view. In: Gessner J, Ritterhoff J (eds) BfN-Schriften 101: International workshop on species differentiation and population identification in the sturgeons *Acipenser sturio* L. and *Acipenser oxyrhinchus*. Bundesamt für Naturschutz, Bonn, pp 83–99

Chapter 29
Post-release Monitoring Techniques

M.L. Acolas, Charles Roqueplo, E. Rouleau, and E. Rochard

Abstract This section deals with the monitoring techniques that have been tested and used in the context of the Gironde population dynamic survey. Very recent methods and results are presented as well as earlier data. Monitoring is currently used mainly to estimate the efficiency of stocking (survival and growth parameters) and spatial and temporal distribution, in order to localize the essential habitat of the fish.

29.1 Introduction

In a fish population restoration program, methods are needed to evaluate the success of stocking (UICN re-introduction specialist group 1998). Despite these recommendations, evaluation of the efficiency of stocking is rarely carried out, and this limits the improvements that can be made in current practice (Williot et al. 2009).

The objectives of monitoring are to assess the efficiency of stocking, which means the actual contribution of stocked fish to the population dynamics. First, this implies localizing individuals spatially and temporally, and catching them in order to estimate their survival and growth (Lochet et al. 2004). For this purpose, size of the fish permitting, each fish can be individually marked in order to monitor its life-history trajectory (see Chap. 24 for more details). Associated species are also sampled to characterise potential intra-species interactions (for food or habitat). Fish are usually stocked at a young stage in rivers, and the choice of monitoring technique depends on the environment and the fish size targeted. To carry out efficient monitoring, the technique used should not cause injuries to the fish; it has to be effective and easily reproducible.

M.L. Acolas (✉) • C. Roqueplo • E. Rouleau • E. Rochard
Cemagref, Estuarine Ecosystems and Diadromous Fish Research Unit, 50 Avenue de Verdun, 33612 Cestas cedex, France
e-mail: marie-laure.acolas@cemagref.fr

P. Williot et al. (eds.), *Biology and Conservation of the European Sturgeon Acipenser sturio L. 1758*, DOI 10.1007/978-3-642-20611-5_29,

In this section, methods and results of recent sampling in the Garonne and Dordogne Rivers will be presented, as well as monitoring of the Gironde estuary based on previous and current experiments.

29.2 Monitoring in the Garonne and Dordogne Rivers

In autumn 2008, the lack of knowledge of juvenile habitat use in freshwater, combined with the recent stocking of the Dordogne and Garonne Rivers with juvenile *A. sturio*, prompted us to initiate sampling in these rivers (Rochard and Williot 2006). The objective was to develop an effective methodology for capturing young of year *A. sturio* in a large river. For this purpose, two techniques were tested: beam trawling and electrofishing.

29.2.1 Materials and Methods

29.2.1.1 Sampling with a Beam Trawl

This gear was tested because it can sample fish near the bottom, where sturgeons are mostly believed to live. To use this technique efficiently, the following conditions must be respected (1) the river bottom has to be as flat as possible, and (2) boat speed must be properly adjusted so that the trawl is pulled (speed higher than the current) without causing injuries to the fish.

Most of these trawling samples were done within the European Water Framework Directive surveys (EU 2000). Sampling areas for beam trawling were located in the tidal part of the Dordogne River, upstream and downstream from the release site of young sturgeons (Fig. 29.1). These sites were chosen among the known *A. sturio* potential spawning areas (Jego et al. 2002). The bottom trawl mouth was 150 cm in width and 50 cm in height, and the stretched mesh size was 28, 20 and 16 mm respectively from top to bottom. Boat ground speed was on average 3.0 km/h; sampling was done against the current either at ebb, flow, or slack water (maximum water current of 1 m/s), and tow duration did not exceed 15 min. Depths explored ranged from 1 to 4 m, and the width of the river was on average 240 m ± 140 (standard deviation).

In the Dordogne River, juveniles of between 3 and 7 g were stocked during the second half of September (Rochard and Roqueplo 2009), and sampling was carried out more than 1 month after release, every day between 21 and 25 October.

29.2.1.2 Electrofishing

Electrofishing is a widely used technique for sampling fish in rivers and shallow waters (Vibert 1967). The principle is to create an electric field between two

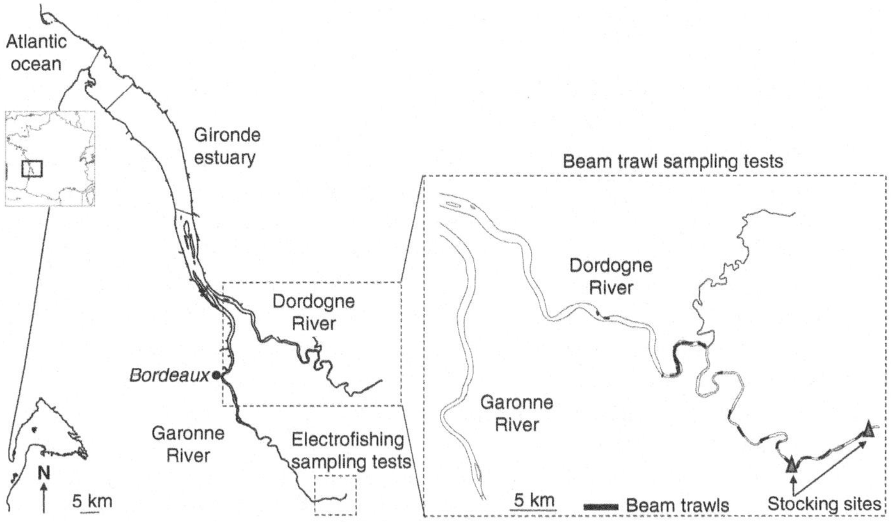

Fig. 29.1 Location of the freshwater sites of sampling tests in the studied area

electrodes in the water; then the fish are attracted towards the anode when the voltage gradient between its head and tail is about 2 V. This technique has been successfully employed in large rivers (Philippart 1979), and young sturgeons (*A. fulvescens*) have been electrofished in a Lake Michigan tributary (Benson et al. 2005).

We used electrofishing in the Garonne River near the young sturgeons' release sites (Fig. 29.1). Sampling was carried out in October from a boat with specific gear (power generator unit 220 V and 2 A direct current (unidirectional flow) output, Dream Electronics, France). This method allowed us to move easily from one point to another and to prospect littoral zones and depths of less than 1.5 m. All the fish species caught were kept in a bucket (less than 30 min) until they were identified and measured before being released.

Nine stations were sampled along a 2.5 km sector upstream and downstream from the fish release site 1 week after stocking. These stations were between 0.2 m and 1.5 m in depth and were representative of different habitat types in terms of substrate, water current and vegetation presence.

In addition to this field sampling, some additional trials were carried out in tanks (2.50 m wide and 0.80 m deep) at our facilities (experimental station at Saint-Seurin-sur-l'Isle). The same electrofishing gear was used, and two types of species behaviour were tested: (1) 15 *A. sturio* 18 months old (40 cm mean fork length) in one tank, and (2) 15 *A. baeri* 20 months old (35–40 cm mean fork length) in another tank. Fish behaviour was observed, such as attraction towards the anode, tetany, recovery and movements in the case of repetitive electrofishing.

29.2.2 Results and Conclusions

29.2.2.1 Sampling with a Beam Trawl

Among 33 trawls carried out in the Dordogne River, two young-of-year *A. sturio* were caught in the same tow. This tow was near the river bank in low water velocity (<25 cm/s) in habitat with fine substrate and vegetation in decomposition. Fish were located 9–20 km downstream from the release sites and 22 days after the last *A. sturio* release. Fish length was 109 and 111 mm for a weight of about 4.8 g. One fish was unscathed and returned to the river, but the other one did not recover from the fishing process. It was probably caught at the beginning of the trawl and stuck under sediment and organic remains. Its gut content was analysed; it had fed only on oligochaetae annelid animal prey (50 *Lumbriculidae* individuals and five *Enchytraeidae* individuals).

This result is in accordance with the recent literature on juvenile sturgeon species, where fine substrates are thought to provide optimal habitat as they have been found to contain large amounts of small-sized benthic prey (Barth et al. 2009). In the near future, this early life stage habitat preference should be considered in order to accurately evaluate recruitment success. This sampling method was relatively successful (one mortality), but other types of gear should be tested and adapted to this large river system where the main constraint is the power of the tide. Passive gear such as gill nets (Barth et al. 2009), fyke nets or active gear such as haul-seine or diving (Benson et al. 2005) could be tested either by day or night.

29.2.2.2 Electrofishing

At the nine survey stations in the Garonne River, no sturgeon was caught. Only eight fish species (*Alburnus alburnus*, *Anguilla anguilla*, *Barbus barbus*, *Cyprinus carpio*, *Leuciscus cephalus*, *Leuciscus leuciscus*, *Perca fluviatilis*, *Silurus glanus*) were caught, which corresponds to very poor fish diversity for this area.

Experimental results provided a potential explanation for this failure. First, young *A. sturio* were sensitive to an electric field, but only 8% of them reached the surface by electrotaxis and were caught with dipnet. It took these sensitive fish a few seconds to recover after the first shock and about 3 min after a second shock. Second, successive electrofishing at 3-min intervals made the young *A. sturio* completely insensitive to the electric field. They stayed motionless on the tank bottom. Third, all *A. baeri* tested swam to the surface by electrotaxis, where it was easy to catch them. In the case of successive electrofishing, 30% of them reached the surface and could still be caught. For these fish, it took them a few seconds to recover, even after a second shock.

In conclusion, electrofishing gave unsatisfactory results for young *A. sturio*, and it is not a technique to consider for sampling them in the future.

29.3 Monitoring in the Estuary

In the Gironde estuary, different sampling methods have been used to survey population dynamics. At first (1981), trammel nets were used with cooperation from professional fishermen (Castelnaud et al. 1991). Next, in addition to trammel nets, scientific trawling with a research vessel was set up until 2000 (Rochard 1992, 2002; Rochard et al. 1997;). This sampling mainly made it possible to localize fish habitat preference (Rochard et al. 2001; Brosse 2003) and to characterise fish diet (Brosse 2003). Moreover, it demonstrated that there were no growth differences between wild fish (Cohort 1988, 1994) and stocked fish (cohort 1995).

Since stocking programmes in recent years (2007–2008–2009), this technique has been used again since spring 2009. The objectives are to estimate fish stocking efficiency (survival, dispersion, growth) and to characterize habitat and juvenile diet.

29.3.1 Materials and Methods

The experimental trawl measures 13 m in width, 2.5 m in height and the bottom stretched mesh is 70 mm. Trawls are conducted by day; the tow duration is 30 min at an average speed of 8.3 km/h in the same direction as the current either by ebb or flood. Before each trawl, physical and chemical characteristics of the water 1 m above the bottom are measured (temperature, salinity, dissolved oxygen, turbidity). Trawl GPS position is recorded with Maxsea software.

Each species captured is determined; 30 fish per species per trawl are measured, and the others are counted. Sturgeon are measured, weighed and individually tagged (Hallprint external tag and PIT tag, see Chap. 24 for more details). To estimate age, a slice of the pectoral fin is sampled (Rochard and Jatteau 1991). In addition, to determine the parents, a genetic sample is taken. Diet is also analysed by gastric lavage (Brosse et al. 2002). Trawling is limited to particular areas in the estuary that are divided into 22 sectors (Fig. 29.2), and one trawl per sector is usually planned at each campaign.

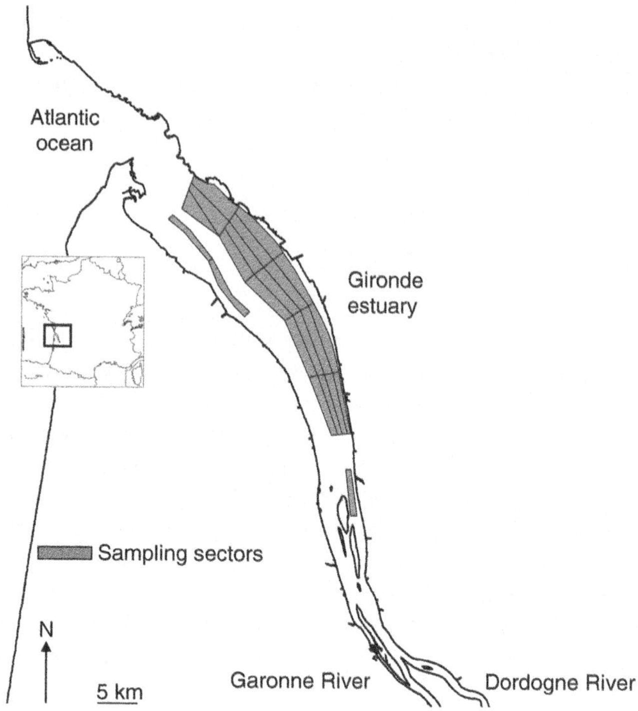

Fig. 29.2 Estuarine sampling areas for trawling

29.3.2 Results and Conclusions

Between mid-2009 and autumn 2010, 128 trawl tows were carried out in 24 days and 41 sturgeons were captured (Table 29.1). About 8.6% of the trawls were successful for sturgeon capture (one trawl may represent single or multiple captures), which is less than the frequency obtained between 1994 and 1996 (about 15%) (Rochard et al. 2001). The mean catch was 0.32 individuals per tow against 0.47 individuals per trawl in 1994–1996. Thirty-four percent of the fish captured were already tagged, and by checking their individual tag number, we knew that they were stocked at over 1 year old. Up to now, very few campaigns have been carried out due to technical and meteorological difficulties, so it is too early to estimate fish survival for the recent stockings.

Monitoring is limited to the middle and the lower part of the right side of the Gironde, because in many areas it is impossible to trawl (navigation channel and wrecks). However, previous telemetry studies (Lepage et al. 2005; Taverny et al. 2002) have demonstrated that the areas most used were in the monitored sectors. In addition to the scientific sampling, voluntary declarations by fishermen or local investigations enabled us to complete our monitoring, because they are fishing throughout the whole estuary and in the river.

Table 29.1 Sturgeon captures between mid-2009 and autumn 2010 by trawl sampling in the Gironde estuary

Sampling dates	Number of trawls	Number of sturgeon captured	Cohort origin	Fish already tagged	Fork length (cm)	Weight (g)
30 June–1 July 2009	10	1	2007	Yes	42.4	450
14–16 September 2009	21	3	Ind	No	45	590
			Ind	No	48.5	690
			Ind	No	60	1,530
28–30 September 2009	13	0				
8–9 December 2009	7	2	Ind	No	52	950
			Ind	No	67	2,800
7 May 2010	5	1	Ind	No	50	860
1–3 June 2010	23	1	Ind	No	50.5	860
23–24 June 2010	12	1	2007	Yes	40.5	330
5–8 July 2010	21	4	2007	Yes	45	425
			Ind	No	40	440
			Ind	No	47.2	660
			Ind	No	49	690
27–30 October 2010	16	28	2009	Yes	36.5	240
			2009	Yes	42	365
			2009	Yes	43	400
			2008	Yes	41	430
			2009	Yes	44.5	465
			2009	Yes	46	515
			2009	Yes	47.5	650
			2009	Yes	48.5	680
			Ind	No	51	875
			Ind	No	53	885
			Ind	No	53.5	970
			2007	Yes	54	1,030
			2007	Yes	55.5	1,060
			Ind	No	57	1,090
			Ind	No	55	1,095
			Ind	No	58	1,195
			Ind	No	59.3	1,210
			Ind	No	57	1,250
			Ind	No	61.3	1,280
			Ind	No	59	1,380
			Ind	No	60	1,400
			Ind	No	59.5	1,410
			Ind	No	60	1,455
			Ind	No	61.5	1,485
			Ind	No	62	1,520
			Ind	No	65.5	1,735
			2007	Yes	71	2,010
			Ind	No	69.5	2,270

"Ind" means that the cohort origin has to be determined. When fish were already tagged externally when captured, "yes" is given in the column "Fish already tagged" and if not, "no" is given

References

Barth CC, Peake SJ, Allen PJ, Anderson WG (2009) Habitat utilization of juvenile lake sturgeon, *Acipenser fulvescens*, in a large Canadian river. J Appl Ichthyol 25(2):18–26

Benson AC, Sutton TM, Elliott RF, Meronek TG (2005) Seasonal movement patterns and habitat preferences of age-0 lake sturgeon in the lower Peshtigo River, Wisconsin. Trans Am Fish Soc 134(5):1400–1409

Brosse L (2003) Caractérisation des habitats des juvéniles d'esturgeon européen, *Acipenser sturio*, dans l'estuaire de la Gironde. Doctorat, Université Paul Sabatié, Toulouse

Brosse L, Dumont P, Lepage M, Rochard E (2002) Evaluation of a gastric lavage method for sturgeons. N Am J Fish Man 22(3):955–960

Castelnaud G, Rochard E, Jatteau P, Lepage M (1991) Données actuelles sur la biologie d'*Acipenser sturio* dans l'estuaire de la Gironde. In: Williot P (ed) *Acipenser*. Cemagref, Antony, France, pp 251–275

EU (2000) Parliament and Council Directive 2000/60/EC of 23 October 2000. Establishing a Framework for Community Action in the field of Water Policy. Official Journal PE-CONS 3639/1/00 REV 1, Brussels

Jego S, Gazeau C, Jatteau P, Elie P, Rochard E (2002) Les frayères potentielles de l'esturgeon européen *Acipenser sturio* L. 1758 dans le bassin Garonne–Dordogne. Méthodes d'investigation, état actuel et perspectives. Bull Fr Pêche Piscic 365–366:487–505

Lepage M, Taverny C, Piefort S, Dumont P, Rochard E, Brosse L (2005) Juvenile sturgeon (*Acipenser sturio*) habitat utilization in the Gironde estuary as determined by acoustic telemetry. In: Spedicato MT, Lembo G, Marmulla G (eds) Aquatic telemetry: advances and applications. Proceedings of the Fifth Conference on Fish Telemetry held in Europe. Ustica, Italy, 9–13 June 2003. FAO/COISPA, Rome, pp 169–177

Lochet A, Lambert P, Lepage M, Rochard E (2004) Growth comparison between wild and hatchery-reared juvenile European sturgeons *Acipenser sturio* (Acipenseridae) during their stay in the Gironde estuary (France). Cybium 28(1):91–98

Philippart JC (1979) Observations concernant l'efficacité de la pêche à l'électricité dans une rivière de la zone à barbeau. Bull Fr Pêche Piscic 273:157–172

Rochard E (1992) Mise au point d'une méthode de suivi de l'abondance des amphihalins dans le système fluvio-estuarien de la Gironde, application à l'étude écobiologique de l'esturgeon, *Acipenser sturio*. Doctorat, Université de Rennes I, Rennes

Rochard E (2002) Restauration de l'esturgeon européen acipenser sturio. Rapport scientifique contrat LIFE n°B-3200/98/460. Etude Cemagref n°80, groupement de Bordeaux

Rochard E, Jatteau P (1991) Amélioration de la méthode de détermination de l'âge de l'esturgeon commun *Acipenser sturio* et premières applications. In: Williot P (ed) Acipenser. Cemagref, Antony, France, pp 193–208

Rochard E, Lepage M, Dumont P, Tremblay S, Gazeau C (2001) Downstream migration of juvenile European sturgeon *Acipenser sturio* L. in the Gironde Estuary. Estuaries 24(1):108–115

Rochard E, Lepage M, Gazeau C (1997) Suivi de la situation de la population relictuelle d'*Acipenser sturio* dans le milieu naturel. In: Elie P (ed) Restauration de l'esturgeon européen *Acipenser sturio*. Contrat Life rapport final du programme d'exécution. Cemagref, Bordeaux

Rochard E, Roqueplo C (2009) Action n° 17: Définition des règles et pratiques d'alevinages. In: Rochard E (ed) Programme de recherche et de conservation de l'esturgeon européen *Acipenser sturio*; bilan scientifique et technique 2008. Etude n°133.Cemagref, Bordeaux, pp 77–81

Rochard E, Williot P (2006) Actions de recherche proposées pour contribuer au plan international de restauration de l'esturgeon européen *Acipenser sturio*. Etude n°103. Cemagref, Bordeaux

Taverny C, Lepage M, Piefort S, Dumont P, Rochard E (2002) Habitat selection by juvenile European sturgeon *Acipenser sturio* in the Gironde estuary (France). J Appl Ichthyol 18 (4–6):536–541

UICN re-introduction specialist group (ed) (1998) Lignes directrices de l'IUCN relatives aux réintroductions. IUCN, Gland, Cambridge

Vibert R (ed) (1967) Fishing with electricity. Its application to biology and management. Fishing News Books, EIFAC, London

Williot P, Rochard E, Kirschbaum F (2009) Acceptability and prerequisites for the successful introduction of sturgeon species. In: Carmona R, Domezain A, Garcia Gallego M, Hernando-Casal J, Rodriguez F, Ruiz Rejon M (eds) Biology, conservation and sustainable development of sturgeons, Fish and Fisheries Series. Springer, New York, pp 369–384, 467p

Chapter 30
Modelling the Future of Stocked Fish

E. Rochard and Patrick Lambert

Abstract Stocking with juvenile sturgeon is a key point of the action plans described below for *A. sturio*. It is essential to assess the expected effects of these stocking measures in order to improve practises. Using a few parameters of population dynamics assessed from monitoring some cohorts, we simulated the survival of the stocked fish. Simulations are obviously highly sensitive to the mortality rate. Based on simulations using the best range of available parameters for the Gironde population, the number of females on spawning grounds constitutes a bottleneck for the dynamic of this population.

30.1 Introduction

Stocking with young fish produced from an ex situ stock of breeders is a key point in the National French and German as well as the Bern Convention action plans (objective 6 "release of *A. sturio* for re-establishment or enhancement") (Rosenthal et al. 2007; Williot et al. 1997; Gessner et al. 2010). In France, juveniles were produced successfully in 1995 (Williot et al. 2000; Elie 1997), 2007 (Rouault et al. 2008), 2008 (Rochard 2009) and 2009 (unpublished data). However, we have a limited number of adults maturing in any given year, and our control of the reproduction processes is still limited; therefore, the number of fish produced was limited. The vast majority of the fish was stocked in the rivers Garonne and Dordogne (Table 30.1). Following the Russian practises (Williot 1984; Charlon and Williot 1978), we stocked most of them at 3–5 g (3–4 months) in the middle of the potential spawning ground areas (Jego et al. 2002).

E. Rochard (✉) • P. Lambert
Cemagref, Estuarine Ecosystems and Diadromous Fish Research Unit, 50 Avenue de Verdun, 33612 Cestas Cedex, France
e-mail: eric.rochard@cemagref.fr

P. Williot et al. (eds.), *Biology and Conservation of the European Sturgeon Acipenser sturio L. 1758*, DOI 10.1007/978-3-642-20611-5_30,

Table 30.1 Number, weight and age of fish stocked from the different cohorts

Cohorts (year of birth)	Number	Weight mean (min, max)	Age
1995	2,000	0.02 g	5 days
	5,000	1.0 g	2 months
	2,000	6.5 g	4 months
2007	6,230	3.4 g (1.3, 5.0)	3 months
	390	353.0 g (21.0, 1,200.0)	27 months
2008	81,935	4.6 g (2.5, 27.8)	3 months
	273	619.0 g (61.0, 2,760.0)	27 months
2009	44,819	5.0 g (3.6, 6.4)	3 months
	1,605	169.0 g (22.0, 655.0)	13 months

It doesn't include fish used for telemetry experiments

We explore the relationships between the number of fish stocked and, based on survival, the number of adults that are likely to return to reproduce each year, by using simple population dynamics simulations. These simulations address the question of the number of spawners we can expect with present stocking practises, and conversely the question of the number of stocked fish needed to reach a certain number of adults on the spawning grounds.

These simulations are obviously limited by our knowledge of the population dynamics parameters (e.g., natural mortality, fisheries-related mortality, maturation rates, etc.) but they are intended to provide a first idea of what could be expected and to help improve current practises.

30.2 Background

The model is simply based on two biological processes: mortality and sexual maturation (i.e., being ready to reach the spawning grounds).

From our observations of the 1995 cohort kept in captivity, we postulated a balanced sex ratio at birth (50% males, 50% females).

We considered an exponential instantaneous mortality coefficient that varies during the life of the fish but not between sexes. It sums up natural mortality M and possibly fishing mortality F by poaching and by-catch.

The number of females $N_{\mathrm{f}}(t)$ and males $N_{\mathrm{m}}(t)$in a cohort at instant t, is obtained using the following equations:

$$N_{\mathrm{f}}(t) = N_{\mathrm{f}}(t-1)\mathrm{e}^{-(M(t)+F(t))}$$
$$N_{\mathrm{m}}(t) = N_{\mathrm{m}}(t-1)\mathrm{e}^{-(M(t)+F(t))}.$$

Total mortality coefficient from age 1 to age 2 was estimated at 0.9 year^{-1} from stock assessment of 1994 and 1995 cohorts during their second autumn in the estuary (Rochard 1992; Rochard et al. 1997).

From monitoring the 1988 and 1994 cohorts, we deduced a total mortality coefficient of at least 0.45 $year^{-1}$ during the marine phase up to their first spawning migration in rivers (Lepage et al. 2002).This value was used to simulate mortality of fish older than 2 years.

Theoretically (Pauly 1980), considering the life span of the species, its observed growth (Lochet et al. 2004), and the temperature in the coastal north-east Atlantic, a natural mortality M of about 0.16 $year^{-1}$ is expected. This value is close to the observed value for *Acipenser oxyrinchus* (Hoenig 1983). Therefore, the 0.45 figure includes an anthropogenic mortality component, probably due to fishing, since we were aware of a large number of by-catch and some poaching events (see Chap. 22). As a consequence of sturgeon recovery plan implementation, we expect fishing mortality to decrease, or even to be completely eliminated. To simulate the present situation and also a potentially improved mortality we built two scenarios, $Z = 0.45$ and $Z = 0.2$, using two values of total instantaneous mortality coefficients of 0.45 and 0.2 $year^{-1}$.

We considered that the most precocious male can reproduce at 10 years and female at 14 years. After that age, we considered that each year 75% of the adult males (half of them mature yearly and half of them only every 2 years) and 33% of the adult females mature (they mature every 3 years).

30.3 Simulations

The aim of the scenarios was to assess the number of adults which could reproduce in the future.

Firstly, we simulated the evolution in the number of fish in the wild with the size of the cohorts during their second autumn (1985, 1988, 1994, 1995) or the number of fish stocked (2007, 2008, 2009) without any reproduction (with no new cohort entering the system) (Figs. 30.1 and 30.2).

The absence of reproduction simulation leads to some decreasing trends which are not necessarily realistic. However, these results do highlight the importance of the recent stocking events (2007–2009), which give us good reason to hope that the disappearance of the species in the wild can at least be delayed. Differences between the two scenarios are not obvious. They become more clearly visible for the oldest fish, which are less numerous than the youngest. We see that even without simulating senescence (i.e., increasing mortality for the older fish), there is a rapid decrease in numbers of mature fish, caused only by the constant mortality.

Secondly, we estimated for each year the possible numbers of males and females from the different cohorts which could be mature.

Simulations presented in Figs. 30.3 and 30.4 give the total number of mature fish in the population. Taking the above-mentioned prerequisites into account, every year 75% and 33% of males and females respectively can be expected to enter the rivers for reproduction. Thus, the simulation indicates that each year we could expect to observe mature adults of both sexes during the simulated period ($Z = 0.2$)

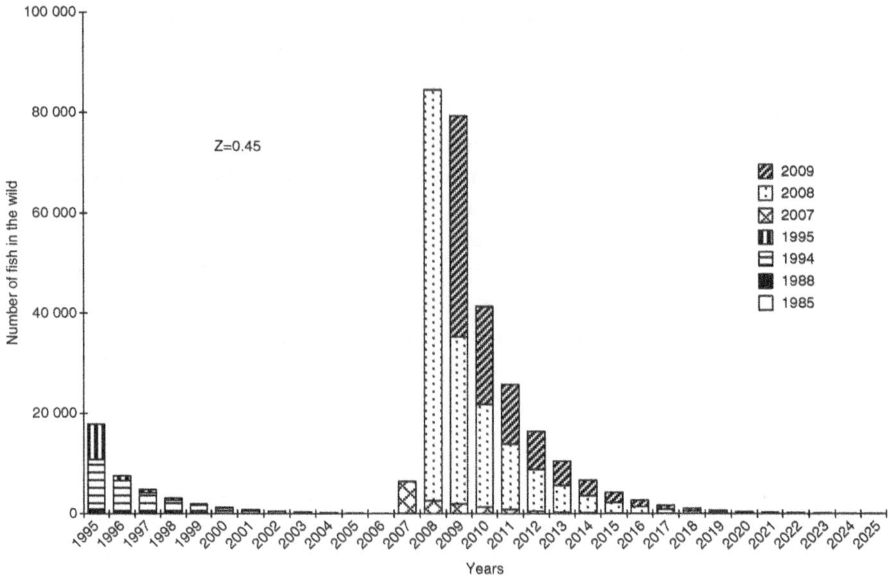

Fig. 30.1 Simulation of the number of *A. sturio* in the wild (1995–2025), considering a constant mortality rate at sea of $Z = 0.45$, and several cohorts

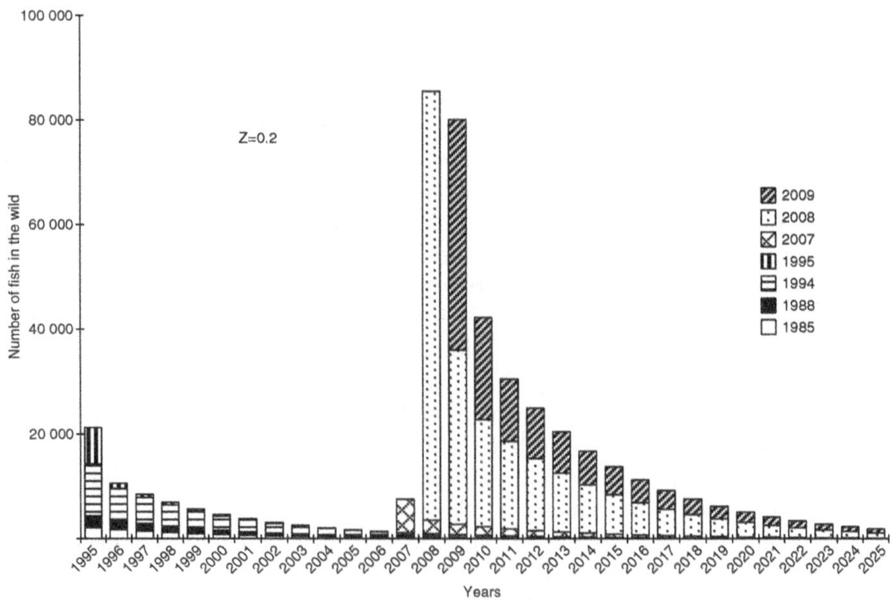

Fig. 30.2 Simulation of the number of *A. sturio* in the wild (1995–2025), considering a constant mortality rate at sea of $Z = 0.2$, and several cohorts

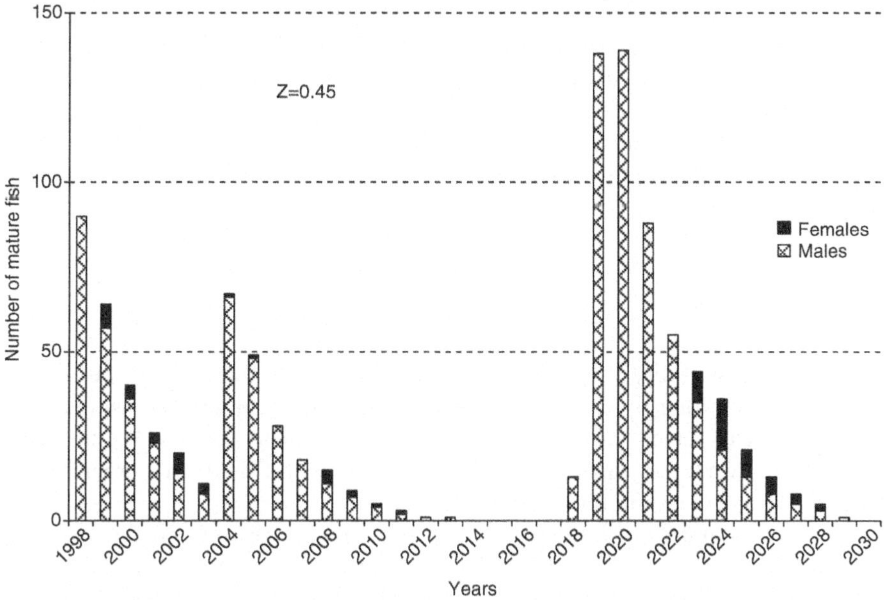

Fig. 30.3 Simulation of the total number of mature adults (males and females) of *A. sturio* in the wild, considering a constant mortality rate at sea of $Z = 0.45$

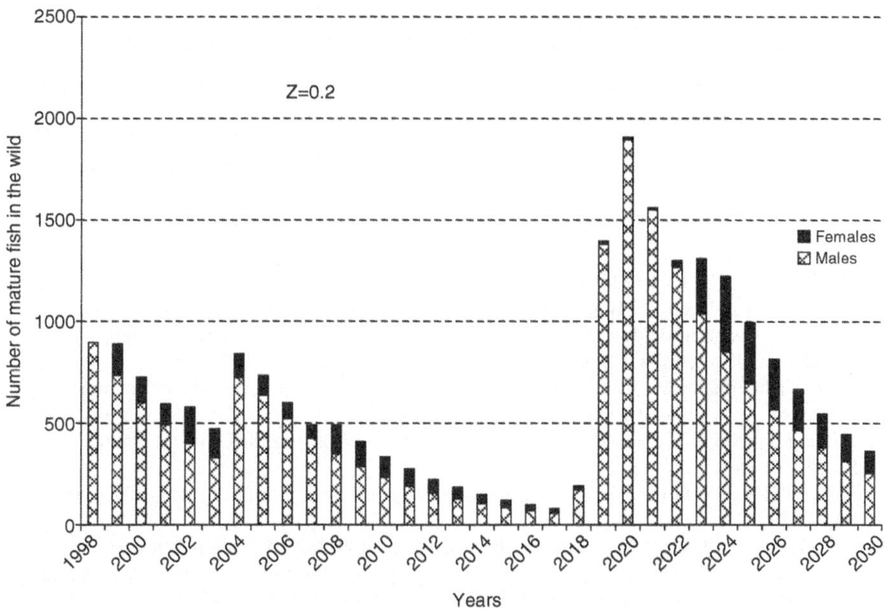

Fig. 30.4 Simulation of the total number of mature adults (males and females) of *A. sturio* in the wild, considering a constant mortality rate at sea of $Z = 0.2$

or only rarely ($Z = 0.45$), before 2012 or after 2023. In the first case we could expect natural reproduction every year, while in the second case only in some years. The presence of mature males does not seem to be a limiting factor, whereas the number of mature females does sometimes appear limiting. Therefore, the number of mature females constitutes a bottleneck for the recovery of the population. Future cohorts will not change this situation, since they will have no impact until after the simulation period.

30.4 Discussion

The assessed mortality rates (cohorts 1988, 1994, 1995) are likely, but do not correspond to a long-lived protected species. Incidental captures at sea and even poaching events have drastically reduced the size of these cohorts. We hope that more recent awareness campaigns and cooperation with fishermen's organisations will lead to a decrease in this mortality.

From our simulation, a highly skewed sex ratio is apparent on the spawning grounds, with many more males than females. This simulated ratio roughly corresponds to historical observations by fishermen and to the number of fish captured incidentally in rivers (Williot et al. 2002). Therefore, situations where there are females without males on the spawning grounds are highly unlikely, while the opposite scenario (male without female) could be frequent at some periods.

This first simulation approach will be improved by explicitly taking the reproduction process into account. The output will then be the probability, according to the level and the duration of the stocking, that a given number of females (the minimum required if reproduction is to be considered successful) will encounter males in a spawning ground.

References

Charlon N, Williot P (1978) Elevage d'esturgeons de repeuplement et de consommation en URSS. Bull Cent Etud Rech Sci Biarritz 12(1):7–156

Elie P (1997) Restauration de l'esturgeon européen *Acipenser sturio*. Contrat Life rapport final du programme d'exécution. Cemagref, Bordeaux

Gessner J, Tautenhahn M, Von Nordheim H, Borchers T (2010) Plan national d'actions pour la protection et la conservation de l'Esturgeon européen (*Acipenser sturio*). Gesellschaft zur Rettung des Störes Acipenser sturio L. e.V. Bundesministerium für Umwelt, Naturschutz und Reaktorsicherheit, Rostock

Hoenig JM (1983) Empirical use of longevity data to estimate mortality rates. Fish Bull 82(1): 898–903

Jego S, Gazeau C, Jatteau P, Elie P, Rochard E (2002) Les frayères potentielles de l'esturgeon européen Acipenser sturio L. 1758 dans le bassin Garonne-Dordogne. Méthodes d'investigation, état actuel et perspectives. Bull Fr Pêche Piscic 365–366:487–505

Lepage M, Rochard E, Brosse L (2002) Suivi de la population estuarienne d'esturgeon. In: Rochard E (ed) Restauration de l'esturgeon européen *Acipenser sturio* Rapport scientifique Contrat LIFE n° B – 3200/98/460. Cemagref Cestas, pp 123–138

Lochet A, Lambert P, Lepage M, Rochard E (2004) Croissance de juvéniles d'esturgeons européens *Acipenser sturio* (Acipenseridae) sauvages et issus d'alevinage, durant leur séjour dans l'estuaire de la Gironde (France). Cybium 28(suppl 1):91–98

Pauly D (1980) On the interrelationships between natural mortality, growth parameters, and mean environmental temperature in 175 fish stocks. Journal du Conseil International pour l'Exploration de la Mer 39:175–192

Rochard E (1992) Mise au point d'une méthode de suivi de l'abondance des amphihalins dans le système fluvio-estuarien de la Gironde, application à l'étude écobiologique de l'esturgeon, Acipenser sturio. Thèse de doctorat, Université de Rennes I, Rennes

Rochard E (ed) (2009) Programme de recherche et de conservation de l'esturgeon européen *Acipenser sturio*; bilan scientifique et technique 2008, vol rapport n° 133. Cemagref, Bordeaux

Rochard E, Lepage M, Gazeau C (1997) Suivi de la situation de la population relictuelle d'Acipenser sturio dans le milieu naturel. In: Elie P (ed) Restauration de l'esturgeon européen Acipenser sturio. Contrat Life rapport final du programme d'exécution, vol 24. Cemagref, Bordeaux, pp 295–333

Rosenthal H, Bronzi P, Gessner J, Moreau D, Rochard E, Lasen C (2007) Draft action plan for the conservation and restoration of the European sturgeon (*Acipenser sturio*). Council of Europe, Convention on the Conservation of European Wildlife and Natural Habitats, Strasbourg

Rouault T, Chèvre P, Rochard E, Jatteau P, Jacobs L, Gonthier P (2008) Programme de recherche et de conservation de l'esturgeon européen *Acipenser sturio*; bilan scientifique et technique 2007. Cemagref, Bordeaux

Williot P (1984) L'expérience soviétique en matière d'exploitation des stocks d'esturgeons en Mer d'Azov et Mer Caspienne. Etude n° 20, Série esturgeon N° 3. Cemagref, Bordeaux

Williot P, Rochard E, Castelnaud G, Rouault T, Brun R, Lepage M, Elie P (1997) Biological characteristics of European Atlantic sturgeon, Acipenser sturio, as the basis for a restoration program in France. Environ Biol Fish 48(1–4):359–370

Williot P, Brun R, Pelard M, Mercier D (2000) Induced maturation and spawning in an incidentally caught adult pair of critically endangered European sturgeon, Acipenser sturio L. J Appl Ichthyol 16(6):279–281

Williot P, Rouault T, Brun R, Pelard M, Mercier D (2002) Status of caught wild spawners and propagation of the endangered sturgeon Acipenser sturio in France: a synthesis. Int Rev Hydrobiol 87(5–6):515–524

Chapter 31
Building a Brood Stock of *Acipenser sturio* in France

Patrick Williot, Thierry Rouault, Rémi Brun, Marcel Pelard, Daniel Mercier, Louis Jacobs, and Frank Kirschbaum

Abstract The aim of this chapter is to describe the strategy which was applied, the changes needed in the programme, and the results of the brood stock building. A five-point process has been set up (when?, how?, which rearing conditions?, which management?, and which rules as reference?). The stock is described, as well as the delay in resuming food intake for wild-originated fish. The only food accepted on a regular basis is shrimps. Three levels of water salinity (freshwater, brackish water and to a lesser extent sea water) have been tested to simulate some of the species' natural environment. At present, brackish water is favoured. Resumption of food intake by wild-originated fish might be delayed. Growth is more or less cyclic, related to water temperature and food consumption. Loss of fish is analysed. Growth between 1 year old and 7.5 years old is similar for both genders.

31.1 Introduction

The restoration–conservation programme in France comprises a three-leg strategy (see Chap. 20): (1) increase knowledge of the remaining population, (2) restocking measures, and (3) integration of a model species to gain insights into the biology of sturgeons in general for later application to *Acipenser sturio*. In the course of the 1980s, the objective of restocking rapidly revealed itself to be an unrealistic plan,

P. Williot (✉)
Sturgeon Consultant, 4 Rue du Pas de madame, 33980 Audenge, France
e-mail: williot.patrick@neuf.fr

T. Rouault • R. Brun • M. Pelard • D. Mercier • L. Jacobs
Cemagref, Estuarine Ecosystems and Diadromous Fish Research Unit, Moulin de Logerie, 33660 Saint-Seurin-sur-l'Isle, France

F. Kirschbaum
Faculty of Agriculture and Horticulture, Biology and Ecology of Fishes, Humboldt University, Philippstr. 13, Haus 16, 10115 Berlin, Germany

P. Williot et al. (eds.), *Biology and Conservation of the European Sturgeon Acipenser sturio L. 1758*, DOI 10.1007/978-3-642-20611-5_31,

due to the difficulty in performing artificial reproduction on wild brood fish. The reasons for this were: (1) the dramatic decline in the wild population, and thus the very low numbers of by-catch wild brood fish, (2) the physiological reproductive status of brood fish, both male and female, which very regularly proved unsuitable, and (3) the frequent non-simultaneous presence of both sexes in spite of some measures to overcome this last constraint, e.g., the appropriate reproductive management of the males (see Chap. 27).

A long-term releasing programme could therefore no longer be based on such uncertainties. Moreover, the conservation of the species emerged as evidence. The only way to tentatively address these two issues was by building a functional brood stock, and this was made possible thanks to the availability of a new research facility at Saint-Seurin-sur-l'Isle (see Chap. 20).

Some limited attempts were made during the 1980s to keep some wild brood fish in freshwater ponds after artificial reproduction, but these failed as the fish refused to feed, and they were finally released after tagging (Williot et al. 2002).

By that time, in the late 1980s and early 1990s, very few data had been published on successful sturgeon farming, especially the anadromous species. The successful reproduction of the Adriatic sturgeon, *Acipenser naccarii*, though not a true anadromous species, using individuals captured in the Po basin and farmed in fresh water for many years (Arlati et al. 1988), was an interesting case. Soon afterwards, a paper by Struffenegger (1992) reported the successful farming, i.e., the completion of the whole biological cycle, of the anadromous White sturgeon *A. transmontanus* in freshwater. However, nothing was published with regard to the Atlantic sturgeon, *A. oxyrinchus*, the sister species of *A. sturio*.

Most of the data presented in this section originate from two previous publications (Williot et al. 2007, 2009), complemented by unpublished data.

The objectives of building the brood stock were: (1) to define the rearing conditions for maintaining the fish alive and making them grow, (2) to find the environmental factors to stimulate gonad maturation (see Chap. 32), (3) to favour non-invasive methods, (4) to set up a management procedure, (5) to promote research for better long-term management of the brood stock, and (6) to gather together all the information gained for a maximal understanding of the biology of the species in the framework of the restoration–conservation programme.

31.2 Material and Methods

31.2.1 The Applied Strategy

It has been necessary to define some general guidelines to address the following issues: (1) how to constitute the brood stock? (2) which rearing conditions have to be provided? (3) what kind of management should be applied? and (4) which general procedures should be applied to deal with unexpected situations?

With regard to the constitution of the brood stock, our main line of action has been to take different possibilities into consideration. The long-term objectives were to increase the rate of rearing success, to obtain the largest potential genetic variability, and to reduce the delay in obtaining mature fish. This indeed is a key issue for conservation. The aim was therefore to integrate the maximum of year classes from the wild, including adults, and to integrate artificially produced progeny as well.

The only examples of a complete biological cycle of sturgeon species in captivity were in the freshwater environment. We therefore decided to apply these rearing conditions ourselves, especially as there are no physiological data available which could help in this choice. As juveniles caught in the Gironde estuary (see Chap. 31) acclimatised better to brackish water than to freshwater, we decided to use brackish water to test its long-term impact on fish.

The main points for management were the type of food, the sampling procedure, and the secure functioning of the rearing tanks. The only food item accepted on a long-term basis was shrimps. A 3-month sampling procedure was applied because this appeared to be the best compromise between getting data on a regular basis and yet limiting possible deleterious effects of more frequent handling. From 2000 onwards, fish were blood-sampled quarterly for hormone determinations, and at the same time a small piece of pectoral fin was sampled for genetic investigations when needed (Ludwig et al. 2004, see Chap. 34). Telephone alerts (when water level and oxygen content fell below a prefixed value) were installed on each of the individual closed water systems.

General rules were established as a reference guide to conservation-related issues. These were: (1) reducing the risk of mortalities, (2) improving our knowledge of husbandry-related issues, (3) adopting a step-by-step procedure, (4) favouring a solution allowing a return (vs one-way route), (5) favouring conservative protocols, (6) referring to the ecology, and (7) favouring non-invasive methods.

31.2.2 Obtaining the Fish and Holding Tanks

The wild-originated fish were captured either by trawl nets by the Cemagref research vessel in the estuary of the Gironde or as a by-catch of brood fish by fishermen. Fish were trucked in a water tank ($2.5 \times 1 \times 0.8$ m; $L \times W \times D$) filled with the original water supplied with oxygen. Transport duration was limited to 30 min to minimize the stress imposed on the fish.

From the early 1990s on, selective keeping of wild-originated brood fish was set up. Most of the by-catch females and the first two males per season were kept in the hatchery. The objective was to use the remaining mature wild fish after tagging to localise the spawning grounds. This is group 1 (Table 31.1), represented initially by nine fish, including the sire and the dam used in the successful artificial reproduction in 1995 (Williot et al. 2000). The year's new wild fish were integrated; however, mortalities also occurred, reducing the total number of fish (see Sect. 31.3.3).

Table 31.1 Composition of the French brood stock of *Acipenser sturio* as of March 2010 updated from Williot et al. (2009)

Origin	Year of birth	Number and sex	Mean weight (kg)
Wild	1981–1988 (group-1)	8 (7 ♂, 1 ♀)	19
	1994 (group–2)	14 ♀	13
		8 ♂	14
		7 ?	
Artificial breeding	1995 (group-3)	23 ♀	13
		17 ♂	11
Total		77	

Twenty-nine fish were caught during spring 1995 in fresh water in the upper Gironde estuary and 13 in summer in brackish water (~15‰) in the middle part of the Gironde estuary. They were held in the same water salinity upon arrival in the hatchery. These fish belong to the last recorded spawning that occurred in the wild in 1994. This is group 2 (Table 31.1). Captures were voluntarily limited in number to give the remainder the opportunity to become adult in the wild. These fish were probably full siblings (Ludwig et al. 2004).

The third group (group 3 fish) was composed of a batch of progenies ($n1 \sim 200$ and $n2 \sim 120$; Williot et al. 2005) from the artificial reproduction achieved in 1995, and they are full siblings (Williot et al. 2000, 2005).

All fish were double-tagged, with one PIT-tag and a Petersen disc tag attached to the base of the dorsal fin (see Chap. 24). The juvenile fish were tagged when they were about 1.5 years old.

The holding tanks were circular (4 × 1 m; $\Phi \times h$ and 4 × 2 m; $\Phi \times h$), each one being part of a recirculation system composed of a pump equipped with a basket to collect refused food, a mechanical sand filter, a biological filter, a UV-device, and a column for oxygen enrichment. All the tanks had a 0.6-m-high wire fence placed on the top of the tanks to prevent the fish from escaping.

31.2.3 Management

Temperature, salinity, pH and oxygen were checked daily. The yearly water quality for all the tanks until autumn 2000 ranged (min–max) from 16 to 24°C, fresh water or 14–16‰ brackish water, pH 6.8–7.5, and O_2 6.8–7.8 mg l^{-1}. Ammonium, nitrite and nitrate were controlled monthly in the inlet and outlet waters from the tanks. From winter 2000–2001 on, water temperature was lowered to 11°C to simulate vernalization. The photoperiod was maintained similar to the natural day length with a low intensity.

Fish were fed frozen shrimps, mainly based on *Palaemonetes varians* and, to a lesser extent, *Crangon crangon*, twice daily ad libitum. Uneaten feed was recovered every day from the tanks, and the pump baskets were weighed in order to adapt the daily food ration, which was reduced when waste feed was >15% of that offered.

The total daily shrimp portion ranged between 1.2 and 3% of the fish's biomass. Other food items were tested such as compound food, mussels, and squid. The compound food was not regularly accepted, and led to malformations (Williot et al. 2004); mussels and squid were refused. Therefore, despite constraints (seasonal availability, heavy costs) and risks (introducing toxic substances and possibly freeze-resistant pathogens), shrimp feeding was applied for the whole period. However, from a long-term perspective, holding brood stock cannot be based on such a feeding procedure, and a modified compound diet has to be developed. In this respect, recent results are promising (see Chap. 33).

With the exception of the first year of the experimental period for the young juveniles (group 2 and group 3 fish) when the fish were weighed every month, current checking was on a 3-month basis. Prior to handling, fish were tranquillised by immersion for about 5 min in a bath containing essential oils of cloves at a concentration of 40 ppm. At low temperatures, dilution in ethanol (1/10, v/v) improves the solubility of the essential oil. Large fish which exhibited continuous growth were biopsied for gender and also for gonad development. Handling of the fish is described in Chap. 27.

31.3 Results and Discussion

31.3.1 Type of Water and Growth Characteristics

Preliminary attempts to maintain wild-originated fish were carried out in 1991 and 1992 under the assumption that fish might become acclimatised and spend all their biological cycle in fresh water. First experiments showed that wild juvenile fish with a total length (TL) of about 105 cm (ca. 8 kg) could be directly acclimatised to fresh water, leading to a delay in food intake of about 6 months and a loss of about 20% of their initial body weight (Williot et al. 1997). This is in agreement with Magnin (1962), who reported that fish over this size were able to move back and forth easily in a very wide range of water salinity. It was decided to tentatively reduce the delay in food intake and to take the original salinity level of the large juveniles into account, thus the growth characteristics of juveniles caught in 1995 (group-2 fish) were tested in parallel either in freshwater or in brackish water of about 15‰. Fish caught in summertime (brackish water) resumed their initial weight faster than those caught in spring, 1.2 ± 1.2 vs 3.6 ± 1.5 months on average. Growth curves (mean weight and SD) of juveniles raised in fresh or brackish water were similar, especially from the middle of 1997 onwards (Fig. 31.1). Prior to this date, the mean weight of the batch of fish kept in brackish water was significantly higher than the mean weight of fish kept in fresh water, because at the start the mean weight was 234 ± 73 g and 503 ± 116 g in freshwater and brackish water respectively. After 5 years, at the end of the experiment, fish (6.5 years old) weighed 6.4 ± 2.9 kg and 5.3 ± 2.5 kg in brackish and fresh

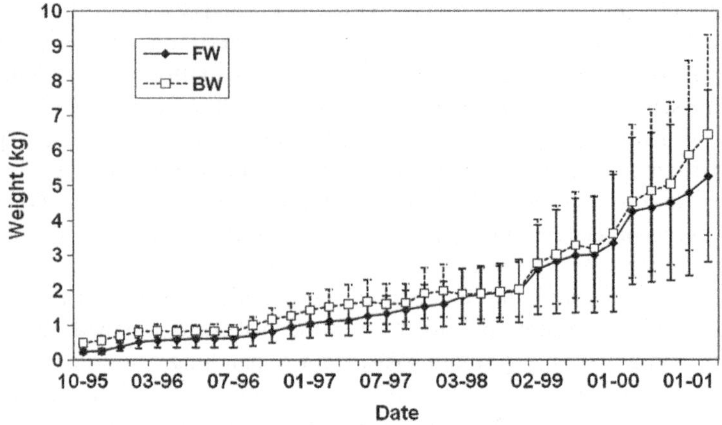

Fig. 31.1 Growth of juvenile *Acipenser sturio* born in the wild in 1994 and held in either fresh water (*FW*) or brackish water (*BW*) (each *dot* is the average weight value for 20 fish till January 1997, 19 fish till April 2000, and 15 and 16 after this date for FW and BW, respectively). *Bars* are standard deviation. Though one third of distributions were not normally distributed, mean ± SD were retained (after Williot et al. 2007)

water respectively, without significant differences ($P < 0.23$) in growth. Interestingly, the range of weight variation was high in both batches, and there was a trend towards better growth in brackish water. The experiment was stopped because the majority (11/15) of the fish kept in fresh water were lost due to an insecure holding system and human error (Williot et al. 2007). Maximum densities were one individual per m^{-2} and 6.4 kg m^{-2}.

A new salinity-growth replicate experiment was set up (with a new seawater batch as the fish spent most of their adult phase in ocean waters) in early 2001 using all the available juveniles. The composition of the different batches (Table 31.2) respected the history of the fish with regard to the type of water in which they had been kept before; changes between different water types were kept to a minimum. The similar number of individuals per group allowed for comparison between the different groups. The only significant difference was in the smaller fish in group 2, kept in fresh water (Fig. 31.2a), which were smaller due to poor rearing conditions of this subsample in the first year of life. Growth curves (Fig. 31.2) fit well in linear regression, and did not differ significantly within the 3-year experimental period (2001–2004) whatever the salinity level. It is worth noting that the growth trend for the two batches in seawater (Fig. 31.2c) was in the lower range, as illustrated by the lowest slopes (Williot et al. 2009). It is also worth noting that despite a good linear fit, yearly cycles are visible in salt and brackish water batches. Unfortunately, this experiment stopped abruptly with the accidental loss of the whole fresh water batch. From then on, we considered the individuals in group-3 fish kept in fresh water in Berlin (IGB) as control fish for this type of water (see Chap. 21). In order to explore in more detail how each fish grew during the 3-year experimental period, weight distributions according to batches were plotted at both the beginning and end of the

Table 31.2 Initial composition in September 2001 with juvenile *Acipenser sturio* of a 3-year experimental study concerning type of water

Salinity (‰)	0		15		33	
Number of fish	36		36		36	
Tank number	1	2	3	4	5	6
Number of fish (cohort)	13 (95)	18 (95)		17 (95)	17 (95)	19 (95)
	2 (94)	3 (94)	16 (94)	3 (94)		
Median weight (g)	4,470	2,020	5,455	3,510	5,260	3,470
Stocking density						
(kg/m^3)	4.5	2.5	6.8	4.5	6	4.7
(ind/m^3)	1	1.4	1.1	1.3	1.1	1.3

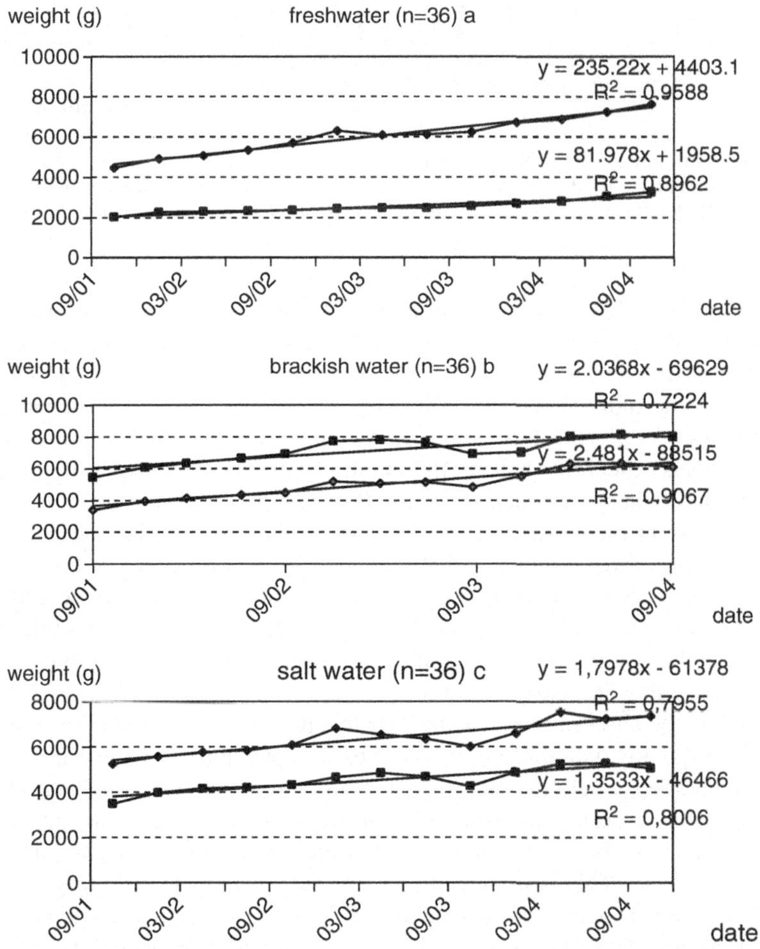

Fig. 31.2 Growth of juvenile Acipenser sturio according to salinity and tanks. Weights represent median values (g) (after Williot et al. 2009)

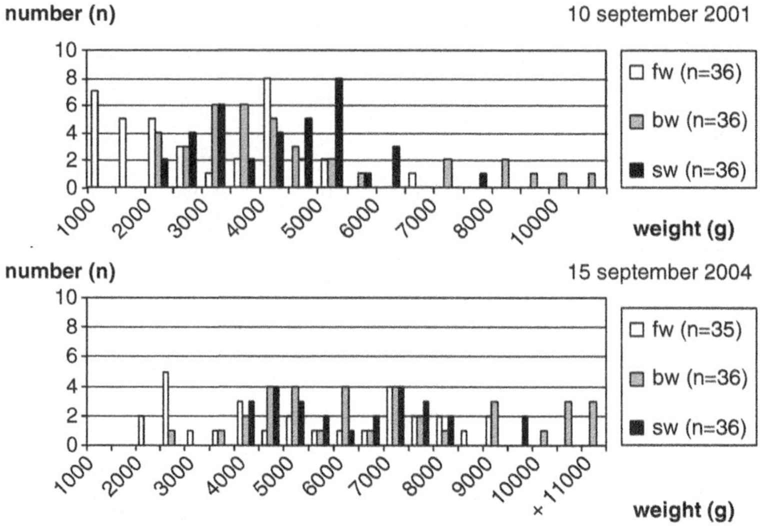

Fig. 31.3 Distribution of weight of juvenile *Acipenser sturio* at the beginning and end of the 3-year experimental period according to the type of water in which they were raised. *FW* freshwater, *BW* brackish water (~15‰), *SW* sea water (~33‰) (after Williot et al. 2005)

period (Fig. 31.3). Due to the selection process, the initial distribution was spread, and the final distribution was spread too. Thanks to individual tagging, the rank correlation coefficient was computed between the two dates. r_s is 0.88, indicating that most of the individuals grew similarly, or at least they did not change their respective rank, i.e., very little significant compensatory growth occurred and no weight loss was observed whatever the salinity level. There was a parallel in growth in both batches kept in brackish water (Fig. 31.2b), indicating similar growth potentials despite their different origin. The results also provide support for the decision to form three batches of individuals from two different year classes and different weights. Some of the fish kept in sea water did not look very healthy, some died and others had reddish-looking skin, and the fish in this batch had the lowest growth rate, though this was not significant. It was therefore decided to move these fish into brackish water early in 2005.

Following the observation made in 1995 that fish in brackish water resumed food intake more quickly than those in fresh water, as a precautionary measure large fish (mainly adults) were held in brackish water throughout the year from early 1996 on. In the growth experiments described above – temperature was about 20°C (range16–24°C) in brackish water – fish were subjected to an irregular cyclic growth (Fig. 31.4). This is in agreement with observations in wild large juveniles during their stay in the estuary, which also exhibited a cyclic growth (Lepage 1995). A batch of juveniles was kept in brackish water and subjected to a cyclical temperature range (10–25°C). The resulting growth was also cyclic, and resumed when temperature decreased (Fig. 31.5).

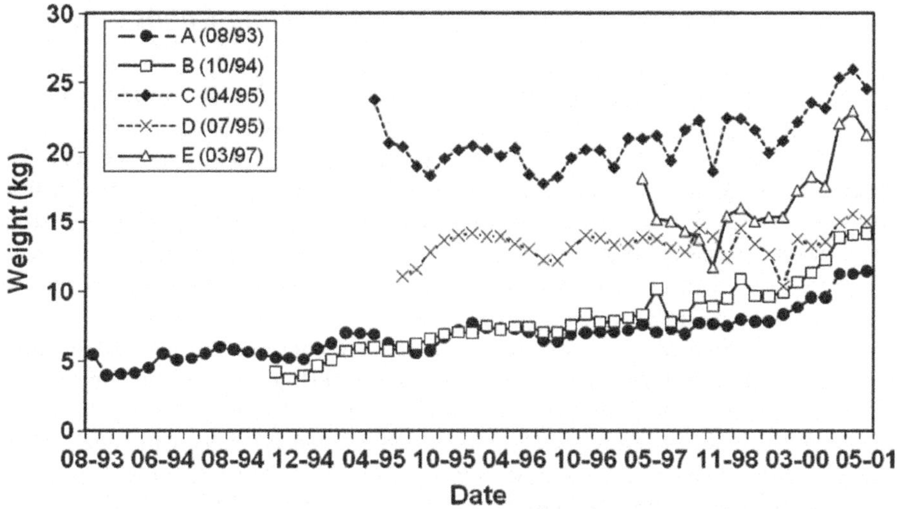

Fig. 31.4 Examples of growth curves of five wild caught *Acipenser sturio* during their first years in captivity (Williot et al. 2007)

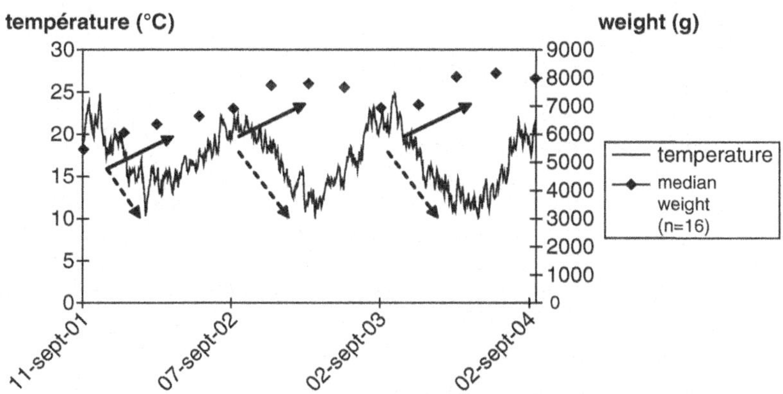

Fig. 31.5 Growth of juvenile *Acipenser sturio* kept in brackish water and corresponding water temperature. Weights are median values (g) and temperatures (°C) are daily values. *Solid arrows* underline increasing growth, and *dotted arrows* signal corresponding decreasing water temperature (Fr) (Williot et al. 2009)

Overall, all the fish were held in brackish water from 2005 onwards, and no other attempt to explore the salinity factor was made with the available fish.

One should note that less than 10 kg wild-originated Atlantic sturgeon (*A. oxyrinchus*) took from 2 to 14 weeks to accept the various passively introduced diets, and an additional 2–23 weeks to successfully train to commercial diet when kept in a water temperature range of 8–25°C and with water salinity in a range of 6–12‰ (Lazur et al. 2010).

31.3.2 Acclimation of Large Fish to the Holding Conditions

These fish were predominantly captured by fishermen, and it was difficult to have them adapt to the holding conditions despite many efforts. During their reproductive migration, brood fish do not take food. Probably as a result, the delay in resuming food intake was long, as illustrated by the time lapse during which fish lost weight (Fig. 31.6). This loss can be up to 30% of the initial weight, or even more in some cases. Current loss was about 20%. One large male, despite his great weight loss (marked by an arrow in Fig. 31.6), recovered and provided good quality sperm some years later.

The weight loss was plotted against the initial weight (Fig. 31.7). Small juvenile fish recovered faster (Fig. 31.7a) than large adult fish (Fig. 31.7c). Three to 5 years were needed for these adult fish to recover their arrival weight. Among the small juveniles, those with the lower weight loss corresponded to fish for which the whole capture process (short tow duration, <30 min, same original water for transportation, and quick transportation into the hatchery) had been quick and efficient. Two

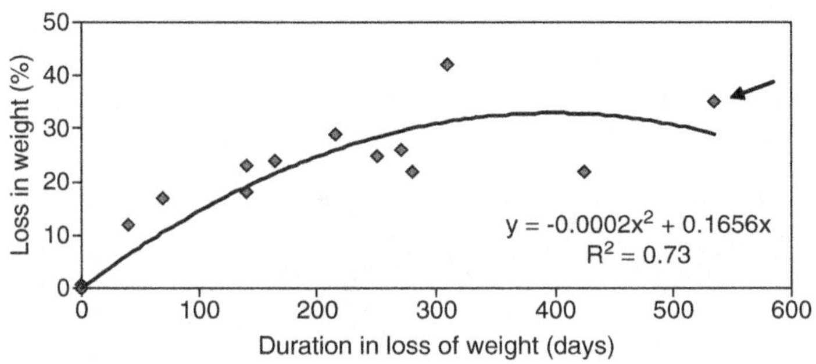

Fig. 31.6 Loss of weight (as percentage of initial weight) in wild *Acipenser sturio* during the adaptation phase in the holding facility (after Williot et al. 2007)

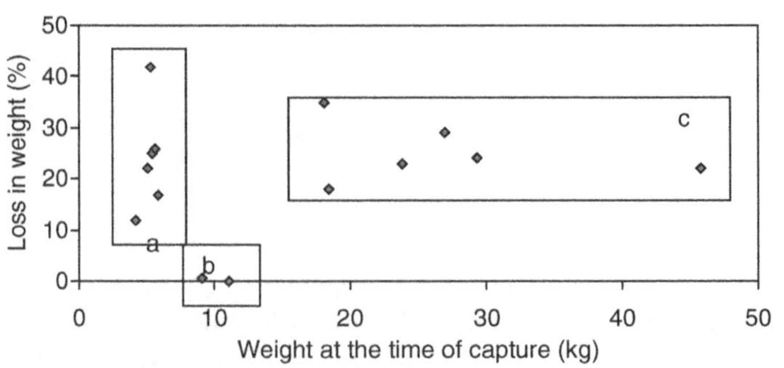

Fig. 31.7 Loss of weight (as percentage of initial weight) of wild *Acipenser sturio* in relation to weight at the time of capture during the adaptation phase in the holding facility (after Williot et al. 2007)

large juveniles did not lose weight at all, and started to feed quickly, soon after capture (Fig. 31.7b). This size most probably corresponds to a very peculiar capacity of the fish to adapt to new conditions (Magnin 1962).

The influence of water depth (1 and 2 m deep tanks) on growth was investigated by transferring fish (during the quarterly sample) from a 1 m tank to a 2 m tank or vice versa. Weight was mainly controlled within 3 months. Ten out of the 12 fish exhibited better growth in deeper tanks ($P = 0.008$). It is worth noting that in one fish there was no weight change after transfer, and in another there was weight loss for both types of transfer. These attempts should be repeated, taking into account the seasonal changes highlighted above.

31.3.3 Overall Management and Mortalities

Patterns of physico-chemical characteristics of the water are very similar in all tanks. Outlet ammonium content varied greatly throughout the year, with peaks in the range of 120–160 mg l^{-1} and minima ranging from 10 to 60 mg l^{-1}, thus revealing a large variation in the yearly metabolic activity of the fish. The good functioning of the biological filter was verified, as the inlet ammonium content remained very low, mainly $\leq$20 mg l^{-1} throughout the year.

From autumn 2006, a water temperature control was set up to maintain the water below 20°C in summer (see Chap. 32).

Consumed food and weight were plotted for the same experiment as described above (Fig. 31.8). The experimental period was divided into five phases. Unmarked phases were similar, as growth remained stable and consumed food displayed sudden changes around a fairly stable trend. Phases a and b were high-growth periods, with a corresponding increase in consumed food. The fish were observed in autumn and winter during 2002–2003 and 2003–2004, i.e., in the period when

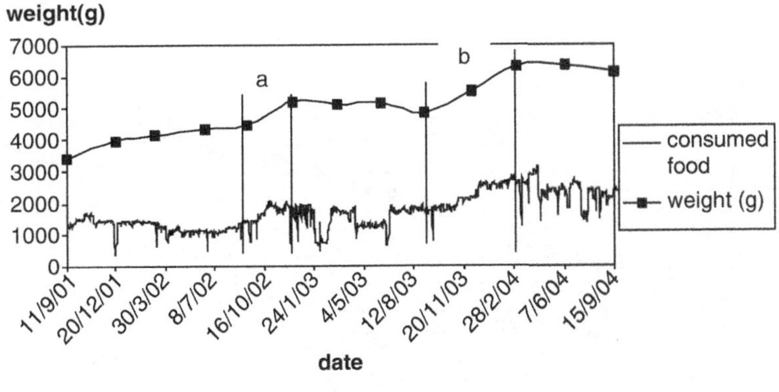

Fig. 31.8 Growth of juvenile *Acipenser sturio* in brackish water and consumed food (g). Weight (g) represents median values. Consumed food (g) is the daily difference between quantities of food delivered and recovered. Phases a and b correspond to the higher growth periods (modified after Williot et al. 2009)

the temperature was rather low. These observations strongly suggest that this is a temperate species (the Atlantic population at least) with preferences for water temperatures around or below 20°C.

Between 1993 and 2006, fish were lost for different reasons (Table 31.3). This account ($n = 78$) includes neither the young specimen lost during feeding experiments (see Chap. 28) nor the two young wild juveniles (group-2 fish) which died soon after arriving in the rearing facility. Despite the security installations, ironically, the majority of losses (~84%) were due to human failure and/or an insecure system (11 mortalities in summer 2001 and 55 late 2004). Those two batches were each full siblings, thus limiting the genetic loss. Unknown causes accounted for nine fish in total, i.e., 11%. Among those fish was the dam which spawned in 1995. The fish never recovered, and 2 years after capture it lost weight abruptly and died. The first three fish which died (Table 31.3) apparently died due to intoxication, as some Siberian sturgeons which were introduced into the tank also died. Observations suggested that both the food and the closed water system were the causes. Altogether, about half of the fish were lost, with 77 fish still alive (Table 31.1).

Table 31.3 *Acipenser sturio* mortalities during the building of the broodstock in the period 1993–2006 (after Williot et al. 2009)

Origin	Year of birth	Number (period)	Causes (number)
Wild	1984–1988	6 (1997–2001)	Intoxication (3) Unknown (3)
	1994	3 (late spring)	Unknown
		11 (summer 2001)	Human error and/or insecure water system
Artificial breeding	1995	55 (late 2004)	Intoxication due to an influx of formaldehyde
		3 (mid 2006)	Unknown
Total		78	

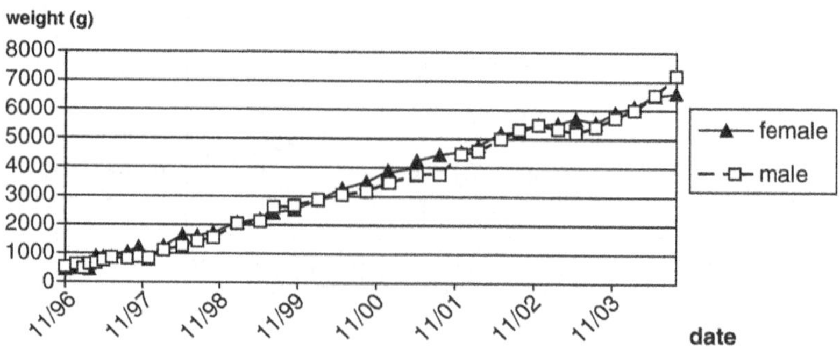

Fig. 31.9 Growth of juvenile male and female *Acipenser sturio* artificially produced in the hatchery in 1995 and which died accidentally in November 2004. Dead fish numbered 55, i.e., the current freshwater fish-group ($n = 36$) and 19 specimens held in three separate freshwater tanks. The database allowed us to reconstruct their growth from late 1996 onwards

After the batch of fish kept in freshwater (see above) died accidentally in 2004, we were able to determine the sex of the fish. The sex ratio was balanced. As these fish were individually tagged, their growth curve could be plotted. This demonstrated that there was no sexual difference in growth rate up to about 7 kg weight (Fig. 31.9); at this size the fish were still immature.

31.4 Conclusions

A general decision-making schema related to building a brood stock of an endangered species, the European sturgeon, *A. sturio*, is presented, based on acquired experience, including the general rules as references to help in case of any unexpected situation.

At present (2010), the brood stock is composed of three groups of fish (Table 31.1): (1) group 1 represents the oldest fish, which are mainly males, (2) group 2 represents wild fish from the natural reproduction of 1994, and (3) group 3 were artificially produced (full siblings) in 1995, Altogether there are less than 80 fish, comprising a restricted genetic variability. It would be very difficult to increase genetic variability, as there are few fish left in the wild. A successful integration of new wild fish into the brood stock depends on the extent to which the fish successfully adapt to captivity conditions. Juvenile fish adapted more easily and started to feed more quickly than adult fish. The comparison of the three types of water (fresh, brackish and salt water) based on the growth of the juveniles did not reveal large differences. The decision to keep most of the fish of the brood stock in brackish water was finally supported by a good compromise with the ecology, quicker resumption of food intake by wild-originated fish, continuous growth, better-looking, maturation of gonads and successful artificial reproduction over several years (see Chap. 32). The environmental factors to stimulate gonad growth were cyclic variations in temperature (25–10°C and 20–10°C respectively) and natural photoperiod. The only food accepted on a regular basis was two kinds of shrimp; however, in the future, compound diets (see Chap. 33) will need to be developed to reduce costs and to match the physiological requirements of this species. There are indications of irregular cyclic growth for large fish and more regular cycles for juvenile fish. Best growth rates apparently occur at temperatures between 15 and 20°C. There is no sexual difference in growth rate in juvenile fish up to 7.5 years of age. Mortalities accounted for about 50% of the fish, most being lost due to human error.

References

Arlati G, Bronzi P, Colombo L, Giovannini G (1988) Induced breeding of the Italian sturgeon (*Acipenser naccarii*) raised in captivity. Riv Ital Acquacolt 23:94–96

Lazur AM, Markin E, Van Heukelem W (2010) Initial evaluation of various foods and diets in feed training wild-caught Atlantic sturgeon (*Acipenser oxyrinchus*) onto a commercial diet. J Appl Ichthyol 26:420–423

Lepage, M (1995) Partie A: Bilan des opérations menées sur l'esturgeon Acipenser sturio de 1984 à 1993. Partie B: Expérience étrangère en matière de repeuplement d'esturgeon. Programme Esturgeon. Cestas, 37 p

Ludwig A, Williot P, Kirschbaum F, Lieckfeld D (2004) Genetic variability of the Gironde population of *Acipenser sturio*. In: Gessner J, Ritterhoff J (eds) Species differentiation and population identification in the sturgeons *Acipenser sturio* L. and *Acipenser oxyrinchus*, vol 101. Bundesamt für Naturschutz, Bonn, pp 54–72

Magnin E (1962) Recherche sur la systématique et la biologie des Acipenséridés *Acipenser sturio* L., *Acipenser oxyrinchus* Mitchill et *Acipenser fulvescens* Raf. Ann Stat Centr Hydrobiol Appl 9:7–24

Struffenegger P (1992) Sturgeon farming in California: a promising new industry. Aquacult Eur 17:6–9

Williot P, Rochard E, Castelnaud G, Rouault T, Brun R, Lepage M, Elie P (1997) Biological characteristics of European Atlantic sturgeon, *Acipenser sturio*, as the basis for a restoration program in France. Environ Biol Fish 48:359–370

Williot P, Brun R, Pelard M, Mercier D (2000) Unusual induced maturation and spawning in an incidentally caught pair of adults of the critically endangered European sturgeon *Acipenser sturio* L. J Appl Ichthyol 16(6):279–281

Williot P, Rouault T, Brun R, Pelard M, Mercier D (2002) Status of caught wild spawners and propagation of the endangered sturgeon *Acipenser sturio* in France: a synthesis. Int Rev Hydrobiol 87:515–524

Williot P, Rouault T, Rochard E, Castelnaud G, Lepage M, Gonthier P, Elie P (2004) French attempts to protect and restore *Acipenser sturio* in the Gironde. In: Gessner J, Ritterhoff J (eds). Status and perspectives, the research point of view. Bundesamt für Naturschutz 101:83–99

Williot P, Brun R, Rouault T, Pelard M, Mercier D (2005) Attempts at larval rearing of the endangered western European sturgeon, *Acipenser sturio* L. (Acipenseridae), in France. Cybium 29:381–387

Williot P, Rouault T, Pelard M, Mercier D, Lepage M, Davail-Cuisset B, Kirschbaum F, Ludwig A (2007) Building a broodstock of the critically endangered sturgeon *Acipenser sturio* L.: problems associated with the adaptation of wild-caught fish to hatchery conditions. Cybium 31:3–11

Williot P, Rochard E, Rouault T, Kirschbaum F (2009) *Acipenser sturio* recovery research actions in France. In: Carmona R, Domezain A, Garcia Gallego M, Hernando-Casal J, Rodriguez F, Ruiz Rejon M (eds) Biology, conservation and sustainable development of sturgeons, Fish & fisheries series. Springer, New York, pp 247–263, 467 p

Chapter 32
Reproduction of the Cultured Brood Fish

Patrick Williot and Patrick Chèvre

Abstract This chapter describes the last management procedure applied to the brood fish. There are five steps, from the staging of sexual maturation advancement to the embryogenesis survey, i.e., in the last 2 months prior to spawning. Results are reported for the few oldest fish (born between 1981 and 1988 in the wild), composed mainly of males (only one female) that have been reconditioned, and the two youngest year classes (1994 and 1995).

The oldest males exhibited very heterogeneous maturation figures, with a 2-year maturation occurring most frequently. Partial maturations in several females are reported, possibly due to inappropriate rearing conditions, namely temperature regime, which was later changed. Precocious reproductive successes were obtained with 13-year-old males and females whatever the year class of younger fish. There are suggestions of a 2-year recurring cycle in females. New indications are given for improving the prediction power of the in vitro maturation test.

32.1 Introduction

Reproduction is the key phase, validating the management techniques applied to the brood fish throughout their growth period from juvenile to puberty, or during the reconditioning period for wild-originated brood fish. In most sturgeon species, this period is long; several years are needed, and *A. sturio* is no exception (Magnin 1962; Holčik et al. 1989). This means that the more advanced the gametogenesis,

P. Williot (✉)
Sturgeon Consultant, 4 Rue du Pas de madame, 33980 Audenge, France
e-mail: williot.patrick@neuf.fr

P. Chèvre
Cemagref, Estuarine Ecosystems and Diadromous Fish Research Unit, Moulin de la logerie, 33660 Saint-Seurin-sur-l'Isle, France

P. Williot et al. (eds.), *Biology and Conservation of the European Sturgeon Acipenser sturio L. 1758*, DOI 10.1007/978-3-642-20611-5_32,

the more the consequences of rearing conditions on the success of reproduction are detectable. The present chapter is therefore closely related to the earlier Chap. 27.

However, the husbandry of the species is still in its infancy, i.e., there is no knowledge that might be considered as a set of reference guidelines containing conclusions with regard to stability in the "rearing conditions–reproduction" combination.

Despite the above-mentioned remarks, the significant progress that has been achieved recently is presented as a preliminary foundation that can be built on for reproduction success.

32.2 Brood Stock Management

The brood stock is composed of three groups (see Table 31.1). Fish in group 1 ($n = 7$) are the oldest fish, and are of wild origin. They were born in 1988 or earlier, were mature when they arrived at the hatchery, and all are male with the exception of the only female. Group 2 consists of 29 individuals, born in the wild in 1994. They are believed to be full siblings (Ludwig et al. 2004), and even two males cannot be excluded (Villazain-Romo, personal communication). Fish in group 3 were the progeny (full siblings) of the artificial reproduction performed in 1995 (Williot et al. 2000a). With the exception of the fish in group 1, the sex ratio is balanced. The fish were passive integrated transponder (PIT)-tagged, and a Petersen disc was fixed at the base of the dorsal fin to which coloured markers were also attached for easy identification of the fish gender within the tanks (see Chap. 31).

Most of the management procedures are similar to those described in the earlier Chap. 27. Rearing conditions – tanks (4 m in diameter and 1 or 2 m depth), water salinity (~15‰), yearly temperature regime (12 to 23–25°C up to mid-2006 and then 12 to 18/20°C (Williot and Rouault 2008) and light regimes (low intensity under regular photoperiod), feeding (frozen shrimps ad libitum), current control in the tanks (cleaning), water controls (physio-chemical characteristics), and fish handling and checking (anaesthesia post-capture in the tanks on a 3-month basis) – are totally similar to those already described (see Chap. 27). It has to be noted that from March 2009 onwards, only 1-yearly controls are to be carried out. Each tank is part of an individual closed water system that includes sand filter, bacteria filter, UV

Table 32.1 Maturation in oldest males over time (group-1 fish born between 1981 and 1988)

Name	1999	2000	2001	2002	2003	2004	2005	2006	2007	2008
Phil		x	x		x					x
Just	x		x	x					x	
Mario		x	x	x	x	x	x	x		x
Ascen		x								
Emel		x	x				x	x		x
Nausi		x	x							

treatment and oxygen enrichment. Temperature, salinity, pH and oxygen are checked daily. Ammonium, nitrite and nitrate are controlled monthly.

The ultimate fish management of brood fish requires five steps: (1) food deprivation of potentially mature fish based on observations made during the preceding controls, about 1 month prior to the start of pre-selection, (2) pre-selection of the fish, (3) simulation of upstream spawning migration by transferring the fish from brackish water to fresh water, (4) definite selection prior to hormonal injection, and (5) reproduction per se, i.e., injection, collection, and treatment of gametes.

Stopping feeding corresponds to *A. sturio's* species-specific trait (as in most anadromous sturgeon species) at the onset of their upstream reproductive migration (Magnin 1962). The food deprivation procedure was set up from 2007 onwards.

Pre-selection consists of staging sexual maturation by sampling a piece of gonad tissue via biopsy. This pre-selection of mature fish is usually performed in May, as this corresponds to the peak period of wild brood fish captures in rivers, when fish swim upstream for potential spawning (Williot et al. 2002, see Chap. 12). Since 2008, the two-step pre-selection has been organised as follows: (1) steroid level plus echography, and (2) biopsy and subsequent observations of ovarian follicles. Males with developed testes are selected and their gonad texture graded as described for the Siberian sturgeon (Williot et al. 2000b). For females, the pre-selection process assesses the size, homogeneity, appearance, and degree of migration of germinal vesicle expressed by the polarisation index (PI) of a sample of ovarian follicles (ovf) (Williot 2002). The lower the PI ($\geq$15 ovarian follicles), the more advanced the maturation.

These pre-selected fish, held in brackish water throughout the year (see Chap. 31), are then transferred into tanks supplied with fresh water for 1 week–10 days to simulate the upstream spawning migration. At the end of this short time lapse, some female ovarian follicles are sampled again. The current value of PI (migration of the nucleus or germinal vesicle to the animal pole) is assessed, and an in vitro maturation competence test of ovarian follicles is performed.

The selection process for ripe females relies on a biotest that evaluates the in vitro maturation competence of ovf. The in vitro maturation competence test is performed as follows: ovf (n = 33) are placed one by one in the standard Siberian sturgeon incubation media (SIS medium) to which progesterone is added (1 μg ml^{-1}) (or not = control) for 24 h at 18°C (Williot et al. 1991; Williot 1997). Each follicle is then evaluated for absence of germinal vesicle envelopes, i.e., germinal vesicle breakdown (GVBD), or presence of germinal vesicle (GV). When GVBD is at least 80–90%, the fish are considered as potentially reactive to hormonal stimulation.

Males and females are injected with mammalian GnRH at a rate of 5 and 10 μg kg^{-1}. Injection is performed directly in the tank without handling the fish, or through the mesh of a sock-like net, allowing the fish to be gently immobilised. In cases where fish are not in an optimal reproductive state (as judged by some of their morpho-physiological characteristics) a two-injection procedure might be applied; an initial priming dose (one-tenth of the dosage) followed by the remainder of the dose 12 h later (Williot et al. 2000a). Males are hormonally injected a few hours before the females, in order to get the best quality sperm when ovulated eggs are collected, as demonstrated in the Siberian sturgeon (Williot et al. 2000b).

Hormonal injections, checking of ovulation, collection of ovulated eggs (stripping and further small abdominal opening) and semen (polypropylene tube and beaker, fertilisation (water-diluted semen), rinsing, removing stickiness (water suspension of clay), and care of brood fish are carried out as previously described for Siberian sturgeon (Williot et al. 1991; Williot 2002). The collection time for the ovulated eggs is delayed for 4–6 h post-onset of ovulation to get the best fertilisable products, as shown for the Siberian sturgeon (Williot et al. 1991). Semen is selected mainly on a motility basis, which takes into account a visual estimation of the number of motile spermatozoa (%), and their speed and type of movement on a 5-grade scale (where rating 5 is the most active semen; Williot et al. 2000b). Semen is stored in covered beakers in the refrigerator (5–8°C) until use.

Incubation is performed in McDonald jars as part of a closed thermo-regulated system at 18.5°C (±0.2°C) equipped with UV treatment. As the quality of the gametes is not consistent, development of *Saprolegnia* sp. was prevented in this case by applying a malachite green treatment[1] (5–10 g m^{-3} for 20–30 min, no water renewal during treatment) in 1995 (Williot et al. 2000a). Later on, other treatments are applied. At the onset of hatching, jars are placed in troughs (2.5 × 0.5 × 0.3 m) where the hatchlings are collected. During incubation, success of embryogenesis is monitored by sampling 90–180 embryos in order to limit potential defects from the monitoring process. This approach is more risk-averse than the recommended monitoring of 300 specimens for survival assessments (Dettlaff et al. 1993).

32.3 Results and Discussion

32.3.1 Oldest Fish (Group-1)

The first signs of maturation (developed testis) in males were observed in 1999 in Group-1 fish, i.e., the oldest fish born in the wild between 1981 and 1988 that had been re-conditioned. Similar observations were made in the following years (Table 32.1). Five out of the six old males matured in 2000 and 2001, with four of them maturing in two consecutive years. In the remaining years, there were only one or two fish that matured. One fish matured every year between 2000 and 2008. Some fish matured for 2 consecutive years, then again 2 years later or vice versa. As a result, the maturation frequency of males is very heterogeneous. The overall pattern of maturation frequency shows a maximum over 2 successive years (Fig. 32.1) which is significantly higher than the other frequencies (U-test, $P = 0.044$). One well-repaired fish (Emel, Table 32.1) was born in 1988 in the

[1]Malachite green has not been allowed to be used in France for some years.

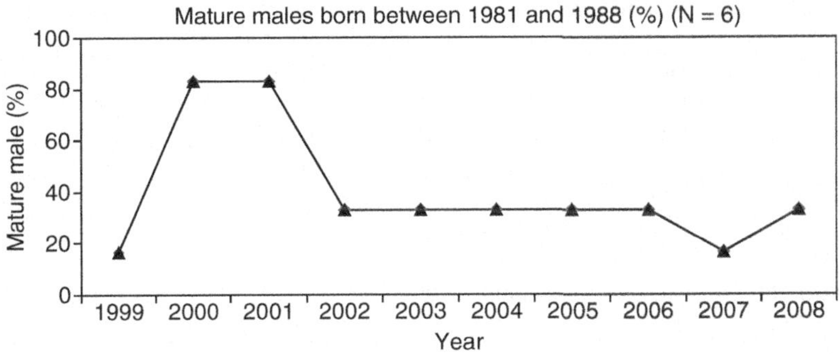

Fig. 32.1 Yearly totals for mature males. Fish are of wild origin, and entered the hatchery as mature fish. Fish were born in 1988 or before. Data are yearly totals from Table 32.1

wild and started to feed very quickly upon entering the hatchery; it matured for the first time in 2000 at 12 years old. From the beginning of maturation, some fish were hormonally injected to test the influence of duration in fresh water (simulation of upstream reproductive migration) on spermiation (Williot et al. 2007, see Chap. 27). Sperm were obtained in the majority of attempts. At least in some cases, sperm showed their fertilisation potential (Kopeika et al. 2000, see Chap. 35). Moreover, these successful attempts in obtaining sperm from reconditioned males enabled us to perform preliminary sperm cryoconservation trials (Kopeika et al. 2000).

The appearance of the first large greyish-dark-coloured ovarian follicles occurred in 2002 in the oldest female. Only a few ovarian follicles could be sampled, as this was the sign of a partial maturation of the ovaries. Moreover, most of the ovarian follicles were brittle and showed unusual patterns, and the abdomen of the fish was not swollen (Williot and Rouault 2008). Similar observations were made in 2005 on the same female and on two group-2 females (11-year-old fish), and in 2006 on three group-1 full sibling females, including the two from the previous year. In most cases, hormonal stimulation increased the rate of GVBD up to the threshold level mentioned above. It is worth noting that, in the majority of the females, PI values exhibited a wide range of variation (e.g., 0.08–0.16) thus illustrating an asynchronous progress in maturation. Further hormonal treatment did not induce ovulation. The oldest female never exhibited complete maturation. Though kept in fresh water (salinity $< 1‰$), male Atlantic sturgeon provided sperm after 6 years of captivity (Mohler and Fletcher 1999).

32.3.2 Youngest Fish (Group-2 and Group-3 Fish Born in 1994 and 1995 Respectively)

First maturations in males born in 1994 (group-2 fish) were observed in 2007 for three fish, i.e., 25% of all the males of the group; two of them were retained for

Table 32.2 Examples of male response to post-hormonal injection: volume, aspect and motility of spermatozoa (spz) in *A. sturio* males in 2007 (after Williot et al. 2009)

Code number (birth year; weight in kg)	Semen volume (ml)	Aspect	Initial motility (%)	Motility post-water addition Percentage of moving spz and intensity of moving (0–5)[a]			Held for mating
				30″	1′	1′30″	
1 (1994; 14.4)	~40	Clear	Very rare spermatozoa				No
2 (1984/1985; 24.3)	~180	Light milky	0	50%, 4	20%, 3	5%, 2	Yes
3 (1994; 17.6)	~30	Light milky	0	95%, 5	90%, 4	50%, 3	Yes
4 (1994; 7.6)	0						No

spz Spermatozoa
[a]Graded from 0 (no movement) to 5 (spz are moving out of the microscope field) and see text

fertilisation (Table 32.2). In 2008, nine fish belonging to the two groups (36% of the total) provided sperm, and ten in 2009, or 40%. The three fish which matured in 2007 exhibited a further maturation in 2008. Two fish matured in both 2008 and 2009. The results obtained in 2007 illustrate the variability in males' response in terms of volume of sperm and of motility of spermatozoa. Whatever the cohort of the males (1994 or 1995), first maturation occurred in 13-year-old fish.

The first complete maturation of cultured females occurred in 2007 with one fish (female 3) out of the four in the group-2 females (born in 1994) (Table 32.3) (Williot et al. 2009). The other three exhibited partial maturation, illustrated by difficulties in ovarian follicle sampling, the brittleness of the ovarian follicles and the high heterogeneity of germinal vesicle migration (PI). This is a possible consequence of underdevelopment due to a slightly smaller size of the ovarian follicles, thus revealing an asynchronous maturation process. Additionally, the fully mature female was the only one of the three tested whose ovarian follicles provided a typical response to the biotest, positive in this case with regard to reproductive potential, with 90% GVBD in presence of hormone and 0% GVBD (or 100% GV) in control. One female provided the same result (80% GVBD) in both presence and absence of the hormone (female 1), and another exhibited a high rate of damaged cells (female 4). As neither of these last two females ovulated, these results underlined the great advantage of carrying out a control of the biotest (In vitro maturation competence) so that the reproductive potential of a given female can be accurately assessed prior to the hormonal injection. In a normal production process, these two females would not have been injected. In the following years, a similar methodology was applied and enabled us to select some mature fish that were also injected, and most of which ovulated (Table 32.4). However some remarks are worth noting. Of the four group-3 females that ovulated in 2008, only one produced progeny. None was obtained from the other three, and the males cannot be held responsible, as they fertilised other batches of eggs from which progeny were obtained in large numbers. The origin of the failure is due to the females, and

Table 32.3 Morpho-physiological characteristics of the cultured pre-selected maturing *A. sturio* females in 2007 (after Williot et al. 2009)

Code number (birth year; weight in kg)	Observations	Diameter of ovarian follicles, mean in mm	Polarisation index (PI), mean (min–max)	In vitro maturation competence of ovarian follicles			Held for hormonal injection
					Hormone	Control	
1 (1988; 19.5)	Abdomen not swollen Difficulties in sampling ovf Ovf delicate	2.3	0.11 (0.07–0.125)	GVBD	80%	80%	Yes
				GV	–	–	
2 (1994; 27.4)	Abdomen not swollen Difficulties in sampling Ovf very delicate	2.3	0.15 (0.10–0.19)	GVBD	Not tested		No
				GV			
3 (1994; 8.5)	Abdomen not swollen Easy to sample	2.5	0.12 (0.07–0.22)	GVBD	90%		Yes
				GV		100%	
4 (1994; 19.2)	Abdomen not swollen Difficulties in sampling Ovf delicate	2.3	0.13 (0.07–0.19)	GVBD	100%		Yes
				GV	(56%)[a]	100%	
				Damaged		(75%)[b]	
				(nb of ovf)	14	7	

Ovf ovarian follicle
[a]GVBD is 100% against the observable ovf, and 56% when taking into account those which were damaged and not observable, and which could have been non-mature (GV)
[b]Similar reasoning for GV as above for GVBD

unfortunately the absence of observation means that no hypothesis can be put forward to avoid such a situation in the future.

Surprisingly, none of the group-3 females matured in 2009. Finally, the only female that ovulated in 2009, a group-2 fish, is the one that had already spawned in 2007. None of the 6–8 maturing females in 2010 ovulated. The considerably higher number of larvae obtained in 2009 compared with 2007 does provide support retrospectively for the changes made in brood fish management throughout 2006 with regard to temperature regime and food deprivation. Plotting the results (Fig. 32.2) of maturing females from group 2 and group 3 suggests a dominant 2-year cycle in oogenesis. Precocious complete maturations in both groups were observed for 13-year-old females. Latency in ovulation is 24 h post-second hormonal injection at ~18°C (2007) (Williot et al 2000a). Later on, latencies close to 30 h were recorded.

Table 32.4 Synthesis of maturation and reproductive results for females born in 1994 and 1995

		Birth year		
Experimental year	Criterion	1994 (group 2)	1995 (group 3)	Mean weight (kg)
2007	Matured	4	–	18.7
	Injected	3	–	15.7
	Ovulated	1[a]	–	–
	Larvae	11,000	–	–
2008	Matured	9	6	15.7 ± 4.1
	Injected	6	6	16 ± 4.9
	Ovulated	3	4	15.1 ± 5.3
	Larvae	38,760 from three ♀	55,784 from one ♀	–
2009	Matured	4	–	12.1 ± 3.8
	Injected	2	–	9.0
	Ovulated	1[b]	–	9.0
	Larvae	52,300	–	–

Data are number of fish (either female or larvae). Injected females selected by in vitro maturation competence test (see text)
[a]There were objective observations to explain the defects between injection and ovulation (see Table 32.3)
[b]This is the same female that had already spawned in 2007

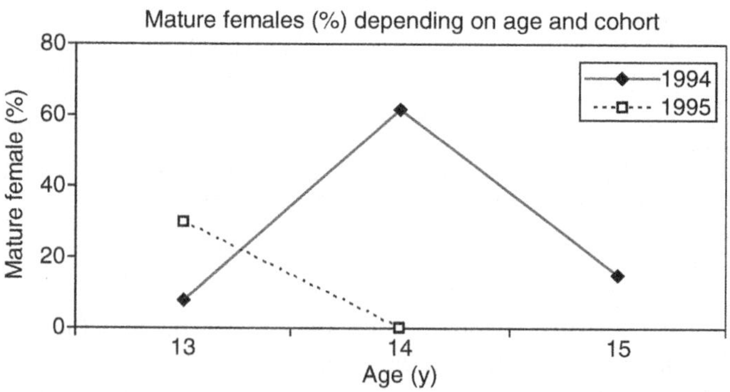

Fig. 32.2 Mature females depending on cohort and age

Embryogenesis was checked carefully in 2007 (Table 32.5). As two sperms exhibited some fertilisation potential (Table 32.2), two batches of eggs from the only ovulating female were prepared, each of them being fertilised by a given sperm in order to increase genetic diversity as far as possible. The fertilisation rate was moderately high (70–80%). There was no significant difference between the two batches during embryogenesis, thus proving that the fertilisation potential of the two sperms was similar. Survival rates decreased by 50–60% during the embryogenesis. As a result, survival was only 40% at the onset of hatching, which is a typical figure for moderate-quality eggs.

Table 32.5 Embryogenesis success of *A. sturio* in 2007 at 18.5°C (after Williot et al. 2009)

Time post-fertilisation (h)	4	24.30	73.50	90.15
Stage name[a]	2nd cleavage, initiation of 3rd cleavage	Early and middle gastrula	Tail tip approaches the heart (onset of heart beating)	Onset of hatching
(Stage number)[a]	(5–6)	(14–15)	(30)	(34–35)
Survival (%)				
♀3 × ♂2	70[a]	66[b]	47[c]	39[d]
♀3 × ♂3	80[a]	71[b]	54[c]	38[d]
	($z = 1.7$; $P = 0.09$)	($z = 0.57$; $P = 0.57$)	($z = 0.92$; $P = 0.36$)	($z = 0.12$; $P = 0.9$)

[a]According to Dettlaff et al. (1993)
[b]Same superscript number in a given column denotes the absence of significant difference

32.4 Conclusions

Though reproduction successes are still far from what is currently expected in sturgeon farming, the present preliminary findings might be considered as a first step in validating the whole management process. This includes on the one hand the long-term rearing conditions, namely water salinity, yearly temperature regime, feeding, and current management, and on the other hand the ultimate five-step procedure described above. Most of the points underlined could be either improved (temperature regime and feeding, among others) or confirmed (water salinity).

For the first time in sturgeon, partial ovary maturation is reported in several fish from different year-classes, and this has sometimes recurred. An inappropriate water temperature regime has been suggested as the cause. Careful observations of future generations should have been performed once it had been changed from 2007 onwards.

In the rearing conditions described and with a rather origin-limited brood fish, maturation patterns in males were extremely variable, thus revealing an absence of a general well-established trend. Additionally, this suggests a concentrated 2-year maturation window. This precocious maturation in both males and females occurred in 13-year old fish. Present findings suggest a dominant 2-year oogenesis cycle. There are indications for carefully setting up the biotest in detail (in vitro maturation competence) to increase its prediction power.

References

Dettlaff TA, Ginsburg AS, Schmalhausen OI (1993) Sturgeon fishes. Developmental biology and aquaculture. Springer, Berlin, 300p

Holčik J, Kinzelbach R, Sokolov LI, Vassilev VP (1989) *Acipenser sturio* Linnaeus, 1758. In: Holčik J (ed) The freshwater fishes of Europe. Aula Verlag, Wiesbaden, pp 367–394

Kopeika E, Williot P, Goncharov B (2000) Cryopreservation of Atlantic sturgeon *Acipenser sturio* L., 1758 sperm: first results and associated problems. Bol Inst Esp Oceanogr 16:167–173

Ludwig A, Williot P, Kirschbaum F, Lieckfeld D (2004) Genetic variability of the Gironde population of *Acipenser sturio*. In: Gessner J, Ritterhoff J (eds) Species differentiation and population identification in the sturgeons *Acipenser sturio* L. and *Acipenser oxyrinchus*, vol 101. Bundesamt für Naturschutz, Bonn, pp 54–72

Magnin E (1962) Recherche sur la systématique et la biologie des Acipensérides *Acipenser sturio* L., *Acipenser oxyrinchus* Mitchill et *Acipenser fulvescens* Raf. Ann Stat Cent Hydrobiol Appl 9:7–242

Mohler JW, Fletcher JW (1999) Induced spermiation in wild Atlantic sturgeons held captive up to six years. North Am J Aquacult 61:70–73

Williot P (1997) Effects of incubation media on maturation of isolated ovarian follicles of Siberian sturgeon (*Acipenser baerii* Brandt) induced by sturgeon gonadotropic preparation or 17α, 20β, dihydroxy progesterone. Comp Biochem Physiol 118(C):285–293

Williot P (2002) Reproduction des esturgeons. In: Billard R (ed) Esturgeons et caviar. Lavoisier Tech et Doc, Paris, pp 63–90

Williot P, Rouault T (2008) Basic management for spawning the cultured sturgeon, *Acipenser sturio* L., 1758, a critically endangered species. Cybium 32(2 suppl):334–335

Williot P, Brun R, Rouault T, Rooryck O (1991) Management of female breeders of the Siberian sturgeon, *Acipenser baeri* Brandt: first results. In: Williot P (ed) Acipenser. Cemagref, Antony, pp 365–379

Williot P, Brun R, Pelard M, Mercier D (2000a) Unusual induced maturation and spawning in an incidentally caught pair of adults of the critically endangered European sturgeon *Acipenser sturio* L. J Appl Ichthyol 16(6):279–281

Williot P, Kopeika EF, Goncharov B (2000b) Influence of testis state, temperature and delay in semen collection on spermatozoa motility in the cultured Siberian sturgeon (*Acipenser baeri* Brandt). Aquaculture 189:53–61

Williot P, Rouault T, Brun R, Pelard M, Mercier D (2002) Status of caught wild spawners and propagation of the endangered sturgeon *Acipenser sturio* in France: a synthesis. Int Rev Hydrobiol 87:515–524

Williot P, Rouault T, Pelard M, Mercier D, Lepage M, Davail-Cuisset B, Kirschbaum F, Ludwig A (2007) Building a broodstock of the critically endangered sturgeon *Acipenser sturio* L.: problems associated with the adaptation of wild-caught fish to hatchery conditions. Cybium 31:3–11

Williot P, Rouault T, Pelard M, Mercier D, Jacobs L (2009) Artificial reproduction of captive endangered European Atlantic sturgeon, *Acipenser sturio*. Endanger Species Res 6:251–257. doi:10.3354/esr00174

Chapter 33
Recent Progress in Larval Rearing of the European Sturgeon, *Acipenser sturio*

Patrick Chèvre, Julien Saint-Sevin, Daniel Mercier, Louis Jacobs, and Patrick Williot

Abstract This chapter deals with updated larvae management techniques, focusing on feed management. A compound diet was produced, and the reasoning behind this, the choice of ingredients, and manufacturing process are described here. Stocking, feeding sequences and results are also given. Co-feeding was based successively on 1-day-old *Artemia salina nauplii* (7–35 dph), *chironomids* (20–49 dph), and the compound diets (from 29 dph onwards). For 65-day-old fish raised at a water temperature range of 18.5–19°C, high survival rates (89%), and average weight of 3 g were recorded. Six-month-old fish weaned on compound diet weighed three times more (90 vs 35 g) than their naturally fed counterparts.

33.1 Introduction

Until recently, very few opportunities have occurred to investigate the larval and juvenile rearing of the sturgeon *Acipenser sturio*. The only published work in the field reported best survival rates (~30%) post-weaning on compound diets for 55 days post-hatch larvae (Williot et al. 2005; see Chap. 28). Larvae were first fed *Artemia salina* nauplii and later chironomids (Williot et al. 2005). Long weaning (3 weeks) provided much better survival than short (3 days). This means that long co-feeding proved to be more effective.

Moreover, the fish weaned on a current commercial compound diet exhibited a fairly good growth rate up to 18–24 months. However, from this age onwards, a

P. Chèvre (✉) • J. Saint-Sevin • D. Mercier • L. Jacobs
Cemagref, Estuarine Ecosystems and Diadromous Fish Research Unit, Moulin de la logerie, 33660 Saint-Seurin-sur-l'Isle, France
e-mail: patrick.chevre@cemagref.fr

P. Williot (✉)
Sturgeon Consultant, 4 Rue du Pas de madame, 33980 Audenge, France
e-mail: williot.patrick@neuf.fr

P. Williot et al. (eds.), *Biology and Conservation of the European Sturgeon Acipenser sturio L. 1758*, DOI 10.1007/978-3-642-20611-5_33,

significant majority (compared with fish fed mostly on natural food) showed deformities (68% vs 19%) which then led to the death of the specimen (Williot et al. 2004). Fish fed on natural food were much less abnormal in shape. This strongly suggested that the regular compound food was involved in the abnormalities observed in juveniles.

Sturgeons are semi-ossified fish species, their axial skeleton being composed of cartilage. The main component of the cartilage is glycosaminoglycan (polysaccharide) of which *keratan* sulphate is one component. It is therefore likely that keratin is an important nutrient to support the cartilaginous part of the sturgeon skeleton.

The main aims of the present investigation were: (1) to define the best weaning procedure regarding survival and growth, (2) to provide the fish with acceptable feed particles, and (3) to produce a palatable experimental compound diet able to improve post-weaning survival.

33.2 Material and Methods

33.2.1 Fish, Rearing Facilities, and Current Management

Larvae were obtained from brood fish reproduction (Williot et al. 2009; see Chap. 32). The swimming larvae were counted, and stocked in 2-m-long troughs (0.5 m wide) at an initial stocking density of 3,000 individuals per trough. Water depth was maintained at close to 15 cm, with sufficient flow rate to maintain an outlet content of oxygen close to or higher than 7 mg l^{-1}. The troughs were made of grey polyester, and were all placed inside a building to avoid direct daylight. The troughs and other rearing structures were supplied with running, degassed and re-aerated (through column filled with plastic units) well water at 18.5–19°C throughout the late spring and early summer. Refusals and faeces were removed twice a day. Dead larvae were removed and counted daily. Fish were kept in the troughs for 2.5 months until they reached a mean weight of 3 g. Next, fingerlings were raised in 2-m-diameter tanks supplied with the same water.

33.2.2 Food and Feeding

The principle of co-feeding was retained, with successive food items consisting of *Artemia salina* nauplii, chironomids, and compound diet.

One-day-old *Artemia salina* nauplii remained a key starter food, well-accepted and effective. They were delivered automatically in controlled quantities at a maximum daily rate of 2,500 nauplii. larvae^{-1} from 7 days post-hatch (dph) until 35 dph. The relative quantity delivered ranged from 20% during the first few days of feeding up to 100% between 22 and 25 dph, and then decreased. Chironomids were also delivered between 20 and 49 dph at a maximum rate of 40% of the fish

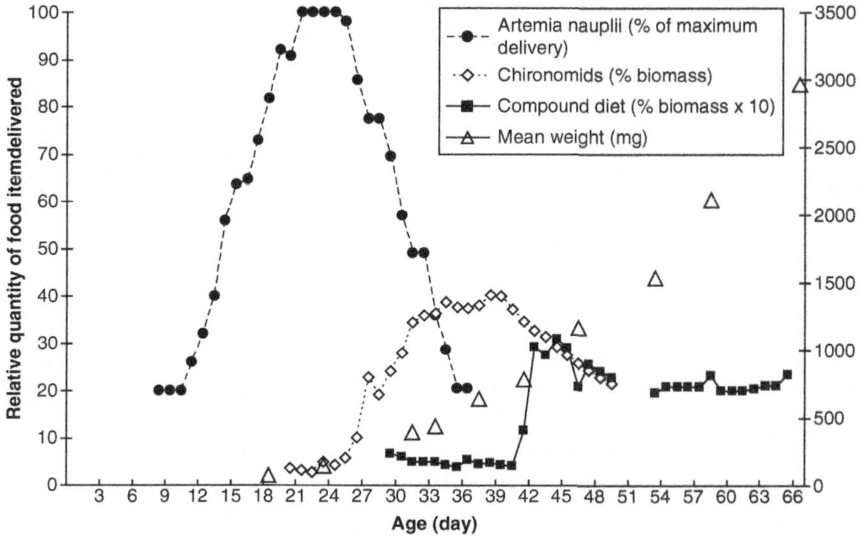

Fig. 33.1 Feeding sequence of *A. sturio* in 2009 until fish are 66 days old with corresponding growth. Weights are mean of three troughs

biomass between 35 and 40 dph. Chironomids were a transition food item between the nauplii and the compound diet. The new prepared compound diet was delivered from 29 dph onwards (Fig. 33.1).

33.2.3 Preparation of a New Compound Food Item

A preliminary test based on a few marine fish foods with a high protein content did not provide satisfactory results due to lack of palatability, which resulted in the sturgeon rejecting the food particles. A commercial food (NRD™ from Ocean Nutrition) was used as a base as it was more porous, contained a high level of protein and krill flour, the latter being rich in keratin. Analysis of the chironomids showed they have a high iron and potassium content.

Thus, to the aforementioned commercial formula (NRD™) were added: (1) blood pork powder rich in iron, (2) shrimp (or krill) powder rich in keratin, (3) fish oil rich in fatty acids, and (4) linseed oil rich in fatty acids (mostly in ω3) and minerals, especially potassium). The components of the compound food are given in Table 33.1. The size of the new particle-pellets started at 0.500 mm, and remained unchanged in composition up to 1.5 mm. Larger sized pellets were composed of only NRD and blood powder.

Overall, composition of the compound diet is now: 65.5% DM (dry matter, percentage of raw material), 62.2% protein (percentage of DM), 19% lipids (percentage of DM), and 22.4 kJ/g DM. Fatty acids (relative content) of sturgeon ovule, as suggested by Gisbert et al. (2002), and different food items, including the

Table 33.1 Components of the compound diet produced for *A. sturio*

Components	Percentage and size of ingredient (μm)	Period (duration)
Chironomid powder	10 (first week) and then 5	First 6 weeks
Krill powder	~7	From second week for 8 weeks
	~5	From ninth week onwards
Shrimp powder	5.2	Last 2 weeks (28/10/09–11/11/09)
NRD™	52 (100–200)	First 3 weeks
	47 (200–300)	One week
	45–46 (300–500)	Two and a half weeks
	46 (300–500)	Two weeks
	55 (300–500 + 500–800)	Three weeks
	57 (500–800)	One week
	56–51 (800–1,200)	Six weeks
Blood powder	9	First week
	13–14.6	Later on
Fish oil	2.1–5.8 (mean = 3.4)	Whole period
Hydrolysable proteins	5.2–3.8	First 4 weeks
	1.9	Later on
Water	20–25.6	

Table 33.2 Relative fatty acid (FA) content in different food items (% of total FA content)

	Food item				
Fatty acid	Ovule	Shrimp (PL)	Shrimp (PV)	Chironomids	Compound diet
Σ Saturated	25.1	21.3	34.1	38.3	26.6
Σ Monosaturated	39.8	27.7	19.3	29.2	34.4
Σ PUFA n-6	8.2	13.5	27.3	19.5	11.0
Σ n-6	6.6	11.1	24.8	16.7	9.7
Σ PUFA n-3	22	28.4	16.3	8.6	25.0
Σ Sat/Σ PUFA	0.8	0.5	0.8	1.4	0.7
n3/n6	3.3	2.6	0.7	0.5	2.6

PL, *Penaeus longirostris* (wild shrimps and preferred food); PV, *Penaeus vannamei* (farmed shrimps)

prepared compound diet, are given in Table 33.2. The fatty acid composition of the prepared compound diet is very similar to that of the ovules.

33.3 Results and Discussion

The mean weight of larvae at 66 days post-hatch is very close to 3 g (Fig. 33.1) which is double the weight observed in a study published earlier (Williot et al. 2005) at a slightly lower water temperature (17.5°C).

The survival rate is 89%, far better than that obtained previously, and similar to that of the Atlantic sturgeon, *Acipenser oxyrinchus* (Mohler 2000).

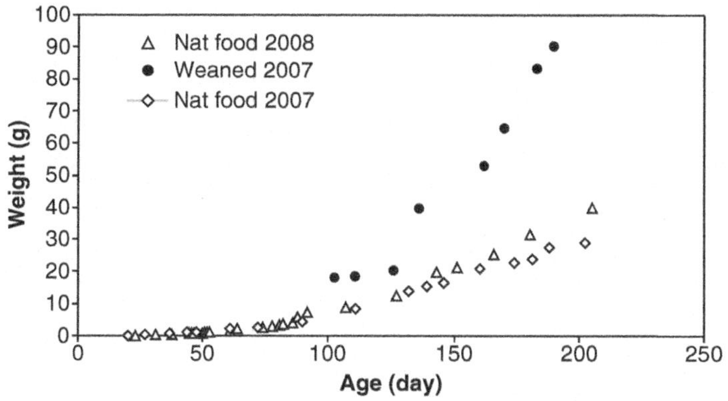

Fig. 33.2 Growth of *A. sturio* larvae fed either natural food items or weaned on compound diet

Mean growth for fish up to 6 months according to food item, natural or compound diet, is shown in Fig. 33.2. Growth curves for fish fed natural food are very similar for both years, 2007 and 2008, with a mean final weight of about 35 g. In contrast, the mean weight of fish weaned on a compound diet is far higher, with a mean weight of about 90 g, and a survival rate of about 85%, thus proving the effectiveness of the prepared food as well as the feeding management technique. It has to be mentioned that in the latter case, fish that were not ready to accept compound food were returned to natural food items for a few days before trying to wean them once again.

Overall, the present findings are promising. They provide confirmation of the advantages of co-feeding, useful reference to ovulated eggs for fatty acid composition, and new inputs in the preparing of a sturgeon-adapted compound diet, all of which are of the utmost importance for the future. Complementary results deal with long-term growth, survival, and potential abnormalities, as revealed in the past for fish about 2 years old that had been successfully weaned on commercial pellets (Williot et al. 2004).

References

Gisbert E, Williot P, Castello-Orvay F (2002) Fully vitellogenic oocyte amino acid profile of Siberian sturgeon (*Acipenser baerii*, Brandt). J Appl Ichthyol 15:347–350

Mohler JW (2000) Early culture of the American Atlantic sturgeon *Acipenser oxyrinchus oxyrinchus* Mitchill, 1815 and preliminary stocking trials. Bol Inst Esp Oceanogr 16(1–4):203–208

Williot P, Rouault T, Rochard E, Castelnaud G, Lepage M, Gonthier P, Elie P (2004) French attempts to protect and restore *Acipenser sturio* in the Gironde: status and perspectives, the research point of view. In: Gessner J, Ritterhoff J (eds) BfN-Schriften 101: International workshop on species differentiation and population identification in the sturgeons *Acipenser sturio* L. and *Acipenser oxyrhinchus*. Bundesamt für Naturschutz, Bonn, pp 83–99

Williot P, Brun R, Rouault T, Pelard M, Mercier D (2005) Attempts at larval rearing of the endangered western European sturgeon, *Acipenser sturio* L. (Acipenseridae), in France. Cybium 29:381–387

Williot P, Rouault T, Pelard M, Mercier D, Jacobs L (2009) Artificial reproduction of captive endangered European Atlantic sturgeon, *Acipenser sturio*. Endangered Species Research 6:251–257

Chapter 34
Genetic Variability of Cultured European Sturgeon *Acipenser sturio*

Ralph Tiedemann, Anja Schneider, Patrick Williot, and Frank Kirschbaum

Abstract From the relict population of the European sturgeon *Acipenser sturio*, a captive breeding stock of specimens has been established in France, in order to prevent extinction of this species and to produce offspring to be released again to the wild. We genetically analyzed 23 wild-caught specimens of this French captive population, as well as 27 specimens of artificially produced offspring, at 12 newly developed microsatellite loci. We quantify the genetic distance among any pair of sturgeons belonging to the captive stock. In order to maintain the genetic variability and minimize inbreeding, we suggest dissortative mating, i.e., mating of any reproducing female with a distantly related male. We explicitly identify such genetically distant suitable mates for any female of the French breeding stock. Forced dissortative mating could be facilitated by cryoconservation of sperm, such that sperm of putatively suitable partners would be available at any time.

34.1 Introduction

Many species face nowadays declines in population sizes. They might experience genetic threats due to an increased likelihood of relatives mating with each other. Such *inbreeding* can cause the expression of recessive maladaptive or even lethal

R. Tiedemann (✉) • A. Schneider
Unit of Evolutionary Biology/Systematic Zoology, University of Potsdam, Karl-Liebknecht-Straße 24-25 (Haus 26), 14476 Potsdam, Germany
e-mail: tiedeman@rz.uni-postdam.de

P. Williot
Sturgeon Consultant, 4 Rue du Pas de madame, 33980 Audenge, France

F. Kirschbaum
Faculty of Agriculture and Horticulture, Biology and Ecology of Fishes, Humboldt University, Philippstr. 13, Haus 16, 10115 Berlin, Germany
e-mail: Frank.kirschbaum@staff.hu-berlin.de; frank.kirschbaum@staff.hu-berlin.de

P. Williot et al. (eds.), *Biology and Conservation of the European Sturgeon Acipenser sturio L. 1758*, DOI 10.1007/978-3-642-20611-5_34,

alleles in homozygous individuals, or reduced resistance against pathogens (reviewed in Charlesworth and Willis 2009). Such *genetic threats* can severely compromise the viability and survival expectancy of an affected population. In addition, if the population size of a threatened species declines below a certain threshold, it is also threatened by *demographic effects*, such as a randomly unfavourable sex ratio (generally, too few reproductive females) or simply the difficulty of two putative mating partners synchronizing in their reproductive activity, and/or actually meeting each other (a demographic Allee effect; Courchamp et al. 1999). The ultima ratio to rescue such a species or population at the verge of extinction is to take the entire population or a substantial fraction of it into captivity. Ultimately, such efforts aim not only at rescuing the particular kept individuals, but also to propagate them in captivity and – ideally – to release their offspring into the natural habitat in order to support or even to re-establish a self-sustaining wild population (Seddon et al. 2007).

A fundamental concept in population genetics is the *effective population size* (N_e). N_e is a population measure which takes into account the expected increase of inbreeding, i.e., the likelihood that genotypes are identical-by-descent, from one to the next generation. N_e can be calculated, according to Hedrick (2000), as follows:

$$N_e = \frac{4N_m N_f}{N_m + N_f},$$

where N_m and N_f are the number of males and females in a population, respectively. N_e is equal to the census population size if (1) the operational sex ratio OSR (i.e., the sex ratio of those specimens participating in reproduction) is even, (2) mating occurs at random throughout the entire population, and (3) fertility in all specimens is equal. In real populations, these conditions are only rarely met, such that N_e is typically lower than the census size. To illustrate this with an extreme example, a population consisting of one male and 99 females would have a census size of $N = 100$, but an effective population size only of about $N_e = 4$. This means that the increase of inbreeding within one generation is the same in (1) a population of one male and 99 females and (2) in a population of two males and two females.

Taking parts of relict populations of threatened species, like the European sturgeon *Acipenser sturio* (Rochard et al. 1990; Williot et al. 1997; Kirschbaum et al. 2006), into captivity has genetic and demographic consequences which need to be considered. First, the establishment of a new captive population causes a founder effect (Barton and Charlesworth 1984), as only a subset of geno- and phenotypes from the wild is captured. These captured specimens are transferred to a new artificial environment. Captivity can cause relaxed selection, as specimens might survive which would have been exterminated in the wild by natural selection. Captivity can even cause unconscious selection towards traits which increase the survival and/or propagation probability in captivity, but might be maladaptive in the wild (Zohary et al. 1998, see Chap. 48). If the captive population is propagated, inbreeding is likely to increase quickly, as effective population sizes N_e of captive populations are typically small. N_e can be further reduced relative to the census size of the captive population, if – for technical reasons – only part of the captive stock reproduces, a complication particularly apparent in the case of the European sturgeon. Here, considerable effort (and sometimes luck) is necessary to bring an individual female into breeding condition. If successful, the best effort should be made to ensure fertilization. The availability of reproductive males at the very same time is a further constraint which – however – might be overcome by cryoconservation of sperm.

From the perspective of maintaining as much genetic variability as possible in the captive sturgeon population, females should be mated to as distantly related males as possible. We here use genetics to suggest suitable (i.e., distantly related) mating partners among the captive sturgeons. Under such a scenario of forced dissortative mating, inbreeding actually increases less than expected at random, such that – in this special case – N_e can exceed the census size, to the benefit of the genetic variability among the offspring.

34.2 Material and Methods

We particularly focus on the French captive breeding population of European sturgeon, from which we genetically assess 23 specimens (nine females, 14 males) captured in the last wild population in the Gironde (Table 34.1). Previous genetic analysis on this captive breeding population indicated that it is not homogeneous, as it is composed both of old (presumably not directly related) fish and full siblings of the last known reproduction in the wild in 1994 (Ludwig et al. 2004). It has also previously been shown that the wild relict population is probably already genetically depleted, relative to the ancestral more widespread population of this species (Ludwig et al. 2000, 2004; Tiedemann et al. 2007). We further include specimens from a successful propagation in 1995 (Williot et al. 2000) into our analysis, i.e., 11 specimens kept still in France and 16 additional specimens transferred to Germany, of which nine are still kept at the Leibniz-Institute of Freshwater Ecology and Inland Fisheries in Berlin.

In order to provide an up-to-date measure of the available genetic variability in these captive sturgeons and to suggest suitable mating partners for captive propagation, we genotyped 50 specimens of the European sturgeon at 12 polymorphic microsatellite loci with proven diploid inheritance. These new microsatellites had been specifically developed for the sturgeon according to the protocol of Paulus and Tiedemann (2003), and will be published in detail elsewhere.

We used our microsatellite data to look for any hidden genetic substructure among the cultured specimens. This was achieved by a genetic clustering algorithm (STRUCTURE, Pritchard et al. 2000) which identifies locus-wise subsets of specimens such that allele distributions within subsets conform to the expectation of random association (so-called Hardy–Weinberg equilibrium, HWE). We performed a principal component analysis (PCA) using the software PCAGEN (available at http://www2.unil.ch/popgen/softwares/pcagen.htm) which stratified the individuals according to the two major variation components. We also

Table 34.1 Characteristics of the *A. sturio* sampled for present genetics investigations

Name	Birth year	Sex
France		
Judes	1970–1972	♂
Justin	1984	♂
Philippe	1984	♂
Ascension	1988	♂
Emeline	1988	♂
Mariette	1988	♂
Bleu	1988	♂
DN	1988	♀
Delphine	1994	♂
Nathalie	1994	♂
Gilles	1994	♂
Fiacre	1994	♀
Aristide	1994	♀
Severine	1994	♀
Jules	1994	♀
Leonce	1994–1995	♀
Martinien	1995	♂
Fulbert	1995	♂
Hervé	1995	♂
Thierry	1995	♀
Julie	1995	♀
Edith	1995	♀
Gautier	n.a.	♂
Astu25 (111)	1995	♀
Astu26 (304)	1995	♀
Astu27 (309)	1995	♀
Astu28 (314)	1995	♂[a]
Astu29 (342)	1995	♂
Astu30 (347)	1995	♀
Astu31(353)	1995	♂
Astu32 (359)	1995	♀
Astu33 (365)	1995	♂
Astu34	1995	n.a.
Astu35	1995	n.a.
Germany		
IGB00-0137-221C	1995	♂
IGB00-0137-81EF	1995	n.a.[b]
IGB00-012D-DEOC	1995	n.a.[b]
IGB00-013C-6EC8	1995	♂
IGB00-012D-DDAE	1995	♀
IGB00-0136-F322	1995	♀
IGB00-0137-8557	1995	♂
IGB00-0136-E8D0	1995	n.a.[b]
IGB00-0136-FAC0	1995	n.a.[b]
IGB00-0137-875C	1995	n.a.[b]

(continued)

Table 34.1 (continued)

Name	Birth year	Sex
IGB00-0137-2A50	1995	♀
IGB00-0136-A3CA	1995	n.a.[b]
IGB00-0137-8138	1995	♂
IGB00-0137-1906	1995	♀
IGB00-0137-7A59	1995	♂
IGB00-0137-7346	1995	n.a.[b]

n.a. data not available; [a]Sex determination not certain; [b]No longer alive

calculated Nei's standard genetic distance (Nei 1972) among any potential mating partners, using the software PHYLIP (Felsenstein 2005).

34.3 Results and Analysis

With the STRUCTURE analysis, we found support for a subdivision of the captive European sturgeons into two groups: Fig. 34.1 provides, for any analyzed specimen, percentages of assignment to the one (blue) or the other (orange) genetic cluster. Interestingly, all individuals of the 1995 propagation, both those in France and in Germany, are assigned to the same orange cluster, all except one (IGB00-013C-6EC8) with high probabilities. The French captive breeding population, on the other hand, is assigned almost entirely to the blue cluster, except for two males, one of which (Justin) is preferably assigned to the orange cluster and a second (Martinien) appears intermediate between the clusters. This result clearly demonstrate that if only a single reproduction (here, in 1995) is considered, further genetic variability is lost such that the offspring appears genetically homogeneous. This finding is, however, to be expected, as this reproduction involved only a single pair of sturgeons (Williot et al. 2000).

If we focus our analysis on the French wild-originated breeding stock only, we were again able to detect a subtle substructure, dividing the specimens into two clusters: A (light blue) and B (dark blue; Fig. 34.1).

Most specimens were assigned with high probability to one or the other cluster, except for three males (Herve, Justin, Gautier) and two females (Edith, Julie) exhibiting intermediate genotypes. As an easy-to-use "rule of thumb", one could attempt to interbreed partners from the two genetic clusters. However, there are only a few cluster-A females (Leonce and Thierry; Table 34.2).

Another approach to visualizing the genetic variability hold in the breeding stock is a stratification of the specimens according to their multilocus genotype by multivariate statistical analysis. We performed a principal component analysis (PCA) using the software PCAGEN (available at http://www2.unil.ch/popgen/softwares/pcagen.htm), which stratified the individuals according to the two major components (Fig. 34.2). The first component (horizontal) which accounts for 18.6% of the variation, separates out exactly the same groups of specimens as

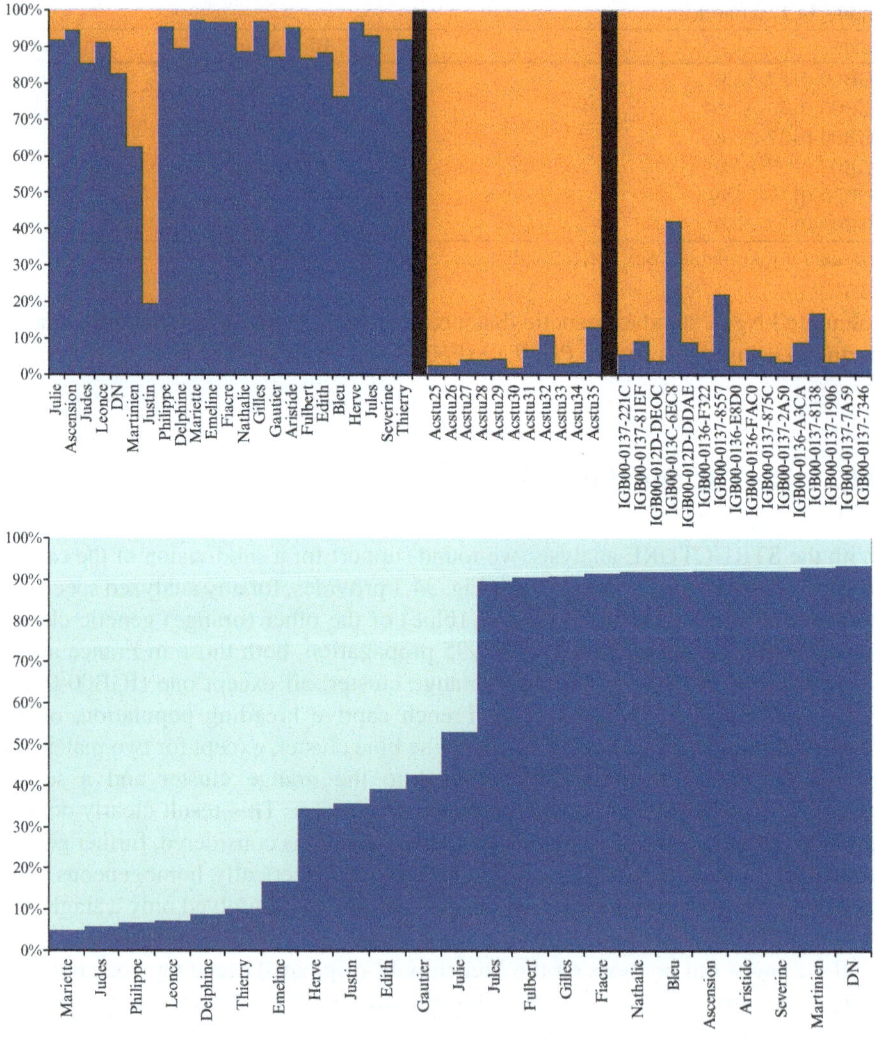

Fig. 34.1 Subdivision of the captive *Acipenser sturio* population according to the genetic assignment analysis (STRUCTURE; Pritchard et al. 2000). Values on the y-axis are percentage of assignment to the first cluster, while $(1 - y)$ provides assignment to the second cluster. *Upper*: All specimens. *Lower*: Wild originated specimens of the French captive breeding population only

the STRUCTURE assignment. The second component (accounting for 15.1% of the variation) adds valuable stratification among individuals, such that one could maximize the distance between pairs of specimens to be mated. For any individual female in the breeding stock, we identified the three most distant males (according to the PCA) which constitute suitable mating partners, in an attempt to promote dissortative mating and to minimize inbreeding (Table 34.2). Two males from cluster A (Philippe and Judes) appeared particularly often distant from the available

Table 34.2 Captive females of the French brood stock of *Acipenser sturio*, mean genetic distance to captive males, and genetically most different males according to the Principal Component Analysis (PCA) and the calculation of pairwise genetic distances

Female	Mean genetic distance to males	Most distant males (principal components)	Most distant males (genetic distance)
		1. Philippe (A)	1. Philippe (A) 0.857
		2. Judes (A)	2. Justin (AB) 0.663
Julie (AB)	0.541	3. Bleu (B)	3. Judes (A) 0.634
		1. Bleu (B)	1. Bleu (B) 0.682
		2. Gilles (B)	2. Philippe (A) 0.631
Leonce (A)	0.449	3. Fulbert (B)	3. Gilles (B) 0.629
		1. Judes (A)	1. Philippe (A) 0.887
		2. Philippe (A)	2. Mariette (A) 0.782
DN (B)	0.461	3. Mariette (A)	3. Gautier (AB) 0.531
		1. Judes (A)	1. Philippe (A) 0.916
		2. Philippe (A)	2. Judes (A) 0.811
Fiacre (B)	0.522	3. Mariette (A)	3. Mariette (A) 0.811
		1. Judes (A)	1. Philippe (A) 1.224
		2. Philippe (A)	2. Judes (A) 1.157
Aristide (B)	0.657	3. Mariette (A)	3. Delphine (A) 0.818
		1. Philippe (A)	1. Philippe (A) 0.990
		2. Judes (A)	2. Emeline (A) 0.945
Edith (AB)	0.613	3. Justin (A)	3. Justin (A) 0.916
		1. Judes (A)	1. Philippe (A) 1.129
		2. Philippe (A)	2. Mariette (A) 0.665
Jules (B)	0.537	3. Mariette (A)	3. Martinien (B) 0.606
		1. Judes (A)	1. Herve (AB) 0.794
		2. Philippe (A)	2. Mariette (A) 0.659
Severine (B)	0.463	3. Bleu (B)	3. Philippe (A) 0.519
		1. Bleu (B)	1. Philippe (A) 0.749
		2. Philippe (A)	2. Martinien (B) 0.661
Thierry (A)	0.461	3. Judes (A)	3. Ascension (B) 0.595

In parentheses, assignment to a genetic cluster is given (A = light blue; B = dark blue) according to Fig. 34.1; intermediate genotypes are termed AB

females and could be – based on this analysis – considered to be particularly valuable for maintaining genetic diversity in the breeding stock. From cluster B, the male Bleu often exhibited the largest distance to females, in particular those assigned to cluster A and those with intermediate genotypes AB.

For every female, we also quantified the mean genetic distance (using Nei's standard genetic distance; Nei 1972) at the analyzed microsatellite loci, averaged over all males. With this measure, one can easily assess for any possible mating combination whether partners are genetically more alike or more unalike than on average (Table 34.2). We explicitly provide data on the three genetically most distant males to illustrate our approach to identify suitable mating pairs to minimize inbreeding. Again, Philippe constituted the male genetically distant to most females, and can be considered a valuable breeder according to this analysis.

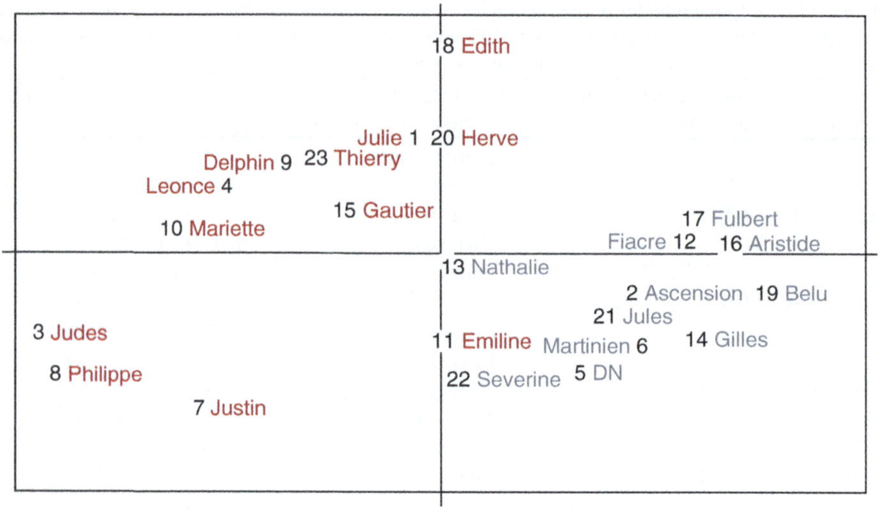

Fig. 34.2 Principal component analysis (PCA) based on multilocus microsatellite genotypes of the wild originated specimens of the French brood stock of *Acipenser sturio*. *Colours* refer to the assignment analysis [*red* refers to cluster A (*light blue* in Fig. 34.1), *blue* refers to cluster B (*dark blue* in Fig. 34.1)]

Our three lines of evidence to identify suitable mating pairs with the aim to minimize inbreeding [i.e., (1) STRUCTURE clustering (Fig. 34.1), (2) PCA stratification (Fig. 34.2), and (3) quantification of pairwise genetic distance (Table 34.2) deliver mostly consistent results: for any given female, suitable males mostly belong to another genetic cluster, and were identified to be distant by both the PCA and the quantitative distance measure.

This study had two objectives, i.e., (1) to give an example of how genetics can assist efforts to minimize inbreeding in a captive population, and (2) to provide an easy-to-use data format for those researchers and managers who actually maintain the captive breeding populations of European sturgeon, and are engaged in their propagation and eventual release into nature (Williot et al. 2009). Typically, the suggested mating partners were known beforehand to be unrelated. Already before any genetic analysis, cross-breeding was performed preferably among such unrelated specimens. The added value of this genetic survey is the possibility to particularly select the genetically most distant partners for mating.

We are well-aware of that logistic constraints might prevent an a priori planned breeding scheme based solely on genetics. Nonetheless, we argue that – if any of the breeding stock females is likely to spawn – our data enable managers to quickly identify suitable mating partners. Cryopreservation of the sperm of all potential male breeders would evidently greatly facilitate the possibility to actually perform those matings suggested by genetics (see Chap. 35). Sperm preservation would additionally allow for multiple paternity of single spawns, such that a single female's offspring would comprise a maximum of genetic variability. By this approach, the increase of inbreeding would be reduced to the minimum inevitable in a small captive population.

One potential caveat of our analysis is the assumed neutrality of the microsatellite variability. Our suggested breeding scheme hence is based solely on the identification of genetically generally unalike individuals, but does not account for specific genotypes of adaptive importance. It would be valuable in future research also to investigate adaptive genetic variation, in particular in the immune system, i.e., at MHC genes (Sommer 2005; Tiedemann et al. 2007). Depletion of this variation has been demonstrated to severely compromise the viability of affected populations. Including information on this marker could hence be of interest in future management.

Acknowledgements Christiane Schröder assisted with the data analysis. Financial support was provided by the German Bundesministerium für Bildung und Forschung (BMBF) project number 0330718.

References

Barton NH, Charlesworth B (1984) Genetic revolutions, founder effects, and speciation. Ann Rev Ecol Syst 15:133–164

Charlesworth D, Willis JH (2009) The genetics of inbreeding depression. Nat Rev Genet 10:783–796

Courchamp F, Clutton-Brock T, Grenfell B (1999) Inverse density dependence and the Allee effect. Trends Ecol Evol 14:405–410

Felsenstein J (2005) PHYLIP (Phylogeny Inference Package) version 3.6. Distributed by the author. Department of Genome Sciences, University of Washington, Seattle

Hedrick PW (2000) Genetics of populations, 2nd edn. Jones and Bartlett, Sudbury, MA

Kirschbaum F, Würtz S, Williot P, Tiedemann R, Arndt GM, Anders E, Bartel R, Gessner J (2006) Prerequisites for the restoration of Atlantic sturgeons, *Acipenser sturio* and *A. oxyrinchus*, in Germany – report on the twelve-year preparatory period. Verhandlungen der Gesellschaft für Ichthyologie 5:79–93

Ludwig AN, Jenneckens I, Debus L, Ludwig A, Becker J, Kirschbaum F (2000) Genetic analysis of archival specimens of the Atlantic sturgeon *Acipenser sturio* L., 1758. Bol Inst Esp Oceanogr 16:221–230

Ludwig A, Williot P, Kirschbaum F, Lieckfeld D (2004) Genetic variability of the Gironde population of *Acipenser sturio*. In: Gessner J, Ritterhoff J (eds) Species differentiation and population identification in the sturgeons *Acipenser sturio* L. and *Acipenser oxyrinchus*. Bundesamt für Naturschutz, Bonn, pp 54–72

Nei M (1972) Genetic distance between populations. Am Nat 106:283–292

Paulus KB, Tiedemann R (2003) Ten polymorphic autosomal microsatellite loci for the Eider duck *Somateria mollissima* and their cross-species applicability among waterfowl species (Anatidae). Mol Ecol Notes 3:250–252

Pritchard JK, Stephens M, Donnelly P (2000) Inference of population structure using multilocus genotype data. Genetics 155:945–959

Rochard E, Castelnaud G, Lepage M (1990) Sturgeons (Pisces: Acipenseridae); threats and prospects. J Fish Biol 37(suppl A):123–132

Seddon PJ, Armstrong DP, Maloney RF (2007) Developing the science of reintroduction biology. Conserv Biol 21:303–312

Sommer S (2005) The importance of immune gene variability (MHC) in evolutionary ecology and conservation. Front Zool 2:16

Tiedemann R, Moll K, Paulus KB, Scheer M, Williot P, Bartel R, Gessner J, Kirschbaum F (2007) Atlantic sturgeons (*Acipenser sturio*, *Acipenser oxyrinchus*): American females successful in Europe. Naturwissenschaften 94:213–217

Williot P, Rochard E, Castelnaud G, Rouault T, Brun R, Lepage M, Elie P (1997) Biological characteristics of European Atlantic sturgeon, *Acipenser sturio*, as the basis for a restoration program in France. Environ Biol Fish 48:359–370

Williot P, Brun R, Pelard M, Mercier D (2000) Unusual induced maturation and spawning in an incidentally caught pair of adults of the critically endangered European sturgeon *Acipenser sturio* L. J Appl Ichthyol 16:279–281

Williot P, Rochard E, Rouault T, Kirschbaum F (2009) *Acipenser sturio* recovery research actions in France. In: Carmona R, Domezain A, Garcia Gallego M, Hernando-Casal J, Rodriguez F, Ruiz Rejon M (eds) Biology, conservation and sustainable development of sturgeons, Fish & fisheries series. Springer, New York, pp 247–263

Zohary D, Techernov E, Kolska Horwitz L (1998) The role of unconscious selection in the domestication of sheep and goats. J Zool Lond 245:129–135

Chapter 35
Sperm Cryopreservation in Sturgeon with a Special Focus on *A. sturio*

Ákos Horváth, Patrick Chèvre, and Béla Urbányi

Abstract Cryopreservation is a low-temperature conservation technique for genetic resources for an extended period of time The cryopreservation of sturgeon sperm has been problematic for many years, due to low fertilization and hatch percentages observed with cryopreserved sperm. During the last decade, a reliable methodology was developed which increased the fertilizing capacity of frozen sturgeon semen. This technique involves the use of methanol as cryoprotectant, instead of the more conventional dimethyl-sulfoxide (DMSO). The method was successfully applied to *Acipenser sturio* by using cyropreserved sperm of *A. sturio* to fertilize the eggs of *A. ruthenus* for the production of hybrids of the two species in 2001–2002 in three successive attempts. Currently, a cryopreserved gene bank is being developed for this species.

35.1 Introduction

Cryopreservation is the cooling and subsequent storage of live cells, tissues and in some cases organisms at very low temperatures, and their use following thawing. The earliest and most common application of this technique is the cryopreservation of sperm. Studies on sperm cryopreservation in aquatic organisms have been conducted since the 1950s (Blaxter 1953), and have become practically applicable since the 1980s. However, few of them have been applied to aquaculture practice. Nevertheless, the advantages of sperm cryopreservation in fish, especially in species conservation, are numerous: it allows the storage of genetic resources of

Á. Horváth (✉) • B. Urbányi
Department of Aquaculture, Szent István University, 2100 Gödöllő, Páter K. u. 1., Hungary
e-mail: Horvath.Akos@mkk.szie.hu

P. Chèvre
Cemagref, Estuarine Ecosystems and Diadromous Fish Research Unit, Moulin de la logerie, 33660 Saint-Seurin-sur-l'Isle, France

P. Williot et al. (eds.), *Biology and Conservation of the European Sturgeon Acipenser sturio L. 1758*, DOI 10.1007/978-3-642-20611-5_35,

particular male individuals or groups, reduction of the male broodstock, and costs associated with its maintenance. A potential application also is a combination of sperm cryopreservation and androgenesis.

The number of studies published on the cryopreservation of *Acipenser sturio* sperm is very limited (Kopeika et al. 2000; Urbányi et al. 2004). In contrast, the experience on sturgeon sperm cryopreservation in general is quite extensive, and due to the great inter-species similarity in sperm biology in sturgeons, it should be possible for cryopreservation methods developed for one species to be applied to other species with minimal changes in the protocols. Therefore, cryopreservation of sturgeon sperm will be discussed in this chapter in detail, and special focus will be placed on the findings for *A. sturio*.

35.2 Principles of Sperm Cryopreservation

The principles of cryopreservation are similar to those of the cooling of water and aqueous solutions. The freezing point of water is 0°C, and that of solutions is generally lower; however, freezing seldom starts at exactly this temperature. Water and aqueous solutions normally supercool beyond the freezing point, and stay in the liquid phase until freezing starts along so-called ice nuclei. Ice nucleation can start along impurities or concentration differences in the solution. As ice nucleation is accompanied by a release of energy (latent heat), the freezing solution warms to the freezing point, which is followed by a slow decrease of temperature as water molecules are incorporated into the crystal structure of ice. Due to this process, concentration of the solution gradually increases until the eutectic point, which is the maximal concentration of solvents before the entire solution freezes. If cooling is accompanied by the formation of crystalline ice, then the process is called freezing. There exists, however, another way of phase transition from liquid to solid as a result of ultra-rapid cooling, which results in the formation of an amorphous glass-like solid state of aqueous solution which is called vitrification (Denniston et al. 2000).

In live cells and tissues, the factor regulating the effects of cooling is the cell membrane, which acts as a semi-permeable barrier that controls osmotic functions. During cooling, cells lose water to compensate for the increase of osmotic pressure in the extracellular environment. Water release is necessary during cooling to avoid the formation of intracellular ice crystals, which typically occurs as a result of very fast cooling rates. On the other hand, cooling should also be sufficiently fast to reduce the toxic effects of concentrated solutes and lethal dehydration. Chemical compounds called cryoprotective agents (CPA-s) or simply cryoprotectants are added to the suspensions of cooled cells. Cryoprotectants have various roles in the cooling process: they reduce the freezing point of the solution, accelerate water release from the cells, stabilize cell membranes, and prevent water molecules from incorporation into the ice crystals by bonding to them. Cryoprotectants can be external or internal, depending on their ability to penetrate the cells. Cryopreserved cells are stored in insulated liquid nitrogen containers called Dewars. During thawing, recrystallization of the cryopreserved cell suspensions can occur again; therefore, sufficiently fast thawing rates must be applied.

Spermatozoa of fish spawning in freshwater are activated upon their release into the aqueous environment. As the period of active movement following activation is relatively short (20 s to 6–10 min depending on the species), sperm must be maintained in a quiescent, immotile state, which is accomplished by means of isoosmotic solutions called extenders. In addition to the isoosmotic nature of the solution, in salmonids and acipenseriform fish an adequate level of potassium in the extender is also required (Gallis et al. 1991). Regarding CPA-s, a wide variety of internal cryoprotectants has been tested (dimethyl-sulfoxide – DMSO, dimethyl-acetamide – DMA, glycerol, methanol, ethylene glycol, etc.) in concentrations typically ranging between 5 and 15%. Dilution of fish sperm with the extender containing the cryoprotectant is done in ratios between 1:1 and 1:9. Fish spermatozoa representing a very simple cell type can tolerate various cooling rates varying between 10 and 80°C/min. Generally speaking, slow cooling rates are not favorable for fish sperm, and solute effects of increasing concentrations of extenders seem to affect cryopreservation success more than intracellular ice formation.

Cooling of fish sperm can be carried out several ways, using dry ice (solid CO_2) or liquid nitrogen (LN_2) as coolant, with the former having a temperature of −79°C and the latter −196°C. Suspended sperm is either frozen in pellets dropped into small indentations in the surface of dry ice, or loaded into cooling containers of various forms and sizes and subsequently frozen on a block of dry ice or in the vapors of liquid nitrogen. Cooling containers can include straws, ampules, cryovials, etc. Following cooling, samples are plunged directly into liquid nitrogen, and stored until thawing and subsequent use for fertilization. Thawing of cryopreserved fish sperm is almost always carried out in a warm (30–40°C) water bath for various periods of time.

35.3 Cryopreservation of Sturgeon Sperm

Sturgeon spermatozoa are unique among bony fish, as they possess several conservative features that were lost in teleosts during evolution, and knowledge of which seems to be crucial for successful cryopreservation. In contrast to teleost sperm cells, sturgeon and paddlefish spermatozoa have an elongated cylindrical head with an approximate length of 5–6 μm, which is covered by an acrosome that contains enzymes such as acrosine and arylsulfatase necessary for the acrosome reaction during the fertilization process. The acrosome reaction consists of the formation of a fertilization filament, which is hypothesized to act as a signal transducer carrying information to the egg on the presence of a fertilizing spermatozoon (Psenicka et al. 2010) The seminal plasma of sturgeons has a low osmolality (30–80 mosmol/kg) in contrast to that of teleost fish (~300 mosmol/kg), which is maintained in spite of the fact that their blood plasma has an osmolality close to 300 mosmol/kg. Sturgeons typically produce a high volume of sperm, up to 1–1.5 l in larger males. However, sperm concentration is lower than that of teleosts, usually being observed in the range of 0.5–2 × 10^9 spermatozoa/ml. Activated spermatozoa maintain their motility for 5–6 min which is significantly longer than what is observed in teleosts (20 s to 2 min). Spermatozoa of different acipenseriform species are remarkably similar, with only minor differences in the morphology, such as the length of the head or position of the postero-lateral projections of the acrosome (for a review, see Psenicka et al. 2007).

All of these above-mentioned characteristics contributed considerably to the failures and successes of sperm cryopreservation protocol development in sturgeons. The first attempts in sturgeon sperm cryopreservation were published by Soviet scientists in the 1960s (Dettlaff et al. 1993). Up to the second half of the 1990s, several attempts were published on the cryopreservation of sturgeon sperm (Drokin et al. 1991, 1993; Cherepanov et al. 1993; Ciereszko et al. 1996; Brown and Mims 1999). Most of the reported techniques and the results were either not repeatable, or resulted in poor fertilization rates. The first positive results on the cryopreservation of sturgeon sperm were reported by Tsvetkova et al. (1996), who used sucrose which was crucial as an extender (23.4 mM sucrose, 118 mM Tris, pH 8.0) for the sperm of *A. baerii*. A study on *A. ruthenus* (Jähnichen et al. 1999) reported remarkably high hatch rates (up to 91%) with sperm cryopreserved with ethylene glycol. However, the method described appeared to be very difficult to repeat. In this latter study, 12.5 and 17.5% ethylene glycol were used in combination with an extender of 25 mM NaCl, 10 mM Tris, pH 8.5 and two cooling profiles. The work of Tsvetkova et al. (1996) was continued by Billard et al. (2000), who tested the effect of cryopreservation on the post-thaw motility of *A. baerii* sperm using stroboscopic illumination, and detected several alterations in the movement of spermatozoa following cryopreservation, along with abnormalities of the flagellum.

The first report of the method we currently use (Horváth and Urbányi 2000) suggested the use of methanol as a cryoprotectant for the sperm of *A. baerii*. This cryoprotectant was further tested by Glogowski et al. (2002) in combination with three basic extenders, a Tris–sucrose–KCl extender (30 mM Tris, 23.5 mM sucrose, 0.25 mM KCl, pH 8.0), Jähnichen's extender (Jähnichen et al. 1999), and a Tris–sucrose extender (20 mM Tris, 400 mM sucrose, pH 8,0). All were combined with a 10% methanol cryoprotectant. Hatch rates of 30 ± 5% were reported with the Tris–sucrose–KCl extender, comparable to the control (18% for the first male and 26% for the second).

The Tris–sucrose–KCl extender (here called modified Tsvetkova's or MT extender) was tested alongside Tsvetkova's original (OT) and modifier Hanks' balanced salt solution (mHBSS), and in combination with 5, 10 and 15% methanol as well as DMSO in two further studies concentrating on the fertilizing capacity (Horváth et al. 2005) and membrane integrity (Horváth et al. 2008) of the sperm of shortnose sturgeon (*A. brevirostrum*) and pallid sturgeon (*Scaphirhynchus albus*). In these experiments, although the use of DMSO as cryoprotectant resulted in the highest post-thaw motility, fertilization percentages were at their highest when using 5% methanol as cryoprotectant. With regard to extenders, the use of MT extender in combination with 5% methanol yielded the highest fertilization and hatch rates. The results were lowest when OT extender was used. These observations suggested that hyperosmotic conditions of the OT extender or those resulting from the addition of DMSO could result in good motility percentages, but will also produce poor fertilization rates. OT extender was hyperosmotic when compared to MT extender (222 mosmol/kg vs 82 mosmol/kg) as well as that of sturgeon seminal plasma. The addition of DMSO resulted in a sharp increase of osmolality (to 719 mosmol/kg when 5% DMSO was added to MT extender), whereas the addition of methanol did not change the osmolality values significantly. These results stimulated a debate as to whether the osmolality measured by a vapor pressure osmometer is the true osmolality value of the extender, or is the result of a systematic measurement error (due to the low molecular weight of methanol). Regardless of the methodological problems described, this observation could be an explanation for the poor fertilization results in DMSO, whereas those of methanol are considerably higher. The described experiment was repeated also on paddlefish (*Polyodon spathula*) sperm (Horváth et al. 2006), using MT and mHBSS extenders in combination with 5 or 10% DMSO or methanol as cryoprotectant. Again, the use of methanol resulted in significantly higher fertilization rates than that of DMSO, regardless of its concentration. The established method was later tested on lake sturgeon (*A. fulvescens*) by Ciereszko et al. (2006), who observed very poor fertilization rates with the use of cryopreserved sperm; however, due to the low fertility of the control, the authors concluded that further studies were needed. Another study on paddlefish sperm (Linhart et al. 2006) reported the use of 2-ml cryotubes and programmable freezers. Subsequent fertilization tests revealed the positive effects of methanol on hatch rates, although the use of both methanol and DMSO yielded good post-thaw motility percentages. A further harmful effect of DMSO as cryoprotectant was discovered by Psenicka et al. (2008), who found that the use of DMSO resulted in premature triggering of the acrosome reaction in sterlet sperm before contact with the eggs, which may be responsible for low fertilizing capacity of sperm samples frozen in the presence of this cryoprotectant. Obviously, premature acrosome reaction did not affect motility parameters of sperm; however, it prevented fertilization. These studies firmly established methanol as a cryoprotectant of preference for the sperm of sturgeon in these research groups.

Thus, crucial parameters for cryopreservation success in sturgeons include the appropriate choice of extender – which should be isoosmotic to the seminal plasma of the species – and the cryoprotectant (methanol).

35.4 Cryopreservation of Sperm of *A. sturio*

The first attempt to cryopreserve the sperm of *A. sturio* was reported by Kopeika et al. (2000). They reported the cryopreservation of sperm collected from a spermiating male individual 4 h after its capture in the Gironde estuary. Motility of collected sperm was 50% upon activation. Sperm was diluted at a ratio of 1:1 in several media containing different concentrations of 56–76% Tris, 14.4–24% DMSO and 9.6–20% egg yolk. Diluted samples were loaded into 1.5-ml ampoules and frozen using a three-step protocol, with a first step of 1–5°C/min from 5°C to −10 to15°C, a second of 20–25°C/min to −70°C, and a third step of plunging the samples into liquid nitrogen. Post-thaw motility of samples ranged between 1 and 15%. Fertilization test were carried out using the eggs of *A. ruthenus* 1 month after cryopreservation. The rate of developing embryos on the second day post fertilization was 23.2–38.3%, with 43% in the control which was fertilized with fresh sperm from *A. ruthenus*. Authors did not specify the developmental stage of eggs when the rate of developing embryos was calculated, and hatch rates were not reported.

On 31 May 2001, sperm samples were shipped to Hungary from two individuals named Emeline and Nausicaa. Samples were collected the same day, with initial motilities of 90% at 30 s, 20% at 60 s and 5% at 90 s post-activation for Nausicaa, and 90% at 30 s and 5% at 60 s post-activation for Emeline (Williot, personal communication). Samples were shipped for 8 h by air in sealed polyethylene bags filled with oxygen, which were placed into a polystyrene box with ice packs. The primary objective was to adopt means for a species-specific technique of cryopreservation. Furthermore, there was an attempt to perform androgenesis using the eggs of sterlet (Table 35.1). The sperm samples were cryopreserved in MT extender (see above) supplemented with 10% methanol. The dilution ratio was 1:1. Diluted sperm samples were loaded into 0.5- and 1.2-ml straws, and frozen in the vapor of liquid nitrogen 3 cm above the level of LN_2. A test cryopreservation with two straws per sample was carried out, to verify the post-thaw motility of samples. Ten straws of each sample were thawed 15 h following freezing, and their motility percentages were determined. Surprisingly, not all straws displayed the same motility percentage, and very low or 0% motility was observed in about half of the samples. This observation was later confirmed with the sperm of *A. baerii* and *A. gueldenstaedtii* to be an effect of very close proximity between individual straws adversely affecting the cooling rates.

Androgenesis was attempted twice during the experimental period using the eggs of *A. ruthenus*, in 2001 and 2002. In each experiment, ten straws of sperm from individual 1 and ten from individual 2 were thawed in a 40°C water bath for 13 s for production of androgenetic offspring and five and five straws respectively for the production of hybrids, in order to check egg quality. As both experiments failed to produce androgenetic progeny, fertilization or hatch rates of the hybrids were not calculated either. Nevertheless, hybrid individuals successfully hatched from eggs fertilized with cryopreserved *A. sturio* sperm. Most hybrid individuals were killed

Table 35.1 Parameters of two *A. sturio* sperm samples shipped to Hungary on 31 July 2001

Sample name	Volume (ml)	Motility before activation (%)	Motility post-activation (%)	Post-thaw motility following test cryopreservation (%)
Nausicaa	60	1–2	90	70
Emeline	80	5	85	60

immediately following hatching to prevent the incidental release of hybrid individuals into natural waters, but several hybrids were grown to the size of 30 cm. None of the fish were grown beyond that size. A series of deformities such as the lack of rostrum were observed in these hybrid fish. The deformities were not uniform, and occurred only with eggs from a single female; thus, we postulate that this deformity can be attributed to egg quality.

An additional experiment was carried out in 2002 to investigate the fertilizing capacity of cryopreserved *A. sturio* sperm along with the frozen sperm of *A. gueldenstaedtii*, *A. baerii* and *A. ruthenus*, using the eggs of *A. ruthenus*. In this experiment, eggs were distributed into batches of 3,000 ova and were fertilized with the contents of six straws per sample in duplicate. In case of *A. sturio*, three and three straws respectively from the two individuals were mixed together and used for fertilization. The post-thaw motility of *A. sturio* sperm was 60–70%, while the hatch rate of eggs fertilized with this sperm was 34%. For comparison, the hatch rate of eggs fertilized with cryopreserved *A. ruthenus* sperm was 30.6%, with *A. baerii* sperm 50% and with *A. gueldenstaedtii* sperm 17.4%. The presence of paternal genes in the offspring was verified using RAPD DNA markers (Urbányi et al. 2004). All hatched hybrid larvae were either killed following hatching or after 5 months of rearing at a commercial hatchery.

After carrying out these experiments, the cryopreserved samples of *A. sturio* sperm stored in the facilities of Szent István University were not used again for fertilization. Individual straws were periodically thawed for verification of sperm survival by the estimation of post-thaw motility, which stayed the same as those reported in Table 35.1 throughout the storage period. Unfortunately, the samples were destroyed in 2005 due to human error.

Thus, the sperm samples stored in Hungary were successfully used to produce viable hybrid progeny using the eggs of *A. ruthenus*. Cryopreserved sperm samples retained their post-thaw motility throughout the storage period.

The Cemagref station in St-Seurin-sur-l'Isle is currently developing a gene bank of cryopreserved *A. sturio* sperm. As of January 2010, the facility had 2,005 straws of cryopreserved sperm from 13 male individuals in its cryo-bank, using either methanol or DMSO as cryoprotectants. Experiments are also carried out on the use of 5-ml macrotubes. Therefore, there is a constant attempt to improve the ex situ conservation of the genetic material of this species, which makes it possible to use these samples in the event of unforeseen fatal incidents.

35.5 Recommended Protocol for Sturgeon Sperm Cryopreservation

Sturgeon sperm is collected following an intramuscular or intraperitoneal injection of different types of hormones. The available hormones and hormone products include carp pituitary with a regular dose of 2–8 mg/kg of body weight, or different GnRH or LHRH analogues at various doses ranging between 5 and 150 μg/kg of body weight, but without the addition of dopamine receptor antagonists. During collection, contact of sperm with urine or feces should be avoided; therefore, the use of an elastic silicone catheter with an external diameter of 4–5 mm inserted into the genital duct of the fish is recommended. The catheter can be attached to a syringe to facilitate collection.

Collected sperm should be stored at 4°C until use. Storage in plastic bags with oxygen allows storage of sperm for several days with only a small loss of quality; however, according to our experience, sperm stored for more than 24 h will be less resistant to cryopreservation. Motility of sperm can be estimated using a dark-field microscope and a 20× objective. There are several ways to estimate motility; the easiest is mixing 20 μl of water with 1 μl of sperm on a microscope slide without coverslip. Motility estimation should be carried out by an experienced person to minimize the subjectivity of measurements. It is recommended that if several samples are collected during one stripping event, then the same person should carry out motility estimation.

For cryopreservation of sturgeon sperm, the following chemicals should be prepared:

Modified Tsvetkova's extender:

- 23.4 mM sucrose
- 2.5 mM KCl
- 30 mM Tris
- pH of the solution should be set to 8.0 using concentrated HCl

The use of methanol in 5 or 10% final concentration as a cryoprotectant is recommended. Final concentration in this case means the concentration of cryoprotectant following dilution of sperm with the mixture of extender and cryoprotectant. The dilution ratio for sturgeon sperm is 1:1. Thus, if 10 ml of sperm are available for cryopreservation and 5% methanol is used as cryoprotectant, the diluted mixture would contain the following:

- 9 ml extender
- 1 ml methanol
- 10 ml sperm

Diluted sperm is loaded into 0.5-ml straws. Alternatively 1.2-ml straws and 5-ml macrotubes are also available for cryopreservation. The use of 5-ml straws seems to have practical advantages, especially for hatchery use in *A. sturio*, since the species can produce up to 1,000 ml of sperm. Up to now, the fertilization with sperm cryopreserved in these macrotubes is largely variable, and further experiments are needed to establish a standard protocol for their use (Horváth et al. 2010).

Straws should be placed horizontally on a 3-cm-high polystyrene frame. According to our experience, a spacing of at least 4–5 mm should be left between individual straws, to avoid undesired contact which can affect the quality of thawed sperm. The frame is placed on the surface of liquid nitrogen. The cooling time for 0.5-ml straws is 3 min, whereas for 1.2-ml straws 5 min is recommended. This results in a cooling rate of approximately 70°C/min for 0.5-ml straws. Following cooling, the straws can be plunged into liquid nitrogen and stored in storage Dewars until needed for fertilization.

Straws should be thawed in a 40°C water bath, while care should be taken that contact of sperm with the thawing water is avoided. The thawing time is 13 s for 0.5-ml straws, 20 s for 1.2-ml straws and approximately 40 s for the 5-ml macrotubes. It is recommended that the motility of thawed samples is checked as described for fresh sperm.

Although sperm/egg ratio is typically used to define the volume of sperm to be used for fertilization of a given amount of eggs, for practical reasons we prefer the use of straw/egg ratio. For fertilization with cryopreserved sperm, we recommend the use of the contents of one 0.5-ml straw to fertilize 5 g of eggs. In the case of sturgeon, the wet fertilization method is used. The sperm is activated in hatchery water and then added to the eggs. For sperm activation, a 200-fold dilution of sperm in hatchery water is recommended (Dettlaff et al. 1993) to avoid polyspermy; however, a recent study by Psenicka et al. (2010) shows that sturgeons employ a very effective system of acrosome reaction that prevents polyspermy. Thus, to ensure fertilization success, lower dilution rates should also be explored.

The methodology described above has successfully been applied to eight species of sturgeon and paddlefish, including *A. ruthenus*, *A. gueldenstaedtii*, *A. baerii*, *A. brevirostrum*, *A. sturio*, *Scaphirhynchus albus*, *Huso huso* and *Polyodon spathula*.

35.6 Conclusions

Sturgeon sperm cryopreservation comprises several crucial factors affecting the successful freezing and subsequent utilization of sperm for fertilization. These factors include the use of extenders isoosmotic to the seminal plasma, the use of methanol as cryoprotectant, and proper exposure to the vapor of liquid nitrogen during cooling. Optimization of these factors strongly relies on the trial-and-error approach. Nevertheless, a reliable protocol for the cryopreservation of *A. sturio* sperm has been developed, and a cryobank for the sperm of this species is currently established. Further optimization of the methodology is currently being carried out, mainly focusing on the development of a protocol for practical-sized amounts of sperm.

Acknowledgments The authors thank the Laboratory of Fisheries of Pannon University, Keszthely, Hungary, especially Dr. Miklós Bercsényi and Tamás Müller, for their kind assistance in incubating the hybrid embryos.

References

Billard R, Cosson J, Linhart O (2000) Changes in the flagellum morphology of intact and frozen/thawed Siberian sturgeon *Acipenser baerii* (Brandt) sperm during motility. Aquacult Res 31:283–287

Blaxter JHS (1953) Sperm storage and cross-fertilization of spring and autumn spawning herring. Nature 172:1189–1190

Brown GG, Mims SD (1999) Cryopreservation of paddlefish *Polyodon spathula* milt. J World Aquacult Soc 30:245–249

Cherepanov VV, Drokin SI, Ochkur SI, Dzuba BB, Chikhachov AS, Kopeika EF (1993) Freezing of sperm of the Azov-Black Sea Acipenserides. In: International symposium on sturgeons, Moscow–Kostroma–Moscow, 6–11 Sept 1993, Abstract Bulletin, p 63

Ciereszko A, Toth GP, Christ SA, Dabrowski K (1996) Effect of cryopreservation and theophylline on motility characteristics of lake sturgeon (*Acipenser fulvescens*) spermatozoa. Theriogenology 45:665–672

Ciereszko A, Dabrowski K, Froschauer J, Wolfe TD (2006) Cryopreservation of semen from lake sturgeon. Trans Am Fish Soc 135:232–240

Denniston RS, Michelet S, Godke RA (2000) Principles of cryopreservation. In: Tiersch TR, Mazik PM (eds) Cryopreservation in aquatic species. World Aquaculture Society, Baton Rouge, pp 59–74

Dettlaff TA, Ginsburg AS, Schmalhausen OI (1993) Sturgeon fishes, developmental biology and aquaculture. Springer, Heidelberg

Drokin SI, Cherepanov VV, Kopeika EF, Shilin NI (1991) Sakhalin sturgeon: how to preserve gene fund. Rybnoe Khoziaistvo 7:38–39 (our translation from Russian)

Drokin SI, Cherepanov VV, Kopeika EF, Shilin NI (1993) Cryopreservation of sperm of Sakhalin sturgeon (Acipenser medirostris micadoi): problems and prospects for cryopreserved sperm collection from rare and endangered sturgeon species. In: International symposium on sturgeons, Moscow–Kostroma–Moscow, 6–11 Sept 1993, Abstract Bulletin, p 64

Gallis JL, Fedrigo E, Jatteau P, Bonpunt E, Billard R (1991) Siberian sturgeon, *Acipenser baerii*, spermatozoa: effects of dilution, pH, osmotic pressure, sodium and potassium ions on motility. In: Williot P (ed) Acipenser. Cemagref, Antony, pp 143–151

Glogowski J, Kolman R, Szczepkowski M, Horváth Á, Urbányi B, Sieczyński P, Rzemieniecki A, Domagała J, Demianowicz W, Kowalski R, Ciereszko A (2002) Fertilization rate of Siberian sturgeon (*Acipenser baeri*, Brandt) milt cryopreserved with methanol. Aquaculture 211:367–373

Horváth Á, Urbányi B (2000) Cryopreservation of sterlet (*Acipenser ruthenus*) sperm. In: Norberg B, Kjesbu OS, Taranger GL, Andersson E, Stefansson SO (eds) 6th International symposium on reproductive physiology of fish, Bergen, 4–7 July 1999, p 441

Horváth Á, Wayman WR, Urbányi B, Ware KM, Dean JC, Tiersch TR (2005) The relationship of cryoprotectants methanol and dimethyl sulfoxide and hyperosmotic extenders on sperm cryopreservation of two North-American sturgeon species. Aquaculture 247:243–251

Horváth Á, Urbányi B, Mims SD, Bean WB, Gomelsky B, Tiersch TR (2006) Improved cryopreservation of sperm of paddlefish (*Polyodon spathula*). J World Aquacult Soc 37:356–362

Horváth Á, Wayman WR, Dean JC, Urbányi B, Tiersch TR, Mims SD, Johnson D, Jenkins JA (2008) Viability and fertilizing capacity of cryopreserved sperm from three North American acipenseriform species: a retrospective study. J Appl Ichthyol 24:443–449

Horváth Á, Urbányi B, Wang C, Onders RJ, Mims SD (2010) Cryopreservation of paddlefish sperm in 5-ml straws. J Appl Ichthyol 26:715–719

Jähnichen H, Warnecke D, Trölsch E, Kohlmann K, Bergler H, Pluta HJ (1999) Motility and fertilizing capability of cryopreserved *Acipenser ruthenus* L sperm. J Appl Ichthyol 15:204–206

Kopeika EF, Williot P, Goncharov BF (2000) Cryopreservation of Atlantic sturgeon *Acipenser sturio* L., 1758 sperm: first results and associated problems. Bol Inst Esp Oceanogr 16:167–173

Linhart O, Mims SD, Gomelsky B, Cvetkova LI, Cosson J, Rodina M, Horváth Á, Urbányi B (2006) Effect of cryopreservation and male on motility parameters and fertilization rate in paddlefish (*Polyodon spathula*) frozen-thawed spermatozoa. J Appl Ichthyol 22:389–394

Psenicka M, Alavi SMH, Rodina M, Gela D, Nebesarova J, Linhart O (2007) Morphology and ultrastructure of Siberian sturgeon (*Acipenser baerii*) spermatozoa using scanning and transmission electron microscopy. Biol Cell 99:103–115

Psenicka M, Dietrich GJ, Wojtczak M, Nynca J, Rodina M, Linhart O, Cosson J, Ciereszko A (2008) Acrosome staining and motility characteristics of sterlet spermatozoa after cryopreservation with use of methanol and DMSO. Cryobiology 56:251–253

Psenicka M, Rodina M, Linhart O (2010) Ultrastructural study on the fertilization process in sturgeon (*Acipenser*), function of the acrosome and prevention of polyspermy. Anim Reprod Sci 117:147–154

Tsvetkova LI, Cosson J, Linhart O, Billard R (1996) Motility and fertilizing capacity of fresh and frozen-thawed spermatozoa in sturgeons *Acipenser baeri* and *A ruthenus*. J Appl Ichthyol 12: 107–112

Urbányi B, Horváth Á, Kovács B (2004) Successful hybridization of *Acipenser* species using cryopreserved sperm. Aquacult Int 12:47–56

Chapter 36
How Non-governmental Stakeholders Have Contributed to the Conservation Programme in France

Didier Moreau

Abstract In spite of considerable scientific progress since the middle of the 1990s and the combined efforts of a growing number of actors in France and Europe during the last decade, the European sturgeon still receives little attention from the general public and the elected officials in France. Despite the various national and international commitments, non-governmental organizations face the problem that the French authorities took action rather reluctantly up to the recent adoption of a National Action Plan. A survey of the steps taken since the end of the two LIFE programmes makes it possible to determine progress towards the international recognition of the species, and French commitment to the activities required.

36.1 Introduction: Conservation Efforts for *Acipenser sturio* Until the Early 2000s

After the first warnings in the early 1970s from fishermen in the Gironde estuary (Rabic, personal communication) and biologists specialised in the species, *Acipenser sturio* has been protected in France since 1982 (Trouvery et al. 1984). At that time, catches of broodfish had already declined to negligible single individuals. Since then, the species has attracted scientific interest in restoration rather than political support: extensive biological knowledge and adaptations of biotechniques for controlled reproduction were needed to support the last population world-wide (Castelnaud et al. 1991; Williot et al. 1997). Such interest from the scientific community and a few stakeholders was particularly increased during two European Union (LIFE) programmes (1994–1997 and 1998–2001). During these programmes, an increasing awareness of the critical situation of the population status was witnessed, with the participation of the river and sea professional fisheries

D. Moreau
WWF-France (consultant), 34 rue des Souchères, 26110 Nyons, France
e-mail: dmoreau@wwf.fr

P. Williot et al. (eds.), *Biology and Conservation of the European Sturgeon Acipenser sturio L. 1758*, DOI 10.1007/978-3-642-20611-5_36,

organizations as well as of national (Ministry of Environment and Adour–Garonne Water Agency, Ministry of Agriculture and Fisheries) and local authorities of the Gironde–Dordogne–Garonne basin (Regional and General Councils, Dordogne and Garonne River Basins Joint Unions – EPIDOR and SMEAG). EPIDOR became the head of the programmes with the Bordeaux-based Cemagref (French Institute for Environmental Technologies). Considerable improved knowledge on many aspects of the biology and rearing technology for the species has been achieved during these years (Elie 1997; Rochard 2002; for results, see Chap.20).

These authorities have enhanced the visibility of local non-governmental associations such as the Association for the Protection of Wild Sturgeon (ADES), which has become a partner in these programmes. The expansion of research activities into the marine environment and a report on the habitat conditions and estuarine environments were considered tools for increasing the involvement of management authorities. The results of this collective emulation through the LIFE programmes correspond to four main points (Guerri, personal communication):

- The launch of a fish breeding project: a repopulation tool had been set up in the Cemagref research station, considering that the sturgeon would certainly not be able to recover on its own, and that it needed help.
- The improvement in knowledge about the natural habitat: despite some problems of toxic pollutants and one drawback concerning the quality of estuary sediments, the Garonne like the Dordogne still possessed a habitat in a satisfactory state, so that the repopulation project would work and sturgeon could reproduce.
- The collaboration with fishermen: the importance of accidental capture risks for the species, especially on the coast, has been realized, as well as the importance of such a collaboration for future developments of the restoration project.
- Information and awareness: there was a huge deficit, not only among the general public, but also in the knowledge of public authorities, even in the different ministries.

Unfortunately, combined efforts during these years have not delivered the expected results in the field of species conservation: the last wild sturgeon population has instead suffered further decline, with numbers of the wild population declining further to a few hundred individuals by the early 2000s (Castelnaud et al. 1991).

Retrospectively, this failure from the conservation point of view can be explained by several reasons: the insufficient resources to achieve the defined task, the restrained area of operation, the lack of mobilization of some stakeholders, the uneven implementation and observation of regulations, a limited access to implementation tools within the framework of LIFE programmes, and so on. And above all, insufficient access to the biological resource: there were not fish available to achieve the aims planned, and there was a large deficit with regard to knowledge concerning the species, which adversely affected the targets to be reached (Gessner, personal communication). In addition, the French national authorities have not been sufficiently involved in the issue: attempts to increase active participation either at administrative or political levels did not result in sufficiently large support, taking into account the international importance of the issue. The above-mentioned experiences made it seem necessary to accelerate and scale-up the project. It was concluded that increased efforts were necessary to mobilize top administrative levels, and also to reach binding national and international commitments. The latter has become increasingly important with respect to the countries of the North-East Atlantic coast, because:

- France was the last country to host a reproductive, though irregular, population, even though the latest natural reproduction was reported in 1994 (Williot et al. 1997). Therefore, the responsibility for the conservation of *Acipenser sturio* mainly rests with France by virtue of its main responsibility for the survival of the last wild specimens. Neighbouring countries also share this responsibility, since marine by-catch can only be dealt with at an international level
- Nevertheless, the responsibility for the species is also shared by several countries throughout Europe, since the goal to restore populations within feasible segments of its historical range had become an accepted view during the last LIFE programme (and was going to become also an accepted view during the continuous process of Action Plan development under the Bern Convention). The successful conservation of the species depends upon a synergy being established between technical, human, and financial resources
- Any project for restoring *Acipenser sturio* in Europe will rely on the sturgeon population of the Gironde, Garonne and Dordogne basin, including the captive stock created from this population (Williot et al. 2004)
- Therefore, those states which have failed to preserve their own sturgeon populations must support France in its moves to safeguard a common European biological heritage

At the international level, despite a favourable environment for the implementation of potential targets, there was a lack of interest to initiate a scaling-up of conservation steps, except from Germany, which proposed to do so in bilateral consultations since 1996 as well as in OSPAR Convention meetings almost annually (Gessner, personal communication). The development of a political commitment within the countries of the historical range was largely missing. In addition, France did not make any attempt to take the lead in endorsing action to save the species from extinction, and always responded very vaguely to Germany's requests. Through the communications to this point, some stakeholders successively began to show increased interest in the protection issue.

36.2 Emergence of a National Working Group That Evolved into a European One in the Early 2000s

A conclusion seminar for the LIFE programme was held in Libourne (Gironde) in 2001 (EPIDOR 2002), bringing together more than 200 scientists, management experts and decision-makers from France and other countries. It outlined future prospects for the conservation of this species. In addition to the conservation of the species and its habitats, the following issues were considered to be of major relevance:

- Education, training and awareness, notably in the fisheries. First attempts were made during the first two LIFE programmes, but an extension of their range to the countries of the North-East Atlantic coastline was considered inevitable.
- A strong enforcement of international, European and national regulations, as well as the transformation of EU standards into national law, appeared to be an essential prerequisite for successful protection, while being lacking throughout the range of the species.
- Drafting a structured plan for protecting the sturgeon nationally and enhancing the consistency between national approaches and harmonizing management programmes in the coastal river basins of the French and European coastlines, where the presence of the species could be envisioned if some stocks were re-established.

The seminar resulted in a guideline to pave the way for a European Action Plan for restoring *Acipenser sturio*. Its implementation would have to be carried out jointly by the involved countries and the EU, as well as national and international organizations.

A range of tools for safeguarding the sturgeon has indeed become available at the national and supra-national levels. The EU regulations for the conservation of the species' habitats, and the quality of the waterbodies from Natura 2000 to the enforcement of the Water Framework Directive, are considered strong instruments. Additionally, the management of fishing activities on the coastline allows the involvement of professional fishermen, while the strategy for the conservation of a captive stock of sturgeon could be carried out under the biodiversity regulations.

In France, the mobilisation of some elected officials, in particular of Senator and EPIDOR's President Bernard Cazeau, who had supported such a European Action Plan alongside the successive French Ministers in charge of Environment attracted new partners expected both to press French national authorities and to draw attention to the matter on the national and supra-national levels. The WWF network, which operates in a few hundred countries and shows a strong presence across Europe, was among the most promising candidates. WWF-France, approached by EPIDOR as early as 2002, had nevertheless stayed aloof from the European sturgeon, while it had been heavily committed to the conservation of the Loire River and the Atlantic salmon for years. Despite a first initiative, which was not continued, from WWF's European Freshwater Team, which tried to put the subject forward inside the network (Gessner, personal communication), WWF did not address this issue until then. The sturgeon indeed stands at the crossroads of a number of different conservation issues which are handled by separate departments at WWF (Freshwater, Oceans, Species), headed by particularly busy project managers. As such, the European sturgeon appears at this time nowhere on the WWF Agenda in countries other than France.

In the past, it was already far from uncommon for the local stakeholders involved in conservation measures in France that over several years, the so-called *Direction de l'Eau* and *Direction de la Nature et des Paysages*, under the authority of the Ministry of the Environment which was in charge of fisheries (and later on of fishes) as well as of protected species, shifted the responsibility for tackling the issue from one to another. The European sturgeon conservation issue, which is related to various sector-based policies, indeed often put public authorities on the spot to decide how to approach it. Moreover, in France as well as in many other countries, the issue of diadromous fish conservation has received very little interest from public bodies responsible for public education, as well as from the majority of organizations promoting recreational fishing. Even in the majority of the environmentalist groups, with a few exceptions, a greater emphasis was placed on land species compared to fishes and other aquatic species (Arnould and Moreau in press). The origin of the lack of interest and sensitivity is based upon the poor media coverage and, consequently, on the poor attention by political actors. The European sturgeon is associated with caviar, but, for many people, is mixed up with its Siberian cousin (*Acipenser baerii*), farmed in France for a luxury production meant to replace the so-called "black gold" formerly extracted from the wild population of *Acipenser sturio*.

This was the situation in 2003 when WWF-France began to address the issue by bringing together a panel of experts and the main stakeholders in the conservation of the European sturgeon, in the framework of an informal working group led by this NGO. One goal was to set in action such a working group, initially at the national level, in order to develop discussion documents and formulate proposals that would open the door for moves towards national and international conservation public policies. The task force initially comprised members from the Bordeaux-based Cemagref, EPIDOR, the ADES and WWF-France. This group was rapidly joined by *Migrateurs Rhône-Méditerranée* (MRM), an organization aiming at the reintroduction of the sturgeon into the Rhône river, as well as the Garonne River Basin Joint Union (SMEAG). Thanks to the collective effort and recognition, this national working group established a counter-balance to the French national authorities. This was considerable progress compared to the past situation when this recognition failed to be achieved, since lobbying was neither within the scope of the Cemagref,as a research institution, nor within that of the locally-based EPIDOR and SMEAG.

Through a consultant (Hervé Lethier, EMC^2-I. See Lethier 2005) engaged by WWF-France, this informal group began its activities by meeting the Nature Protection and Biodiversity's Unit Head of the European Commission's DG Environment in Brussels. This department was very familiar with the case, as a result of the two LIFE programmes it had funded. DG Environment was about to integrate the reflection carried out by the OSPAR Convention (Convention for the Protection of the Marine Environment of the North-East Atlantic) on the conservation of marine biodiversity into its work on Special Areas of Conservation defined by the European Union's Habitats Directive (92/43/EEC) inside the territorial seawaters, and the Exclusive Economic Zone. It also had already called on the national authorities in the Member States where the European sturgeon was still present (France) or occasionally reported at sea (Channel and North Sea) to provide detailed information about the legal status and the possible considered conservation public policies for this species (especially under Natura 2000 policy). The majority of the European Commission's expectations concerning the conservation of the species rested with France, as the Commission was going to examine a case filed by *France Nature Environnement* and backed up by WWF-France, about the large-scale navigation channel to be planned in the Gironde estuary for the transport of the Airbus A380 parts, nearby the nursery areas for sturgeon juveniles. This meeting with the European Commission was the first step to try to make international conservation public bodies put pressure on French national authorities.

Progressively, the working group took on an international dimension, with German NGOs, namely the Society to Save the Sturgeon (GRS) and the World Sturgeon Conservation Society (WSCS), as well as the Berlin-based cooperation partner of the Cemagref (the IGB or Leibniz Institute of Freshwater Ecology and Inland Fisheries) joining it, while also securing the support of the IUCN's Sturgeon Specialist Group. It was poised to push ahead with a European-wide Action Plan (Rosenthal et al. 2007).

36.3 A Marathon Across the World to Get Political Tools Operated

At the end of 2004, despite positive feedback from the European Commission and the Council of Europe, no straightforward political commitment had been made yet by either of the two institutions. During the negotiations, the Bern Convention on the Conservation of European Wildlife and Natural Habitats was identified as a prime instrument to progress with preserving the threatened species. With this aim in mind, WWF-France and the WSCS presented the issue at the Steering Committee meeting of the Convention (with participation from France, Germany, the Netherlands, Romania, Ireland, Cyprus and the United Kingdom as well as Cemagref, EPIDOR and EMC^2-I). In response to the presentation, an agreement based on a British proposal was reached. This agreement suggested that the principle of a European-wide action plan for the species should be applied. To communicate the ideas, it was suggested thata special side event devoted to sturgeon issues be held in the following year in Strasbourg, at the 25th anniversary of the Convention. The Council of Europe supported the plan with some funds for WWF-France to commence with the current work.

During the first half of the year 2005, significant progress was made in bringing the case to notice publicly:

- An additional meeting with the Nature Protection and Biodiversity Unit of the European Commission's DG Environment, which called on WWF-France to give an oral presentation later in the autumn as a side-event to the Habitats Committee meeting

- A meeting with the DG Fish of the European Commission, which invited WWF-France to approach the new European Fisheries Fund. EFF at that time projected proposals for awareness-oriented campaigns focusing on the interactions between fishing activities, environment and society
- At a meeting in Dubai in the framework of its European Sub-Committee, WWF International tried to obtain a rating of the sturgeon as a priority species as part of the global strategy of the organization in the Europe–Middle East area. The project received a favourable reception for the North-East Atlantic ecoregion. This step was devised as a way of enhancing the legitimacy of the network's lobbying, mainly at the high-profile international meetings due to be held in the end of the year (Steering Committees of the Bern and Bonn Conventions)
- The OSPAR Convention had added the sturgeon to the list of priority species after considerable efforts of the German Federal Ministry for the Environment in 2005. A meeting with the Secretariat also brought up the opportunity for WWF to give a presentation later in the autumn at the MASH task force (Working Group on Marine Protected Areas, Species and Habitats). From 2006 onwards, on their own, Cemagref and IGB jointly completed the MASH yearly forms
- A meeting with the Secretary-General of the Convention on Migratory Species (CMS or Bonn Convention), indicated the possibility of adding the European sturgeon to Annex I of the Convention, providing that one member state requested it before the 20th of June for consideration at the next Conference of the Parties

WWF-France, supported by the working group in collaboration with the French Fisheries Committee, heavily lobbied the French national authorities in an attempt to make a proposal within the requested timeframe. Such a proposal was presented to the Environment Minister, who responded in favour of the suggestion. The communications which followed with the ministry's relevant departments gave the opportunity to observe their awareness to put the sturgeon conservation issue on EU's agenda. However, they did not want to support the proposal to add the species to the CMS' Annex I, since a number of protective national and European regulations already existed. Rumours also indicated that some people involved in the process viewed such a move unnecessary, since the species was considered to be threatened with extinction anyway. At last, after continued discussions, agreement was reached that the Ministry eventually ended up handing in the requested proposal and asking the other relevant ministries (Foreign Affairs as well as Agriculture and Fishing) to follow this line. Finally, the proposal was handed over to the Secretariat of the Bonn Convention within the allotted time.

In the summer of 2005, WWF-France continued to establish contacts, in particular within the framework of a task force meeting held in Petronell (Austria), to draw up an Action Plan for the Danube sturgeons. The meeting was organized by the Austrian Ministry of Environment and the National Park Donau–Auen, the Swiss Federal Institute for Environmental Science and Technology and the International Association for Danube Research, WWF Austria and WWF's

Danube–Carpathian programme, with the support of the Bern Convention and the WSCS.

In September, presentations of the outline for the European Action Plan were given at the MASH task force of the OSPAR Convention and the European Commission's Habitats Committee. Exchange between WWF-France and the French Fisheries Committee continued at the organization's Interprofessional Commission for Migratory Fish and Estuaries. With fisheries and conservationists starting to cooperate, a potential option opened up to respond to the invitation of DG Fish, to renew the efforts to mobilize all the fishing stakeholders (professional fishermen, fishery organizations and relevant public authorities) with regard to the protection needs of the European sturgeon and to existing regulations, in order to decrease the mortality among wild population caused by incidental catch. The campaign was targeted on French coastal and marine waters and subsequently on the whole North-East Atlantic.

In November 2005, following the 13th meeting of the Convention on Migratory Species in Nairobi (Kenya), held between the 21th and the 25th of this month, and thanks to the support of WWF International and a few other national WWF offices in Europe, the European sturgeon's status increased from Annex II to Annex I of the Convention. On 30 November, at the 27th meeting of the Steering Committee of the Bern Convention in Strasbourg, WWF-France, on behalf of the working group, carried out a special side-event devoted to the sturgeon, as proposed by the Secretariat of the Convention in the previous year. About 30 delegates and experts from the member states of the Convention, coming from 14 countries of the Council of Europe, took part in the side-event. The vast majority of the participants showed a genuine interest in setting up a European task force intended to draw up the proposed Action Plan, which it was suggested should be presented at the meeting in November 2006. Moreover, France, Germany, and Great Britain also committed themselves to contributing financially to the preparation of the plan. WWF-France was appointed to chair the task force comprising most of the participants in the previous informal working group, plus the French Fisheries Committee, an expert from the University of Padova (Italy), the WSCS, and the Secretary of the Bern Convention.

The establishment of the framework of the European Action Plan was based upon the results of a series of meetings held in France (Bordeaux) in 2006. But the European Action Plan could not be completed by that year's meeting of the Steering Committee of the Bern Convention. Consequently, its deveopment was prolonged until September 2007, with WSCS taking over the compilation and editing of the plan after a final project meeting in Germany (Berlin) in June 2007. Thanks to effective lobbying of the French and German delegations, the plan was unanimously adopted on at the 27th meeting of the Steering Committee, in late November in Strasbourg (Rosenthal et al. 2007).

36.4 Lost Momentum: A Smooth Development Process for the European Action Plan, a Tortuous Way Towards a Successful Implementation

The Bern Convention Action Plan was adopted at a time when the awareness campaign targeted at professional fishermen, launched by the French Fisheries Committee, was beginning to expand to the other countries in the North-East Atlantic. It also stemmed from the decision by the French state to impose a ban on gravel extraction in the Gironde estuary. This decision was a result of the battle fought by a local elected official, Philippe Plisson, who was serving as President of the Local Water Commission of the Water Resources Planning and Management Scheme, as well as being Vice-President of the River Basin Joint Union for Sustainable Development of the Gironde Estuary (SMIDDEST). More importantly, it took place a few months after the Cemagref's successful attempt to conduct the first artificial reproduction of brood fish which had grown in captivity (Williot et al. 2009). The enthusiasm stirred up by the event was reflected in the announcement made in November by the French Secretary of State in charge of Ecology, who announced a National Action Plan to be started in the next 6 months. Such a decision was consistent with the commitments made a couple of months earlier in the framework of the "Grenelle Environment" Round Table process (a multiple-partner conference instigated by the French President Nicolas Sarkozy in the summer of 2007 to define the key points of environment public policy and draw up an Action Plan for the following 5 years), as well as with priority status of the species obtained in the National Strategy for Biodiversity, a process to which WWF-France contributed significantly. Despite a number of worrying facts appearing in the few months after the Secretary of State's announcement (for example the authorization to introduce and farm non-native species of sturgeon in French hatcheries for caviar production), this series of governmental decisions seemed to be a fresh start for ambitious concrete conservation and restoration measures of the European sturgeon.

However, more than three years after the Secretary of State's announcement were necessary so that the French Action Plan is finalized and validated. But at the time of finalizing the present chapter (June 2001), and thus soon four years after this announcement, questioning remains on the governance of the issue, and no publication still went out in order to make visible the French governmental strategy. In this context, the moves of each partner fail to be really coordinated, it remains partly unclear what the guidelines, the decision-making process, and the technical or strategic tools will be. What is more, the fact that the French national authorities fail up to now to take the lead in the matter at a European level, despite suggestions and requests from Germany and the Netherlands which remain without clear response, weakens any effort to launch a long-term ambitious programme for restoring the European sturgeon within its historical range.

36.5 Conclusion

Since the early 2000s, the grouping of partners involved in the sturgeon conservation has grown in size, as a sign that the present evolution is hopeful for a successful restoration of the species. The general approach to the sturgeon issue is more optimistic than in the past: there is some evidence of the species surviving in its natural environment, that attempts to curb the mortality caused by incidental catch can come to fruition, and, more importantly, that techniques for its artificial reproduction, ahead of the reintroduction in its natural environment are, in the main, successful (in spite of the failure of the artificial reproduction attempts in 2010). Each of these partners, like the other stakeholders campaigning for a patrimonial and integrated management of the Gironde (gathered into the so-called *Collectif Estuaire*), are expecting the French State to make a strong and constant commitment, consistent with its national and international commitments.

The Water Resources Planning and Management Scheme of the Gironde Estuary approved in September 2010, the project of a second acclimatization structure in Saint-Seurin for the European sturgeon, the management of which will be given to the association MIGADO (Migrateurs Garonne Dordogne), as well as new strategies to share ex situ conservation actions with other organizations than Cemagref, are positive points to achieve the expected results in term of reinforcement and restoration of the wild population. But current or future collaborative efforts between the State departments, research bodies, local authorities and professional or non-governmental organizations can indeed quickly deliver poor results in the event of the French government not showing more conviction in working in partnership in the sturgeon conservation issues, as it has so far taken action more by obligation than by devotion, failing to provide the required emphasis to give rise to a genuine synergy with all its partners at the local, national, and European levels. The risk of the whole project losing momentum indeed exists, as some stakeholders, such as professional fishermen, can be discouraged from involving themselves again later if the efforts they already made or the actions they implement do not have sufficient feedback from the governmental authorities, especially as time is against the conservation of the species: despite the considerable achievements in the past decade, the next few years are crucial, making it necessary to cash in on the established results of these partnerships and to reverse the evolutionary as well as the demographic trend of the species in the long run.

A more ambitious collective action than that deployed by a single European-wide working group will surely be needed to give a fresh impetus to the current situation. The main problem is that the majority of political actors still pay scant regard to the European sturgeon conservation issue, in spite of information documents (booklets, film) produced and focused on this stakeholder category. Other species, less threatened but the conservation issues of which are conflictual, such as Brown bear and Grey wolf among mammals or European eel and Atlantic bluefin tuna among fishes, arouse much more interest among the general public, the media and so the political actors, as these issues have social and economic side-effects. Conditions of a true balance of power between governmental decision-makers, local authorities and civil society are therefore still missing to encourage the French government to more actively support the conservation programme, enhance the visibility of its policy, and open windows of opportunity for its French and European partners, for the development of a national and European strategy, not only verbally but factually too.

Acknowledgements This chapter is derived from numerous WWF-France working documents and minutes of meetings that have not been published. A lot of people contributed to these documents, among whom the present author and: Cyrille Deshayes, Mohend Mahouche, Stéphane Ringuet, Benoît Guérin (WWF-France), Paul Gonthier, Eric Rochard, Patrick Williot (Cemagref), Olivier Guerri (EPIDOR), Jacqueline Rabic, Nicolas Michelet (French Fisheries Committee), Gilbert Miossec (ADES), Jean-Yves Ménella (Rhône-Migrateurs-Méditerranée), Hervé Lethier (EMC^2-I), Joern Gessner, Frank Kirschbaum (IGB) and Harald Rosenthal (WSCS), who are gratefully recognized for their assistance.

References

Arnould M, Moreau D (2010) Contradictions in public policies on management of aquatic species and habitats, the case of policies for diadromous fish. In: Bergès L, Brun JJ, Gosselin M, Martin C, Mauz I, Rochard E (eds) Public policy and biodiversity. Scientific topics, political issues and local actions. Sciences Eaux and Territoires 3bis:56–62. On-line paper. URL: http://ww.set.revue.fr/contradictions-public-policies-management-aquatic-species-and-habitats-case-policies-diadromous-fis-0

Castelnaud G, Rochard E, Jatteau P, Lepage M (1991) Données actuelles sur la biologie d'*Acipenser sturio* dans l'estuaire de la Gironde. In: Williot P (ed) Acipenser. Cemagref, Antony, pp 251–275

Elie P (ed) (1997) Restauration de l'esturgeon européen *Acipenser sturio*. Contrat Life B4-3200/94/754. Rapport final du programme d'exécution, Etude Cemagref (Bordeaux, GMA, RAC) n 24

EPIDOR (2002) Quel avenir pour l'esturgeon européen? Actes du séminaire, Libourne, 3–4 Octobre 2001. Presses France Quercy, Cahors, 183 p

Lethier H (2005) Vers un plan international de restauration de l'esturgeon, *Acipenser sturio*, document préliminaire d'orientation. EMC^2-I pour WWF-France, Paris, 58 p

Rochard E (coord) (2002) Restauration de l'esturgeon européen *Acipenser sturio*. Rapport scientifique Contrat LIFE n B-3200/98/460, Cemagref, Etude n 80, Groupement de Bordeaux, 224 p

Rochard E, Castelnaud G, Lepage M (1990) Sturgeons (Pisces: Acipenseridae); threats and prospects. J Fish Biol 37(suppl A):123–132

Rosenthal H, Bronzi P, Gessner J, Moreau D, Rochard E (2007) Action plan for the conservation and restoration of the European sturgeon. Convention on the Conservation of European Wildlife and Natural Habitats (Bern Convention), Nature and Environment No. 152. Council of Europe Publishing, Strasbourg, 116 p

Trouvery M, Williot P, Castelnaud (1984) Biologie et Ecologie d'*Acipenser sturio*. Etude de la pêcherie. Cemagref, Etude n 17, Série Esturgeon n 1, 79 p

Williot P, Rochard E, Castelnaud G, Rouault T, Brun R, Lepage M, Elie P (1997) Biological characteristics of European Atlantic sturgeon, *Acipenser sturio*, as the basis for a restoration program in France. Environ Biol Fish 48:359–370

Williot P, Rouault T, Rochard E, Castelnaud G, Lepage M, Gonthier P, Elie P (2004) French attempts to protect and restore *Acipenser sturio* in the Gironde: Status and perspectives, the research point of view. In: Gessner J, Ritterhoff J (eds) Species differentiation and population identification in the sturgeons Acipenser sturio L. and Acipenser oxyrinchus, vol 101. Bundesamt für Naturschutz, Bonn, pp 83–99

Williot P, Rouault T, Pelard M, Mercier D, Jacobs L (2009) Artificial reproduction of captive endangered European Atlantic sturgeon, *Acipenser sturio*. Endanger Species Res 6:251–257. doi:10.3354/esr00174

Chapter 37
Why, How and Results from an Awareness Campaign Within Marine Professional Fishermen for the Protection of Large Migratory Fish, the European Sturgeon *Acipenser sturio*

Nicolas Michelet

Abstract Mortalities arising from accidental catches of European sturgeon at sea represent one of the main threats to the diadromous species. The information and awareness-raising among fishermen, and more generally the maritime sector on an European scale, is of the highest importance for the success of the species' restoration programmes. The French national committee for marine fisheries and sea farming (CNPMEM) has been addressing French professional fishermen in that way since 2006, as they are the most likely to be involved in accidental catching of the fish. The communication campaign has been extended since 2008 over the area of marine distribution of the species, with the creation and mobilisation of an international information network. Several communication tools are nowadays available (leaflets, posters, video documentary, website, etc.) and contribute to the dissemination of the campaign messages. In addition to the proactive assistance provided by the European maritime sector, the accidental catches of six adults and 39 juveniles, released alive since the end of 2006, have been reported through the information network. Taking into account the progressive dispersion into the natural environment of the young sturgeons issued from recent artificial reproductions, communication campaigns must be continued in France and abroad.

N. Michelet (✉)
French National Committee for Marine Fisheries and Sea Farming (CNPMEM), 134 avenue de Malakoff, 75116 Paris, France
e-mail: nmichelet@comite-peches.fr

P. Williot et al. (eds.), *Biology and Conservation of the European Sturgeon Acipenser sturio L. 1758*, DOI 10.1007/978-3-642-20611-5_37,

37.1 Introduction

37.1.1 The Watchword: Save the European Sturgeon!

Even though the species is since the early 1970s no longer the target of organised commercial fishing (Trouvery et al. 1984), the European sturgeon frequents, throughout its life cycle, the estuaries and shallow coastal waters that are also inhabited by numerous other fished species, and where most of the professional fleets and amateur fishing activities of the countries concerned are concentrated (Rochard et al. 1997). Mortalities arising from accidental catches are now considered to be one of the main threats to this great diadromous species.

Victims not only of their physical appearance and sometimes extraordinary length, but also due to the exceptional and symbolic mystique that surrounds the catching of a specimen, sturgeons are sometimes kept on board boats, unloaded and even sold.

These observations and conclusions have been obtained from different work programmes carried out since the beginning of the 1980s on the future of the sturgeon in its natural maritime habitat. These studies have, in particular, enabled an area of the marine distribution of the species to be mapped out and the impact of by-catches at sea on the size of the remaining population to be assessed. Furthermore, they have highlighted the profound ignorance of the situation and legal status of the species by those involved in maritime affairs, both in France and throughout the European Community as a whole (Rochard et al. 1997).

In 2001, the necessity of curbing the decline in numbers of the population in the wild was reaffirmed through a new action plan that gave emphasis to an international approach to conserving the species. So it was decided to carry out wide-ranging information, training and awareness campaigns throughout the Atlantic coast countries that are on the boundary of or close to the epicentre of the current population, in an endeavour to reduce the risk of mortality through accidental catch in the species' natural habitat.

37.2 The Urgency of Increased Awareness Within the French Fisheries Sector

37.2.1 Relaunching the Awareness-Raising Campaign

As part of a group initiative coordinated by WWF-France, and in order to meet new targets set for the protection and restoration of the species, the French National Committee for Marine Fisheries and Sea Farming (CNPMEM) undertook, in 2006[1],

[1]The 2006 campaign was carried out with the financial support of the European Union, the Agence de l'Eau Adour–Garonne, the WWF-France foundation, the French Ministry of Ecology and

major communication campaigns aimed at the French fisheries sector to raise awareness of the problems surrounding the conservation of the European sturgeon. The involvement of the profession followed upon the sales of two adult sturgeons in France and Wales in 2004 (Guth and Laurent 2004). As French fishermen are the most likely to be involved in accidental catching of the European sturgeon, regular communication with the sector should enable mobilisation and levels of vigilance to be maintained commensurate with the stakes and risks of the extinction of the species. Generally speaking, this communication strategy is aimed at increasing the chances of an individual specimen being able to return from the sea to its spawning grounds, and at improving our understanding of the species in its maritime environment.

This campaign, also called the "Sturio Programme", was based on teachings and recommendations drawn from previous communication programmes, and more especially the "Atlantic Sturio" operation (Mayer and Lepage 2001) conducted in France on the Atlantic and Channel coasts up until the year 2000. In the same way, different communication tools (information pamphlets, posters, stickers, voluntary declaration forms, desk pads, mouse mats, and slide shows) were drawn up and disseminated (Michelet 2006):

- Either directly to the campaign's priority targets, i.e., the professional French fishing fleets using benthic gear, fixed nets or bottom trawlers in the species' marine distribution area
- Or through the intermediary of partners in an information network comprising special contacts in the fisheries social and occupational environment, i.e., representative organisations within the fishing sector with the most relevant role to play in passing on (disseminating and collecting) information: local and regional maritime fishing committees, maritime affair departments, fish auctions and markets, network of IFREMER technical researchers, fish producer organisations, professional maritime schools, and French amateur or sports fishing federations as well as maritime cooperatives and educational organisations established along the coast

In addition to general information on the biology and legal status of the species, communication tools emphasised the attitude to adopt in the event of an accidental catch, the necessity of mobilising those involved in the sector, and hopes placed in the possible solution of captive breeding stock. A procedure has been put into place in order to follow up each by-catch declaration through a standardised response that thanks the fishermen and partners involved.

This work has enabled the French network of partners and actors in the field wishing to make a proactive contribution to the protection of the sturgeon to be

Sustainable Development (MEDD), the French Ministry of Agriculture and Fishing (MAP), and with the technical participation of the Cemagref, the river basin administrative departments EPTB Gironde (SMIDDEST), Garonne (SMEAG) and Dordogne (EPIDOR), the Ecole de la Mer at La Rochelle and the Wild Sturgeon Conservation Association (ADES).

reactivated and strengthened. With the support of some 300 professional, institutional or association representatives from both the maritime and fluvial world, from over 60 ports in the 17 departments on the French Atlantic coastline, the Channel and North Sea, this campaign has reached a fleet of over 2,000 trawlers and gillnetters and nearly 6,000 sea fishermen. Conducted for the most part along the Atlantic Arc coast, it was relayed in the media through the local and regional press, and was written about in articles published in different magazines and newsletters. Issues faced by the species and associated regulations have been included in initial training programmes for personnel working in maritime affairs departments and recorded in fishing inspection manuals.

37.2.2 An On-Going Process

Through a partnership with WWF-France, the CNPMEM awareness-raising campaign[2] continued in 2007 and 2008 (Michelet and Moreau 2009).

Apart from the dynamic created by the 2006 campaign and its initial feedback, the information network has been boosted by the regular broadcasting of current news presenting the state of progress in the communication campaign, and by specific actions carried out on a nationwide or European scale. Different meetings and events organised along the coast have been the occasion for new partners to join the movement and for the communication work promoting the species to be endorsed. Participation of professional partners is optimised by the dissemination of simple and clear instructions, adapted to situations faced by actors in the field, and the efforts made to limit the time taken to respond to by-catch declarations.

This work proved to be even more important, insofar as the first artificial reproduction of European Sturgeon from stock bred in captivity at the Cemagref research institute facilities was concurrently successful (Williot et al. 2009). The release of hatchery-produced alevins and juvenile stock into the Gironde–Garonne–Dordogne river system began in 2007, and required particular communication efforts with regard to the fishing fleets and others using the repopulation basin. Instructions and by-catch declaration procedures, which are more adapted to the species at the earlier stages in their life cycle, were defined, including thanking the participants.

At the same time, information systems were set up concerning the two main species present in open French waters, the protected European sturgeon and the Siberian sturgeon (*Acipenser baerii*) reared on fish farms. A leaflet enabling the two species to be differentiated was distributed to fishermen fishing the waters where the two species might coexist.

[2]The 2007–2008 phase was carried out thanks to the financial support of the European Commission (FEP), the French Ministry of Ecology and Sustainable Development (MEDD), the Agence de l'Eau Adour–Garonne and the regional councils of Aquitaine and Poitou–Charentes.

Since the end of 2008, CNPMEM has carried out monitoring work[3] on by-catch declarations, and endeavours to reply to requests made. The growth and dispersion of alevins and juvenile stock into the Garonne basin have increased the potential of accidental catches, particularly since 2010, in the Gironde estuary and the surrounding coastal waters. In order to support local incentives and optimise the acquisition of catch data, new awareness campaigns have been conducted aimed at fisheries and local managers, and a form for multiple declarations, more adapted to the situation faced by local fishermen, is now being used.

37.3 The European Dimension of the Campaign

Given the migration of the species and the prospects of reintroducing specimens into their natural environment, protection of the sturgeon requires the proactive participation of fisheries throughout Europe. The recent unloading and sale of two European sturgeons in Wales (2004) and The Netherlands (2007), which demonstrate the lack of information outside of France, only go to prove that this initiative is heading in the right direction.

In order to be consistent with the targets of the International Action Plan for the Protection and Restoration of the European sturgeon and to contribute to its implementation in France, the CNPMEM has offered to promote the dissemination of messages from the French campaign to the main European countries whose fishing fleets are implicated in the issue of European sturgeon by-catches (with priority being given to the United Kingdom, Ireland, Denmark, Germany, The Netherlands and Belgium). This work involves the identification and mobilisation of relay structures that can then continue the communication work with the maritime sector of each of the countries concerned, and provide support for the implementation of the initiative (Michelet and Moreau 2009).

In order that these targets may be met, the CNPMEM counts on the assistance of the "Regional Advisory Council" (RAC) networks, which are European Commission executive agencies and mainstays of the common Fisheries Policy reform, and are the appropriate discussion platforms from where campaign messages are broadcast and constructive contact is established with professional representatives. Their field of expertise extends over the area of marine distribution of the European sturgeon: the South Western Waters RAC for the Bay of Biscay, the North Western Waters RAC for the Channel and Irish Sea, and the North Sea RAC for the sector that bears that same name.

A European information network (Table 37.1) has been gradually built up, with European, scientific, administrative associations and partners joining forces and the setting-up of the international action plan to protect the European sturgeon. Within

[3]Work carried out in 2009 and 2010 by the CNPMEM is financially supported by the Ministry of Ecology, Energy and Sustainable Development and the Sea (MEEDDM).

Table 37.1 National intermediaries that are part of the international information network

State	National intermediaries belonging to the international information network
UK and Ireland	CEFAS – Centre for Environment, Fisheries and Aquaculture Science
Belgium	RBINS – Royal Belgian Institute of Natural Sciences
Netherlands	DFO – Dutch Fishermen Organisation
Germany	GRS – Gesellschaft zur Rettung des Störs
Denmark	ZMUC – Zoological Museum University of Copenhagen
Spain	AZTI Technalia Foundation and IEO – Spanish Institute of Oceanography
Portugal	Institute of Oceanography
Norway	NINA – Norwegian Institute for Nature Research

the member states concerned, as well as in Norway, one or two key structures have been mobilised in order to organise the dissemination of campaign messages along the coast, and to collect any information about the species, with the support of local networks.

Different communication aids (leaflets, posters and declaration forms), based on the French models but adapted to an international context and translated into five languages (English, German, Dutch, Spanish and Danish), have been drawn up and then offered to partners in the European network. These partners, in order to enable them to appropriate the messages and tools employed in this campaign, were included in the adaptation of these aids to the particular requirements of their local context. Eleven thousand leaflets and 2,000 posters have been disseminated by the national networks to fisheries concerned in the North-East Atlantic. The instructions to be followed in the event of an accidental catch are identical to those given to fisheries on the French coasts: the fishermen are invited to note the date and place that the specimen was caught, and to measure or estimate its size and weight before carefully releasing the fish and notifying the catch. If the sturgeon has an external tag, this should be left in place and the tag number noted.

Specialised press and the main international organisations in the sector have also been mobilised. Furthermore, the CNPMEM remains available to provide any assistance that may be required to set up the campaign; advice on the general procedure used to process and reply to by-catch declarations is also offered.

37.4 Making of a Documentary Film and the Creation of a Website

In addition to the actions carried out on the North European maritime coasts, and in order to widen the impact of the awareness-raising campaign, a video documentary lasting 25 min was made by WWF-France with the participation of the CNPMEM and widely broadcast in France and Europe, especially to institutional structures and professional and amateur fishing sectors, in 2008. This film, recorded in French with English sub-titles, takes up the issue of protection of the species, and talks about the hopes that have been kindled thanks to the national and international efforts that are now underway. The documentary recounts the ground that has already been covered, throws the spotlight on the efforts to be continued and developed, and highlights the importance of the partnerships and mobilisation of all those involved in the protection and restoration of the sturgeon.

Furthermore, the CNPMEM has recently developed a multi-language Internet website that is specific to the European sturgeon. This site is a European-scale, up-to-date information platform, and an interactive tool allowing exchanges and cooperation between French and foreign partners who are part of the national and international protection and restoration action plans for the European sturgeon. This tool, which is currently available in French and English, comprises an operational interface for structures seeking proactive involvement in the protection of the species, and is online at the following address: http://www.sturio.eu. Among other elements, it contains:

- Information for the general public (historical background, taxonomy, biological characteristics, threats, protection of the species, conservation programmes and news about actions underway, etc.)
- Communication aids and tools made available to those who wish to contribute to the saving of the European sturgeon (especially managers, schools and other educational establishments)

It also enables accidental catch declarations to be made online, which facilitates the processing and feedback of information, and provides a space in which partners, from the historical distribution area of the European sturgeon, may communicate.

37.5 Results from and Prospects for Communication Campaigns Since 2006

The accidental catches of six European sturgeons, spawners and sub-adults, have been reported since the end of 2006 by French and Belgian fishermen through the information network (Fig. 37.1). All the specimens survived capture (and were released alive or transported to the tanks at the Cemagref research institute facilities so that they may participate in future artificial reproduction programmes) and bear witness to the strong resistance of these fish when caught. Two other specimens, captured in Dutch and Norwegian coastal waters in 2007 and 2008, were dead when unloaded before being notified by national partners. These events took place before the implementation of the awareness-raising campaigns in these countries, but local dissemination of information was facilitated by making the most of the media coverage resulting from these by-catches.

In addition, 39 juveniles, from the Cemagref research institute facilities, have been accidently caught since 2008 (28 of them in 2010) on the lower part of the Gironde–Garonne–Dordogne system and in peripheral coastal waters (see Sect. 36.4). These young European sturgeons were all released in good condition.

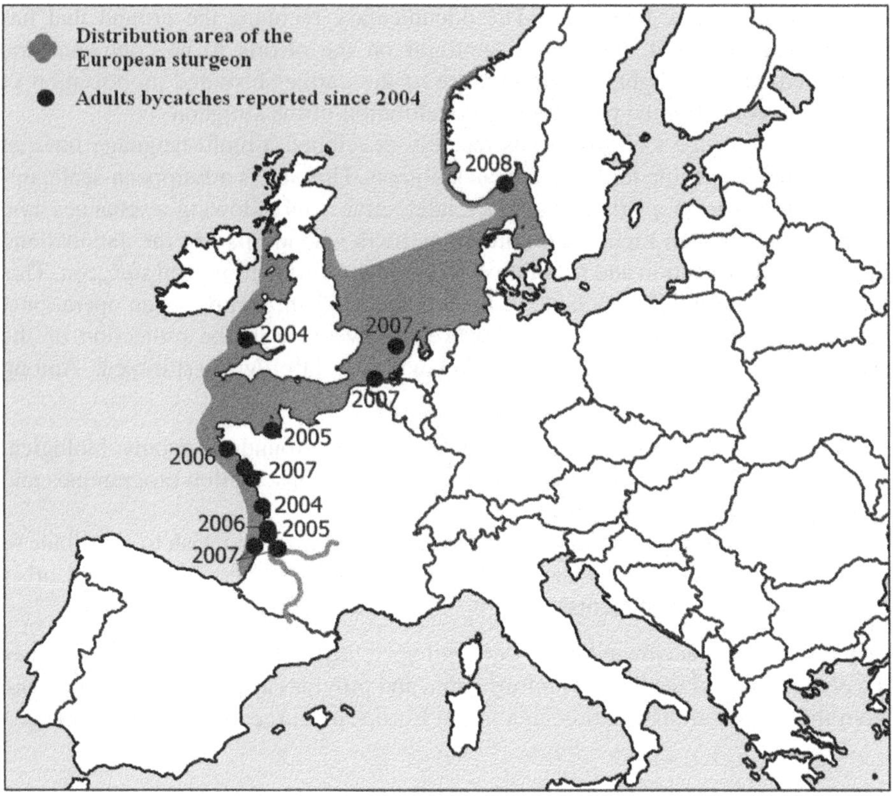

Fig. 37.1 By-catch position of European sturgeon adults reported since 2004

These recordings can be added to the specimen that was found dead on the Bay of Biscay coast and occasional observations gathered in the catchment basin. In addition, a juvenile, born in France and then released in 2009 in the Elbe basin in Germany, was caught, released alive and then reported by a professional fisherman in Danish coastal waters in 2010.

Information is circulated, and it is appropriate to consider that the awareness-raising campaigns have reached their objectives as far as an effective alert network is concerned, amongst others. These voluntary declarations demonstrate the proactive assistance provided by the maritime sector in the protection of this migratory fish, and the good performance of the information networks. The precious contribution to this programme made by professional fisheries in the Gironde–Garonne–Dordogne river system, in which the European sturgeon enjoys a high heritage factor, must be especially mentioned.

Meetings and discussions with the maritime sector have been the occasion to take count of by-catches that have not been declared since the previous campaign, and to specify the geographic location at sea of the different multiple catch areas off the French coast. Furthermore, nine catches of sturgeon, kept alive before being formally identified as not belonging to the European species, have been notified in French and foreign freshwater zones. In the same way, awareness-raising programmes carried out for the European sturgeon have also benefitted another protected species, *Acipenser oxyrinchus*, that was reintroduced into several German and Polish catchment basins in the Baltic Sea some years earlier (see Sects. 6.4 and 6.5). In addition to the information gathered from recent by-catches in Danish coastal waters, some specimens have been successfully released alive.

While there has been clear progress in the mentality within the maritime sector over the last 10 or so years, the interest and readiness to listen to how this migratory species can be saved demonstrated by those working in the field shows the extent to which fishermen are committed to a responsible attitude towards the protection of the species. The commitment to such an action by the professional bodies themselves is in itself an undertaking towards a brighter future for the European sturgeon. This responsible attitude enhances the image of the profession, and also enables campaign messages to be more easily appropriated by the sector. There must be no let-up in these efforts. Optimisation of the rate of voluntary by-catch declarations can be achieved through the creation and maintaining of a trusting relationship with fishermen. Fears evoked by some, linked to the consequences that might occur from such voluntary declarations (closure of fishing zones, in particular), require however that these results be put into perspective, and it should be considered that not all by-catches at sea have been declared.

Communication campaigns must be continued in both France and abroad in order to maintain and develop the mobilisation of the maritime sector and the information networks. From the point of view of greater efficiency of the voluntary declaration system and the way to prepare for the dispersion of young sturgeon to repopulate the distribution area of the species, it is essential that trusting relationships with the professional fishers be conserved, and that awareness be developed among amateur fishermen and other sea users. One of the campaign slogans reflects this image: "the preservation of the sturgeon shall not be made to the detriment of the fisherman, nor can any preservation be effective without the fisherman's help".

Not only has the setting-up of this information network throughout the distribution area of the European sturgeon, and the creation of new multi-language communication tools, provided the foundations upon which future cooperation can be built, it also maintains the degree of mobilisation within a more general context of the protection of biodiversity, with which the professional fisheries are involved in terms of the responsible management and sustainable development of fishing resources. Moreover, the Sturio information and awareness programme is emblematic and "trailblazing" in respect of the environmental issues raised by France in the Grenelle Environnement and the Sea round-table debates.

References

Guth MO, Laurent JL (2004) Retour d'expérience sur la capture et la vente illicite d'un esturgeon en criée aux Sables-d'Olonne (Vendée). Ministère de l'Ecologie et du Développement durable, Rapport de l'Inspection générale de l'Environnement, N IGE/04/2004, 57 p

Mayer N, Lepage M (2001) Sauvegarde et restauration de l'esturgeon européen. Action de communication et de sensibilisation, Opération Atlantique Sturio. Programme Life Nature 98, Contrat Epidor – Agedra, Epidor, Castelnaud La Chapelle, 43 p + anx

Michelet N (2006) Campagne d'information et de sensibilisation du secteur des pêches maritimes relative aux captures accidentelles d'esturgeon européen en mer. Contribution à la mise en œuvre d'un plan d'action international de restauration de l'esturgeon européen. CNPMEM, Paris, 23 p + anx

Michelet N, Moreau D (2009) Campagne d'information et de sensibilisation du secteur des pêches maritimes relative aux captures accidentelles d'esturgeon européen. Mise en œuvre du plan international de restauration et contribution à l'élaboration du plan national de restauration de l'esturgeon européen. CNPMEM et WWF-France, Paris, 11 p + anx

Rochard E, Lepage M, Meauzé L (1997) Identification et caractérisation de l'aire de répartition marine de l'esturgeon européen *Acipenser sturio* à partir de déclarations de captures. Aquat Living Resour 10:101–109

Trouvery M, Williot P, Castelnaud G (1984) Biologie et écologie d'*Acipenser sturio*, étude de la pêcherie. Etude Cemagref/AGEDRA, Bordeaux, 79 p

Williot P, Rouault T, Pelard M, Mercier D, Jacobs L (2009) Artificial reproduction of captive endangered European Atlantic sturgeon, *Acipenser sturio*. Endanger Species Res 6:251–257. doi:10.3354/esr00174

Chapter 38
The French–German Cooperation: The Key Issue for the Success of the Preservation and Restoration of the European Sturgeon, *Acipenser sturio*, and Its Significance for Other Sturgeon Issues

Patrick Williot and Frank Kirschbaum

Abstract The first elements of the French–German cooperation arose in 1993 at a sturgeon meeting in Moscow, and the actual initiation represented the transfer of 40 juvenile *Acipenser sturio* from France to Germany in 1996. This resulted in an extensive exchange of ideas, material, and data concerning the biology and restoration of *A. sturio* between the two institutions Cemagref and Leibniz-Institute of Freshwater Ecology and Inland Fisheries, and in addition stimulated many studies on sturgeon biology in general. This will be outlined in this chapter, describing in short the main results and adding the appropriate references. We believe that the French–German cooperation was the key point for the success of the restoration efforts both in France and Germany.

38.1 Starting Point

During the Second International Symposium on Sturgeon in Moscow (1993), we (PW and FK) met for the first time and started to discuss issues concerning the restoration of *Acipenser sturio* in Europe. On the occasion of the Aquaculture Conference in Bordeaux in 1994 we enlarged these discussions, taking into consideration the first experience in the establishment of the brood stock in France and the rising interest in the restoration issue in Germany, exemplified by the establishment in 1994 of the German Society to Save the Sturgeon, *Acipenser sturio*.

P. Williot (✉)
Sturgeon Consultant, 4 Rue du Pas de madame, 33980 Audenge, France
e-mail: williot.patrick@neuf.fr

F. Kirschbaum
Faculty of Agriculture and Horticulture, Biology and Ecology of Fishes, Humboldt University, Philippstr. 13, Haus 16, 10115 Berlin, Germany
e-mail: Frank.kirschbaum@staff.hu-berlin.de; frank.kirschbaum@staff.hu-berlin.de

P. Williot et al. (eds.), *Biology and Conservation of the European Sturgeon Acipenser sturio L. 1758*, DOI 10.1007/978-3-642-20611-5_38,

The key point of the beginning of the French–German cooperation was the first successful artificial reproduction and rearing of larvae in Saint-Seurin-sur-l'Isle in summer 1995. After this tremendous success, we had a crucial meeting in Bordeaux in December 1995.

Our analysis of the situation was as follows:

French partner: The French Ministry of Environment did not financially support either the building of the rearing facility for the future brood stock in Saint-Seurin-sur-l'Isle, or general scientific research on the species. Additionally, the Ministry criticized our strategy of working on farming of the model species, the Siberian sturgeon. Thus, the Cemagref was in lack of official support of a Ministry which was reluctant to help in sturgeon conservation (see Chap. 36, Moreau). A promising way was to establish cooperation with other European countries to create pressure on the French administration and to enlarge the range of potential fields of research.

German partner: Despite tremendous interest in the re-introduction of *A. sturio* in Germany, the lack of fishes was the main cause for the lack of activities and support in this field. On the other hand, the Leibniz-Institute of Freshwater Ecology and Inland Fisheries (IGB) in Berlin, newly founded in 1992, had the potential to work on freshwater ecosystems comprising restoration issues, and to perform experimental and molecular genetic work on sturgeon.

Therefore, a French–German cooperation would allow the French partner to get involved in general research on *A. sturio*, to reduce the risk of losing specimens of the very valuable future brood stock, and to influence German activities concerning restoration. For the German partner, the advantages would be a substantial support of the re-introduction efforts concerning *A. sturio* in Germany, and the opportunity to perform research on various aspects of sturgeon biology.

38.2 Initiation

The initiation of the project was made possible thanks to a scientific approach proposed by the German partner in studying the thermal preferendum of juvenile fish, as this phase of the biological cycle was totally unknown. The transfer of 40 juvenile *A. sturio* (born in Saint-Seurin-sur-l'Isle in 1995) to the IGB in Berlin in May 1996 therefore represented the actual start of the French–German cooperation, initiated many common activities in the framework of restoration activities of both countries, and finally led to a successful restoration programme with first experimental releases of juveniles both in France and Germany in 2008; it yielded at the same time many scientific data of sturgeon biology in general. A description of this complicated process will be given below.

38.3 Investigations on *A. sturio*

38.3.1 Thermal Preference Behaviour

The experimental device at the IGB made it possible to investigate thermal preference behaviour and activity rhythms of the juvenile *A. sturio* transferred to the IGB in 1996. Interestingly, the fish did not seem to have a thermal preference; they are active for 24 h, with the highest swimming activity occurring around dawn (Staaks et al. 1999).

38.3.2 Keeping Strategy and Feeding Behaviour of Juveniles

One aspect of the transfer of fish from France to Germany was the chance to apply different strategies to the two groups of fish, those kept in France and Germany respectively, with regard to feeding, growth and brood stock development.

In France, fresh, brackish, and salt water were used for rearing and maturing the fish; at the IGB only freshwater was applied. In France, for years very low light intensities, a natural photoperiod and an annual temperature variation between 25 and 10°C were applied, whereas at the IGB the fish were raised at constant temperatures around 20°C at bright light intensities and with a natural photoperiod. The weighing intervals were several months in France, and 2 to 4 weeks at the IGB. In France, the fish were fed on different species of shrimp only, whereas at the IGB a variety of natural and composed food items was tested.

The 40 juveniles transferred to IGB had reached a mean total length of 58 cm and a mean weight of 838 g at an age of 29 months (Kirschbaum et al. 1998); this growth performance was slightly slower than for wild fish of the same age.

At the end of the third year, the juveniles had reached a median length of 68 cm (56–76 cm) and a median length of 1,281 g (512–2,097 g). The range was quite high, and the largest fish were as large as wild fish of the same age. Small frozen chironomids were fed for more than 2 years, and to this food were then added frozen large chironomids, krill, and small marine fish (Kirschbaum et al. 2000a).

During the following 2½ years, these fish were used for feeding experiments in the course of a PhD study (E. Hensel), applying short weighing intervals of 2 weeks in general. The four different food components fed simultaneously before were tested as single food items. The large chironomids resulted in the highest growth rates; in addition, the fish showed pronounced compensatory growth and large individual differences with regard to food preference. Weaning to dry food was not successful (Hensel et al. 2002).

In additional feeding experiments, we found out that it takes a long time to accustom the fish to different food items. Large tanks and a low stocking density led to a better growth than small tanks and high stocking densities (Kirschbaum et al. 2006a, b).

Decreasing growth and even mortalities led us to analyse the feed offered for many years. We found that the large chironomids contained pesticides which were transferred to the fish. We therefore accustomed our fish over a period of several months to two kinds of shrimps, the same food that had been successfully used in Saint-Seurin-sur-l'Isle for many years.

Though we only had a limited number of fish available at IGB, we learned during the 10 years of experimental approach many aspects about growth and feeding behaviour in *A. sturio* in general.

The growth experiments performed with juvenile fish in Saint-Seurin-sur-l'Isle showed that there was no difference in growth performance between fish kept in fresh or salt water (Williot et al. 2007).

38.3.3 Adaptation, Growth, and Maturation of Brood Fish

Future captive brood fish caught either as juveniles or adults in the Gironde estuary were difficult to adapt to captive conditions; some fish lost up to 30% of their initial weight, but regained it after a long recovery process. Analysis of steroid hormones allowed to distinguish between males and females, and to follow the maturation process. Maturation and successful artificial reproduction could finally be obtained under natural photoperiod, vernalization (10–25°C) and imitation of spawning migration (Williot et al. 2007).

The fish at IGB were kept in fresh water at constant temperature around 20°C and under natural photoperiod. In 2005 a biopsy yielded a ripe female (12.6 kg, 1.36 m) and males in late maturation state. This result showed that under these keeping conditions at least some fish can obtain maturity. Artificial reproduction was not attempted with these fish.

More general aspects concerning brood stock management were presented during a workshop in China (Williot et al. 2004a).

38.3.4 Transfer of Juvenile A. sturio from France to Germany

In addition to the 40 juveniles (9 months old; 27 cm; 72.5 g) transferred in May 1996, further juveniles were transported to Germany after successful artificial reproduction in Saint-Seurin-sur-l'Isle (Fr) at three timepoints: (1) in autumn 2007 (300 specimens, 11 cm, 7.1 g) reproduced in late spring 2007, (2) in spring 2009 (1,000 specimens; 16.6 cm; 19.8 g) reproduced in late spring 2008, and (3) in spring 2010 (300 specimens; 16.6 cm; 19.8 g) after successful reproduction in late spring 2009. These juveniles were used both for first experimental restocking in the North Sea tributaries from 2008 on, and in part for enlarging the German brood stock.

38.3.5 *First Experimental Releases and Reports on Progress of Restoration*

Fifty-one of the *A. sturio* juveniles obtained in 2007 (see above) were taken in 2008 for stocking in the River Elbe in Germany. Thus, *A. sturio* was re-introduced in the River Elbe about 50 years after its disappearance from this river, a truly historic event (Kirschbaum et al. 2009a, b).

Reports on the advancement of the restoration efforts were published regularly (Kirschbaum and Gessner 2000, 2002; Kirschbaum 2002; Kirschbaum et al. 2004, 2006a, b, 2009a, b; Williot et al. 2004b, 2009), and also more general aspects of restoration and management (Kirschbaum et al. 2000b; Williot et al. 2002).

38.3.6 *Population Genetics and Cytogenetic Data*

38.3.6.1 North Sea and Garonne Populations

For restoration measures, it is very important to know the genetic status of the specimens to be used for re-introduction. The French partner failed to obtain the involvement of a French laboratory working on molecular genetics. Therefore, at the IGB the genetic status of *A. sturio* originating from the North Sea tributaries (museum samples) was compared with specimens from the Garonne system. Interestingly, the fishes of the two origins were genetically practically identical (Ludwig et al. 2000), and therefore the fishes from the Garonne river system can be used for re-introduction into the tributaries of the North Sea. However, the genetic variability of the remaining *A. sturio* population is quite limited (Ludwig et al. 2004).

38.3.6.2 Baltic Sea Population

The situation in the Baltic Sea represents a complicated scenario which is far from being fully understood. mtDNA studies revealed that ten archived specimens from the Baltic and one from the Oste River (North Sea) carried the *A. oxyrinchus* haplotype A, typical for the most northern populations of *A. oxyrinchus* inhabiting the east coast of North America. This finding further supported the investigation of the differential morphology of the scutes of the two species *A.sturio* and *A.oxyrinchus*, and indicated the colonization of the Baltic by *A. sturio* about 3,000 years ago, the presence of *A. oxyrinchus* in the Baltic about 1,200 years ago, and – after the sympatric occurrence of both species for several hundred years – a dominance of the *A. oxyrinchus* mitochondrial haplotype A (Ludwig et al. 2002). The investigation of one microsatellite did not reveal any sign of hybridization in the sympatric populations of the two sturgeon species.

Based on these results, an international workshop in 2002 (International workshop on "Species differentiation and population identification in the sturgeons *Acipenser sturio* L. and *Acipenser oxyrinchus*". Blossin, Germany, 27.-28.07.2002) proposed to change the restoration strategy: re-introduction of *A. oxyrinchus* into the Baltic Sea (Gessner et al. 2004a; Gessner et al. 2004b; Gessner et al. 2006; Gessner et al. 2007a; Gessner et al. 2008a; Gessner et al. 2008b) and re-introduction of *A. sturio* into the North Sea tributaries, in particular the rivers Elbe and Rhine (Kirschbaum 2002).

However, extensive hybridization was also reported between *A. sturio* and *A. oxyrinchus* in the Baltic Sea, based on investigations of nuclear MHCII genes and on microsatellite studies (Tiedemann et al. 2007). In the light of these new findings, the current German restoration strategy, stocking of the Baltic with *A. oxyrinchus*, needs to be reconsidered. Therefore, an international workshop was organized in Poland to discuss this issue. Unfortunately, there is no document available to the scientific community to see the outcome of this discussion.

Cytogenetically, *A.sturio* and *A.oxyrinchus* are quite similar (Fontana et al. 2008).

38.3.7 French–German Cooperation Concerning the OSPAR Convention Response (J. Gessner and PW) with Regard to the Status of A. sturio *and Its Conservation (Preservation)*

Yearly documents (MASH) that have to be provided by each country were filled in jointly by the French and German scientists, to ensure that the documents arrived from both France and Germany to the OSPAR convention secretary.

38.4 Investigations of Other Sturgeon Species

38.4.1 Control of Vitellogenesis in the Sterlet, Acipenser ruthenus

An important aspect of ex situ measurement is the understanding of vitellogenesis. In *A. sturio* many aspects are not well-understood in this respect. We therefore undertook a PhD study (S. Wuertz) investigating the role of paracrine hormones in controlling first vitellogenesis, using the model species *A. ruthenus*. It could be shown that the IGF-I system plays an important role as paracrine mediator in vitellogenesis, indicating it as a candidate for the vitellogenesis-inducing factor (VIF) (Wuertz et al. 2006c, 2007a, b). In the final phase of oocyte maturation, C21 steroids act on germinal vesicle breakdown (GVBD). Out of three C21 steroids (progesterone, 11-deoxycortisone, 11-deoxycortisol) tested, progesterone proved to be the most potent steroid in provoking GVBD in short exposure experiments (Semenkowa et al. 2008).

38.4.2 *Hermaphroditism in the Sterlet,* Acipenser ruthenus

One aspect of the reproductive biology of sturgeons is the occurrence of hermaphroditic specimens at a very low percentage. This phenomenon seems to hold for many species, though it is not well-studied. In the Sterlet, *A. ruthenus*, we were able to show, applying various methods, that even simultaneous hermaphrodites occur, with the surprising potential of self-fertilisation (Williot et al. 2005).

38.4.3 *Chemical Characteristics of Caviar*

In the course of an EU-Project (Inco-Copernicus IC 15 CT 96-1005), various aspects of the chemical characteristics of caviar were investigated, comprising the analysis of fatty acids including a comparison of caviar from farmed and wild fish (Kirschbaum et al. 1999; Wirth et al. 2002; Gessner et al. 2008c). Interestingly, but not surprisingly, the caviar of wild fish may be polluted with pesticides.

38.4.4 *Genetic Sex Determination*

Genetic sex determination has been demonstrated in several species of fish. In sturgeon, the understanding of this phenomenon would be useful both in the context of applied research (caviar production) and fundamental research, as this would make it possible to determine early in ontogeny the gender of fish and the sex ratio of potential brood fish. In an EU-Project (CRAFT 1999-72183) involving three very experienced laboratories from France, Italy, and Germany, we did not succeed in demonstrating the existence of genetic sex determination mechanisms in four species of sturgeon – *A. baerii*, *A. naccarii*, *A. gueldenstaedtii*, and *A. ruthenus* (Wuertz et al. 2006a).

38.4.5 *Sturgeon Farming*

A review concerning the significance of sturgeon farming in Western Europe was presented at the Conference of Nice (2000) and published subsequently (Williot et al. 2001).

In farmed Shortnose sturgeon, *Acipenser brevirostrum,* an investigation on rearing density was carried out, with a focus on cortisol-stress measurement (Wuertz et al. 2006b).

38.4.6 Further PhD Studies

In addition to the two PhD studies (E. Hensel, S. Würtz) mentioned before, two other PhD projects were undertaken partly or totally related to restoration issues. E. Gisbert studied early ontogeny-related issues on a model species *A. baerii* and found interesting behaviours at early stages of life; he described the initiation of first feeding, and the enzymatic changes in the digestive tract (Gisbert and Williot 1997; Gisbert et al. 1998a, b, 1999a, b).

J. Gessner was very efficient in particular in obtaining funds for restoration issues from the Ministry of Environment via the German Society to save the Sturgeon. FK integrated him on his demand in 1996 into his Department of Biology and Ecology of Fishes of the IGB. Here he worked for years on restoration issues, concentrating on the remediation of *A. oxyrinchus* in the Baltic (Gessner and Arndt 2006; Gessner et al. 2006, 2007b, 2009a, b; see Chap. 41, Gessner et al. for more references).

38.5 Additional Issues

38.5.1 Gravel Extraction

An important threat to *A. sturio* represents gravel extraction in the Gironde estuary in France. The German partner wrote to the French administration in favour of stopping the continuation of the activities comprising the extraction of gravel in zones adjacent to or superposed with nursery grounds. This was finally successful, probably in part due to this international action.

38.5.2 Action Plan

The idea and the necessity for an action plan for *A. sturio* arose in 2001 at the restitution seminar for the second Life program devoted to *A. sturio* (B-3200/98/460) held in October 2001 at Libourne (Fr). This idea was later on followed by WWF-France (see Chap. 36, Moreau). In the course of the first meeting organized by WWF-France, both authors of the present chapter were active contributors. Subsequently, the efforts were put on a more international level by integration of the secretariat of the Bern Convention (see Chap. 18, Rochard) and several subsequent international meetings [e.g., PW participated in meetings in Strasbourg (European Council) and Brussels (European Commission)], and these activities finally led to the establishment of a European Action Plan (Rosenthal et al. 2007) and subsequently to a German Action Plan (Gessner et al. 2010).

38.5.3 Sturgeon Societies

38.5.3.1 German Society to Save the Sturgeon

FK was one of the foundation members of the German Society to save the Sturgeon established in 1994. Both authors were for many years members of the Scientific Advisory Committee of the Society, which plays an active role in remediation issues in Germany.

38.5.3.2 World Sturgeon Conservation Society (WSCS)

The idea of the establishment of an International Sturgeon Conservation Society was launched in 1989 at the first International Sturgeon Symposium (ISS1, Williot (ed) Acipenser 1991) held in Bordeaux. Both authors are foundation members of the WSCS finally founded in Germany in 2003. In the meantime, the Society plays an important role in organizing important sturgeon meetings and supports conservation issues.

38.5.3.3 Representations at Workshops and Symposia

The related talks and conferences are listed below, each meeting presented in chronological order.

3rd International Symposium on Sturgeon (Acipenseriformes), Piacenza, Italy, 1997
Staaks G, Kirschbaum F, Williot P (1997) Experimental studies on thermal behavior and diurnal activity rhythms of juvenile European Sturgeons (*Acipenser sturio*). Third International Symposium on Sturgeons (Acipenseriformes), Piacenza, Italy, Book of abstracts, A3-O
Grüne Woche Berlin, 1997, Berlin, Germany
Kirschbaum F (1997) Möglichkeiten und Grenzen nachhaltiger Fischproduktion am Beispiel des Atlantischen Störs *Acipenser sturio*. 21.01.1997, Berlin, Germany
AQUAROM "98 – Fisheries management in the Danube river basin. Galati, Romania, 1998
Kirschbaum F, Gessner J, Williot P (1998) Growth performance of *Acipenser sturio* reared under experimental indoor conditions. AQUAROM '98 – Fisheries management in the Danube river basin. Galati, Romania, 18.–22.5.98 Extended abstracts: 227–228

(continued)

Meeting of the French Academy of Agriculture, Paris, France, 1999

Kirschbaum F, Wirth M, Gessner J, Krüger A, Patriche N, Williot P, Billard R (1999) Les caractéristiques physico-chimiques de caviars d'esturgeon d'élevage et sauvages, Acad Agric Fr, Paris, Frankreich, 24.11.99

Symposium on Conservation of the Atlantic Sturgeon Acipenser sturio in Europe. Madrid and Sevilla, 1999

Kirschbaum F, Gessner J (1999) Re-establishment programme for *Acipenser sturio* L., 1758: the German approach. Symposium on Conservation of the Atlantic Sturgeon *Acipenser sturio* in Europe. Madrid and Sevilla 06.–1.09. 99

Kirschbaum F, Gessner J, Williot P (1999) Restoration of *Acipenser sturio* L., 1758 in Germany, I: growth characteristics of juvenile fish reared under experimental indoor conditions. Symposium on Conservation of the Atlantic Sturgeon *Acipenser sturio* in Europe. Madrid and Sevilla 06.–1.09.99

AQUA 2000, European Aquaculture Society, Nice, France, 2000

Kirschbaum F, Williot P, Arlati G (2000) Restoration of the sturgeon species in the European Union with emphasis on restocking programs. AQUA 2000, European Aquaculture Society, Nice

Williot P, Sabeau L, Gessner J, Arlati G, Bronzi P, Gulyas T, Berni P (2000) Main characteristics, most recent evolution and perspectives in Western Europe sturgeon farming.

Caviar Production and Closing Session of the EU-INCO Copernicus Project (IC 15 CT 96-1005), Berlin, Germany, 2000

Hensel E, Kirschbaum F, Williot P (2000) Effects of different food items on specific growth rates of *Acipenser sturio*. Symposium Caviar Production and Closing Session INCO Copernicus Project (IC 15 CT 96-1005), Berlin, 13.–5.05.2000, Abstract book:23

Kirschbaum F, Gessner J, Hensel E, Williot P (2000) Brood stock management: the case of *Acipenser sturio*. Symposium Caviar Production and Closing Session INCO Copernicus Project (IC 15 CT 96-1005), Berlin, 13.–15.05.2000, Abstract book:21

Symposium Fische in der Elbe. Havelberg, Germany, 2001

Kirschbaum F (2001) Perspektiven der Wiedereinbürgerung des Störs in der Elbe. Havelberg, 07.04.2001

Fourth International Symposium on Sturgeon, Oshkosh, USA. 2001

Gessner J, Arndt GM, Williot P, Rochard E (2001) Environmental impact of farming and trade of sturgeons in Western Europe. Fourth International Symposium on Sturgeon, Oshkosh, USA. 2001

Williot P, Rouault T, Pelard, Mercier D, Davail B, Kirschbaum F, Ludwig A. Setting up of a farmed broodstock of the critically endangered sturgeon, *Acipenser sturio* with special emphasis on large fish. Fourth International Symposium on Sturgeon, Oshkosh, WI, USA. 2001

Kirschbaum F, Hensel E, Williot P. Restoration of *Acipenser sturio* in Germany. III. A new approach of measuring food preference in juvenile fish. Fourth International Symposium on Sturgeon, Oshkosh, WI, USA. 2001

Restitution seminar for the LIFE programme, Libourne, France, 2001

Kirschbaum F (2001) La dimension européenne de la stratégie de sauvegarde de l'esturgeon. Séminaire sur l'esturgeon, 04.–05.10.2001, Libourne, France

Man-made aquatic Habitats Alterations and their effects on sturgeon populations, Beijing, China, 2004

Williot P, Doroshov S, Ireland, Kirschbaum F, Rouault T (2004) Brood stock management of sturgeon to secure reproduction for stock enhancement, aquaculture and ex situ conservation.

Fifth International Symposium on Sturgeon, Ramsar, Iran, 2005

Kirschbaum F, Hensel E, Williot P, Gessner J. Restoration of *Acipenser sturio* L., 1758 in Germany: long term feeding to accustom large juveniles to a new food item and the influence of stocking density on growth. Fifth International Symposium on Sturgeon, Ramsar, Iran, 9.–13-05.2005, extended Abstracts, LH12, 270

Kirschbaum F, Hensel E, Wirth M, Gessner J, Tiedemann R, Rouault T, Williot P (2005) Ex situ

(continued)

measure of *Acipenser sturio* – the basis for the restoration of the species. Fifth International Symposium on Sturgeon, Ramsar, Iran, 9.–13.05.2005, extended Abstracts, LH14, 276

First International Workshop on the Biology, Conservation and Sustainable Development of Sturgeons in Southern Europe. Bioresturgeon, Granada, Spain, 2005

Williot P, Rochard E, Rouault T, Lepage M, Kirschbaum F, Bouju, Elie P, Gonthier P (2005) *Acipenser sturio* recovery in France. First International Workshop on the Biology, Conservation and Sustainable Development of Sturgeons in Southern Europe. Bioresturgeon, Granada, Spain

Williot P, Rochard E, Lepage M, Elie P, Kirschbaum F (2005) Acceptability and necessary conditions for the successful introduction of an aquatic species: the example of sturgeon. First International Workshop on the Biology, Conservation and Sustainable Development of Sturgeons in Southern Europe. Bioresturgeon, Granada, Spain

Kirschbaum F, Arndt GM, Anders E, Tiedemann R, Bartel R, Williot P, Gessner J (2005) Restoration of Atlantic Sturgeons in Germany – background, current situation, and perspectives. First International Workshop on the Biology, Conservation and Sustainable Development of Sturgeons in Southern Europe "BIORESTURGEONS", Granada, Spain, 26. –30.11.2005

Symposium zur Ökologie, Ethologie und Systematik der Fische, Zoologisches Forschungsinstitut und Museum Alexander Koenig, Bonn, Germany, 2005

Kirschbaum F, Arndt GM, Anders E, Tiedemann R, Gessner J (2005) Wiedereinbürgerung der Atlantischen Störe in Deutschland. Bonn, 24.09.2005

International workshop on fish elevators: a tool for overcoming barriers for large migratory fish, Piacenza, Italy, 2006

Williot P, Kirschbaum F (2006) *Acipenser sturio*: Recovery projects in Europe. International workshop on fish elevators: a tool for overcoming barriers for large migratory fish, Piacenza, Italy, 2006

Gessner J, Arndt GM, Anders E, Kirschbaum F, Bartel R, Zdanowski B (2006) Sturgeon remediation in the tributaries to the Baltic Sea. International workshop on fish elevators: a tool for overcoming barriers for large migratory fish, Piacenza, Italy, 2006

In addition to the presentations mentioned above, we organized several workshops in Bordeaux and Berlin comprising discussions on the progress of the activities and preparing further common activities. In addition, at a meeting in Szcezin, Poland, both authors presented the restoration efforts of their appropriate countries in front of Polish scientists working on sturgeon remediation in the Baltic.

38.6 Conclusion

French–German cooperation has resulted in an extensive exchange of ideas, material, and data concerning the biology and restoration of *Acipenser sturio*, and in addition stimulated many studies on sturgeon biology in general and led to common attendance at symposia and congresses and organisation of workshops; we therefore believe that the French–German cooperation was the key point for the success of the restoration efforts both in France and Germany. It is also a valuable example of a fruitful cooperation at the European level. Last not least, this book is a good example of fruitful common activity.

References

Fontana F, Lanfredi M, Kirschbaum F, Garrido-Ramos MA, Robles F, Forlani A, Congiu L (2008) Comparison of karyotypes of *Acipenser oxyrinchus* and *A. sturio* by chromosome banding and fluorescent *in situ* hybridisation. Genetica 132:281–286

Gessner J, Arndt GM (2006) Modifications of gill nets to minimize by-catch of sturgeons. J Appl Ichthyol 22(Suppl 1):166–171

Gessner J, Arndt G-M, Anders E, Kirschbaum F (2004a) Arterhaltung und Wiedereinbürgerung der Atlantischen Störe (*A. oxyrinchus*) in Nord- und Ostsee. 13. Meeresumweltsymposium, Bundesministerium für Umwelt, Naturschutz und Reaktorsicherheit, Bundesamt für Schiffahrt und Hydrographie 3. –4. 6. 2003, Meeresschutz, 139–151

Gessner J, Arndt G-M, Anders E, Kirschbaum F (2004b) Perspektive für die Wiedereinbürgerung der Atlantischen Störe (*Acipenser sturio* und *A. oxyrinchus*) in Nord- und Ostsee. Handb Angew Limnol 19:1–12, Erg.Lfg. 7/04

Gessner J, Arndt GM, Tiedemann R, Bartel R, Kirschbaum F (2006) Remediation measures for the Baltic sturgeon: status review and perspectives. J Appl Ichthyol 22(Suppl 1):23–31

Gessner J, Arndt GM, Ludwig A, Kirschbaum F (2007a) Remediation of Atlantic sturgeon in the Baltic Sea: background, status, and perspectives. Am Fish Soc Sympos 56:301–317

Gessner J, Arndt GM, Ludwig A, Kirschbaum F (2007b) Remediation of Atlantic sturgeon in the Baltic Sea: background, status, and perspectives. Am Fish Soc Symp 56:301–317

Gessner J, Arndt GM, Anders E, Wuertz S, Bartel R, Zdanowski B (2008a) The development of a broodstock and rearing of *Acipenser oxyrinchus* between 1998 and 2007 as a prerequisite for stocking the tributaries to the Baltic Sea. In: Kolman R, Kapusta E (eds) Actual status and active protection of sturgeon populations endangered by extinction. Wydawictnictwo Instytutu Rybactwa Srodalowengo, Olsztyn, Poland, pp 19–30

Gessner J, Midgalska B, Tautenhahn M, Domagala J, Freedrich F, Bartel R (2008b) Migration analysis of juvenile sturgeons (*Acipenser oxyrinchus*) in the Odra River catchment as determined by catch data. In: Kolman R, Kapusta E (eds) Actual status and active protection of sturgeon populations endangered by extinction. Wydawictnictwo Instytutu Rybactwa Srodalowengo, Olsztyn, Poland, pp 151–162

Gessner J, Würtz S, Wirth M, Kirschbaum F (2008c) Biochemical composition of caviar as a tool to discriminate between aquaculture and wild origin. J Appl Ichthyol 24(Suppl 1):52–56

Gessner J, Horvath A, Arndt GM, Urbanyi B, Anders E, Hegyi A, Wuertz S (2009a) Interrcontinental transfer of adult *Acipenser oxyrinchus* – impact assessment of aviation transport conditions on blood parameters. J Appl Ichthyol 25:365–371

Gessner J, Würtz S, Kamerichs CM, Kloas W (2009b) Substrate related behavioural response in early life stages of American Atlantic sturgeon *A. oxyrinchus*. J Appl Ichthyol 25(suppl 2):83–90

Gessner J, Tautenhahn M, von Nordheim H, Borchers T (2010) Nationaler Aktionsplan zum Schutz und zur Erhaltung des Europäischen Störs (*Acipenser sturio*). Bundesministerium für Umwelt, Naturschutz und Reaktorsicherheit (BMU), Bundesamt für Naturschutz (BfN). Silber Druck, Niestetal

Gisbert E, Williot P (1997) Larval behaviour and effect of the timing of initial feeding on growth and survival of Siberian sturgeon larvae under small scale hatchery production. Aquaculture 156:63–76

Gisbert E, Rodriguez A, Castellò-Orvay F, Williot P (1998a) A histological study of the development of the digestive tract of Siberian sturgeon (*Acipenser baeri* Brandt) during early ontogeny. Aquaculture 167(3/4):195–209

Gisbert E, Williot P, Castello-Orvay F (1998b) Morphological development of Siberian sturgeon (*Acipenser baeri* Brandt) during prelarval and larval stages. Riv Ital Acquacolt 33:121–130

Gisbert E, Williot P, Castellò-Orvay F (1999a) Behavioural modifications in the early life stages of development of Siberian sturgeon (*Acipenser baeri* Brandt). J Appl Ichthyol 15:237–242

Gisbert E, Sarasquette MC, Williot P, Castello-Orvay F (1999b) Histochemistry of the development of the digestive system of Siberian sturgeon during early ontogeny. J Fish Biol 55:596–616

Hensel E, Kirschbaum F, Williot P, Wirth M, Gessner J (2002) Restoration of the European sturgeon, *Acipenser sturio* L., 1758 in Germany: effect of different feed items on specific growth rates of large juvenile fish. Int Rev Hydrobiol 87:539–551

Kirschbaum F (2002) La dimension européenne de la stratégie de sauvegarde de l'esturgeon. Quel avenir pour l'esturgeon européen? Actes du séminaire. Libourne EPIDOR 4/5 octobre 2001, 118–132

Kirschbaum F, Gessner J (2000) Re-establishment programme for *Acipenser sturio* L., 1758: the German approach. Bol Inst Esp Oceanogr 16(1–4):149–156

Kirschbaum F, Gessner J, Williot P (1998) Growth performance of *Acipenser sturio* reared under experimental indoor conditions. AQUAROM "98 – Fisheries management in the Danube river basin. Galati, Romania, 18.–22.5.98 Extended abstracts, pp. 227–228

Kirschbaum F, Wirth M, Gessner J, Krüger A, Patriche N, Williot P, Billard R (1999) Les caractéristiques physico-chimiques de caviars d'esturgeons d'élevage et sauvages. CR Acad Agric Fr 85(8):85–96

Kirschbaum F, Gessner J, Williot P (2000a) Restoration of *Acipenser sturio* L., 1758 in Germany: Growth characteristics of juvenile fish reared under experimental indoor conditions. Bol Inst Esp Oceanogr 16(3):157–165

Kirschbaum F, Williot P, Arlati G (2000b) Restoration of the sturgeon species in the European Union with emphasis on restocking programs. AQUA 2000, Nice (2000), European Aquaculture Society, Special publication no. 28: 336

Kirschbaum F, Gessner J (2002) Perspektiven der Wiedereinbürgerung des Störs, *Acipenser sturio* L., in der Elbe – Perspectives for the re-introduction of the European sturgeon, *Acipenser sturio* L., in the Elbe River. Z Fischk Suppl Bd 1:217–232

Kirschbaum F, Ludwig A, Hensel E, Würtz S, Kloas W, Williot P, Tiedemann R, Arndt GM, Anders E, von Nordheim H, Gessner J (2004) Status of the project on protection and restoration of Atlantic sturgeon in Germany: Background, current situation, and perspectives. In: Gessner J, Ritterhoff J (eds) Species differentiation and population identification in the sturgeons *Acipenser sturio* L. and *Acipenser oxyrinchus*. BfN-Skripten 101. Bundesamt für Naturschutz, Bonn, pp 36–53

Kirschbaum F, Hensel E, Williot P (2006a) Feeding experiments with the European Atlantic sturgeon, *Acipenser sturio* L., 1758 to accustom large juveniles to a new feed item and the influence of tank size and stocking density on growth. J Appl Ichthyol 22(suppl 1):307–315

Kirschbaum F, Würtz S, Williot P, Tiedemann R, Arndt GM, Anders E, Bartel R, Gessner J (2006b) Prerequisites for the restoration of Atlantic sturgeons, *Acipenser sturio* and *A. oxyrinchus*, in Germany – Report on the twelve-year preparatory period. Verh Ges Ichthyol 5:79–93

Kirschbaum F, Fredrich F, Williot P, Gessner J (2009a) Wiedereinbürgerung des Europäischen Störs, *Acipenser sturio,* in Deutschland – Vorbereitende Maßnahmen und erster Besatz. Naturschutz Landschaftspflege Brandenburg 18(3):76–82

Kirschbaum F, Wuertz S, Williot P, Tiedemann R, Arndt G-M, Anders E, Krüger A, Bartel R, Gessner J (2009b) Prerequisites for the restoration of the European Atlantic sturgeon, *Acipenser sturio* and the Baltic sturgeon (*A. oxyrinchus* ♀ x *A. sturio* ♂) in Germany. In: Carmona R, Domezain A, Garcia Gallego M, Hernando-Casal J, Rodriguez F, Ruiz Rejon M (eds) Biology, conservation and sustainable development of sturgeons, vol 29, Fish and Fisheries Series. Springer, New York, pp 385–401

Ludwig A, Debus L, Lieckfeldt D, Wirgin I, Benecke N, Jenneckens I, Williot P, Waldman JR, Pitra C (2002) When the American sea sturgeon swam east. Nature 419:447–448

Ludwig A, Williot P, Kirschbaum F, Liekfeld D (2004) Genetic variability of the Gironde sturgeon population. Proceedings of the International workshop on species differentiation and population identification in the common sturgeon *Acipenser sturio* L. Blossin, Germany, 27. –28.07.2002. BfN Skripten 101:54–72

Ludwig AN, Jenneckens I, Debus L, Ludwig A, Becker J, Kirschbaum F (2000) Genetic analyses of archival specimens of the Atlantic sturgeon *Acipenser sturio* L., 1758. Bol Inst Esp Oceanogr 16:221–230

Rosenthal H, Bronzi P, Gessner J, Moreau D, Rochard E (2007) Action plan for the conservation and restoration of the European sturgeon. Convention on the Conservation of European wildlife and natural habitats (Bern Convention), Council of Europe Publishing, Nature and Environment no. 152

Semenkowa T, Bayunova L, Wuertz S, Kirschbaum F, Barannikova I (2008) Effect of C21 steroids on germinal vesicle break down in sturgeon follicles in vitro. J Appl Ichthyol 32(Suppl 2):2 68–269

Staaks G, Kirschbaum F, Williot P (1999) Experimental studies on thermal behaviour and diurnal activity rhythms of juvenile European sturgeons (*Acipenser sturio*). J Appl Ichthyol 15:243–247

Tiedemann R, Moll K, Paulus KB, Scheer M, Williot P, Bartel R, Gessner J, Kirschbaum F (2007) Atlantic sturgeons (*Acipenser sturio*, *Acipenser oxyrinchus*): American females successful in Europe. Naturwiss 94:213–217. doi:10.1007/s0014-006-0175-1

Williot P (ed) (1991) *Acipenser*. Proceedings of the first International Symposium on Sturgeon. Cemagref, Antony, France, 518p

Williot P, Sabeau L, Gessner J, Arlati G, Bronzi P, Gulyas T, Berni P (2001) Sturgeon farming in Western Europe: recent developments and perspectives. Aquat Living Resourc 14:367–374

Williot P, Arlati G, Chebanov M, Gulyas T, Kasimov R, Kirschbaum F, Patriche N, Pavlovskaya L, Poliakova L, Pourkazemi M, Kim Yu, Zhuang P, Zholdasova IM (2002) Status and management of Eurasian sturgeon: an overview. Int Rev Hydrobiol 87:483–506

Williot P, Doroshov SI, Chebanov M, Kirschbaum F, Rouault T, Zhuang P (2004a) Brood stock management of sturgeon to secure reproduction for stock enhancement, aquaculture and ex situ conservation. (Workshop "Man-made aquatic habitats alterations and their effects on sturgeon populations" Beijing 2004). In: Proceedings of the Sixth International Congress of Zoology, China Zoological Society, Beijing, China, pp. 227–228

Williot P, Rouault T, Rochard E, Castelnaud G, Lepage M, Gonthier P, Elie P (2004b) French attempts to protect and restore *Acipenser sturio* in the Gironde: Status and perspectives, the research point of view. In: Gessner J, Ritterhoff J (eds) Species differentiation and population identification in the sturgeons *Acipenser sturio* L. and *Acipenser oxyrinchus*. BfN-Skripten 101. Bundesamt für Naturschutz, Bonn, pp 83–99

Williot P, Brun R, Rouault T, Pelard M, Mercier D, Ludwig A (2005) Artificial spawning in cultured sterlet sturgeon, *Acipenser ruthenus* L., with special emphasis on hermaphrodites. Aquaculture 246:263–273

Williot P, Rouault T, Pelard M, Mercier D, Lepage M, Davail-Cuisset B, Kirschbaum F, Ludwig A (2007) Building a broodstock of the critically endangered sturgeon *Acipenser sturio* L.: problems associated with the adaptation of wild-caught fish to hatchery conditions. Cybium 31:3–11

Williot P, Rochard E, Rouault T, Kirschbaum F (2009) *Acipenser sturio* recovery research actions in France. In: Carmona R, Domezain A, Garcia Gallego M, Hernando-Casal J, Rodriguez F, Ruiz Rejon M (eds) Biology, conservation and sustainable development of sturgeons, vol 29, Fish and Fisheries Series. Springer, Berlin, pp 247–263

Wirth M, Kirschbaum F, Gessner J, Williot P, Patriche N, Billard R (2002) Fatty acid composition of sturgeon caviar: comparison of wild and aquaculture specimens. Int Rev Hydrobiol 87:629–636

Wuertz S, Gaillard S, Barbisan E, Carle S, Congiu L, Forlani A, Aubert J, Tosi E, Kirschbaum F, Zane L, Grillasca J (2006a) Extensive screening of sturgeon genomes by random screening techniques revealed no sex-specific markers. Aquaculture 258:685–688

Wuertz S, Gessner J, Kirschbaum F, Kloas W (2007a) Expression of IGF-I and IGF-I receptor in male and female sterlet, *Acipenser ruthenus* Linnaeus, 1758 – evidence for an important role in gonad maturation. Comp Biochem Physiol A 147:223–230

Wuertz S, Lutz I, Gessner J, Loeschau P, Hogans B, Kirschbaum F, Kloas W (2006b) The influence of rearing density as environmental stressor on cortisol response of Shortnose sturgeon (*Acipenser brevirostrum*). J Appl Ichthyol 22(Suppl 1):269–273

Wuertz S, Nitsche A, Gessner J, Kirschbaum F, Kloas W (2006c) IGF-I and its role in maturing gonads of female sterlet, *Acipenser ruthenus* Linnaeus, 1758. J Appl Ichthyol 22(Suppl 1): 346–352

Wuertz S, Nitsche A, Jastroch M, Gessner J, Klingenspor M, Kirschbaum F, Kloas W (2007b) The role of the IGF-I system for vitellogenesis in maturing female sterlet, *Acipenser ruthenus* Linnaeus, 1758. Gen Comp Endocrinol 150(1):140–150

Part IV
Restoration: Conservation Programmes of *Acipenser oxyrinchus*

Chapter 39
Conservation and Restoration of *Acipenser oxyrinchus* in the USA

John Waldman

Abstract The US East and Gulf Coast subspecies of Atlantic sturgeon *Acipenser oxyrinchus* have undergone severe population size reductions from a combination of overfishing, blockages by damming of spawning runs, and (possibly) the effects of water pollution. The life history characteristics of these fishes (late age of maturity, low lifetime fecundity, low straying rates) render their abundances susceptible to low levels of fishing mortality (such as by-catch) and low rates of natural recolonization where extirpated. Some experimental restocking has occurred, but to little effect. Present management rests mainly on decades-long cessation of directed fishing. A formal petition to list Atlantic sturgeon as a federally endangered species is under consideration. If designated, even longer-term protection might be offered.

39.1 Biology

Acipenser oxyrinchus is a large anadromous sturgeon species of Eastern North America. Populations of *Acipenser oxyrinchus oxyrinchus* (Atlantic sturgeon) occur in rivers from the St. Lawrence, Canada, to the St. Johns, in Florida (Fig. 39.1). *Acipenser oxyrinchus desotoi* (Gulf sturgeon) occurs in Gulf of Mexico drainages from the Apalachicola River, Florida, to the Mississippi River, Louisiana (Fig. 39.1). Both subspecies can grow large, with Atlantic sturgeon approaching 300 kg; they also mature late (10–25 years, showing a positive relationship with latitude) and can live ~60 years.

Atlantic and Gulf sturgeon are morphologically similar (with minor differences; Wooley and Crateau 1985) but differ genetically, with fixed differences seen in

J. Waldman (✉)
Biology Department, Queens College, 65–30 Kissena Boulevard, Flushing, New York, NY 11367, USA
e-mail: john.waldman@qc.cuny.edu

P. Williot et al. (eds.), *Biology and Conservation of the European Sturgeon Acipenser sturio L. 1758*, DOI 10.1007/978-3-642-20611-5_39,

Fig. 39.1 Map of locations of mouths of rivers and other geographic features mentioned in text

mitochondrial DNA haplotypes (Ong et al. 1996). Atlantic and Gulf sturgeon also have divergent life histories: Atlantic sturgeon are highly migratory, spending considerable time in marine waters as subadults and in between spawning years as adults, whereas thermal constraints restrict Gulf sturgeon to rivers except for cooler seasons. This discrepancy results in substantial differences in movement patterns – the need to annually seek thermal refuges does not allow Gulf sturgeon to make extensive marine migrations. However, Atlantic sturgeon appear to range widely, with subadults remaining largely but not completely restricted to the marine zoogeographic provinces of their natal rivers, often moving from estuary to estuary. Fewer data exist for adult migrations, but the discovery that Atlantic sturgeon had colonized some northern European waters in recent millennia (Ludwig et al. 2002) indicates that some individuals of primordial stocks of Atlantic sturgeon may have made broad oceanic movements.

Prior to the advent of molecular techniques, little was known of the stock structure of Atlantic and Gulf sturgeon, nor if they home to their natal rivers. Subsequent investigations that employed mitochondrial DNA and microsatellite nuclear DNA markers have shown that both subspecies exhibit pronounced stock structure (e.g., Stabile et al. 1996; Wirgin et al. 2000). Concordant with this strong stock structure are high levels of homing fidelity, as deduced from gene flow estimates. Whereas both subspecies showed low gene flow in comparison with other anadromous fishes, estimates for Gulf sturgeon were 3× lower than for Atlantic sturgeon. Waldman et al. (2002) interpreted the lower values for Gulf sturgeon as a consequence of their lower vagility in marine waters (and lower probability of straying), due to their need for seasonal thermal refuges.

39.2 Status in US Waters

39.2.1 Historical

39.2.1.1 Atlantic Sturgeon

Atlantic sturgeon were fished by Native Americans, with utilization traced to 2190 BC in New England (Ritchie 1969). In Colonial times, there were populations of *A. oxyrinchus* in approximately 24 East Coast watersheds (treating Chesapeake Bay as a single system) (Waldman and Wirgin 1998). Colonial fisheries, focused on spawning adults moving up rivers, became pronounced during the late 1800s, especially during the international "caviar craze" of the 1890s (Waldman 1995; Limburg and Waldman 2009). US landings peaked in 1890 at approximately 3,350 mt (shortnose sturgeon *A. brevirostrum* was not distinguished in these landings, but probably represented a minor component). Rampant, unregulated overfishing led to collapse of many populations within a decade; by 1901 US landings had fallen to 295 mt (<10% of the peak; Smith and Clugston 1997).

Harvests during the late 1800s centered on mid-Atlantic rivers but focused most strongly on the lower Delaware River, where decades of accumulated biomass were rapidly removed through the efforts of 26 fishing camps. In 1888, Delaware Bay contributed 2,700 mt of the 3,200 mt of Atlantic sturgeon landed coastwide. Secor and Waldman (1999) estimated a decline rate of 11% annually for Delaware Bay over 1890–1892. Ninety percent of the harvest consisted of (mostly large) females; mean female weight landed in 1897 was 115 kg, and male fractions of the harvests ranged between 2.6 and 4.3% by weight. By 1901, the fishery had crashed, with landings in Delaware and New Jersey only 6% of peak levels observed in 1889.

Atlantic sturgeon harvests were not nearly as intense in the 1900s, partly due to the reduction in population sizes. By the late 1970s, southern states (North

Carolina, South Carolina, Georgia) that had not had their stocks decimated in the 1800s accounted for about 80% of U.S. landings (Smith and Clugston 1997). However, in the 1990s, the pattern shifted to preeminent catches being made in the Hudson River and coastal New York and New Jersey, with these regions producing 93% of US landings over the period 1990–1992.

39.2.1.2 Gulf Sturgeon

Gulf sturgeon were known from at least 12 Gulf Coast watersheds (Waldman and Wirgin 1998). Aperiodic commercial landing statistics are available from 1887 to 1985 for Gulf sturgeon (USFWS and GSMFC 1995). Directed fisheries for Gulf sturgeon occurred only in Florida and Alabama. The first commercial fishery on the Florida Gulf Coast opened in Tampa Bay in 1886, and operated effectively for only 3 years, until annual sturgeon catches decreased from approximately 2,000 to 7 individuals (US Commission of Fish and Fisheries 1902). Landings in metric tons in the Gulf peaked in the early 1900s at about half of peak Atlantic coast harvests, and dwindled to low levels by 1920 (Smith and Clugston 1997).

Gulf sturgeon fishing was most extensive and best documented in the Apalachicola River. The fishery declined from approximately 9,000–27,000 kg annually before 1917 to low levels thereafter, effectively ending in 1970 when only five individuals were taken (Wooley and Crateau 1985).

39.2.2 Present

39.2.2.1 Atlantic Sturgeon

All Atlantic sturgeon populations (*Acipenser oxyrinchus oxyrinchus*) are severely reduced from historical levels, largely because of overfishing and due to the blockage of spawning runs by dams (Smith and Clugston 1997). Contamination has often been proposed as another potential contributor to decline of sturgeons, and worrisome levels of xenobiotics have been seen in Atlantic sturgeon (USFWS and GSMFC 1995), but their direct effects are not well-understood. However, it is clear that in some systems, such as the Chesapeake Bay, lower dissolved oxygen levels from eutrophication caused by sewage and other nutrient inputs have created "habitat squeezes" due to the combined constraints of oxygen and temperature (Niklitschek and Secor 2005).

Waldman and Wirgin (1998) reviewed the status of Atlantic sturgeon populations, and found they ranged between those probably extirpated (e.g., St. Marys River, Georgia–Florida) to rivers that may not have had breeding populations in historical times, but from which occasional specimens were caught (St. Johns River, Florida), to rivers presently lacking breeding populations but seasonally occupied with individuals from other populations (e.g., Merrimack River, Massachusetts), to rivers with small to moderate reproducing populations (e.g., Savannah River, Georgia), to rivers with ongoing commercial fisheries (Shubenacadie River, Nova Scotia).

39.2.2.2 Gulf Sturgeon

Gulf sturgeon (*Acipenser oxyrinchus desotoi*) have also suffered from similar environmental insults and overfishing. Generally, information on Gulf sturgeon populations is less complete than for Atlantic sturgeon populations. Gulf sturgeon populations do show a similar range of statuses to Atlantic sturgeon populations (Waldman and Wirgin 1998). In the St. Lucie River, Florida, sturgeon are occasionally seen in by-catch, but there is no historical evidence of spawning there. Low numbers were seen in recent surveys of the Pascagoula River, Mississippi, and the Yellow River, Florida. More robust numbers are known from the Apalachicola River, Florida (<200 adults in late 1990s) and the Suwanee River, Florida (numbered in low thousands in late 1990s).

39.3 Conservation and Restoration Initiatives

Concern for Atlantic and Gulf sturgeons became pronounced in the late twentieth century. Along the Atlantic coast, modest sturgeon fisheries persisted in several rivers, but management was the responsibility of the individual states. Almost a century had elapsed since the overfishing of the caviar craze, with no individual populations returning to robust abundances. In 1990, the Atlantic States Marine Fisheries Commission (ASMFC; a compact of coastal states) completed a Fishery Management Plan (FMP) for Atlantic sturgeon. The goal of the plan was to provide the framework to allow restoration of Atlantic sturgeon to fishable abundance throughout its range, with fishable abundance defined as an annual harvest of 317 M or about 10% of peak landings in 1890. However, the value of 10% was criticized for being arbitrary and lacking a sound scientific basis (Waldman 2000).

Management objectives of the FMP included: (1) protection from further stock depletion, (2) expansion of knowledge concerning the stocks, and (3) coordination of research and management activities throughout the species' range. Thirteen specific management recommendations were made; most importantly, that each state limit harvest: (1) by establishing a fishing moratorium, or (2) by establishing a minimum harvest size of 213 cm total length, coupled with a monitoring program, or (3) by developing a conservation equivalency plan acceptable to ASMFC.

Around the time the FMP was issued, some small fisheries sharply expanded to worrisome levels: landings in New York's coastal waters and the Hudson River increased from 7,700 kg in 1993 to almost 16,000 in 1994; while landings in New Jersey's coastal waters (in a putative by-catch fishery) went from 5,900 kg in 1988 to 100,000 kg in 1990. Subsequent genetic analysis showed that approximately 99% of these catches were originating from the Hudson River (Waldman et al. 1996).

With evidence of a decline in the Hudson River Atlantic sturgeon stock, and with continued low abundances in some rivers and possible extirpations in others, in the 1990s stocking of hatchery-produced specimens was strongly considered. Beginning in 1993, the US Fish and Wildlife Service (USFWS) hatchery in Lamar, Pennsylvania experimented with culturing Atlantic sturgeon brood stock captured from the Hudson (Mohler 2000). In 1995, an ASMFC sturgeon subcommittee considered the role of hatcheries in the conservation biology of Atlantic sturgeon (St. Pierre 1999). A chief concern was the maintenance of effective population size, given the difficulty of procuring sufficient numbers of ripe females. A protocol was developed to yield a generational effective population size of 100 that would have allowed flexibility in annual numbers of females spawned, but which required on average six brood stock per year (preferably three of each sex) for many consecutive years.

Annually during 1993–1996, the USFWS hatchery obtained 1–3 mature females and several ripe males by gill net for breeding experiments (St. Pierre 1999). In autumn 1994, progeny from a single Hudson River female became available from the hatchery in Lamar, and 4,929 of them were marked and stocked into the Hudson, despite their lack of genetic diversity, through a "research exception." In 1995, Peterson et al. (2000) captured 15 marked and 14 unmarked (wild) age-1 Atlantic sturgeon. Using these data, they estimated there were 4,314 wild age-1 Atlantic sturgeon in the Hudson River in 1995, a decline of about 80% from the similarly conducted population estimate of 1977.

Because of these dramatic increases in landings from the Hudson River, combined with evidence of decreasing recruitment (Bain et al. 2000), in 1996 the New York State Department of Environmental Conservation declared a moratorium on the harvest of Atlantic sturgeon in New York waters, and in 1997 New Jersey reduced its allowable harvest of Atlantic sturgeon as by-catch to zero. This was followed in 1998 by a major, proactive management action – the institution by ASMFC of a moratorium of directed harvest of Atlantic sturgeon in US waters for as many as 40 years. This duration was selected to establish 20 protected year classes of females in each spawning stock.

Gulf sturgeon was afforded protection through separate but similar means. The USFWS and the US National Marine Fisheries Service jointly listed the Gulf sturgeon as threatened under the authority of the Endangered Species Act of 1973, but this designation did not become official until 1991. The Gulf States Marine Fisheries Commission (GSMFC) began in1991 to draft a fishery management plan for the Gulf sturgeon. The USFWS and the Gulf States Marine Fisheries Commission (a compact of states) completed a Gulf Sturgeon Recovery/Management Plan (GSRMP) 4 years later (USFWS and GSMFC 1995).

The major goal of the GSRMP is to provide a strategy to remove Gulf sturgeon from threatened status by 2023. The recovery plan lists three objectives, and recommends the first two be met before delisting occurs. The first objective is to prevent further reduction of existing wild populations, to be monitored on catch per unit bases compared with baseline levels. The second objective is to establish population levels that would allow delisting on a management unit basis by 2023.

These management units would need to be self-sustaining, and efforts would also be required to restore Gulf sturgeon habitat. The third objective is to establish self-sustaining populations robust enough to withstand directed fishing.

39.4 Prognosis

Waldman and Wirgin (1998) believed that in the mid-1990s, efforts to restore both subspecies of *Acipenser oxyrinchus* were at a critical juncture, with only two realistic management options; to reduce or eliminate harvests to allow stocks to rebuild naturally, or to stock Atlantic sturgeon. It's clear that subsequent management has favored the first option, what Waldman and Wirgin (1998) termed a "laissez faire" approach. There has not been any assessment of whether the Hudson stocking produced any meaningful population augmentation, nor whether its limited genetic representation did any harm to the population. In 1996, an additional 3,300 yearling Atlantic sturgeon cultured at the USFWS hatchery were stocked into the Nanticoke River of Chesapeake Bay, a relatively small system which did not have a reproducing population at that time (St. Pierre 1999). But to date there is no evidence of homing and reproduction by these fish in the Nanticoke. Thus, the only stocking that has occurred has been on an ad hoc basis, with little promise demonstrated with which to justify additional efforts of the kind.

Waldman and Wirgin (1998) did suggest that stocking might be warranted in systems where populations appeared to be extirpated, such as the Delaware River. Perpetuation of the original Delaware stock since the crash of the late 1800s was in doubt, with no clear evidence of ongoing reproduction. Given the century-long gap since the crash, and the finding by Waldman et al. (2002) that gene flow rates for Atlantic sturgeon are low in comparison with other anadromous fishes (using gene flow as a proxy for the likelihood of reproductive straying), stocking would have been better justified in the Delaware than in the Hudson. However, in recent years, telemetry tracking of adults tagged in Delaware Bay and the lower Delaware River and the capture of young-of-the year and yearlings indicate either that a relict stock has begun to increase or that a recolonization from elsewhere occurred. Genetic analysis suggested that the Atlantic sturgeon seen today in the small but recovering Delaware River, and James and York Rivers (of Chesapeake Bay) populations include expansion from relict stocks (Wirgin et al. 2007).

A contextual effort that could affect Atlantic sturgeon management was an effort to designate distinct population segments (DPS), management units often larger than populations that show discreteness from other such units, and that have some special significance to the species to which they belong. In 1997 representatives from the National Marine Fisheries Service, US Fish and Wildlife Service, and US Geological Service, with input from state and regional biologists (Atlantic Sturgeon Status Review Team), conducted an Atlantic sturgeon status review. A five-DPS scheme was proposed by ASSRT (2007), which included Gulf of Maine, New York Bight, Chesapeake Bay, Carolina, and South Atlantic regions. At that time, the

ASSRT suggested the New York Bight, Chesapeake Bay, and Carolina DPS deserved threatened status, and that there was too little information for the other two DPS to make a recommendation. They noted that the chief recovery issues included the presence of dams, poor water quality, dredging effects, by-catch in other fisheries, vessel strikes, and inadequate regulatory protection.

Grunwald et al. (2007) advocated a higher number of DPS units based on their genetic findings from their own study, one that included more individuals and populations. However, they also argued for focusing management and restoration on management units (individual populations), and against the general utility of DPS units for a highly philopatric species that exhibits strong structure, such as Atlantic sturgeon.

The prognoses for both *Acipenser oxyrinchus* subspecies are good, though threats exist. No significant directed illegal fishing activity has been uncovered since legal harvests were banned. But by-catch remains a concern. The life history characteristics of these fishes, with their late maturity and low lifetime fecundities, render their population abundances highly vulnerable to by-catch mortality (Boreman 1997). Along the US east coast, Atlantic sturgeon range widely, and are caught accidently in many kinds of fishing gear (Stein et al. 2004). Enough perish to be of concern; Stein et al. (2004) estimated that immediate mortality rates are 22% in sink gill nets and 10% in drift gill nets, and that annual mortality from these sources alone may reach 1,500 individuals. Secor and Waldman (1999) estimated that for the Delaware Bay, based on a scenario of no directed fishing and constant and low recruitment of yearlings but with continued by-catch of subadults, restoration to historical levels would require well over a century.

Climate change remains a threat, in particular warming experienced by Gulf sturgeon and southern populations of Atlantic sturgeon that already exist under marginal thermal conditions, as has been seen for another partly sympatric anadromous species, rainbow smelt *Osmerus mordax*, which has retreated from the southern portion of its Atlantic coast range (Waldman et al. 2006).

Though a movement is growing to remove dams in the US, obstacles targeted are usually smaller dams, not the large mainstem dams that typically affect sturgeon. But exceptions exist. In 1999, the Edwards Dam on Maine's Kennebec River was removed, allowing renewed access by anadromous fishes (including Atlantic sturgeon) to 20 miles of historical spawning grounds.

Atlantic sturgeon was proposed for federal endangered species status in September 2009 by the Natural Resources Defense Council (an environmental advocacy organization), and the proposed DPS listings were released in October 2010 and are under review. These include threatened status for the Gulf of Maine DPS and endangered status for the other four DPS. Comments were due by January 2011. But regardless of that outcome, the absence of directed fishing alone provides a stark contrast to the dire prospects for some populations of the species in the 1990s. Periodic monitoring is required to assess the status of Atlantic sturgeon populations under this management regime, and it is likely, given the slow intrinsic population growth rates, that monitoring will not show dramatic increases. Certainly, a long-term perspective is required (Bain et al. 2000). Nonetheless, recoveries are likely, and there are reasons to be optimistic. Some of the lessons learned in the US may be applicable to restoration of European sturgeons (Waldman 2000).

References

ASSRT (Atlantic Sturgeon Status Review Team) (2007) Status review of Atlantic sturgeon (*Acipenser oxyrinchus oxyrinchus*). Report to the National Marine Fisheries Service, Northeast Regional Office. Gloucester, MA, p 187

Bain M, Haley N, Peterson D, Waldman J, Arend K (2000) Harvest and habitats of Atlantic sturgeon *Acipenser oxyrinchus* Mitchill, 1815 in the Hudson River estuary: lessons for sturgeon conservation. Boletín Instituto Español de Oceanografía 16:43–53

Boreman J (1997) Sensitivity of North American sturgeon and paddlefish populations to fishing mortality. Environ Biol Fish 48:399–405

Grunwald C, Maceda L, Waldman J, Stabile J, Wirgin I (2007) Conservation of Atlantic sturgeon *Acipenser oxyrinchus oxyrinchus*: delineation of stock structure and distinct population segments. Conserv Gen 8:1566–1572

Limburg KE, Waldman JR (2009) Dramatic declines in North Atlantic diadromous fishes. Bioscience 59:955–965

Ludwig A, Debus L, Lieckfeldt D, Wirgin I, Benecke N, Jenneckens I, Williot P, Waldman JR, Pitra C (2002) When the American sea sturgeon swam east. Nature 419:447–448

Mohler JW (2000) Early culture of the American Atlantic sturgeon *Acipenser oxyrinchus oxyrinchus* Mitchill, 1815, and preliminary stocking trials. Biol Inst Esp Oceanogr 16:203–208

Niklitschek EJ, Secor DH (2005) Modeling spatial and temporal variation of suitable nursery habitats for Atlantic sturgeon in the Chesapeake Bay. Est Coast Shelf Sci 64:135–148

Ong TL, Stabile J, Wirgin I, Waldman JR (1996) Genetic divergence between *Acipenser oxyrinchus oxyrinchus* and *A. o. desotoi* as assessed by mitochondrial DNA sequencing analysis. Copeia 1996:464–469

Peterson DL, Bain MB, Haley N (2000) Evidence of declining recruitment of Atlantic sturgeon in the Hudson River. N Am J Fish Manage 20:231–238

Ritchie WA (1969) The archaeology of Martha's Vineyard. The National History Press, Garden City, NY, p 253

Secor DH, Waldman JR (1999) Historical abundance of Delaware Bay Atlantic sturgeon and potential rate of recovery. Am Fish Soc Symp 23:203–216

Smith TIJ, Clugston JP (1997) Status and management of Atlantic sturgeon, *Acipenser oxyrinchus*, in North America. Environ Biol Fish 48:335–346

St. Pierre RA (1999) Restoration of Atlantic sturgeon in the northeastern USA with special emphasis on culture and restocking. J Appl Ichthyol 15:180–182

Stabile J, Waldman JR, Parauka F, Wirgin I (1996) Stock structure and homing fidelity of Gulf sturgeon (*Acipenser oxyrinchus desotoi*) based on restriction fragment length polymorphism and sequence analyses of mitochondrial DNA. Genetics 144:767–775

Stein AB, Friedland KD, Sutherland M (2004) Atlantic sturgeon marine bycatch and mortality on the continental shelf of the northeast United States. N Am J Fish Manage 24:171–183

US Commission of Fish and Fisheries (1902) Report of the Commissioner (Part XXVII) for the year ending June 30, 1901. US Government Printing Office, Washington

USFWS and GSMFC (US Fish and Wildlife Service and Gulf States Marine Fisheries Commission) (1995) Gulf Sturgeon Recovery Plan, Atlanta, Georgia, p 170

Waldman JR (1995) Sturgeons and paddlefishes: a convergence of biology, politics, and greed. Fisheries 20(9):20–21

Waldman JR (2000) Restoring *Acipenser sturio* L., 1758 in Europe: lessons from the *Acipenser oxyrinchus* Mitchill, 1815 experience in North America. Bol Instit Esp de Oceanogr 16:237–244

Waldman JR, Wirgin II (1998) Restoration options and prospects for Atlantic sturgeon *Acipenser oxyrinchus* in North America. Conserv Biol 12:631–638

Waldman JR, Hart JT, Wirgin II (1996) Stock composition of the New York Bight Atlantic sturgeon fishery based on analysis of mitochondrial DNA. Trans Am Fish Soc 125:364–371

Waldman JR, Grunwald C, Stabile J, Wirgin II (2002) Impacts of life history and biogeography on the genetic stock structure of Atlantic sturgeon *Acipenser oxyrinchus oxyrinchus*, Gulf sturgeon *A. o. desotoi*, and shortnose sturgeon *A. brevirostrum*. J Appl Ichthol 18:509–518

Waldman JR, Lake T, Schmidt RE (2006) Biodiversity and zoogeography of Hudson Estuary fishes. Am Fish Soc Symp Ser 51:129–150

Wirgin II, Stabile JE, Waldman JR (1997) Molecular analysis in the conservation of sturgeons and paddlefish. Environ Biol Fish 48:385–398

Wirgin I, Waldman JR, Rosko J, Gross R, Collins MR, Rogers SG, Stabile J (2000) Genetic structure of Atlantic sturgeon populations based on mitochondrial DNA control region sequences. Trans Am Fish Soc 129:476–486

Wirgin I, Grunwald C, Stabile J, Waldman J (2007) Genetic evidence for mid-Atlantic coast relict Atlantic sturgeon stocks. N Am J Fish Manage 27:1214–1229

Wooley C, Crateau EJ (1985) Movement, microhabitat, exploitation, and management of Gulf of Mexico sturgeon, Apalachicola River, Florida. N Am J Fish Manage 5:590–605

Chapter 40
Atlantic Sturgeon (*Acipenser oxyrinchus oxyrinchus*) Fishery Management in the St. Lawrence Estuary, Québec, Canada

Guy Verreault and Guy Trencia

Abstract Atlantic sturgeon (*Acipenser oxyrinchus oxyrinchus*) have been commercially harvested in the St. Lawrence estuary (Quebec, Canada) for decades. After a severe crash between 1966 and 1976, this non-regulated fishery recovered, and landings peaked at record level in early 1990s with signs of overexploitation. Annual monitoring was set up in the mid-1990s in conjunction with fishing restrictions. Since then, length size distributions in annual landings has moved in accordance with sturgeon sizes selected by fishermen. Fishermen preferably select sturgeons larger than 105 cm, but this behavior is counterbalanced by decreasing gillnet selectivity over this threshold. Regulation enforcement and fishermen's compliance has helped to stabilize the landings, and our analyses suggest that Atlantic sturgeon population abundance is increasing and can sustain the fishery. The results from annual monitoring show that management measures are efficient, and that large sturgeons can escape the fishery and contribute to the spawning stock.

40.1 Introduction

The American Atlantic sturgeon (*Acipenser oxyrinchus oxyrinchus*) has been fished commercially in major tributaries and coastal areas of the Atlantic drainages in United States and Canada, with annual landings peaking at 3,350 tons in the nineteenth century (Smith and Clugston 1994). Commercial harvesting has been

G. Verreault (✉)
Ministère des Ressources Naturelles et de la Faune, 186 rue Fraser, Rivière-du-Loup, QC, Canada, G5R 1C8
e-mail: Guy.verreault@mrnf.gouv.qc.ca

G. Trencia
Ministère des Ressources Naturelles et de la Faune, 8400 Sous-le-Vent, Charny, QC, Canada, G6X 3S9

P. Williot et al. (eds.), *Biology and Conservation of the European Sturgeon Acipenser sturio L. 1758*, DOI 10.1007/978-3-642-20611-5_40,

closed in the United States since 1996 (Petersen et al. 2000) but still persists in the St. Lawrence estuary, the northernmost location in the species distribution range (Trencia et al. 2002). Unlike other sturgeon fisheries, the St. Lawrence estuary fishery has always targeted juveniles and sub-adult fish for flesh, not the caviar.

The St. Lawrence estuary is a very large and dynamic system in North America, characterized by tidal activity with amplitude up to 5 meters, strong currents, and salinities ranging between 0 and 30 ppm on a 175 km stretch from Quebec City to Rivière-du-Loup (Fig. 40.1). An estuarine transition zone (ETZ) where freshwater from upstream mixes with saltwater from the ocean leads to conditions favoring the development and retention of sediment, zooplankton and benthos assemblages (Nellis et al. 2007). These physical and biological characteristics provide an important feeding potential for Atlantic sturgeon, and represent a prime habitat for this species (Hatin et al. 2007).

Fisheries statistics have shown large variations in landings during the last 70 years, and can be divided into five periods (Fig. 40.2). Although the quality and precision of the data varies widely between these periods, they provide an approximation of the underlying numbers, and can be used as references when applied with due care. The first period lasts from 1940 to 1960 and is characterized by a relative stability with a mean harvest of 35 tons annually; the second period between 1960 and 1967 shows a gradual decline from 58 to 24 tons, followed by the third period (1967–1975) with a complete crash of the fishery. The fishery recovered during the fourth period (1976–1992), with record landings reaching 142 tons although underestimated, which were followed by the last period (1990–present) of

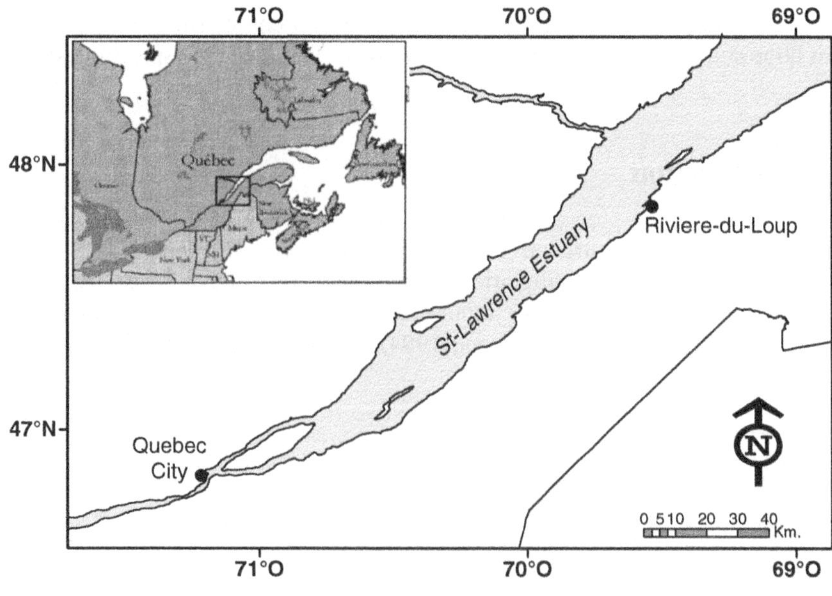

Fig. 40.1 Map of the St. Lawrence estuary where Atlantic sturgeon fishery is permitted

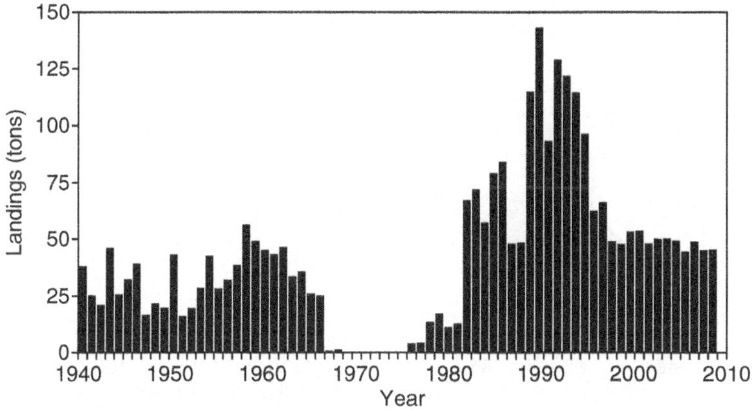

Fig. 40.2 Atlantic sturgeon landings for the St. Lawrence estuary from 1940 to 2009 (*source*: Quebec Min. Agr., Fish. & Food; and Quebec Min. Nat. Res. & Wildl.)

progressive fishery restrictions and consequently, decreased landings (and more accurate records).

The last two periods raised conservation concerns, and prompted managers to adopt a conservative approach towards the fishery and to initiate major measures for ensuring a sound exploitation of this valuable resource. The management measures included: (1) to introduce a robust annual monitoring of the sturgeon population and the fishery, (2) to gather population dynamics key parameters, and (3) to progressively implement regulation measures to protect the fish resource while permitting data collection from the fishery. These measures will be discussed and assessed in this paper.

40.2 Methods

Starting in 1994, the sturgeon population and the fishery were monitored annually using the fishermen's statistics. This fishery evolved then from a non-regulated activity to the enactment of strict commercial fishing rules (Trencia et al. 2002). At the same time, enforcement by fishery officers on the fishing grounds increased. Since 1997, regulation in place includes a maximum size limitation imposed for every landed sturgeon at 150 cm (FL), a standardized gillnet mesh size (20.4 cm stretched mesh), a fixed fishing season and individual annual quota for each fisherman based on their historical statistics. The quota (number of sturgeon) is adjusted yearly with the aim to keep overall fishing mortality under 60 mt, the highest value reported before the crash. Every landed Atlantic sturgeon must be individually registered and tagged under current regulations. Registration includes date of capture, fishing zone, and legal length (LL) (length (cm) from the opercular margin to the back end of the dorsal fin). LL is transformed into fork length (FL)

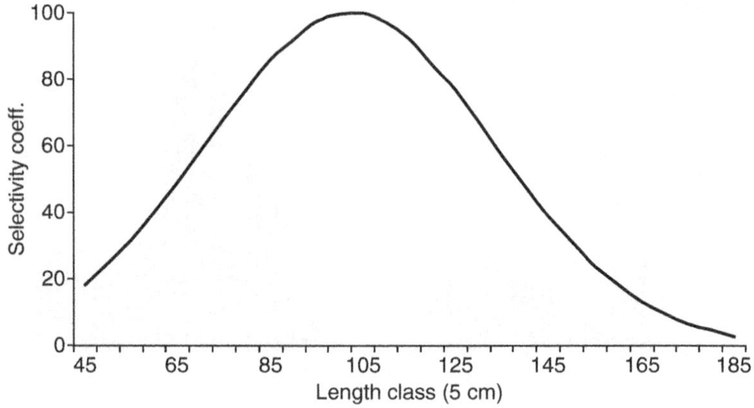

Fig. 40.3 Gillnet selectivity curve (20.4 cm mesh size) for Atlantic sturgeon

using: FL(cm) = 5.118 + 1.674 × LL(cm) (r = 0.98; n = 10,788) and total weight is estimated using length–weight relationship, in Weight(kg) = −12.402 + 3.116 × ln LF(cm) (r = 0.93; n = 840). Fishing effort for standard gillnets (length 50 m; 20.4 cm mesh size) is estimated with data collected through an Index Fishermen program implemented in the fishing zones. Index fishermen were selected for their large part of the total catch and their reliability. Annual catch per unit of effort (CPUE) is calculated using total catch of Atlantic sturgeons divided by fishing effort. Index fishermen recorded the length and fate of every sturgeon caught in gillnets, even those discarded on account of their small size or being larger than the size limit (150 cm FL). Total instantaneous mortality rates (Z) is based on catch curves corrected for gillnet selectivity. Gillnet selectivity is assessed using the method proposed by Hamley (1975), and peak selectivity for sturgeon in the 20.4 cm meshed gillnet is reached at the 105–110 cm length class (Fig. 40.3).

All statistical analyses were performed with Statview for Windows®, version 5.0 from SAS Institute Inc.

40.3 Results

Figure 40.4 summarizes the length size characteristics of landed Atlantic sturgeon over the last 13 years. During this time, median fork length size increased and decreased steadily following a bell shape, and varied from a minimum of 108.9 cm in 2009 to a maximum of 132.3 cm in 2004. Associated mean fork length size varied from a minimum of 114.2 (SD = 17.3) in 2009 to a maximum of 130.7 (SD = 12.2) in 2004. Over this period, fishery regulation related to length size limit or gillnet mesh size did not change, while annual fishing quota evolved from 6,015 to 3,735 Atlantic sturgeons to adjust to the predicted mean weight and the overall landings under 60 mt. Although the annual landed sturgeon statistic gives accurate

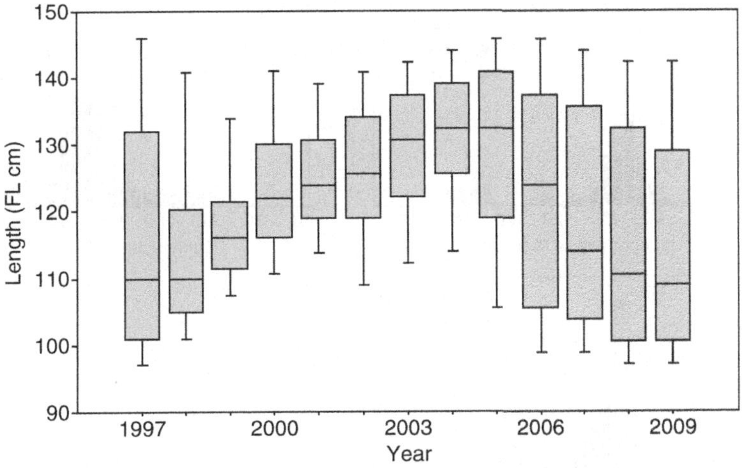

Fig. 40.4 Box plot of landed Atlantic sturgeon length in the St. Lawrence estuary fishery 1997–2009. The box boundaries indicate the 25th and 75th percentiles and *line* within the box marks the median. *Bars* indicate the 90th and 10th percentiles

data on the exploited population segment, it lacks information on the overall sturgeon population on the fishing grounds. Because Atlantic sturgeons entangled in gillnets can stay alive for hours and days, fisherman can choose and keep those they wish among fishes caught, below the maximum length size limit, and discard others.

Data gathered by index fishermen are more accurate and help to assess the fish population. Discarded sturgeons usually exceed landed sturgeons in numbers. For example, 56.5% of the total catch ($n = 3{,}942$) in index fishermen gillnets was discarded alive in 2009. Length frequency distributions from index fishermen changed over the period, and evolved towards a wider distribution of discarded fish (Fig. 40.5). These discarded fish represent a great opportunity to predict and assess the impact of fishing on the sturgeon population. At first hand, it makes it possible to see the arrival of strong cohorts on the fishing grounds a few years before they appear in the landings. These young recruits can then be tracked in successive years to estimate their contribution and vulnerability to the fishery. In the St. Lawrence estuary fishery, following the growth over the years 1997 to 2001 of the smaller sturgeons without visible recruitment, new cohorts began to appear in the catch starting from 2000, while the 100–105 cm FL still supported the fishery; it can be estimated that an Atlantic sturgeon cohort is exploited for approximately 15 years. It can also be useful to evaluate relative numbers of maturing individuals and spawners escaping the fishery when they finally exceed 150 cm (FL). Depending on the year, discarded Atlantic sturgeon can represent between 35 and 60% of total catch. Moreover, discarded fish over 150 cm varied between 6.1% (2009) and 15.1% (2003). This observation is already interesting, but it can be more accurate if those numbers are corrected for gillnet selectivity.

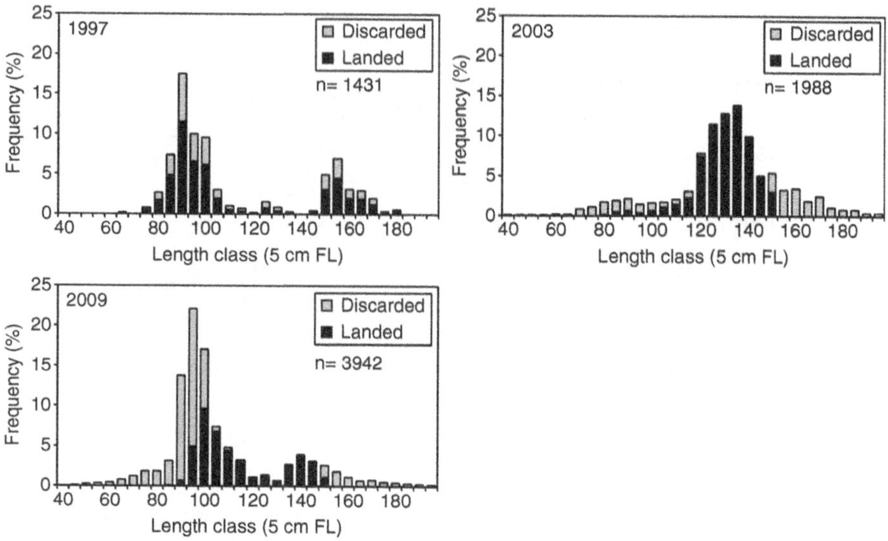

Fig. 40.5 Landed and discarded Atlantic sturgeon length frequency distribution in the St. Lawrence estuary fishery in 1997, 2003 and 2009

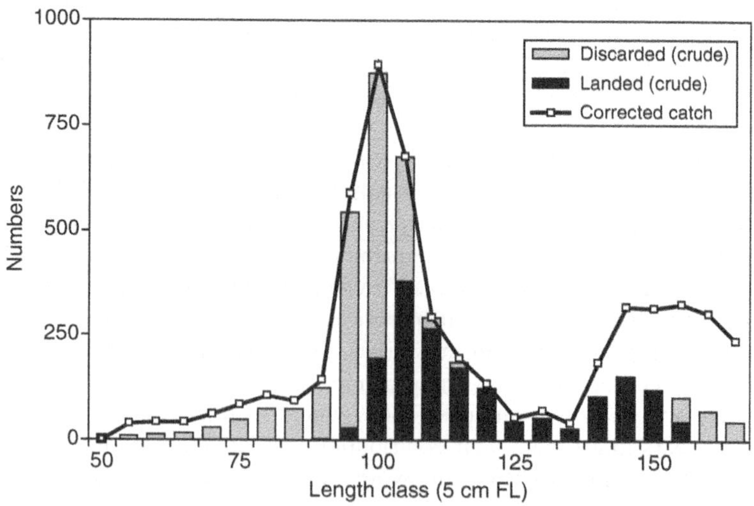

Fig. 40.6 Atlantic sturgeon length class distribution for crude discarded and landed catches, with numbers corrected for gillnet selectivity, in the St. Lawrence estuary fishery, 2009

As shown in Fig. 40.3, selectivity coefficient for gillnet used in this fishery for Atlantic sturgeon is at its maximum for the 105–110 cm length class, and is over 85% for length size comprised between 85 and 120 cm. For those length classes, distribution corrected for gillnet selectivity modifies the numbers slightly, but increases progressively for other length sizes (Fig. 40.6). Vulnerability for Atlantic sturgeons decreases rapidly in proportion as their length size increases; from 140 cm, selectivity coefficient is less than 50%, and at 150 cm it is only 31%. Corrected numbers allows a better evaluation for this stock, especially when data from index fishermen are used. This dataset gives accurate information referring to length size and CPUE for sturgeon population, landed and discarded, on the fishing grounds. When corrected for gillnet selectivity, those two parameters lead to an interesting opportunity to estimate population dynamics.

Instantaneous mortality rates (*Z*) can be achieved using the annual moving of the unique modal length class from year to year along with corrected CPUE. This procedure makes it possible to track the decreasing CPUE of the modal class, and we used data from 1997 to 2004 to construct a catch curve for estimating *Z*. Over this time period, modal class moved from 100 to 140 cm, and its abundance decreased from 42.5 fish/gillnet/day to 12.7 fish/gillnet/day for instantaneous mortality rates of −0.122 (Fig. 40.7). Annual survival rates for Atlantic sturgeons ranging from 100 to 140 cm in this population is then estimated at 88.5%. The arrival of new cohorts after 2004 made it risky to use modal length class to detect any trends in the fishery CPUE.

Along with gillnet selectivity, fishermen behavior must be taken in account by managers to measure the performance of the fishery regulations applied on this population. The adaptation of fishermen to changes in regulations implemented since the late 1990s resulted in unintended biological effects. Shortened season, maximum length size and individual quota are key components in controlling exploitation, and influence fishers' decision-making. This could be illustrated by the fate of fish caught in gillnets following selection by fishers to be kept and landed or discarded. In our analysis based on 25,107 sturgeons caught by index fishermen, the selection process applied on the catch depending on the size of individual fish can be seen as a sigmoid-shaped cumulative distribution curve (Fig. 40.8). Sturgeons under 95 cm are neglected and almost systematically discarded. Between

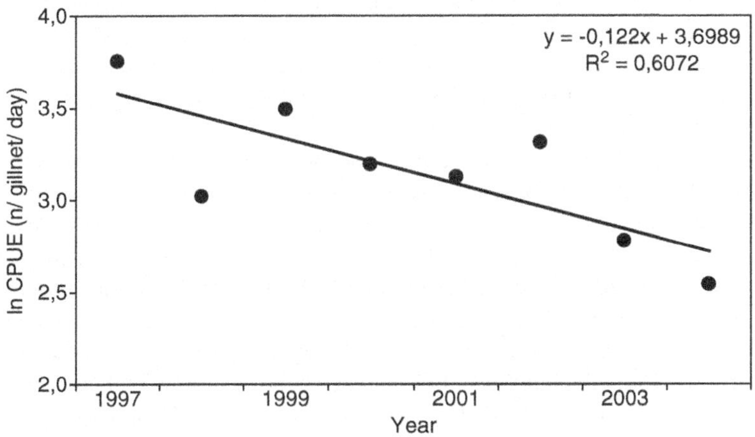

Fig. 40.7 Catch curve for Atlantic sturgeon corrected for gillnet selectivity

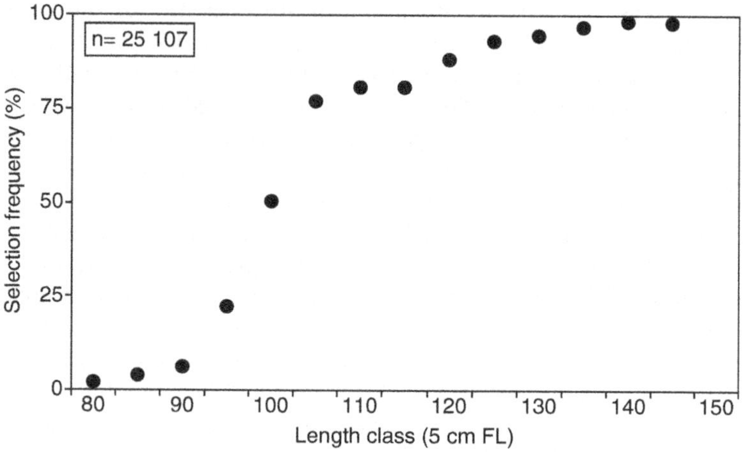

Fig. 40.8 Fishers' selection for Atlantic sturgeon in the St. Lawrence estuary fishery

95 and 105 cm, selection increases sharply, from 22.2 to 76.7% of the fish caught being kept and landed. Afterward, the selection process continues to increase until maximum length size (150 cm FL) is reached.

40.4 Discussion

Since the implementation of restrictive regulations for Atlantic sturgeon in the St. Lawrence estuary in 1997, the registration of every landed sturgeon and total catch statistics by index fishermen allow a robust monitoring of the fishery. Moreover, data collected are very useful to gain insight into the sturgeon population dynamics and to evaluate the impact of the fishery on the stock. Maintaining active fishery operations while implementing new regulations has made it possible to gain precious data and control over an unregulated fishery. Landings and fishing mortality have been lowered to 60 mt, and kept under this value for the last 12 years by way of regulations and their strict enforcement.

Registered landing statistics shows a moving annual length size distribution over time. The increasing modal length from 1997 to 2003 is consistent with the exploitation of one strong cohort over this timeframe. In 2004, length distribution starts to widen, a sign that young and small individuals are recruited to the fishery at the same time that larger fish continue to be exploited. The subsequent decrease in modal length size in the following years reflects the gradual and increasing appearance of smaller and younger sturgeons replacing larger and older ones gradually being absent from the landings. For the near future, modal length distribution should begin to shorten and modal length to decrease until the arrival of another strong cohort. This expectation is largely based on index fishers monitoring results. With this monitoring, it is possible to detect a strong recruiting cohort 5 years before they begin to enter in landings statistics. It is seen as a good predictor for future exploitation and a meaningful relative estimator of larger fish escaping the fishery. This latter can be seen as a performance measure for managers. In fact, those large sturgeons escaping fishing pressure are the guarantee that the reproductive stock can be maintained, ensuring spawning potential and ultimately new young cohorts for sustaining the Atlantic sturgeon population and the fishery.

In this small-scale fishery, fishers must adapt their strategy to regulations. Among these, gillnet mesh size, season, sturgeon maximum length size, and individual quota are the most stringent measures influencing their activity. Individual quotas are set in numbers of fish and vary among individual fishers based on their fishing history. With their quota, only a limited number of individual sturgeon can be kept and registered for landings during the short fishing season. This is an incentive for fishers to develop a strategy by selecting larger sturgeons during a fishing trip. It is well-known that resource scarcity and strict quotas provoke a stringent focus on optimizing profit (Gaertner et al. 1999; Nielsen and Mathiesen 2003). But selecting towards larger fish is not an easy task. In fact, this behavior is counterbalanced by gillnet selectivity. It was shown that fishers preferably select sturgeons larger than 105 cm, but at the same time, gillnets' selectivity coefficient is decreasing. Interaction between these two opposite trends gives a vulnerability coefficient (fishers' behavior × gillnet selectivity) as illustrated in Fig. 40.9. Vulnerability coefficient for Atlantic sturgeon peaks at 0.77 for the 105–110 cm length class, then decreases slightly to 125–130 cm length class and declines sharply afterward. This phenomenon ensures a relative protection to the stock and, moreover, a lowering fishing pressure for sturgeons as they become more valuable (e.g., maturing individuals and spawners).

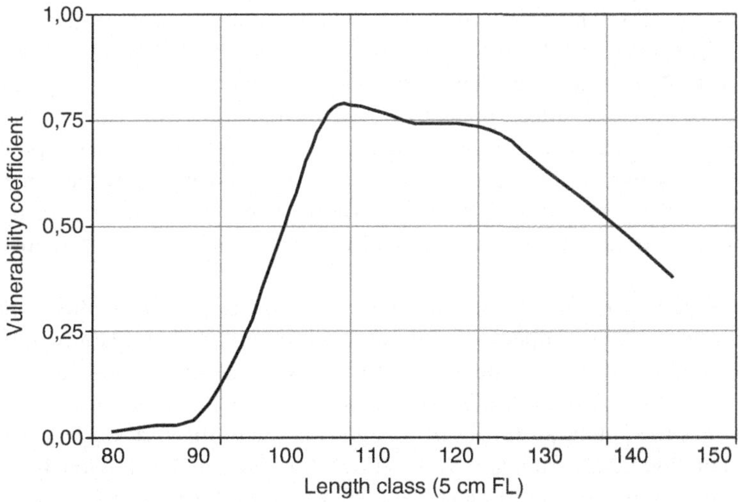

Fig. 40.9 Vulnerability coefficient (gillnet selectivity × fishers' selection) for Atlantic sturgeon in the St. Lawrence estuary fishery under current regulation

Research in compliance behavior has identified several factors influencing rule compliance with fishermen (Nielsen and Mathiesen 2003; van Oostenbrugge et al. 2008). Among these factors, fishermen's involvement in decision-making appears to be a stronger incentive to comply. During the implementation of regulation for the Atlantic sturgeon fishery, fishers were involved in decision-making, on an annual basis, to identify best options in compliance with managers' objective of reducing annual landings under 60 mt. Implementation of fishery regulations was not an easy task, but 13 years later, fishers are still involved in management for improving compatibility between fishing exercises and regulations. Over the last few years, young Atlantic sturgeons have been very abundant among discarded fish. This observation is seen as a result of management practices, especially the exploitation ban on fish larger than 150 cm. These large unexploited sturgeons are building up an important spawners' reserve that will contribute to the future recruitment of the population and ensure a sustainable fishery.

Under current regulations, our observations and analyses suggest that this Atlantic sturgeon stock can sustain the fishery. Nevertheless, with an actual annual total mortality rate of 0.12 and approximately 15 years of exploitation, we estimate that 34.1% of a recruited cohort is escaping the fishery after maximum length size (150 cm) is reached. This estimate should not be seen as an opportunity to increase exploitation, it rather encourages the maintenance of current management practices and carrying on annual monitoring for landings and discarded fish. The gain in confidence in annual monitoring results, and the perception of efficiency in managing this small-scale fishery by fishers, legitimize the current regulations. This may have a positive impact with regulation compliance and fish resource conservation.

While the current regulations have been implemented for the last 14 years, we think that its impact on population demographics is only beginning to be seen in monitoring results. This northern slow-growth population reach sexual maturity at 200 cm FL for female (Vladykov 1955) and 163.5 cm FL for male (Caron et al. 2002), well above maximum length size limit set for the fishery. With spawning periodicity spaced by more than 2 or 3 years (Smith 1985; Bain 1997), many spawners are requested to ensure annual spawning and strong cohort production. In our case, a gap of 11–12 years has been observed between the appearances of cohorts strong enough to sustain the fishery. Thus, the challenge for managers and fishers is to ensure that enough pre-spawners and spawners can survive the fishery.

But fishing mortality is not the only issue that can threaten this population. Among other issues, impact of habitat alteration and degradation are largely unknown and probably very significant for the species (Smith and Clugston 1994). Recent research on habitat use by juveniles highlights the importance of ETZ as a nursery, and the potential impact of dredged sediment disposal on feeding habitat (Hatin et al. 2002; McQuinn and Nellis 2007). More research is obviously needed to identify and protect essential habitat that can sustain this valuable resource at the northern edge of the species' distribution range.

40.5 Perspectives

The St. Lawrence estuary Atlantic sturgeon fishery recovered from a severe crash followed by an almost complete disappearance in the 1970s, to record high landings never seen before in early 1990s. Overexploitation was suspectedm and regulation measures were progressively implemented to prevent another crash. This process was not an easy task, and it left fishers, researchers, and managers on a tightrope at the start. The results have shown that those measures put in place at an appropriate time were sound and efficient: large fish are escaping the fishery in numbers, and young recruits are appearing and filling the space left by their growing parents. Fishers are still operating, and are as optimistic as the managers regarding the future of the fishery. Atlantic sturgeon is a distinctive species in the St. Lawrence estuary, and its smoked flesh is used as a fine local product. The fishery has always targeted juveniles and sub-adult fish for flesh. We did not intend to change this traditional fishery towards caviar exploitation. On the contrary, we strongly believe that protection of spawners is the best option for ensuring fishery sustainability for the future. Fishermen are also wary observers of any habitat problem that could affect the species. Atlantic sturgeon can be seen as an umbrella species for the St. Lawrence estuary, and the protection of its habitat is covering much more as it is also extended to other species of concern in the same area.

Acknowledgements Annual monitoring was financially supported by the Quebec Ministry of Natural Resources and Wildlife. We wish to thank commercial fishermen for their cooperation in data collection for this study. We are also indebted to the Fish and Wildlife enforcement officers for their continuing efforts in maintaining strict but equitable conditions in the fishery.

References

Bain M (1997) Atlantic and shortnose sturgeons of the Hudson River: common and divergent life history attributes. Environ Biol Fish 48:347–358

Caron F, Hatin D, Fortin R (2002) Biological characteristics of adult Atlantic sturgeon (*Acipenser oxyrinchus*) in the St Lawrence River estuary and the effectiveness of management rules. J Appl Ichtyol 18:580–585

Gaertner D, Pagavino M, Marcano J (1999) Influence of fishers' behaviour on the catchability of surface tuna schools in the Venezuelan purse-seiner fishery in the Caribbean Sea. Can J Fish Aquat Sci 56:394–406

Hamley JM (1975) Review of gillnet selectivity. J Fish Res Board Can 32:1943–1969

Hatin D, Fortin R, Caron F (2002) Movements and aggregation areas of adult Atlantic sturgeon (Acipenser oxyrinchus) in the St. Lawrence River estuary, Québec, Canada. J Appl Ichtyol 18:586–594

Hatin D, Munro J, Caron F, Simons RD (2007) Movements, home range size, and habitat use and selection of early juvenile Atlantic sturgeon in the St Lawrence estuarine transition zone. Am Fish Soc Symp 56:129–155

McQuinn I, Nellis P (2007) An acoustic-trawl survey of middle St Lawrence estuary demersal fishes to investigate the effects of dredged sediment disposal on Atlantic sturgeon and lake sturgeon distribution. Am Fish Soc Symp 56:257–271

Nellis P, Munro J, Hatin D, Desrosiers G, Simons RD, Guilbard F (2007) Macrobenthos assemblages in the St Lawrence estuarine transition zone and their potential as food for Atlantic sturgeon and lake sturgeon. Am Fish Soc Symp 56:105–128

Nielsen JR, Mathiesen C (2003) Important factors influencing rule compliance in fisheries: lessons from Denmark. Mar Policy 27(5):409–416

Petersen DL, Bain MB, Haley N (2000) Evidence of declining recruitment of Atlantic sturgeon in the Hudson River. N Am J Fish Manag 20:231–238

Smith TIJ (1985) The fishery, biology, and management of Atlantic sturgeon, *Acipenser oxyrhynchus*, in North America. Environ Biol Fishes 14:61–72

Smith TIJ, Clugston JP (1994) Status and management of Atlantic sturgeon, *Acipenser oxyrinchus*, in North America. In: International Conference on Sturgeon Biodiversity and Conservation, New York, 28–30 July 1994, 32 p

Trencia G, Verreault G, Georges S, Pettigrew P (2002) Atlantic sturgeon (*Acipenser oxyrinchus oxyrinchus*) fishery management in Québec, Canada, between 1994 and 2000. J Appl Ichthyol 18:455–462

Van Oostenbrugge HJAE, Powell JP, Smith JPG, Poos JJ, Kraak SBM, Buisman EFC (2008) Linking catchability and fisher behaviour under effort management. Aquat Living Resour 21(3):265–273

Vladykov VD (1955) Fishes of Québec: album no 5, sturgeons. Department of Fisheries, Québec, 11 p

Chapter 41
Remediation of Atlantic Sturgeon *Acipenser oxyrinchus* in the Oder River: Background and First Results

Jörn Gessner, Gerd-Michael Arndt, Frank Fredrich, Arne Ludwig, Frank Kirschbaum, Ryszard Bartel, and Henning von Nordheim

Abstract At the end of the nineteenth century, sturgeon populations in the southern Baltic Sea drainages started to decrease dramatically. The last captures were reported from Lake Ladoga in 1984 and off Estonia in 1996. Today, sturgeons are considered extinct or missing in the Baltic. In 1996, the pilot phase of a federally funded remediation program was initiated. In parallel, population genetic analyses of historic samples of the species were carried out at the IGB, revealing the

J. Gessner (✉)
Leibniz-Institute of Freshwater Ecology and Inland Fisheries, Müggelseedamm 310, 12587 Berlin, Germany
e-mail: sturgeon@igb-berlin.de

G.-M. Arndt
Fisch und Umwelt Mecklenburg-Vorpommern e.V., Fischerweg 408, 18069 Rostock, Germany

F. Fredrich
Leibniz-Institut für Gewässerökologie und Binnenfischerei, Müggelseedamm 310, 12587 Berlin, Germany

A. Ludwig
Department of Evolutionary Genetics, Institute for Zoo and Wildlife Research, Alfred-Kowalke-Str. 17, 10315 Berlin, Germany

F. Kirschbaum
Faculty of Agriculture and Horticulture, Biology and Ecology of Fishes, Humboldt University, Philippstr. 13, Haus 16, 10115 Berlin, Germany
e-mail: Frank.kirschbaum@staff.hu-berlin.de; frank.kirschbaum@staff.hu-berlin.de

R. Bartel
Instytut Rybactwa Śródlądowego, Zakład Ryb Wędrownych, ul. Synów Pułku 37, 80-298 Gdańsk, Poland

H. von Nordheim
Fachgebiet Meeres- und Küstennaturschutz, Bundesamt für Naturschutz, Außenstelle Insel Vilm, 18581 Putbus, Germany

P. Williot et al. (eds.), *Biology and Conservation of the European Sturgeon Acipenser sturio L. 1758*, DOI 10.1007/978-3-642-20611-5_41,

succession of two different species in the Baltic Sea drainage area. As a result, broodstock development is being carried out using Canadian populations of *A. oxyrinchus*. The evaluation of habitat status for early life stages has been performed in the River Oder. Alternative fisheries techniques, based on by-catch data of exotic sturgeons, have been developed to increase juvenile survival upon release. Since 2006, experimental releases of juvenile *A. oxyrinchus* have been carried out annually to determine habitat use, migration patterns and risks for population development.

41.1 Introduction

Sturgeons exhibit an unusual combination of size, life-cycle duration and behavior, which make them highly vulnerable to anthropogenic impact resulting from such elements as fisheries, hydroconstruction, and habitat degradation (Rochard et al. 1990; Beamesderfer and Farr 1996; Boreman 1996; Gessner 2000), which are the reasons why most of the 27 species of sturgeons worldwide are endangered (Birstein 1993; Birstein et al. 1998). However, the combinations of the various anthropogenic impacts vary for each species and sometimes from river system to river system. The return of diadromous fish species into some of the Central European rivers seems feasible under certain prerequisites, since water quality is not the limiting factor anymore (Sych et al. 1996; Schmidt 1996). To mediate rehabilitation of such fish populations, restoration and connection of critical habitats for sturgeon as well as for other diadromous fish species is essential, in order to overcome the most pressing bottleneck and permit the species to efficiently reproduce under natural conditions (Beamesderfer and Farr 1996). Nevertheless, the historical reasons for the decline of the species in the Oder River (Fig. 41.1) have to be understood for each of the river systems potentially utilized for re-establishment, in order to assess the requirements for future remediation measures, including pressure on fisheries (Gessner and Bartel 2000; Arndt et al. 2006). Therefore, data obtained from monitoring experimental releases will provide the means to improve our understanding of the habitat-related deficits and pressures acting upon the species.

41.2 Historic Range and Decline

41.2.1 Population Genetics of the Historic Baltic Sea Population

Historically, the European sturgeon *Acipenser sturio* (Linnaeus 1758) ranged from the Black Sea via the Mediterranean Sea and the Eastern North Atlantic, the North Sea, to the Baltic Sea (Holčik et al. 1989). Based on archaeological records,

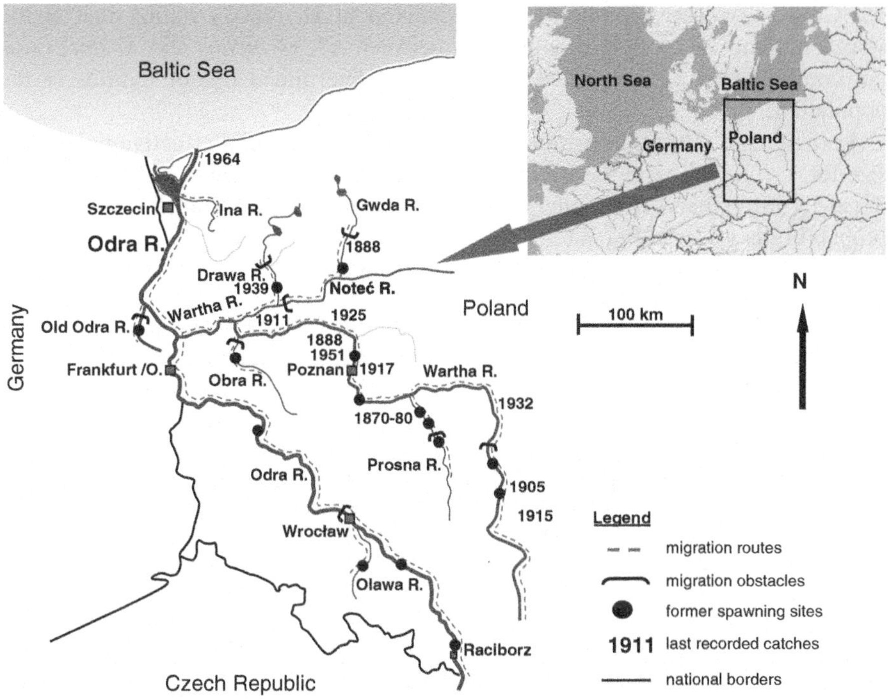

Fig. 41.1 Historic spawning sites, migration obstacles and last records of sturgeon catches in the Oder River system

European sturgeon had colonized the Baltic Sea by around 3,000 years BP (Benecke 1995).

The historical presence of a geographical subpopulation of *A. sturio* in the Baltic was discussed for decades, based on morphological and ecological characteristics (Tikhiy 1929; Magnin 1963; Holčik et al. 1989; Gröger and Debus 2000; Artyukhin and Vecsei 1999; Holčik 2000; Elvira and Almodovar 2000).

Genetic differences were observed in preliminary studies between North and Baltic Sea sturgeons (Birstein et al. 1998), thus enhancing the need for characterization of the genetic structure of the European Atlantic sturgeon from the North and Baltic Seas to determine suitability of *A. sturio* and other potential sources for remediation attempts. Ludwig et al. (2002) identified the sturgeon in the Baltic to be closely related to *Acipenser oxyrinchus oxyrinchus* (Mitchill 1815) based on genetic as well as morphological characteristics. Archaeological records suggest an initial post-Pleistocene colonization of Baltic waters by Atlantic sturgeon (*A. oxyrinchus*) not earlier than 1,300 years BP (Ludwig and Gessner 2007), which is questioned by Stankovic et al. (see Chap. 42), who postulate that first evidence for the species dates back to 300 BC. *A. oxyrinchus* became the dominant species in the Baltic during the early Middle Ages between 1,200 and 800 years BP, with only limited occurrence of *A. sturio* in the samples (Ludwig et al. 2008, 2009;

see Chap. 42). The hypothesis by Tiedemann et al. (2007) stating that Baltic sturgeon comprised a hybrid population between *A. oxyrinchus* and *A. sturio* was not confirmed by Ludwig et al. (2008), who determined 6% of hybrids in the population.

Ludwig et al. (2008) determined that the a posteriori initial effective population size for *A. oxyrinchus* in the Baltic Sea drainage system would have been between 3 and 18 individuals. The model runs are summarized in Table 41.1. The resulting probability is given in Fig. 41.2.

Table 41.1 *Demographic modeling* estimated size of the founding population (N_F) to the Baltic Sea at the Early Middle Ages

Model	Posterior density (N_F)		Assumptions					
run	Mean	95% HPD	N_A	N_B	T_F	T_{bot}	Source	Remark
1	3.8	2–10	2,000	2,000	60	1	Ca + Mid	Baseline
2	3.0	2–6	10,000	2,000	60	1	Ca + Mid	
3	20.4	2–82	2,000	2,000	60	1	Ca	
4	3.6	2–10	2,000	2,000	50	1	Ca + Mid	
5	18.6	2–38	2,000	2,000	60	10	Ca + Mid	
6	10.4	2–26	2,000	2,000	60	–	Ca + Mid	Exponential growth from N_F to N_B

The ABC method was applied to 1,000,000 simulated genetic data sets (mtDNA control region and seven microsatellite loci). The following population history was assumed as a baseline scenario 1: a small part of the source (Canadian, Ca, and Mid-Atlantic, Mid) populations colonized the Baltic Sea at 1,200 years or 60 generations before present (T_F), experienced single-generation bottleneck (T_{bot}), then the populations of both sides of the Atlantic (N_A, N_B) kept a constant size (effective population size = 2,000) until the Baltic population became extinct. Modified population assumptions were tested in the scenarios 2–6; 95% HPD (highest probability density) intervals are listed

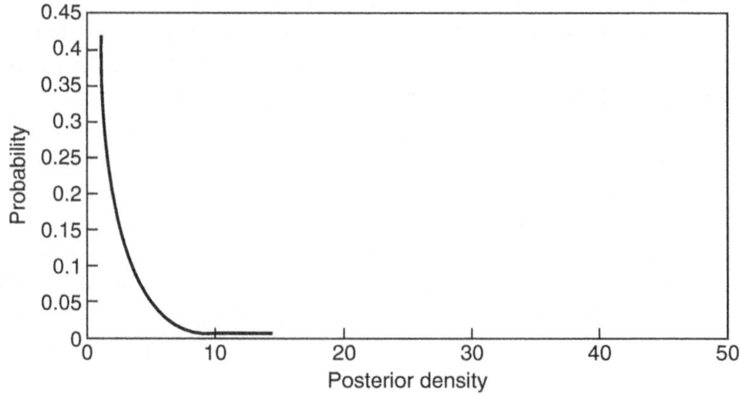

Fig. 41.2 Estimated effective size of the founding population to the Baltic Sea at the Early Middle Ages. The ABC method was applied to the simulated genetic data sets (mtDNA and seven microsatellite loci). This estimates includes only females and the 95% HPD (highest posterior density) interval is [1, 7]

41.2.2 Founder Population

According to the data on mtDNA and microsatellites, the Saint John and Saint Lawrence Rivers populations of Atlantic Canada resemble the recent Baltic sturgeon haplotype (Ludwig et al. 2002). It is therefore evident that restoration of sturgeon populations should focus on these populations. The decision to use *A. o. oxyrinchus* for the re-introduction into the Baltic Sea was recommended by an international expert workshop in 2002 (Gessner and Ritterhoff 2004), and is further supported by the recent studies of Ludwig et al. (2008, 2009). A joint expert workshop on July 27th 2009 in Warszawa revealed a common position of the experts involved in this research.

41.2.3 Decline

A comprehensive overview on the fisheries and population development of sturgeons in the Baltic are given by Kolman (see Chap. 15). The reasons for the decline of the species cannot be attributed to overfishing alone (Seligo 1931; Mamcarz 2000), since the comparison of different case studies reveal different timelines in the trends between stocks, and these are linked to the differences in impacting factors (Bauch 1958; Gessner 2000; Kirschbaum and Gessner 2000, 2002; Gessner et al. 2006). In the Oder River, the combination of impacts comprising the conversion of the river valley to agriculture and pasture land was intensified after 1740. In particular, the drainage of the bifurcation area between the city of Frankfurt/Oder and Hohensaaten (see Fig. 41.1) resulted in large-scale habitat loss. River corrections reduced the river length locally by 30–60%. During the nineteenth century, the alteration of the river habitats for navigation purposes and caused by the increasing waste water discharge were further intensified by industrialization of the upper and middle reaches of the river (Schiemenz 1905; Schräder and Czensny 1930). The associated losses of floodplains, side channels, and meanders reduced the available habitat as well as its productivity both for reproduction and for the juveniles, and were associated with the decrease in population size in the Oder River. This becomes evident also from the data by Przybyl (1976), who showed that at least since the late nineteenth century the main proportion of sturgeon catches in the drainage basin originated from the Warta River and its tributaries. Similar developments were observed on the Vistula Rivers at different time scales (Mamcarz 2000).

The convergence of industrialization and the disappearance of the sturgeon were noted in both the Warta and Oder Rivers (Kulmatycki 1932). In particular, the intensive development of paper mills, sugar refineries, glass and iron smelting plants, leather production and the resulting load of chemical waste and sewage in the river drastically increased from the seventeenth century onwards (Maleczynski 1963). Consequently, habitat alterations were changing the dominance in the fish communities drastically. The reduced abundance of typical river fish species, including the decline of the sturgeon populations and increased abundance of generalists like roach and perch, was considered typical for waterways (Albrecht 1964; Wolter and Vilcinkas 1998). The second main impact upon the populations resulted from high fishing pressure, which has been detrimental for the remaining sturgeon specimen throughout the river basin as described for other catchments too (Jego et al. 2002; Jatteau 1998; Castelnaud et al. 1991). The sturgeon in the Oder River had already become rare at the beginning of the twentieth century. The last catches were observed from the 1950s on until 1968, when the last specimen was caught near Szczecin (Przybyl 1976), and in 1971 in the Rega River (Bartel and Pender 2007).

41.3 Remediation Attempts

41.3.1 Remediation Strategy

During the last four decades of the twentieth century, no catches of endemic sturgeons were reported in the German part of the Baltic (Arndt et al. 2000). The very last autochthonous sturgeon was caught in the Baltic Sea in 1996 (Paaver 1996).

Enhancement of various fish populations through stocking has been common practice since the nineteenth century (Bennecke 1881), since natural recruitment of the released species (salmonid, coregonids, alewife) suffered from degraded environmental conditions.

Increasing attention for sturgeon is represented by the remediation plan for anadromous species in Poland (Sych et al. 1996), while in Germany the interest in restoring the sturgeon increased in the early 1990s (Arndt and Anders 1997; Kirschbaum and Gessner 2000). The need to work on this subject in a coordinated manner, to combine and focus national and international activities, was recognized as early as 1992, leading to the foundation of the Society to Save the Sturgeon as a means to unify the interested parties (Elvira and Gessner 1996). From 1996 on, the Federal Agency for Nature Conservation began to support these activities, while also providing substantial funding for the current program in Germany (von Nordheim et al. 2001). In addition, cooperation in sturgeon remediation between German and Polish scientists was initiated in the second half of the 1990s, being accompanied by a HELCOM project group that attempted an early coordination role in the Baltic Sea area, at that time still focusing on the utilization of *A. sturio*.

41.3.2 Broodstock Development of Acipenser oxyrinchus *in Germany*

The first prerequisite for the remediation project was the attempt to identify the remaining population of the species throughout its historic range. According to genetic analysis of the mitochondrial (mt) D-Loop sequences of sturgeons from the Baltic, only haplotype A was present in the Baltic (Ludwig et al. 2002; see also Chap. 42). For this reason, restoration efforts utilizing similar genetic characteristics for the donor population are focused on fish from the northernmost rivers on the Atlantic coast of North America. Broodstock development was carried out in cooperation with two Canadian research facilities in New Brunswick (Arndt and Gessner 2004). Collection of spawners and sub-adults from the St. John River was undertaken by commercial fisheries (Fig. 41.3).

In order to establish the broodstock for the reproduction and subsequent rearing and release of stocking material in Germany, transfer of adult fish seemed the only means to reduce the time-span for having reproductive fish available (Gessner et al. 2006). This decision resulted in the involvement of the Working Group on International Transfers of Marine Organisms of the International Council for the Exploration of the Sea in the possibility of application of the ICES Code of Practice (ICESCOP) for the transfer of broodstock from North America to Germany.

Following the adaptation of air transport techniques – which initially lead to mortalities, due to blood pH drop and ammonium toxicity – the fish were maintained in boxes equipped with polyethylene bags with buffered water and a pure oxygen atmosphere (Gessner et al. 2008a). As a result, 36 individuals of adult *A. oxyrinchus* of the St. John population were transferred successfully to Germany in 2006 and 2007 (Gessner et al. 2009a). After transfer, quarantine rearing took place in a facility of the Regional Fishery Research Institute of Mecklenburg–Vorpommern according to the criteria of the ICES COP for an additional year. Following quarantine, fish are held in brackish water (2–6‰) under controlled conditions to allow monitoring for growth, gonad maturation, parasite infestations

Fig. 41.3 Catch of *A. oxyrinchus* adult during spawning migration in St. John River, NB, Canada July 2004

and diseases. Rearing conditions provide semi-natural conditions in ponds with sand and gravel substrate. The water supply utilizes brackish surface waters with high turbidity, and stocking densities are maintained at 2 kg/m^2, which in combination is designed to minimize the selection in the broodstock towards artificial rearing conditions.

Weaning of the wild-caught broodstock was carried out during the first year in quarantine. Fish were offered oligochaetes, squid and commercial semi moist diet for striped bass. Weaning took between 2 and 6 months, when feeding of striped bass pellets was supplemented by commercial salmon broodstock diet. Upon transfer, feeding with pelleted feed commenced. Broodstock reveals good growth under captive conditions, although the variability of individual growth is extremely large (Fig. 41.4). Compared to wild-caught fish of the St. John population reaching maturity at 1.6 m in males and 1.8 m in females (Arndt and Gessner 2004) at ages of 20–25 years (Dadswell 2002), fish in captivity increased body weight by 50% in 3 years following transfer. In juveniles, sexual maturity is reached after 9 years in males at ambient (water temperature and light regime) conditions (Gessner unpubl.)

According to the genetic requirements imposed on the program for effective broodstock size for reproduction, a large, diverse gene-pool is required for propagation (St Pierre 1996; see also Chap. 42). In total, 170 individuals of different age-classes and maturity stages are needed to meet the aim of providing an effective population size of ten, under the assumption of a generation time of 15 years. The population size of the broodstock is considered necessary to cope with the obstacles of extended maturation cycles, occasional losses from the stock, long-term restocking requirements, and late maturation. Since microsatellite data in *A. oxyrinchus* show intraspecific differences in allelic distribution (King et al. 2001), they are valuable tools for future breeding plans. Further fish have been integrated

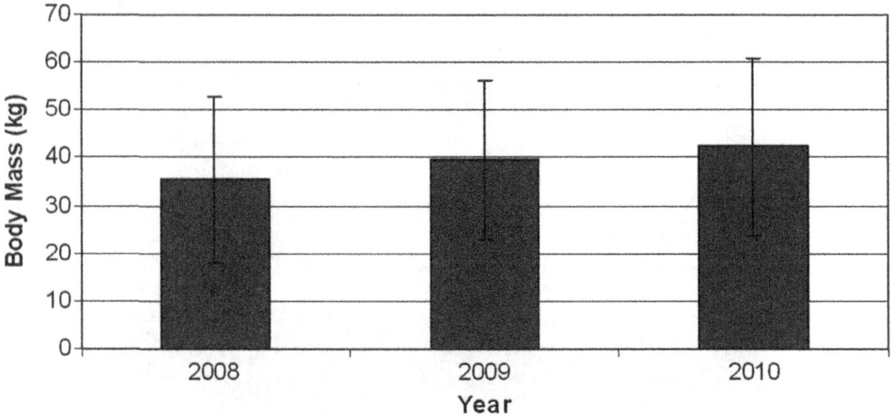

Fig. 41.4 Mean body mass ± sd of *A. oxyrinchus* broodstock ($N = 36$) following intercontinental transfer 2006

into the future broodstock from imported juveniles reproduced on site since 1997 (Arndt and Gessner 2004).

Fish from the 1997 and 2001 yearclasses have been provided to Polish counterparts since 2006 in support of the attempts to develop an autonomous broodstock (Gessner et al. 2008a). It is envisaged to further increase collaboration to maintain sufficient-sized broodstocks and to exchange gametes and offspring to increase the risk split over time. Currently, joint genetic management of broodstock is established to prevent adverse effects arising from both inbreeding and outbreeding depressions (see Chap. 42).

In 2008, reproduction from the captive broodstock in the Mecklenburg facility was not successful since the female did not respond to the GnRH injection. Two females matured in 2009 but did spawn naturally in the rearing tank. In 2010, three females were ovulated in a controlled approach and 400,000 fertilized eggs were incubated. As a backup, reproductions from wild-caught spawners were carried out in Canada until 2009.

41.3.3 Habitat/Ecology of the Oder River

The Oder system has a catchment area of 120,000 km^2, of which 89% belong to Poland, 6% to the Czech Republic and 5% to Germany. The river is 865 km long, with an average slope of 0.07‰. The river is navigable from Raciborz (see Fig. 41.1), at km 63, to its mouth near Szczecin. The runoff throughout the year varies between 250 and 1,500 m^3/s. The main tributary is the Warta River, which contributes approximately 40% of the total flow.

The Oder in its lower part provides floodplains, backwaters and side channels. Large stretches of the river-system are freely accessible (in the Oder River 400 km and in the Warta River 500 km). The decreasing pollution resulting from the application of international agreements (HELCOM 1993) provides an appropriate water quality for the sensitive life-cycle stages. Spawning runs of *Salmo trutta*, *Vimba vimba* and *Coregonus lavaretus* are still observed in the river, and for *Salmo salar* a re-establishment project was initiated in Poland for the Oder tributaries in the beginning of the 1990s (Bartel 2004). In addition, there are numerous recent studies on the fish communities of large rivers which indicate successful reproduction and recruitment of various riverine species (Bischoff and Wolter 2001; Thiel et al. 1998). These results suggest ecological conditions (Prange 2000) that would also allow successful reproduction and recruitment of sturgeons.

The assessment of critical habitat has mainly focused on spawning and nursery habitat for sturgeons because of its key role for recruitment, addressing the most vulnerable life stages. Since habitat requirements for sturgeons are poorly known, experimental verification of the importance of substrate for the development of the early life stages have been conducted (Gessner et al. 2009b). Sturgeon spawning grounds comprise river habitat with rocky-gravel substrate, sufficient water exchange in interstitial water, and absence of sedimentation (Sulak and Clugston

1999; Elie 1997; Jatteau 1998; Jego et al. 2002). The evaluation was performed on historic spawning sites (Przybyl 1976; Sych et al. 1996).

Historical sturgeon spawning (Fig. 41.1) is reported as far upriver as Raciborz, only 63 km below the spring of the Oder River. However, these sites have been inaccessible since the construction of weirs at Wroclaw (Breslau) and upstream since the late nineteenth century.

Today, excessive transport of fine sediments due to construction of groyne fields can be observed throughout the river bed in the middle and lower part. Thereby, the main river channel of Oder River is largely excluded as a potential spawning site for sturgeon. Only very limited areas of suitable substrate have been identified in groyne fields and along river banks. In the Warta River, the main tributary to the Oder, migration obstacles are solely located in the upper reaches of the river as well as in some of its tributaries (Notec, Obra, Prosna). Two historical spawning sites are readily accessible in Warta River near the city of Poznan (Arndt et al. 2006). To assess the status of potential spawning sites, migration obstacles and water quality (i.e., oxygen availability and BOD), dynamic of the discharge and substrate were evaluated following the criteria listed in Table 41.2. Water quality data were obtained from the monitoring data of the regional state authorities. Substrate composition was determined by video-picture analysis linked to an echo-sounder and GPS positioning system. These data were included in a GIS map to allow easier access and control for subsequent verification of status. In the Warta River, the high nutrient and organic load originating from insufficient wastewater treatment, increasing the potential for fungal and bacterial pressure during egg incubation, is still the main obstacle for the suitability for natural reproduction. Adverse effects of high BOD and organic load have been observed for egg and embryonic survival (Gessner unpubl.). In the Drawa River – sturgeon spawning sites are reported to have been used until 1939 – potentially suitable habitat was determined to be readily available.

41.3.4 Habitat Alterations

The removal of migration obstacles is one key option for improving the situation for diadromous species in some rivers of the system, according to the EU

Table 41.2 Criteria for spawning habitat as a prerequisite for the verification of the present status of documented historic sturgeon spawning sites in the Oder River (after Gessner et al. 1999)

Habitat criterion	Requirements
Substrate	Pebble >3 cm or rock 10–30 cm
Sedimentation	Absent
Oxygenation	>5.0 mg/l in the interstitial
Current velocity	0.4–2.0 m/s
Temperature	17–20°C
Depth	1–12 m
Biological oxygen demand (BOD)	Low (no threshold values established)

Water Framework Directive. Nevertheless, continuous pressure to increase navigability of the Oder River conflicts with the need for structural diversity as a key element for the long-term effectiveness of natural reproduction (Beamesderfer and Farr 1996). Therefore, the future development of the rivers for navigation as outlined in the "Oder 2006" plan is of significant concern. The alterations to river structure required to accommodate increasing size of ships will further decrease availability of spawning habitat for sturgeon and other psammophilic species. In particular, increased water depth is accompanied by the development of uniform bank structures with high exposure to wave action, steep bank profiles, uniform current patterns, and construction of additional dams in order to provide navigable waters for bigger ships, leading to canal-like structures (Hermel 1995). This significantly contrasts with the aim of the management body for the connection of backwaters and floodplains to the river system in order to prevent flooding events and thereby increase natural habitat (IKSO 2004), and would increase the pressure to develop functional habitat-rehabilitation in the mainstem Oder River.

41.3.5 Fisheries Impact

Overfishing was one important factor significantly contributing to the decline of the species, and it is still considered a major threat for ongoing restoration measures (Gessner et al. 2010). Making use of reported catches of sturgeon in the Oder and coastal waters of the western Baltic, adverse fisheries techniques were identified (Arndt et al. 2000). Based on this assessment, the main risk is associated with gill netting.

Therefore, development of adapted fisheries techniques to reduce the probability of sturgeon by-catch in the coastal waters of the Oder was performed. The design of the gill-nets was adopted in accordance with the results from trials that were carried out with Siberian sturgeons in ponds. Data on catch probability for different distances of the net over the bottom were used as the basis for re-designing gillnets for perch (*Perca fluviatilis*) and Sander (*Sander lucioperca*). The modifications were tested with sturgeon under culture conditions in earthen ponds. Nets with only 30 cm space between bottom and the net revealed a highly significant decrease of 98% in catch rate for pond-reared sturgeon (Fig. 41.5). Secondly, the modified nets were utilized in the target fishery in comparison to common gillnets, to assess the effectiveness of the modified gear for the main targeted species (Gessner and Arndt 2006).

41.3.6 Experimental Releases

The initial stocking trials were designed for scientific research only, to verify the conditions for subsequent mass release. They rely on telemetry as well as tagging

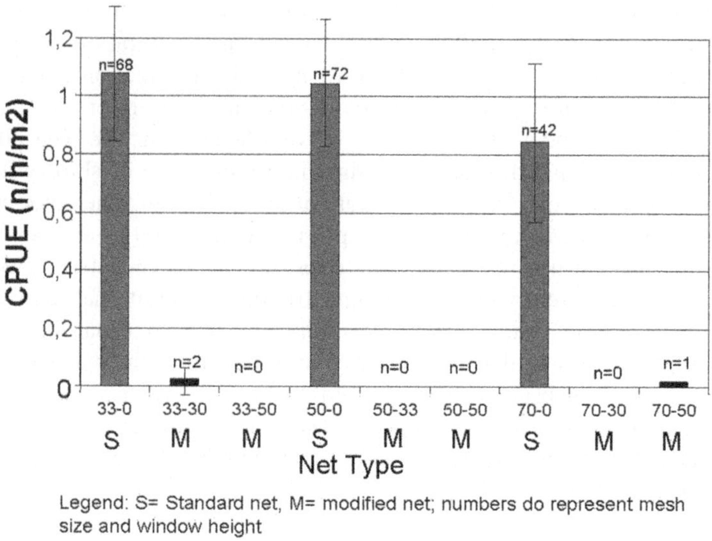

Fig. 41.5 By-catch of sturgeon in experimental fishing trials in an earthen pond utilizing standard nets and nets with 30 and 50 cm space between bottom and net

studies to identify the preferred habitat, food selectivity, and migration patterns, as well as the main impacts on the stocked fish.

Experimental release and monitoring of habitat utilization using telemetry is performed with only small numbers of up to 20 individuals per year. For tagging studies (for details on methodology see Chap. 29) to assess survival, potential threats, migration patterns, habitat use, and feed selectivity, the release of 5,000–60,000 fish annually is carried out in various places (Fig. 41.6).

In total, 6,250 fish were released in 2007 into the Oder River and its tributaries, of which 1,250 were tagged. Tagged fish varied in size between 20 and 85 cm TL according to group. A total of 624 tagged recaptures were reported in the same year. The catch reports increased significantly after the release in October, with more than 500 reports made available. Relatively, the reported catch rate over the first 6 months following release increased from 15.1% in summer releases to 64.2% in fall releases. In subsequent years, releases focused mainly on spring and summer for 1+ fish (TL 25–45 cm) as well as on 0+ fish (TL 3–12 cm) from September to October. Reported by-catch of tagged fish was reduced from an average of 46% in 2007 to 28% in 2008 and 16% in 2009. Due to large numbers of untagged fish released at 1.2–4 cm of length, the by-catch rate under current release conditions is much lower.

The gear employed for sturgeon bycatch mainly comprises fyke- and gill-nets. Despite the fact that approximately 56% of the reports originate from gill nets that are checked only once a day, the overall mortality of the reported catches was low. In 2007, out of the 527 captured fish only three mortalities were reported.

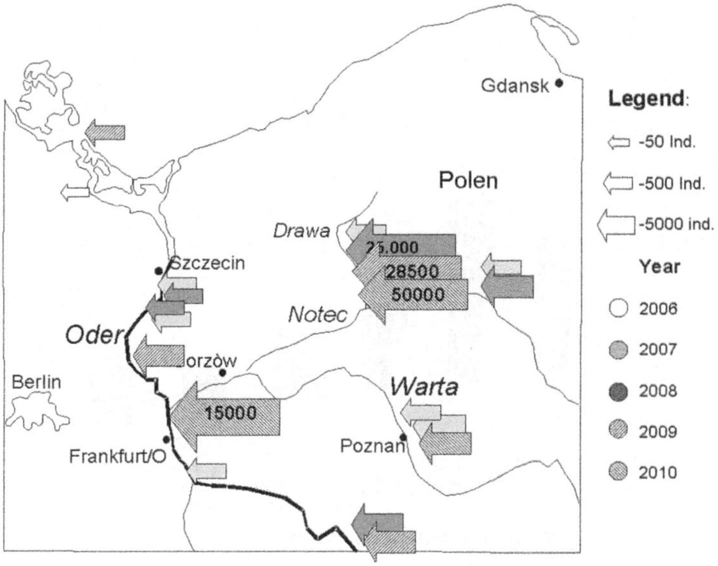

Fig. 41.6 Releases of *A. oxyrinchus* in the Oder River system between 2006 and 2009 (see *shaded circles*); the size of the *arrows* indicates numbers released (see legend)

Reporting was most intensive in areas with increased communication prior to the release and close contacts upon initial catches. In 2007, less than 10% of the catches were reported from waters downstream of Szczecin, whereas the majority of reports originated from the fishermen along the river. With the onset of tagging studies in the Pomeranian Bight and the release of fish with more than 4 kg and 1 m TL, the reports from Scandinavian fishermen increased drastically. The fish that left the Pomeranian Bight were predominantly caught near Bornholm. Two months catch reports were provided from the Belt Sea, while single individuals were caught during the winter in the Arkona Deep off Rügen Island again. In fish of this large size, reported by-catch totals 20% of the released numbers in the first year. Tag loss of Floy tags, observed in large fish predominantly, makes it difficult to assess the accurate number of fish recaptured over time.

41.3.7 Migration

In May 2007, 20 juveniles (29.1 ± 2.2 cm TL, 89.9 ± 18.9 g WW) were released in two groups on 7 May (group 1) and 24 May (group 2) into the Drawa River below Kamienna dam. With the onset of dawn, they started to move downstream over longer distances. On the first day after release, all fish had left the release site and

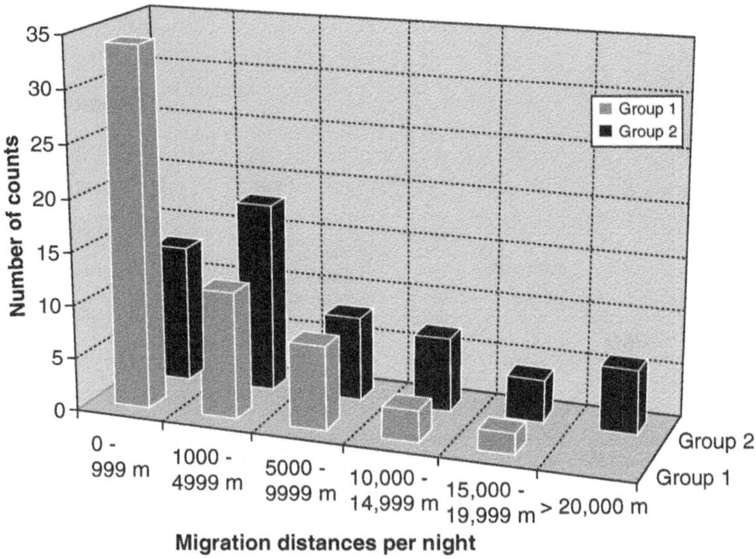

Fig. 41.7 Daily downstream migration distances in two subsequent releases (group 1 and 2 on May 7th and May 24th 2007 with $N = 10$ fish respectively) 1 + *A. oxyrinchus* during 14 days following release in Drawa River

were tracked between 1,050 and 18,815 m downstream. This heterogenic behavior continued over the following 3 weeks of observation (Fig. 41.7).

Throughout the observation time, the fish were never observed to leave the main channel to move into tributaries or backwaters. No differences were observed in utilization of bank structures between both tracking groups. During daylight hours, sturgeon were largely immobile, and utilized habitats associated to structures (sunken trees, pools) in the river. Approximately 60% of the tracking data show sturgeon of both groups to prefer deep holes in the river bed for prolonged stationary phases during daylight (Fig. 41.8) (Fredrich et al. 2008).

Migration speed was assessed by the intervals between release and captures at various sites. Migration speed of fish released 2007 in the Drawa River with an average body length of 56.3 ± 2.4 cm during the first days post-release was similar (2.5 km/h) to that of fish released in the tributaries Gwda and Warta. Single individuals were observed to reach maximum speeds of 3.4 km/h. One fish was reported from the vicinity of Kołobrzeg on the Pomeranian coast east of the Oder River mouth 10 days after release, revealing a continuous migration speed of 1.6 km/h over this time. Specific migration speed equaled 0.7 body lengths per second (bl/s) at an average water temperature of 9°C. Only fish released at water temperatures of below 4°C revealed a considerably slower migration (0.5 km/h or 0.2 bl/s), despite the fact that elevated water levels and increased flow velocity were observed during this time.

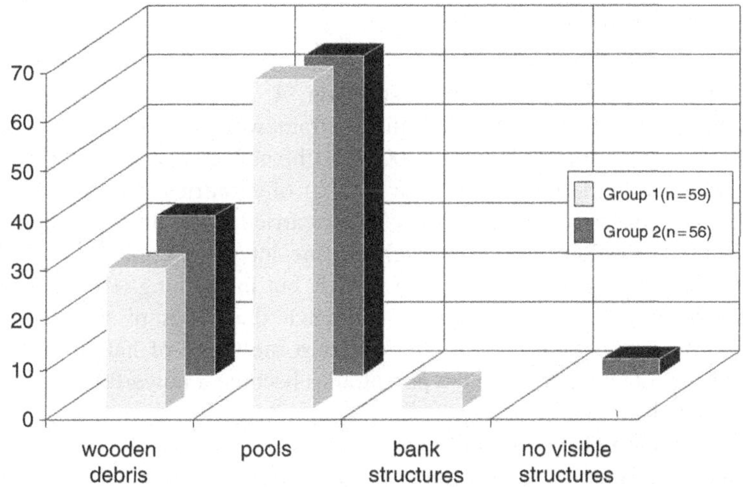

Fig. 41.8 Observation rate for utilization of habitat structures by two groups of juvenile *A. oxyrinchus* during downstream migration in Drawa River

Out of the 527 recorded catches, 12% were repeated recaptures over intervals of up to 3 months, indicating that a substantial proportion of the fish was utilizing the lower Oder River between Krajnik Dolny and Gryfino over an extended period of time (Gessner et al. 2008b). This also became evident from the time elapsed between the release of the fish and the time of catch for single individuals, which revealed a maximum of 2.5 years.

41.3.8 Growth

Growth of the fish in the river was fast. Increases in length of 100% in 60 days were determined from the catch data in August and September with a mean specific growth rate in length of 0.8%/day. With colder water in fall, the SGR decreased to a mean of 0.5%/day (Gessner et al. 2008b).

In September and October 2007, 20 fishes of 53–80 cm and 0.6–1.5 kg were used to determine stomach contents by pulse flushing upon narcosis (Brosse et al. 2000). Eighty percent of the stomachs investigated contained chironomids only. Numbers varied significantly, between four and 63 individuals per stomach. No other organisms were recovered from the stomachs. In November, with water temperatures below 7°C, the stomachs of 20 individuals sampled were completely devoid of food (Gessner et al. 2008b).

Releases of YOY sturgeon and habitat utilization is currently being investigated through recapture experiments both with feeding larvae and 2.5–5 cm fingerlings.

41.3.9 International Support

The re-establishment of the Atlantic sturgeon, *A. o. oxyrinchus*, is a long-term project that will only be successful in the framework of close and continuous international cooperation (Gessner 2000; Kirschbaum and Gessner 2002). This is a first step towards the integration of a number of countries sharing the responsibility for an ecosystem, the measures to be carried out, and the resources that derive from it. The starting point might be the intensive collaboration between the researchers trying to cope with the matter, but in the long term an integrated management system will be required to match the needs of the users of the resources. As emblematic species that utilize a multitude of habitats, sturgeons, being highly protected, do have the potential to become a powerful tool to support river catchment management.

The cooperation program with Poland largely results from the country's accession to the EU. Since sturgeons migrate for thousands of kilometers (Castelnaud et al. 1991; Williot et al. 1997), the need for the integration of the project into a management body governing the Baltic in total was apparent. This integration was established under HELCOM Habitat – as a working group, providing the opportunity to aim at a basinwide conservation approach. Political support by Federal and regional authorities is considered essential for the program. Therefore, it is creating an opportunity for not only a release program but to increase its outreach by utilizing the species as an umbrella to improve habitat management in riverine and coastal areas.

41.3.10 Public Information Campaigns

To ensure maximum public support, an increase in the public awareness on the restoration of Atlantic sturgeons in Germany is an important element of the project. The effort is aimed at the general public, but has a special focus on fishermen and other users of the resources. Exhibitions on the subject, such as the one recently presented in the German Marine Museum in Stralsund, are a major step towards this aim. Additionally, the effectiveness and the responses are to be investigated by a team of sociologists to continuously improve the communication strategy. A common flyer has been developed by HELCOM EC Nature in all languages of the countries surrounding the Baltic Sea, to increase awareness in the fishery. First results have been extremely positive, especially in the Danish and Swedish fisheries, which continuously report catches of sturgeons through national contacts.

International collaboration for the program, having been initiated primarily for the protection of released fish in the commercial fishery today, reflects increasing interest in participation in the remediation of the species in various rivers throughout the Baltic.

41.4 Perspective

Based upon the first controlled reproductions of broodstock in 2009 and 2010, the expansion and genetic improvement of the broodstock has highest priority (see Chap. 42).

Provided continuous support and successful subsequent reproductions are available, the onset of the mass releases for remediation of *A. oxyrinchus* in Baltic tributaries would be in 2012. Upon the first years of returning adults, an assessment of natural reproduction will prove the efficiency of the measures so far. This will determine the second step to be taken. Depending on the results, long-term stocking might be required to compensate for fisheries related losses.

Up to now, international harmonization of activities and adoption of international agreements in the various countries are still lacking to protect the species effectively during its marine phase (Rochard et al. 1997).

It is strongly recommended that the inclusion of viable sturgeon populations as one of the main criteria for successful habitat restoration in European rivers should be adopted by the managing authorities.

Natural populations, however, can only be established if the most important factors that contributed to the decline of the species more than 100 years ago are no longer prevalent or can be overcome in the rivers selected for the re-mediation. This will require the continuous effort of all stakeholders, based upon an insight into the necessity of the required alterations.

Acknowledgments The authors thank the Federal Agency of Nature Conservation (BfN) and the Federal Ministry for Environment, Nature Protection and Nuclear Safety (BMU) for their continuous and substantial support and funding of this program. J.G. was employed from 1996 to 2009 through grants Nr. Az.: Z1.2-892 11-4/96, -7/99, and 9/05, and Z1.3-892 11 8/09. The Regional Research Institute for Agriculture and Fisheries has been a major supporter and partner in the work over the past 20 years. The Federal Ministry for Education and Research (BMBF) supports the program by institutional and project (FKZ 0330532, FKZ 0330718) funding. Without this support, the work would not have been possible.

References

Albrecht ML (1964) Die Oder als Fischgewässer. Z Fisch 12(6/7):479–506

Arndt GM, Anders E (1997) Der Stör (*Acipenser sturio* L.) und seine Wiedereinbürgerung in Mecklenburg–Vorpommern und Deutschland. Fischerei in Mecklenburg–Vorpommern 4:14–19

Arndt GM, Gessner J (2004) Aufbau eines Elterntierbestandes als Grundlage für die Wiedereinbürgerung des Störs im Ostseeeinzugsgebiet. Wissenschaftliche Schriftenreihe Fisch und Umwelt Mecklenburg–Vorpommern e.V.:5–22

Arndt GM, Gessner J, Anders E, Spratte S, Filipiak J, Debus L, Skora K (2000) Predominance of exotic and introduced species sturgeons among sturgeons captured from the Baltic and North Seas and their watersheds, 1981–1999. Bol Inst Oceanogr 16(1–4):29–36

Arndt GM, Gessner J, Bartel R (2006) Characteristics and availability of spawning habitat for Baltic sturgeon in the Oder River and its tributaries. Appl Ichthyol 22(suppl 1):172–181

Artyukhin E, Vecsei P (1999) On the status of Atlantic sturgeon: conspecificity of European *Acipenser sturio* and North American *Acipenser oxyrinchus*. J Appl Ichthyol 15:35C–37C

Bartel R (2004) The Polish attempts in Salmon restoration and their effects. BfN Skripten, 104. Bundesamt für Naturschutz, Bonn, pp 111–118

Bartel R, Pender R (2007) Prospects and perspectives of fisheries in the coastal zone of the southern Baltic. In: Kubiak J (ed) Inland fisheries in Western Pomerania in the light of new regulations of European Union Policies. Prospects and Perspectives of Fisheries in coast vol II. Szczecin, pp 103–110

Bauch G (1958) Untersuchung über die Gründe für den Ertragsrückgang der Elbfischerei zwischen Elbsandsteingebirge und Boizenburg. Z Fisch 7:161–437

Beamesderfer RCP, Farr RA (1996) Alternatives for the protection and restoration of sturgeons and their habitat. Environ Biol Fish 18:407–417

Benecke N (1995) Neue archäozoologischer Forschungen am Burgwall von Lossow, Ortsteil von Frankfurt/Oder – einige vorläufige Ergebnisse. Acta Praehist et Archaeol 26/27:14–23

Bennecke B (1881) Fische, Fischerei und Fischzucht in Ost- und Westpreußen. Hartungsche Verlagsdruckerei, Königsberg, 331 pp

Birstein VJ (1993) Sturgeons and Paddlefishes: Threatened Fishes in Need of Conservation. Conservation Biology 7:773–787

Birstein VJ, Betts J, De Salle R (1998) Molecular identification of *Acipenser sturio* specimens: a warning note for recovery plans. Biol Conserv 84:97–101

Bischoff A, Wolter C (2001) The flood of the century on the River Oder: effects on the 0+ community and implications for floodplain restoration. Regul Rivers 17:171–190

Brosse L, Rochard E, Dumont P, Lepage M (2000) First results on the diet of the European sturgeon, Acipenser sturio, in the Gironde estuary. Comparison with the benthic fauna. Cybium 24:49–61

Castelnaud G, Rochard E, Jatteau P, Lepage M (1991) Données actuelles sur la biologie d'*Acipenser sturio* dans l'estuaire de la Gironde. In: Williot P (ed) Acipenser. Cemagref, Antony, pp 251–275

Dadswell MJ (2002) A review of the status of Atlantic sturgeon in Canada, with comparisons to populations in the United States and Europe. Fisheries 31(5):218–229

Elie P (ed) (1997) Restauration de l'esturgeon Européen *Acipenser sturio*. Rapport final du programme execution, Operations: I-III, 1994–1997, vol 24. Cemagref, Antony, 381 pp

Elvira B, Almodovar A (2000) Further observations on the morphological characters of *Acipenser sturio* L., 1758 from the Iberian Peninsula: a comparison with North and Adriatic Sea populations. Bol Inst Esp Oceanogr 16:89–97

Elvira B, Gessner J (1996) The society to save the sturgeon *Acipenser sturio*. Sturgeon Q 4(1/2):7

Fredrich F, Kapusta A, Ebert M, Duda A, Gessner J (2008) Migratory behaviour of young sturgeon, *Acipenser oxyrinchus* in the Oder (Odra) River catchment – preliminary results of a radio telemetric study in the Drage (Drawa) River, Poland. Arch Pol Fish 16(2):105–117

Gessner J (2000) Reasons for the decline of *Acipenser sturio* L., 1758 in central Europe, and attempts at its restoration. Symposium on conservation of the Atlantic sturgeon *Acipenser sturio* L., 1758 in Europe, Madrid. Bol Inst Esp Oceanogr 16(1–4):117–126

Gessner J, Arndt GM (2006) Modifications of gill nets to minimize by-catch of sturgeons. J Appl Ichthyol 22(suppl 1):166–171

Gessner J, Bartel R (2000) Is there still suitable habitat for sturgeons in the Oder River tributary? Symposium on conservation of the Atlantic sturgeon *Acipenser sturio* L., 1758 in Europe, Madrid. Bol Inst Esp Oceanogr 16(1–4):127–137

Gessner J, Debus L, Filipiak J, Spratte, S, Skora KE, Arndt G-M (1999) Development of sturgeon catches in German and adjacent waters since 1980. J Appl Ichthyol 15:136–141

Gessner J, Ritterhoff J (eds) (2004) Species differentiation and population identification in the common sturgeon Acipenser sturio L. In: Proceedings of the International Expert Workshop Blossin, Germany, 27–28th June 2002. BfN Skripten 74. Bundesamt für Naturschutz, Bonn

Gessner J, Arndt G M, Tiedemann R, Bartel R, Kirschbaum F (2006) Remediation measures for the Baltic sturgeon: status review and perspectives. Journal of Applied Ichthyology 22:23–31

Gessner J, Arndt GM, Anders E, Wuertz S, Bartel R, Zdanowski B (2008a) The development of a broodstock and rearing of *Acipenser oxyrinchus* between 1998 and 2007 as a prerequisite for stocking the tributaries to the Baltic Sea. Pp. In: Kolman R, Kapusta A (eds) Actual status and active protection of sturgeon populations endangered by extinction. Wydawictnictwo Instytutu Rybactwa Srodalowengo, Olsztyn, pp 19–30

Gessner J, Midgalska B, Tautenhahn M, Domagala J, Fredrich F, Bartel R (2008b) Migration analysis of juvenile sturgeons (*Acipenser oxyrinchus*) in the Oder River catchment as determined by catch data. In: Kolman R, Kapusta A (eds) Actual status and active protection of sturgeon populations endangered by extinction. Wydawictnictwo Instytutu Rybactwa Srodalowengo, Olsztyn, pp 151–162

Gessner J, Horvath A, Arndt GM, Urbanyi B, Anders E, Hegyi A, Wuertz S (2009a) Intercontinental transfer of adult *Acipenser oxyrinchus* – impact assessment of aviation transport conditions on blood parameters. J Appl Ichthyol 25:365–371

Gessner J, Würtz S, Kamerichs CM, Kloas W (2009b) Substrate related behavioural response in early life stages of American Atlantic sturgeon *A. oxyrinchus*. J Appl Ichthyol 25 (suppl 2):83–90

Gessner J, Tautenhahn M, von Nordheim H, Borchers T (2010) Nationaler Aktionsplan zum Schutz und zur Erhaltung des europäischen Störs (*Acipenser sturio*). Bundesministerium für Umwelt, Naturschutz und Reaktorsicherheit und Bundesamt für Naturschutz, Bonn, 84 pp

Gröger J, Debus L (2000) Morphometric comparison of *Acipenser sturio* L. populations based on mixed estimation and morphometric measurements. Arch Fish Mar Res 48:175–193

HELCOM (1993) Second Baltic Sea pollution load compilation. Baltic Sea Environ Proc 39:1–162

Hermel UR (1995) Möglichkeiten eines ökologischen Wassermanagements unter Berücksichtigung der Belange der Schifffahrt, des Hochwasserschutzes, der Landwirtschaft und der Fischerei. Thesis, European Postgraduate Programme 1994/1995 in Environmental Management, Berlin, Germany

Holčik J (2000) Major problems concerning the conservation and recovery of the Atlantic sturgeon *Acipenser sturio*. Bol Inst Esp Oceanogr 16:139–148

Holčik J, Kinzelbach R, Sokolov LI, Vasil'ev V (1989) *Acipenser sturio* Linnaeus, 1758. General introduction of fishes. Acipenseriformes. In: Holcik J (ed) The freshwater fishes of Europe, vol 1. AULA Verlag, Wiesbaden, pp 367–394

IKSO (2004) Aktionsprogramm Hochwasserschutz im Einzugsgebiet der Oder. Internationale Kommission zum Schutz der Oder gegen Verunreinigung, Wroclaw, 37 pp

Jatteau P (1998) Etude bibliographique des principales caractéristiques de l'écologie des larves d'Acipenséridés. Bull Fr Pêche Piscic 350–351:445–464

Jego S, Gazeau C, Jatteau P, Elie P, Rochard E (2002) Les frayères potentielles de l'esturgeon Européen *Acipenser sturio* L. 1758 dans le bassin Garonne–Dordogne. Méthodes d'investigation, état actuel et perspectives. Bull Fr Pêche Piscic 365(366):487–505

King TL, Lubinski BA, Spidle AP (2001) Microsatellite DNA variation in Atlantic sturgeon (*Acipenser oxyrinchus oxyrinchus*) and cross-species amplification in the Acipenseridae. Conserv Genet 2:103–119

Kirschbaum F, Gessner J (2000) Re-establishment program for *Acipenser sturio* L., 1758: the German approach. Symposium on conservation of the Atlantic sturgeon *Acipenser sturio* L., 1758 in Europe, Madrid. Bol Inst Esp Oceanogr 16:1–4, 149–156

Kirschbaum F, Gessner J (2002) Perspectives for the re-introduction of the European sturgeon Acipenser sturio in the Elbe River system. In: Thiel R (ed) Ökologie der Elbefische Z Fischk Suppl, vol 1, pp 217–232

Kulmatycki W (1932) About the protection of the sturgeons in Polish Rivers. Ochrona Przyrody 12:8–28 (In Polish)

Ludwig A, Gessner J (2007) What makes the difference – sea sturgeon on both sides of the Atlantic Ocean. Am Fish Soc Symp 56:285–300

Ludwig A, Debus L, Lieckfeldt D, Wirgin I, Benecke N, Jenneckens I, Williot P, Waldman JR, Pitra C (2002) When the American sea sturgeon swam east. Nature 419:447–448

Ludwig AN, Arndt U, Lippold S, Benecke N, Debus L, King TL, Matsamura S (2008) Tracing the first steps of American sturgeon pioneers in Europe. BMC Evol Biol 8:221–236

Ludwig A, Makowiecki D, Benecke N (2009) Further evidence of trans-Atlantic colonization of Western Europe by American Atlantic sturgeons. Archaeofauna 18:185–192

Magnin E (1962) Recherches sur la systématique et la biologie des Acipensérідés *Acipenser sturio* L., *Acipenser oxyrhynchus* Mitchill et *Acipenser fulvescens* Raf. Ann Stat Cent Hydrobiol Appl 9:3–242

Maleczynski K (1963) History of Silesia. Part III: from the end of the 16th century to 1763. Ossolineum, Wroclaw, 626 pp

Mamcarz A (2000) Decline of the Atlantic sturgeon *Acipenser sturio* L., 1758 in Poland: an outline of problems and prospects. Bol Inst Esp Oceaogr 16:191–202

Paaver T (1996) A common or Atlantic sturgeon, *Acipenser sturio*, was caught in the Estonian waters of the Baltic Sea. Sturgeon Q 4(3):7

Prange A, Furrer R, Einax JW (2000) Die Elbe und ihre Nebenflüsse – Belastung, Trends, Bewertung, Perspektiven, ATV-DVWK-Forschungsberichte. GfA, Hennef, p 168

Przybyl A (1976) Historic distribution of common sturgeon *Acipenser sturio* in Warta River. OT: Wystepowanie jesiotra zachodniego *A. sturio* w dorceczu Warty. Ochrona Przyrody 32(5): 5–12

Rochard E, Castelnaud G, Lepage M (1990) Sturgeons (Pisces: Acipenseridae); threats and prospects. Journal of Fish Biology 37:123–132

Rochard E, Lepage M, Meauze L (1997) Identification et caractérisation de l'aire de répartition marine de l'esturgeon européen *Acipenser sturio* à partir de déclarations de captures. Aquat Living Res 10:101–109

Schmidt GW (1996) Wiedereinbürgerung des Lachses Salmo salar in Nordrhein-Westfalen - Allgemeine Biologie des Lachses sowie Konzeption und Stand des Wiedereinbürgerungsprogramms unter besonderere Berücksichtigung der Sieg. Landesanstalt für Ökologie, Bodenordnung und Forsten/ Landesdamt für Agrarordnung NRW. LÖBF-Mitt 11:1–194

Schiemenz P (1905) Weitere fischereiliche Studien über organische Abwässer. Z f Fischerei 13:49–81

Schräder T, Czensny R (1930) Die Untersuchung der oberschlesischen Oder am 2. und 3. Oktober 1928. Z Fisch 28:191–220

Seligo A (1931) Die Seefischerei von Danzig. Handbuch der Seefischerei Nordeuropas VIII (7):25–28

St Pierre R (1996) Breeding and stocking protocol for cultured Atlantic sturgeon. Final report from the Atlantic Sturgeon Aquaculture and Stocking Committee to the Atlantic States Marine Fisheries Commission Atlantic Sturgeon Management Board, 16 pp

Sulak KJ, Clugston JP (1999) Early life history and population structure of Gulf sturgeon, *Acipenser desotoi*, in the Suwanee River, Florida. J Appl Ichthyol 15:116–128

Sych R, Bartel R, Bieniarz K, Mastynski J (1996) Project for the restoration of migratory fish species in Poland. OT: Projekt Restytucji Ryb Wedrownych w Polsce. Opracowanie Zespolowe, 40 pp

Thiel R, Nellen W, Ergenzinger P, Kirschbaum F, Knösche R, Larink O, Winkler HM (1998) Ökologische Zusammenhänge zwischen Fischgemeinschafts- und Lebensraumstrukturen der Elbe. Z Ökol Naturschutz 7:67–70

Tiedemann R, Moll K, Paulus KB, Scheer M, Williot P, Bartel R, Gessner J, Kirschbaum F (2007) Atlantic sturgeons (*Acipenser sturio*, *Acipenser oxyrinchus*): American females successful in Europe. Naturwissenschaften 94:213–217

Tikhiy MI (1929) German sturgeon in the Rioni River. OT: Nemetskiy osetr v Rione. Priroda 4:369

von Nordheim H, Gessner J, Kirschbaum F, Anders E, Arndt G-M (2001) Das Wiedereinbürgerungsprogramm für A. sturio – Hintergründe und Konzeption. In: Der Stör Acipenser sturio –

Fisch des Jahres 2001. Hrsg. Verband Deutsche Sportfischer. Verlag M. Faste, Offenbach, pp 30–49

Williot P, Rochard E, Castelnaud G, Rouault T, Brun R, Elie P (1997) Biological and ecological characteristics of European Atlantic sturgeon, *Acipenser sturio*, as foundations for a restoration programme in France. Environ Biol Fishes 48:359–370

Wolter C, Vilcinkas A (1998) Fish community structure in lowland waterways: fundamental and applied aspects. Pol Arch Hydrobiol 45(2):137–149

Chapter 42
The Past and Future of Sturgeons in Poland: The Genetic Approach

Ana Stanković

Abstract To support the sturgeon reestablishment program in Poland, DNA analyses of the extinct sturgeon populations of Odra and Vistula rivers were performed. According to the data obtained, the dominant sturgeon species since the third century BC was *Acipenser oxyrinchus*, rather than *A. sturio* as previously stated. Two ancient populations of *A. oxyrinchus* from Poland, one dated for the third century BC and one for the ninth to thirteenth centuries AD, were genetically characterized based upon mtDNA region (used here) and microsatellite analysis. The two populations are genetically close to *A. oxyrinchus* populations from the St. John River population, NB, Canada. Starting from 2004, fertilized eggs obtained from sturgeons in the St. John River were transferred to Poland. In 2006/2007, the first experimental stocking was performed. The fish maintained as a brood stock, as well as those used for stocking, were genetically characterized to monitor the level of genetic polymorphism.

42.1 Introduction

As in the other parts of Northern Europe, Polish rivers were inhabited by sturgeons after the last glaciation. From the Baltic Sea they migrated upstream in the Oder and Vistula Rivers for spawning. Archaeological studies have permitted an estimation that in the tenth to eleventh century the sturgeons made up over 50% of the total fish catch in the Gdansk area (Urbanowicz 1965), while in the

A. Stanković (✉)
Department of Biology, Institute of Genetics and Biotechnology, University of Warsaw and Institute of Biochemistry and Biophysics, Polish Academy of Sciences, Pawińskiego 5A, 02-106 Warsaw, Poland
e-mail: anams@ibb.waw.pl

P. Williot et al. (eds.), *Biology and Conservation of the European Sturgeon Acipenser sturio L. 1758*, DOI 10.1007/978-3-642-20611-5_42,

eighteenth century a drastic reduction of the sturgeon population in Eastern Baltic was observed due to anthropogenic impacts (Mamcarz 2000 and references therein). The total ban on sturgeon catches, introduced in 1936, did not stop the decline of the sturgeon and the subsequent extinction of the species from Polish waters after 1965 (Debus 1996). In all of Europe, sturgeons are on the verge of extinction, with *A. sturio* being listed in the Red Books of the European Union (Gessner 2000), while in Poland it is recognized as an extinct species.

In 1998, the program of sturgeon restitution in Poland was launched (Sych et al. 1998). All species restitution programs should have a strong genetic backbone (IUCN Species Survival Commission 1996). In the case of the Baltic sturgeon, it was important to answer two questions: (1) which species, *A. sturio* or *A. oxyrinchus,* was the native species in the Baltic prior to extinction? and (2) which of the sturgeon populations in Europe or North America is genetically closest to the one which inhabited the Vistula and Oder tributaries and could serve as a donor population? Once these questions were answered (Ludwig et al. 2002; Ludwig et al. 2008), the reestablishment process should be carefully monitored to secure the proper selection of breeding material and sufficient genetic polymorphism of the future population.

42.2 Materials and Methods

The material for DNA analysis of Hudson and St. Lawrence *A. oxyrinchus* populations, as well as for the *A. oxyrinchus desotoi* population, was obtained by courtesy of Dr. Timothy L King. The material for analysis of *A. sturio* population came from IGB in Berlin and Cemagref in Bordeaux thanks to Dr. Jörn Gessner and Dr. Patrick Williot. The remaining material (*A. oxyrinchus* specimens from St. John and from Polish hatcheries) was supplied by Dr. Ryszard Kolman. Sturgeon samples from archaeological sites were provided by Dr. Daniel Makowiecki and Dr. Krzysztof Stefaniak. The dating of twentieth century samples was obtained from museum files, while samples excavated were dated by accelerator mass spectrometry (^{14}C AMS). Bony scutes served as a material for DNA isolation in the case of both archival and excavated samples.

The work on ancient sturgeon DNA was carried out in the special ancient DNA laboratory of the Institute of Genetics and Biotechnology, University of Warsaw. Special care was taken to avoid contamination with modern sturgeon DNA. Recent samples were never processed in the ancient DNA lab. The details of the amplification and sequencing procedures were applied as described by Pääbo et al. (2004).

The references to software used for population analyses are given in figure and table legends.

42.3 Results

42.3.1 A. sturio *or* A. oxyrinchus*?*

Until 2002 it was common understanding that *A. sturio* was the only sturgeon species which inhabited Baltic and European Atlantic coastal waters, despite the differences in meristic characteristics described previously (Magnin 1962; Artyukhin and Vecsei 1995). Ludwig et al. (2002), based upon an analysis of ancient DNA extracted from archival sturgeon samples, found that since 800 years BP, *A. oxyrinchus* – a species today inhabiting North American rivers along the Atlantic Coast from Labrador to Florida – was the dominant species in the samples analyzed from the Baltic region. According to Ludwig et al. (2002), the Atlantic sturgeons might have benefited from the so-called Little Ice Age in Europe and, due to their ability to reproduce in lower temperatures, replaced "European" *A. sturio* in the Baltic drainage system.

To verify the results of Ludwig et al. (2002) and to identify the Polish sturgeon species prior to their extinction, we extracted DNA from sturgeon samples obtained from museums and from archeological excavation sites. Sampling sites were located in both Vistula and Oder drainage systems, and the origin of the samples analyzed covered well over 2,000 years, from the third century BC to the end of the twentieth century.

The DNA sequences obtained were compared with sequences of *A. sturio* and *A. oxyrinchus* deposited in GenBank. As shown in Table 42.1, only one specimen dated for fourteenth century was classified as *A. sturio*. All the remaining 97 samples, including those from the third century BC, were determined as *A. oxyrinchus*. These results clearly indicate that *A. oxyrinchus* appeared in the Baltic region much earlier than the postulated dominance by Ludwig et al. (2002, 2008). Therefore, it should not have been connected with climatic changes during the Little Ice Age. Desse-Berset (2009), studying fish bones and scutes from several

Table 42.1 The list of sturgeon fossil and archival specimens analyzed

Locality	Dating of samples	Number of individuals tested	Cytochrome *b* sequences obtained (n)	Cr mtDNA sequences obtained (n)
Vistula River (Elbląg site)	3rd BC	40	22	18
Vistula River (Nieszawa site)	9th–13th	44	33	43
	20th	5	2	2
	4th–5th	5	–	1
	9th–14th	40	27	24
	14th		1*	–
Oder River (Estuary site)	17th–18th	28	–	3
	20th	2	1	1

Species attribution to *A. oxyrinchus* or *A. sturio* (*) was performed based upon cytochrome *b* and D-loop mtDNA sequences (Ludwig et al. 2000)

French archeological sites, determined that remains of *A. oxyrinchus* were present at the Atlantic coast 5,000 years ago and until approximately 200 AD. She assumed that *A. oxyrinchus* probably lived in sympatry with *A. sturio* because of the abundant remains of the latter species in the archaeological sites.

Genetic analyses of sturgeon populations from two medieval sites at Ralswick, Isle of Rugia on the German Baltic coast performed by Ludwig et al. (2002) indicated that between 1,200 and 800 years ago there had been a remarkable species shift from *A. sturio* to *A. oxyrinchus*. Ludwig et al. (2008), in an excellent example of a population study based on ancient DNA analysis, managed to determine mtDNA control region haplotypes of 227 historic specimens, and to designate profiles of seven polymorphic microsatellite loci in 29 specimens. The genetic structure of several North American populations was compared with the structure of the ancient Baltic population. The results demonstrated a high similarity of allele frequencies of the Baltic population with the Canadian one. According to the authors, these results proved that the Baltic *A. oxyrinchus* population was founded by immigrants from Canada. The same authors, using ancient and contemporary DNA data for eight genetic loci (seven autosomal microsatellites and mtDNA), and employing the approximate Bayesian computation (ABC) method (Beaumont et al. 2002), estimated the size of the founding population to be less than 20 individuals.

42.3.2 Genetic Polymorphism and Structure of Historic and Contemporary Sturgeon Populations

A. oxyrinchus populations still exist in the major North American Atlantic coastal rivers from the St. Lawrence River to the St. Johns River (Wirgin et al. 2000, 2002). During the last 2 centuries, the abundance of American Atlantic sturgeons has dramatically declined for the same reasons as in Europe (overfishing, damming of rivers and water pollution) (for details see Chap. 39, Waldman). The endangered status of *A. oxyrinchus oxyrinchus* in North America triggered a series of genetic studies of the remaining populations. Analyses of both mitochondrial DNA (mtDNA) haplotypes (Wirgin et al. 2000) and microsatellite loci (King et al. 2001) have shown that all populations are highly structured and that they differ significantly from each other, indicating a strong homing effect. The two northernmost populations (St. Lawrence and St. John) are exceptional for their monomorphism of mtDNA haplotypes (Wirgin et al. 2002), the same monomorphic haplotype which is characteristic for the extinct Baltic population (Ludwig et al. 2002).

We have chosen three North American populations of *A. oxyrinchus* (St. John, St. Lawrence and Hudson) for comparison of genetic polymorphism with two historic sturgeon populations from Poland represented by the relatively high number of specimens. One of these populations was dated for the ninth to thirteenth century, and was excavated from archaeological sites close to the Vistula Estuary (Elbląg), the second was dated for the third century BC and originated from a site located about 200 km upstream on the Vistula River (Nieszawa). Additionally, the genetic polymorphism of two populations of closely related sturgeon species was studied: *A. oxyrinchus desotoi* from the Choctawatchee River and *A. sturio* from a mating pair of a controlled reproduction of 1995 from the Gironde River. For each of these specimens, microsatellite profiles were obtained for eight loci: *Aox*C45, *Aox*D170, *Aox*D234, *Aox*45, *Aox*D54, *Aox*D161, *Aox*188 and *Aox*D297 (King et al 2001; Henderson-Arzapalo and King 2002). Despite some problems with amplification of longer alleles in the historic material, no evidence of large allele dropouts and presence of false alleles were found except for the *Aox*188 (351 BP) locus, which was used in calculations only when contemporary material was analyzed. The number of alleles per locus, allelic richness and other indicators of polymorphism are given in Table 42.2.

A high degree of genetic diversity was detected, as 13.75 alleles were observed across the seven loci, ranging from three alleles at *AoxC45* to 23 at loci *AoxD234* and *AoxD297*. The *AoxC45* locus was the only one which in the *A. sturio* population appeared to be monomorphic. The number of alleles (Na) and allelic richness (R) was much higher in all *A. oxyrinchus* populations than in *A. sturio* and *A. desotoi* populations. The size of alleles in one locus (*AoxD*188) was different in the *A. sturio* population (245–250 BP) compared with *A. oxyrinchus* populations (270–356 BP). The heterozygosity values were at similar level for all *A. oxyrinchus* populations and lower for *A. sturio* and *A. desotoi*. The F_{IS} values ranged from 0.092 to −0.099.

Most of the genetic diversity indices of the extinct Baltic populations appeared to be similar to those obtained for the contemporary *A. oxyrinchus* populations from

Table 42.2 Allelic variation and genetic diversity of sturgeon populations based on analysis of seven microsatellite loci

Species	*A. oxyrinchus*			*A. desotoi*	*A. oxyrinchus*		*A. sturio*
Populations	1	2	3	4	5	6	7
N	83	31	20	19	23	38	12
Na	10.86	6.71	8.00	4.57	8.14	9.43	3.14
Ne	4.32	3.16	4.58	2.77	4.33	4.35	2.13
R	5.25	4.47	5.78	3.58	5.53	5.50	2.96
Np.	1.43	0.71	0.86	0.71	1.29	1.71	1.29
Ho	0.64	0.66	0.78	0.55	0.68	0.64	0.63
He	0.66	0.65	0.76	0.51	0.73	0.70	0.57
Fis	0.028	−0.013	−0.032	−0.064	0.068	0.092	−0.099
HWE	–	–	*	–	*	*	–

N – number of individuals, Na – number of alleles, Ne – effective number of alleles, R – allelic richness, Np. – number of private alleles, H_O – observed heterozygosity, H_E – expected heterozygosity, F_{IS} – inbreeding coefficient, HWE – Hardy–Weinberg Equilibrium Test.

* means significant deviation from HWE ($P < 0.05$). Analyzed populations: (1) *A. oxyrinchus* (St.John), (2) *A. oxyrinchus* (St. Lawrence), (3) *A. oxyrinchus* (Hudson), (4) *A. desotoi* (Choctawatchee River), (5) *A. oxyrinchus* (Poland 3rd BC), (6) *A. oxyrinchus* (Poland-9th–13th AD), (7) *A. sturio* (Gironde)

North America, and different from the *A. sturio* population from the Gironde. This was expected, since although there is some evidence that *A. oxyrinchus* and *A. sturio* hybridized with each other (Tiedemann et al. 2007), they are considered to be separate species, easily distinguishable by morphological traits (Peng et al. 2007). The samples from the Gironde population, as expected, exhibit very low genetic polymorphism, much smaller than all *A. oxyrinchus* populations studied. This is not surprising considering the origin of the samples (see above).

In the case of the St. John population, we have detected a bottleneck effect. This population is not in Hardy–Weinberg equilibrium when eight msDNA loci are taken into account (slightly positive F_{IS} value), and shows an average genetic polymorphism. The values obtained for the number of alleles per locus and heterozygosity were similar to those obtained for the St. John population by Henderson-Arzapalo and King (2002). It seems that this population is in a relatively good genetic condition, although the effective population size was estimated at 98.4, which is below the brood-stock size proposed for the reestablishment of populations proposed by the Atlantic Sturgeon Status Review Team (2007).

42.3.3 Population Structure and Genetic Distance Between Sturgeon Populations

The population structure was tested using the Bayesian-clustering approach implemented in STRUCTURE 2.3.3 (Pritchard et al. 2000). Simulations were performed assuming admixture and correlated allele frequencies (F-model), using no population information. An evident structure was detected. All analyzed sturgeons were assigned to five clusters ($\Delta K = 5$) (Fig. 42.1). The cluster Q1 consisted mainly of St. John individuals, cluster Q2 of Hudson and St. Lawrence

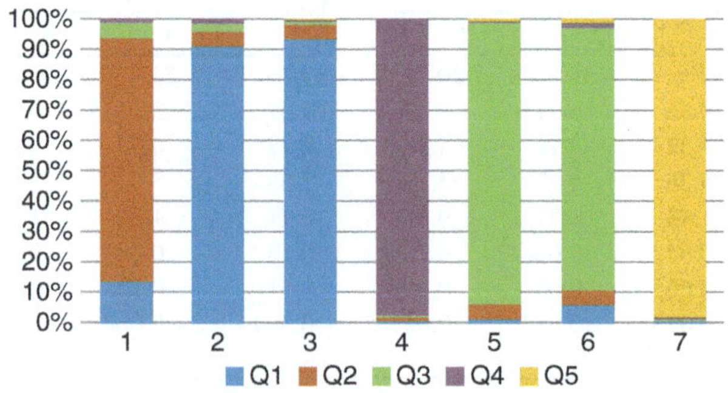

Fig. 42.1 Proportion of membership of each population in each of the defined clusters (Q1–Q5) based on analysis of seven microsatellite loci. Calculations were performed with the STRUCTURE software (Pritchard et al. 2000). (1) *A. oxyrinchus*, St. John ($n = 83$); (2) *A. oxyrinchus*, St. Lawrence ($n = 31$); (3) *A. oxyrinchus*, Hudson ($n = 20$); (4) *A. desotoi*, Choctawatchee River ($n = 19$), (5) *A. oxyrinchus*, Poland, 3rd BC ($n = 23$); (6) *A. oxyrinchus*; Poland, 9th–13th AD ($n = 38$); (7) *A. sturio*, Gironde ($n = 12$)

individuals, cluster Q3 of sturgeons from ancient Baltic populations, clusters Q4 and Q5 of *A. oxyrinchus desotoi* and *A. sturio* populations respectively. It can be seen that the fraction of individuals from St. John population showing genetic similarity to Baltic sturgeons is much greater than the fraction of Baltic sturgeons showing similarity to *A. oxyrinchus desotoi* or *A. sturio.*

The values of pairwise F_{ST} indicate that the genetic distance between St. John and Baltic populations is low (0.075 and 0.067) compared to the distance between Baltic populations and populations of *A. sturio* and *A. desotoi* (Table 42.3 and Fig. 42.2). The genetic distance between Baltic and St John populations is even smaller than between populations of *A. oxyrinchus* from St. Lawrence and Hudson, and similar to that obtained by King et al. (2001) for the St. John and St. Lawrence

Table 42.3 Pairwise estimates of genetic differentiation (F_{ST}) between sturgeon populations

	A.oxyrinchus St. John	*A.oxyrinchus* St. Lawrence	*A.oxyrinchus* Hudson	*A. desotoi* Choctawatchee	*A.oxyrinchus* Poland, 3rd BC	*A.oxyrinchus* Poland, 9th–13th AD
A. oxyrinchus St. Lawrence	0.096					
A. oxyrinchus Hudson	0.134	0.151				
A. desotoi Choctawatchee	0.283	0.302	0.300			
A. oxyrinchus Poland, 3rd BC	0.075	0.131	0.140	0.278		
A. oxyrinchus Poland, 9th–3th AD	0.067	0.096	0.144	0.249	0.042	
A. sturio Gironde	0.300	0.352	0.288	0.455	0.244	0.266

Analyses of F_{ST} values were conducted in software FSTAT. All values are significant ($P < 0.05$)

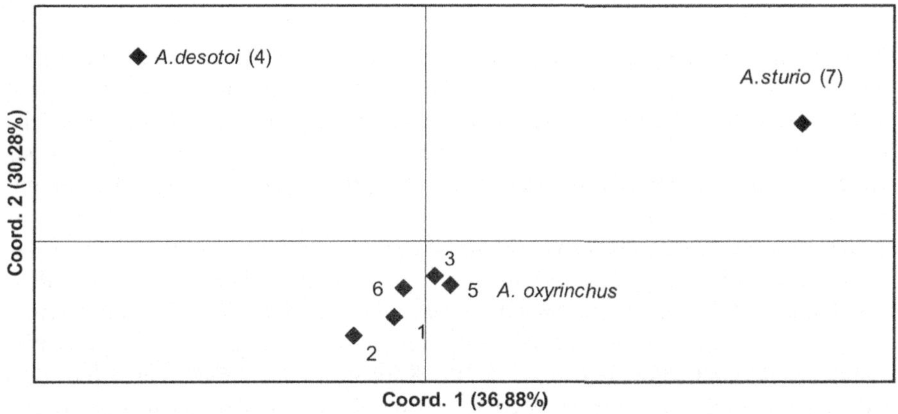

Fig. 42.2 PCA of genetic divergence (F_{ST}) values (see Table 3) obtained with GeneAlEx software (Paekall and Smouse 2001) (1) *A. oxyrinchus,* St. John; (2) *A. oxyrinchus,* St. Lawrence; (3) *A. oxyrinchus,* Hudson; (4) *A. desotoi;* (5) *A. oxyrinchus,* Baltic, 3rd BC; (6) *A. oxyrinchus,* Baltic 9th–13th AD; (7) *A. sturio,* Gironde

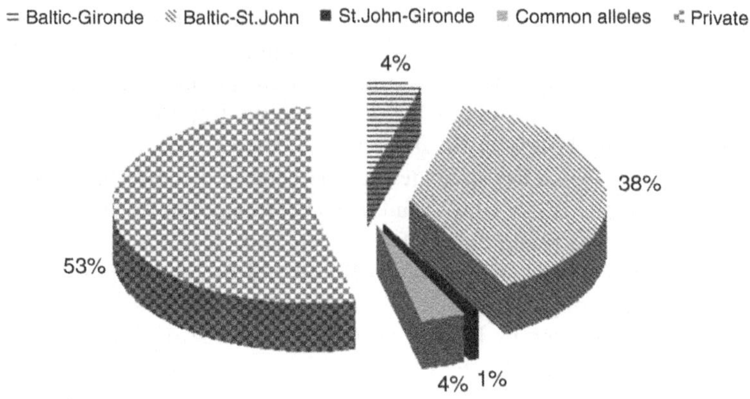

Fig. 42.3 Distribution of seven microsatellite loci among three populations: Gironde (*A. sturio*), St. John (*A. oxyrinchus*) and combined Baltic *A. oxyrinchus* populations (3rd BC and 9th–13th AD). Common alleles are those shared by all populations. Private alleles are characteristic for only one population

populations (=0.086). Importantly, the historic Baltic population is evidently genetically close to the contemporary North American *A. oxyrinchus* populations.

The genetic proximity of the St. John and Baltic populations is also clearly visible when distribution of alleles is examined. Figure 42.3 shows allele distribution between Gironde, St. John and two combined Baltic populations. It can be seen that Baltic and St. John populations share 38% of alleles, much more than Baltic and Gironde (4%) or St. John and Gironde (1%).

42.3.4 Sturgeon Restitution Program in Poland

From the genetic analyses of sturgeon population samples, three main conclusions important for starting the sturgeon restitution program in Poland may be drawn: (1) the species to be restituted should be *A. oxyrinchus* rather than *A. sturio*, (2) the St. John *A. oxyrinchus* population is genetically similar to the sturgeon populations which populated Polish Rivers before their extinction, and (3) the St. John population is in a relatively good genetic shape, i.e., it is of sufficient size and reveals sufficient polymorphism to serve as a donor population for stocking Polish rivers.

Between 2004 and 2009, representatives of the Inland Fisheries Institute (IRS) from Olsztyn, Poland, and the Society to Save the Sturgeon from Germany carried out the controlled reproduction of adult sturgeons caught in the St. John River (Arndt et al. 2006; Gessner et al. 2008; Kapusta et al. 2008). Fertilized eggs and eleutheroembryos were shipped to various hatchery stations in Poland and Germany where hatchlings were reared. The fry obtained were kept either for building the future brood stock or to carry on experimental stocking of Polish rivers (for details see Chaps. 43 and 41)

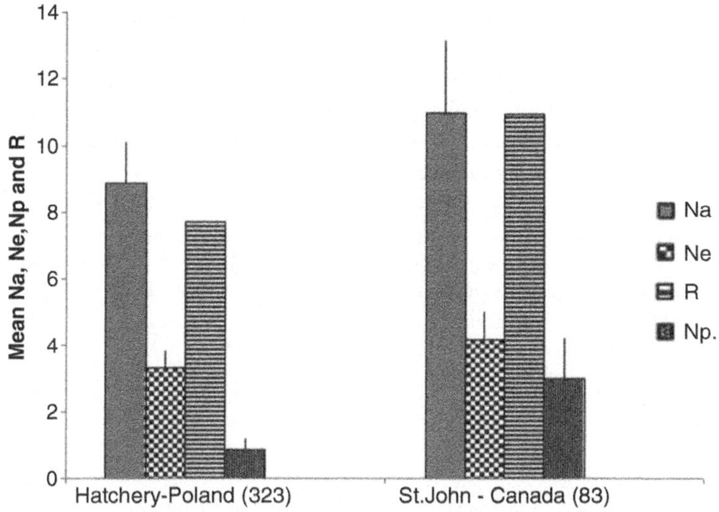

Fig. 42.4 Polymorphism indices of the population founded in Poland and of the St. John founder population. Eight microsatellite loci were analyzed. *N* – number of individuals, *Na* – number of alleles, *Ne* – effective number of alleles, *R* – allelic richness, *Np.* – number of private alleles

To verify the polymorphism of the future brood stock maintained in Poland and to select specimens which should be included into the brood stock as a basis for future stocking of Polish rivers, genetic analyses of fishes used for reproduction in Canada as well as their progeny were performed. Profiles of eight microsatellites were obtained for 323 specimens in total, comprising all batches of eggs hatched in Poland between 2004 and 2008. The results shown in Fig. 42.4 indicate that the polymorphism indices of fishes reared in Poland today are still lower than those calculated for the native population. This low polymorphism is based on the low number of parental fish used for spawning from the St. John River. For example, in 2004 one ♀ and two ♂ participated in spawning, in 2005 one ♀ and four ♂ but only two males had a mating success, in 2007 three ♀ and four ♂ with male mating success 37.5, 29, 21 and 12.5%. The mating success of males was calculated by microsatellite analysis on parent fishes and the progeny. The expected and observed heterozygosity values for Polish hatchery populations were 0.640 and 0.639, while for the St. John population they were 0.665 and 0.663. This result again indicates that the differences between the native St. John population and the broodstock kept in Poland are rather low.

42.4 Conclusions

It is much too early to speculate on the perspective of the sturgeon restitution in Poland. The genetic analyses presented in this paper indicate that the choice of *A. oxyrinchus* population from St. John River in Canada is well-substantiated.

The results of preliminary stocking of Polish rivers indicate that sturgeon from Canada may adapt to local environmental conditions. An additional effort should be made to build up a broodstock of greater genetic variability than observed among fishes already kept in Poland. One can, however, expect that joint efforts of Polish and German breeders will result in building up a sturgeon founding stock of high genetic quality and diversity in future years.

Acknowledgements The fish tissues for DNA analyses were kindly provided by D. Makowiecki and K. Stefaniak (antique samples) and by T. L. King, R. Kolman, P. Willot and J. Gessner (contemporary samples). The experimental work was supported by NCBiR grant No. ZPB/62/72/380/IT2/10 and by the CePT program.

References

Arndt GM, Gessner J, Bartel R (2006) Characteristics and availability of spawning habitat for Baltic sturgeon in the Odra river and its tributaries. J Appl Ichthyol 22(suppl 1):172–181

Artyukhin EN, Vecsei P (1995) On the status of Atlantic sturgeon: conspecificity of European *Acipenser sturio* and North American *Acipenser oxyrinchus*. J Appl Ichthyol 15:35–37

Beaumont MA, Zhang W, Balding DJ (2002) Approximate Bayesian computation in population genetics. Genetics 162:2025–2035

Debus L (1996) The decline of the European sturgeon *Acipenser sturio* in the Baltic and North Sea. In: Kirchofer A, Hefti D (eds) Symposium, advances in life sciences: conservation of endangered freshwater fish in Europe. Birkhaeuser, Basel, pp 147–156

Desse-Berset N (2009) First archaeozoological identification of Atlantic sturgeon (*Acipenser oxyrinchus* Mitchill 1815). CR Palevol 8:717–724. doi:10.1016/j.crpv.2009.06.001

Gessner J (2000) Reasons for the decline of *Acipenser sturio* L., 1758 in central Europe, and attempts at its restoration. Boletin Instituto Espanol De Oceanografia 16:117–126

Gessner J, Arndt GM, Anders E, Würtz S, Bartel R, Zdanowski B (2008) The development of a broodstock and rearing of Acipenser oxyrinchus between 1998 and 2007, as a prerequisite for stocking the tributaries to the Baltic Sea. Wyd. IRS, Olsztyn, pp 19–30

Henderson-Arzapalo A, King TL (2002) Novel microsatellite markers for Atlantic sturgeon (*Acipenser oxyrinchus*) population delineation and bloodstock management. Mol Ecol Notes 2:437–439

Kapusta A, Duda A, Kolman R (2008) Movement of juvenile American Atlantic sturgeon, (Acipenser oxyrhynchus Mitchill) in the Drwęca River (central Poland) in Actual status and active protection of sturgeon fish populations endangered by extinction. Wyd. IRS, Olsztyn, 310 pp

King TL, Lubinski B, Spidle AP (2001) Microsatellite DNA variation in Atlantic sturgeon (*Acipenser oxyrinchus oxyrinchus*) and cross-species amplification in the Acipenseridae. Conserv Genet 2:103–119

Ludwig A, Debus L, Lieckfeldt D, Wirgin I, Benecke N, Jenneckens I, Williot P, Waldman JR, Pitra Ch (2002) When the American sea sturgeon swam east. Nature 419:447–448

Ludwig A, Arndt U, Lippold S, Benecke N, Lutz Debus L, Timothy L, King TL, Matsumura S (2008) Tracing the first steps of American sturgeon pioneers in Europe. BMC Evol Biol 8:221

Mamcarz A (2000) Decline of the Atlantic sturgeon *Acipenser sturio* L., 1758 in Poland: an outline of problems and prospects. Bol Inst Esp Oceanogr 16:191–200

Magnin E (1962) Recherches sur la systématique et la biologie des Acipenséridés: *Acipenser sturio* L. *Acipenser oxyrinchus* Mitchell et *Acipenser fulvescens* Raf. Ann Stat Cent Hydrobiol Appl 9:7–242

Paekal R, Smouse PE (2001) GenAlEx V5: Genetic analysis in Excel. Population genetic software for teaching and research. http://www.anu.ed.au/BoZo/GenAlEx/

Peng Z, Ludwig A, Wang D, Diogo R, Wei Q, He S (2007) Age and biogeography of major clades in sturgeons and paddlefishes (Pisces: Acipenseriformes). Mol Phylogenetic Evol 42:854–862

Pääbo S, Poinar H, Serre D, Jaenicke-Despres V, Hebler J, Rohland N, Kuch M, Krause J, Vigilant L, Hofreiter M (2004) Genetic analyzes from ancient DNA. Ann Rev Genet 38:645–679

Pritchard JK, Stephens M, Donnelly P (2000) Inference of population structure using multilocus genotype data. Genetics 155:945–959

Sych R, Bartel R, Bieniarz K, Mastyński J (1998) Project for the restoration of migratory fish species in Poland. OT: Projekt Restytucji Ryb Wędrownych w Polsce. Opracowanie Zespołowe, 40

Tiedemann R, Moll K, Paulus KB, Scheer M, Williot P, Bartel R, Gessner J, Kirschbaum F (2007) Atlantic sturgeons (*Acipenser sturio*, *Acipenser oxyrinchus*): American females successful in Europe. Naturwissenschaften 94:213–217

Wirgin I, Waldman JR, Rosko J, Gross R, Collins MR, Rogers SG, Stabile J (2000) Genetic structure of Atlantic sturgeon populations based on mitochondrial DNA control region sequences. Trans Am Fish Soc 129:476–486

Wirgin I, Waldman J, Stabile J (2002) Comparison of mitochondrial DNA analysis in estimating population structure and gene flow rates in Atlantic sturgeon *Acipenser oxyrinchus*. J Appl Ichthyol 18:313–319

Chapter 43
Re-establishing the Atlantic Sturgeon (*Acipenser oxyrinchus oxyrinchus* Mitchill) in Poland

Ryszard Kolman, Andrzej Kapusta, and Arkadiusz Duda

Abstract The objective of this chapter is to present the aquaculture-based restitution program for Atlantic sturgeon *Acipenser oxyrinchus* Mitchill in Poland. Broodstock development using *A. oxyrinchus* population from St. John River (Canada) is currently under way. The juvenile sturgeon obtained from eggs imported from Canada was reared in recirculating aquaculture system (RAS). Currently, there are two rearing facilities in Poland where selected fish from 1998 and 2001–2010 are held. To examine the feasibility of the reestablishing program, 63,660 hatchery-produced juveniles were released to the historical sturgeon rivers in the Vistula and Oder drainages. Radio-telemetry studies of 0+ and 1+ sturgeon were conducted in the Drwęca River between 2007 and 2009. Sturgeon moved downstream during the night, and no schooling behavior was noted. The mean swimming speed was 1.18 km h^{-1} (0.73 BL s^{-1}), while the maximum was 8.73 km h^{-1} (9.51 BL s^{-1}). Sturgeon preferred deep holes in the river bed for prolongated stationary phases during the daytime.

43.1 Introduction

The Atlantic sturgeon, *Acipenser oxyrinchus* Mitchill was only one diadromous representative of the genus *Acipenser* inhabiting the Baltic Sea basin. *A. oxyrinchus* was historically widely distributed in the rivers flows into this sea. Mature individuals undertook spawning migrations in the rivers of the eastern and southern Baltic such as the Neva, Volkhov, Daugava, Nemunas, Pregola, Vistula, and Oder, where they would migrate to spawning grounds located in the upper reaches of

R. Kolman (✉) • A. Kapusta • A. Duda
Department of Ichthyology, Inland Fisheries Institute, Oczapowskiego 10, 10-719 Olsztyn-Kortowo, Poland
e-mail: kolrys@infish.com.pl

P. Williot et al. (eds.), *Biology and Conservation of the European Sturgeon Acipenser sturio L. 1758*, DOI 10.1007/978-3-642-20611-5_43,

these rivers or in tributaries (Wałecki 1864; Kulmatycki 1932; Berg 1948; Kuderskii 1983). The Atlantic sturgeon was a commercially important species in the Baltic Sea basin (Kolman et al. 2011). Continuously decreasing abundance of all European *A. oxyrinchus* populations was noted at the end of the nineteenth century (Kirschbaum and Gessner 2000; Mamcarz 2000; Kolman et al. 2008). The decline of the sturgeon has been considered for a long time to have occurred as a result of anthropogenic impact (Gessner et al. 2006). The last Atlantic sturgeon was caught in the Vistula River in 1965 (Mamcarz 2000) and in 1996 in the Baltic Sea (Paaver 1996).

The objective of this chapter is to present the aquaculture-based restitution program for Atlantic sturgeon in Poland. To examine the feasibility of this program, hatchery-produced juvenile *A. oxyrinchus* were released to the historical sturgeon rivers in the Vistula and Oder drainages. Radio telemetry was used to track juvenile Atlantic sturgeon and habitat use in the Drwęca River.

43.2 Restoring the Atlantic Sturgeon in Poland

Positive environmental changes occurred in the late twentieth and early twenty-first centuries in Polish rivers, and consequently in the southern Baltic Sea basin. Furthermore, the clarification of the species classification of the Atlantic sturgeon prompted the Inland Fisheries Institute in Olsztyn to undertake introductory steps to restore the *A. oxyrinchus*, based upon the restoration plan by Sych et al. (1996). This reintroduction strategy included the following steps:

- To use aquaculture systems to rear juveniles that will, in the future, form a broodstock to provide stocking material for the reestablishment project in the future
- Pilot stocking, with the aim of testing the adaptive abilities and behavior of sturgeon juveniles reared under controlled conditions

43.3 Broodstock Development

Beginning in 2004, various materials were imported from Canada to Poland annually, including fertilized eggs, hatchlings, and fingerlings. In the past 4 years, only fertilized eggs have been imported. This material was obtained from wild-caught spawners of the St. John River population (Photo 43.1). Artificial reproduction and the incubation of eggs to the advanced gastrula stage was performed at the hatchery of Acadian Sturgeon and Caviar, Inc., which is located on the St. John River about 30 km upstream from the river mouth to the Atlantic Ocean.

Tissue samples from the spawners as well as from the offspring were subjected to genetic testing. The results will be used after the broodstock has been

Photo 43.1 Catches of Atlantic sturgeon spawners in the St. John River

Photo 43.2 Closed aquaculture systems used for rearing sturgeon stocking material

established to maintain a high degree of genetic diversity in the progeny produced (see Chap. 42).

The hatch obtained from the sturgeon eggs imported from Canada was reared in recirculating aquaculture system (RAS) tanks to a mean weight of 5–7 g (Photo 43.2). These fry were used for stocking, since current studies indicated they were sufficiently resistant and had the ability to adapt to natural conditions (Kapusta et al. 2008; Kolman et al. 2008). About 200 individuals from each batch of these fry were retained for further rearing. Some of the fish that are reared to a mean body weight exceeding 30 g are used in experimental stocking and migration studies, while the rest are reared to the select stage for use as the subsequent year of the broodstock. Currently, there are two rearing facilities in Poland where selected fish from 1998 and 2001–2010 are held. These fish are tagged with PIT tags to allow individual identification after genetic characterization.

43.4 Habitat Use and Behavior

Since 2006, pilot stocking has been performed in the Drwęca River in the Vistula drainage and in the Oder River drainage of the Drawa, Warta, and Gwda rivers, all of which were previously inhabited by sturgeon. The results of these studies indicate that currently all of these rivers meet the requirements for normal fry growth and, in the future, for sturgeon spawning. Both stocking and experimental work is being conducted in the Oder River basin jointly with German scientists under the auspices of a co-operative agreement between the Inland Fisheries Institute in Olsztyn and the Leibniz Institute of Freshwater Ecology and Inland Fisheries in Berlin. Within the framework of this co-operation, more than 63,000 individuals of various sizes of American Atlantic sturgeon stocking material have been released in the Vistula (51,650 individuals) and Oder (12,010 individuals) catchment areas (Table 43.1). Stocking has been performed at various times of the year and with material of varying sizes. Some of these fish had been reared under natural conditions in flow-through ponds that provided access to natural food, while others had been reared in tanks on commercial diets. Some of the juvenile released into the rivers were fitted not only with external Carlin tags, but also with internal telemetry tags (micro radio transmitters) (Table 43.1). This made it possible to follow the movements of the fish. With species threatened with extinction, discovering what kinds of habitats they prefer can be decisive when implementing conservation measures.

Radio-telemetry studies of sturgeon were conducted in the Drwęca River between 2007 and 2009. Sturgeon were tagged with ATS F1805, F1810 and F1815 transmitters (respectively, weight: 3, 6, 7 g and operational life: 19, 34, 114 days), and Holohill BD-2 transmitters (weight of 0.62 g, operational life 21 days). A total of 101 juvenile sturgeon fitted with telemetry transmitters were released, and information was collected for 86 of these fish (Table 43.1). No natural mortality was confirmed among the fish fitted with the transmitters. The juvenile sturgeon released in spring remained at the release site from 1 to 21 days, and then moved downstream (Kapusta et al. 2008). In the fall, all of the fish released left the site within 2 days. The rate at which they descended the river was highly variable. Periods of slow movement altered with periods of rapid movement. The mean swimming speed was 1.18 km h^{-1} (0.73 BL s^{-1}), while the maximum was

Table 43.1 Characteristics of the Atlantic sturgeon stocking material released into Polish rivers

River basin	River stocked	Date	Type of material age/weight (g)	Quantity (ind)	Tag type
	Drwęca	9.10.2006	0+/7–9	1,500	–
	Drwęca	12.06.2007	1+/400–500	12	Carlin + T-M
	Drwęca	15.06.2007	1+/400–500	200	Carlin
	Drwęca	29.10.2007	0+/7–9	700	–
	Drwęca	30.10.2007	0+/20–40	250	Carlin
	Drwęca	30.10.2007	0+/30–50	20	Carlin + T-M
	Drwęca	30.10.2007	1+/500–650	10	Carlin + T-M
	Drwęca	3.06.2008	1+/300–400	370	Carlin
	Drwęca	3.06.2008	1+/300–400	30	Carlin + T-M
	Drwęca	15.10.2008	1+/400–500	30	Carlin + T-M
	Drwęca	19.06.2009	1+/26	1,900	Carlin
	Drwęca	3.07.2009	1+/350–500	430	Carlin + T-M
	Drwęca	20.10.2009	0+/7–9	10,500	–
	Drwęca	21.10.2009	0+/3–5	22,000	–
	Drwęca	26.10.2009	0+/5	1,630	–
	Drwęca	10.11.2009	0+/7–9	10,000	–
	Drwęca	30.04.2010	1+/80–150	1,900 (120)	Carlin
	Wisłoka	06.10.2009	1+/450–600	26	Carlin + T-M
Vistula	Wisłoka	06.10.2009	0+/5–7	150	–
	Drawa	10.05.2007	0+/150–250	10	Carlin + T-M
	Drawa	28.10.2007	2+/1,600–1,800	200	Carlin
	Gwda	29.10.2007	2+/1,600–1,800	238	Carlin
	Warta	29.10.2007	2+/1,600–1,800	200	Carlin
	Warta	29.10.2007	0+/5–7	4,000	–
	Gwda	19.03.2008	1+/200–400	550	Floy Tag + T-M
	Wisłoka	05.05.2008	0+/100–200	103	Carlin
	Barycz	21.10.2008	0+/7–12	2,000	–
	Warta	05.11.2008	0+/7–12	500	–
	Warta	05.11.2008	0+/30–50	100	–
	Warta	05.11.2008	0+/300–500	100	–
	Barycz	04.05.2009	1+/40–70	2,100(600)	Carlin + Floy Tag
	Warta	22.10.2009	0+/3–6	3,175	–
Oder	Warta	14.04.2010	0+/100	835	Carlin

Carlin, Floy Tag – external identification tags; T-M – radio telemetry tags

8.73 km h^{-1} (9.51 BL s^{-1}). In fall 2007, the migration speed of two groups of juvenile sturgeon was compared. The migration rates of fish aged 1+ was statistically significantly slower ($P < 0.05$) than that of the fish aged 0+ (0.83 and 2.94 BL s^{-1} respectively). The overall migration rate in the Drwęca was similar to that noted in 0+ sturgeon in the Drawa River (Fredrich et al. 2008), while it differed substantially from that of 0+ sturgeon in the Nanticoke River, which descended to the estuary at rates ranging from 0.04 to 0.4 km day^{-1} (Secor et al. 2000).

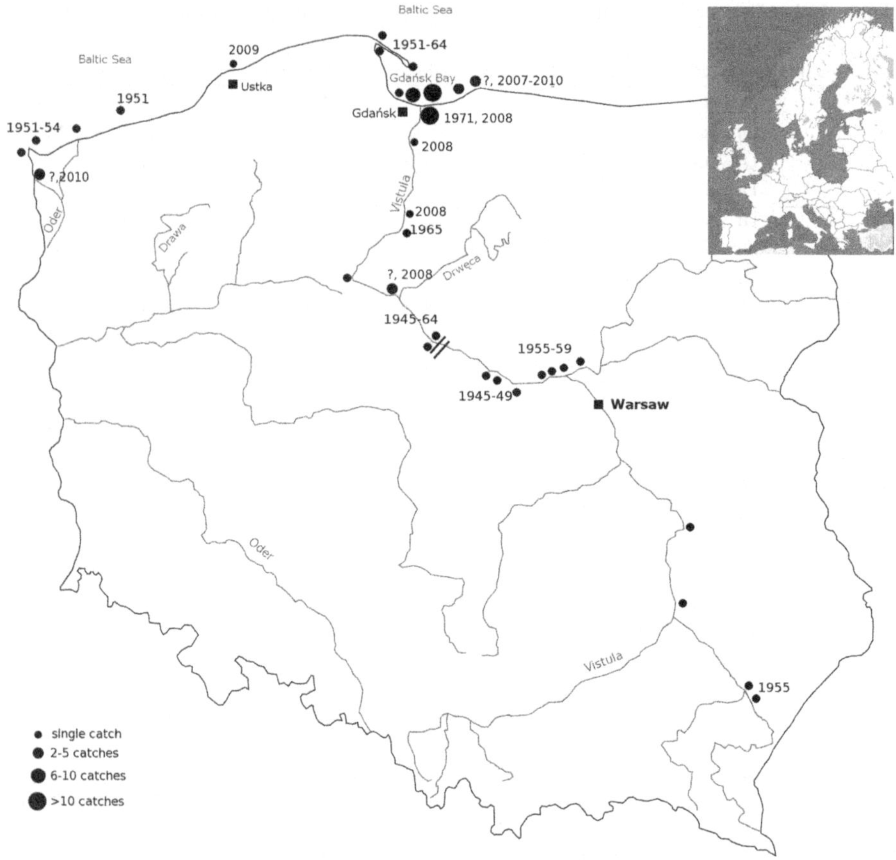

Fig. 43.1 Historic occurrence of Atlantic sturgeon *Acipenser oxyrinchus* in the second half of the twentieth century (according to Rolik 1959; Jaskowski 1962; Anon. 1965; Rudnicki 1966; Grabda 1968) and reported recapture of sturgeon released into the Drwęca River in the 2006–2009

After being released into rivers, sturgeons descend the rivers into the estuaries (Fig. 43.1). All of the tagged sturgeon exhibited a tendency to migrate downstream to the mouths of the rivers where they remained for about 2 weeks feeding intensively, as is evidenced by their increased body lengths and weights (Table 43.2).

Fredrich and Gessner (2007) reported that it is possible for the fish to migrate upstream; one specimen migrated a distance of 7 km in the Peene River, Germany. No such cases were noted in the Polish studies. During migration, some individuals remained in a given area for lengths of time ranging from a few days to over 2 weeks (10% of fish localizations), while sometimes they moved upstream by 50–200 m (5% of fish localizations); nevertheless, these movements were within areas characterized by a single, distinct type of bottom structure (Kapusta et al. 2008).

Table 43.2 Characteristics of sturgeon individuals caught in the Gdańsk Bay in the Vistula catchment

	Total length (cm)		Body length (cm)		Body weight (g)		
Tag number	Stocked	Caught	Stocked	Caught	Stocked	Caught	Site caught
P620AC	66.5	67.5	52.0	54.5	925	931	Vistula mouth
P891AC	58.0	59.0	45.5	48.5	465	502	Vistula mouth
P934AC	53.0	53.5	42.0	42.5	400	396	Vistula mouth
P862AC	66.0	–	51.5	–	589	–	Vistula–Grudziądz
P808AC	59.0	81.0	46.5	–	536	2,450	Gdańsk Bay near Mikoszewo
–	–	71.0	–	57.0	About 520	1,700	Gdańsk Bay near Junoszyno
P675AD	59.0	65.0	40.5	–	312	870	Vistula mouth

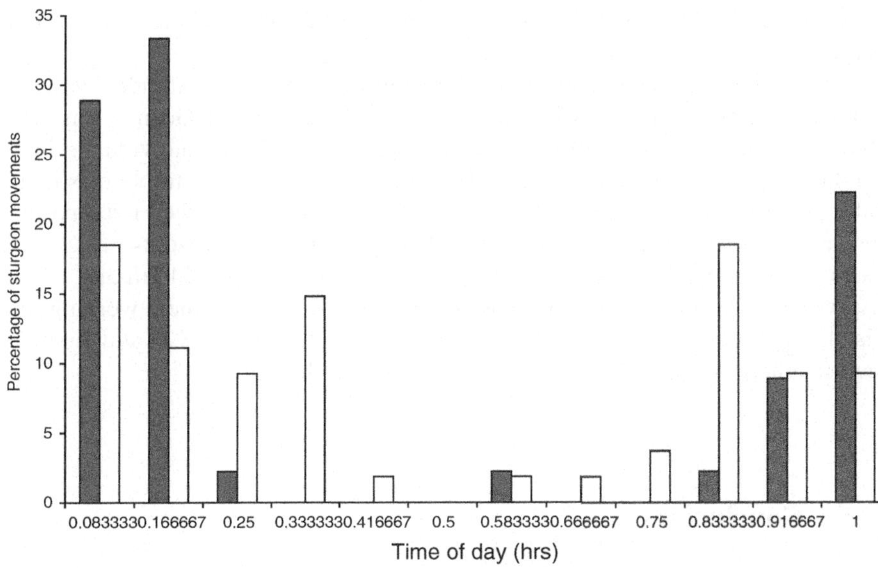

Fig. 43.2 Percentage of juvenile Atlantic sturgeon movements at different time of day in the Drwęca River registered by an automatic data-logging telemetry receiver. Spring and autumn movements are represented by solid and open bars respectively

The sturgeon released from rearing facilities moved in the Drwęca river at night (Kapusta et al. 2008). Substantial differences were noted in daily migration cycles depending on the seasons of the year (Fig. 43.2). The peak of migration occurred at night, and only single individuals were noted during the days. In fall, changes in the numbers of individuals registered throughout the day was less clear. Fish that moved at night dominated, but because this part of the day was longer, migration was also further and was more extended over time. Despite these differences, the general pattern of daily migration was similar. The first sturgeon migrations were registered about 1 h before sunset, and only single individuals continued to migrate during the day, usually during the morning hours.

There is little information available regarding the habitat preferences of young sturgeon during the river phase of their lives (Fredrich et al. 2008). In the Drwęca River, sturgeons were noted most frequently at deep holes, and no records were made of any fish in any shallower spots. The most fish were noted in smaller or larger channels that ran parallel to the river bed (Kapusta et al. 2008; Kolman et al. 2008). The share of sturgeon in the other habitats was similar (Fig. 43.2). The most common type of habitat in the segment of the Drwęca studied was lengths of even bottom, formed of sand and organic material. The habitat preferred in the Drawa River differed only slightly (Fredrich et al. 2008). Similarly to the Drwęca, most fish were registered in the channels. Nearly threefold more sturgeon occurred in the vicinity of partially submerged, dead trees in the Drawa River than were recorded at similar habitats in the Drwęca. Taking into consideration, however, the fact that depressions in river beds always form beneath submerged trees, the differences detected do not diverge from the overall pattern of habitat choice by juvenile sturgeon.

A particularly spectacular example of high growth rate is the individual with tag number P808AC, which was caught in mid November in the Gdańsk Bay near Mikoszewo at a depth of 30 m. Over a 6-month period, this individual's body weight increased by about 450%. Another sturgeon was caught in turbot nets in the Baltic Sea near Ustka on August 20, 2009. This specimen had been released into the Drwęca in the fall of 2008, and immediately prior to its release its total length was 55.5 cm and body weight was 528 g, and 10 months later 74 cm and 1.8 kg respectively. According to information obtained from fishermen working the Vistula Mouth and the Gdańsk Bay, by December 2008 a total of 27 sturgeon had been caught, most of which were released back into the water.

43.5 Conclusions

Our program will be regarded as a success when the released fish survive and return as adults. However, the results obtained to date make it possible to conclude that there are indications that future restoration programs will succeed. Atlantic sturgeon fry exhibit the ability to adapt to natural conditions, and they are finding good environmental conditions, including food supply, which not only permit them to remain in good health, but also to attain fast growth rates. Better fisheries monitoring is needed, along with enforcement to ban illegal catches. This study provides the first detailed information on the migratory behavior and habitat use by juvenile *A. oxyrinchus* in the Drwęca River. Our results establish that deep areas in river are important for juvenile sturgeon. Further studies designed to answer some of the questions are necessary for conservation and sustained management of this threatened species.

References

Anon (1965) Sturgeon caught. Gosp Ryb 12:21 (in Polish)

Berg LS (1948) Ryby presnykh vod SSSR i sopredelnykh stran, vol 1. AN SSSR, Moskwa

Fredrich F, Gessner J (2007) Ultrasonic study of downstream migration and habitat use of yearling sturgeon in the lower part of River Peene, Germany. In: 7th Conference on Fish Telemetry held in Europe, Book of Abstracts, Silkeborg, 17–21 June 2007

Fredrich F, Kapusta A, Ebert M, Duda A, Gessner J (2008) Migratory behavior of young sturgeon, *Acipenser oxyrinchus* Mitchill, in the Oder River drainage. Preliminary result of radio telemetric study in the Drawa River, Poland. Arch Pol Fish 16:105–117

Gessner J, Arndt GM, Tiedemann R, Bartel R, Kirschbaum F (2006) Remediation measures for the Baltic sturgeon: status review and perspectives. J Appl Ichthyol 22(suppl 1):23–31

Grabda E (1968) Endangered sturgeon. Ochrona Przyrody 33:177–191 (in Polish)

Jaskowski J (1962) Instructive materials on the ichthyofauna of the Warta River and its tributaries. Fragmenta Faunistica 28:22–23 (in Polish)

Kapusta A, Duda A, Fredrich F, Gancarczyk J, Raczkowski M, Kolman R (2008) Applying biotelemetry in studies of juvenile specimens of *Atlantic sturgeon, Acipenser oxyrhynchus* Mitchill, in the Drwęca and Drawa rivers. Komun Ryb 1:11–14 (in Polish)

Kirschbaum F, Gessner J (2000) Re-establishment program for *Acipenser sturio* L., 1758: the German approach. Bol Inst Esp Oceanogr 16:149–156

Kolman R, Kapusta A, Szczepkowski M, Duda A, Bogacka-Kapusta E (2008) Baltic sturgeon *Acipenser oxyrhynchus oxyrhynchus* Mitchill. Wyd. IRS, Olsztyn (in Polish)

Kolman R, Kapusta A, Morzuch J (2011) History of the sturgeon in the Baltic Sea and Lake Ladoga. In: Williot P et al (eds) Biology and conservation of the European sturgeon Acipenser sturio L. 1758. Springer, Heidelberg

Kuderskii LA (1983) Osetrovye ryby v bassejnach onezhskogo i ladozhskogo ozer. Ryby onezhskogo ozera i ikh khoziajstvennoe ispol'zovanie. Sborn Nauch Trud GosNIOPKh 128–148

Kulmatycki W (1932) About protection of the sturgeon in Polish rivers. Ochrona Przyrody 12:8–28 (in Polish)

Mamcarz A (2000) Decline of the Atlantic sturgeon *Acipenser sturio* L., 1758 in Poland: an outline of problems and prospects. Bol Inst Esp Oceanogr 16:191–202

Paaver T (1996) A common or Atlantic sturgeon, *Acipenser sturio*, was caught in the Estonian waters of the Baltic Sea. Sturgeon Q 4:3–7

Rolik H (1959) Giant sturgeon from the Bugo–Narew. Gosp Ryb 6:19–20 (in Polish)

Rudnicki A (1966) Sturgeon, conservation, and poaching. Gosp Ryb 1:21–22 (in Polish)

Secor DH, Niklitschek EJ, Stevenson JT, Gunderson TE, Minkkinen SP, Richardson B, Florence B, Mangold M, Skjeveland J, Henderson-Arzapalo A (2000) Dispersal and growth of yearling Atlantic sturgeon, Acipenser oxyrinchus, released into Chesapeake Bay. Fish Bull 98:800–810

Sych R (1996) About the project of migratory fish restoration in Poland. Zool Pol 41(suppl):47–59

Wałecki A (1864) Materials about the ichthyological fauna of Poland. II Systematic review of Polish fish. Drukarnia Gazety Polskiej, Warszawa (in Polish)

Part V
Perspectives and Prospectives

Chapter 44
Population Conservation Requires Improved Understanding of In Situ Life Histories

M.L. Acolas, J. Gessner, and E. Rochard

Abstract This section focuses on future research that should be carried out to contribute to the success of the sturgeon restoration programme. We consider both *A. sturio* and *A. oxyrinchus* data to support proposals for future experiments. Three main research topics are proposed: young-of-year behaviour and habitat use in rivers, downstream migration patterns of juveniles, environmental factors influencing migration.

44.1 Introduction

Knowledge of the in situ status and behaviour of *Acipenser sturio* populations mainly focuses on the time the fish spend in the estuary. Investigations have addressed migration and habitat use, population size, mortality, prey selectivity and growth of fish between 3 and 7 years old (Rochard et al. 2001; Brosse 2003). Life at sea, migration tactics, either juvenile downstream migration or reproduction migration, and the associated threats are yet not documented, and nor is the ecology of less than 1-year-old-fish.

In order to understand critical stages in the species life cycle and to protect its essential habitats, in situ experiments and observations are needed. For the near future, three topics have a high priority:

M.L. Acolas (✉) • E. Rochard
Cemagref, Estuarine Ecosystems and Diadromous Fish Research Unit, 50 Avenue de Verdun, 33612 Cestas cedex, France
e-mail: marie-laure.acolas@cemagref.fr

J. Gessner
Leibniz-Institute of Freshwater Ecology and Inland Fisheries, Müggelseedamm 310, 12587 Berlin, Germany

P. Williot et al. (eds.), *Biology and Conservation of the European Sturgeon Acipenser sturio L. 1758*, DOI 10.1007/978-3-642-20611-5_44,

44.2 Young-of-Year Behaviour and Habitat Use in Rivers

Ecology of the early juvenile phases of sturgeons in rivers is largely unknown, apart from some ancient reports (Jacobson 1889). Up to now, only laboratory experiments have been carried out (Jatteau 1998; Charles et al. 2009; Gessner et al. 2009), while only a few attempts to localise fish in rivers have recently been made, in the Dordogne River (Fig. 44.1) (see Chap. 29) for *A. sturio* and in the Odra River basin (Fig. 44.1) for *A. oxyrinchus*.

To assess migration patterns and habitat use, as well as feed preferences, in situ recapture and mesocosm experiments were carried out with *A. oxyrinchus* in the Odra River basin in 2009–2010. Larvae at first feeding, fingerlings (2.5 cm and 4.5 cm, total length TL) were released in the lower reaches of the Drawa River in summer. Fixed nets (2-m^2 opening equipped with plankton nets of 6 m length and 0.5 and 1 mm mesh size) were then used to recapture fish during downstream migration. In addition, releases of young-of-year fish of 10–15 cm and 1 year-old tagged fish of 25–35 cm TL were used to verify the methodology in the Barycz River. In general, during first feeding fish larvae utilise to a large extent gravel substrate and obstacles in the river bed, which minimises predation (Fuiman et al. 2010; McCormick and Meekan 2010). Also, feeding seems to take place concealed

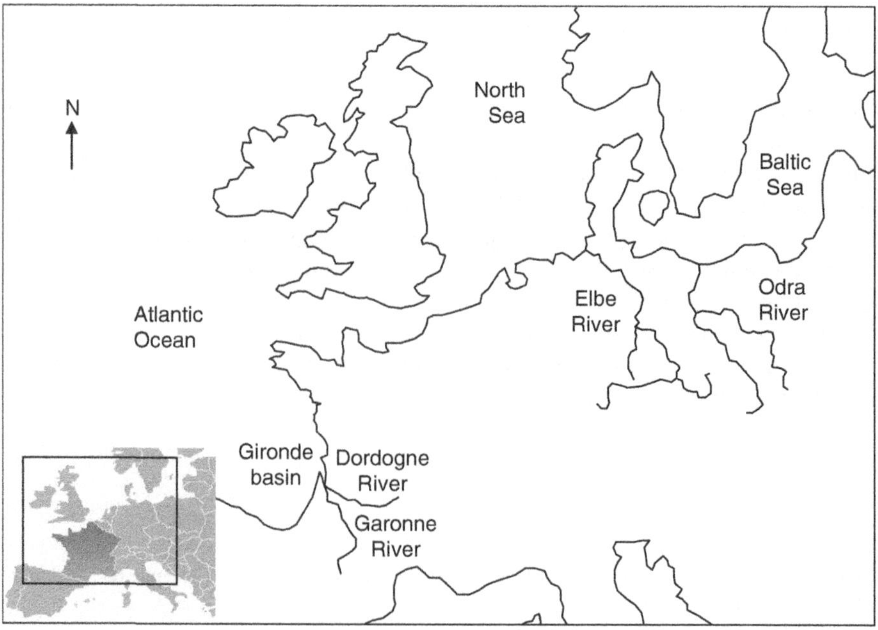

Fig. 44.1 European rivers with recent stocking of *A. sturio* (Gironde basin, France; Elbe River, Germany) and *A. oxyrinchus* (Odra River flowing through Germany, Poland and Czech Republic)

in the gravel and on substrate surfaces, reducing exposure to currents and thereby minimizing the drift speed. In larvae released at onset of feeding, recapture rates 5 and 12 km downstream of the release site over 4 days did not exceed 0.002% of the numbers released (75,000 in total). This could indicate the capacity of early life stages to maintain in suitable places for prolonged times despite their limited swimming ability. However, it may also suggest that the recapture technique is inefficient. For the onset of feeding, small (<300 μm) zooplankton, including rotifers, are the main prey items. In mesocosm experiments, feed uptake, growth, and a survival above 95% were confirmed with gravel substrate in river water. Both in situ data and the results of the laboratory experiments on *A. oxyrinchus* indicate that eleutheroembryos and larvae preferred gravel for resting (Gessner et al. 2009).

In *A. oxyrinchus* fingerlings, stomach analyses mainly comprise benthic invertebrates (Chironomidae and Oligochaetae). As a result of changing feeding preferences, substrate choice and thereby mobility should change also. Feeding habitats seem to be mainly sandy or include deep muddy areas. Mobility largely increases due to downstream migration, and recapture rates by fixed nets positioned at the edge of the main current increase to 1.5–2.5% of the fish released over a 3-day period. In larger fish exceeding 25 cm TL, habitat utilisation was determined using radio-telemetry. Due to safety issues, telemetry by boat was possible only during the day. Habitats used were predominantly deep holes or back-eddies beneath sunken trees, comprising more than 80% of the relocations (Fredrich et al. 2008).

For *A. sturio*, the Dordogne River (Fig. 44.1) samplings suggest that fish of 10 cm TL (about 5 months old) maintained positions near banks, in low velocity areas (< 25 cm s^{-1}) with fine sediment, while feeding on Oligochaetae (Acolas et al. 2009). However, very few fish were sampled in the river, and the techniques for sampling young fish in large rivers (width >300 m) in a non-invasive way still need to be improved (see Chap. 29). Laboratory experiments carried out on the same fish size confirm the preference for fine substrate. When different substrates were offered, individuals much preferred sand (0.8–1.4 mm) rather than gravel (8–16 mm) or stones (50–95 mm) (Charles et al. 2009). These authors also demonstrate that activity increases at night, which could be considered as anti-predatory behaviour of high importance, which fish have to acquire before being stocked. A higher level of activity at night has also been described in 1-year-old fish (Staaks et al. 1999). To what extent this behaviour is influenced by visibility or turbidity of the water body as well as water depth is still unknown.

Within the frame of stocking programs, data on the habitat preferences of juveniles are urgently needed to determine the optimal stocking site for their development. Moreover, knowledge of their habitat preferences, in terms of water current for example, would help to train fish before stocking. To answer these questions, both in situ and enclosure experiments are needed, and a comparative approach between the two species could also help to determine the adaptive significance of their behaviour.

44.3 Downstream Migration Patterns of Juveniles

Downstream migration from the river to the sea has not been documented in full detail. Up to now, knowledge of the duration that fish spend in the river before reaching the estuary, as well as their downstream migration patterns, is extremely limited.

In addition to the river sampling presented above, recent telemetry experiments on *A. sturio* were implemented in the Gironde (Fig. 44.1) (Acolas, unpublished data) and the Elbe (Fig. 44.1) (Gessner et al. 2010) to understand downstream migration from the river to the estuary and assess the impacting factors. Fish younger than 2 years were tagged (acoustic transmitters), and their movements were monitored between 1 and 6 months. The analysis of these data will provide a first description of downstream migration patterns. In addition, fishermen's declarations and a sampling survey in the Gironde estuary will help to understand fish movement thanks to individual tagging (PIT tags and external tags) (see Chap. 29).

44.3.1 Downstream Migration in Rivers

In the Odra River basin (Fig. 44.1), for instance, first data on *A. oxyrinchus* fingerlings (2.5–4.5 cm in total length) reveal a crepuscular to nocturnal migration pattern in rather shallow, clear water rivers. Downstream migration is very slow, not reaching 4 km in 3 days if suitable habitat is available. Main migration was observed after sunset and until 1 h after sunrise, while only very few fish were observed in the drift during the day. Similar patterns were observed in telemetry studies for fish of 25–65 cm TL in the same area. Those fish revealed migrations of up to 20 km downstream at night, whereas daytime was mainly used for very short distance migrations of up to 500 m (Fredrich et al. 2008). During night time, in general, no upstream migrations were observed in the river sections. This behaviour changed when the fish reached the lower part of the river, where current is reduced due to wind drift and increased river width. In these areas, limited upstream movements and daytime activity were observed (Gessner et al. 2008). From the telemetry as well as from the fisheries data, there are considerable inter-individual differences in migration speed and pattern (Gessner et al. 2008). Some fish leave the river within 10 days to migrate into coastal waters (fish with TL >40 cm), and some stay in the lower river section for more than 2 years (fish up to 92 cm TL).

In the Elbe River (Fig. 44.1), *A. sturio* were released in the middle reach, and downstream migration from the release site was rapid at speeds of up to 40 km/day, and initially also took place during daytime. As for *A. oxyrinchus*, in the lower area of the river with reduced current, migration slowed and alternating upstream and downstream movements of up to 2 km took place (Gessner et al. 2010).

44.3.2 Downstream Migration in Tidal Area

In the Dordogne River, 10 cm TL *A. sturio* were captured downstream of their release site (9–20 km) about 22 days after having being stocked, which is within the range of the distance covered by *A. oxyrinchus* in non-tidal rivers. Moreover, thanks to reports from fishermen in the Dordogne River, fish between 10 and 25 cm TL were localised 35 to 55 km downstream of the release site between 3 and 6 months after stocking. These data suggest that fish frequent the freshwater tidal area for at least 6 months before reaching the estuary, but this hypothesis has still to be confirmed.

In the Gironde, 94 *A. sturio* were tagged in 2008–2009, and their movements were recorded between 1 and 6 months over an area of 270 km in length (freshwater and saline estuary) (Acolas, unpublished data). Preliminary results highlight that several patterns of downstream migration exist (straight and rapid downstream migration, downstream migration with lots of back and forth movements, settlement in the estuary and settlement in freshwater), and that daytime does not seem to influence displacements. The influence of tide and home range characterisation are currently being tested on this data set.

During a telemetry study, one *A. sturio* of 35 cm TL was actively tracked in freshwater tidal waters of the Elbe River, upstream of the city of Hamburg. This individual's movements predominantly followed the main current patterns making use of the tide and utilizing the available habitat over 10 days before the fish became inaccessible in the Hamburg harbour area (Gessner et al. 2010). These preliminary results have been verified in additional telemetry experiments, also addressing the impact of oxygen deficiency in the harbour on the migration pattern. In these habitats, comprising deep and rather turbid waters with high suspended sediment load, differences between daytimes still exist, as observed in clear rivers, but are less pronounced.

44.3.3 Outmigration to Sea

To date, only incidental captures have provided information about the fishes' migration to sea (Rochard et al. 1997). These data are crucial in understanding the ecology at sea but they are incomplete, as they are limited to occasional reports from the main fishing areas, and in addition they depend on voluntary declarations.

First, it is important to evaluate movements between the sea and the estuary, focusing on depth, salinity, temperature, and substrate preferences. This would allow a characterization of essential coastal habitats. It is therefore, planned to equip fish with data storage tags (DST), in both the Gironde and Elbe estuaries, to record temperature, depth and salinity that the fish encounter. With these data, it should be possible to assess movements between estuary and sea. The main drawback is the fact that the DSTs are expensive, the fish have to be caught and the storage tag is to be removed by the fisherman. To further refine tracking technology, pop-up tags (i.e., archival tags permitting satellite transmission of the data) could be applied, in order to characterise fish trajectory at sea without being dependent upon the fish being caught.

To answer questions concerning the criteria for habitat selection and migration stimuli, more detailed individual movement analyses are required, in conjunction with habitat assessments and experimental work to assess overall denominators (Brown et al. 2005). These data will help to understand migration routes (are they common to many individuals or is there individual variability?) and to identify essential habitat (areas that should be protected, for instance).

44.4 Environmental Factors Influencing Sturgeon Migrations

It might be postulated that several drivers of animal behaviour interact, and that the resulting behaviour can be considered as a "personality"-based integration of anticipated risks and opportunities. Among the triggers of migration behaviour, food availability, preferences for temperature, oxygen, and salinity as well as depth, thigmotaxis, and illumination, which are involved with predator avoidance, are often mentioned in the literature of migratory species (McCleave et al. 1984). Depending on the "personality" of the individual, the resulting migration occurs with variable speed and distances (Stamps 2007). Finally, habitat choice is the result of the integration of the various factors impacting with the personality of the fish (Jones and Godin 2009). Nevertheless, individual variability in the different responses makes it difficult to clearly determine essential traits and threshold levels for habitat selection.

The environmental factors that account for fish leaving or using a specific habitat are largely associated with the developmental stage of the individual. During embryogenesis, important developmental requirements are the availability of prey, the avoidance of predators, and suitable oxygen and temperature ranges, as well as the dilution of metabolic end products and the avoidance of bacterial and fungal or toxic contamination. The provision of these needs would reflect the functional integrity of the habitat. The optimum way to provide these characteristics would be to use substrate with large crevices and low load (e.g., gravel banks with no siltation). Similar conditions should be available for the development of the eleutheroembryos. To reduce predation risk, and fungal and bacterial pressure through decaying eggs, it becomes necessary to leave the spawning site for a suitable substrate, swimming with the current (swimming against the current is not an option due to physical limitations). After the yolk sac has been resorbed, feed uptake and predation avoidance contribute equally to habitat choice. Also, since these life phases are unable to regulate their ion balance, habitats have to be located in freshwater. But what trigger dominates behaviour at this stage or at subsequent stages, and what is considered suitable by the individual?

Considering the drivers mentioned, the question as to when it is beneficial to change habitat remains unanswered. For example, the tolerance of osmotic pressure does increase gradually with size. The other drivers certainly include feed availability, oxygen, temperature, and substrate preferences in order to optimise energy budget.

44.5 Conclusion

In addition to these in situ experiments planned for the near future, it is still important to document current estuarine survival and occurrence, in order to assess stocking efficiency and to compare current locations (after 2010) with past locations (1981–2000). Moreover, reproduction migration is largely unknown for *A. sturio,* although some attempts have been made to verify adult utilisation of spawning grounds (Taverny and Piefort 2002), and there have been studies on other sturgeon species (Paragamian et al. 2002; Benson et al. 2007; Heublein et al. 2009). The results might be of special importance, not only in confirming the utilisation of potential spawning grounds but also to determine the key drivers for migration and maturation in river habitats. This knowledge could also be used to discriminate ecological adaptations to specific sites and thereby optimize in situ stock reproduction strategy. Since in situ experiments suggest that marine water is not essential to reach sexual maturity (Gessner, unpublished data), other drivers must be involved in reproduction migration. Here also the determination of genetic differences should be taken into consideration. Considering the population size and the age structure of the remaining population, this is a more long-term action to consider. Additionally, a comparative approach to *A. sturio* and *A. oxyrinchus* life histories would help our understanding of the relative occurrence of these two species in European basins.

References

Acolas ML, Rouleau E, Roqueplo C, Le Barh R, Rochard E (2009) Action n°10 Localisation et caractérisation des habitats fluviaux fréquentés par les jeunes. In: Rochard E (ed) Programme de recherche et de conservation de l'esturgeon européen *Acipenser sturio* ; bilan scientifique et technique 2008. Cemagref de Bordeaux, étude n 133, pp 64–78

Benson RL, Turo S, McCovey BW (2007) Migration and movement patterns of green sturgeon (*Acipenser medirostris*) in the Klamath and Trinity rivers, California. USA Environ Biol Fish 79(3–4):269–279

Brosse L (2003) Caractérisation des habitats des juvéniles d'esturgeon européen, *Acipenser sturio*, dans l'estuaire de la Gironde. Doctorat, Université Paul Sabatié, Toulouse

Brown C, Jones F, Braithwaite V (2005) In situ examination of boldness–shyness traits in the tropical poeciliid, *Brachyraphis episcopi*. Anim Behav 70:1003–1009

Charles K, Roqueplo C, Jatteau P (2009) Action n°9: identification expérimentale des préférences d'habitat des jeunes stades. In: Rochard E (ed) Programme de recherche et de conservation de l'esturgeon européen Acipenser sturio; bilan scientifique et technique 2008. Cemagref de Bordeaux, étude n°133, pp 54–64

Fredrich F, Kapusta A, Ebert M, Duda A, Gessner J (2008) Migratory behavior of young sturgeon, *Acipenser oxyrinchus* Mitchill, in the Oder River drainage. Preliminary results of a radio telemetric study in the Drawa River, Poland. Pol Arch Fish 16(2):105–117

Fuiman LA, Meekan MG, McCormick MI (2010) Maladaptive behavior reinforces a recruitment bottleneck in newly settled fishes. Oecologia 164:99–108

Gessner J, Migdalska B, Tautenhahn M, Domagala J, Fredrich F, Bartel R (2008) Migration analysis of juvenile sturgeon (*Acipenser oxyrinchus*) in the Odra River catchment as determined by catch data. In: Kolman R, Kapusta A (eds) Actual status and active protection of sturgeon populations endangered by extinction. Wydawictnictwo Instytutu Rybactwa Srodalowengo, Olsztyn, Poland, pp 151–162

Gessner J, Kamerichs CM, Kloas W, Wuertz S (2009) Behavioural and physiological responses in early life phases of Atlantic sturgeon (*Acipenser oxyrinchus* Mitchill 1815) towards different substrates. J Appl Ichthyol 25:83–90

Gessner J, Fredrich F, Williot P, Kirschbaum F (2010) Preparatory measures and initial release as a prerequisite for the remediation of the European sturgeon, *Acipenser sturio*, in Germany. Bull Fish Biol 11(1/2):21–36

Heublein JC, Kelly JT, Crocker CE, Klimley AP, Lindley ST (2009) Migration of green sturgeon, *Acipenser medirostris*, in the Sacramento River. Environ Biol Fish 84(3):245–258

Jacobson H (1889) The artificial propagation of sturgeon in Schleswig-Holstein, Germany. Bulletin of the United States Fish Commission:81–90

Jatteau P (1998) Etude bibliographique des principales caractéristiques de l'écologie des larves d'Acipenséridés. Bull Fr Pêche Piscic 350–51:445–464

Jones KA, Godin J-GJ (2009) Are fast explorers slow reactors? Linking personality type and anti-predator behaviour. Proc Royal Soc B 277(1681):625–632

McCleave JD, Arnold GP, Dodson JJ, Neill WH (1984) Mechanisms of migration in fishes. Plenum, New York

McCormick MI, Meekan MG (2010) The importance of attitude: the influence of behaviour on survival at an ontogenetic boundary. Mar Ecol Prog Ser 407:173–185

Paragamian VL, Wakkinen VD, Kruse G (2002) Spawning locations and movement of Kootenai River white sturgeon. J Appl Ichthyol 18(4–6):608–616

Rochard E, Lepage M, Meauzé L (1997) Identification et caractérisation de l'aire de répartition marine de l'esturgeon européen *Acipenser sturio* à partir de déclarations de captures. Aquat Living Resour 10(2):101–109

Rochard E, Lepage M, Dumont P, Tremblay S, Gazeau C (2001) Downstream migration of juvenile European sturgeon *Acipenser sturio* L. in the Gironde Estuary. Estuaries 24(1):108–115

Staaks G, Kirschbaum F, Williot P (1999) Experimental studies on thermal behaviour and diurnal activity rhythms of juvenile European sturgeon (*Acipenser sturio*). J Appl Ichthyol-Z Angew Ichthyol 15(4–5):243–247

Stamps JA (2007) Growth–mortality tradeoffs and personality traits in animals. Ecol Lett 10:355–363

Taverny C, Piefort S (2002) Validation des zones de fréquentation des zones de frayères par *Acipenser sturio*. Etude Cemagref n°80. In: Rochard E (ed) Restauration de l'esturgeon européen Acipenser sturio. Rapport scientifique Contrat LIFE n°B-3200/98/460. Groupement de Bordeaux, pp 65–78

Chapter 45
Potential Effects of Forthcoming Climate Change and Biological Introductions on the Restoration of the European Sturgeon

Géraldine Lassalle, M. Béguer, and E. Rochard

Abstract The restoration of threatened species is taking place in a changing global environment, creating additional problems for stakeholders and decision-makers to face. Here, the potential influences on the restoration measures developed for the European sturgeon (*Acipenser sturio* L., 1758) of both climate change and, to a lesser extent, biological introductions are assessed and discussed. Both evaluations were based on intensive literature searches and species distribution models. Most of the former European sturgeon distribution range was calculated as being strongly affected by global warming, especially in basins along its southern margin. Five basins where re-introductions are envisaged were projected to remain highly suitable for the species under climate change assumptions. In addition, some large French basins appeared to be suitable for other sturgeon species that share many ecological similarities with *A. sturio*.

45.1 Introduction

In the context of the changing global environment, the restoration of threatened species is a complex process involving interdisciplinary and multi-scale analysis as well as long-term perspectives. On the one hand, recent climate change has

---

G. Lassalle (✉)
Cemagref, Estuarine Ecosystems and Diadromous Fish Research Unit, 50 Avenue de Verdun, 33612 Cestas Cedex, France

UMR 6250 LIENSs, University of La Rochelle – CNRS, Institut du Littoral et de l'Environnement, 2 rue Olympe de Gouges, 17000 La Rochelle, France
e-mail: geraldine.lassalle@univ-lr.fr

M. Béguer • E. Rochard
Cemagref, Estuarine Ecosystems and Diadromous Fish Research Unit, 50 Avenue de Verdun, 33612 Cestas Cedex, France
e-mail: melanie.beguer@cemagref.fr; eric.rochard@cemagref.fr

P. Williot et al. (eds.), *Biology and Conservation of the European Sturgeon Acipenser sturio L. 1758*, DOI 10.1007/978-3-642-20611-5_45,

been demonstrated to have significant effects on both plant and animal species (Parmesan and Yohe 2003; Walther et al. 2002). Consequently, the future distribution of various organisms was assessed using different temperature rise scenarios for the coming decades, e.g., trees (Thuiller 2003), amphibians (Araujo et al. 2006), and freshwater fishes (Buisson et al. 2008). When European diadromous fishes were assessed in their turn, they appeared particularly sensitive to climatic disruption in the surrounding environment (Lassalle et al. 2009). It has been calculated that in the near future, a large proportion of the hydrographic basins where a species is or was known to occur will decrease in favourability or become totally unsuitable for the given species. Even with the wide spectrum of uncertainties inherent in modelling species distribution, these results could be useful to stakeholders and decision-makers to enable them to take more informed management measures. They should be regarded as the first step in a decision-making process combining information on local habitat quality and on the social perception of restoration efforts (Lassalle et al. 2010).

On the other hand, current and future global change could raise some other questions on the restoration of threatened species. One major issue concerns the potential interactions with allochthonous species that could proliferate in the niches left vacant by the autochthonous endangered taxa or compete with the remaining weakened populations, leading in some cases to competitive exclusion (Bohn et al. 2008). Most of the time, non-native fish species were introduced accidentally, or intentionally through recreational and commercial activities such as aquaculture and pet fish industries. Nevertheless, they can also colonize new places that become suitable following human-induced environmental modifications. This latter case is illustrated when species extend their range dozens of kilometres, mainly northward, due to current global warming (Thomas and Lennon 1999). In this context, several anadromous and non-anadromous sturgeon species coexist with the European sturgeon (*Acipenser sturio* L., 1758) in the Eastern part of its distribution range (Holcik et al. 1989). Many of these species are currently farmed in Western European countries, e.g., *A. baerii* in France, *Huso huso* in Italy or even *A. gueldenstaedtii* in Belgium and Netherlands (Williot et al. 2009). In 1999, farmed specimens of *A. baerii* accidentally escaped into the Gironde system, survived and have occasionally been recaptured by fishermen (Maury-Brachet et al. 2008). Similarly, *A. gueldenstaedtii* individuals, originally limited to the Ponto-Caspian region, were regularly caught along the shores of the North and Baltic seas after escapes in the Oder estuary (Gessner et al. 1999; Arndt et al. 2002). No introduction of sturgeon in Europe was followed by the installation of a functional population in the foreign environment. Nevertheless, these examples suggest that there is a high risk of potentially interactive species being introduced into the area where practical measures are being undertaken to restore European sturgeon. Such introductions could expose the protected species to competition for habitat and food resources, and to new parasites and diseases, and could reduce its reproductive success because of interbreeding and the subsequent production of fertile or unfertile hybrids. It has not yet been proven that there have been deleterious effects on the European sturgeon population following the introduction of non-indigenous counterparts, but the potential must be recognised and evaluated.

Consequently, in the context of the conservation and restoration of the European sturgeon at European level (Rosenthal et al. 2007), the aim of this chapter is to discuss re-introduction opportunities in the light of two major environmental changes that could be linked, i.e., climate change and to a lesser extent biological introductions. Firstly, potential future habitat favourability for *A. sturio* was assessed under climate change assumptions over its entire former distribution range. Secondly, the potential habitat favourability of French basins was estimated for three non-native sturgeon species, which could be of interest for aquaculture diversification (Williot and Rochard 2007). It seemed particularly relevant to consider France, because this country hosts the last functional Atlantic sturgeon population. Both sub-sections were based on a large-scale modelling approach.

45.2 Material and Methods

45.2.1 Future Habitat favourability for A. sturio *in a Changing Global Environment*

Details of the material and methods are given in Lassalle et al. (2010). A three-step procedure was applied as follows: (1) the native 1750–1850 distribution of the species was assessed, as described extensively in Chap. 6, (2) the distribution was then modelled using biogeographical, topographical and climatic variables such as temperature and precipitations, and (3) the distribution was finally projected using potential future climatic conditions as new inputs into the niche model. Projections were performed at two time-steps which are biologically relevant for the European sturgeon: 2020–2049 and 2070–2099. In accordance with the view of French meteorologists and the recent evaluation of greenhouse gas concentrations (www.esrl.noaa.gov/gmd/ccgg/trends), the A2 and A1FI (Fossil Intensive) emission scenarios were retained (IPCC 2000). Potential distribution maps were produced showing the probable suitability for each basin in the new climatic conditions. Three probability classes were defined as follows: low, less than 0.15; medium, between 0.15 and 0.50; high, greater than 0.50.

45.2.2 Potential Habitat favourability for Three Non-Native Sturgeon Species

The species for consideration were selected according to four criteria: (1) fish farmers are interested to introduce these species, (2) their realized niche overlaps with that of the European sturgeon, indicating ecological similarities between taxa,

(3) they had been recorded several times outside their native range, i.e., in the present study, in France and/or in neighbouring countries, and (4) their distribution is clearly delineated and precise information about it can be gathered. The fourth criterion is a prerequisite to niche-based modelling. However, since the range margins and biogeographical history of the sister species, the Atlantic sturgeon (*Acipenser oxyrinchus* Mitchill, 1815), are still being revised, e.g., remains were recently discovered in French sites (Desse-Berset 2009), this species was not taken into account here. The three species finally retained were *A. gueldenstaedtii*, *A. stellatus* and *H. huso*, for which distributional boundaries are fairly well-defined.

A three-step procedure was applied as follows: (1) native distribution for the end of the nineteenth century was established over 34 basins flowing into the Black, Azov and Caspian Seas for the first two species, and over 52 basins including systems ending into the Ionian, Aegean and Adriatic Seas for the Beluga (information was obtained through literary searches), (2) distribution was then modelled using a common set of variables describing the abiotic and biotic environment, and (3) the model was finally applied to 39 French basins that cover the entire national territory. Potential distribution maps were produced, and depict, for each French basin, its probable suitability for the investigated species. Full details of the statistical technique are given in Béguer et al. (2007), and the general methodology is presented in Appendix 10 by Williot and Rochard (2007).

45.3 Results and Discussion

45.3.1 Future Habitat favourability for A. sturio *in a Changing Global Environment*

It is important to note that the niche model had statistical performances sufficient to allow projections into a new spatio-temporal frame, i.e., in this case, a new climatic context. Added to this, it integrated two general climatic descriptors, i.e., air temperature and precipitation, underlining the key role of climate in determining the distribution of this species. Moreover, the corresponding response curves are in accordance with the species' thermal preferences and hydrological habitat requirements (see Lassalle et al. 2010 for illustrations).

For the period "2020–2049", less than 55% of basins where the sturgeon historically reproduced (called latter spawning basins and indicated with a red scale in Fig. 45.1) have a high probability of being suitable for the species, i.e., a value greater than 0.5 (Fig. 45.1a). For the end of the twenty-first century, this number decreases sharply to 21%, and the mean probability decreases by 16% (Fig. 45.1b). Suitability values are low, lower than the threshold of 0.15, for the spawning basins at the southern distributional edge, i.e., from West to East: the Tiber (Italy), Ebro (Spain), Rhône (France), Adige (Italy), Buna (Albania), Inguri (Georgia) and Rioni (Georgia) basins. In contrast, five basins bordering the North

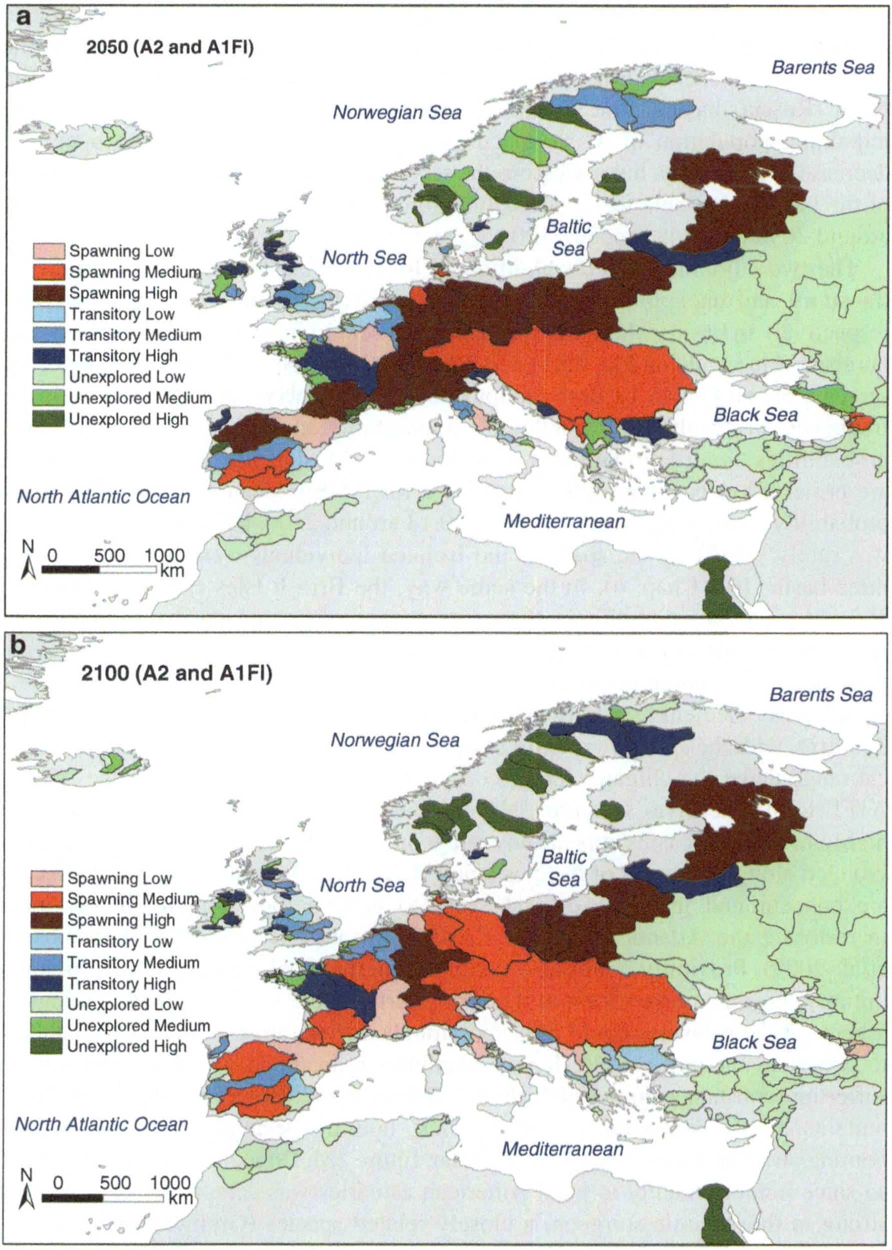

Fig. 45.1 Mean probabilities of providing potential future basin suitability for the European sturgeon, using final niche model and A2 and A1FI (Fossil Intensive) climate change scenarios for middle [(**a**), 2050] and end of the twenty-first century [(**b**), 2100]. The three functional groups of basins are shown with different colour scales, with the *darker shade* corresponding to higher probability. Spawning, transitory and unexplored basins are in *red*, *blue* and *green* respectively

and Baltic seas remain particularly favourable throughout the century, with a mean probability of 0.69 ± 0.04 around 2100, i.e., from West to East: the Rhine (Netherlands), Oder (Germany/Poland), Vistula (Poland), Neman (Lithuania) and Neva (Russia) basins. The Gironde–Garonne–Dordogne system, where the last remaining population of *A. sturio* currently survives, is calculated to operate a decrease in suitability, but not before the second half of the century. The probability of the basin being able to support a functional population is calculated to be 0.87 around 2050 and 0.48 around 2100.

The two adjoining patterns highlighted for spawning basins are also noted for transitory and unexplored basins. In the historical distribution, they correspond respectively to basins where the species was observed infrequently and in very low numbers (blue scale on Fig. 45.1) and basins where the species was never recorded (green scale on Fig. 45.1). Both groups of basins cited above show a decrease in the frequency of "high" and "medium" favourability classes and a shift of high probabilities to northern latitudes (Fig. 45.1a, b). Most of the Scandinavian basins are projected to become highly favourable for the European sturgeon. The mean probability over the peninsula is 0.80 ± 0.14 around 2100. Historically, this region was rarely visited by the species, and isolated individuals were recorded in only three basins (see Chap. 6). In the same way, the British Isles constitute the third zone of great favourability in the context of climatic upheaval. Originally, no spawning shoals entered these basins, only regular single individuals straying from nearby spawning populations (see Chap. 6).

All these elements strongly suggest a northward shift of suitable habitats for *A. sturio*, with the species virtually disappearing from southern parts of the historical range under the climate change conditions calculated according to the A2 and A1FI (fossil intensive) scenarios. Nevertheless, the future spatial configuration of highly favourable basins appears advantageous for the species, i.e., they are mainly grouped along the shores of the North and Baltic seas. Indeed, most of these basins are concentrated in countries that seem to have a long-standing involvement in restoring the Atlantic sturgeon (Kirschbaum and Gessner 2000; Kirschbaum et al. 2009). Proximity could also facilitate international agreement and common initiatives for the preservation of *A. sturio*. Furthermore, considering the biological aspects, these measures could offer individuals a wide range of alternative habitats to complete their life cycle. It is particularly relevant in a global environment suffering constant abiotic and biotic disturbances. However, this hypothesis remains highly dependent on the species' homing behaviour and the way its homing/straying ratio evolves in the near future (McDowall 2001), all the more so since homing fidelity to natal American estuaries was demonstrated to be very strong in the Atlantic sturgeon, a closely related species (Grunwald et al. 2008). However, one major drawback is the geographical location of this hotspot of favourability. It greatly overlaps the reintroduction area of the Atlantic sturgeon, which has been released in both the Oder and Vistula basins (see Chaps. 42 and 44). From a practical point of view, restoration of two different sturgeon species in the same province could blur the conservation message, and from a biological point of view it could lead to various interferences (genetic, epidemiological, ecological etc) that might make the action plan even more complex. Indeed, in past history, extensive hybridization between the two targeted species was strongly suspected in the Baltic Sea region (Tiedemann et al. 2007).

45.3.2 Potential Habitat Favourability for Three Non-Native Sturgeon Species

For the three species investigated, 95–100% of the presences recorded in the Ponto-Caspian and Mediterranean biogeographical regions are correctly fitted by the models. When applied to 39 French basins, the models indicate that a large majority would be favourable for these non-native sturgeon species, the Adour, Garonne, Loire, Rhône and Seine basins being the most suitable (Fig. 45.2). The lowest probabilities were observed in the small basins of Brittany and the Mediterranean region. Key conclusions emerging from this large-scale study are: (1) the homogeneous response pattern to the physical constraints of the environment shared by

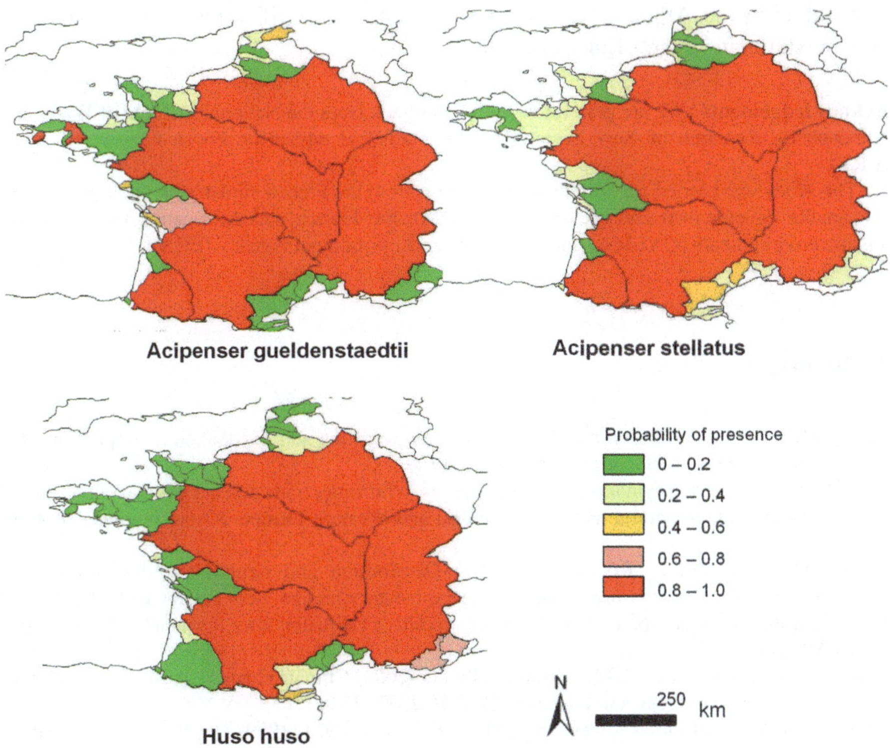

Fig. 45.2 Probabilities of providing potential French basin suitability for three non-native sturgeon species using the niche model based on their historical Ponto-Caspian and Mediterranean distribution

sturgeon species, and (2) the extent to which the general characteristics of large French basins are suitable for these potential exotic species. These new insights strengthen the need for further measures to reduce deliberate releases by anglers and aquarists and also accidental releases or escapements from fish farms.

45.4 Conclusion

This study brings concrete elements to the plan for the conservation and restoration of the European sturgeon, where both climate change and introductions of allochthonous species are considered as risks and threats for the natural *A. sturio* population (see Sects. 4.5 and 4.6 of Rosenthal et al. 2007). Following the present results, two points in particular should be assessed to further develop an integrated recovery process for this species: (1) scenarios or general evolution trends for the major causes of rarefaction such as damming, fishing activities and pollution, and (2) a specific evaluation of the potential consequences of a massive re-introduction of the natural ecological counterpart, the Atlantic sturgeon. As already mentioned, reintroducing the Atlantic sturgeon will depend on clarification of its population genetic status in several European sites.

Acknowledgements We are grateful to John Leathwick from the National Institute of Water and Atmospheric Research of New Zealand for his statistical advice on boosted regression trees (BRT).

This study was carried out with financial support from the French ministry for ecology, energy, sustainable development and the sea (MEEDDM), the French national agency for water and aquatic environments (ONEMA) and the European Environment Agency (EEA).

References

Araujo MB, Thuiller W, Pearson RG (2006) Climate warming and the decline of amphibians and reptiles in Europe. J Biogeogr 33(10):1712–1728

Arndt G-M, Gessner J, Raymakers C (2002) Trends in farming, trade and occurrence of native and exotic sturgeons in natural habitats in Central and Western Europe. J Appl Ichthyol 18(4–6): 444–448

Béguer M, Beaulaton L, Rochard E (2007) Distribution and richness of diadromous fish assemblages in Western Europe: large scale explanatory factors. Ecol Freshw Fish 16:221–237

Bohn T, Amundsen PA, Sparrow A (2008) Competitive exclusion after invasion? Biol Invasions 10:359–368

Buisson L, Thuiller W, Lek S, Lim P, Grenouillet G (2008) Climate change hastens the turnover of stream fish assemblages. Glob Change Biol 14:2232–2248

Gessner J, Debus L, Filipiak J, Spratte S, Skora KE, Arndt GM (1999) Development of sturgeon catches in German and adjacent waters since 1980. J Appl Ichthyol 15:136–141

Grunwald C, Maceda L, Waldman J, Stabile J, Wirgin I (2008) Conservation of Atlantic sturgeon *Acipenser oxyrinchus oxyrinchus*: delineation of stock structure and distinct population segments. Conserv Genet 9(5):1111–1124. doi:10.1007/s10592-007-9420-1

Holcik J, Kinzelbach R, Sokolov LI, Vassilev VP (1989) *Acipenser sturio* Linnaeus, 1758. In: Holcik J (ed) The freshwater fishes of Europe Part II. General introduction to fishes, Acipenseriformes, vol 1. Aula Verlag, Wiesbaden, pp 367–394

IPCC (2000) Emissions scenarios. Special Reports of the IPCC. Cambridge University Press, Cambridge, UK

Kirschbaum F, Gessner J (2000) Re-establishment programme for *Acipenser sturio* L., 1758: The German approach. Bol Inst Esp Oceanogr 16(1–4):149–156

Kirschbaum F, Wuertz S, Williot P, Tiedemann R, Arndt G-M, Anders E, Krüger A, Bartel R, Gessner J (2009) Prerequisites for the restoration of the European Atlantic sturgeon, *Acipenser sturio* and the Baltic sturgeon (*A. oxyrinchus* × *A. sturio*) in Germany. In: Carmona R, Domezain A, Garcia Gallego M, Hernando-Casal J, Rodriguez F, Ruiz Rejon M (eds) Biology, conservation and sustainable development of sturgeons, Fish & fisheries series. Springer, Berlin, pp 385–401

Lassalle G, Béguer M, Beaulaton L, Rochard E (2009) Learning from the past to predict the future: responses of European diadromous fish to climate change. In: Haro AJ, Smith KL, Rulifson RA et al (eds) Challenges for diadromous fishes in a dynamic global environment, vol 69, AFS Symposium. American Fisheries Society, Bethesda, MD, pp 175–193

Lassalle G, Crouzet P, Gessner J, Rochard E (2010) Global warming impacts and conservation responses for the critically endangered European Atlantic sturgeon. Biol Conserv 143: 2441–2452

Maury-Brachet R, Rochard E, Durrieu G, Boudou A (2008) The "storm of the century" (December 1999) and the incidental escape of Siberian sturgeons (*Acipenser baerii*) in the Gironde estuary (SW France): an original bioindicator for metal contamination. Environ Sci Pollut Res 15(1): 89–94

McDowall RM (2001) Anadromy and homing: two life-history traits with adaptive synergies in salmonid fishes? Fish Fish 2(1):78–85

Parmesan C, Yohe G (2003) A global coherent fingerprint of climate change impacts across natural systems. Nature 421:37–42

Rosenthal H, Bronzi P, Gessner J, Moreau D, Rochard E (2007) Action plan for the conservation and restoration of the European sturgeon. Council of Europe Publishing, Strasbourg, Nature and Environment N° 152

Thomas CD, Lennon JJ (1999) Birds extend their ranges northwards. Nature 399(6733):213

Thuiller W (2003) BIOMOD – optimizing predictions of species distributions and projecting potential future shifts under global change. Glob Change Biol 9:1353–1362

Tiedemann R, Moll K, Paulus KB, Scheer M, Williot P, Bartel R, Gessner J, Kirschbaum F (2007) Atlantic sturgeons (*Acipenser sturio*, *Acipenser oxyrinchus*): American females successful in Europe. Naturwissenschaften 94:213–217

Walther GR, Post E, Convey P, Menzel A, Parmesan C, Beebee TJC, Fromentin JM, Hoegh-Guldberg O, Bairlein F (2002) Ecological responses to recent climate change. Nature 416:389–395

Williot P, Rochard E (2007) Biologie, exigences environnementales et éléments d'évaluation à priori du risque relatifs à plusieurs espèces d'esturgeon susceptibles d'être élevées en pisciculture en France: phases 1 and 2. Etude Cemagref-EPBX pour le Ministère de l'Environnement et du Développement durable

Williot P, Rochard E, Kirschbaum F (2009) Acceptability and prerequisites for the successful introduction of sturgeon species. In: Carmona R, Domezain A, Garcia Gallego M, Hernando-Casal J, Rodriguez F, Ruiz Rejon M (eds) Biology, conservation and sustainable development of sturgeons, Fish & fisheries series. Springer, Berlin, pp 369–84

Chapter 46
Population Viability Analysis of the European Sturgeon (*Acipenser sturio* L.) from the Gironde Estuary System

Ivan Jarić, J. Knežević-Jarić, G. Cvijanović, and Mirjana Lenhardt

Abstract This study presents a population viability analysis (PVA) conducted on the Gironde population of *Acipenser sturio* using the Vortex software package. As identified by the model, the most important objects of the future research efforts related to *A. sturio* life history should be the determination of the mean fecundity, age at which females reach maturity, female spawning frequency and population sex ratio, as well as the intrinsic population growth rate and the age-specific natural mortality, especially of the youngest age classes. The model has confirmed a high population susceptibility to unsustainable fishery, and a slow recovery potential that can span over a number of decades. The detected lag between the stocking activities and the initiation of the actual population recovery should be taken into account in the planning of restoration projects. PVA should be recognized as an important tool, and integrated within future research, management and policy development efforts.

I. Jarić (✉)
Institute for Multidisciplinary Research, Kneza Višeslava 1, 11000 Belgrade, Serbia
e-mail: ijaric@imsi.rs

J. Knežević-Jarić
Ecological Society "Endemit", Oračka 42, 11080 Belgrade, Serbia
e-mail: jelak@isp.b92.net

G. Cvijanović
Institute for Multidisciplinary Research, Kneza Višeslava 1, 11000 Belgrade, Serbia
e-mail: mitrandir@imsi.rs

M. Lenhardt
Institute for Biological Research, Despota Stefana 142, 11000 Belgrade, Serbia
e-mail: lenhardt@ibiss.bg.ac.rs

P. Williot et al. (eds.), *Biology and Conservation of the European Sturgeon Acipenser sturio L. 1758*, DOI 10.1007/978-3-642-20611-5_46,

46.1 Introduction

European sturgeon (*Acipenser sturio* L., 1758) is considered to be the most endangered sturgeon species in the world (Paaver 1999). It was once one of the most widely spread sturgeon species, but due to a number of different anthropogenic impacts, such as overfishing and pollution, it became extirpated from the majority of the European watersheds (Williot et al. 2002, 2009a; Jarić et al. 2009a). It is believed that the only remaining populations are the ones inhabiting the Gironde–Garonne–Dordogne Basin in France and the Rioni River in Georgia (Williot et al. 2002). Since 1982, following the collapse of *A. sturio* stocks in the Gironde Estuary, it has become a fully protected species in France (Trouvery et al. 1984; Williot et al. 2009a). The population, however, continued to decline, mainly due to by-catch mortality (Rochard et al. 1997). These negative impacts have been further aggravated by the lack of knowledge concerning its demography and life history.

Due to the above-described problems, the development of population models that are able to assess viability and the extinction risks of a population could be a tool that would bridge the existing gaps in understanding of sturgeon ecology and life history (Jarić et al. 2010). Population viability analysis (PVA) represents a method that is able to cope with uncertainties present in ecological data, and it is considered to be an indispensable tool in conservation biology (Boyce 1992; Akçakaya and Sjögren-Gulve 2000). PVA program packages have become very popular, since they are able to provide sufficient reliability and replication ability, as well as having low programming skill requirements. Vortex simulation software has been routinely used by the World Conservation Union (IUCN) to establish quantitative classification of endangered species (Brook et al. 1997).

Within the present study, we developed a model of the Gironde *A. sturio* population by means of the Vortex software package. We assessed the sensitivity of the model with regard to changes in different life-history parameters, and also conducted simulations using different fishing pressures and stocking dynamics, to assess patterns of their influence on population viability. Outcomes of the present study could be important both for the scientific community and for *A. sturio* management and policy development authorities.

46.2 Model Parameterization and Scenario Development

Population viability analysis of the *A. sturio* population from the Gironde Estuary system was conducted using the Vortex software package (version 9.72). The package consists of a computer model that runs an individual-based Monte Carlo simulation of a population through the action of deterministic factors and environmental, demographic, and genetic stochasticity (Lacy 1993). For a more detailed description of Vortex, see Lacy (1993, 2000) and Miller and Lacy (2005).

As a first step in model parameterization, a literature survey was conducted in order to define the population parameter values that are necessary for scenario-building in Vortex. For parameters that have been provided by different authors as ranges of values the most probable mean value was determined, and the ranges were included in the model as an environmental variation. If the ranges provided by different authors proved to be inconsistent, the values that were given in the majority of the publications were applied, and the other values were included in the sensitivity analysis. The values of all life history parameters used in the model are presented in Table 46.1.

In general, the available data on the deterministic growth rate (the *r* value) and the natural mortality for sturgeon populations are very poor. Based on the limited available information, we applied a number of different population growth rates ranging between 0.05 and 0.15 (see Jarić et al. 2010). Other authors have applied either a uniform mortality in their population models across all age classes following the age 0+ (Jager 2005, 2006; Heppell 2007; Beamesderfer et al. 2007; Kennedy and Sutton 2007; Doukakis et al. 2010), or fitted the mortality to reach the expected population growth, by placing a higher mortality on the younger age classes and significantly lower mortality on adults (Jarić et al. 2010). We applied both approaches in this study, but placed a greater emphasis on the latter approach. This decision was made because the former one was shown to be less ecologically realistic when applied in Vortex. Namely, a uniform distribution of age-specific mortality allowed the existence of only a small number of adults in a population, and prevented them from reaching the old ages that have been reported for this species.

To define a starting point for the determination of age-specific natural mortality, we used Rikhter and Efanov's method for the estimation of natural mortality for fish species in temperate regions (FISAT II 2000), which requires age at maturity of females as the input variable. This approach yielded 3.8% as a natural mortality value. Thereafter, we applied three different population growth rates (i.e., 0.05, 0.10 and 0.15) and fitted the age-specific natural mortality in such a way that each of the *r* values was met in different scenarios. We assumed that the younger age groups are characterized by higher annual variability in natural mortality, which decreases towards the adult age groups. We have applied the general rule for the standard deviation of the natural mortality of each of the groups for ages 1–3 to be equal to 40% of the natural mortality of those age groups. With regard to the age groups from age 4 to adulthood it was equal to 25%, while for adults it was 10% of the age-specific natural mortality. However, if the mean mortality of a certain age plus two standard deviations resulted in a higher value than 100 (mean + 2x SD >100), the standard deviation was reduced to the level where the outcome was equal to 100. Variation in the natural mortality of the age 0–1 was incorporated through the standard deviation of the mean number of the offspring.

For the uniform mortality scenarios, we determined a starting value based on those applied in population models of sturgeon species that share similar life history with *A. sturio* (Beamesderfer et al. 2007; Heppell 2007). Thereafter, the value was modified to make it possible to reach the expected Gironde *A. sturio* life expectancy

Table 46.1 Population parameter estimates used in a Vortex simulation model of the European sturgeon from the Gironde Estuary

Variable	Value	Source
Age of first offspring for females	17	Magnin 1962 cited in Rochard et al. (1991), Magnin 1963 cited in Gessner et al. (2007)
Age of first offspring for males	14	Magnin 1962 cited in Rochard et al. (1991), Magnin 1963 cited in Gessner et al. (2007)
Maximum age of reproduction	40	Magnin 1962 cited in Williot personal communication
Maximum number of progeny per year	160[a]	–
Sex ratio at birth – in % males	50	–
Percent adult females breeding	29.15%	(Every 3–4 years) Williot personal communication
Environmental variation in % adult females breeding	2 percentage units	–
Mean number of offspring per year	70[b]	–
Standard deviation in number of offspring per year	10	–
Percent males in breeding pool	33.3%	(Every 3 years) Williot personal communication

Natural mortality fitted to the three different values of intrinsic population growth rate, and values of the uniform mortality applied:

Age	$r = 0.05$	$r = 0.10$	$r = 0.15$	Uniform mortality ($r = 0.129$)
age 0–1	0.0 ± 0.0[c]	0.0 ± 0.0[c]	0.0 ± 0.0[c]	0.0 ± 0.0[c]
age 1–2	85.0 ± 7.5	69.0 ± 15.5	40.0 ± 16.0	10.0 ± 2.5
age 2–3	60.0 ± 20.0	45.0 ± 18.0	30.0 ± 12.0	10.0 ± 2.5
age 3–4	35.0 ± 8.75	25.0 ± 6.25	20.0 ± 5.0	10.0 ± 2.5
age 4–5	10.0 ± 2.5	10.0 ± 2.5	10.0 ± 2.5	10.0 ± 2.5
age 5–17	3.8 ± 0.95	3.8 ± 0.95	3.8 ± 0.95	10.0 ± 2.5
age 17+	3.8 ± 0.38	3.8 ± 0.38	1.0 ± 0.1	10.0 ± 2.5

[a] $400{,}000 \times 0.0004$: maximum fecundity 400,000 (Williot personal communication); 0.0004 is the maximum natural survival of age 0–1 (Pine et al. 2001; Jager 2005, 2006; Bajer and Wildhaber 2007; Kennedy and Sutton 2007)

[b] $350{,}000 \times 0.0002$: average fecundity 300,000–400,000 (Williot personal communication); 0.0002 is the average natural survival of age 0–1 (Pine et al. 2001; Jager 2005, 2006; Bajer and Wildhaber 2007; Kennedy and Sutton 2007)

[c] Age 0–1 natural mortality is incorporated in the mean and maximum number of offspring; see model parameterization and scenario development for clarification

and, at the same time, to produce the r value that falls within the range 0.05–0.15. As a result, a uniform mortality rate of 0.10 was reached, resulting in an r value of 0.129. In this case, we applied uniform standard deviation of the natural mortality across all age classes.

The age 0–1 natural mortality was incorporated in the model through a proportional reduction of the number of offspring produced. As a result, the young-of-the-year that did not survive up to the age of 1 were not included in the simulation (see Miller and Lacy 2005). Therefore, it is important to note that in the presented results of the model, the population size (*N*) does not include embryos and young-of-the-year that do not survive up to the age of 1.

As proposed by a number of authors (Beissinger and Westphal 1998; Akçakaya 2000), the pseudo-extinction criterion was applied in this study, instead of scoring the risk of reaching absolute extinction, and the pseudo-extinction threshold was set to ten individuals. Sex ratio was set to be equal, and effects of an unequal sex ratio were tested in the sensitivity analysis. Age distribution at the beginning of each simulation was set to be stable, since the stochastic fluctuations in age distribution and year-class strength already represent an integral part of simulations in Vortex. The percentage of individuals of each sex that participate annually in spawning was established as $S = 100/I$, where I represents the time between two spawning migrations. For the sake of simplicity, as well as to reduce the number of assumptions already made in model parameterization, parameters such as density dependence in reproduction and the existence of catastrophic events were not included in the model. While demographic stochasticity is incorporated in the Vortex model, the problem of other Allee effects was partly resolved by introducing pseudo-extinction (Akçakaya 2000).

To increase the precision of the results, the number of iterations was set at 1,000 for all simulations and, occasionally, at 10,000. Excluding sensitivity analysis and the stocking scenarios, carrying capacity was set to be equal to the initial population size, (see Reed et al. 2003 for similar approach).

Four basic groups of scenarios were developed within this study: (1) assessment of the minimum viable population size (MVPs), (2) sensitivity analysis, (3) harvest scenarios, and (4) reintroduction and supportive stocking. The MVPs was assessed by carrying out a series of simulations with varying initial population size. The simulations covered 500 years, in order to assess long-term viability, while carrying capacity was retained at the level of the initial population size.

Sensitivity analysis was conducted to test the dependence of the model behaviour on changes in each of the life-history parameters, as well as to span the ranges of life-history parameter values provided by different authors. As stated by Miller and Lacy (2005), sensitivity analysis can reveal the effect of uncertainty in the applied parameters, as well as the relative influence of different parameters on the population dynamics and model projections. To test the effect of change in any single parameter, only one parameter was varied at a time. The age distribution of the initial population was kept constant in all scenarios, to avoid the effect of change in age distribution when life-history parameters are being modified. The tested changes in individual parameters either reflected the ranges provided for that parameter by different authors, or a particular range of change was applied (e.g., ±50%).

The results of the sensitivity analysis were assessed through the use of a ranking approach. Absolute changes in output variables (i.e., pseudo-extinction probability, deterministic and stochastic population growth rates, expected heterozygosity, and the mean time to recovery) that were caused by changes in each of the life history parameters were ranked from the highest to the lowest. The ranking was performed

Table 46.2 Results of the sensitivity analysis in a Vortex simulation model, presenting the influence of each life-history parameter through the total rank value (where higher value signifies higher influence) for each of five resulting parameters – pseudo-extinction probability (*P*), deterministic population growth rate (r_{det}), stochastic population growth rate (r_{stoch}), heterozygosity level (*H*) and the population recovery time (*R*)

Parameter[a]	*P*	r_{det}	r_{stoch}	*H*	*R*	Total
Mean offsp[b]	143	237	240	208.5	241.5	1,070
Fem age	176.5	221.5	225.5	170.5	232.5	1,026.5
Sex ratio	113.5	237	224.5	175	169	919
1–4 mort	123	200.5	202	141	169	835.5
Fem breed	91.5	186	184	116.5	174.5	752.5
N	192	69	92	216	177.5	746.5
4-adult mort	119	164	161	139	145.5	728.5
Adult mort	125.5	165	166.5	135	135.5	727.5
Max age	129.5	149	139	124	102	643.5
K EV	166.5	69	77	142.5	114.5	569.5
Mort EV	159.5	69	98.5	164	74.5	565.5
Male age	128	69	72	93	91	453
Mean offsp EV	142.5	69	52	73.5	68	405
Male breed	109	69	58	82	68	386
Max offsp[b]	111	69	60	65	74.5	379.5
Fem breed EV	82	69	60	66.5	74.5	352

[a]Mean offsp – mean number of offspring (also represents egg-to-age-1 natural mortality); fem age – time needed for females to reach maturity; 1–4 mort – natural mortality of year 1–4 age groups; fem breed – female spawning frequency; N – initial population size; 4-adult mort – natural mortality of year 4 to adulthood age groups; adult mort – natural mortality of adult individuals; max age – maximum reproductive age; K EV – environmental variation in carrying capacity; mort EV – environmental variation in natural mortality of all age groups; male age – time needed for males to reach maturity; mean offsp EV – environmental variation in mean number of offspring; male breed – male spawning frequency; max offsp – maximum fecundity; fem breed EV – environmental variation in female spawning frequency

[b]Age 0–1 natural mortality is incorporated in the mean and maximum number of offspring; see model parameterization and scenario development for clarification

separately for each of the four fitted age-specific natural mortalities (i.e., for three population growth rates applied and uniform mortality fitting – see Table 46.2) and the ranks were thereafter summed up for each life-history parameter. The life-history parameters were ranked separately for each output variable to reveal the relative influence of each parameter on the output variables, and then summed up across the output variables to determine the life-history parameters that have the highest overall influence on population viability.

The basic question addressed in the harvest scenarios was the determination of the threshold level of the harvest pressure below which population viability would still be preserved. The harvest was introduced in the model as the percentage of adults taken every year.

Additional scenarios were developed to test the effectiveness of reintroduction because of the ongoing restoration efforts across its former range (Kirschbaum and Gessner 2000; Williot et al. 2009a, b). The main goal was to test different approaches in supportive stocking, and their influence on the population viability. In general, the stocking dynamics varied between different scenarios along three main dimensions: the total number of individuals that were released, age of released individuals (whether they were juveniles or adults), and the sex ratio of released individuals.

Output variables tracked in simulations were pseudo-extinction probability, population size, deterministic and stochastic population growth rates, expected heterozygosity, and the time to recovery (i.e., the mean time needed for population to reach half of the carrying capacity). As suggested by other authors (Ebenhard 2000; Beissinger 2002; Hanski 2002; Reed et al. 2002), the results of all scenarios were predominantly used to compare their relative effects, for instance to compare different management scenarios, rather than to be employed for absolute predictions. An extinction risk of $P < 0.05$ was applied as the threshold to evaluate viability (Beissinger and Westphal 1998).

46.3 Results and Discussion

Estimated population age structure of the undisturbed population is presented in Fig. 46.1. The ratio of adult individuals in a population was generally low for all population growth rates applied, and ranged from 3.5 to 7.8%. As depicted in

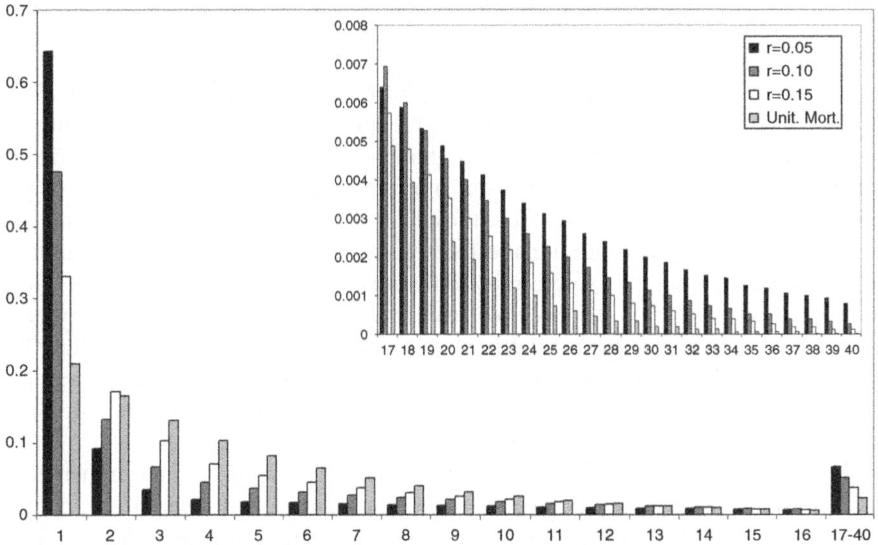

Fig. 46.1 Age structure of the undisturbed *A. sturio* population in the Gironde system, estimated by the Vortex population model according to four different population growth rates (r) applied. Small plot in the *upper right* corner presents the age structure within adult age groups. *Unif. mort.* –10% mortality rate applied across all age groups ($r = 0.129$)

Fig. 46.1, the largest ratio of adult individuals was reached at the lowest population growth rate ($r = 0.05$), while the lowest value was recorded when the uniform mortality across all age groups was applied ($r = 0.129$). This was caused by a high juvenile mortality applied at the lower population growth rates, thus resulting in a larger relative number of adults. The mean female generation time varied across different scenarios between 22 and 25.5 years, and was the longest at low population growth rates. These values are in accordance with findings of other authors (e.g., the range of 20–25 years according to Gessner 2000).

As expected, the scenarios for the assessment of the MVPs showed a high sensitivity of the population persistence probability to the population size. A long-term viability ($P < 0.05$) became more likely only for the population sizes of 200–500 individuals (representing all age classes at the stable age distribution, but without eggs and young-of-the-year that do not survive to age 1). However, the loss of genetic variability was still very high at such population sizes, and only at a population size of 2,000 or more individuals was the loss of heterozygosity lower than 10%. It is important to emphasize that the presented MVPs were estimated under zero negative impact scenarios. Consequently, pressure from any negative impact (e.g., pollution, by-catch mortality) would result in an increase of the MVPs.

Results of the sensitivity analysis are presented in Table 46.2. It revealed similar results across different population growth rates applied. The most influential parameters on the model output were the mean number of offspring, egg-to-age-1 natural mortality, sex ratio, and the age at which females reach maturity. The changes in juvenile and adult natural mortality, female spawning frequency and the initial population size also resulted in high model sensitivity. The results are similar to the population models of other sturgeon species (Pine et al. 2001; Beamesderfer et al. 2007; Kennedy and Sutton 2007), which have determined highest model sensitivity to egg mortality, overall natural mortality, age at which females reach maturity, and the spawning frequency. When all output variables are ranked together, the most influential parameters overall were the mean number of offspring and the egg-to-age-1 natural mortality. However, since these two life history parameters were combined within the model, it was not possible to make a distinction between their individual effects. If the influence on pseudo-extinction is considered separately, the single most influential parameter was the age at which females reach maturity. In general, male-related traits and the environmental variation of parameters had a relatively weak effect on the model behaviour. The low influence of male-related traits has also been recognized by other authors, and as a result, many sturgeon population models have been restricted to females only (e.g., Bajer and Wildhaber 2007; Kennedy and Sutton 2007).

The weaker influence of adult mortality, when compared with the influence of other life-history parameters, could probably be explained by a low ratio of adults in the population. The lower ratio of adult individuals means that the total number of individuals affected by a change in adult mortality would be also low, especially when compared with the ratio of youngest age classes in a population (Fig. 46.1). Findings of other authors have significantly varied with regard to the sensitivity of their models to adult mortality. While Kennedy and Sutton (2007) found a lower influence of adult mortality in comparison to egg and juvenile mortality, Pine et al. (2001), Boreman (1997) and Heppell (2007) found the opposite.

The relationship between the deterministic growth rate and the other model output variables (pseudo-extinction probability, expected heterozygosity and the population recovery time), based on the results of sensitivity analysis, is presented in Fig. 46.2. A strong correlation between these variables was also confirmed by the

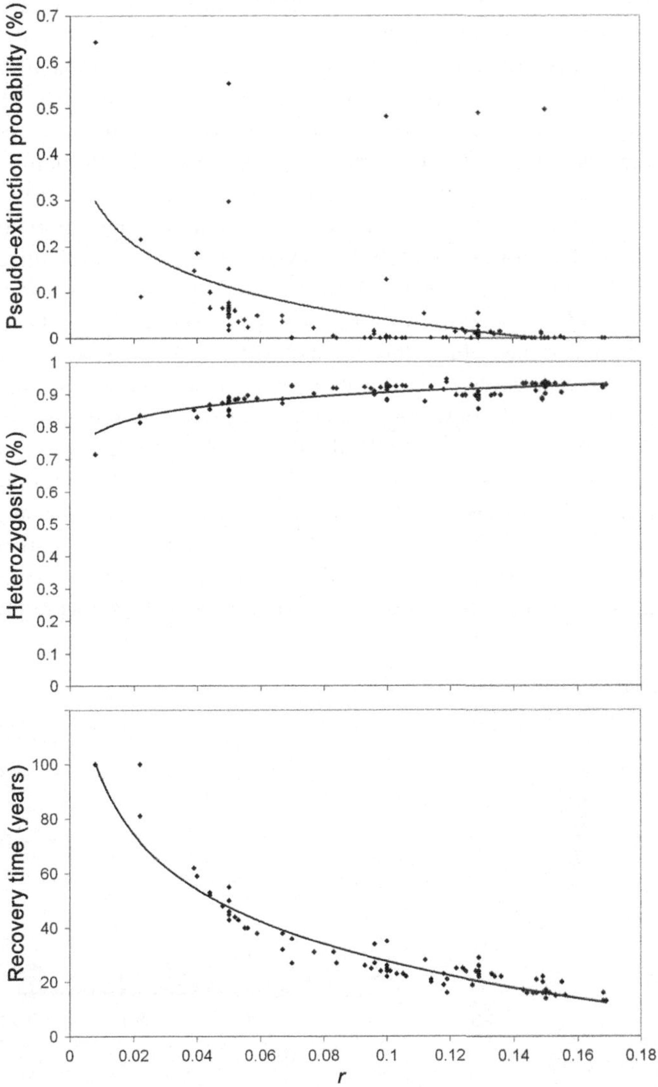

Fig. 46.2 Relationship between the deterministic population growth rate (r) and the pseudo-extinction probability, expected heterozygosity and population recovery time, based on the results of sensitivity analysis

use of Spearman's non-parametric correlation test (correlation coefficients ranging between 0.641 and 0.928, $P < 0.001$). Although the deterministic population growth rate does not consider stochasticity, male-related traits, and genetically linked demographic traits, it has a potential to be used as a predictor or indicator of the population state (Jarić et al. 2010).

The life-history parameters that are the most influential upon the population viability are usually proposed for the highest management and/or research priority (Mills and Lindberg 2002). Although fecundity, female age at maturity, female spawning frequency, sex ratio, and egg-to-age-1 and juvenile natural mortality represented the most influential parameters in this study, they have been poorly studied in the Gironde *A. sturio* population. There is a virtually complete lack of data on *A. sturio* spawning frequency in the Gironde Estuary system (Williot personal communication), and the same is true for age-specific natural mortality and the intrinsic population growth rate. The data on age at maturity and mean fecundity provided by different authors is inconsistent, and the reports on sex ratio have been often biased for different reasons (Williot personal communication). As a result, given the importance of these parameters and the current lack of knowledge, they should be widely recognized by the scientific community as future research priorities. Another issue which has been poorly addressed within the present literature on sturgeon ecology is the existence of senescence in sturgeons, the decline in fecundity and spawning success before the maximum longevity has been reached (Jarić et al. 2010). While the senescence has been detected in aquaculture (Williot et al. 2005), there are no data available for wild populations.

In harvest scenarios, unsustainable fishing pressures (i.e., where pseudo-extinction probability was $P > 0.05$) have always reduced population to the half of its initial size for no more than 2 or 3 decades, and resulted in a completely depleted stock within a single century of fishing (Fig. 46.3). The caviar industry within the Gironde

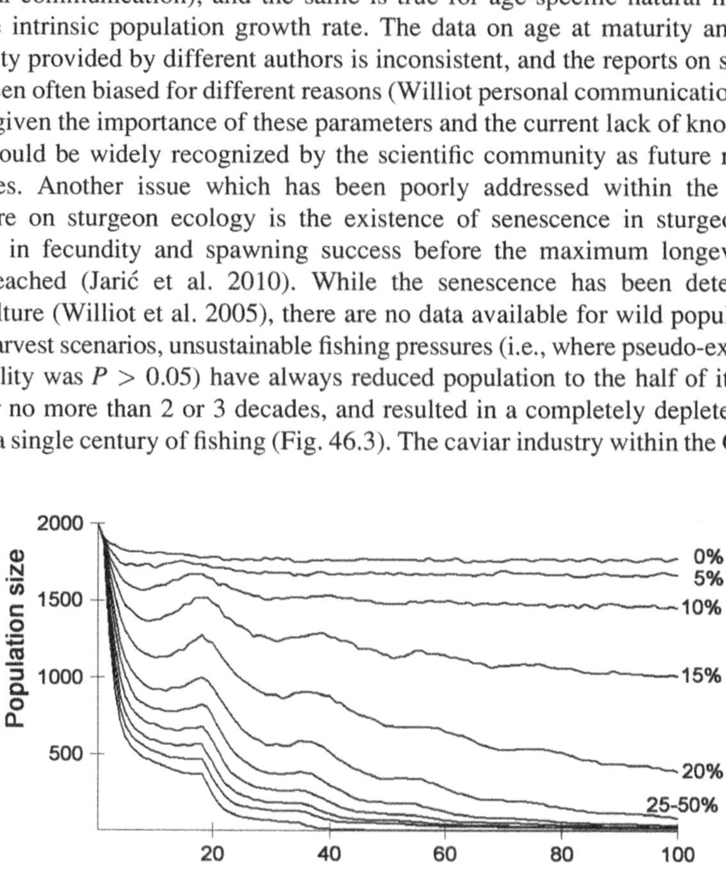

Fig. 46.3 Effects of different fishing pressures applied on the Gironde *A. sturio* population ($r = 0.05$). *Y*-axis – average population size among all iterations in a simulation; *X*-axis – years of the simulation; percentage values represent different fishing pressures. Initial population size and carrying capacity in all scenarios were set to 2,000 individuals

system was established in 1920, and by the early 1970s, *A. sturio* was no longer an economically important resource (Williot et al. 1997; Holčík 2000). Such a time frame for complete stock depletion (i.e., within 50 years) would correspond in this model to fishing pressures far above the sustainable ones (i.e., 15–20% units higher). The maximum sustainable harvest rates at the population growth rates of 0.05 and 0.10 were, respectively, 0.15 and 0.40, signifying percent of adults taken from the population each year. These threshold levels are similar to those determined for other sturgeon species, which ranged from 0.05 to 0.35 (Beamesderfer et al. 1995; Quist et al. 2002; Colombo et al. 2007; Kennedy and Sutton 2007). However, these values are somewhat higher than those determined for the historical *A. sturio* population in the Danube River basin at the same population growth rates (0.10 and 0.25, respectively; Jarić et al. 2010), which can be explained by different parameter values used in the model, due to significant differences in their life history among different river basins. The threshold harvest level determined at the population growth rate of 0.15 was rather high (0.60), which could be an indication that such a population growth rate might be unrealistically high for such slow-maturing species. A total fishing ban in France was introduced in 1982 (Williot et al. 2009a), but the accidental by-catch continues to threaten the species (Limburg and Waldman 2009). According to Gessner (2000), these catches alone might be sufficient to extirpate the population in the course of approximately 15 years. The harvest scenarios described thus have a potential to be applied for the evaluation of harvest through by-catch.

Results of some of the reintroduction scenarios are presented in Fig. 46.4. At lower population growth rates (0.05 and 0.10), the number of juveniles that needed to be released to ensure population viability ($P < 0.05$) had to be in the range from a few thousands to a few tens of thousand specimens. When compared with the number of adults needed to produce the same model outcomes, the juvenile/adult ratio ranged approximately from 10/1 to 100/1, being on average 42/1 across all scenarios. The juvenile/adult ratio generally decreased towards higher population growth rates. The optimal sex ratio of released individuals (i.e., the one that produces quickest recovery and smallest extinction risk and heterozygosity loss) was shifted towards a greater female ratio. The best outcomes, when juveniles were used for stocking, were reached with 65–70% of individuals being females.

In both adult and juvenile stocking scenarios, there was a clearly defined lag before the actual population recovery, which clearly corresponded to the time needed for females to reach maturity (Fig. 46.4). At lower numbers of stocked individuals, there was another lag in population recovery following the first one, prolonging the phase of a more rapid population recovery to twice the female age at maturity (17×2 in present scenarios). Such a lag, caused by the time needed for juveniles to reach maturity and start sustaining population through reproduction, has been found in other sturgeon species as well (Jarić et al. 2010). This should be taken into account in all restoration project planning, and confirms the claim that the recovery of sturgeon populations is a multi-decadal affair (Lenhardt et al. 2006). As stated by Gessner (2000), management programmes might prove their efficiency

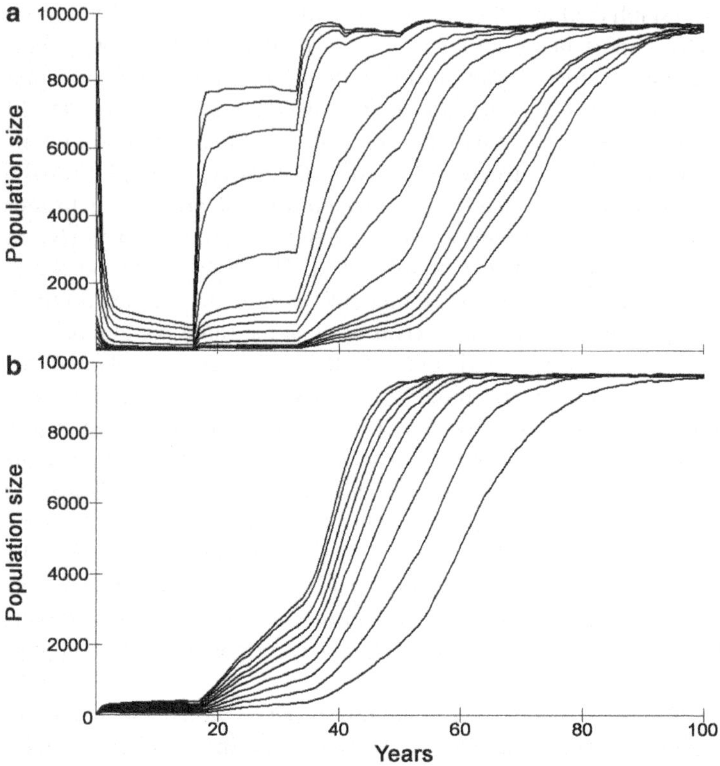

Fig. 46.4 Recovery of the *A. sturio* population after its reintroduction ($r = 0.10$). From the lowest to the highest one, respectively, full lines in each graph represent (**a**) the release of 20, 40, 60, 80, 100, 200, 400, 600, 800, 1,000, 2,000, 4,000, 6,000, 8,000 and 10,000 juvenile (age 1) individuals, and (**b**) the release of two, four, six, eight, ten, 12, 14, 16, 18 and 20 adult individuals; *Y*-axis – average population size among iterations with surviving populations in a simulation; *X*-axis – years of the simulation

only after 15–25 years, which unfortunately represents a time frame hard to cover by consistent administrative policies.

The decision to use the Vortex simulation model as a suitable tool for the assessment of *A. sturio* population viability was based on a number of its advantages: (1) it is able to perform sensitivity analysis and identify key life-history parameters, (2) it has options for simulating both harvest and reintroduction scenarios, and (3) as an individual-based model, it is able to track population genetic diversity and include demographic and environmental variation within the model. However, there are a number of other available PVA and fishery assessment software packages that might be applied for the assessment of *A. sturio* population viability and the major anthropogenic impacts. For instance, RAMAS software is able to handle very large population sizes, which is an advantage when it is applied to species with high fecundities, such as fish (Brook et al. 1997). It is also suitable for cases when there are spatial and habitat suitability factors that should be incorporated in the model. On the other hand, RAMAS is not an individual-based model, but an age- or stage-structured model that therefore lacks genetic considerations and more complex demographic patterns. Furthermore, while some fishery assessment program packages have been successfully used for the assessment of sturgeon fishery sustainability (e.g., Doukakis et al. 2010), they are less suitable for the *A. sturio* population, since it is no longer an object of commercial fishery. In the end, none of the existing PVA software packages can be recognized as a superior one, since each one has its advantages and drawbacks, and studies that involve comparison of efficiency of different models are common (Brook et al. 1997, 1999, 2000; Lindenmayer et al. 1995).

Ongoing *A. sturio* restoration efforts are and will be hindered to an extent by the lack of knowledge regarding its life history and demography. As stated by Waldman (2000), currently there is a limited species-specific knowledge and experience available regarding *A. sturio* and, given its scarcity, there are few specimens to be used for scientific purposes. However, due to an expressed need for urgent conservation measures, the development of *A. sturio* PVA models should not wait for better data. On the contrary, PVA can provide directions to investigators with regard to the research activities of greatest importance (Jarić et al. 2009b). Therefore, PVA should be recognized as an important tool to be included in the process of research and management planning and policy development. As has been stated by Williot et al. (2009b), PVA studies can also help managers dealing with broodstock and supportive stocking management to choose the main lines for future breeding programmes. Figure 46.5 presents general principles of the integration of PVA within these processes. The greatest potential of PVA models is their ability to provide insights regarding the parameters that could have the strongest impact on population persistence and thus define research priorities (Jarić et al. 2009b), and to compare the efficacy of different management scenarios (Ebenhard 2000; Reed et al. 2002). At the same time, new insights acquired through research and monitoring could be used to evaluate and refine the models, as well as to define further modelling directions. A productive interaction among the PVA developers, scientific community,

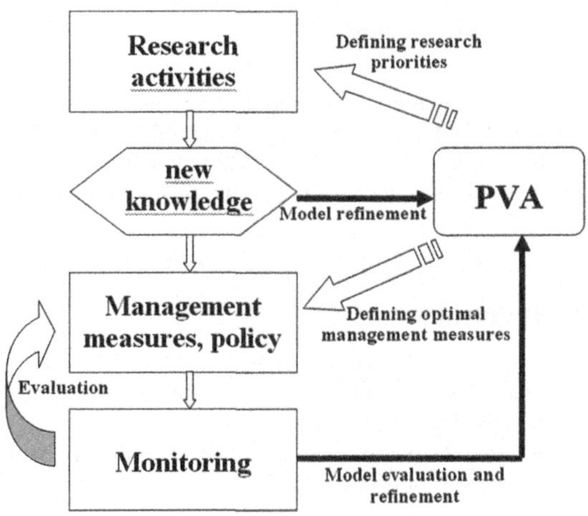

Fig. 46.5 Principles of population viability analysis (PVA) integration within the research and management planning and policy development processes

management authorities, policy developers and the general public will be essential for the general effectiveness of this approach. At the same time, the availability of PVA program packages such as VORTEX, which have already proved their usefulness and reliability (Brook et al. 2000), could diminish the need for experience in programming and developing PVA models (Jarić et al. 2009b).

46.4 Conclusions

The population viability analysis presented in this paper is one of the few models that have been developed for the Gironde *A. sturio* population. While there is an urgent need to establish effective recovery measures for the population, the existing knowledge regarding its life history and demography is still severely limited. In addition to some commonly recognized factors, such as political will and public awareness, improving knowledge with regard to this species could be one of the key factors for its successful preservation and recovery. The most important direction of future research efforts related to *A. sturio* life history (as was identified by the model) should be the determination of the mean fecundity, age at which females reach maturity, female spawning frequency and population sex ratio, as well as intrinsic population growth rate and age-specific natural mortality, especially of the youngest age classes. The question regarding the existence of senescence in wild populations should be also resolved.

The model confirmed a high population susceptibility to unsustainable fishery, and a slow recovery potential, which could span over a number of decades. The lag between stocking activities and the initiation of actual population recovery, which corresponds to the time the juveniles need to reach maturity, should be taken into account when planning the restoration projects.

The development of *A. sturio* PVA models should not wait for better data. On the contrary, they should be recognized as an important tool, and be integrated within the research, management, and policy development efforts. If used with necessary caution, PVA can provide invaluable help for the successful recovery of this endangered species.

Acknowledgements This study represents a part of the activities within Project No. 173045, funded by Ministry of Science of Republic Serbia. The authors would like to thank Dr Patrick Williot for his help in resolving different questions of *A. sturio* life history, as well as Dr Torbjörn Ebenhard for his guidance regarding use and model development in the Vortex simulation model.

References

Akçakaya HR (2000) Population viability analyses with demographically and spatially structured models. Ecol Bull 48:23–38

Akçakaya HR, Sjögren-Gulve P (2000) Population viability analyses in conservation planning: an overview. Ecol Bull 49:9–21

Bajer PG, Wildhaber ML (2007) Population viability analysis of Lower Missouri River shovelnose sturgeon with initial application to the pallid sturgeon. J Appl Ichthyol 23:457–464

Beamesderfer RCP, Rein TA, Nigro AA (1995) Differences in the dynamics and potential production of impounded and unimpounded White Sturgeon populations in the Lower Columbia River. T Am Fish Soc 124:857–872

Beamesderfer RCP, Simpson ML, Kopp GJ (2007) Use of life history information in a population model for Sacramento green sturgeon. Environ Biol Fish 79:315–337

Beissinger SR (2002) Population viability analysis: past, present, future. In: Beissinger R, McCullough DR (eds) Population viability analysis. University of Chicago Press, Chicago, pp 5–17

Beissinger SR, Westphal MI (1998) On the use of demographic models of population viability in endangered species management. J Wildl Manage 62(3):821–841

Boreman J (1997) Sensitivity of North American sturgeons and paddlefish to fishing mortality. Environ Biol Fish 48:399–405

Boyce MS (1992) Population viability analysis. Annu Rev Ecol Evol S 23:481–506

Brook BW, Lim L, Harden R, Frankham R (1997) Does population viability analysis software predict the behaviour of real populations? A retrospective study on the Lord Howe Island woodhen *Tricholimnas sylvestris* (Sclater). Biol Conserv 82:119–128

Brook BW, Cannon JR, Lacy RC, Mirande C, Frankham R (1999) Comparison of the population viability analysis packages GAPPS, INMAT, RAMAS and VORTEX for the whooping crane (*Grus americana*). Anim Conserv 2:23–31

Brook BW, O'Grady JJ, Chapman AP, Burgman MA, Akçakaya HR, Frankham R (2000) Predictive accuracy of population viability analysis in conservation biology. Nature 404:385–387

Colombo RE, Garvey JE, Jackson ND, Brooks R, Herzog DP, Hrabik RA, Spier TW (2007) Harvest of Mississippi River sturgeon drives abundance and reproductive success: a harbinger of collapse? J Appl Ichthyol 23:444–451

Doukakis P, Babcock EA, Pikitch EK, Sharov AR, Baimukhanov M, Erbulekov S, Bokova Y, Nimatov A (2010) Management and recovery options for Ural River beluga sturgeon. Conserv Biol 24(3):769–777

Ebenhard T (2000) Population viability analyses in endangered species management: the wolf, otter and peregrine falcon in Sweden. Ecol Bull 48:143–163

FISAT II (2000) FAO-ICLARM Fish stock assessment tools. Version 1.2.2. FAO, Rome, Italy

Gessner J (2000) Reasons for the decline of *Acipenser sturio* L., 1758 in central Europe, and attempts at its restoration. Bol Inst Esp Oceanogr 16(1–4):117–126

Gessner J, Van Eenennaam JP, Doroshov SI (2007) North American green and European Atlantic sturgeon: comparisons of life histories and human impacts. Environ Biol Fish 79:397–411

Hanski I (2002) Metapopulation of animals in highly fragmented landscapes and population viability analysis. In: Beissinger R, McCullough DR (eds) Population viability analysis. University of Chicago Press, Chicago, pp 86–108

Heppell SS (2007) Elasticity analysis of green sturgeon life history. Environ Biol Fish 79:357–368

Holčík J (2000) Major problems concerning the conservation and recovery of the Atlantic sturgeon *Acipenser sturio L.*, 1758. Bol Inst Esp Oceanogr 16(1–4):139–148

Jager HI (2005) Genetic and demographic implications of aquaculture in white sturgeon (*Acipenser transmontanus*) conservation. Can J Fish Aquat Sci 62:1733–1745

Jager HI (2006) Chutes and ladders and other games we play with rivers I. Simulated effects of upstream passage on white sturgeon. Can J Fish Aquat Sci 63:165–175

Jarić I, Lenhardt M, Cvijanović G, Ebenhard T (2009a) *Acipenser sturio* and *Acipenser nudiventris* in the Danube – extant or extinct? J Appl Ichthyol 25:137–141

Jarić I, Lenhardt M, Cvijanović G, Ebenhard T (2009b) Population viability analysis and potential of its application to Danube sturgeons. Arch Biol Sci 61(1):123–128

Jarić I, Ebenhard T, Lenhardt M (2010) Population viability analysis of the Danube sturgeon populations in a VORTEX simulation model. Rev Fish Biol Fish 20(2):219–237

Kennedy AJ, Sutton TM (2007) Effects of harvest and length limits on shovelnose sturgeon in the upper Wabash River, Indiana. J Appl Ichthyol 23:465–475

Kirschbaum F, Gessner J (2000) Re-establishment programme for *Acipenser sturio* L., 1758: the German approach. Bol Inst Esp Oceanogr 16(1–4):149–156

Lacy RC (1993) VORTEX: a computer simulation model for population viability analysis. Wildl Res 20:45–65

Lacy RC (2000) Structure of the VORTEX simulation model for population viability analysis. Ecol Bull 48:191–203

Lenhardt M, Jaric I, Kalauzi A, Cvijanovic G (2006) Assessment of extinction risk and reasons for decline in sturgeon. Biodivers Conserv 15:1967–1976

Limburg KE, Waldman JR (2009) Dramatic declines in North Atlantic diadromous fishes. BioSci 59(11):955–965

Lindenmayer DB, Burgman MA, Akçakaya HR, Lacy RC, Possingham HP (1995) A review of the generic computer programs ALEX, RAMAS/space and VORTEX for modelling the viability of wildlife metapopulations. Ecol Model 82:161–174

Miller PS, Lacy RS (2005) VORTEX: A stochastic simulation of the extinction process. Version 9.50 user's manual. Conservation Breeding Specialist Group (SSC/IUCN), Apple Valley, MN, USA

Mills LS, Lindberg MS (2002) Sensitivity analysis to evaluate the consequences of conservation actions. In: Beissinger R, McCullough DR (eds) Population viability analysis. University of Chicago Press, Chicago, pp 338–366

Paaver T (1999) Historic and recent records of native and exotic sturgeon species in Estonia. J Appl Ichthyol 15:129–132

Pine WE III, Allen MS, Dreitz VJ (2001) Population viability of the Gulf of Mexico sturgeon: inferences from capture–recapture and age-structured models. T Am Fish Soc 130:1164–1174

Quist MC, Guy CS, Pegg MA, Braaten PJ, Pierce CL, Travnichek VH (2002) Potential influence of harvest on shovelnose sturgeon populations in the Missouri River system. N Am J Fish Manag 22:537–549

Reed DH, O'Grady JJ, Brook BW, Ballou JD, Frankham R (2003) Estimates of minimum viable population sizes for vertebrates and factors influencing those estimates. Biol Conserv 113:23–34

Reed JM, Mills LS, Dunning JB Jr, Menges ES, McKelvey KS, Frye R, Beissinger SR, Anstett MC, Miller P (2002) Emerging issues in population viability analysis. Conserv Biol 16(1):7–19

Rochard E, Williot P, Castelnaud G, Lepage M (1991) Elements de systematique et de biologie des populations sauvages d'esturgeons. In: Williot P (ed) Acipenser. Cemagref, Antony, pp 475–507

Rochard E, Lepage M, Meauzé L (1997) Identification et caractérisation de l'aire de répartition marine de l'esturgeon européen *Acipenser sturio* à partir de déclarations de captures. Aquat Living Resour 10:101–109

Trouvery M, Williot P, Castelnaud G (1984) Biologie et ecologie d'*Acipenser sturio*. Etude de la pêcherie. Cemagref, Etude no 17, Série Esturgeon no 1, p 79

Waldman JR (2000) Restoring *Acipenser sturio* L., 1758 in Europe: lessons from the Acipenser oxyrinchus Mitchill, 1815 experience in North America. Bol Inst Esp Oceanogr 16(1–4):237–244

Williot P, Rochard E, Castelnaud G, Rouault T, Brun R, Lepage M, Elie P (1997) Biological characteristics of European Atlantic sturgeon, *Acipenser sturio*, as the basis for a restoration program in France. Environ Biol Fish 48:359–370

Williot P, Arlati G, Chebanov M, Gulyas T, Kasimov R, Kirschbaum F, Patriche N, Pavlovskaya LP, Poliakova L, Pourkazemi M, Kim Y, Zhuang P, Zholdasova IM (2002) Status and management of Eurasian sturgeon: an overview. Int Rev Hydrobiol 87(5–6):483–506

Williot P, Brun R, Rouault T, Pelard M, Mercier D, Ludwig A (2005) Artificial spawning in cultured sterlet sturgeon, *Acipenser ruthenus* L., with special emphasis on hermaphrodites. Aquaculture 246:263–273

Williot P, Rochard E, Rouault T, Kirschbaum F (2009a) *Acipenser sturio* recovery research actions in France. In: Carmona R et al (eds) Biology, conservation and sustainable development of sturgeons. Springer, Dordrecht, pp 247–263

Williot P, Rochard E, Kirschbaum F (2009b) Acceptability and prerequisites for successful introduction of sturgeon species. In: Carmona R et al (eds) Biology, conservation and sustainable development of sturgeons. Springer, Dordrecht, pp 369–384

Chapter 47
One Alternative to Germ Cells Cryopreservation: Cryobanking of Somatic Cells in Sturgeon

Catherine Labbe, Alexandra Depince, Pierre-Yves Le Bail, and Patrick Williot

Abstract Cryobanking of somatic cells is one option for preserving both parental genomes when gametes are not available or are unsuitable for cryopreservation. Somatic cells can be obtained from skin or fin biopsies, which are easy to collect long before sexual maturation is reached. This chapter develops a strategy to be considered for cryobanking of *Acipenser sturio* somatic cells, with the example of fin as a cell reservoir. The collected fins should be appropriately stored to allow shipping to laboratory facilities where they will be either frozen as a whole, or cultured to grow out cells which will be cryopreserved thereafter. Fish reconstruction from somatic cells will require the use of nuclear transfer technology, where the nucleus of the cryopreserved cell is transferred into the enucleated oocyte of a closely related species. The promises and drawbacks of this reconstruction technology are developed.

47.1 Introduction

It should be emphasized that although cryobanking of reproductive and somatic cells permits an efficient and dependable storage of genetic resources, cryobanking should always be regarded as a backup strategy. Cryobanking should never hinder strategies more dedicated to the in situ preservation of habitat and populations, as already rightfully claimed by Corley-Smith and Brandhorst (1999). Cryobanks allow cell storage over tens of years, and stored cells are easy to handle for diffusion. Cryobanks, however, are completely dependent on institutional regime

C. Labbe (✉) • A. Depince • P.-Y. Le Bail
INRA, UR1037 SCRIBE, Cryopreservation and Regeneration, F-35000, Rennes, France
e-mail: Catherine.Labbe@rennes.inra.fr

P. Williot
Sturgeon consultant, 4, Rue du pas de madame, 33980 Audenge, France

P. Williot et al. (eds.), *Biology and Conservation of the European Sturgeon Acipenser sturio L. 1758*, DOI 10.1007/978-3-642-20611-5_47,

and politics: collection management and liquid nitrogen supplying require funding which should under no circumstances be cut back. Cryobanking also relies on cell availability and quality, and on a cryopreservation method which preserves at best the cell quality and functionality. Finally, cryobanking is only one half of the process. The availability of a regeneration method which will allow the reconstitution of an animal from the cryopreserved cells is the essential second half of the process.

Somatic cell and tissue cryopreservation is becoming a compelling option for fish genetic resources management, as a complement to sperm cryopreservation. Indeed, somatic cells bear both paternal and maternal genome. This compensates for the inability of oocytes and whole embryo to withstand cryopreservation (Robles et al. 2009), a feature specific to fish because of the difficulty of protecting the yolk compartment (Hagedorn et al. 1998). In the case of European Sturgeon *Acipenser sturio*, whose sexual maturity is reached after 12–20 years, somatic cells can be collected very early in the life of the animal, well before sexual maturity is achieved, thereby allowing an early preservation of the genome of a given population. It is therefore important to control the quality of the collected tissues so that they can produce proliferating cells in culture condition. It is also of major importance that a simple procedure be set up to reliably cryopreserve samples of different types such as isolated cells or tissue fragments. Finally, cryobanking of somatic cells or tissue requires that efficient reconstruction biotechnologies are established in order to bring back to life the genetically inherent information of the cell nucleus. In this section, all three technical steps of cryobanking will be discussed with regard to sturgeon somatic cell cryopreservation and its use for animal reconstruction. Examples taken in other acipenserids and in teleost species will help to illustrate the global picture which could be drawn and adapted to *A. sturio*.

47.2 Primary Cell Culture from Fin Explants

47.2.1 A. sturio *Fin Sampling*

In the case of valuable animals such as in *A. sturio*, the easiest tissue to sample and the least damaging to the fish is the fin. The healing ability and regenerative capacities of fins in most species prevent any long-term disabling of the fish (Akimenko et al. 2003), and such samples are considered to be reliable cell reservoirs when fin explants are cultured in suitable conditions.

Rearing facilities or natural habitats where the fish are to be sampled are most often far from the laboratory facilities required for the conservation and reconstruction technologies. Therefore, it is essential that the fin samples can be collected and stored in suitable conditions, so that they can be shipped to the laboratory facilities. To our knowledge, such a procedure has not been reported in *A. sturio*, although it is obvious that many groups do carry out the sending of refrigerated samples (Fontana et al. 2008). Such procedure was experimentally explored on goldfish (*Carassius auratus*) fins (Mauger et al. 2006; Moritz and Labbe 2008) to test whether any delay between fin sampling and fin processing for explant culture could affect culture success. It was demonstrated that fin pieces can be stored and shipped for at least 4 days without any conditioning medium, provided that they are kept under humid ambiance at 4°C (Moritz and Labbe 2008). Moreover, fin pieces can be kept for at least 3 days in culture medium with antibiotics at ambient temperature (Mauger et al. 2006).

In June 2006, we successfully applied the method to *A. sturio* pectoral fin collected at the Cemagref hatchery (Saint-Seurin-sur-l'Isle, France):

1. Pectoral fin pieces (2 × 0.5 cm) were collected on three 8-year-old females of about 5 kg body weight. The samples were at least 4 mm thick, and displayed blood vessels and flesh.
2. Samples were plunged in the culture medium described in Mauger et al. (2006), containing L15 buffered culture media, 2 mM L-glutamine, 10% (v:v) fetal bovine serum (FBS) and antibiotics, and shipped at 4°C in an insulated box.
3. After the 2-day shipping, the medium was discarded and the fin pieces were prepared for culture.

47.2.2 Pectoral Fin Culture in A. sturio

In acipenserids, several works report the production of cell lines after primary culture of fin explants (Fontana et al. 1995; Tagliavini et al. 1998; Wang et al. 2003; Zhou et al. 2008), as cell lines are routinely used in these species for karyotype studies or virus infection studies. Prior to the primary culture, fin pieces are wiped, minced in very small explants, and washed in culture medium containing high antibiotics concentrations. Dulbecco's modified Eagle medium (DMEM) or Leibovitz 15 medium (L15) are the most common media, and high FBS concentrations (20%) are used at the beginning of the culture. Fin explants are usually plated into 25 cm^2 flasks and cultured at 25°C. Although fin explant behavior with regard to cell proliferation ability was little described by the authors, and little information was given on culture duration, no specific difficulties with these primary cell cultures were ever reported.

The protocols described in the literature for acipenserids are very close to those which were set up for teleost fins (Mauger et al. 2006). The following culture conditions were applied to the *A. sturio* fins retrieved after shipping (see Sect. 47.2.1.):

1. Fin pieces were thoroughly minced and reduced to 1-mm^2 explants. Explants were mildly digested in collagenase (C-2674, Sigma) 0.2 mg/ml for 30 min at 25°C, to allow a better release of the cells. Explants were then plated in 12-well plates at 25°C.

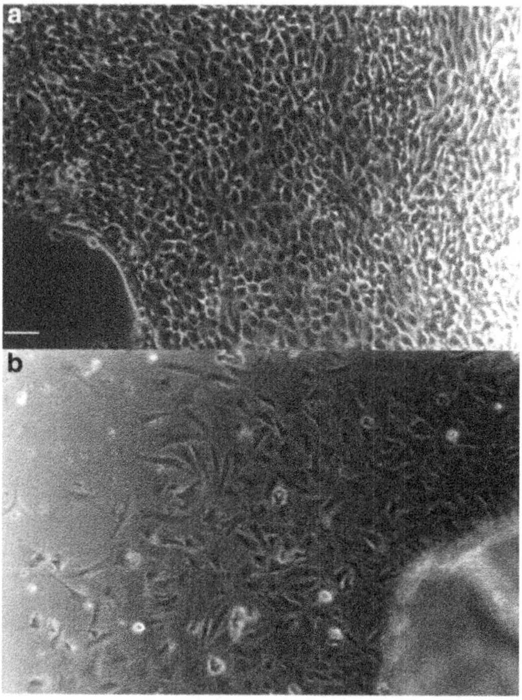

Fig. 47.1 Cell outgrowing from *Acipenser sturio* fin explants in culture. **a** Fresh explants after 2 days in culture; note the tightly packed cells with a cubic epithelial-like morphology. **b** Frozen-thawed explants 13 days after thawing and culture. Cells showed a looser density. *Bar* = 50 μm

2. Cells outgrew from the explants at the end of the first day and proliferated rapidly (Fig. 47.1a). Confluence was reached after 4–5 days, and more than 2×10^6 cells were obtained from each initial 2×0.5 cm fin piece.

The fast cell growing obtained in primary culture with the shipped samples indicates that no major adverse effect of the sampling and storage condition was evident at this stage. Therefore, *A. sturio* fin explants can be used to outgrow fin cells in short-term primary culture. In these conditions, however, although explant adhering and cell outgrow were good during the first 5 days, it was observed that many cells became loose and detached from the culture plate after 5 days. We suspect that, unlike culture started with fins from very small individuals (1-g fish in Wang et al. (2003) on *Acipenser transmontanus*), the fin pieces used in our *A. sturio* experiment contained a large variety of cell types (vessel cells, muscle). Because the culture conditions were set up for skin cells, most of the other cell types died after 5 days. To confirm the assumption that fins taken on big fish are more heterogeneous than those taken on smaller fish, we carried out the same experiment on anal fin samples of 20–25 cm *Acipenser baerii*. In this case, confluence was also reached after only 5 days, but we did not face losing cells afterwards. This time, the cell maintenance in culture may be due to the fact that *A. baerii* explants taken on small fish were much less heterogeneous than the *A. sturio* ones. *A. baerii* explants were mainly made of fin rays and skin whose cells proliferated in our culture conditions. We observed the same phenomenon with rainbow trout

(*Oncorhynchus mykiss*) explants (unpublished data): explants taken from thick fins of 2-kg fish were difficult to maintain in culture, whereas explants taken on 20-g fish showed a good cell proliferation in our culture conditions.

We conclude from the literature and from our preliminary trials on *A. sturio* that cultured cells were easily obtained after fin explant plating, and that the procedure where samples were collected in situ and sent to laboratory facilities worked well. However, when large batches of cells are required, for distribution between several cryobanks for example, the efficiency of the culture has to be improved when fin samples are very thick and heterogeneous. One possibility could be to isolate the epidermis from the fin pieces, as described for example by Lamche et al. (1998) for rainbow trout skin. These authors explain that skin samples digestion in dispase II solution removes the hemisdesmosomal junctions which connect the keratinocytes to the basal lamina. Thereafter, dermal and muscle tissue remnants can be scraped off easily from the epidermis. Plating of cell suspension after thorough digestion of the explants might be another way out. Indeed, skin cells will adhere very rapidly (within 2 h), which may not be the case with other cell types, which may therefore be washed off after the initial culture seeding. In addition, a longer culture time which might enrich the culture with skin epithelial cells well-characterized in teleosts (Mauger et al. 2009) is another possibility. Overall, sampling of very young fish helps to circumvent the problems described above.

47.3 Cryopreservation

47.3.1 Cryopreservation of Fin Cultured Cells

Cryopreservation of cultured cells has been successfully achieved in fish for more than 40 years (Wolf and Quimby 1969). Fontana et al. (1995) in *Acipenser naccarii*, *Acipenser ruthenus*, and *Acipenser gueldenstaedtii*, and Wang et al. (2003) in *A. transmontanus* proposed very simple cryopreservation protocols in culture medium using dimethyl sulfoxide (DMSO) as a cryoprotectant and a slow cooling rate. Slow cooling is easily achieved when the cryotubes containing the cells are wrapped within several layers of absorbent paper before they are placed into a −70°C chest freezer for 2–24 h. After this slow cooling phase, cryotubes can be plunged into liquid nitrogen for long-term storage. In our experiment on *A. sturio* using the protocol set up for goldfish cells (Mauger et al. 2006), 50–80% of the cryopreserved cells were recovered after thawing and washing, and cell viability of the recovered cells was more than 90%. The same high viability was reported for *A. transmontanus* cells (Wang et al. 2003). Little information is given, however, on cell proliferation more than 2–3 days after plating. The cells probably perform well, as they are routinely used for karyotyping by the author. In our conditions with *A. sturio* cells, almost all thawed cells were lost after 4 days in culture. The fact that *A. sturio* cells were viable at thawing, but did not survive after a few days of culture, might be explained by the already mentioned heterogeneity in cell types when thick fin pieces are used. It cannot be ruled out, either, that some cryodamage within the fin piece may have affected cell ability to proliferate after explant plating. Although the thawed cells still can be readily used for reconstruction experiments by nuclear transfer, as in teleosts (Le Bail et al. 2010), this lack of proliferation hinders the possibility to put the cells back in the cryobank collection. Moreover, the origin of the halt in proliferation should be explored further, to make sure that no apoptosis or necrosis was induced in the cultured cells. Such damage, which will ultimately affect DNA, would indeed dramatically impair zygote formation after nuclear transfer.

47.3.2 Cryopreservation of Fin Pieces

Because cryopreservation of somatic tissues for genetic resources management in cryobank is a very recent perspective, only a few works have addressed this issue either in mammals (Silvestre et al. 2002, 2003, 2004) on skin and ear pieces, or in fish (Cardona-Costa et al. 2006; Moritz and Labbe 2008) on fin pieces. One advantage inherent to cryopreservation of fin pieces, in contrast to that of cultured cells is that, at the time of sample collection, the whole procedure can be handled with very little equipment, and neither cell culture facilities nor sterile conditions are required. Therefore, tissue sampling and cryopreservation can be performed first on many samples, after which thawing, culture for cell growing, and nuclear transfer for fish regeneration can be performed only on those samples whose interest is high at a given time.

The simple procedure set up for Goldfish (Moritz and Labbe 2008) was tested on *A. sturio* samples during the same June 2006 experiment as above. The pectoral fin pieces collected and shipped as described in sub-section 47.1.1. were minced into thin stripes, placed into 2-ml cryovials and covered with 1.5-ml cryoprotective media containing buffered L15, 2 mM L-glutamine, 10% (v/v) FBS, 10% (v/v) DMSO and 166 mM sucrose. In order to ensure a very slow freezing rate, the vials were place in a polystyrene box into a −20°C conventional chest freezer. After 3 h, the vials were dropped into liquid nitrogen. After several days of storage at −196°C, the vials were thawed in a water bath at 37°C. The cryoprotective media was discarded, and culture media was added. Stripes were reduced to 1-mm^2 explants, and treated as fresh samples (see Sect. 47.2.1). We observed that the frozen-thawed explants adhered very well to the culture plate, but cell outgrowth from the frozen-thawed explant was very slow. After 13 days in culture (Fig. 47.1b), only 3,000 cells per plated explant could be recovered. The slow cell outgrowth from thawed fin explants from *A. sturio* is similar to the results obtained with Goldfish thawed fin explants (Moritz and Labbe 2008), but in this latter case this was compensated for by a large cell proliferation after several days. In the case of *A. sturio,* we can only suspect that, again, the procedure was not optimal for the complex fin pieces that were collected. More attention should be paid to the size of the pieces in the cryovials, and to the freezing rate.

We believe that cryopreservation of *A. sturio* fin pieces is possible, but the procedure described above should be optimized in order to increase the number of recovered cells. The cryopreservation of skin pieces isolated from the fin after dispase II digestion (see Sect. 47.2.2), instead of whole fin pieces, may help to improve the procedure. Another issue should be addressed, related to the fact that cryovials are not easy to handle in most animal cryobanks, because most samples are sperm in straws, and the tanks used are designed for straw storage. We believe that it would be more suitable to set up a procedure in which the fin or skin pieces will be inserted in the conventional straws used for sperm cryopreservation. This will require that samples be chopped into very thin pieces prior to cryopreservation.

47.4 Reconstruction by Nuclear Transfer

47.4.1 Principle of nuclear transfer

Nuclear transfer, also called cloning, is a reconstruction method which does not rely on conventional fertilization. A diploid nucleus is transferred either by injection or by electrofusion into a recipient oocyte which has previously been enucleated (Fig. 47.2). Therefore, the so-called reconstructed embryo will bear the nuclear genome of the injected nucleus. Early development, however, will rely on the

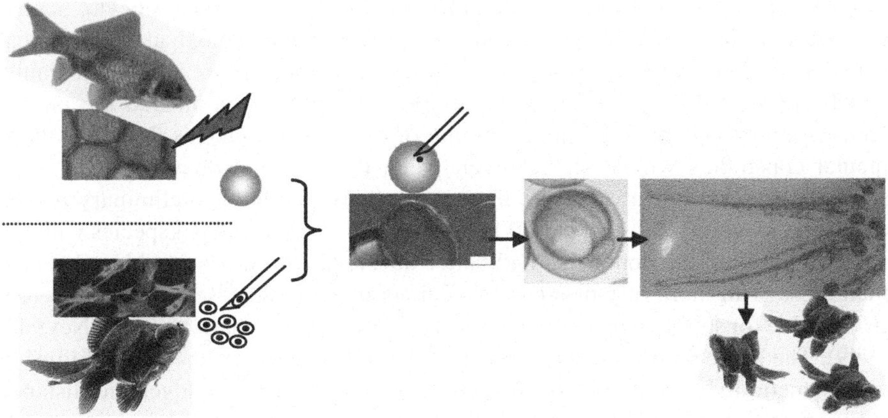

Fig. 47.2 Description of the nuclear transfer steps, using goldfish as an example. Fins from a valuable individual (*lower part*) are used as somatic cell reservoir. After culture, one fin cell is aspirated into the nuclear transfer microcapillary. Recipient oocyte from more common individuals (*upper part*) is enucleated, either by irradiation or by aspiration of the oocyte DNA, to become the recipient ooplasm. The donor cell is then injected through the micropyle at the animal pole of the ooplasm. Egg activation, in water for goldfish, initiates embryo development. The reconstructed nucleocytoplasmic hybrid bears the nuclear genome of the donor cell

cytoplasmic components of the recipient enucleated oocyte (mitochondria, maternal RNA, proteins). This is why reconstructed embryos are also called nucleo-cytoplasmic hybrids. The nuclear transfer method is divided into different categories which take into account the nucleus cell type, embryonic or somatic, and the species of the donor nucleus and that of the recipient oocyte. When the donor nucleus is taken from an embryonic cell, nuclear transfer is called embryonic cell nuclear transfer (ECNT), whereas when the donor nucleus is taken from a differentiated somatic cell, such as fin cells, nuclear transfer is called somatic cell nuclear transfer (SCNT). When the donor cells are from a species which is different from that of the recipient oocytes, nuclear transfer is referred as to interspecific or cross-subfamily nuclear transfer. None of these reconstruction methods have ever been applied to Acipenserids, although some experiments with variable success have been reported in teleosts (see Siripattarapravat et al. 2009 and Hattori et al. 2011 in zebrafish *Danio rerio* for the latest), as discussed later in this chapter.

47.4.2 The case of interspecific nuclear transfer

In some cases, recipient oocytes from the species deemed for reconstruction by nuclear transfer will no longer be available. This can be the case when the species is extinct or if it is no longer reared and reproduced in fish farms. In this case, nuclear transfer will be performed by the donor nucleus being injected into the enucleated oocyte from another close species. The latter is usually chosen because it is easily reared. In the case of *A.sturio*, injection of the fin cell nucleus could be considered either using A. sturio or in the second case *A. baerii* as a recipient oocyte. Such a procedure, which is very simple to state, bears in fact many biological and technical constraints to a proper embryonic development taking place. Some constraints which are shared by intra and interspecific nuclear transfer include the oocyte enucleation procedure and reprogramming of the injected nucleus. More developmental constraints will arise, however, when the recipient oocyte belongs to a species different from that of the injected nucleus. Based on preliminary results obtained in Cyprinids, Liu et al. (2002) proposed that oocytes from species with the closer relation to the donor fish had better development rates up to blastula stage than those with more distant relations. This issue was addressed from a different perspective for sturgeon in androgenesis experiments (Grunina et al. 2006; Neyfakh 1999). The authors raised the hypothesis of nucleocytoplasmic incompatibilities in androgenetic hybrids, which are not necessarily related to phylogenetic distance between the sperm donor and the oocyte recipient. For example, androgenetic hybrids were viable when enucleated eggs from Stellate (*Acipenser stellatus*) were fertilized with Beluga (*Huso huso*) sperm, but not viable when fertilized with Russian sturgeon (*Acipenser queldenstaedtii*) sperm. How the variable chromosome numbers between these different species interfered with the so-called nucleocytoplasmic incompatibility might be addressed, although egg DNA inactivation should have resolved this issue. After interspecific nuclear transfer, Sun et al.

(2005) and Yan et al. (1991) showed that the tempo of embryo development reflected that of the recipient species. These authors propose that mortalities after hatching might be due to the burning out of all vitelline reserves before the reconstructed fry has developed enough to start feeding. An additional origin of the nucleocytoplasmic incompatibility may be that specific cytoplasmic components brought by the recipient ooplasm do not meet the needs of the developing embryo. Maternal RNA and ooplasmic mitochondria are the best candidates. For instance, the initial number of mitochondria in the oocyte, which is of major importance to ensure proper ATP production necessary to cell division (see the review in mammals by St John et al. 2004), may be inadequate. Regulation of the mitochondrial genome, which results from a precise intercommunication with the nucleus genome, could also be affected, at least in the mitochondria brought by the ooplasm. A very thoughtful strategy is therefore necessary when interspecific nuclear transfer is required, to ensure that the right recipient species will have been identified and will be available at the time the reconstruction is planned.

47.4.3 How operative is nuclear transfer in fish

In all, nuclear transfer is still far from being used routinely for reconstruction strategy. Several groups, however, have developed promising research on this technology, the best results having been obtained after embryonic cell nuclear transfer. The first report of a successful fish reconstruction using interspecific embryonic cell nuclear transfer was published on teleosts by the leading group in nuclear transfer at the Academia Sinica in Wuhan, China (Tung et al. 1965). In this work, the fry resulting from nuclear transfer between two sub-families (*Carassius auratus* × *Rhodeus sinensis*) died, however, few days after hatching. Fertile adults were obtained afterwards by the same group with *Carassius auratus* and *Cyprinus carpio* (Tung and Tung 1980; Yan et al. 1984). These successes followed the first intraspecific nuclear transfers with embryonic donor nucleus, which were published more than 60 years ago in amphibians (Briggs and King 1952) and which were adapted to fish (Niwa et al. 1999; Wakamatsu et al. 2001; Hongtuo and Chingjiang 2001; Gasaryan et al. 1979; Sun et al. 2005; Tung et al. 1963). Although this was several years before the first successful report in mammals (McGrath and Solter 1983), it is in mammals, nevertheless, that the first success of nuclear transfer with somatic donor nucleus was reported, with the birth of the sheep Dolly (Wilmut et al. 1997). This indicated that the difficulty of reprogramming the fully differentiated somatic nuclei to allow embryo development could be overcome. Today, somatic cell nuclear transfer is almost routinely used in mammals to produce fertile adults, especially in sheep, cattle and horses. In fish, somatic cell nuclear transfer is still the objective of active research, as very little development success has been reported so far (Lee et al. 2002; Liu et al. 2002; Siripattarapravat et al. 2009; Le Bail et al. 2010; Bubenshchikova et al. 2007, 2008; Kaftanovskaya et al. 2007; Wakamatsu 2008; Ju et al. 2004; Huang et al. 2003). Nucleo-cytoplasmic interaction after nuclear transfer should be studied further, especially when interspecific nuclear transfer is at stake. Such interspecific study is already being developed by the Wuhan group on fish (Pei et al. 2007, 2008, 2009; Sun et al. 2005), and the number of trials in mammals, aiming to reconstruct endangered animals thanks to domestic species, shows a constant increase. In addition, extensive research is being developed in mammals about nuclear reprogramming, which should benefit fish. Fish species bear several advantages compared to mammals, the first one being the lack of extra embryonic tissues whose alteration in mammalian clones is responsible for many abortions. Moreover, the zygotic genome activation is triggered much later in most fish (around the tenth mitosis) than in mammals (first to second mitosis), a specificity which should be favorable to experimentally induced reprogramming of the embryo. An unknown parameter should be raised in Acipenserids, whose embryos display holoblastic cleavage (Colombo et al. 2007). This cleavage over the entire egg volume is similar to that of mammals, at least during the first cleavages, which is in contrast to the meroblastic cleavage in teleosts, at the top of the yolk material.

47.5 Conclusion

From now on, cryobanking of *A. sturio* somatic cells is at hand, and only minor improvements are needed to fix this issue. Fish reconstruction by somatic cell nuclear transfer, or by the more successful embryonic cell nuclear transfer, is however still at a level of basic research, which is mostly carried out on teleost species. The statement of Corley-Smith and Brandhorst (1999) several years ago is still appropriate for fish: restoration of species and populations by transfer of cryopreserved cell nuclei to eggs must be considered a promising but unproven technology. Other reconstruction strategies should still be explored with the same tenacity. Among them, the perspective of producing interspecific androgenotes using enucleated oocytes from a fish-farmed species which has undergone dispermic fertilization with cryopreserved sperm (Grunina et al. 2006) might be another approach for production of heterozygotic individuals from valuable species using cryobanked material. Such work proves that no single reconstruction strategy should be explored. The reader is referred to the review of Corley-Smith and Brandhorst (1999) for a synthetic description of the strength and pitfalls of these different strategies. From now on, our responsibility is to make sure that several cell types are cryobanked, so that at the time reconstruction is needed, the best reconstruction strategy can be chosen.

Finally, beyond the technical and biological issues of the reconstruction technologies, the utilization of the reconstructed fish should be carefully thought over. In the context of population restoration, we believe that utilization of the reconstructed fish should be restricted to offspring production in controlled facilities. Reconstructed males and females produced by intraspecific nuclear transfer can be crossed, and the genetic identity of the offspring is maintained. Such offspring could therefore be released for population restoration. The issue is very different in the case of broodstock produced by interspecific nuclear transfer, or interspecific androgenesis, because heteroplasmy at the mtDNA level (see the review in mammals by St John et al. 2004) will affect the genomic identity of the animal. Since mtDNA is transmitted by the female, reconstructed fish will transmit mtDNA originating from the recipient genome to their offsprings. To close the loop with the introduction on cryobanking, this may become acceptable only if the original broodstock is dramatically threatened. In this case, it may become preferable to allow some mitochondrial introgression rather than suffer the irremediable loss of the whole genome.

Acknowledgements The study on *Acipenser sturio* fin culture and cryopreservation was funded by the INRA AIP CRB-Bioressources.

References

Akimenko MA, Mari-Beffa M, Becerra J, Geraudie J (2003) Old questions, new tools, and some answers to the mystery of fin regeneration. Dev Dyn 226:190–201

Briggs R, King TJ (1952) Transplantation of living nuclei from blastula cells into enucleated frogs'eggs. Proc Natl Acad Sci USA 38:455–463

Bubenshchikova E, Kaftanovskaya E, Motosugi N, Fujimoto T, Arai K, Kinoshita M, Hashimoto H, Ozato K, Wakamatsu Y (2007) Diploidized eggs reprogram adult somatic cell nuclei to pluripotency in nuclear transfer in medaka fish (*Oryzias latipes*). Dev Growth Differ 49:699–709

Bubenshchikova E, Kaftanovskaya E, Hattori M, Kinoshita M, Adachi T, Hashimoto H, Ozato K, Wakamatsu Y (2008) Nuclear transplants from adult somatic cells generated by a novel method using diploidized eggs as recipients in medaka fish (*Oryzias latipes*). Clon Stem Cell 10:443–452

Cardona-Costa J, Roig J, Perez-Camps M, Garcia-Ximenez F (2006) Vitrification of caudal fin explants from zebrafish adult specimens. Cryo Lett 27:329–332

Colombo RE, Garvey JE, Wills PS (2007) A guide to the embryonic development of the shovelnose sturgeon (*Scaphirhynchus platorynchus*), reared at a constant temperature. J Appl Ichtyol 23:402–410

Corley-Smith GE, Brandhorst BP (1999) Preservation of endangered species and populations: a role for genome banking, somatic cell cloning, and androgenesis? Mol Reprod Dev 53:363–367

Fontana F, Manfredi M, Rossi R, Bronzi P, Arlati G (1995) Established cell lines from three sturgeon species. Sturgeon Quarterly 3:6–7

Fontana F, Lanfredi M, Kirschbaum F, Garrido-Ramos MA, Robles F, Forlani A, Congiu L (2008) Comparison of karyotypes of *Acipenser oxyrinchus* and *A. sturio* by chromosome banding and fluorescent in situ hybridization. Genetica 132:281–286

Gasaryan KG, Hung NM, Neyfakh AA, Ivanenkov VV (1979) Nuclear transplantation in teleost *Misgurnus fossilis* L. Nature 280:585–587

Grunina AS, Recoubratsky AV, Tsvetkova LI, Barmintsev VA (2006) Investigation on dispermic androgenesis in sturgeon fish. The first successful production of androgenetic sturgeons with cryopreserved sperm. Int J Refrig 29:379–386

Hagedorn M, Kleinhans FW, Artemov D, Pilatus U (1998) Characterization of a major permeability barrier in the zebrafish embryo. Biol Reprod 59:1240–1250

Hattori M, Hashimoto H, Bubenshchikova E, Wakamatsu Y (2011) Nuclear Transfer of Embryonic Cell Nuclei to Non-enucleated Eggs in Zebrafish, Danio rerio. Int J Biol Sci 7:460–468

Hongtuo F, Chingjiang W (2001) Nuclear transfer in loach (*Paramisgurnus dabryanus* Sauvage) by cell-to-cell electrofusion. Aquacult Res 32:267–275

Huang H, Ju B, Lee KY, Lin S (2003) Protocol for nuclear transfer in zebrafish. Clon Stem Cell 5:333–337

Ju B, Huang H, Lee KY, Lin S (2004) Cloning zebrafish by nuclear transfer. Meth Cell Biol 77:403–411

Kaftanovskaya E, Motosugi N, Kinoshita M, Ozato K, Wakamatsu Y (2007) Ploidy mosaicism in well-developed nuclear transplants produced by transfer of adult somatic cell nuclei to nonenucleated eggs of medaka (*Oryzias latipes*). Dev Growth Differ 49:691–698

Lamche G, Meier M, Suter M, Burkhardt-Holm P (1998) Primary culture of dispersed skin epidermal cells of rainbow trout *Oncorhyncus mykiss* Walbaum. Cell Mol Life Sci 54:1042–1051

Le Bail PY, Depince A, Chenais N, Mahe S, Maisse G, Labbe C (2010) Optimization of somatic cell injection in the perspective of nuclear transfer in goldfish. BMC Dev Biol 10:64

Lee KY, Huang H, Ju B, Yang Z, Lin S (2002) Cloned zebrafish by nuclear transfer from long-term-cultured cells. Nat Biotechnol 20:795–799

Liu TM, Yu XM, Ye YZ, Zhou JF, Wang ZW, Tong JG, Wu CJ (2002) Factors affecting the efficiency of somatic cell nuclear transplantation in the fish embryo. J Exp Zool 293:719–725

Mauger PE, Le Bail PY, Labbe C (2006) Cryobanking of fish somatic cells: optimizations of fin explant culture and fin cell cryopreservation. Comp Biochem Physiol B Biochem Mol Biol 144:29–37

Mauger PE, Labbe C, Bobe J, Cauty C, Leguen I, Baffet G, Le Bail PY (2009) Characterization of goldfish fin cells in culture: some evidence of an epithelial cell profile. Comp Biochem Physiol B Biochem Mol Biol 152:205–215

McGrath J, Solter D (1983) Nuclear transplantation in the mouse embryo by microsurgery and cell fusion. Science 220:1300–1302

Moritz C, Labbe C (2008) Cryopreservation of goldfish fins and optimization for field scale cryobanking. Cryobiology 56:181–188

Neyfakh AA (1999) Nucleo-cytoplasmic incompatibility of androgenetic hybrids in sturgeons. J Appl Ichthyol 15:318–319

Niwa K, Ladygina T, Kinoshita M, Ozato K, Wakamatsu Y (1999) Transplantation of blastula nuclei to non-enucleated eggs in the medaka, *Oryzias latipes*. Dev Growth Differ 41:163–172

Pei DS, Sun YH, Chen SP, Wang YP, Hu W, Zhu ZY (2007) Identification of differentially expressed genes from the cross-subfamily cloned embryos derived from zebrafish nuclei and rare minnow enucleated eggs. Theriogenology 68:1282–1291

Pei DS, Sun YH, Chen CH, Chen SP, Wang YP, Hu W, Zhu ZY (2008) Identification and characterization of a novel gene differentially expressed in zebrafish cross-subfamily cloned embryos. BMC Dev Biol 8:29

Pei DS, Sun YH, Zhu ZY (2009) Identification of a novel gene K23 over-expressed in fish cross-subfamily cloned embryos. Mol Biol Rep 36:1375–1380

Robles V, Cabrita E, Herraez MP (2009) Germplasm cryobanking in zebrafish and other aquarium model species. Zebrafish 6:281–293

Silvestre MA, Saeed AM, Escriba MJ, Garcia-Ximenez F (2002) Vitrification and rapid freezing of rabbit fetal tissues and skin samples from rabbits and pigs. Theriogenology 58:69–76

Silvestre MA, Saeed AM, Cervera RP, Escriba MJ, Garcia-Ximenez F (2003) Rabbit and pig ear skin sample cryobanking: effects of storage time and temperature of the whole ear extirpated immediately after death. Theriogenology 59:1469–1477

Silvestre MA, Sanchez JP, Gomez EA (2004) Vitrification of goat, sheep, and cattle skin samples from whole ear extirpated after death and maintained at different storage times and temperatures. Cryobiology 49:221–229

Siripattarapravat K, Pinmee B, Venta PJ, Chang CC, Cibelli JB (2009) Somatic cell nuclear transfer in zebrafish. Nat Meth 6:733–735

St John JC, Lloyd REI, Bowles EJ, Thomas EC, El Shourbagy S (2004) The consequences of nuclear transfer for mammalian foetal development and offspring survival. A mitochondrial DNA perspective. Reproduction 127:631–641

Sun YH, Chen SP, Wang YP, Hu W, Zhu ZY (2005) Cytoplasmic impact on cross-genus cloned fish derived from transgenic common carp (*Cyprinus carpio*) nuclei and goldfish (*Carassius auratus*) enucleated eggs. Biol Reprod 72:510–515

Tagliavini J, Williot P, Congius L, Chicca M, Lanfredi M, Rossi R, Fontana F (1998) Molecular cytogenetic analysis of the karyotype of the European Atlantic sturgeon, *Acipenser sturio*. Heredity 83:1–6

Tung TC, Tung YYF (1980) Nuclear transplantation in teleosts. I. Hybrid fish from the nucleus of carp and the cytoplasm of crucian. Sci Sin XXIII:518–523

Tung TC, Wu SC, Tung YYF, Yen SS, Tu M, Lu TY (1963) Nuclear transplantation in fishes. Sci Sin 14:1244–1245

Tung TC, Tung YFY, Luh TY (1965) Transplantation of nuclei between two subfamilies of teleosts (domesticated goldfish *Carassius auratus*, and Chinese bitterling *Rhodeus sinensis*). Acta Zool Sin 19:210–212

Wakamatsu Y (2008) Novel method for the nuclear transfer of adult somatic cells in medaka fish (*Oryzias latipes*): use of diploidized eggs as recipients. Dev Growth Differ 50:427–436

Wakamatsu Y, Ju B, Pristyaznhyuk I, Niwa K, Ladygina T, Kinoshita M, Araki K, Ozato K (2001) Fertile and diploid nuclear transplants derived from embryonic cells of a small laboratory fish, medaka (*Oryzias latipes*). Proc Natl Acad Sci USA 98:1071–1076

Wang G, LaPatra S, Zeng L, Zhao Z, Lu Y (2003) Establishment, growth, cryopreservation and species of origin identification of three cell lines from white sturgeon, *Acipenser transmontanus*. Meth Cell Sci 25:211–220

Wilmut I, Schnieke AE, McWhir J, Kind AJ, Campbell KH (1997) Viable offspring derived from fetal and adult mammalian cells. Nature 385:810–813

Wolf K, Quimby MC (1969) Fish cell and tissue culture. In: Hoar WS, Randall DJ (eds) Fish physiology Vol III. Academic, New York, pp 253–305

Yan SY, Lu DY, Du M, Li GS, Lin LT, Jin GQ, Wang H, Yang YQ, Xia DQ, Liu AZ (1984) Nuclear transplantation in teleosts. Hybrid fish from the nucleus of crucian and the cytoplasm of carp. Sci Sin Ser B Chem Biol Agric Med Earth Sci 27:1029–1034

Yan SY, Mao ZR, Yang HY, Tu MA, Li SH, Huang GP, Li GS, Guo L, Jin GQ, He RF (1991) Further investigation on nuclear transplantation in different orders of teleost: the combination of the nucleus of Tilapia (*Oreochromis nilotica*) and the cytoplasm of Loach (*Paramisgurnus dabryanus*). Int J Dev Biol 35:429–435

Zhou GZ, Gui L, Li ZQ, Yuan XP, Zhang QY (2008) Establishment of a Chinese sturgeon *Acipenser sinensis* tail-fin cell line and its susceptibility to frog iridovirus. J Fish Biol 73:2058–2067

Chapter 48
Some Ex-Situ-Related Approaches for Assessing the Biological Variability of *Acipenser sturio*

Patrick Williot

Abstract Intra-specific biological variability is the core subject of this chapter, the aim being to improve the effectiveness of the long-term restoration programme. This programme is divided into five steps (from brood-stock building to post-release monitoring), the first four being more exclusively part of the ex-situ actions. Each step comprises different key tasks. The following two sections of the chapter deal with the known biological variability of the brood stock, both genetic and behavioural. With regard to genetic variability, recent approaches based on genomics and proteomics are presented, with a brief analysis of the factors that may impact variability. Next, some proposals are made to meet short-term targets, i.e., those which can potentially be applied immediately, then medium- or long-term proposals which need additional research.

48.1 Introduction

Maintaining the fitness of wild individuals and populations is (or should be) the objective of any species management programme. "Populations have demographic and genetic thresholds below which non-adaptive, random forces prevail over adaptive, deterministic forces" (Soulé 1986). A population needs biological variability to ensure its long-term function and thus its survival, so that it is able to produce an evolutionary response (e.g., Bernatchez 2004). This necessity is more acute in endangered species, which require an ex-situ brood stock to circumvent total extinction; and in such cases, wise management of the resource is essential. With regard to *A. sturio*, the question has to be addressed urgently, as the brood stock is limited both in numbers and in its composition. This assessment of

P. Williot (✉)
Sturgeon Consultant, 4 Rue du pas de madame, 33980 Audenge, France
e-mail: williot.patrick@neuf.fr

P. Williot et al. (eds.), *Biology and Conservation of the European Sturgeon Acipenser sturio L. 1758*, DOI 10.1007/978-3-642-20611-5_48,

biological variability should be primarily conducted during the ex-situ phases of the restoration programme, to better organize its preservation.

The objective of this short chapter is to outline tentatively some of the fields which should be investigated to optimise the management of the brood stock and thus produce the best quality progenies for adaptation to the wild, while maximizing their chances of returning to spawn.

In this chapter we (1) outline the different phases of the hatchery-stocking process, (2) summarize the experience of the ex-situ measures for the species with a special focus on its biological diversity, (3) present briefly the tools available for biological variability assessment, and (4) propose some directions for both short- and long-term future work to improve our knowledge of biological variability and its potential applications.

48.2 Context of the Hatchery-Stocking Process

The term hatchery stocking, coined by Araki and Schmid (2010), is understood to also include monitoring of the released fish. The five steps in the hatchery-stocking process are summarized in Table 48.1. The process can be logically divided into five homogeneous phases. The first is building the brood stock, which includes the constitution of the brood stock (number, origin, age, etc.), rearing conditions in term of structures (tanks or ponds, size, etc.), water quality (temperature range,

Table 48.1 Outline of the five-step process for hatchery-stocking with the aim of either sustaining, enhancing, restoring, or conserving a fish species, with special reference to *Acipenser sturio*

Step	Building a brood stock 1	Reproducing 2	Rearing of fingerlings 3	Releasing 4	Monitoring 5
Related task	Development of brood stock, Rearing conditions (tanks, water, etc.), Current management, (density, feeding, etc.), Long-term management (replacement, sex ratio, origin, etc.) Current genetic variability	Managing brood fish (last months prior to reproduction, fasting), Preselecting and selecting brood fish fit for reproduction, Reproduction procedure, Mating, Gamete managing	Rearing conditions (tanks, water, light, etc.), Feeding (food items , sequences), Density	Preparing (ensuring disease-free, counting, etc.), Marking/ tagging, Transporting	Movements, Survival, Growth
	Current genetic variability, Transcriptome, proteome behaviour	*Reproduction and crossing guides*	*Current genetic variability, Behaviour, Transcriptome, proteome*	*Current genetic variability (in case step 3 not achieved)*	*Current genetic variability, Transcriptome, proteome*

In italics are suggested tasks to consider in the further restoration programme salinity level, oxygen etc.), environmental conditions (lighting, isolation from noise, inside/outside the building), current management strategies (density, feeding, cleaning, monitoring of fish), and long-term management (strategy for improving knowledge, research promotion, risk assessment). All the above-mentioned stages have already been reported and detailed in Williot et al. (2007) and in Williot et al. (Chaps. 31 and 32).

The second step deals with the management of brood-fish from the few months before reproduction until reproduction itself, selection of the most suitable brood-fish for reproduction, management of gametes, and the organization of reproductions.

The third step consists of fingerling rearing, with the main sub-items of rearing conditions (structures, water, and environmental characteristics), feeding (type of food and delivery), and managing (cleaning, removing mortalities, monitoring for health and growth).

Step four deals with release. Preparing (counting, health status, weaning), marking and/or tagging, and transporting the fish to be released are the three main tasks.

Finally, the fifth step focuses on monitoring released fish in the wild. This is a key step, as it is the only way to obtain data on the effectiveness of the preceding steps (Williot et al. 2009a, b).

This very classical scheme will be used as a reference to assess biological variability, which ideally should be known at each step of the hatchery-stocking process.

48.3 Status of Knowledge on Brood Stock Building and Managing

48.3.1 Brood Stock Development and Genetic Characteristics

The *A. sturio* brood stock consists of three groups of fish; group 1 are the oldest fish, with only one female, group-2 fish were born in 1994 in the wild (last recorded natural spawning), and group-3 fish are the progenies of a couple of wild brood fish artificially reproduced in 1995. Thus, they are full siblings. The few fish that are still held in Berlin (Germany) are exclusively group-3 fish. As a result, the present brood stock includes a very limited number of unrelated specimens (see Chap. 31, Williot et al., for details). The current known genetic variability of the brood stock is summarized in Table 48.2. Genetic variability of *A. sturio* was assessed in the late 1990s and the early 2000s (Ludwig et al. 2000, 2004). Eleven primer pairs (Afu-19, Afu-22, Afu-23, Afu-34, Afu-39, Afu-54, Afu-57, Afu-58, Afu-62, Afu-68, Afu-69) were used in the previous study and six (Afu-19, Afu-39, Afu-54, Afu-68, Aoxy-23, Aoxy-45) in the second study. The microsatellites were designed for the Lake sturgeon, *Acipenser fulvescens*, and for the Atlantic sturgeon, *Acipenser oxyrinchus*.

Table 48.2 Genetic variability of *Acipenser sturio* brood stock by the early 2000s in France (Fr) and Germany (D)

Criterion	Archival specimen	Artificially-reproduced in 1995 (group 3)	Wild-originated fish (group 1)	Born in the wild in 1994 (group 2)
		Nuclear DNA		
No. of alleles	19 ($n = 38$)[a]	10 ($n = 87$, Fr)	24 ($n = 28$)[b] (Fr)	14 ($n = 38$) (Fr)
Level of homozygosity		0.72	0.45	0.46
		Mitochondrial DNA		
No. of haplotypes	3 (A,B[d]) North Sea C[e] (Tiber River)	1 (A[c])	1 (A)	1 (A)
		← Ludwig et al. (2004) →		
No. of alleles		8 ($n = 27$, D)		
	← Ludwig et al. (2000) →			

Level of homozygosity and number of alleles from six microsatellite loci. Alleles were amplified using primer pairs designed for *Acipenser fulvescens* (Afu-19, Afu-39, Afu-54, Afu-68) and *Acipenser oxyrinchus* (Aox-23, Aox-45) (Ludwig et al. 2004)

[a]Archival specimens were analysed with five primer pairs designed for *A.fulvescens* (Afu-19, Afu-34, Afu-39, Afu-54, Afu-68) (Ludwig et al. 2000)

[b]Live and dead specimens combined

[c]The sequence is saved in EMBL as AJ249673 (Ludwig et al. 2004)

[c, d & e]The A, B, and C for the haplotypes do not refer to any previous spelling

With reference to the allelic diversity of archival specimens ($n = 19$) (Table 48.2), live specimens exhibited either a higher richness ($n = 24$, group-1 fish) or lower ($n = 14$, group-2 fish and $n = 10$, group-3 fish). As a result, heterozygosity is greater in both group-1 and -2 fish which are of wild origin, in contrast to the group-3 fish which are full sibling progeny of a pair of wild-caught brood fish which reproduced in 1995 (Williot et al. 2000).

Three haplotypes were isolated from *A. sturio* (living and archival specimens) (Ludwig et al. 2000). Two of these were observed in present collapsed populations, one in the Tiber River (Italy) and the other from one sample in the North Sea. The third haplotype, and the most common, is present in the French Atlantic coast and North Sea. Recent palaeogenetic investigations (see Chap. 8) on *A. sturio* at the European scale revealed a more diverse past status of genetic variability with regard to haplotypes.

Average relatedness, determined by Ludwig et al. (2004) with 0.62 in group-1 and 0.81 in group-2 fish, reveals that these specimens are closely related, although specimens with rare alleles were listed (Ludwig et al. 2004). Unfortunately, the great majority were sequenced by sampling dead fish, and "no rare alleles were observed in group-2 fish born in the wild in 1994" (Ludwig et al. 2004). Three fish left showing rare alleles in the survey belonged to group 1 (wild originated fish).

Based upon the data for relatedness and allelic distribution, Ludwig et al. (2004) suggested that it was most likely that only one pair of fish was involved in the last natural reproduction in 1994 (group-2 fish). However, as the segregation of the alleles of three of the markers does not follow a Mendelian mode of inheritance, it has been suggested that two females and one male or two males and one female could have been involved in the 1994 spawning (Ortega-Villaizàn Romo, personal communication).

A last result is worth noting with regard to the genetic variability of brood fish, thanks to a series of 12 new microsatellites and a further three lines of analysis (Chap. 34). The sampled brood fish are all the group-1 fish (the oldest), most of the group-2 fish, and a batch of group-3 fish. The authors showed that brood fish could be separated mainly into two groups, and drew up a table with the most distant recommended crossing of reproductive partners.

All the aforementioned deal with brood fish, i.e., with step 1 of the table, and provide some points to consider with regard to step 2, i.e., how to better organize crossing reproduction while maintaining genetic variability.

48.3.2 *Inter- and Intra-Individual Variability in Fish Behaviour*

Throughout the ex-situ rearing of brood stock within the restoration programme, major variability was observed in the response of the fish. Inter-individual differences were observed in adaptive characteristics towards rearing (delay in resuming food intake), in growth curves of wild-originated fish, in growth performance following a change in tank size (1 m versus 2 m depth) (Williot et al. 2007), and in food intake preference (Henzel et al. 2002). Hensel et al. also showed that fish may change their food preference in a very short time span, indicating their ability to switch their behaviour rapidly. It has further been demonstrated that specific growth rates varied greatly, thus limiting the interpretation of the response using this criterion (Kirschbaum et al. 2006).

For much of the growth data, the variation range increased rapidly, so that significant differences were not statistically supported, e.g., growth curves of batches held in fresh and brackish water (Williot et al. 2007). This means that, even within closely related fish [at least half sibs in the present case (group-2 fish born in the wild in 1994)], sturgeon may exhibit very variable responses of unknown origin. Also, it has been shown that the maturation of males exhibited very different cycles, such as repeated yearly maturation and a recurring 2-year cycle followed by many years of sexual rest, among others (Williot et al. 2007, Williot and Rouault 2008).

48.3.3 *What Influences Biological Variability? How Can It Be Measured?*

It has been shown that the known available genetic diversity of the brood stock is somewhat low, mainly due to a loss of two haplotypes for the species. However, a great uncertainty remains in the field. Low haplotype diversity indicates that the species (or population) is experiencing or has experienced a bottleneck, or has undergone recent expansion or selection. Therefore, some questions arise: (1) what was the level of genetic diversity of the Garonne-basin population at a time when it was still functional, for instance one century ago? (2) is it possible to address the question? (3) is there any chance of increasing the genetic variability of the present brood stock by incorporating newcomers with a rare genetic background?

What is to be done with the available data? The data on rare alleles observed in the previous brood stock constitutes preliminary information on genetic variability that should be preserved. However, "rarer alleles are more likely not to be transmitted to any offspring in the next generation than are more common alleles" (Lacy 1995). This means that the heritability of rare characteristics is not guaranteed. This is also illustrated in the case of the sterlet sturgeon (*Acipenser ruthenus*), as it has been shown that fertilization rates depend also on the interaction of both gametes and not only on each gamete (Williot et al. 2005).

Are the current genetic techniques the only means to detect biological variability? Recent new genetic approaches, genomics and proteomics have proved their effectiveness, as illustrated by a synthesis to explore the quality of fish progenies (Cerdà et al. 2008). Functional genomics and proteomics are the expression of the genes and proteins respectively at tissue level. The technologies of genome sequencing have recently been extended to partially sequence the genome and transcriptome, i.e., the mRNA repertoire of a tissue or the expressed sequence tags (ESTs) from normalized or subtracted cDNA[1] libraries (Cerdà et al. 2008). The power of the proteomic studies has been shown in other fields such as crowding, which impacts on muscle and blood plasma proteomes of Atlantic salmon, *Salmo salar* (Veiseth-Kent et al. 2010). Several significant variations in abundance of some proteins and their isoforms in the muscle of gilthead sea bream, *Sparus aurata*, were detected, related to growth and environmental factors (Addis et al. 2010). Proteomics made possible the discrimination of spermatozoa of five sturgeon species and one paddlefish (Li et al. 2010). Among the other many useful aspects of proteomics, we should point out the distinction of sperm proteins responsible for infertility in humans (Ainsworth 2005).

Is genetics the only source of variability? As described briefly above, in many ways, *A. sturio* have exhibited a remarkable heterogeneity in their response to a given situation. As mentioned by Shumway (1999), "behaviour diversity needs to be explicitly conserved, as the stability of a population can be affected by individual variability in reproductive behaviour, dominance, and spatial distribution". Moreover, the author stated that maximizing genetic diversity may not adequately conserve behavioural diversity, as some traits of life are not necessarily detectable with current genetic techniques. Also, part of behavioural diversity is learned or determined environmentally (Shumway 1999; Frost et al. 2007; Magnhagen 2007; Oosten et al. 2010). In support of this statement is a reported experiment on the

[1]cDNA are complementary DNA realised from ARMm thanks to an enzyme.

effects of changing environments in juvenile cod that generate results in terms of more flexible fish, probably more ready to adapt in the wild after release (Braithwaite and Salvannes 2005). Learning might be a likely explanation for a reduced variability in self-feeder adaptation in amago salmon (*Oncorhynchus masou masou*) in duoculture with rainbow trout (*Oncorhynchus mykiss*) in contrast with monoculture (Flood et al. 2010). Frost et al. (2007) showed that rainbow trout (*Onchorynchus mykiss*) changed its boldness according to its previous experience, thus demonstrating that learning might be a key factor for future adaptive (or maladaptive) behaviours.

The impact of the environment on a fish – either brood fish or juvenile for release – is rarely assessed systematically. The impact of stocking density (6 fish m^{-2} and 3 fish m^{-2}) in ponds of stellate sturgeon offspring (*Acipenser stellatus*) on characteristics of rearing, as well as on genetic trait loci associated with fitness, was reported by Ryabova et al. (2006). One of the main conclusions was that genetic variability increased with decreasing densities. Similarly, it has been demonstrated that the behaviour of young perch (*Perca fluviatilis*) might be modified according to previous social interactions (Magnhagen 2007; Oosten et al. 2010).

Indeed, minimizing adaptation to captive conditions requires increasing the number of individuals per generation kept in captivity (Lacy 2000; Williams and Hoffman 2009). If this is unfeasible, then Williams and Hoffman suggest delaying reproduction and cryopreserving the germplasm. Potential changes caused by artificial selection through farming have been reported in salmon, by changes in transcription profile genes in progenies (larvae prior to exogenous feeding) of farmed and wild specimens (Roberge et al. 2006).

As a result of the brief overview, known genetic variability has been focused on brood fish with primers sequenced from other sturgeon species. There are many signs of biological variability through behaviours, and physiological responses that should also be preserved. Recent genetic-related approaches are promising for assessing new aspects of variability, as well as providing tools to trace the different options of the hatchery-stocking tasks. Finally, there is support for interactions between genetic characteristics and the environment that might be of the utmost importance, either in detecting the consequences of rearing conditions or in further assessing adaptation and survival in the long term. The abovementioned comments deal either with the brood fish (i.e., steps 1 and 2 of Table 48.1), or with young fish, i.e., step 3, in giving examples of how biological variability might be approached.

48.4 Proposals

Two levels of recommendation are proposed. The first deals with the short-term application of known and available materials and methods. The second level identifies some lines of thought for future improvements.

48.4.1 Short-term Suggestions

A key prerequisite for future tasks, including investigations on biological variability, lies with the effective management of brood stock reproduction. Special attention must therefore be paid from the few months preceding reproduction until reproduction per se. The methods have often been described (Williot et al. 2009b), including in the present volume (see Chap. 32); therefore, any failure should be analysed in depth to ensure the presence of yearly progenies.

The composition of the brood stock has changed from when the current preliminary genetic screening was performed. This is due to some mortalities and the arrival of a small number of newcomers in the group-1 fish. The description of the genetic variability of the present brood stock now needs to be updated, using similar methods and tools as for former studies (Ludwig et al. 2000, 2004). After this, a synthesis should be established by combining our results with those of Tiedemann et al. (see Chap. 34), who showed that thanks to 12 new loci, brood fish could for the most part be separated into two groups, and also proposed the most distant crossing pairs. In the same field, more preliminary work has been carried out by building an enriched library which is ready for testing 6–8 loci in *A. sturio* (Congiu, personal communication). There is no doubt that the achievements of the studies mentioned above should be promoted to provide the first homologous microsatellites in *A. sturio,* which will be so useful in promoting further species genetic variability assessment, as illustrated by a repatriation programme on Galapagos tortoises (Milinkovitch et al. 2004).

Even though the chances are very low with regard to the status of the species, every opportunity of including a newcomer (from a genetic point of view) must be seized.

In the present situation, where two out of the three groups (group 2 and group 3) of brood stock are either half- or full siblings respectively, crossing fish within the two groups has to be avoided to minimize kinship and thus limit inbreeding (Lacy 2000; Bernatchez 2004; Ludwig et al. 2004). In addition, "every attempt should be made to breed animals with rare alleles which are at risk of being lost to the next generation, because of lack of prior opportunity to breed, poor behavioural, physiological, and/or genetic adaptation to the captive environment, or just bad luck" (Lacy 2000). And this is particularly true for small populations, as illustrated by *A. sturio*. According to the simulation, "To be effective, supplemental breeding programs require crossing strategies that maximize offspring production while maintaining genetic diversity within each supplemental year class" (Fiumera et al. 2004). The authors demonstrated that a factorial crossing design was better than monogamous crossing. This is to be applied in cases where every specimen is of equal genetic value.

The actions suggested above, in italics in the table, are targeting steps 1and 2 of the hatchery-stocking process (Table 48.1).

So that time will not be lost, it is strongly recommended that building the framework for medium- and long-term studies be set in motion as soon as possible.

48.4.2 Medium- and Long-term Suggestions

The only chance of gaining a broader knowledge of past genetic variability of the population is to carry out a systematic investigation of the available material from excavations of sturgeon remains in France (Desse-Berset 2009; Chap. 7). The present volume presents preliminary results for such an investigation (see Chap. 8), and this work should now be extended. This would give a better representation of the current genetic variability available in the brood stock (step 1 of the hatchery-stocking process).

We would then be able to compare former (wild) genetic variability with the present brood stock, as Chebanov et al. (2002) did for Stellate sturgeon (*Acipenser stellatus*) and Russian sturgeon (*Acipenser gueldenstaedtii*) populations in the Azov sea by using the allelic frequency in two loci (albumin and transferrin) and computing the average heterozygosity.

It has been recommended that genetic characteristics in wild and captive populations be compared by developing hyper-variable microsatellites and amplified fragment polymorphism markers, to ensure that genetic diversity in the hatchery offspring approximates that of the wild population (Waldman et al. 2008). This is of the utmost importance for sturgeon restoration programmes because of the very limited number of brood stock, as illustrated by *A. sturio*. Moreover, the above-quoted authors mentioned that in the case of fewer loci, the techniques may even enable familial origin of hatchery-released offspring to be identified.

One of the deleterious consequences of managing an ex-situ brood stock comes from a domestication selection that favours some traits in captivity, which are most likely not favoured by natural conditions (Araki and Schmid 2010). It was mentioned above that a limitation on the number of generations kept in captivity should be the rule, as well as the cryopreservation of sperm (Lacy 2000; Williams and Hoffman 2009). In the absence of any present precise reference in the means for measuring changes, how can we apply these general rules? A consistent and wise management plan should be set up. In practice, some basic points could be suggested: (1) holding similar rearing conditions to those described for the French brood stock, which were established to mimic natural conditions as far as possible and which proved effective (Williot et al. 2007; Chaps. 31 and 32), (2) maintaining a low stocking density to limit close constant relationships between fish, (3) maintaining the current monitoring of brood fish to accumulate data on the species, and (4) mixing the sexes. Further potential changes would be programmed with part of the next generation of brood fish.

As previously mentioned, our uncertainty about the transmission of genetic characteristics from parents to progenies represents a major problem. Genetic screening of released juveniles and recaptures is therefore a key feature that should be established to determine the genetic characteristics and the associated performance of released fish. This might be helpful in interpreting their long-term survival. This involves steps 1, 3, 4, and 5 of the hatchery-stocking process. It was this type of screening that made it possible to detect skew in natural crossing of the endangered Giant Galapagos tortoise, and facilitated a change in the breeding programme to alleviate this problem (Milinkovitch et al. 2004).

Environmental conditions and hatchery practices, from brood stock to fingerlings (when fingerlings are the size targeted for release) and management (feeding, density, behaviour, etc.), can influence the quality of offspring (Shumway 1999; Le Vay et al. 2007). Extensive screening should be undertaken of the transcriptome of both parents and progenies (steps 1, 2, and 3 of the hatchery-stocking process) according to rearing conditions, to further detect the influence of the hatchery-stocking process on fingerlings to be released (step 4) and to assess the impact of larval and rearing production. The study should then be extended to recaptured stocked fish (step 5) to assess the characteristics of those that survived.

With regard to the abovementioned proposals, it is recommended that the various patterns of fish behaviour be characterized. This should be done at brood fish and progeny levels, i.e., at steps 1 and 3 of the hatchery-stocking process. Obviously, this is dependent on rearing conditions.

With the exception of suggestions/recommendations that deal with archaeological remains, suggestions are given in italics in Table 48.1.

48.5 Conclusions

There are many indicators for biological variability through genetics and behaviour that should be preserved. Recent genetic-related approaches are promising for assessing new aspects of variability, as well as providing tools to determine different options for the hatchery-stocking process. Finally, there is support for interactions between genetic characteristics and the environment which might be of the utmost importance, either to detect the consequences of rearing conditions or to further assess adaptation, and later survival, in a long-term perspective.

The abovementioned broad understanding of variability should be applied in any restoration programme. In the case of *A. sturio* in particular, this is becoming crucial, as the present population is now exclusively dependent upon stocking activities. A programme must be started as soon as possible; even changes are likely to be difficult to assess, due to the low level of genetic changes in sturgeon. A restoration programme such as this will need constant revision and the inclusion of innovative approaches for many years, even decades (Peterson et al. 2006).

This chapter has not taken into account all the potential factors that may impact biological variability. The purpose was rather to provide methods supported by examples that could be applied to improve the restoration programme and ensure that it becomes successful. However, as for brood-fish management, as a precaution, a low stocking density and natural food delivery should be favoured for present larval rearing. At the same time, a thorough analysis and further tests should be carried out, as recommended in the paragraph on short-term suggestions.

References

Addis MF, Cappucinelli R, Tedde V, Pagnozzi D, Porcu MC, Bonaglini E, Roggio T, Uzzau S (2010) Proteomic analysis of muscle tissue from gilthead sea bream (*Sparus aurata*, L) farmed in offshore floating cages. Aquaculture 309:245–252

Ainsworth C (2005) The secret of sperm. Nature 436:770–771 (11 August)

Araki H, Schmid C (2010) Is hatchery stocking a help or harm? Evidence, limitations and future directions in ecological and genetic surveys. Aquaculture 308:S2–S11

Bernatchez L (2004) Considérations génétiques et protocole de reproduction relatifs au plan de rétablissement du chevalier cuivré (*Moxostoma hubbsi*). Rapport technique 16–22, viii + 38p. Ministère des Ressources naturelles, de la Faune et des Parcs. Direction de l'aménagement de la faune de Montréal, de Laval et de la Montérégie

Braithwaite VA, Salvannes AGV (2005) Environmental variability in the early rearing environment generates behavioural flexible cod: implications for rehabilitating wild populations. Proc R Soc B. doi:10:1098/rspb.2005.3062

Cerdà J, Bobe J, Babin PJ, Admon A, Lubzens E (2008) Functional genomics and proteomic approaches for the study of gamete formation and viability in farmed finfish. Rev Fisheries Sci 16(S1):54–70

Chebanov MS, Karnaukhov GI, Galich EV, Chmir YuN (2002) Hatchery stock enhancement and conservation of sturgeon, with an emphasis on the Azov Sea populations. J Appl Ichthyol 18:463–469

Desse-Berset N (2009) First archaeozoological identification of Atlantic sturgeon (*Acipenser oxyrinchus* Mitchill 1815) in France. CR Palevol 8:717–724

Desse-Berset N, Williot P (2011) Emerging questions from the discovery of the long term presence of *Acipenser oxyrinchus in France*. J Appl Ichthyol 27:263–268

Fiumera AC, Porter BA, Looney G, Asmussen MA, Avise JC (2004) Maximizing offspring production while maintaining genetic diversity in supplemental breeding programs of highly fecund managed species. Conserv Biol 18:94–101

Flood MJ, Noble C, Kagaya R, Damsgård B, Purser GJ, Tabata M (2010) Growing amago and rainbow trout in duoculture with self-feeding systems: implications for production and welfare. Aquaculture 309:137–142

Frost AJ, Winrow-Giffen A, Ashley PJ, Sneddon LU (2007) Plasticity in animal personality traits: does prior experience alter the degree of boldness? Proc R Soc B 274:333–339

Henzel E, Kirschbaum F, Williot P, Wirth M, Gessner J (2002) Restoration of the European sturgeon, *Acipenser sturio* L., 1758 in Germany: effect of different feed items on specific growth rates of large juvenile fish. Int Rev Hydrobiol 87:539–551

Kirschbaum F, Hensel ECK, Williot P (2006) Feeding experiments with the European Atlantic sturgeon, *Acipenser sturio* L., 1758 to accustom large juveniles to a new feed item and the influence of tank size and stocking density on growth. J Appl Ichthyol 22:307–315

Lacy R (1995) Clarification of genetic terms and their use in the management of captive populations. Zoo Biol 14:565–578

Lacy RC (2000) Should we select genetic alleles in our conservation breeding programs? Zoo Biol 19:279–282

Le Vay L, Carvalho GR, Quinitio ET, Lebata JH, Ut VN, Fushimi H (2007) Quality of hatchery-reared juveniles for marine fisheries stock enhancement. Aquaculture 268:169–180

Li P, Hulak M, Rodina M, Sulc M, Li ZH, Linhart O (2010) Comparative protein profiles: potential molecular markers from spermatozoa of Acipenseriforms (Chondrostei, Pisces). Compar Biochem Physiol Part D 5:302–307

Ludwig AN, Jenneckens I, Debus L, Ludwig A, Becker J, Kirschbaum F (2000) Genetic analysis of archival specimens of the Atlantic sturgeon *Acipenser sturio* L., 1758. Bol Inst Esp Oceanogr 16:221–230

Ludwig A, Williot P, Kirschbaum F, Lieckfeld D (2004) Genetic variability of the Gironde population of *Acipenser sturio*. In: Gessner J, Ritterhoff J, (eds) Species differentiation and population identification in the sturgeons *Acipenser sturio* L. and *Acipenser oxyrinchus*. Bundesamt für Naturschutz 101: 54–72

Magnhagen C (2007) Social influence on the correlation between behaviours in young-of-the-year perch. Behav Ecol Sociobiol 61:525–531

Milinkovitch MC, Monteyne D, Gibbs JP, Fritts TH, Tapia W, Snell HL, Tiedemann R, Caccone A, Powell JR (2004) Genetic analysis of a successful repatriation programme: giant Galápagos tortoises. Proc Roy Soc Lond B271:341–345

Oosten JE, Magnhagen C, Hemelrijk CK (2010) Boldness by habituation and social interactions: a model. Behav Ecol Sociobiol 64:793–802

Peterson DL, Vescei P, Jennings CA (2006) Ecology and biology of the lake sturgeon: a synthesis of current knowledge of a threatened North American *Acipenseridae*. Rev Fish Biol Fisheries. doi:10:1007/s11160-006-9018-6

Roberge C, Einum S, Guderley H, Bernatchez L (2006) Rapid parallel evolutionary changes of gene transcription profiles in farmed Atlantic salmon. Mol Ecol 15:9–20

Ryabova GD, Klimonov VO, Afanas'ev KI, Vyshkvartsev DI, Moskaleichik FF, Rubtsova GA (2006) Variation in morphometric and genetic characteristics of stellate sturgeon juveniles raised at different densities. Russ J Genet 42:182–191

Shumway CA (1999) A neglected science: applying behaviour to aquatic conservation. Environ Biol Fishes 55:183–201

Soulé M (1986) (ed) Conservation Biology. The science of scarcity and diversity. Sinuauer Associates, Inc. Publishers, Sunderland, Massachusetts. 584p

Veiseth-Kent E, Grove H, Færgestad EM, Fjæra SO (2010) Changes in muscle and blood plasma proteomes of Atlantic salmon (*Salmo salar*) induced by crowding. Aquaculture 309:272–279

Waldman JR, Doukakis P, Wirgin I (2008) Molecular analysis as a conservation tool for monitoring the trade of North American sturgeons and paddlefish. J Appl Ichthyol 24(suppl 1):20–28

Williams SE, Hoffman EA (2009) Minimizing generic adaptation in captive breeding programs: a review. Biol Conserv 142:2388–2400

Williot P, Rouault T (2008) Basic management for spawning the cultured sturgeon, *Acipenser sturio* L., 1758, a critically endangered species. Cybium 32(2 suppl):334–335

Williot P, Brun R, Pelard M, Mercier D (2000) Unusual induced maturation and spawning in an incidentally caught pair of adults of the critically endangered European sturgeon *Acipenser sturio* L. J Appl Ichthyol 16(6):279–281

Williot P, Brun R, Rouault T, Pelard M, Mercier D, Ludwig A (2005) Artificial spawning in cultured sterlet sturgeon, Acipenser ruthenus L., with special emphasis on hermaphrodites. Aquaculture, 246:263–273

Williot P, Rouault T, Pelard M, Mercier D, Lepage M, Davail-Cuisset B, Kirschbaum F, Ludwig A (2007) Building a broodstock of the critically endangered sturgeon *Acipenser sturio* L.: problems associated with the adaptation of wild-caught fish to hatchery conditions. Cybium 31:3–11

Williot P, Rochard E, Kirschbaum F (2009a) Acceptability and prerequisites for the successful introduction of sturgeon species. In: Carmona R, Domezain A, Garcia Gallego M, Hernando-Casal J, Rodriguez F, Ruiz Rejon M (eds) Biology, conservation and sustainable development of sturgeons. Fish & fisheries series. Springer, Heidelberg, pp 369–384

Williot P, Rouault T, Pelard M, Mercier D, Jacobs L (2009b) Artificial reproduction of captive endangered European Atlantic sturgeon, *Acipenser sturio*. Endang Species Res 6:251–257. doi:10.3354/esr00174

Part VI
General Conclusions

Chapter 49
What Is the Future for the European Sturgeon?

Patrick Williot, Eric Rochard, Nathalie Desse-Berset, Jörn Gessner, and Frank Kirschbaum

49.1 An Overview of the Content

The present book appears 20 years after the peer-reviewed proceedings of "Acipenser", the first International Symposium on sturgeon (ISSI) (Williot 1991) where two preliminary investigations on the European sturgeon were presented (Castelnaud et al. 1991; Rochard et al. 1991).

What is the present assessment after such a lapse of time?

There are 49 chapters written by almost 70 collaborators with 27 different first authors from nine different countries; seven are European and two North American. Half of the first authors are French, one fifth are German. This illustrates the relative importance of these two countries in research into this species. It also reflects on the one hand the fact that France is the last country where the species is still present, and on the other hand the fruitful French–German scientific cooperation (see Chaps. 20 and

P. Williot (✉)
Sturgeon Consultant, 4 Rue du Pas de madame, 33980, Audenge, France
e-mail: williot.patrick@neuf.fr

E. Rochard
Cemagref, Estuarine Ecosystems and Diadromous Fish Research Unit, 50 Avenue de Verdun, 33612 Cestas Cedex, France

N. Desse-Berset
Université de Nice-Sophia Antipolis – CNRS-CEPAM-UMR 6130, SJA3, 24 avenue des Diables Bleus, 06357 NICE Cédex 4, France

J. Gessner
Leibniz Institute of Freshwater Biology and Inland Fisheries, Müggelseedamm 310, 12587 Berlin, Germany

F. Kirschbaum
Faculty of Agriculture and Horticulture, Biology and Ecology of Fishes, Humboldt University, Philippstr. 13, Haus 16, 10115 Berlin, Germany
e-mail: Frank.kirschbaum@staff.hu-berlin.de; frank.kirschbaum@staff.hu-berlin.de

P. Williot et al. (eds.), *Biology and Conservation of the European Sturgeon Acipenser sturio L. 1758*, DOI 10.1007/978-3-642-20611-5_49,

38). All European countries that have in the past been home to significant populations of European sturgeon are represented by overviews of their sturgeon fisheries, with the exception of Spain. It is worth noting that Spanish European sturgeon-related issues were the subject of a workshop in Madrid and Seville in 1999 which led to a specific publication (Elvira 2000), and to a lesser extent in Carmona et al. (2009). Additionally, the sturgeon-species issue in Spain is discussed in Chap. 9.

In addition, the large representation of European contributors, whatever the past-historical importance of the species in the different European countries, is a good indicator of what European scientists are able to initiate when faced with a conservation question. It is also an indication of the potential of this emblematic fish to become a European umbrella species for fish and habitat protection. This contrasts with the absence of any appropriate management of fish populations, especially migratory fish such as the European sturgeon.

The main objective of this book is to gather together all the knowledge that can usefully contribute to safeguarding the species, and thus provide a basis on which to build the future of the species. In so doing, it is worth noting that more than 30 disciplines or sub-disciplines have been mobilised, which only serves to illustrate the difficulties with which the European sturgeon conservation programme is faced. These disciplines and sub-disciplines are: archaeology, archaeozoology, behavioural sciences, biology, reproduction biology, biometry, communication sciences, conservation biology, cooperation, cryoconservation, cytogenetics, ecology, feeding sciences, fisheries, genetics, geography, geomatics, history, husbandry, hydrobiology, hydrology, iconography, legislation, management, modelling, morphology, ontogenesis, osteometry, palaeogenetics, palaeogeography, physiology, policy, population dynamics, and sedimentology.

49.2 What Are the Main Inputs of This Book Dedicated to European Sturgeon Conservation?

49.2.1 Historic Changes in European Sturgeon Fisheries

For the first time, historic changes in European sturgeon fisheries are described, especially in France (see Chaps. 13 and 20), Germany (see Chap. 14), and Italy (see Chap. 16). The present status of the species complements previous analysis for Poland (see Chap. 15), or provides information for the very first time, as for Georgia (see Chap. 17).

A few conclusions can be drawn. Fisheries regulations or species conservation restrictions, if any, have almost never been in tune with the ecology of the species. Most year-classes could be legally fished (Fr), but regulations were never effectively implemented, as we saw in the late 1990s, when the recaptures of tagged fish could not be explained unless illegal captures of large juveniles had taken place at sea in French shelves close to the Gironde estuary (see Chap. 22). This lowered overall survival considerably. More recently (2003), protected fish captured at sea were auctioned in several countries (Fr, UK, Nl, Be, De, It), and this was even with the primary agreement of the fisheries administration (Fr) (see Chap. 20). Proposals from scientists to protect the resource and then to preserve long-term income for fishermen were never heeded (see Chap. 20). Some attempts were made in Germany at the turn of the twentieth century, with early reproduction of wild-originated brood fish, and further larval stocking and spawning in protected areas (see Chap. 14). Unfortunately, pressures from other economic sectors, together with the past inability to provide evidence of the effectiveness of restocking and the cost of this action, led to these efforts being rapidly halted. It is worth noting that the lack of management in sturgeon fisheries, with no secure long-term capture statistics (nor any studies on past population dynamics when populations were commercially exploited) led to the impossibility of obtaining any references on the past status of the functioning of sturgeon populations.

This absence of management that has been highlighted for fisheries is also true for species conservation in France, where no effective policy exists, as shown by the extremely late ban that was implemented, and also the lack of support for those engaged in this type of actions.

49.2.2 New Insights

Archaeozoology is omnipresent in the book, either directly (see Chaps. 3 and 7) or indirectly by providing material for genetic studies of past sturgeon populations (see Chaps. 8, 9, 42). The most prominent result deals with the former presence in France of another sturgeon species, the Atlantic sturgeon (*Acipenser oxyrinchus*), the so-called sister species of the European sturgeon, *Acipenser sturio* (see Chap. 7). This confirms and extends recently published field data (Desse-Berset 2009a, b; Desse-Berset and Williot 2011). With the exception of the French Mediterranean coast, both species – the European and Atlantic sturgeons – were present simultaneously from the Neolithic until the seventeenth century on the French Atlantic façade, the Channel and even close to a river (Scarpe) in the Scheldt basin, i.e., the North Sea basin. In many cases, sturgeon remains were excavated from sites far upstream, thus confirming the very recent past native status of the Atlantic sturgeon in France, especially as remains correspond to large fish like brood fish (see Chap. 3 and 7). These findings stimulated the search for the former distribution of both species over Europe, which led to the first study in the field focusing on the European sturgeon (see Chap. 8). This study looked particularly at genetic variability (mtDNA) assessed from both archaeozoological remains and museum samples which were compared to extant specimens. There is a far greater genetic variability in the Mediterranean French façade, through a large set of remains, than for the Black Sea, the Atlantic façade, the North Sea and Baltic Sea populations.

This low variability confirms previous findings by Ludwig et al. (2000, 2004), summarised in Chap. 48. European species probably originated in the Mediterranean region, and extended gradually to the east (Black Sea) and to the west

(Atlantic Ocean, North Sea and Baltic Sea). This dispersion can itself explain the low variability of the outlying populations as being a regional adaptation, and thus this restricted genetic variability would in fact be representative rather than being considered as a genetically deteriorated population. Apart from the preliminary result itself, the interest of this contribution (see Chap. 8) is that the genetic tools are available to extend the study to more samples in order to provide the best possible updated overview in the subject. It was also the analysis of archaeozoological remains that allowed confirmation of the only presence of European sturgeon in Spain (see Chap. 9).

Building a brood stock of an endangered species is ideally a simple process, with the main principle being to hold "a representative subset" of the known variability of the species that is dealt with, and to take care in reproductive mating. However, most of the time, the building of a brood stock is begun too late to satisfy these conditions. Additionally, there are very few or no guidelines on rearing conditions and management. It has been necessary to set up all the steps for the European sturgeon, including some general rules for coping with unexpected situations (see Chap. 31). It is worth noting that, for the first time, a brood stock of an anadromous sturgeon species has been maintained for years in brackish water, allowing the fish to develop a normal gametogenesis which confirms previous data (Williot et al. 2007, 2009).

Based on a set of species-specific microsatellites, a new analysis of the brood stock is proposed in Chap. 34. Most brood fish can be separated into two groups, and a table of best mating is provided.

Larval rearing was under-documented with only one study (Williot et al. 2005; see Chap. 28), due to the very few successful artificial breeding occurrences. Long-term maintenance of brood fish was in need of an appropriate compound diet, as preliminary results did not provide attractive results. Based on a similar weaning sequence to that used previously, a "home-made" formula produced extremely good results in terms of growth and survival (see Chap. 33).

Despite great interest, there was no description of the early ontogenic development of the species, Chap. 5 provides the data references.

An unusual chapter dealing with the iconography of sturgeon, illustrated mainly with ancient pictures taken in France, and also with some stamps is included (see Chap. 19). This is to remind us of the large specimens that occurred in the past (TL $\geq$ 2.40 m) along most French sea shores and river systems. Main sturgeon uses in the early twentieth century, and the beginnings of population dynamic studies, are described. This illustrates the fact that in current fisheries surveys, the disappearance of large-sized specimens from landings would be the first sign of a deteriorating population.

49.2.3 Many Updated or New Methods Are Given

These methods are basic "standardized" key points to go further in conservation–restoration programmes. They are age determination (see Chap. 23), marking and tagging the fingerlings to be released (see Chaps. 25 and 24 respectively), sex determination, and staging (see Chap. 26). In most cases, conditions for use and limitations of the methods are provided.

Species discrimination keys are given based on the bone morphology and osteometry of three species: European, Atlantic, and Adriatic sturgeons (see Chap. 3). In addition to the well-known criteria (surface morphology of dermal bones in *A. sturio* and *A. oxyrinchus*), the morphology of scutes and of the bones of the splanchnocranium (maxillary, dentary, and palatopterygoid) strongly suggest differences between *A. sturio* and *A. naccarii* on the one hand, and between *A. sturio* and *A. oxyrinchus* on the other hand. Interestingly, thanks to highly correlated measurements between TL and bone size (dentary and pectoral spine), it is possible to back-calculate the size of ancient specimens, some of them being extremely large [5.63 m from remains excavated in Arles (South–East Fr)], and then providing an overview of past populations in cases where there is a large data set (Desse-Berset 1994; see Chap. 3).

The previous morphological distinction between large juveniles and adults of *A. sturio* and *A. oxyrinchus* (Magnin and Beaulieu 1963) is confirmed for small juvenile fish (see Chap. 4).

Reproduction methods and results for the European sturgeon, either wild-originated (see Chap. 27) or farmed (see Chap. 32), are described. The necessary management procedures are provided from about 6 months prior to reproduction until reproduction actually occurs, to ensure good quality gametes, with special focus on the final preparation phase (a few weeks at least) and final selection of the best potential reproductive female brood fish.

49.2.4 Thematic Overviews (Other Than Those Already Described Above)

The geographical distribution of the species was revisited based on past descriptions (see Chap. 6). The European catchments were divided into two groups; those where spawning most likely occurred, and those (the majority) where sturgeon were vagrants. Spawning basins were generally separated by hundreds or even thousands of km of coast.

As an anadromous species that spends part of its juvenile period and most of its adulthood in Atlantic continental shelves, the European sturgeon in the French Atlantic façade exhibits another peculiarity at the juvenile stage, with yearly back and forth movements into the Gironde estuary (see Chap. 10). Special emphasis is placed on reproductive cycles (see Chap. 12). As a consequence of some of the above statements, juveniles spend long periods in the Gironde estuary by eating intensively (see Chap. 11).

In the context of species conservation, sperm cryoconservation might be of considerable help in various situations; a synthesis is given of the different methods used with guidelines for sturgeon (see Chap. 35).

49.2.5 Actions in Favour of the Management or Conservation: Restoration of the Atlantic Sturgeon in Northern America and Northern Europe

The North American situation is very different from that in Europe as the status of Atlantic sturgeon, represented by several distinct population segments (DPS), although highly threatened in some areas, is not extirpated from the majority of its natural catchments in the US (see Chap. 39). The DPS are now managed separately. In spite of previous limitations set on sturgeon fisheries, none recovered their previous level, thus illustrating the very long timescale required in sturgeon management. The general umbrella of the federal Endangered Species Act (ESA) prompted state agencies (or interstate bodies such as the Atlantic States Marine Fisheries Commission, ASMFC) to complete a Fishery Management Plan for the Atlantic in the 1990s (see Chap. 39).

In Canada, there are two populations, one in the Saint John River basin (NB, Ca), still exploited for brood fish under a quota system, and the other in the Saint Lawrence River Basin (QC, Ca) where a very strict management process was set up (see Chap. 40). Two characteristics are worth noting in the latter. First, the exploited part of the population is only a fraction of the juveniles, thus allowing all fish over a certain size to become adults and to spawn. Second, instead of banning the fishery, the challenge is to allow juveniles to be exploited, but this is strictly controlled by size and quota, and in this way information can be obtained from fishermen who are closely associated with the management process. In this way, the collection of fishery information is free. This small fishery does not produce caviar but presents some encouraging signs in terms of maintenance.

As a consequence of the primary evidence of the unexpected presence of the Atlantic sturgeon in the Baltic Sea (Ludwig et al. 2002), decisions were rapidly taken to act to re-introduce the species into the area (see Chaps. 41, 42, and 43). Following the genetic investigation findings, Canadian specimens were imported into both Poland and Germany to initiate such actions, which led to the first experimental stocking in 2006 (see Chap. 41).

49.2.6 Importance of Actions That Are Not Strictly Scientific

As mentioned above (Sect. 49.2.1) the shortcomings of the French policy for fish species conservation (see Chap. 20) has prompted NGOs to become involved, especially WWF-France, which after taking some time to be convinced eventually agreed to organize lobbying actions in favour of European sturgeon conservation at both national and European levels (see Chap. 36). This proved to have been a determining factor at the French level in obtaining official support for preservation of the species. It also resulted in a European action plan endorsed by the European Council (Rosenthal et al. 2007).

With the aim of protecting large juveniles and adults on the continental shelves at both French and European levels, a constant awareness campaign among marine professional fishermen is now having an effect (see Chap. 37). The significant positive impact of the French–German cooperation in the European sturgeon programme is described in Chap. 38. This proved to have been a key factor for safeguarding the species.

49.2.7 Some Current and Future Perspectives

Part V gathers together work to be developed for a better knowledge of the in-situ life history (see Chap. 44). Two simulations are included. One deals with the potential settlement in France of non-native sturgeon species (*Acipenser stellatus*, *Acipenser gueldenstaedtii*, and *Huso huso*) when considering future climate change (see Chap. 45). This is part of a larger study (Williot and Rochard 2007) undertaken at the request of French sturgeon farmers supported by the Ministry of the Environment. The main conclusions are that it is highly probable that the three species will settle in all the large French river basins, and that the southern part of the European sturgeon's previous distribution area will probably become unsuitable for the species. The second simulation is a population viability analysis carried out on the European sturgeon in the Garonne basin (Chap. 46). Among the findings is a list of the most crucial biological parameters to be documented: mean fecundity, age at which females reach maturity, female spawning frequency, sex ratio, intrinsic population growth rate, and the age-specific natural mortality, especially of the youngest age classes. Also of importance is an assessment of the minimum viable population needed. In the case that the species continues to deteriorate, an extreme safeguard solution was explored by cryobanking somatic cells (see Chap. 47). The study suggests that results depend on the size of the fish donor. The last chapter of part V is a broad overview of the biological variability that needs to be conserved to ensure the long-term survival of the species. Based on current knowledge, suggestions are made with new complementary genetic approaches, and highlight behaviour as a field to tackle.

49.3 What Should Be Done?

49.3.1 Traditional Scientific Queries Regarding Stocking

Due to the crucial importance of the brood stock, a sound physiological knowledge is needed for optimal long-term management of the fish. Two fields should be favoured, reproduction and osmoregulatory functions. The former has received some attention (Davail-Cuisset et al. 2008), and should be extended over several years. The latter is totally under-documented. Given the ecology of the species and present French brood stock management (see Chap. 31), a better knowledge of this physiological compartment would be very helpful, not only for management but also for a better understanding of the ecology of the species, illustrated by the short migration of juveniles (see Chap. 10). This should be performed in relation to suggestions provided in Chap. 48 (peculiarly proteomic and genomic) and applied simultaneously to the brood fish held in France and in Germany. Similarly, a common appropriate reproduction protocol based on methods described in Chap. 32 should be set up, again in relation to suggestions in Chap. 48, one of the stakes being the definite completion of rearing conditions, especially salinity.

Thanks to the "package of methods" presented in the book, from brood-fish management to post-release monitoring, new questions might be explored: for instance, the rearing conditions of fish to be stocked and the optimum size for release. There are complex and long-term actions which are still poorly documented in spite of past stocking programmes with sturgeon worldwide (see Williot et al. 2009 for a review). Some suggestions are given regarding rearing conditions (see Chap. 48).

49.3.2 New Scientific Questioning with Applied Consequences

There is now evidence for the former simultaneous presence of both sister species, European and Atlantic sturgeons, in northern and western parts of Europe (Ludwig et al. 2002; Tiedemann et al. 2007; Desse-Berset 2009a, b; Desse-Berset and Williot 2011; see Chaps. 7 and 42). Not surprisingly, the presence of Atlantic sturgeon in the North Sea is strongly suggested, in the Scheldt basin (Be) and Rhine basin (see Chap 7), and at the mouth of the Elbe River (De) (the Oste River) (Ludwig et al. 2002).

In France, with the exception of the Mediterranean façade where the only species found was the European sturgeon (Desse-Berset 1994; Desse-Berset et al. 2008; Pagès et al. 2009; see Chaps. 7 and 8), the relevant period ranged from the Neolithic (5,000 BP) on the Atlantic coast to the seventeenth century in the Seine River, which is a Channel catchment. The colonisation of the Baltic Sea occurred in the Middle Ages (1200–800 BP) according to Ludwig et al. (2002), and earlier (300 BC) (Chap. 42). Altogether, there is evidence to suggest that Baltic Sea colonisation by the Atlantic sturgeon was from Western European specimens, as suggested by Desse-Berset and Williot (2011) rather than from the direct migration of North American specimens as previously proposed (Ludwig et al. 2002, 2008; Ludwig and Gessner 2007). The later hypothesis disagrees with the fact that sturgeons are not known to migrate in deep waters (Bemis and Kynard 1997); the coastal migration of the anadromous species (the few sturgeon species which spend long periods at sea) is the rule.

As a result, the dynamics of the distribution of the Atlantic sturgeon in Europe is on the table, as well as its origin. Palaeogenetic studies of both European remains and American samples (ancient and actual) would be fruitful, as suggested by Desse-Berset and Williot (2011). A similar approach to the colonisation of Europe by the European sturgeon is possible, based on preliminary analysis in Chap. 8. This would give a more consistent appreciation of the past genetic variability of the Western Atlantic population of the species, and thus provide references for managing the present brood stock.

It has been suggested that the explanation for the substitution of the European sturgeon by the Atlantic sturgeon in the Baltic Sea was that this was a better adaptation on the part of the Atlantic sturgeon to spawn at lower temperatures, compared with the European sturgeon; one of our team, PW, was at least partly responsible for this in assuming that the species was not known to spawn at water temperatures below 20°C. Indeed, Spillmann (1961) reported that incubation lasted 5 days at 19.5°C to 7 days at 14°C which suggests that the species is able to reproduce at 14°C. A wider range for the incubation duration is mentioned by Buslé and Quignard (2001), with 10–18°C and even 14 days at 7.7°C. In his synthesis, Holčik et al. (1989) reported that spawning may occur within the range of 7.7–22°C. These data appear more ecologically compatible for a temperate species such as the European sturgeon. Therefore, the temperature effect as a potential reason for the substitution of the European sturgeon by the Atlantic sturgeon in the Baltic Sea, at least, has to be revisited. In addition, several sturgeon species in sympatry have been observed in several basins, and this cannot automatically explain the disappearance of a species, even for very close species like the European and Atlantic sturgeon, where a competitive exclusion mechanism could be considered. It remains true that interbreeding is possible between sturgeon, as hybrids have been described (e.g., Kozhin 1964; review by Rochard et al. 1991). Indeed, hybrids between European and Atlantic sturgeons have been genetically described in the Baltic Sea (Tiedemann et al. 2007; Ludwig et al. 2008), with the former giving a significantly higher percentage of hybrids (z test = 2;96, $p = 0.003$) but with a different set of more recent samples (18th–20th centuries) and a different method. Tiedemann et al. (2007) also showed that there was no significant difference between the relative number of assignments to each pure line (Khi2 $= 12, p = 0.21$).

Fontana et al (2008) reported that both species presented fairly similar karyotypes and cytogenetic features. Based on fluorescent in-situ hybridization of two satellite DNA sequences, the authors detected no hybridization signal between the two species, while between stellate and Russian sturgeons signals were detected.

The authors concluded that both species appear very similar from a cytogenetic point of view, and therefore represent a sister clade compared with all sturgeon species studied at present. Fontana's contribution (see Chap. 2) totally confirms previous conclusions suggesting that both species are probably the most primitive sturgeon species that started to differentiate about 90 MY ago, and thus the presence of an interbreeding population, as suggested by Tiedemann et al. (2007), in which species-specific markers do not significantly deviate from Hardy–Weinberg equilibrium, apparently indicates the absence of mating barriers between the two species.

All the data sets show that both species were present in sympatry from ancient times until recently, thus making the Atlantic sturgeon a potential native species to French Atlantic and Channel façades, and most likely to the North Sea.

The situation in France has therefore become more complicated with regard to the restoration of the sturgeon species. Until recently, only the European sturgeon was considered. Indeed, the primary objective of the book was to synthesize and update what had been achieved in restoring the species. And this might be promoted in other European basins than those presently concerned, where conditions might be judged acceptable (quality of habitats, concurrent fisheries activity, post-release monitoring, involvement of all stakeholders), just as has been developed in France, Germany, and partly in Poland (see Chaps. 20, 21, 41, 43). A preliminary enquiry was conducted some years ago at European level (Rochard 2002) and could be updated. One of the difficulties was to find competent accredited bodies in the field.

Given the new situation with the two species, many actions should be set up to address the matter. Taking advantage of the rearing facility in France, it is strongly suggested that work on the Atlantic sturgeon begin as soon as possible.

With regard to restoration programmes, preliminary work was recently carried out in France to explore the potential consequences of introducing six new sturgeon species for farming alongside the European sturgeon (Williot and Rochard 2007). The work served as a background for Chap. 45. An updated version of the guide to decision-making should be produced. In the first version, as a precaution, it was suggested that the Garonne basin should be kept, at least for a while, as a sanctuary for the European sturgeon. In practice, this meant that any new introduction of a non-native sturgeon species into a fish farm would be refused in the Garonne River Basin as well as in the nearby north Charente River basin, to avoid any installation in case of escape.

49.3.3 Governance of Sturgeon Status for the Future

Thanks to the present example of a conservation–restoration programme, the time needed for recovery is very long, all the more so since there are not yet any certainties as to the recovery of the species. This long time dimension has been outlined in similar sturgeon cases (e.g., Secor et al. 2002; Williot et al. 2002; Legg and Nagy 2006) and for all species by Støttrup and Sparrevohn (2007). This is reinforced by the population viability analysis with the European sturgeon in the Garonne basin (Fr) (see Chap. 46). Such long-term actions need adequate management to support and guide the scientific actions. This dimension is totally or almost totally absent from the sturgeon action plan already published (Bloesch et al. 2006; Rosenthal et al. 2007). It is worth noting that the French version of the European action plan is still to be produced by the Ministry of Environment. Management plans should contain objectives and the means to evaluate the objectives (Pullin et al. 2004; Legg and Nagy 2006). It is the responsibility of the State to lay down the main lines for action or to delegate specific agencies to do so. There is a French research agency for biodiversity but not for species conservation. A good example of this dysfunction in species conservation is given by Caramel (2010) who reported that out of 55 French conservation plans only two have been completed and 15 are still ongoing. A political–administrative management process should be set up so that long-term actions such as this one for sturgeon, but also for other species, can be carried out. Indeed the fate of most endangered species should not depend solely on the voluntary actions of a few people. As far as we know, such structures are absent in Italy and Spain as well.

49.4 Conclusions

The legal status of the species is under severe threat (see Chap. 18). For the most part, the species is on the verge of extinction. This is unfortunately a good example of the extinction vortex described by Gilpin and Soulé (1986).

At present, a breakthrough (reproduction of farmed brood fish, see Chap. 32) has been successfully completed, and therefore a primary challenge has been accepted. However, the successful restoration of the European sturgeon is facing one of several new challenges with the long-term management of brood stock, as this represents the last biological unit of the species (see Chap. 48). Apart from the scientific and technical issues developed in the book, the long-term success of the species restoration project depends on setting up long-term scientific monitoring which has proved to be a key issue (Williot et al. 2009) among all those presented here (see Chaps. 29 and 30), operational business management of the overall restoration plan, and the reduction of pressures on the available habitats.

References

Bemis WE, Kynard B (1997) Sturgeon rivers: an introduction to acipenseriform biogeography and life history. Environ Biol Fishes 48:167–183

Bloesch J, Jones T, Reinartz R, Striebel B (2006) Action plan for the conservation of sturgeons (Acipenseridae) in the Danube River Basin. Nature and Environment N 144. Council of Europe Publishing, Strasbourg

Truslé J, Quignard JP (2001) Biologie des Poissons d'eau douce européens. Editions Lavoisier Tec et Doc, Paris, p 625
Caramel L (2010) Face à l'érosion de la biodiversité, la France tâtonne faute d'une stratégie globale et cohérente sur tout le territoire, les objectifs fixés pour 2010 n'ont pas été tenus. Le Monde 27/02/2010 page 4
Carmona R, Domezain A, Garcia Gallego M, Hernandol JA, Rodriguez F, Ruiz Rejon M (eds) (2009) Biology, conservation and sustainable development of sturgeons, Fish & Fisheries Series. Springer, Heidelberg, 467p
Castelnaud G, Rochard E, Jatteau P, Lepage M (1991) Données actuelles sur la biologie d'*Acipenser sturio* dans l'estuaire de la Gironde. In: Williot P (ed) *Acipenser*. Cemagref, Antony, France, pp 251–275
Davail-Cuisset B, Lacomme S, Viaene E, Williot P, Lepage M, Gonthier P, Davail S, Rouault T (2008) Hormonal profile in adults of Atlantic European sturgeon, *Acipenser sturio*, adapted to hatchery in France. Cybium 32(2 Suppl):169–170
Desse-Berset N (1994) Sturgeons of the Rhône during Protohistory in Arles (6th–2nd century BC). In: Fish exploitation in the past, Proceedings of the 7th meeting of the ICAZ Fish Remains Working Group (Louvain, Sept. 1993). Annales du Musée royal de l'Afrique centrale, Tervueren, 274: pp. 81–90
Desse-Berset N (2009a) First archaeozoological identification of Atlantic sturgeon (*Acipenser oxyrinchus* Mitchill 1815) in France. CR Palevol 8:717–724. doi:10.1016/j.crpv.2009.06.001, 8 décembre 2009
Desse-Berset N (2009b) La pêche dans l'économie de subsistance des sites de Ponthezières et de la Perroche (île d'Oléron, Néolithique final). In : Laporte L (ed) Des premiers paysans aux premiers métallurgistes sur la façade atlantique de la France (3500–2000 av. J.-C.). Édition Association des publications Chauvinoises, Mémoire XXXIII Chauvigny: pp 584–609
Desse-Berset N, Williot P (2011) Emerging questions from the discovery of the long term presence of *Acipenser oxyrinchus* in France. J Appl Ichthyol 27:263–268. doi:10.1111/j.1439-0426.2010.01649.x
Desse-Berset N, Pages M, Brosse L, Tougard C, Chassaing O, Hänni C, Berrebi P (2008) Specific identification of the extinct population of sturgeon from the Rhône River by mtDNA analysis from bone remains (Jardin d'Hiver, Arles, France, 6th–2nd century BC. In: Béarez P., Grouard S, Clavel B (eds), Archéologie du poisson. 30 ans d'archéo-ichtyologie au CNRS. Hommage aux travaux de Jean Desse et Nathalie Desse-Berset, XVIIIe rencontres internationales d'archéologie et d'histoire, Editions APDCA, Antibes, France, pp 195–200
Elvira B (2000) Symposium on conservation of the Atlantic sturgeon *Acipenser sturio* L., 1758 in Europe. Bol Inst Esp Oceanogr 16:253p
Fontana F, Lanfredi M, Kirschbaum F, Garrido-Ramos MA, Robles F, Forlani A, Congiu L (2008) Comparison of karyotypes of *Acipenser oxyrinchus* and *A. sturio* by chromosome banding and fluorescent in situ hybridization. Genetica 1342:281–286
Gilpin ME, Soulé ME (1986) Minimum viable populations : processes of species extinction. In: Soulé ME (ed) Conservation Biology. The science of scarcity and diversity. Sinauer Associates, Inc. – Publishers, Sunderland, Massachusets (USA), pp 19–34
Holčik J, Kinzelbach R, Sokolov LI, Vassilev VP (1989) *Acipenser sturio* Linnaeus, 1758. In: Holčik J (ed) The freshwater fishes of Europe. Aula Verlag, Wiesbaden, pp 367–394
Legg CJ, Nagy L (2006) Why most conservation monitoring is, but need not to be, a waste of time. J Environ Manag 78:194–199
Ludwig A, Gessner J (2007) What makes the difference? – Sea sturgeon on both sides of the Atlantic Ocean. Am Fish Soc Symp 56:285–300
Ludwig A, Arndt U, Lippold S, Benecke N, Debus L, King TL, Matsumura S (2008) Tracing the first steps of American sturgeon pioneers in Europe. BMC Evol Biol 8:221. doi:10.1186/1471-2148-8-221

Ludwig AN, Jenneckens I, Debus L, Ludwig A, Becker J, Kirschbaum F (2000) Genetic analysis of archival specimens of the Atlantic sturgeon *Acipenser sturio* L., 1758. Bol Inst Esp Oceanogr 16:221–230

Ludwig A, Williot P, Kirschbaum F, Lieckfeld D (2004) Genetic Variability of the Gironde population of *Acipenser sturio*. In: Gessner J and Ritterhoff J (eds) Species Differentiation and Population Identification in the sturgeons *Acipenser sturio* L. and *Acipenser oxyrinchus*. Bundesamt für Naturschutz 101:54–72

Ludwig A, Makowiecki D, Benecke N (2009) Further evidence of trans-Atlantic colonization of Western Europe by American Atlantic sturgeons. Archaeofauna 18:185–192

Ludwig A, Debus L, Lieckfeldt D, Wirgin I, Benecke N, Jenneckens I, Williot P, Waldman JR, Pitra C (2002) When the American sea sturgeon swam east. Nature 419:447–448

Magnin E, Beaulieu G (1963) Etude morphométrique comparée de l'*Acipenser oxyrinchus* Mitchill du Saint Laurent et de l'*Acipenser sturio* Linné de la Gironde. Le Naturaliste Canadien XC (1):5–38

Pagès M, Desse-Berset N, Tougard C, Brosse L, Hänni C, Berrebi P (2009) Historical presence of the sturgeon *Acipenser sturio* in the Rhône basin determined by the analysis of ancient DNA cytochrome *b* sequences. Conserv Genet 10:217–224. doi:10.1007/s10592-008-9549-6

Pullin AS, Knight TM, Stone DA, Charman K (2004) Do conservation managers use scientific evidence to support their decision-making? Biol Conserv 119:245–252

Rosenthal H, Bronzi P, Gessner J, Moreau D, Rochard E (2007) Action plan for the conservation and restoration of the European sturgeon. Nature and Environment N° 152. Council of Europe Publishing, Strasbourg

Rochard E, Williot P, Castelnaud G, Lepage M (1991) Eléments de systématique et de biologie des populations sauvages d'esturgeons. In: Williot P (ed) Acipenser. Cemagref, Antony, France, pp 475–507

Rochard E (coord) (2002) Restauration de l'esturgeon européen *Acipenser sturio*. Rapport scientifique Contrat LIFE n°B-3200/98/460, Cemagref, Département Gestion des Milieux Aquatiques, UR Ressources Aquatiques Continentales, Etude n° 80. 224 p

Secor DH, Anders PJ, Van Winkle W, Dixon DA (2002) Can we study sturgeons to extinction? What we do and don't know about the conservation of North American Sturgeons. Am Fish Soc Symp 28:3–10

Spillmann CL (1961) Faune de France, 65 poissons d'eau douce. Lechevallier, Paris, 303 p

Støttrup JG, Sparrevohn CR (2007) Can stock enhancement enhance stocks? J Sea Res 57:104–113

Tiedemann R, Moll K, Paulus KB, Scheer M, Williot P, Bartel R, Gessner J, Kirschbaum F (2007) Atlantic sturgeons (*Acipenser sturio, Acipenser oxyrinchus*): American females successful in Europe. Naturwissenschaften 94:213–217. doi:10.1007/s00014-006-0175-1

Williot P (ed) (1991) ACIPENSER. Actes du premier colloque international sur l'esturgeon. Cemagref, Antony, France, 519 p

Williot P, Arlati G, Chebanov M, Gulyas T, Kasimov R, Kirschbaum F, Patriche N, Pavlovskaya L, Poliakova L, Pourkazemi M, Kim Yu, Zhuang P, Zholdasova IM (2002) Status and management of Eurasian sturgeon: an overview. Int Rev Hydrobiol 87:483–506

Williot P, Rochard E (coord) (2007) Biologie, exigences environnementales et éléments d'évaluation a priori du risque relatif à plusieurs espèces d'esturgeons susceptibles d'être élevées en pisciculture en France: phases 1 & 2. Etude Cemagref EPBx pour le Ministère de l'Environnement et du Développement durable, 231 p

Williot P, Rochard E, Kirschbaum F (2009) Acceptability and prerequisites for the successful introduction of sturgeon species, Chap. 23. In: Carmona R, Domezain A, Garcia Gallego M, Hernando-Casal J, Rodriguez F, Ruiz Rejon M (eds) Biology, conservation and sustainable development of sturgeons, Fish & Fisheries Series. Springer, Heidelberg, pp 369–384, 467p

Index

P. Williot et al. (eds.), *Biology and Conservation of the European Sturgeon Acipenser sturio L. 1758*, DOI 10.1007/978-3-642-20611-5,

The manufacturer's authorised representative in the EU is Springer Nature Customer Service Centre GmbH, Europaplatz 3, 69115 Heidelberg, Germany. If you have any concerns regarding our products, please contact ProductSafety@springernature.com

Printed and bound by CPI Group (UK) Ltd, Croydon, CR0 4YY
22/09/2026
02220956-0017